2020台达杯国际太阳能建筑设计竞赛获奖作品集

Awarded Works from International Solar Building Design Competition 2020

U0159598

阳光 · 稚梦

SUNSHINE & CHILDLIKE DREAM

中国可再生能源学会太阳能建筑专业委员会　编

Edited by Special Committee of Solar Buildings, CRES

执行主编：张　磊　鞠晓磊

Chief Editor: Zhang Lei　Ju Xiaolei

中国建筑工业出版社

CHINA ARCHITECTURE & BUILDING PRESS

图书在版编目（CIP）数据

阳光·稚梦：2020台达杯国际太阳能建筑设计竞赛获奖作品集 = Sunshine & childlike dream：Awarded Works from International Solar Building Design Competition 2020 / 中国可再生能源学会太阳能建筑专业委员会编．—北京：中国建筑工业出版社，2021.3
ISBN 978-7-112-25958-8

Ⅰ.①阳… Ⅱ.①中… Ⅲ.①太阳能住宅－建筑设计－作品集－世界－现代 Ⅳ.①TU241.91

中国版本图书馆CIP数据核字（2021）第039324号

本书汇集了2020台达杯国际太阳能建筑设计竞赛的获奖作品，作品以阳光·稚梦为主题，分别选取福建省南平市建阳区景龙幼儿园、新疆巴音郭楞州和静县建设兵团牧场幼儿园及服务中心两个赛题，面向全球征集作品，以幼儿园为平台，将生态、绿色的理念融入到学前教育中。本次作品集通过不同的设计方式，贯通同一个环保理念，展现了当代设计师对太阳能建筑设计和技术的创新思维。不仅传播了太阳能建筑理念，加强了公众对建筑低碳、节能技术应用的认知度，更促进建筑从高耗能向产能方向迈进。本书适用于建筑师、高校师生，以及从事太阳能技术的相关从业者参考阅读。

责任编辑：吴 绫 唐 旭 张 华
文字编辑：李东禧
责任校对：王 烨

2020台达杯国际太阳能建筑设计竞赛获奖作品集
Awarded Works from International Solar Building Design Competition 2020
阳光·稚梦
SUNSHINE & CHILDLIKE DREAM
中国可再生能源学会太阳能建筑专业委员会 编
Edited by Special Committee of Solar Buildings, CRES
执行主编：张 磊 鞠晓磊
Chief Editor: Zhang Lei Ju Xiaolei
*
中国建筑工业出版社出版、发行（北京海淀三里河路9号）
各地新华书店、建筑书店经销
北京雅盈中佳图文设计公司制版
临西县阅读时光印刷有限公司印刷
*
开本：787毫米×1092毫米 1/12 印张：$24\frac{1}{3}$ 插页：1 字数：732千字
2021年3月第一版 2021年3月第一次印刷
定价：198.00元
ISBN 978-7-112-25958-8
(37164)

版权所有 翻印必究
如有印装质量问题，可寄本社图书出版中心退换
（邮政编码 100037）

随着我国生育政策从“单独二孩”到“全面二孩”的调整，对基础教育资源的配置提出了新的需求。党的十九大报告明确提出“优先发展教育事业”，强调“办好学前教育”，如何按照“发展中保障和改善民生”基本方略，满足“幼有所育”的要求也成为本次竞赛的关注点。为此，竞赛以“阳光·稚梦”为名，将幼儿园设计作为竞赛主题。“稚梦”是稚子之梦，幼儿园是孩子们启蒙时期的第一课堂，汲取知识也为他们种下梦想的种子；“稚梦”也是织梦，希望参赛者用建筑的语言为孩子们编织一对寻梦的翅膀，将生态、绿色的理念融入到学前教育中。

感谢台达集团资助举办2020台达杯国际太阳能建筑设计竞赛！

谨以本书献给致力于学前教育与绿色发展的同仁们！

The adjustment from “selective two-child policy” to “universal two-child policy” in China has proposed new requirements for the allocation of basic education resources. It’s clearly noted in the report at the 19th CPC National Congress to “place education development as priority” for improved preschool education, so how to meet the demands of “education accessible to young children” in the basic strategy to “ensure and improve people’s livelihood during development” has also become the focus of the competition. As a result, the kindergarten design is taken as the competition theme named with “Sunshine & Childish Dream”. “Childish Dream” represents children’s dream. The enlightenment in kindergartens symbols the first step in children’s life, and acquiring knowledge will help them scatter the seed of dreams as well; “Childish Dream” also means creating dreams for children. It’s hoped that contestants will create more opportunities for children to realize their dreams through the construction of kindergartens, and integrate ecological and green concepts into preschool education.

Sincere gratitude goes to Delta Group for sponsoring the International Solar Building Design Competition 2020!

This book is written for paying tribute to those committed to preschool education and green development!

目　录
CONTENTS

阳光·稚梦　SUNSHINE & CHILDLIKE DREAM

2020台达杯国际太阳能建筑设计竞赛过程回顾
General Background of International Solar Building Design Competition 2020

2020台达杯国际太阳能建筑设计竞赛评审专家介绍
Introduction to Jury Members of International Solar Building Design Competition 2020

获奖作品　Prize Awarded Works　001

综合奖·一等奖　General Prize Awarded · First Prize

积木·乐园（南平）　Kindergarten Built by Tetris（Nanping）　002

沐日（新疆）　Step into the Sunshine（Xinjiang）　008

综合奖·二等奖　General Prize Awarded · Second Prize

追光者（南平）　Sun Chaser（Nanping）　012

林间风过（南平）　Rambling in the Wind（Nanping）　018

稚梦·星盒（新疆）　Dreamy Star Box（Xinjiang）　024

江格尔与日出（新疆）　Janger and the Sunrise（Xinjiang）　030

综合奖·三等奖　General Prize Awarded · Third Prize

风·跳跃空间（南平）　Wind · Jumping Space（Nanping）　036

漫步园景（南平） Walking in the Garden（Nanping） 042
林海风语（南平） Feel the Wind（Nanping） 048
层峦・沐光行（南平） Layers of Mountains・Bathe in Light（Nanping） 054
竹巷・风生（南平） Bamboo Alley・Wind Form（Nanping） 060
太阳升起的地方（新疆） Follow the Sun（Xinjiang） 066

综合奖·优秀奖 General Prize Awarded · Honorable Mention Prize

光之谷（南平） The Valley of Light（Nanping） 072
律动・魔方（南平） Rotating Cubes（Nanping） 078
阳光之下・田园之上（南平） Sunshine Pouring・Children Geting（Nanping） 084
晨・星（南平） Sun & Star（Nanping） 090
童之梦・光之巢（南平） The Dream of Children・the Nest of Light（Nanping） 096
光・遇（南平） Meet the Light in the Corner（Nanping） 102
山行・竹荫里（南平） Mountain Trip・In the Shade of Bamboo（Nanping） 108
光合积木（南平） Building Blocks of Sunlight（Nanping） 114
童年・里巷・巢下（南平） Childhood・Cold Lane・Under the Nest（Nanping） 120
稚梦・风光（南平） Childlike Dream・Wind and Light（Nanping） 126
树下幼儿园（南平） Under the Tree（Nanping） 130
向阳・梦园（南平） Sunny・Dream Garden（Nanping） 136

“小世界”环游记（南平）Dream Quest Travel（Nanping）	142
趣园（南平） Interesting Garden（Nanping）	148
院・积木之家（南平） Courtyard・House of Building Blocks（Nanping）	152
巷院・阡陌（南平） Lane・Compound（Nanping）	158
稚巷・拾光（南平） Children Alley & Chasing Childlight（Nanping）	164
楼台・织梦（南平） Platforms of Dream（Nanping）	170
巷・间（南平） Cold Lane・Interspace（Nanping）	176
瑶域・疆来（新疆） Yao Yu・Xin Jiang Future（Xinjiang）	180
晨兴日起巴伦台（新疆） Bring the Sunrise to BLUNTAI（Xinjiang）	184
风・院（新疆） Wind・Yard（Xinjiang）	188
向阳而生（新疆） Grow in the Sun（Xinjiang）	194
童旭・土生（新疆） “Raising” Children｜Vernacular Homeland（Xinjiang）	200
朝晖・光场（新疆） Haven（Xinjiang）	204
光之幼儿园（新疆） Collecting Sunshine（Xinjiang）	210
阳光托马斯（新疆） Sunny Thomas（Xinjiang）	216
杲杲日光（新疆） High in the Sun（Xinjiang）	222
煦阳・和风（新疆） Warm Sunshine・Gentle Wind（Xinjiang）	228
腔间光语（新疆） The Melody of Sunlight in the Chamber（Xinjiang）	234

专项奖 · 技术奖　Special Prize Awarded · Technology Prize

记忆中的糖块（南平）　Candy in Memory（Nanping）　240

稚梦飞扬（南平）　Childlike Dreams Flying（Nanping）　246

有效作品参赛团队名单
Name List of All Participants Submitting Valid Works　250

2020台达杯国际太阳能建筑设计竞赛办法
Competition Brief on International Solar Building Design Competition 2020　260

2020台达杯国际太阳能建筑设计竞赛过程回顾

General Background of International Solar Building Design Competition 2020

主题：阳光·稚梦

2020 台达杯国际太阳能建筑设计竞赛由国际太阳能学会、中国可再生能源学会、中国建筑设计研究院有限公司主办，国家住宅与居住环境工程技术研究中心、中国可再生能源学会太阳能建筑专业委员会承办，台达集团冠名。在社会各界的大力支持下，竞赛组委会先后组织了竞赛启动、媒体宣传、云讲堂、作品注册与提交、作品评审、现场答辩等一系列活动。这些活动得到了海内外业界人士的积极响应和参与。

一、赛题设置

随着我国生育政策从“单独二孩”到“全面二孩”的调整，对基础教育资源的配置提出了新的需求。党的十九大报告明确提出“优先发展教育事业”，强调“办好学前教育”，并在“坚持在发展中保障和改善民生”基本方略中，提出了“幼有所育”的要求。幼儿园建设是城镇公共服务设施建设的重要内容，是扩大普惠性学前教育资源的重要途径，更是保障和改善民生的重要举措。在这样的政策背景下，2020 台达杯国际太阳能建筑设计竞赛以“阳光·稚梦”为主题，落实国家推动城镇居住区幼儿园建设、着力补齐农村学前教育短板的政策，以幼儿园建筑为平台，将生态、绿色的理念融入到学前教育中。通过组织专家进行实地考察，确定了福

2020台达杯国际太阳能建筑设计竞赛
INTERNATIONAL SOLAR BUILDING DESIGN COMPETITION 2020
阳光·稚梦
启动会
STARTUP MEETING
.03.22
ISBD

中国可再生能源学会太阳能建筑专业委员会秘书长张磊主持会议
Zhang Lei, Secretary General of Special Committee of Solar Buildings, CRES presided over the conference

Theme: Sunshine & Childlike Dream

This competition is hosted by the International Solar Energy Society (ISES), China Renewable Energy Society (CRES) and China Architecture Design & Research Group (CAG), organized by China National Engineering Research Center for Human Settlements (CNERCHS) and the Special Committee of Solar Buildings, CRES, with the title sponsor of the Delta Group. With great support from all sectors of society, the Organizing Committee for this competition has organized a series of activities ranging from competition start up, media campaigns, cloud lectures, registration and submission, work evaluations to on-site statement. These activities have received positive responses and active participation from industry experts at home and abroad.

I. Competition Preparation

The adjustment from "selective two-child policy" to "universal two-child policy" in China has proposed new requirements for the allocation of basic education resources. It's clearly noted in the report at the 19th CPC National Congress to "place education development as priority" for improved preschool education, and the demands of "education accessible to young children" are put forward in the basic strategy to "ensure and improve people's livelihood during development" . The construction of kindergartens constitutes an integral part of the construction of urban public service facilities, an important way to expand inclusive preschool education resources, and also an essential measure to ensure and improve people's livelihood. In the context of such policies, the International Solar Building Design Competition 2020, themed with "Sunshine & Childlike Dream" , aims to implement the national policy of promoting the construction of kindergartens in urban residential areas and offsetting weakness in rural preschool education. The competition attempts to integrate ecological and green concepts into preschool education through the platform of kindergartens. After on-site inspections, experts determine two competition topics, namely Jinglong Kindergarten in Jianyang District, Nanping

建省南平市建阳区景龙幼儿园和新疆巴音郭楞州和静县建设兵团牧场幼儿园及服务中心两个赛题，并编制了设计任务书。

二、竞赛启动

2020 年 3 月 22 日，2020 台达杯国际太阳能建筑设计竞赛在北京启动。在疫情依然严峻的情势下，竞赛首次采用线上启动，开启竞赛云端作品征集、展示、传播的新模式。中国建设科技有限公司党委书记、董事长文兵，中国工程院院士、中国可再生能源学会理事长谭天伟，中国建筑设计研究院有限公司副总建筑师、中国可再生能源学会太阳能建筑专委会主任委员仲继寿，台达集团创办人暨荣誉董事长郑崇华作为嘉宾参加本次竞赛线上启动仪式并分别致辞。

2020 台达杯
国际太阳能建筑设计竞赛
阳光·稚梦
竞赛题目
福建南平市建阳区景龙幼儿园
新疆巴音郭楞州和静县建设兵团牧场
幼儿园及服务中心
奖项设置
奖项 数量 奖金（含税）
一等奖 2名 人民币 100000 元
二等奖 4名 人民币 20000 元
三等奖 6名 人民币 5000 元
优秀奖 30名
时间安排
2020年3月1日-6月30日 → 注册时间
2020年9月15日前 → 提交作品
2020年9月-10月 → 作品评审
国际太阳能建筑设计竞赛组委会
联系人：鞠晓磊、张星儿、郑晶茹
联系电话：86-010-88983377、86-010-88983383
E-MAIL :isbdc2020@126.com
QQ 交流群：49266054
微信公众号：国际太阳能建筑设计竞赛
网址：www.isbdc.cn
微信公众号
官方微博

竞赛海报
Competition poster

City, Fujian Province and Construction Corps Pasture Kindergarten and Service Center in Hejing County of Bayingol Mongolian Autonomous Prefecture, Xinjiang, and compile task books of design.

II. Competition Start up

On March 22, 2020, the International Solar Building Design Competition 2020 was launched in Beijing. Against the severe COVID-19 epidemic, the competition was initiated online for the first time, so the works were collected, displayed and disseminated online. Wen Bing, President of China Construction Technology Consulting Co., Ltd. (CCTC), Tan Tianwei, Academician of Chinese Academy of Engineering and Chairman of CRES, Zhong Jishou, Deputy Chief Architect of CAG and Chief Commissioner of Special Committee of Solar Buildings, CRES, Zheng Chonghua, Founder and Honorary Chairman of Delta Group, participated in the online opening ceremony as guests and delivered speeches respectively.

In response to the national call to promote the construction of kindergartens in urban residential areas and offset weakness in rural preschool education, the Organizing Committee sets "Sunshine & Childlike Dream" as the competition theme, and integrates ecological and green concepts into preschool education with kindergartens as a platform. Finally, the two competition topics of Jinglong Kindergarten in Jianyang District, Nanping City, Fujian Province and Construction Corps Pasture Kindergarten and Service Center in Hejing County of Bayingol Mongolian Autonomous Prefecture, Xinjiang were selected for global calls on works, aimed at creating low-carbon green kindergartens featuring childlike interest through the optimization of architectural design and utilization of such renewable energy technologies as solar energy. Therefore, kindergartens and related public facilities characterized by less energy consumption and more comfort are required in the competition topics in combination with the construction needs and applied characteristics of projects, use of active and passive solar technologies, and surrounding natural environment and climate of projects.

"迈向产能建筑"系列云上讲堂
2020"台达杯"
国际太阳能建筑设计竞赛
宣讲会
直播时间：2020年4月25日 10:00
宣讲主题
"阳光·稚梦"竞赛赛题解析
太阳能建筑技术综合应用与案例分析
建筑一体化设计与案例分析
直播入口
"迈向产能建筑"系列云上讲堂
2020"台达杯"
国际太阳能建筑设计竞赛
竞赛获奖作品实地建设分享——灾后重建专题
绿色生态幼儿园建设案例分析
直播时间：2020年5月16日 14:30
宣讲主题
竞赛问题讲解&答疑
直播入口
"迈向产能建筑"系列云上讲堂
"深挖建筑节能潜力"之
建筑能效提升与智能化控制
2020年6月19日 14:30
宣讲主题
直播入口
"迈向产能建筑"系列云上讲堂
"建筑光伏一体化"
技术标准解读与分析
2020年7月23日14:30
宣讲主题
直播入口
"迈向产能建筑"系列云上讲堂
2020"台达杯"
国际太阳能建筑设计竞赛
宣讲会
2020年8月20日14:00
宣讲主题
直播入口
网络宣讲海报
Poster on online publicity

为响应国家推动城镇居住区幼儿园建设、着力补齐农村学前教育短板的政策号召，组委会将本届竞赛主题设为"阳光·稚梦"，以幼儿园为平台，将生态、绿色的理念融入到学前教育中。甄选福建省南平市建阳区景龙幼儿园和新疆维吾尔自治区巴音郭楞州和静县建设兵团牧场幼儿园及服务中心为赛题，面向全球征集作品，力求通过优化建筑设计，利用太阳能等可再生能源技术，打造绿色、低碳、童趣的幼儿园。赛题要求结合实际项目的建设需求和使用特点，充分利用主、被动太阳能技术，结合项目周边自然环境和气候特点，建设实现低建筑能耗、高舒适性的幼儿园及相关公共设施。

三、网络宣讲

自 2005 年第一届竞赛举办以来，竞赛组委会已先后前往清华大学、天津大学、东南大学、重庆大学、山东建筑大学等 50 多所建筑院校开展国际太阳能建筑设计竞赛巡讲活动，受到了高校师生的积极响应和好评。本届竞赛受到疫情影响，组委会将线下校园巡讲活动转至线上，主题不仅包括竞赛介绍、答疑，优秀竞赛

III. Online Publicity

Since the first competition was held in 2005, the Organizing Committee has paid a visit to more than 50 architectural colleges and universities including Tsinghua University, Tianjin University, Southeast University, Chongqing University and Shandong Jianzhu University to deliver lectures on the competition, receiving positive response and praise from both teachers and students. Affected by the COVID-19 epidemic, the Organizing Committee changed offline campus tours to online activities. The event consisted of the introduction to the competition, Q&A, interpretations on excellent works and site construction, and industry experts in solar building were also invited to deliver cloud lectures on the topics such as design of solar building integration, comprehensive application of building energy-saving technologies and improvement of building energy efficiency. As a result, it has advanced the knowledge acquisition of audience and the dissemination of relevant

2020 台达杯国际太阳能建筑设计竞赛
INTERNATIONAL SOLAR BUILDING DESIGN COMPETITION 2020
庄惟敏：中国工程院院士、全国工程勘察设计大师、梁思成建筑奖获得者
清华大学建筑学院院长、教授、博士生导师、清华大学
研究院院长、总建筑师。

网络宣讲直播截图
Live Broadcast Screenshot of Online Publicity

Bai百度
2020台达杯国际太阳能建筑设计竞赛
百度一下
建筑设计竞赛-【匠人匠品网】
2020国际太阳能建筑设计竞赛
2020台达杯国际太阳能建筑设计竞赛-建筑与设计学院
2020台达杯国际太阳能建筑设计竞赛-索比光伏网
其他人还在搜
2020台达杯国际太阳能建筑设计竞赛获奖名单在京揭晓

竞赛信息网络发布
Online competition information release

2020台达杯国际 太阳能建筑 设计竞赛
International Solar Building Design Competition 2020
首页
竞赛办法
竞赛简介
作品实施
评审专家
竞赛论坛（维护中）
关注我们
新闻公告
2020台达杯国际太阳能建筑设计竞赛获奖作品公布[2020-12-19]
2020台达杯国际太阳能建筑设计竞赛入围最终评审名单[2020-11-24]
2020"迈向产能建筑"太阳能建筑设计作品"创能奖"网络评选活动[2020-11-11]
2020竞赛主题
2020台达杯国际太阳能建筑设计竞赛的主题为"阳光·稚梦"。
分别选取福建省南平市建阳区景龙幼儿园和新疆巴音郭楞州和静县建设兵团牧场幼儿园及服务中心进行设计，面向全球征集作品。贯彻落实国家推动城镇居住区幼儿园建设、着力补齐农村学前教育短板的政策，以幼儿园为平台，将生态、绿色的理念融入到学前教育中。
延承了往届的做法，本届竞赛将实地建造，使得太阳能建筑成为可供观摩、可经受使用检验的建筑实体，推动清洁能源等绿色建筑技术在建筑上的应用。
竞赛启动会直播
国际太阳能建筑设计竞赛
微信公众号
2020竞赛宗旨

竞赛网站
Official Website of the Competition

technologies, and the publicity lecture on the competition has become an important platform to popularize and exchange industry knowledge, technologies and ideas in the field of solar building.

IV. Media Campaign

The Organizing Committee has carried out media publicity through multiple channels since the launch of the competition, such as real-time bilingual report of the competition progress and scientific popularization of knowledge in solar building on the official competition website, and keywords set in Baidu to facilitate public inquiry and access to the official website. Besides, relevant information of the competition has been released and linked on more than 50 websites

2020年竞赛线上启动仪式视频回顾及报名方式

国际太阳能建筑设计竞赛 2020-03-23

2020年3月22日，2020台达杯国际太阳能建筑设计竞赛线上启动，面向全球正式展开幼儿园作品征集活动。该项大赛由国际太阳能学会、中国可再生能源学会、中国建筑设计研究院有限公司主办，国家住宅与居住环境工程技术研究中心、中国可再生能源学会太阳能建筑专业委员会承办，台达集团冠名。

本届竞赛以"阳光·稚梦"为主题，幼儿园为平台，将生态、绿色的理念融入到学前教育中。竞赛选择福建省南平市建阳区景龙幼儿园和新疆巴音郭楞州和静县建设兵团牧场幼儿园及服务中心为赛题，力求打造绿色、低碳、童趣的幼儿园。

目前竞赛已经开放注册，请大家登录竞赛官方网站 www.isbdc.cn 。点击右上角"注册" 按钮进行注册。欢迎大家积极参与。

English
2020台达杯国际太阳能建筑设计竞赛
International Solar Building Design Competition 2020
注 册

爱之园幼儿园IBG School见学点滴

国际太阳能建筑设计竞赛 2020-06-22

IBG爱之园幼儿园位于北京市西城区西直门内北礼士路，是由具有国际幼儿园设计经验丰富的日比野设计工作室完成，建筑面积3600平方米，包含12个班级和艺术教室、创客空间、多功能厅、图书馆、开放式餐厅等儿童活动区域。这个幼儿园项目是由既有建筑改造而成，虽然原有的建筑结构及空间为改造带来了一定的难度，但改造后的幼儿园还是向人们呈现出了特有的魅力。下面就建筑与教育、细部设计以及既有建筑改造等几方面，谈谈参观后的个人感受。

感触1：色彩

形体与色彩决定了建筑给人的第一印象，通常幼儿园建筑给人的感觉大多是色彩丰富，造型可爱，而"爱之园幼儿园"采用的是基本统一的色系。设计师的初衷是"幼儿园的主体是小孩子，而不是五颜六色的装饰。简洁的色彩更容易启发孩子们的思维，在对外界的探索中，把决定权交给孩子自己。"整个幼儿园，仅仅在地面的一角点缀了一组鲜艳的小色块。引导孩子们自己去探索未知，是比强加给他们对于世界的理解更高级的给予。

室内色彩也是以简洁的白色和温暖原木色为主，营造出轻松亲切的氛围。

竞赛公众号宣传
WeChat public account promotion

作品讲解与实地建设等内容，还邀请了领域专家开展包括太阳能建筑一体化设计、建筑节能技术综合应用、建筑能效提升等内容的云上讲堂，提高了竞赛关注群体和观众的知识积累及相关技术素养的传播，使竞赛宣讲会成为太阳能建筑领域行业知识、技术、理念的重要科普和交流平台。

四、媒体宣传

组委会自竞赛启动以来通过多渠道开展媒体宣传工作，包括：竞赛官方网站（双语）实时报道竞赛进展情况并开展太阳能建筑的科普宣传；在百度设置关键字搜索，方便大众查询，从而更快捷地登陆竞赛网站。在新华网、腾讯网、新浪网等 50 余家网站上报道或链接了竞赛的相关信息；同时，组委会与多所国外院校和媒体取得联系并发布竞赛信息与动态。通过微信公众号、微博实时发布竞赛进展、云讲堂预告等动态，并提供竞赛相关资料下载与案例介绍等，有效提高了竞赛的影响力及参赛团队的技术能力。

台达杯太阳能国际建筑竞赛 6351 推广 阅读
20-4-30 11:33 来自公开课 · 视频社...

2020年4月25日上午十点，"迈向产能建筑"系列云上讲堂开启！本次宣讲会邀请到了国家住宅工程中心太阳能建筑技术研究所所长鞠晓磊、国家住宅工程中心副总建筑师曾雁和龙焱能源科技（杭州）有限公司BIPV事业部总经理刘志钱为大家带来2020年竞赛赛题分析、太阳能建筑技术应用案例及建筑一体化设计的相关内容。

特别提示：在本次宣讲会结尾处，有专家对参赛者们的提问进行答疑哦~千万不要错过！
视频回顾见下方： 台达杯太阳能国际建筑竞赛的微博视频

竞赛微博宣传
Weibo Promotion

五、竞赛注册及提交情况

本次竞赛的注册时间为 2020 年 3 月 1 月至 2020 年 7 月 15 日，共 1200 个团队通过竞赛官网进行了注册，其中，包括来自中国港澳台地区以及美国、澳大利亚、英国、意大利、瑞典、日本等国家的注册团队共 18 个。截至 2020 年 9 月 15 日，竞赛组委会收到英国等国家和中国港澳台地区提交的有效参赛作品 235 份。

六、作品初评

2020 年 9 月 16 日，组委会组织评审专家对通过形式筛查的全部有效作品开展初评工作。专家根据竞赛办法中规定的评比标准对每一件作品进行评审，将各位评审专家作品评分汇总后由高到低排序，筛选出 100 份作品进入中评。

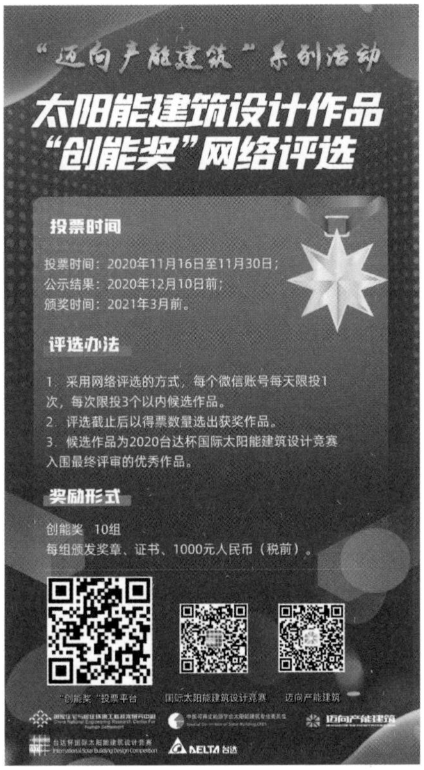
"迈向产能建筑"系列活动
太阳能建筑设计作品
"创能奖"网络评选
投票时间
投票时间：2020年11月16日至11月30日；
公示结果：2020年12月10日前；
颁奖时间：2021年3月前。
评选办法
1. 采用网络评选的方式，每个微信账号每天限投1次，每次限投3个以内候选作品。
2. 评选截止后以得票数量选出获奖作品。
3. 候选作品为2020台达杯国际太阳能建筑设计竞赛入围最终评审的优秀作品。
奖励形式
创能奖 10组
每组颁发奖章、证书、1000元人民币（税前）。
"创能奖"投票平台
国际太阳能建筑设计竞赛
迈向产能建筑
迈向产能建筑
台达杯国际太阳能建筑设计竞赛
International Solar Building Design Competition
DELTA 台达

活动海报
Poster of "Energy Innovation Award"

including xinhuanet.com, qq.com and sina.com. Meanwhile, the Organizing Committee has also contacted with a number of foreign universities and media to release information of the competition. Through the WeChat public account and Weibo, the Organizing Committee will release the progress of the competition and the preview of the cloud lecture, and provide the download of competition-related materials and case introductions, which effectively improve the influence of the competition and the technical capabilities of the participating teams.

V. Registration and Submission

The registration time of the competition ranged from March 1 to July 15, 2020, and a total of 1,200 teams made registration via the official competition website. Among them, there were 18 registered teams from China's Hong Kong, Macao and Taiwan regions, the United States, Australia, Britain, Italy, Sweden, Japan, and other countries. As of September 15, 2020, the Organizing Committee had received 235 valid works from Britain and other countries, as well as China's Hong Kong, Macao and Taiwan regions.

VI. Preliminary Evaluation

On September 16, 2020, the Organizing Committee organized jury members to make a preliminary evaluation of all valid works that have passed the screening of forms. After experts reviewed all the works according to the appraisal standard stipulated in the competition brief for International Solar Building Design Competition 2020, the works were ranked from high to low based on the scores, and 100 works were selected for mid-review.

From September 30 to November 10, 2020, the Organizing Committee organized jury members to review the above-mentioned 100 works. After rigorous review, the Organizing Committee screened out 60 works based on the scores for final review. From November 15 to November 30, 2020, the jury members selected 15 works from the 60 works for on-site statement after another round of review.

2020 年 9 月 30 日 ~ 11 月 10 日，组委会组织评审专家对 100 份进入中评的作品开展评审工作。经过专家组的严格评审，组委会根据评分结果筛选出 60 份作品进入终评阶段。2020 年 11 月 15 日 ~ 11 月 30 日，经过再一轮评审，评审专家从进入终评的 60 份作品中筛选 15 份作品进入现场答辩环节。

同时，为了传播太阳能建筑理念、加强公众对建筑低碳、节能技术应用的认知度，促进建筑从高耗能向产能方向迈进，中国可再生能源学会太阳能建筑专业委员会以进入终评阶段的 60 份作品为评选对象，开展 2020〝迈向产能建筑〞太阳能建筑设计作品〝创能奖〞网络评选活动，在为期两周的投票期内，共收到有效票数 286015 票，太阳能建筑及太阳能建筑设计竞赛得到了全社会的更广泛关注。

七、作品终评

竞赛现场评审会于 2020 年 12 月 18 日在北京召开。为了让学生更好地展示作品设计内容，同时便于评审专家综合评价参赛团队，本届竞赛终评会首次设置了现场答辩环节。由崔愷院士领衔的 10 名国内外知名专家组成了评审组。上午，15 个团队的成员分别通过图像、文字、语言与视频等方式向评审专家展示设计作品，并回答专家提问。下午，国内外评审通过视频会议的方式连线，历经 3 轮评选和讨论。最终评选出 44 件获奖作品，其中一等奖 2 名、二等奖 4 名、三等奖 6 名、优秀奖 30 名、技术专项奖 2 名。

Meanwhile, to spread the concept of solar building, strengthen public awareness to apply low-carbon and energy-saving technologies in buildings, and promote the development from intensive energy consumption to more production capacity in buildings, the 60 works in final review were set as the selection target by the Special Committee of Solar Buildings, CRES for the Online Selection Event of ˝Energy Innovation Award˝ in 2020 ˝Towards Capacity-oriented Buildings˝ Solar Building Design Works. A total of 286015 valid votes were received during the two-week period. The concept of solar building and the competition received extensive attention from the society.

VII. Final Review

The on-site review was conducted in Beijing on December 18, 2020. The on-site statement was set for the first time for students' better display of their works, and comprehensive evaluations of the participating teams by jury members. Academician Cui Kai served as the leader of the ten Well-known jury members.In the morning, members of 15 teams presented their works through such ways as image, text, language and video, and answered questions from the experts. In the afternoon, Domestic and foreign reviewers made three rounds of selection and discussion through a video conference. In the end, 44 prize awarded works were selected, that is, two winners for First Prize, four winners for Second Prize, six winners for Third Prize, 30 winners for Honorable Mention Prize, and two winners for Prize for Technical Excellence Works.

终评会现场　Scenes of final evaluation conference

终评专家组与答辩师生合影　Members of final evaluation juries and on-site statement teachers and students

2020台达杯国际太阳能建筑设计竞赛评审专家介绍

Introduction to Jury Members of International Solar Building Design Competition 2020

评审专家

Jury Members

杨经文，马来西亚汉沙杨建筑师事务所创始人、**2016** 梁思成建筑奖获得者

King Mun YEANG: President of T. R. Hamzah & Yeang Sdn. Bhd (Malaysia), 2016 Liang Sicheng Architecture Prize Winner

Deo Prasad，澳大利亚科技与工程院院士、澳大利亚勋章获得者、新南威尔士大学教授

Deo Prasad: Academician of Academy of Technological Sciences and Engineering, Winner of the Order of Australia, and Professor of University of New South Wales, Sydney, Australia

Peter Luscuere，荷兰代尔伏特理工大学建筑系教授

Peter Luscuere: Professor of Department of Architecture, Delft University of Technology

林宪德，台湾绿色建筑委员会主席、台湾成功大学建筑系教授

Lin Xiande: Chairman of Taiwan Green Building Committee and Professor of Faculty of Architecture of Cheng Kung University, Taiwan

Cui Kai

崔愷，中国工程院院士、全国工程勘察设计大师、中国建筑设计研究院有限公司总建筑师

Cui Kai: Academician of China Academy of Engineering, National Engineering Survey and Design Master and Chief Architect of China Architecture Design & Research Group (CAG)

仲继寿，中国可再生能源学会太阳能建筑专业委员会主任委员、中国建筑设计研究院有限公司副总建筑师

Zhong Jishou: Chief Commissioner of Special Committee of Solar Buildings, CRES, and Deputy Chief Architect of CAG

宋晔浩，清华大学建筑学院建筑与技术研究所所长、教授、博士生导师，清华大学建筑设计研究院副总建筑师

Song Yehao: Director, Professor and Doctoral Supervisor of Institute of Architecture and Technology, School of Architecture, Tsinghua University, and Deputy Chief Architect of Architectural Design and Research Institute of Tsinghua University

钱 锋

钱锋，同济大学建筑与城市规划学院教授、博士生导师，高密度人居环境生态与节能教育部重点实验室主任

Qian Feng: Professor and Doctoral Supervisor of College of Architecture and Urban Planning Tongji University (CAUP), Director of Key Laboratory of Ecology and Energy-saving Study of Dense Habitat (Tongji University), Ministry of Education

黄秋平，华东建筑设计研究总院总建筑师

Huang Qiuping: Chief Architect of East China Architectural Design & Research Institute (ECADI)

冯雅，中国建筑西南设计研究院顾问总工程师

Feng Ya: Chief Engineer of China Southwest Architectural Design and Research Institute Corp. Ltd

获奖作品
Prize Awarded Works

综合奖 · 一等奖
General Prize Awarded · First Prize

注 册 号：7851

项目名称：积木·乐园（南平）
Kindergarten Built by Tetris (Nanping)

作　　者：谢星杰、高嘉婧、徐　涵

参赛单位：重庆大学、华侨大学

指导教师：黄海静、周铁军、张海滨、欧达毅

专家点评：

该作品空间分割与布局有创意，流线设计合理，采用预制装配式模块的建造手法，显著提高项目的可行性和成本效益，形体空间丰富，主动和被动式太阳能策略及集成技术运用巧妙合理，太阳能构件与建筑构件的结合有创意。

The work, innovative in space division and layout, and reasonable in streamline design, significantly improves the feasibility and cost-effectiveness of the project through the construction method of prefabricated modules. It shows abundant space, ingenious and reasonable application of active and passive solar strategies and integrated technologies, and creative ideas through the combination of solar components and building components.

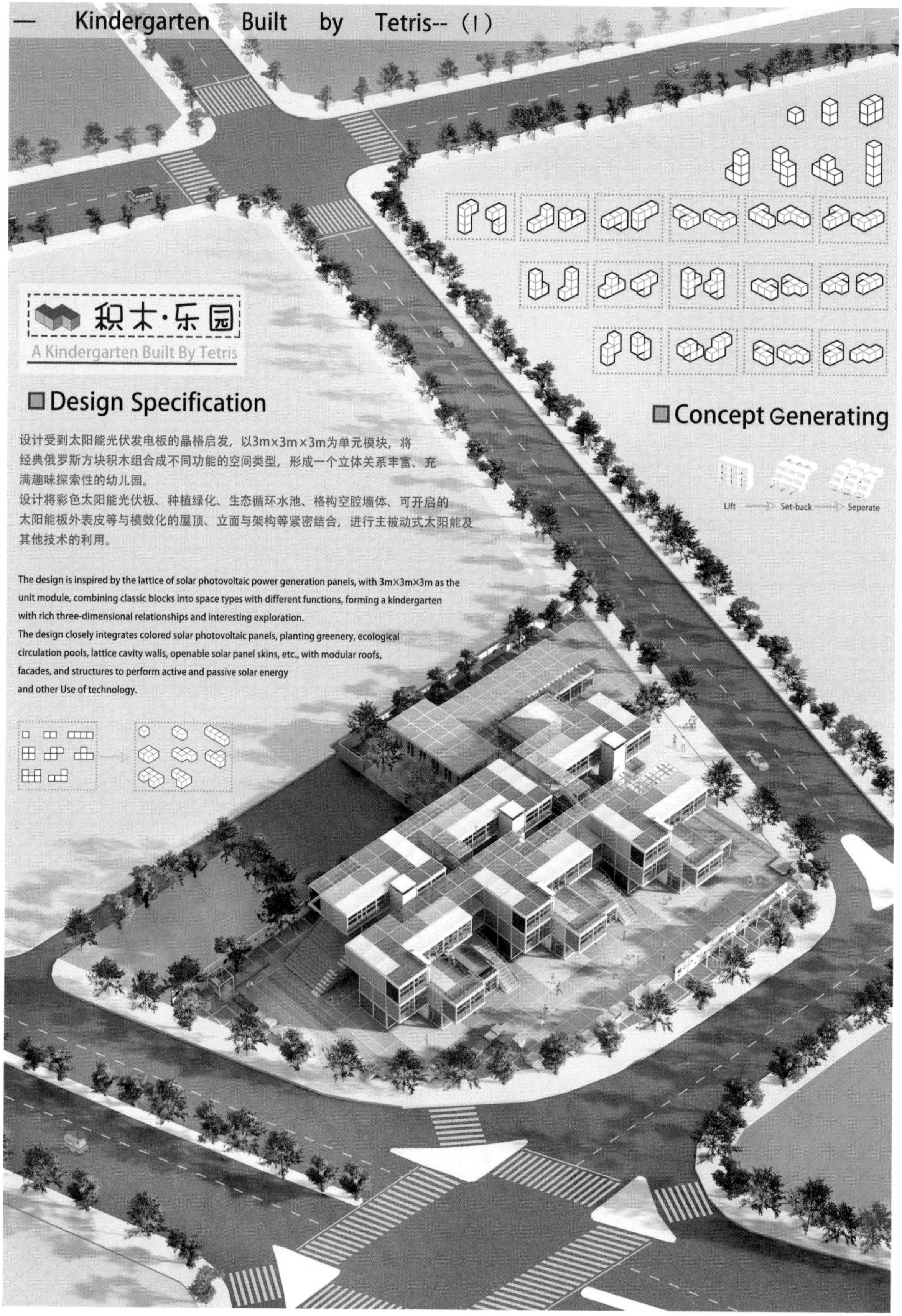
— Kindergarten Built by Tetris-- (I)
积木·乐园
A Kindergarten Built By Tetris
Design Specification
设计受到太阳能光伏发电板的晶格启发，以3m×3m×3m为单元模块，将经典俄罗斯方块积木组合成不同功能的空间类型，形成一个立体关系丰富、充满趣味探索性的幼儿园。
设计将彩色太阳能光伏板、种植绿化、生态循环水池、格构空腔墙体、可开启的太阳能板外表皮等与模数化的屋顶、立面与架构等紧密结合，进行主被动式太阳能及其他技术的利用。
The design is inspired by the lattice of solar photovoltaic power generation panels, with 3m×3m×3m as the unit module, combining classic blocks into space types with different functions, forming a kindergarten with rich three-dimensional relationships and interesting exploration.
The design closely integrates colored solar photovoltaic panels, planting greenery, ecological circulation pools, lattice cavity walls, openable solar panel skins, etc., with modular roofs, facades, and structures to perform active and passive solar energy and other Use of technology.
Concept Generating
Lift
Set-back
Seperate

积木 · 乐园（2） — Kindergarten Built by Tetris--

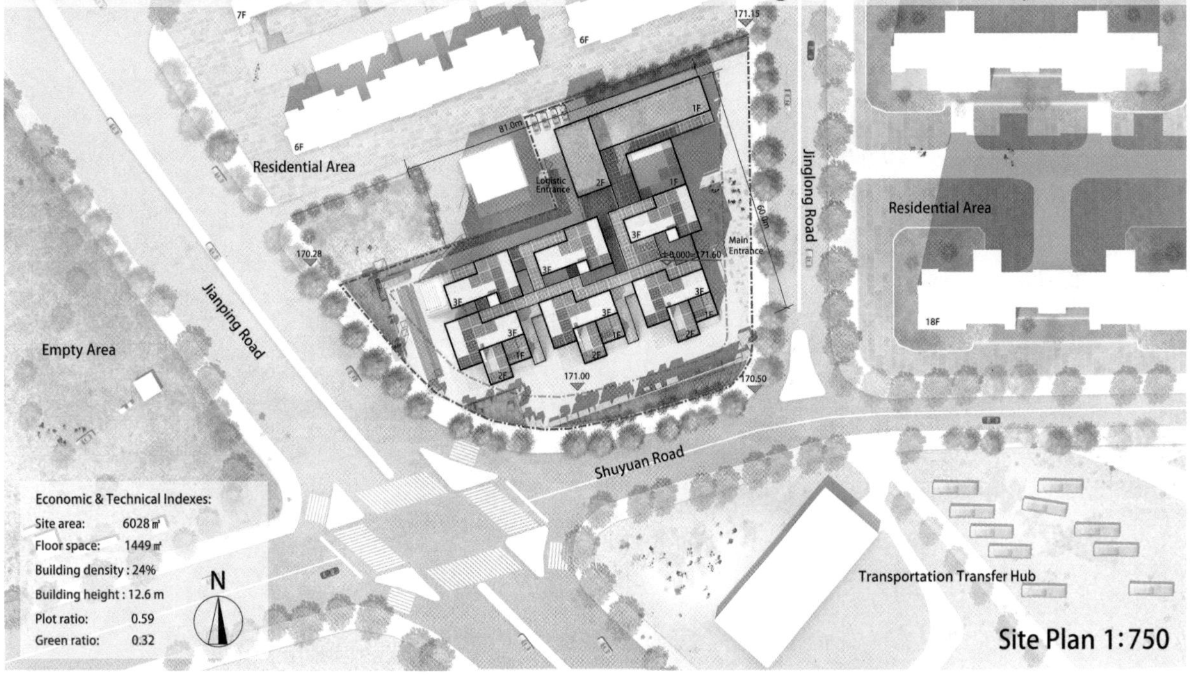
Residential Area
Residential Area
Jinglong Road
Jianping Road
Shuyuan Road
Empty Area
Transportation Transfer Hub
Logistic Entrance
Main Entrance
81.0m
60.0m
±0.000=171.60
171.15
170.28
171.00
170.50
7F
6F
6F
18F
1F
2F
3F
N
Economic & Technical Indexes:
Site area: 6028㎡
Floor space: 1449㎡
Building density : 24%
Building height : 12.6 m
Plot ratio: 0.59
Green ratio: 0.32
Site Plan 1:750

Site Analysis

The base is located in the center of the city, with convenient transportation, and the land is high in the northeast and low in the southwest. There are no existing buildings and trees that need to be preserved in the site, and the site is relatively open. Jianyang District of Nanping City is located on the southeast side of the northern section of the Wuyi Mountains. It has a subtropical monsoon climate, rich in light and heat resources, short winters and long summers, pleasant climate, quiet winds, and concentrated rainy seasons. The annual average temperature is 18.1℃, the frost-free period is 322 days, and the average annual rainfall The volume is 1742 mm, the annual sunshine averages 1802 hours, and the average annual relative humidity is 82%.

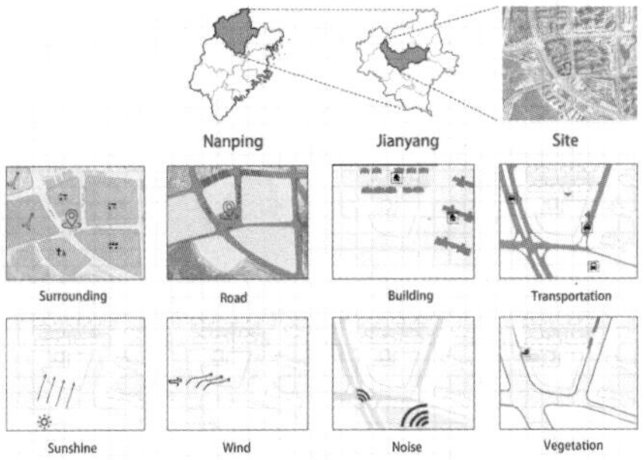
Nanping
Jianyang
Site
Surrounding
Road
Building
Transportation
Sunshine
Wind
Noise
Vegetation

Diagram of Design Process

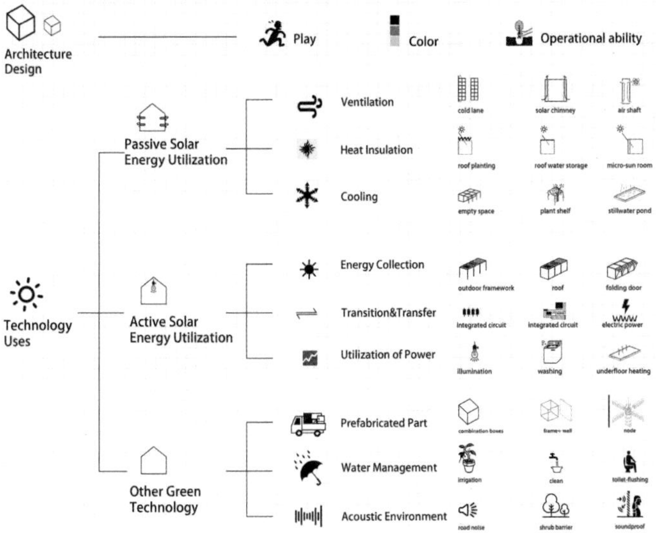
Architecture Design
Play
Color
Operational ability
Technology Uses
Passive Solar Energy Utilization
Ventilation
cold lane
solar chimney
air shaft
Heat Insulation
roof planting
roof water storage
micro-sun room
Cooling
empty space
plant shelf
stillwater pond
Active Solar Energy Utilization
Energy Collection
outdoor framework
roof
folding door
Transition&Transfer
integrated circuit
integrated circuit
electric power
Utilization of Power
illumination
washing
underfloor heating
Other Green Technology
Prefabricated Part
combination boxes
frame+ wall
node
Water Management
irrigation
clean
toilet flushing
Acoustic Environment
road noise
shrub barrier
soundproof

Climatic Simulation

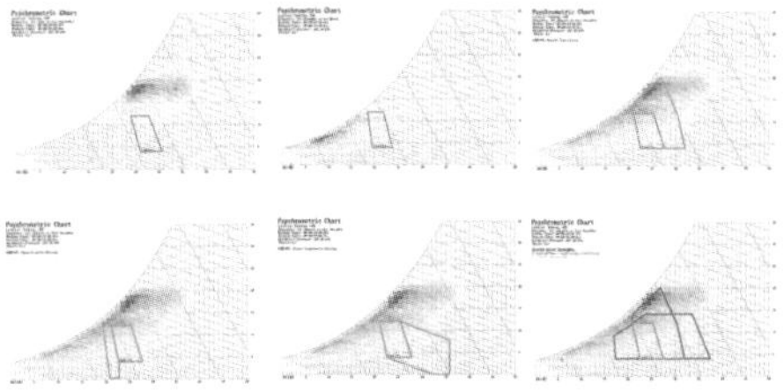

Through the simulation of different passive building strategies, it is concluded that thesuitable passive strategies for this area are: evaporation + natural ventilation; passive solar heating

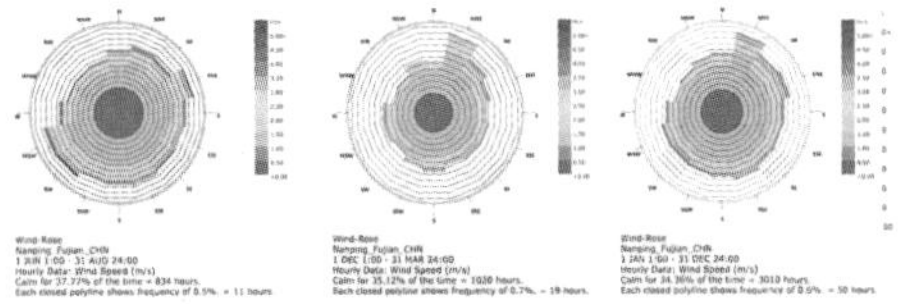

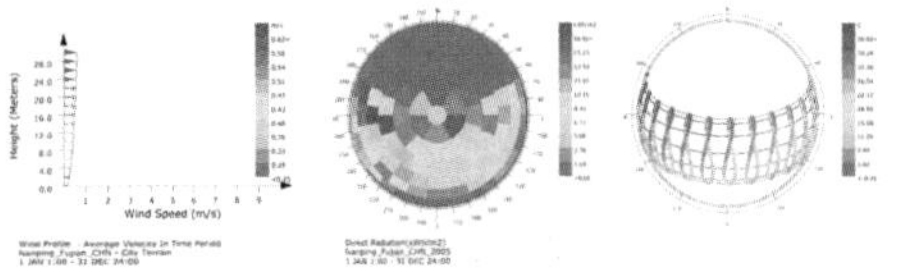
Wind Speed (m/s)

According to software analysis and simulation, the dominant wind direction and annual solar radiation situation in Nanping City, Xiamen are obtained, so as to arrange the kindergarten class and logistics service block

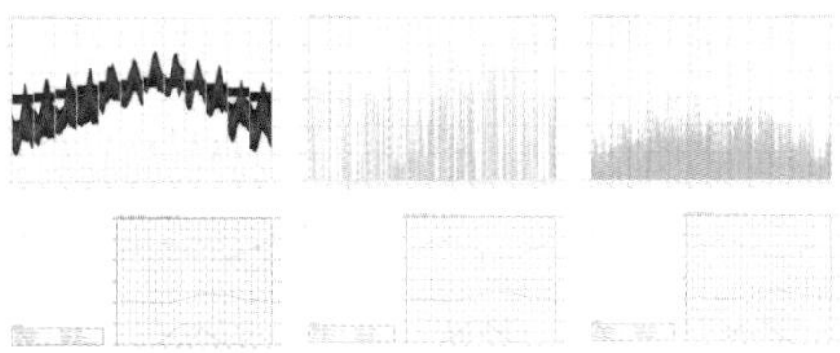

According to the annual average daily temperature and direct/indirect solar radiation conditions, consider the distribution of solar panels

积木·乐园（?） — Kindergarten Built by Tetris--

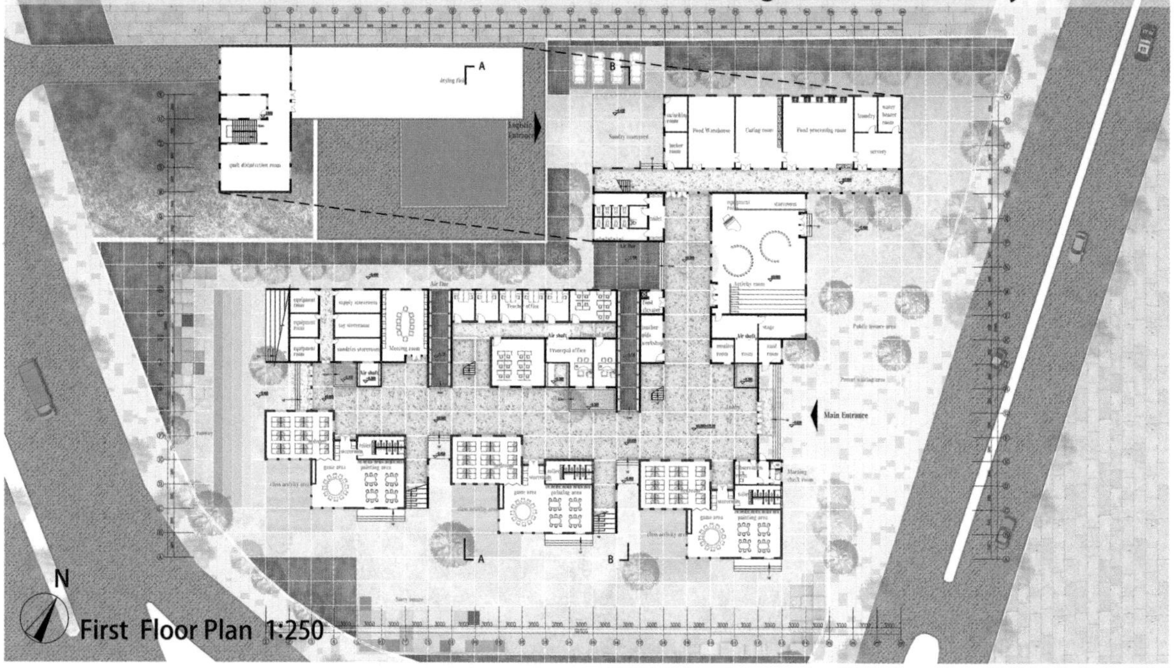
Main Entrance
N
First Floor Plan 1:250

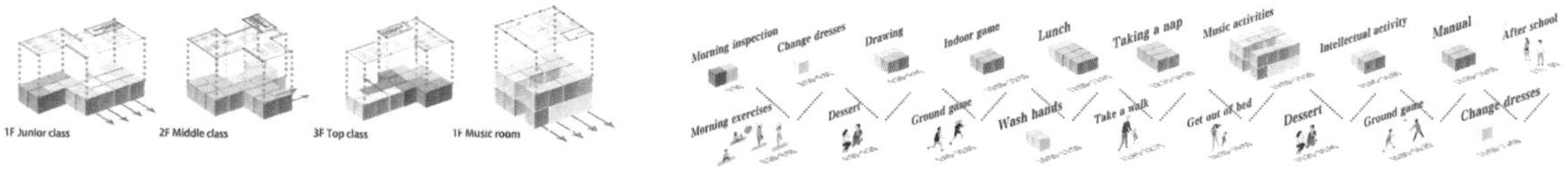
1F Junior class
2F Middle class
3F Top class
1F Music room
Morning inspection
Change dresses
Drawing
Indoor game
Lunch
Taking a nap
Music activities
Intellectual activity
Manual
After school
Morning exercises
Dessert
Ground game
Wash hands
Take a walk
Get out of bed
Dessert
Ground game
Change dresses

Function Streamline

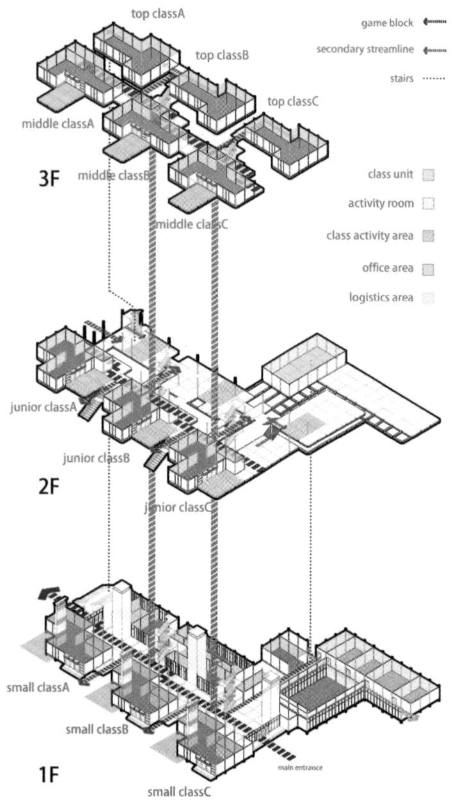
top classA
top classB
top classC
middle classA
middle classB
middle classC
3F
junior classA
junior classB
junior classC
2F
small classA
small classB
small classC
1F
main entrance
game block
secondary streamline
stairs
class unit
activity room
class activity area
office area
logistics area

Second Floor Plan 1：250

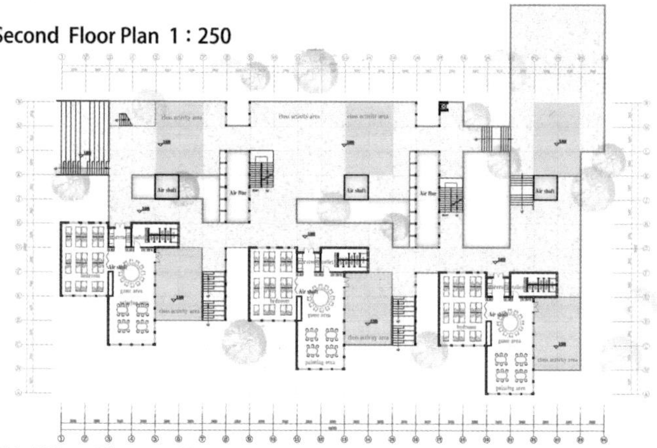

Third Floor Plan 1：250

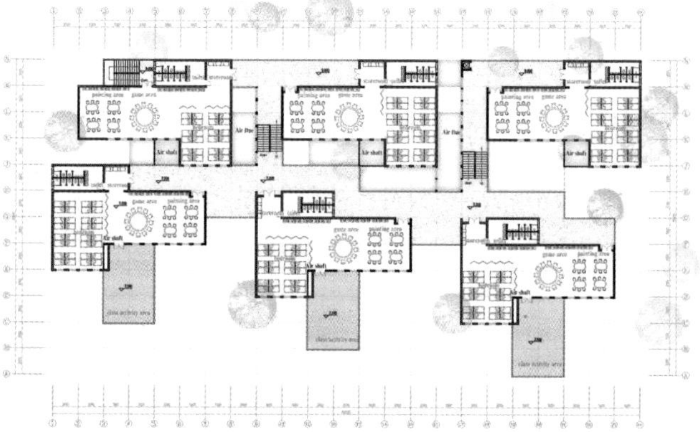

积木·乐园（4） — Kindergarten Built by Tetris--

Active Solar Technology

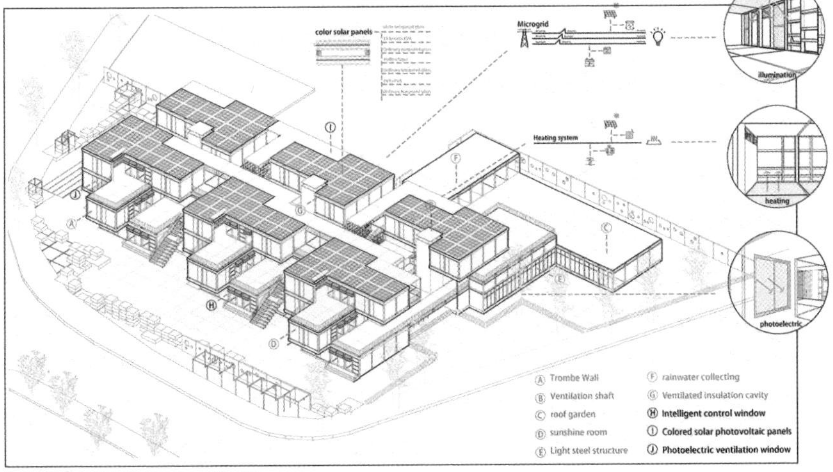
color solar panels
Microgrid
illumination
Heating system
heating
photoelectric
Ⓐ Trombe Wall
Ⓑ Ventilation shaft
Ⓒ roof garden
Ⓓ sunshine room
Ⓔ Light steel structure
Ⓕ rainwater collecting
Ⓖ Ventilated insulation cavity
Ⓗ Intelligent control window
Ⓘ Colored solar photovoltaic panels
Ⓙ Photoelectric ventilation window

Energy Calculation

Kindergarten power consumption

According to the electricity consumption index of civil buildings, the kindergarten′ s hourly electricity consu mption: 20W/m² （1h）

Based on an area of 3000 square meters and an average electricity consumption of 8 hours, the kindergarten's daily electricity consumption:

3000m²×20w/m²×9h=540000w/h=540kw/h

Solar panel power generation

The 1KW module has effective sunshine for 6 hours, con sidering the 30% loss du generate 4.2 electricity per day

Solar panel power generation a day:

The area (roof) required for 1kw solar panels to generate electricity is about 6.5m²

918m²÷6.8m²×4.2kw/h=567kw/h>540kw/h

Conclusion:

The current power generation capacity of solar panels can basically meet the electricity demand of kindergartens

Plant Footprint

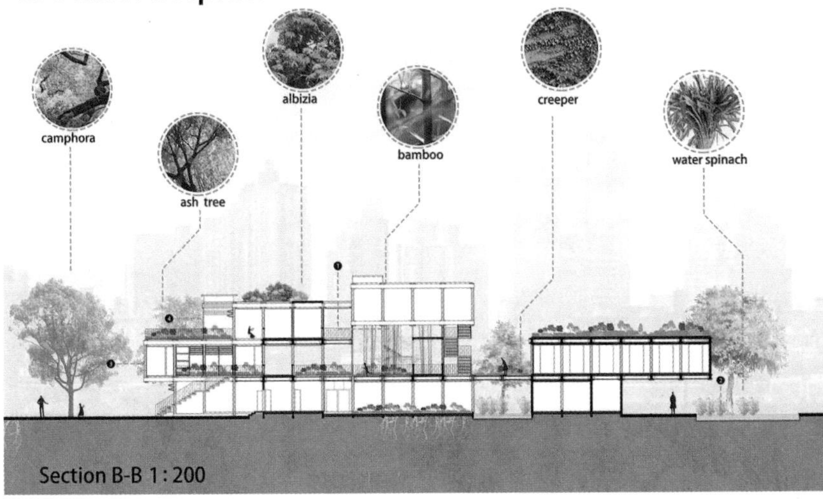
camphora
ash tree
albizia
bamboo
creeper
water spinach
Section B-B 1 : 200

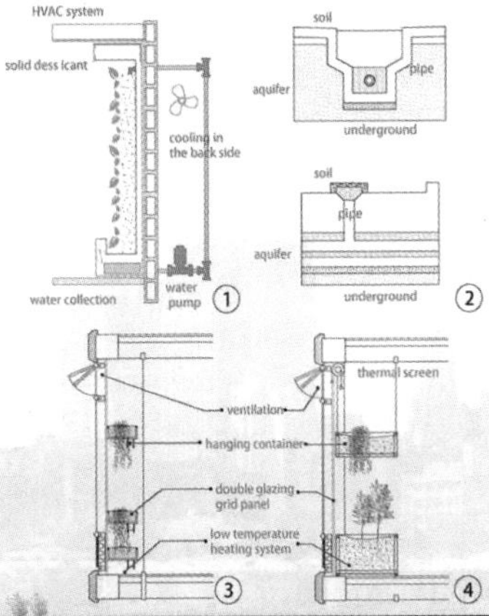
HVAC system
solid dess icant
cooling in the back side
water collection
water pump
soil
pipe
aquifer
underground
ventilation
hanging container
double glazing grid panel
low temperature heating system
thermal screen

积木·乐园（5） — Kindergarten Built by Tetris--

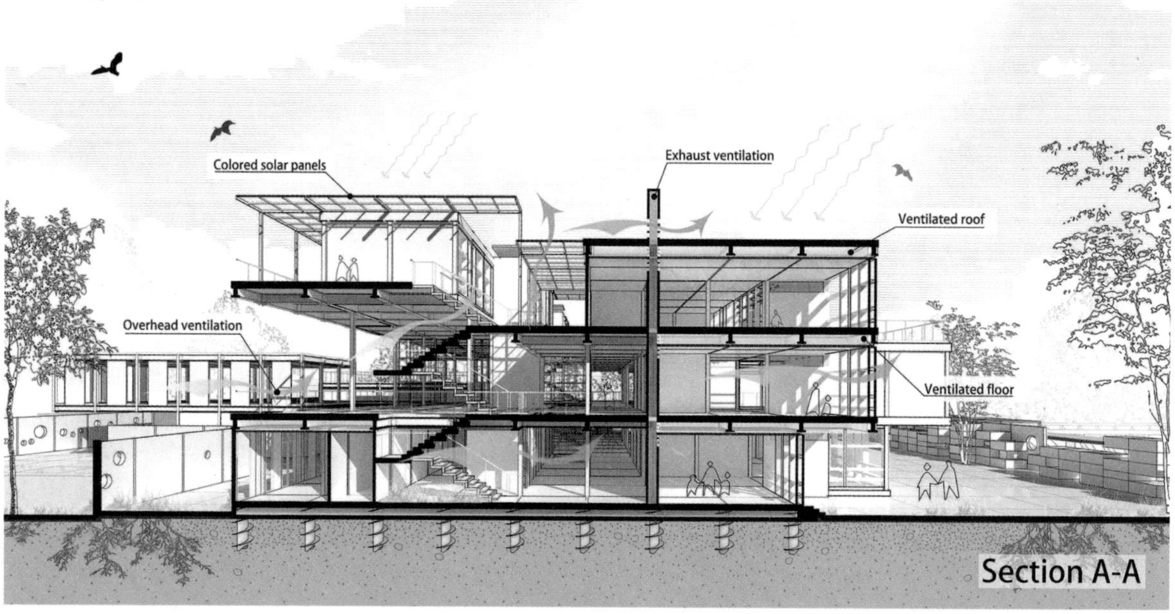
Colored solar panels
Exhaust ventilation
Ventilated roof
Overhead ventilation
Ventilated floor
Section A-A

Wind Shaft Details

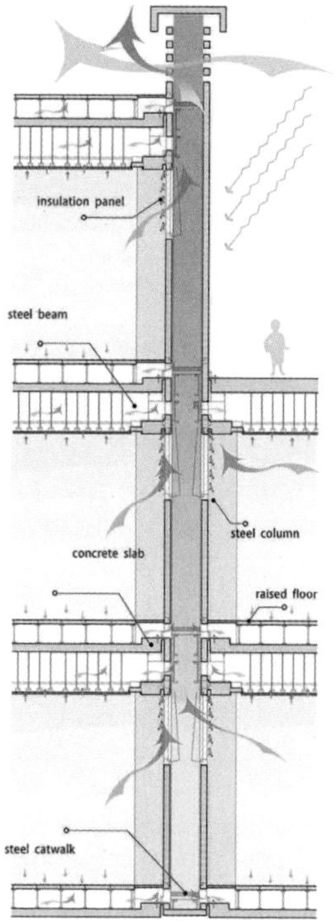
insulation panel
steel beam
steel column
concrete slab
raised floor
steel catwalk

Wind Simulation

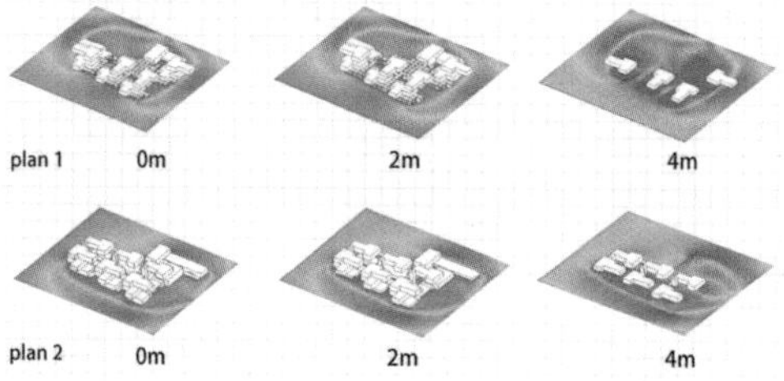
plan 1
0m
2m
4m
plan 2
0m
2m
4m

Sunshine Simulation

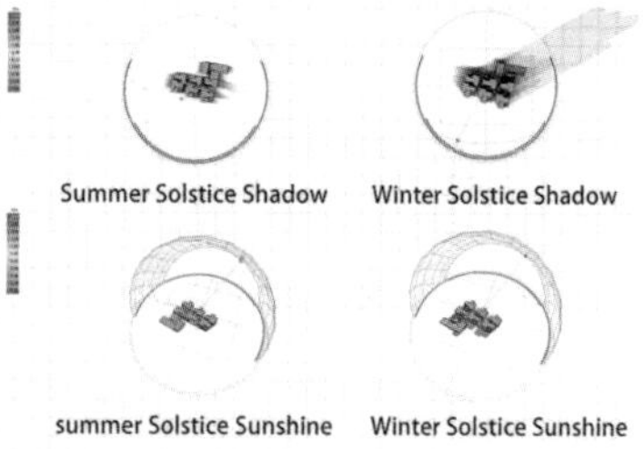
Summer Solstice Shadow
Winter Solstice Shadow
summer Solstice Sunshine
Winter Solstice Sunshine

Section Strategy

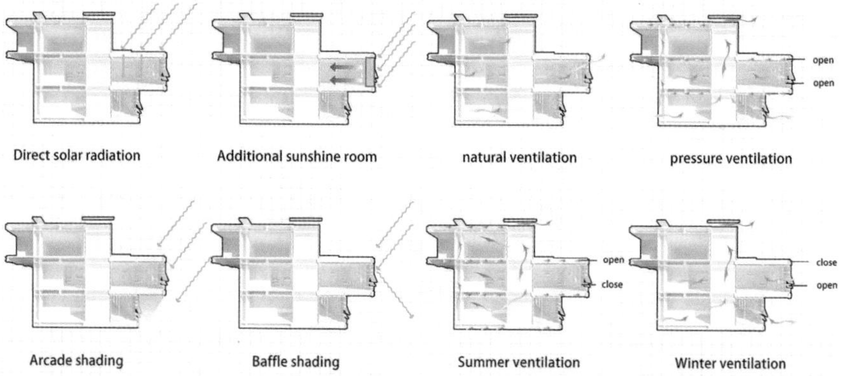
open
open
open
close
close
open
Direct solar radiation
Additional sunshine room
natural ventilation
pressure ventilation
Arcade shading
Baffle shading
Summer ventilation
Winter ventilation

积木 · 乐园（6）— Kindergarten Built by Tetris--

Model Photos

3m×3m×3m blocks

A combination of Tetris

Built on Stilts

Solar Absorption and Indoor Lighting

Natural Ventilation

Note Details & Materials

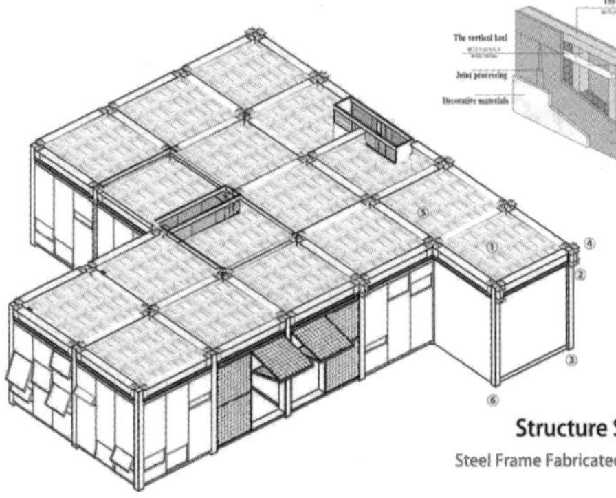

Structure System :

Steel Frame Fabricated Structure

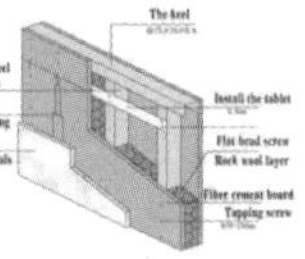

The keel
The vertical keel
Joint processing
Decorative materials
Install the tablet
Flat head screw
Rock wool layer
Fiber cement board
Tapping screw

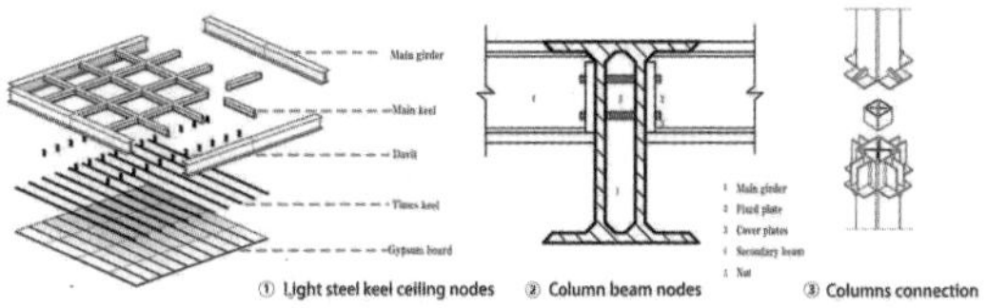

Main girder
Main keel
Davit
Times keel
Gypsum board

① Light steel keel ceiling nodes ② Column beam nodes ③ Columns connection

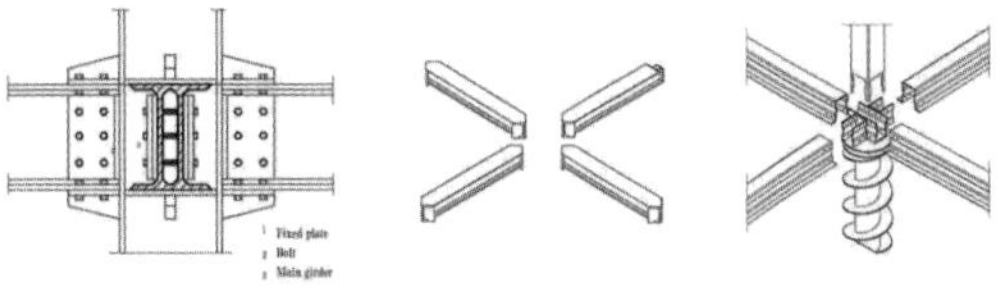

④ Primary and secondary beam connection ⑤ Secondary beams connection ⑥ Joint of column, floor beam and pile foundation

Concrete Steel Maple Oak Plastic

Water Footprint

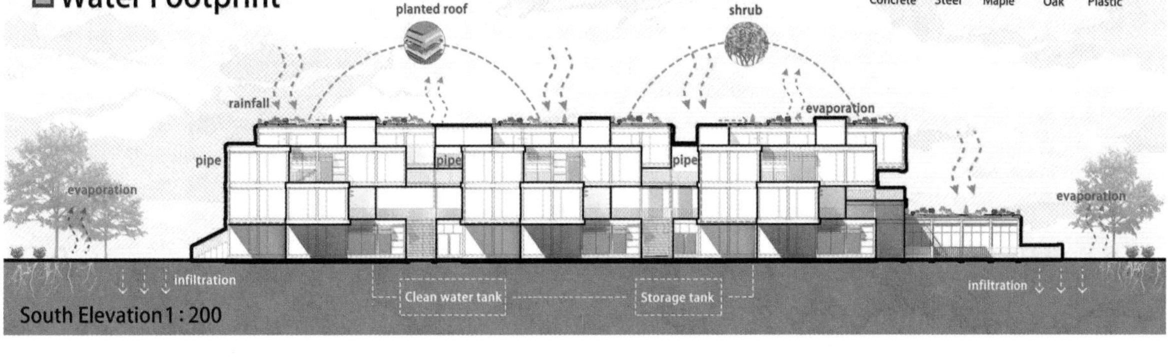

planted roof
shrub
rainfall
evaporation
pipe
pipe
pipe
evaporation
evaporation
infiltration
Clean water tank
Storage tank
infiltration

South Elevation 1 : 200

综合奖 · 一等奖
General Prize Awarded · Frist Prize

注 册 号： 7063

项目名称： 沐日（新疆）
Step into the Sunshine
（Xinjiang）

作　　者： 吴昊天、徐建军、文冬霞

参赛单位： 昆明理工大学

指导教师： 谭良斌

专家点评：

该作品的建筑朝向采用与场地成一定角度的布置方式，有利于对太阳能的充分吸收利用，建筑布局简洁清晰，建筑平面紧凑，功能分区合理，通风坪和暖房的组合使建筑有地域的特色。将建筑的肌理以挡风墙的形式进行延续，实现形式语言的创新，对被动式太阳能的应用合理且有新意。

The certain angle of the building orientation to the site is conducive to the full absorption and utilization of solar energy. The work, concise and clear in layout, compact in building plan, and reasonable in functional zoning, presents the building's characteristics suitable for the region through the combination of ventilating flats and green houses. The texture of the building is continued in the form of a wind-shield wall to innovate the language of forms. Moreover, the work is reasonable and creative in the application of passive solar energy.

沐日 Step into the Sunshine
新疆和静乡村幼儿园设计 01
本次设计从地域气候及文化出发，旨在设计出符合地域特色的绿色舒适性建筑。在对气候回应方面，本方案以充分运用当地丰富的太阳辐射资源来解决冬季热舒适问题为出发点，通过对建筑形态控制以获得更多的有利朝向面，并通过设置阳光间、特朗勃墙、蓄热体等被动式策略来提升室内舒适性。在主动式运用方面，通过计算得出的最佳的太阳能光伏板角度，并通过调整建筑形态以达到太阳能与建筑设计的一体化，并通过主动式与被动式结合的方式，保障建筑在冬季使用时的舒适性。
在对地域性及场地回应中，通过提取当地民居原型，并通过插入墙的语言对其进行变形和组合，形成自然的地域建筑形态，与周边草场与远处山景形成对景。在对地域材料的考虑中，充分利用当地丰富的土和石资源，发挥地域材料自身优秀的物理性能运用在建筑的地域表达和建造技术之中。
同时，通过对幼儿及家长心理及行为的分析和观察，在建筑设计中融入趣味性空间和符合儿童行为的室内活动空间，满足不同季节、场所、时间的幼儿活动需求，最终以达到舒适、绿色、趣味的设计目标。
This design starts from the regional climate and culture, aiming to design a green and comfortable building in line with the regional characteristics. In terms of climate response, the scheme starts from making full use of the local abundant solar radiation resources to solve the problem of thermal comfort in winter. By controlling the building form, more favorable orientation surfaces can be obtained, and passive strategies such as sunshine space, Trumbo wall and heat storage body can be set up to improve indoor comfort. In terms of active application, the optimal solar photovoltaic panel Angle is calculated, the integration of solar energy and architectural design is achieved by adjusting the building form, and the comfort of the building in winter is guaranteed through the combination of active and passive.
In response to the regionality and site, the prototype of local folk houses was extracted, and the language inserted into the wall was used to transform and combine them, so as to form a natural regional architectural form, which was opposite to the surrounding grassland and distant mountain scenery. In the consideration of regional materials, the rich local soil and stone resources should be fully utilized, and the excellent physical properties of regional materials should be given play to in the regional expression and construction technology of architecture.
At the same time, through the analysis and observation of the psychology and behavior of children and their parents, interesting space and indoor activity space conforming to children's behavior are integrated into the architectural design to meet children's activity needs in different seasons, places and times. Finally in order to achieve comfortable, green, fun design goals.
Location Situation
新疆/Xinjiang
新疆/Xinjiang
巴音郭楞蒙古自治州
Bayingolin Mongol Autonomous Prefecture
和静县/Hejing County
amount of solar radiation
Climate Analysis
The location of the project is cold and long in winter and cooler in summer.The heat preservation and heating of buildings in winter is an important problem to be solved.
As can be seen from the figure above, the wind speed throughout the year is relatively moderate Relatively rich resources, can make full use of wind power generation and other energy The source added.
The annual precipitation of the project location is only 111-289mm mainly from June to August, so it is necessary to make full use of local water resources and adopt the reclaimed water treatment system.
Optimum orientation of building
perspective

沐日 Step into the Sunshine
新疆和静乡村幼儿园设计 02

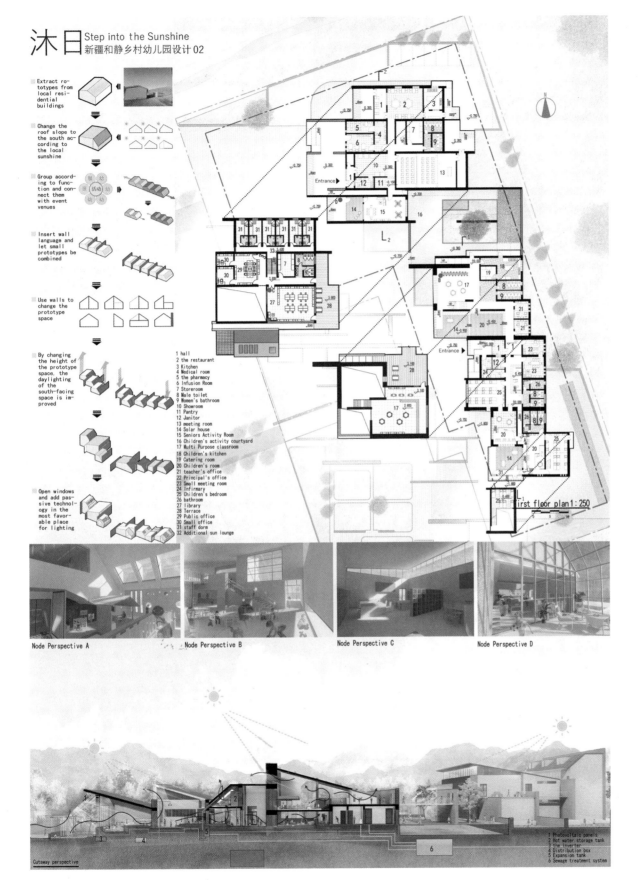

Extract rototypes from local residential buildings

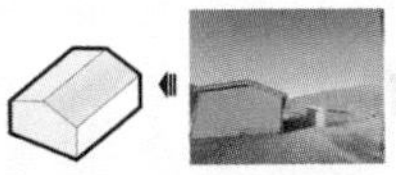

Change the roof slope to the south according to the local sunshine

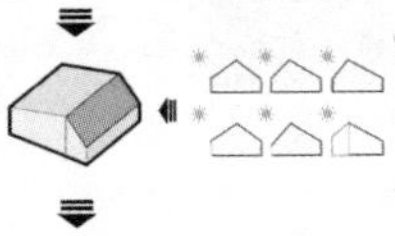

Group according to function and connect them with event venues

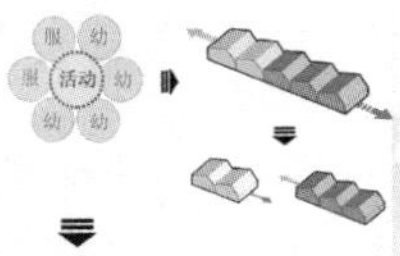

活动

Insert wall language and let small prototypes be combined

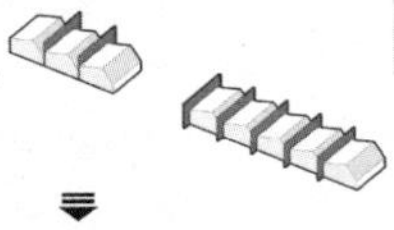

Use walls to change the prototype space

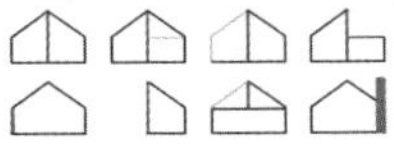

By changing the height of the prototype space, the daylighting of the south-facing space is improved

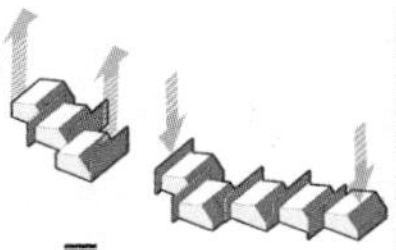

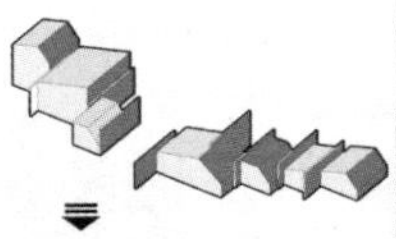

Open windows and add passive technology in the most favorable place for lighting

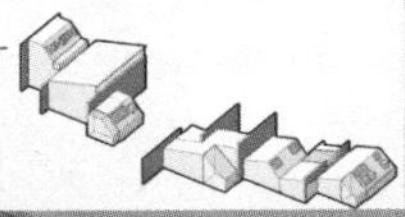

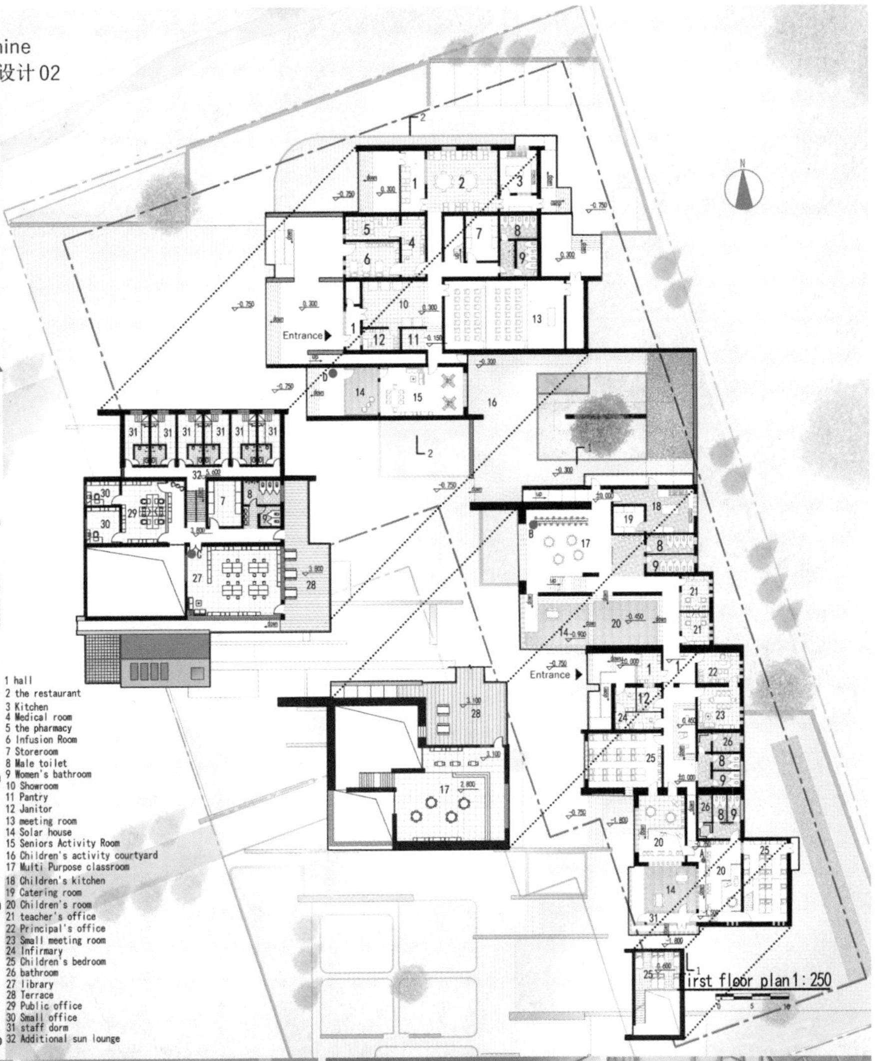

Entrance
Entrance
First floor plan 1: 250

1 hall
2 the restaurant
3 Kitchen
4 Medical room
5 the pharmacy
6 Infusion Room
7 Storeroom
8 Male toilet
9 Women's bathroom
10 Showroom
11 Pantry
12 Janitor
13 meeting room
14 Solar house
15 Seniors Activity Room
16 Children's activity courtyard
17 Multi Purpose classroom
18 Children's kitchen
19 Catering room
20 Children's room
21 teacher's office
22 Principal's office
23 Small meeting room
24 Infirmary
25 Children's bedroom
26 bathroom
27 library
28 Terrace
29 Public office
30 Small office
31 staff dorm
32 Additional sun lounge

Node Perspective A

Node Perspective B

Node Perspective C

Node Perspective D

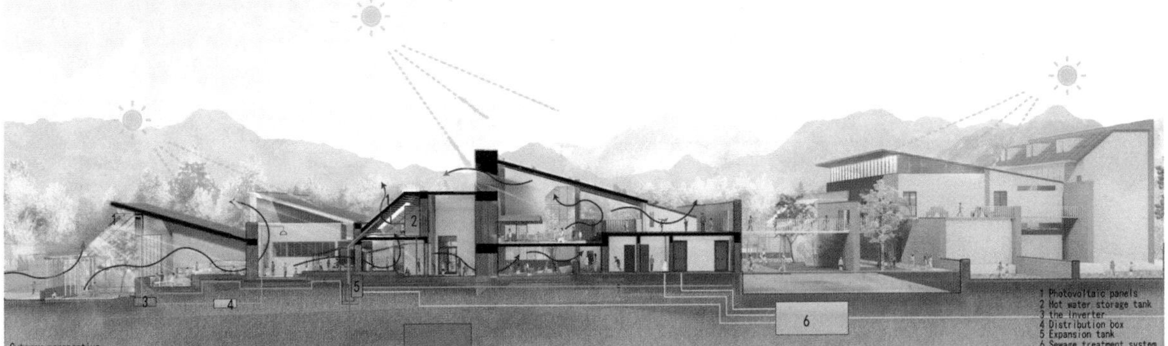

1 Photovoltaic panels
2 Hot water storage tank
3 the inverter
4 Distribution box
5 Expansion tank
6 Sewage treatment system

Cutaway perspective

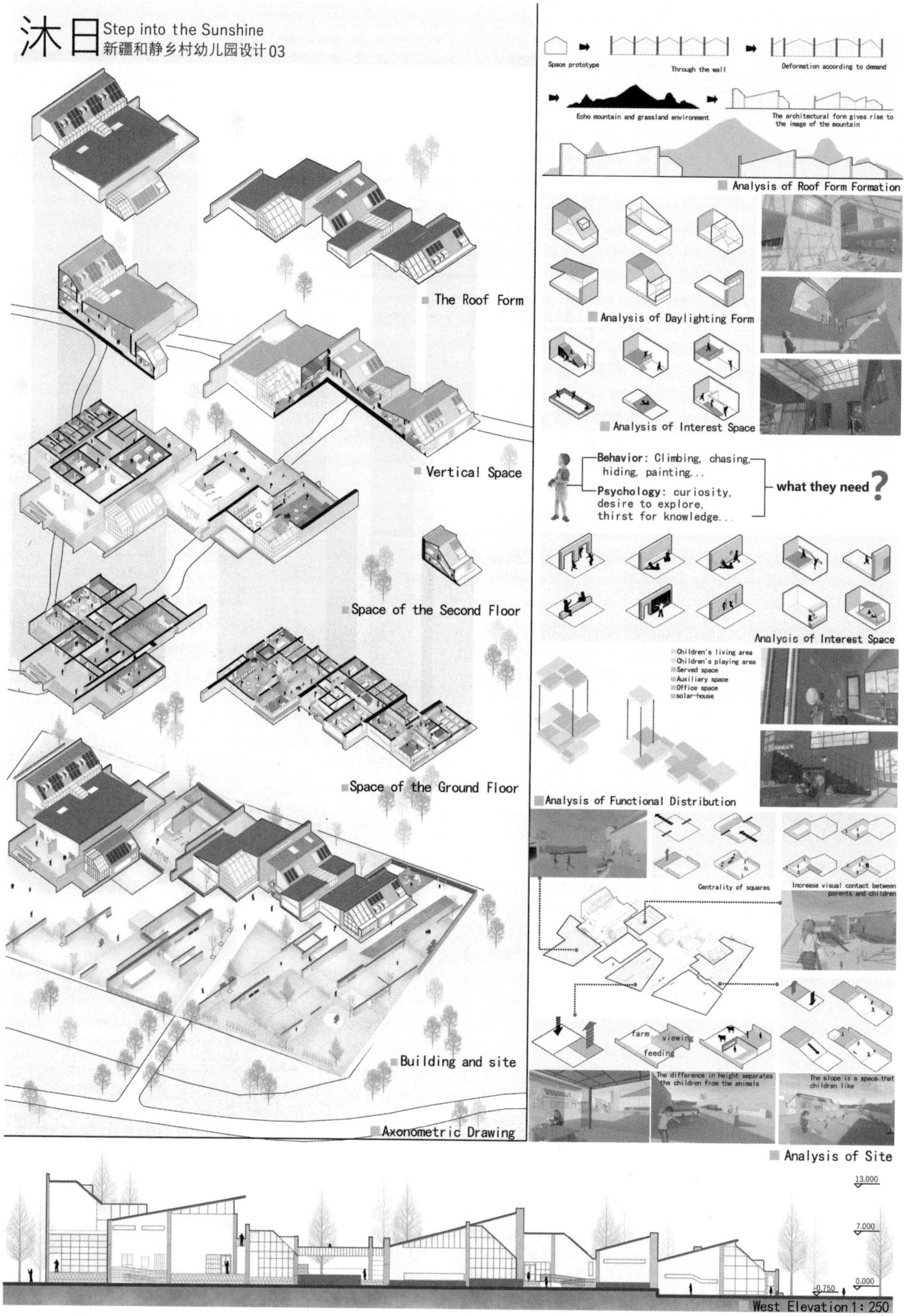

沐日 Step into the Sunshine
新疆和静乡村幼儿园设计03
The Roof Form
Vertical Space
Space of the Second Floor
Space of the Ground Floor
Building and site
Axonometric Drawing
Space prototype
Through the wall
Deformation according to demand
Echo mountain and grassland environment
The architectural form gives rise to the image of the mountain
Analysis of Roof Form Formation
Analysis of Daylighting Form
Analysis of Interest Space
Behavior: Climbing, chasing, hiding, painting...
Psychology: curiosity, desire to explore, thirst for knowledge...
what they need?
Analysis of Interest Space
Children's living area
Children's playing area
Served space
Auxiliary space
Office space
solar-house
Analysis of Functional Distribution
Centrality of squares
Increase visual contact between parents and children
farm
viewing
feeding
The difference in height separates the children from the animals
The slope is a space that children like
Analysis of Site
13.000
7.000
0.000
-0.750
West Elevation 1: 250

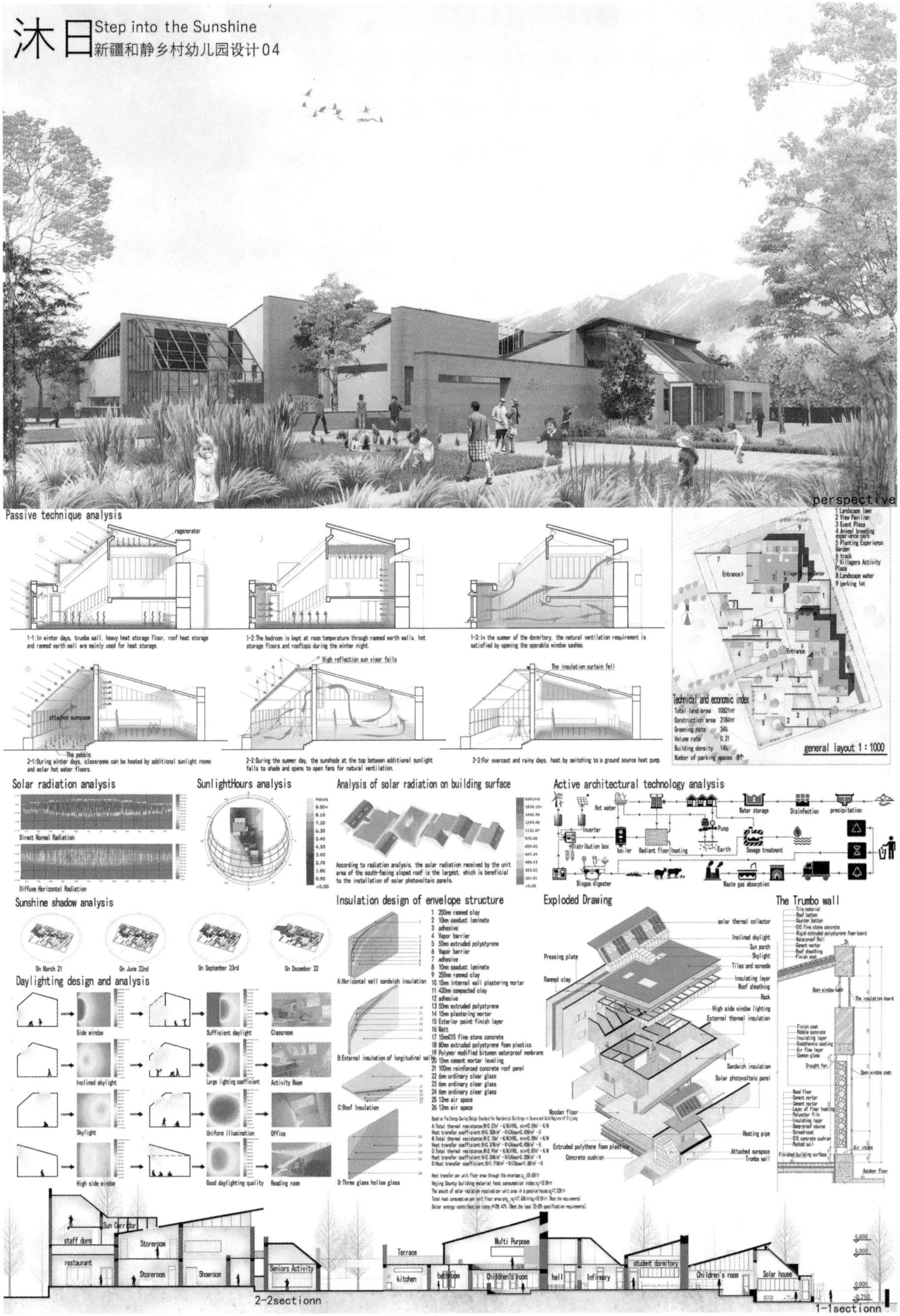
沐日 Step into the Sunshine
新疆和静乡村幼儿园设计 04
perspective
Passive technique analysis
general layout 1:1000
Technical and economic index
Solar radiation analysis
SunlightHours analysis
Analysis of solar radiation on building surface
Active architectural technology analysis
Sunshine shadow analysis
On March 21
On June 22nd
On September 23rd
On December 22
Daylighting design and analysis
Insulation design of envelope structure
Exploded Drawing
The Trumbo wall
2-2sectionn
1-1sectionn

综合奖 · 二等奖
General Prize Awarded · Second Prize

注 册 号：6900
项目名称：追光者（南平）
Sun Chaser（Nanping）
作　　者：肖正天、王　琦、王亚鑫、高鸣飞
参赛单位：天津大学、华北水利水电大学
指导教师：冯　刚、杨　崴

专家点评：

作品吸纳福建土楼的传统建筑形式，以弧形应对场地四周的城市道路，空间友好度高，场地流线清晰，各功能空间相互独立。建筑内部空间丰富、旋转坡道的设计富有童趣。对动态表皮的设计富有创意，且进行了细致的分析。但是在被动式太阳能利用的方面略显不足。

The work adopts the traditional architectural form of Fujian earth buildings, united with urban roads around the site in an arc shape, thus creating a large space, clear site streamline and independent functional spaces. The building interior is spacious, and the design of rotated ramps is full of childlike interest. The work shows creative design of the dynamic surface and detailed analysis on it. However, it's slightly insufficient in passive solar energy utilization.

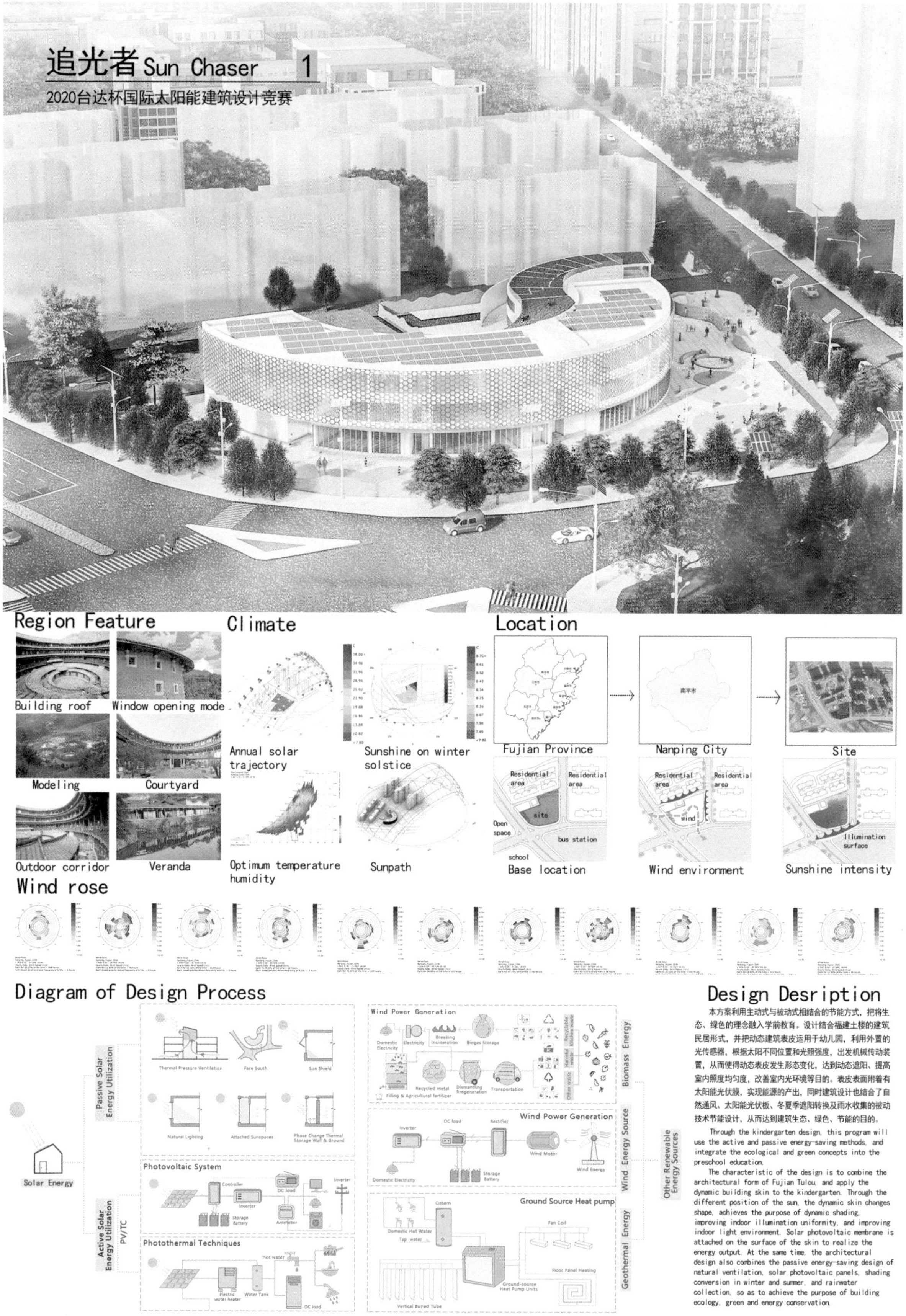
追光者 Sun Chaser 1
2020台达杯国际太阳能建筑设计竞赛
Region Feature
Building roof
Window opening mode
Modeling
Courtyard
Outdoor corridor
Veranda
Climate
Annual solar trajectory
Sunshine on winter solstice
Optimum temperature humidity
Sunpath
Location
Fujian Province
Nanping City
Site
Base location
Wind environment
Sunshine intensity
Wind rose
Diagram of Design Process
Solar Energy
Passive Solar Energy Utilization
Active Solar Energy Utilization
Photovoltaic System
Photothermal Techniques
Wind Power Generation
Ground Source Heat pump
Biomass Energy
Wind Energy Source
Geothermal Energy
Other Renewable Energy Sources
Design Desription
本方案利用主动式与被动式相结合的节能方式，把将生态、绿色的理念融入学前教育。设计结合福建土楼的建筑民居形式，并把动态建筑表皮运用于幼儿园，利用外置的光传感器，根据太阳不同位置和光照强度，出发机械传动装置，从而使得动态表皮发生形态变化，达到动态遮阳、提高室内照度均匀度，改善室内光环境等目的。表皮表面附着有太阳能光伏膜，实现能源的产出，同时建筑设计也结合了自然通风、太阳能光伏板、冬夏季遮阳转换及雨水收集的被动技术节能设计，从而达到建筑生态、绿色、节能的目的。
Through the kindergarten design, this program will use the active and passive energy-saving methods, and integrate the ecological and green concepts into the preschool education.
The characteristic of the design is to combine the architectural form of Fujian Tulou, and apply the dynamic building skin to the kindergarten. Through the different position of the sun, the dynamic skin changes shape, achieves the purpose of dynamic shading, improving indoor illumination uniformity, and improving indoor light environment. Solar photovoltaic membrane is attached on the surface of the skin to realize the energy output. At the same time, the architectural design also combines the passive energy-saving design of natural ventilation, solar photovoltaic panels, shading conversion in winter and summer, and rainwater collection, so as to achieve the purpose of building ecology, green and energy conservation.

追光者 Sun Chaser 2

2020台达杯国际太阳能建筑设计竞赛

Logical generation

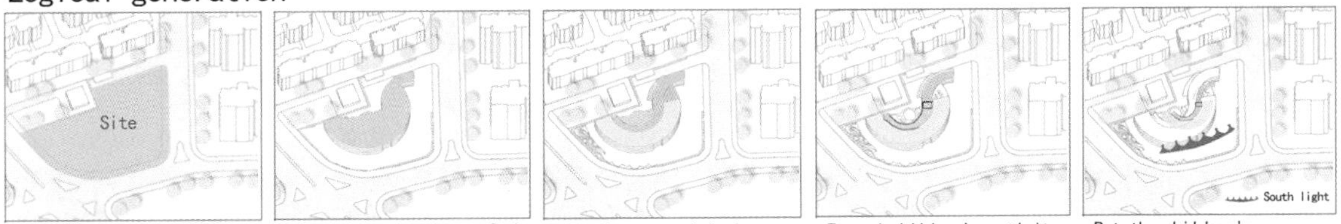
Site
South light

The base area is 6028 m², close to the residential area.

According to the shape of the site, the shape of the building is designed as curve.

In order to enrich the architectural form and meet the functional requirements.

Expand children's activity area and make the building platform interesting.

Put the children's activities and activities in the south.

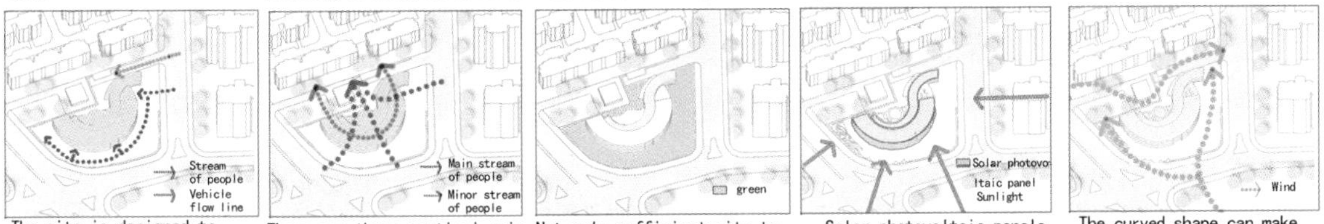
Stream of people
Vehicle flow line
Main stream of people
Minor stream of people
green
Solar photovoltaic panel
Sunlight
Wind

The site is designed to separate people and vehicles.

There are three vertical axis of pedestrian flow, one is along the architectural .

Not only sufficient site to increase the green area, but also designed the roof greening.

Solar photovoltaic panels are installed on the roof and dynamic skin.

The curved shape can make the site ventilation more smooth.

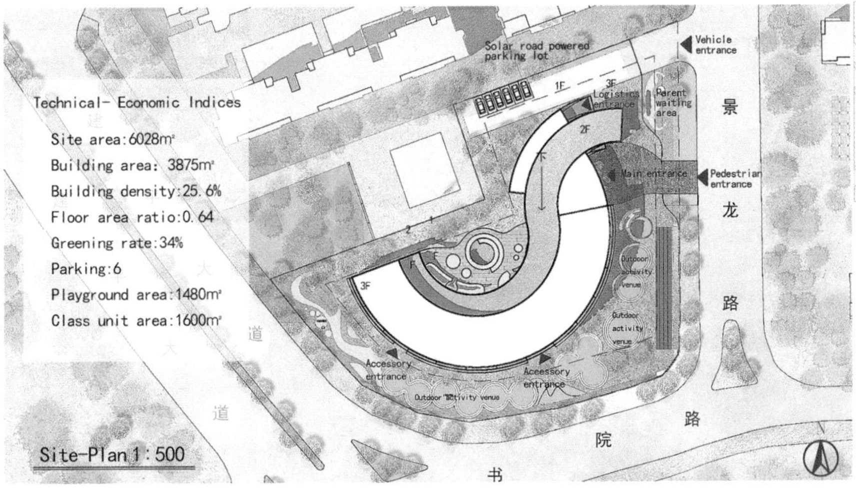
Technical- Economic Indices
Site area:6028m²
Building area: 3875m²
Building density:25.6%
Floor area ratio:0.64
Greening rate:34%
Parking:6
Playground area:1480m²
Class unit area:1600m²
Solar road powered parking lot
Vehicle entrance
Logistics entrance
Parent waiting area
Main entrance
Pedestrian entrance
Outdoor activity venue
Outdoor activity venue
Accessory entrance
Accessory entrance
Outdoor activity venue
1F
2F
3F
3F
景
龙
路
院
路
书
道
道

Site-Plan 1:500

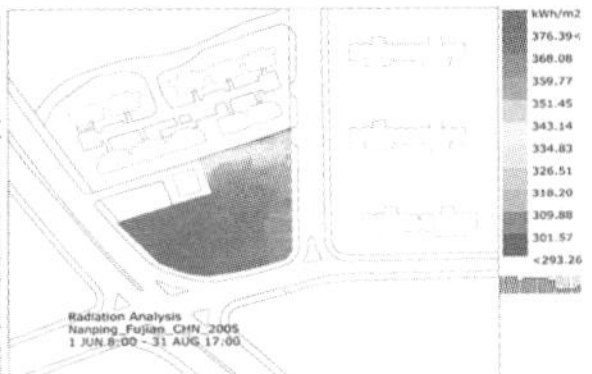
kWh/m2
Radiation Analysis
Nanping_Fujian_CHN_2005
1 JUN 8:00 - 31 AUG 17:00

Analysis of solar radiation in winter

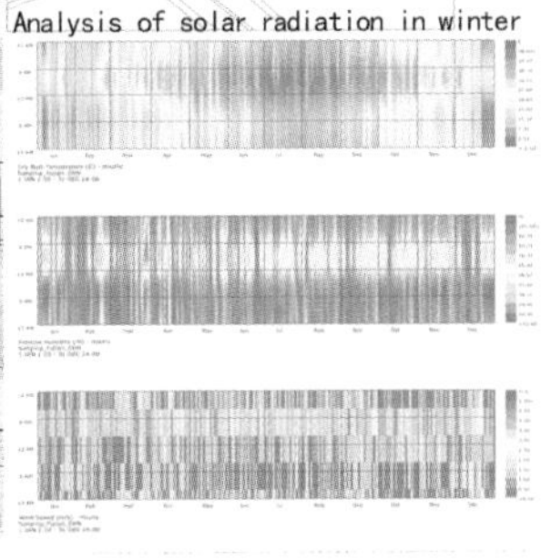

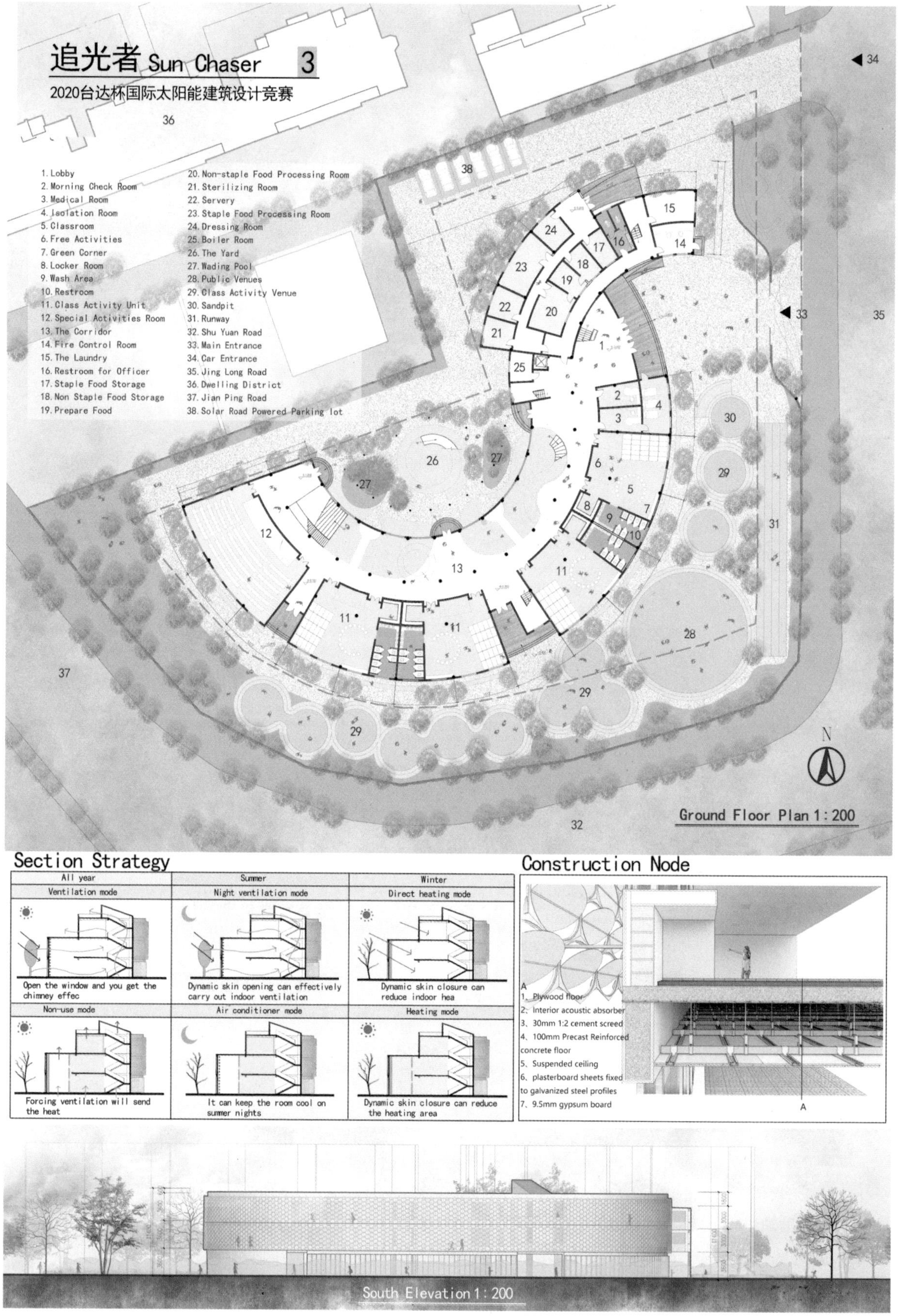

追光者 Sun Chaser 3
2020台达杯国际太阳能建筑设计竞赛
1. Lobby
2. Morning Check Room
3. Medical Room
4. Isolation Room
5. Classroom
6. Free Activities
7. Green Corner
8. Locker Room
9. Wash Area
10. Restroom
11. Class Activity Unit
12. Special Activities Room
13. The Corridor
14. Fire Control Room
15. The Laundry
16. Restroom for Officer
17. Staple Food Storage
18. Non Staple Food Storage
19. Prepare Food
20. Non-staple Food Processing Room
21. Sterilizing Room
22. Servery
23. Staple Food Processing Room
24. Dressing Room
25. Boiler Room
26. The Yard
27. Wading Pool
28. Public Venues
29. Class Activity Venue
30. Sandpit
31. Runway
32. Shu Yuan Road
33. Main Entrance
34. Car Entrance
35. Jing Long Road
36. Dwelling District
37. Jian Ping Road
38. Solar Road Powered Parking lot
Ground Floor Plan 1:200
Section Strategy
All year
Summer
Winter
Ventilation mode
Night ventilation mode
Direct heating mode
Open the window and you get the chimney effec
Dynamic skin opening can effectively carry out indoor ventilation
Dynamic skin closure can reduce indoor hea
Non-use mode
Air conditioner mode
Heating mode
Forcing ventilation will send the heat
It can keep the room cool on summer nights
Dynamic skin closure can reduce the heating area
Construction Node
1. Plywood floor
2. Interior acoustic absorber
3. 30mm 1:2 cement screed
4. 100mm Precast Reinforced concrete floor
5. Suspended ceiling
6. plasterboard sheets fixed to galvanized steel profiles
7. 9.5mm gypsum board
South Elevation 1:200

追光者 Sun Chaser 4

2020台达杯国际太阳能建筑设计竞赛

Floor Plan 1 : 200

5. Classroom
6. Free Activities
7. Green Corner
8. Locker Room
9. Wash Area
10. Restroom
11. Class Activity Unit
12. Special Activities Room
13. The Corridor
16. Restroom for Officer
25. Boiler Room
38. Outdoor Platform
39. Teaching AIDS workshop
40. Storeroom
41. Director Room
42. Accounting Office
43. Library
44. Staircase

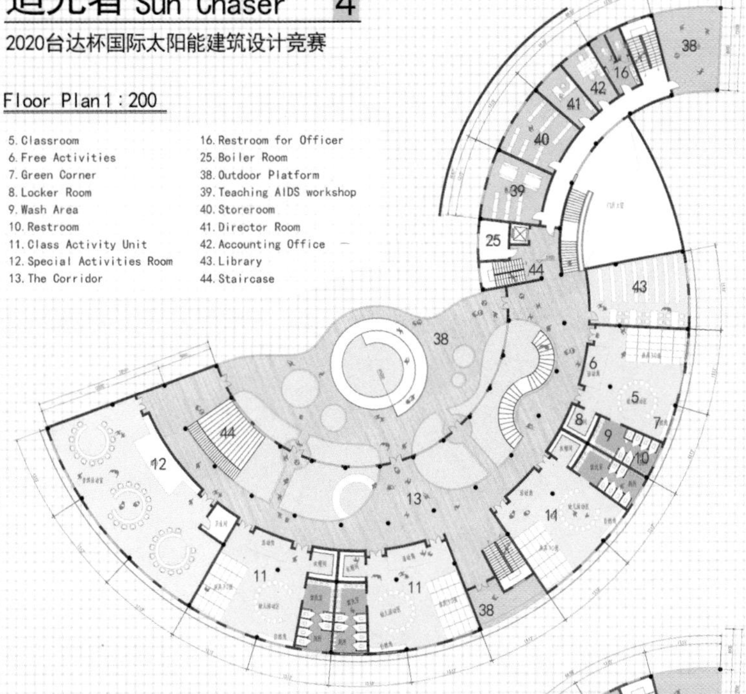

Three-floor Plan 1 : 200

5. Classroom
6. Free Activities
7. Green Corner
8. Locker Room
9. Wash Area
10. Restroom
11. Class Activity Unit
12. Special Activities Room
13. The Corridor
16. Restroom for Officer
25. Boiler Room
38. Outdoor Platform
43. Library
44. Staircase
45. Roofdeck
46. Administration Office

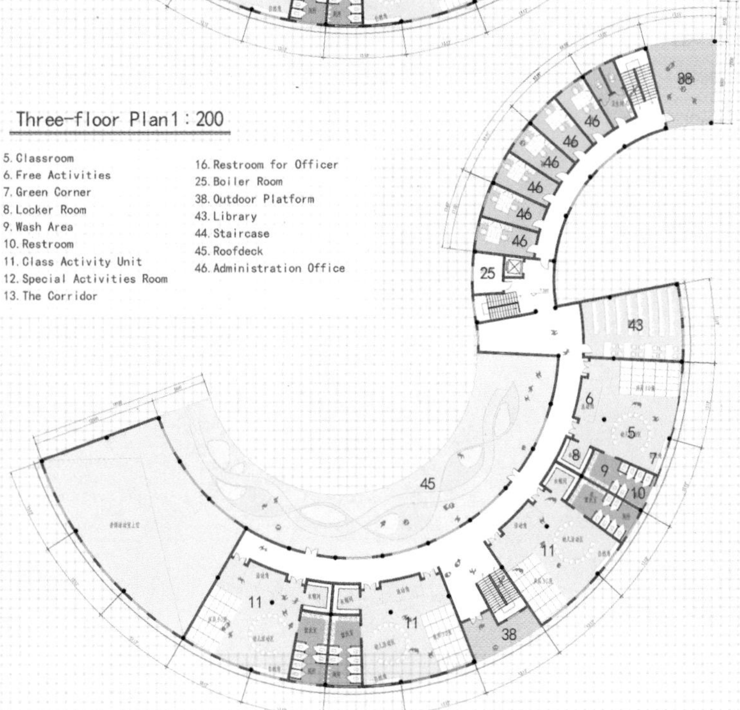

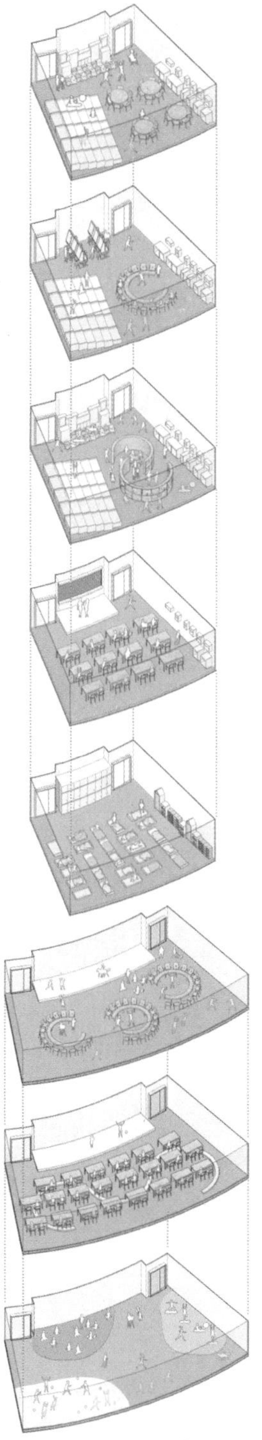

Behavioral Activity Analysis

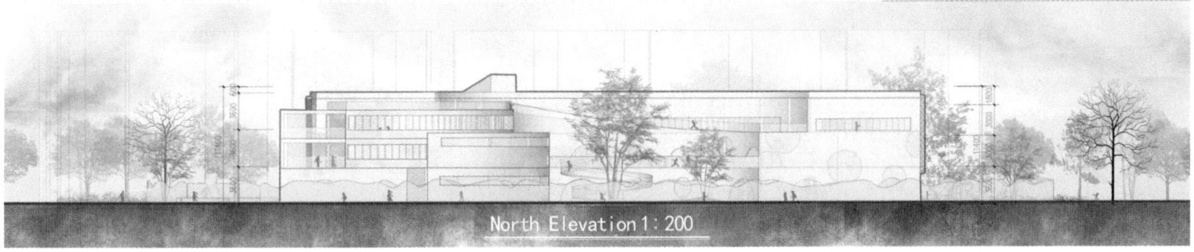

North Elevation 1 : 200

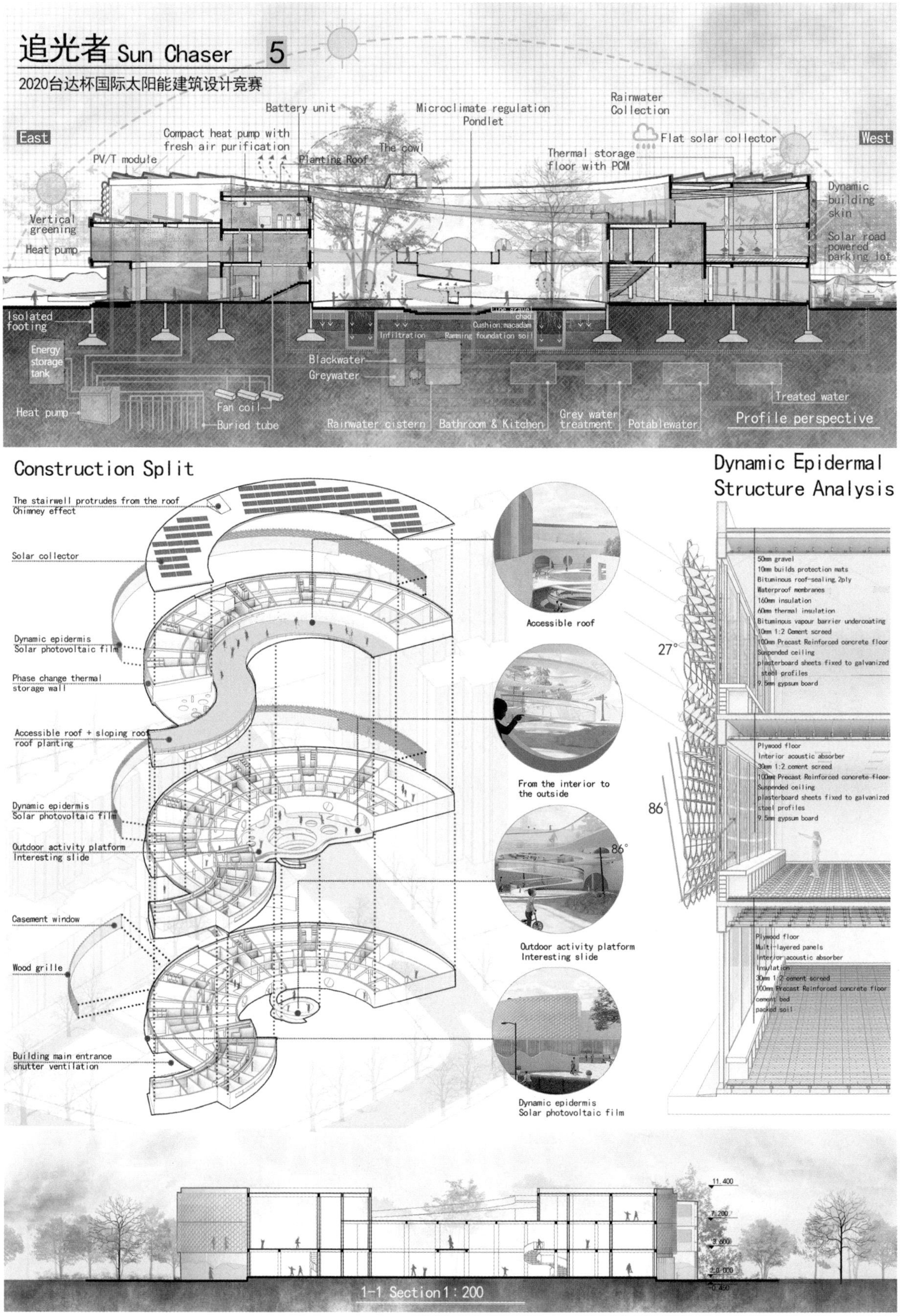
追光者 Sun Chaser 5
2020台达杯国际太阳能建筑设计竞赛
East
West
Battery unit
Microclimate regulation
Pondlet
Rainwater
Collection
Compact heat pump with
fresh air purification
PV/T module
Planting Roof
The cowl
Flat solar collector
Thermal storage
floor with PCM
Dynamic
building
skin
Solar road
powered
parking lot
Vertical
greening
Heat pump
Isolated
footing
Infiltration
Ramming foundation soil
Cushion:macadam
Energy
storage
tank
Blackwater
Greywater
Fan coil
Heat pump
Buried tube
Rainwater cistern
Bathroom & Kitchen
Grey water
treatment
Potablewater
Treated water
Profile perspective
Construction Split
The stairwell protrudes from the roof
Chimney effect
Solar collector
Dynamic epidermis
Solar photovoltaic film
Phase change thermal
storage wall
Accessible roof + sloping roof
roof planting
Dynamic epidermis
Solar photovoltaic film
Outdoor activity platform
Interesting slide
Casement window
Wood grille
Building main entrance
shutter ventilation
Accessible roof
From the interior to
the outside
86°
Outdoor activity platform
Interesting slide
Dynamic epidermis
Solar photovoltaic film
Dynamic Epidermal
Structure Analysis
27°
86°
50mm gravel
10mm builds protection mats
Bituminous roof-sealing 2ply
Waterproof membranes
160mm insulation
60mm thermal insulation
Bituminous vapour barrier undercoating
10mm 1:2 Cement screed
100mm Precast Reinforced concrete floor
Suspended ceiling
plasterboard sheets fixed to galvanized
steel profiles
9.5mm gypsum board
Plywood floor
Interior acoustic absorber
30mm 1:2 cement screed
100mm Precast Reinforced concrete floor
Suspended ceiling
plasterboard sheets fixed to galvanized
steel profiles
9.5mm gypsum board
Plywood floor
Multi-layered panels
Interior acoustic absorber
Insulation
30mm 1:2 cement screed
100mm Precast Reinforced concrete floor
cement bed
packed soil
11.400
7.200
3.600
±0.000
-0.450
1-1 Section 1：200

追光者 Sun Chaser 6
2020台达杯国际太阳能建筑设计竞赛

Dynamic Skin Unit Model

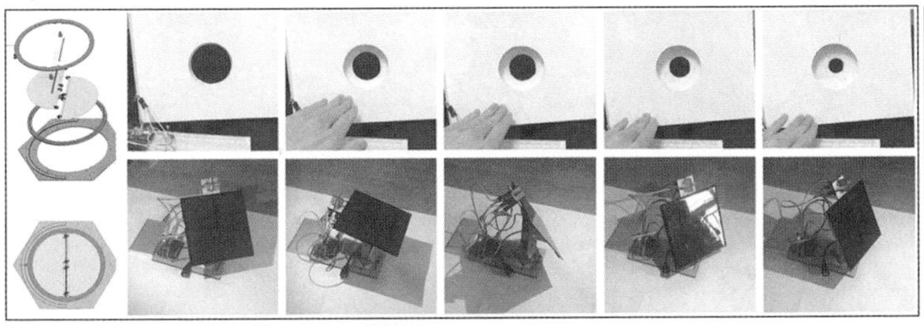

Dynamic Shading

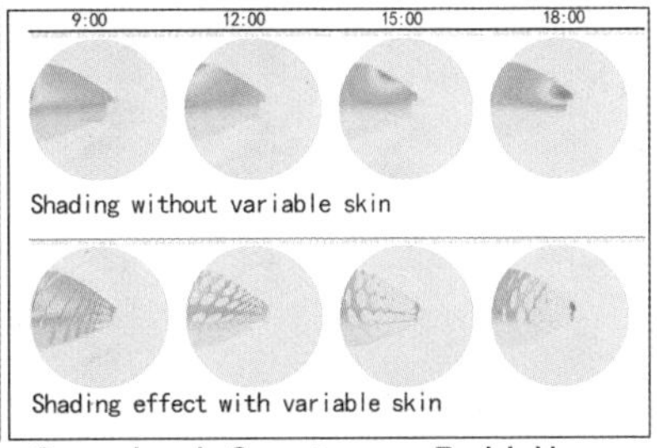
9:00
12:00
15:00
18:00
Shading without variable skin
Shading effect with variable skin

Analysis of Shading in Different Seasons

Winter
Summer

Standard for Green Building

Energy Consumption Comparison

Type of energy consumption	Designed building		Reference building	
Heating set (kWh)	E	29080.28	E	28211.12
Air-conditioner set (kwh)	E	118042.18	E	203512.13
Lighting (kWh)	E	128964.58	E	102514.22
All-year energy consumption (kWh)	B	276087.04	B	334237.47
Reduction of energy consumption	17.40%			

Renewable Energy Calculation

Parameter		Unit	Designed building
Annual solar nadiation in Nan Ping	H	MJ/ m²	4843
Solar panel area	A	m²	980(only roof)
Comprehensive correction	K		0.115
Generation energy of solar panel	E	MJ	32438

Thermal Paramerers of the Enclosure Structure

Hear Transfer Coeffcient			Unit	Designed building	Reference budilding
Roof			W/(m²·K)	0.71	0.8
Exterior wall			W/(m²·K)	1.52	1.5
exterior window	K	East	W/(m²·K)	2.3	3
		South	W/(m²·K)	2.4	2.7
		West	W/(m²·K)	2.5	2.7
		North	W/(m²·K)	2.4	3
	SHGC	East		0.32%	0.35%
		South		0.35%	0.35%
		West		0.32%	0.35%
		North		0.32%	0.45
Roof light		K	W/(m²·K)	2.4	3
		SHGC	SHGC	0.29	0.5
		Area proportion		0.03	0.2

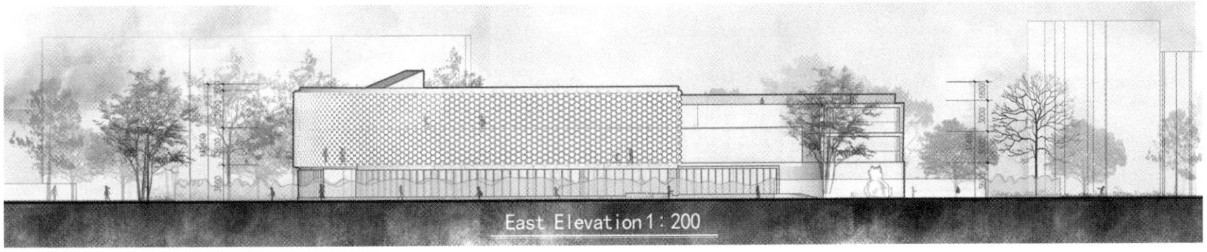
East Elevation 1：200

综合奖 · 二等奖
General Prize Awarded · Second Prize

注 册 号：7594

项目名称：林间风过（南平）
Rambling in the Wind
(Nanping)

作　　者：常逸凡、陆雨瑶、曹志昊

参赛单位：南京工业大学

指导教师：胡振宇、薛春霖、董　凌

专家点评：

作品以树林立意，引入第六立面的设计概念，有创意。立体构成的空间丰富且适度，树状木结构集成了通风、遮阳、隔热、保温等被动式功能且有趣。充分利用当地丰富的竹子资源作为建筑材料且进行合理应用。然而，树状木结构与建筑室内空间的联系方面有所欠缺。

The work, based on the woods, is creative by introducing the design concept of the Sixth Elevation. The three-dimensional space is abundant and moderate, and the tree-like timber structure integrates passive functions such as ventilation, sun shading, heat insulation and thermal insulation, and is interesting as well. Bamboos, abundant local resources, are used as building materials for reasonable application. However, it lacks certain connection between the tree-like timber structure and interior space of the building.

林间风过 Rambling in the Wind
Tree shadows, water currents, floral fragrances, butterfly dances... Those wild interests that should have been deeply rooted in childhood are gradually disappearing in high-density cities. How should the new generation of children find a sunny deciduous forest and experience this unique gift given by nature in the reinforced concrete city?
Starting from natural memory, this design integrates the symbolic expression of the forest, the creation of a childlike space, the new use of local building materials, and green building elements such as light, heat and ventilation to stimulate the children's nature of exercise and exploration. At the same time, a basic experimental framework is provided for future improvement and promotion.
树影、水流、花香、蝶舞……那些本该深深扎根于童年生活的森野之趣在高密度的城市中逐渐消弭，新一代的儿童又该如何在钢筋混凝土林立的都市之中觅得片阳光灿烂的落叶林，体验这一份自然赋予的独特礼物？
本设计从自然记忆出发，将森林的符号化表达、童趣空间的塑造、本土建材的新运用与光热通风等绿建元素相融合，以激发儿童运动与探索的天性。同时提供了一种基本的试验性构成骨架，以期未来完善与推广。
Concept Analysis
Get close to nature and explore the world
Children's behavior patterns
Low-point care
Sixth facade
More possibilities
Concept extension
Form and function
Result
Ⅰ. Looking up is the norm for children
1. Communicate with adults.
2. Adapt to things designed according to adult human scale.
3. Looking up at the starry sky and nature.
Ⅱ. "The sixth facade" concept
While paying attention to the facade and roof design of the building, consider the sixth facade design for indoor and outdoor.
For example
1. Domer in indoor atrium.
2. Overhanging underside
3. Indoor suspended ceiling.
Ⅲ. Architectural vocabulary and natural elements
1. The natural bottom like green plants makes the space more vivid.
2. The overhead way makes the building more flexible.
3. The combination of actual function implantation and energy saving.
Ⅳ. Scheme presentation
1. Simulate nature and create a sense of lush bamboo forest.
2. Imitate the principle of chimney ventilation, use pillars for ventilation.
3. The weave of the secondary beams enrichs the bottom surface.
Site Analysis
1. Tight site environment
2. Lack of children's playground
3. Rich in bamboo and wood resources
Scheme Generation
STEP1:Functional Division
ACTIVITIES
STEP2:Column Insertion
STEP3:Dislocation
STEP4:Generate Volume
STEP5:Overhang&Ramp
STEP6:Roof Deformation
Site Plan 1 : 500
Economic technology index
Cover Area: 6028 m^2
Overall floorage: 4507 m^2
Building Area: 1679 m^2
Building density: 0.28
Building height: 15.2m
Volume Fraction: 0.75
Greening Rate: 0.30
Car Entrance
Logistic Entrance
Office Entrance
Main Entrance
1F
2F
3F
N

林间风过 Rambling in the Wind

Wow!

Awesome!

Enjoy myself !

Amazing!

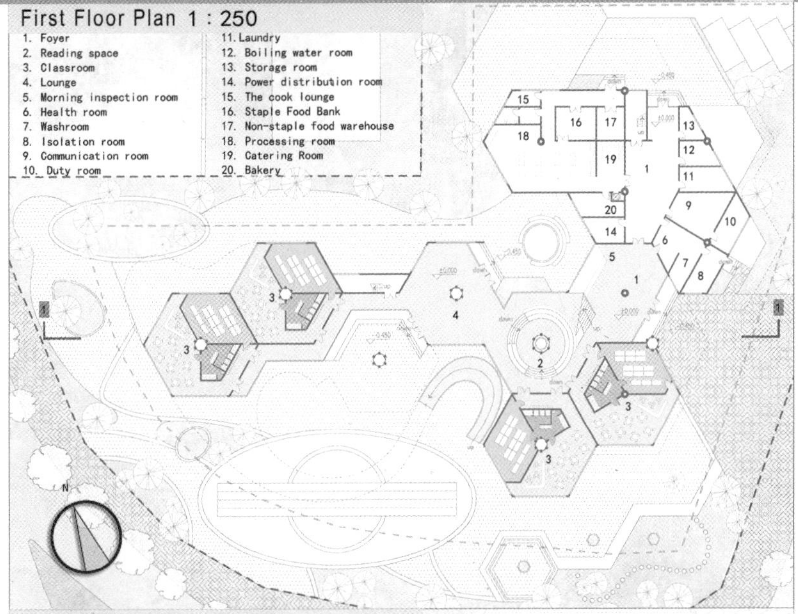
First Floor Plan 1：250
1. Foyer
2. Reading space
3. Classroom
4. Lounge
5. Morning inspection room
6. Health room
7. Washroom
8. Isolation room
9. Communication room
10. Duty room
11. Laundry
12. Boiling water room
13. Storage room
14. Power distribution room
15. The cook lounge
16. Staple Food Bank
17. Non-staple food warehouse
18. Processing room
19. Catering Room
20. Bakery

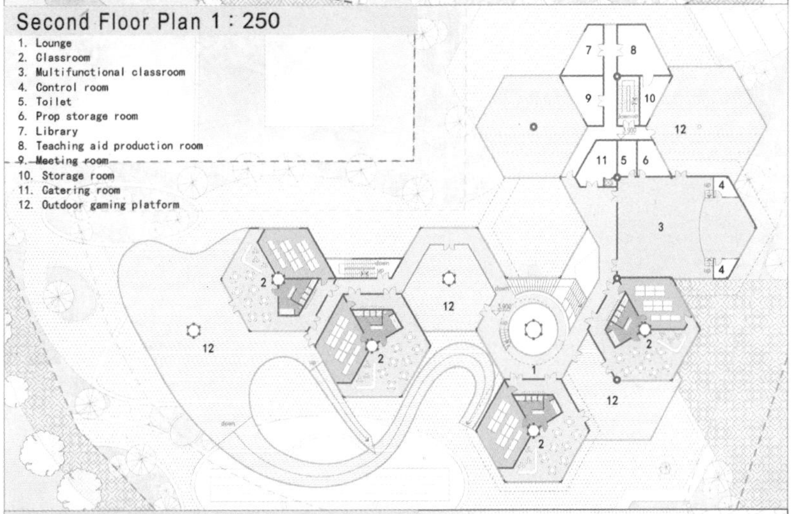
Second Floor Plan 1：250
1. Lounge
2. Classroom
3. Multifunctional classroom
4. Control room
5. Toilet
6. Prop storage room
7. Library
8. Teaching aid production room
9. Meeting room
10. Storage room
11. Catering room
12. Outdoor gaming platform

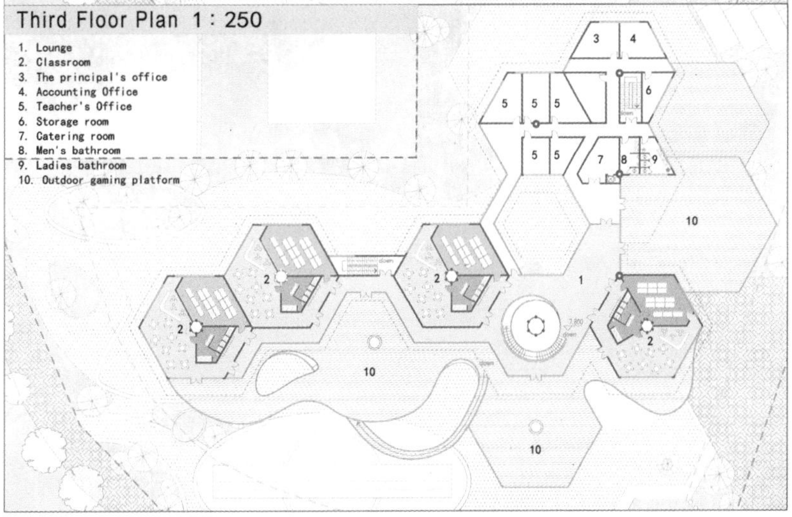
Third Floor Plan 1：250
1. Lounge
2. Classroom
3. The principal's office
4. Accounting Office
5. Teacher's Office
6. Storage room
7. Catering room
8. Men's bathroom
9. Ladies bathroom
10. Outdoor gaming platform

林间风过 Rambling in the Wind

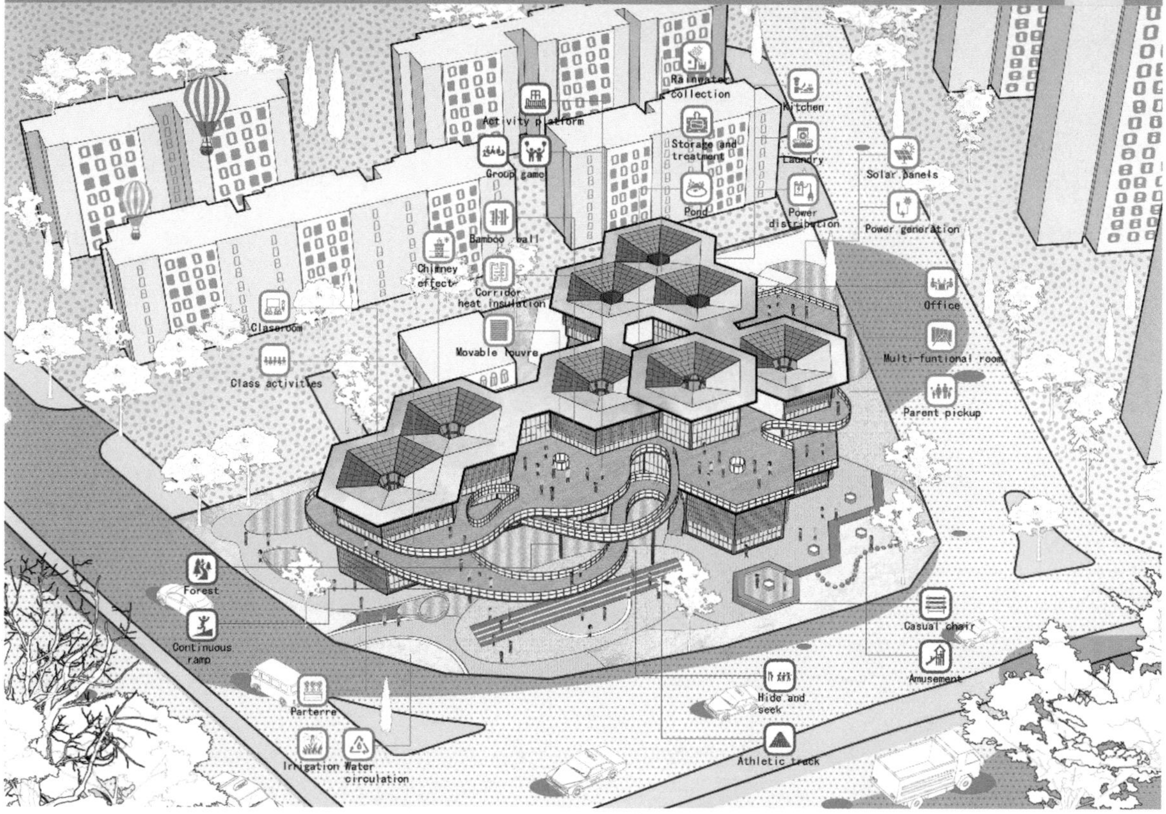
Rainwater collection
Kitchen
Activity platform
Storage and treatment
Laundry
Group game
Solar panels
Pond
Power distribution
Power generation
Bamboo wall
Chimney effect
Corridor heat insulation
Office
Classroom
Movable louvre
Multi-funtional room
Class activities
Parent pickup
Forest
Casual chair
Continuous ramp
Amusement
Hide and seek
Parterre
Irrigation
Water circulation
Athletic track

Site Climate Analysis

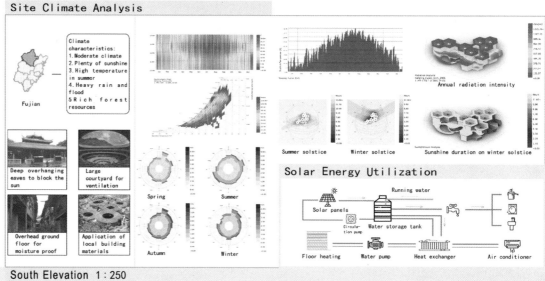

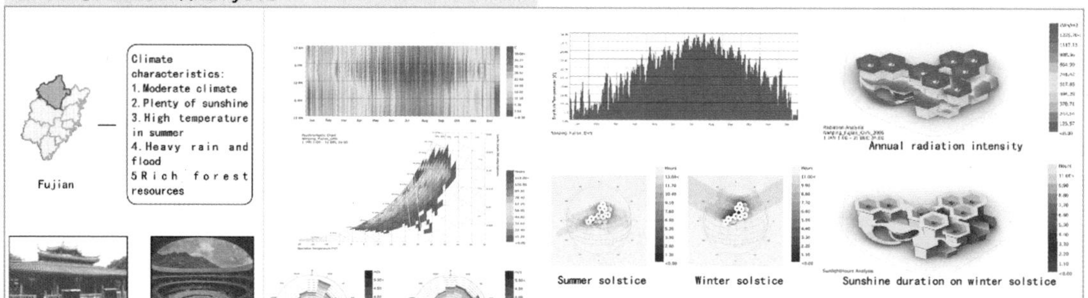
Fujian
Climate characteristics:
1.Moderate climate
2.Plenty of sunshine
3.High temperature in summer
4.Heavy rain and flood
5Rich forest resources
Deep overhanging eaves to block the sun
Large courtyard for ventilation
Overhead ground floor for moisture proof
Application of local building materials
Spring
Summer
Autumn
Winter
Annual radiation intensity
Summer solstice
Winter solstice
Sunshine duration on winter solstice

Solar Energy Utilization

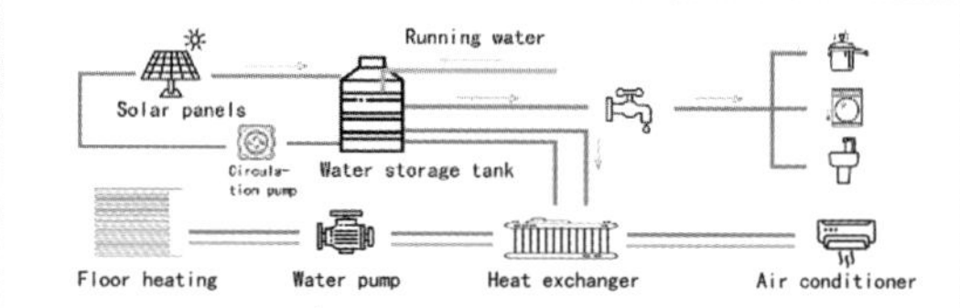
Running water
Solar panels
Water storage tank
Circulation pump
Floor heating
Water pump
Heat exchanger
Air conditioner

South Elevation 1：250

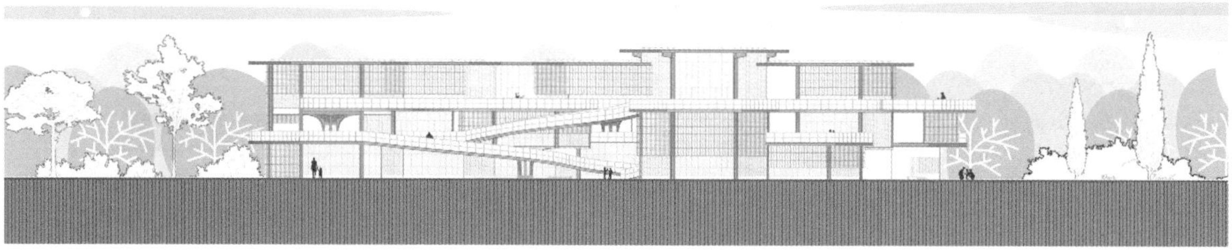

林间风过 Rambling in the Wind

Ⅳ

Classroom Unit Analysis

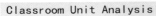

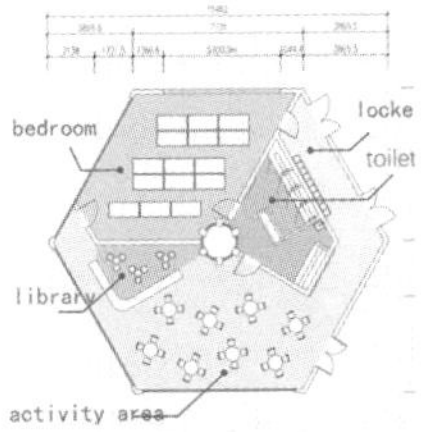
bedroom
locke
toilet
library
activity area

Bamboo Louvers Analysis

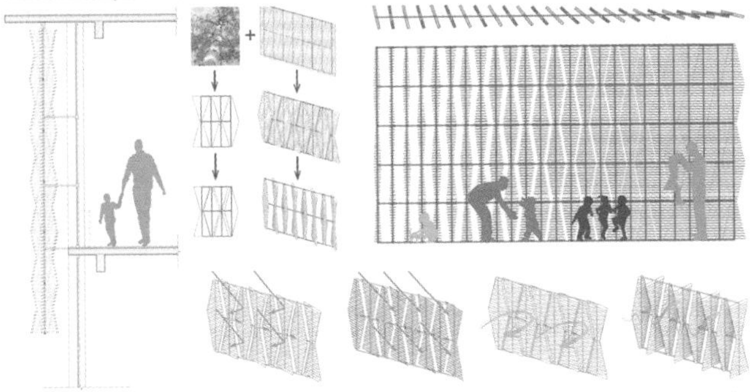

Detailed Structure

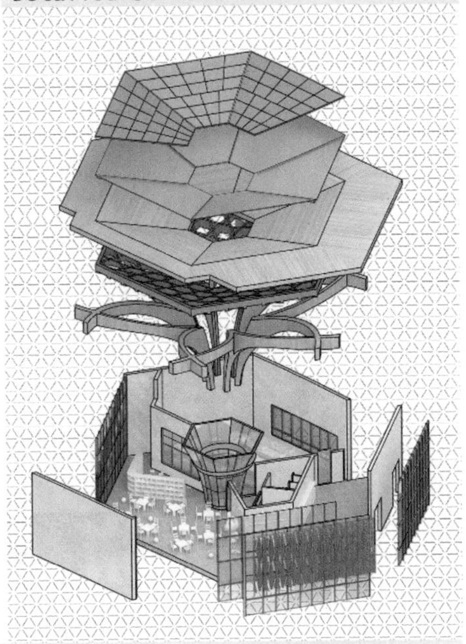

Section 1-1 1 : 250

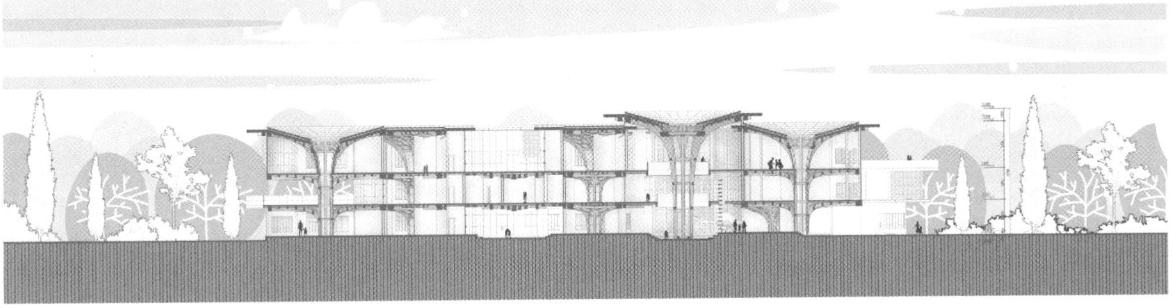

林间风过 Rambling in the Wind

V

Bamboo Louvers

Knittingtexture

Detailed Structure

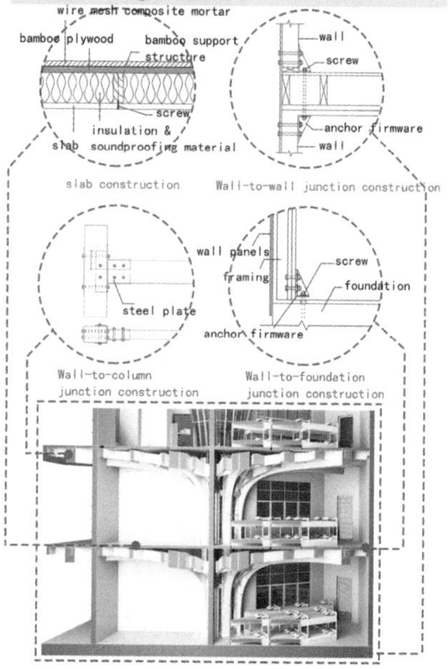

wire mesh composite mortar
bamboo plywood
bamboo support structure
screw
insulation & soundproofing material
slab
slab construction
wall
screw
anchor firmware
wall
Wall-to-wall junction construction
steel plate
Wall-to-column junction construction
wall panels
framing
screw
foundation
anchor firmware
Wall-to-foundation junction construction

Energy Saving Analysis

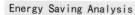

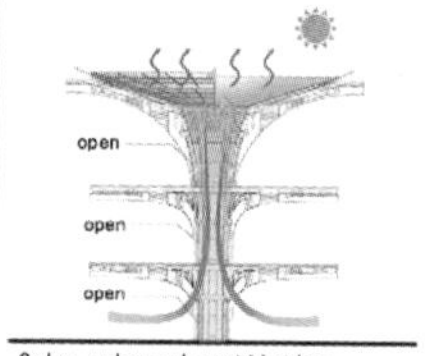

Summer
open
open
open

Solar enhanced ventilation

In summer, solar radiation is used to heat the air in the bowl-mounted roof space to provide more buoyancy for the natural circulation system.

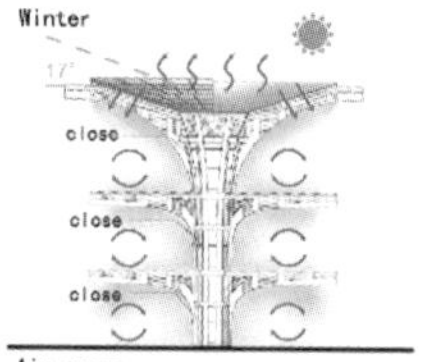

Winter
17°
close
close
close

Air space

In winter, the air space and the double-glazed windows achieve good thermal insulation effect. Meanwhile, it is calculated that the best local solar panel inclination angle is 17°.

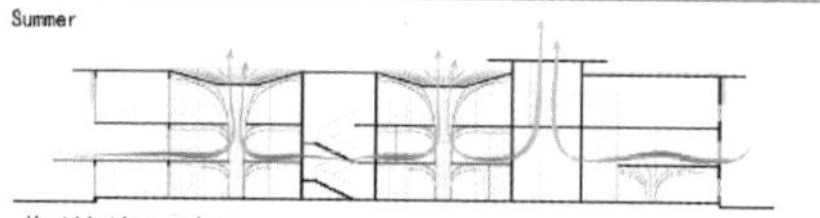

Summer

Ventilation column

This structural design solves the problem that in low-rise buildings, ordinary atriums cannot efficiently use the chimney effect to achieve air circulation due to the insufficient height and diameter ratio, and at the same time enrich the space

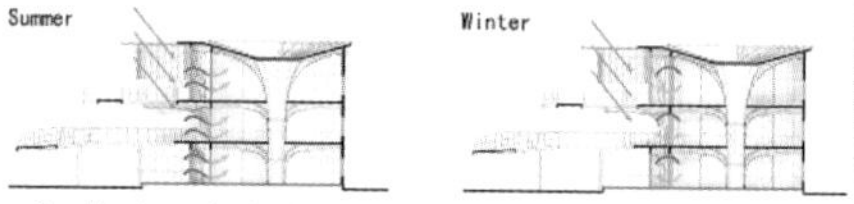

Summer
Winter

Corridor heat insulation

The layout of the south-facing corridor forms a good thermal insulation buffer space, achieving thermal insulation in summer and thermal insulation in winter.

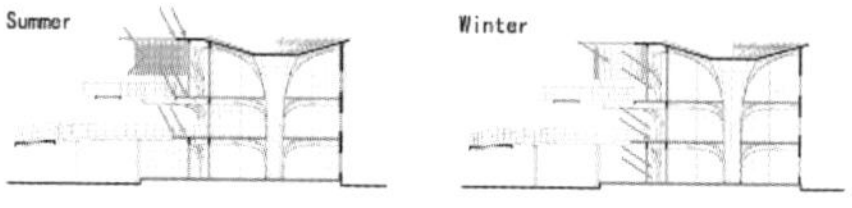

Summer
Winter

Movabel louvers and deep overhanging eaves

The louver design, which can be changed in both horizontal and vertical directions, can freely and flexibly respond to various lighting environments and provide comfortable lighting in the room. The deep overhanging eaves also adapt to local climate characteristics.

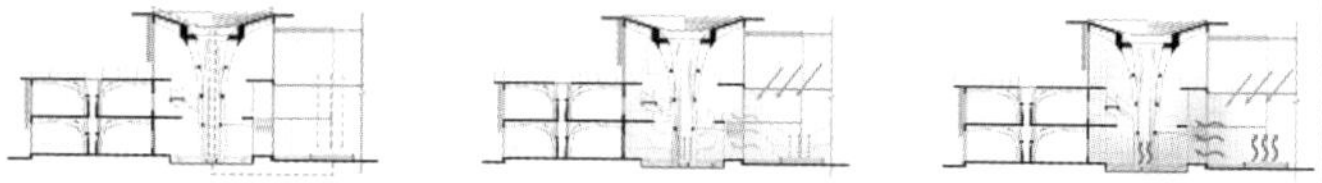

Rain water collection　Summer　Winter

East Elevation 1 : 250

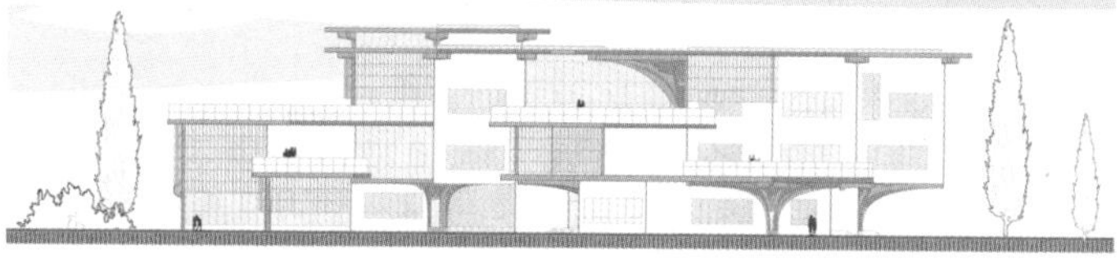

Game Circulation System

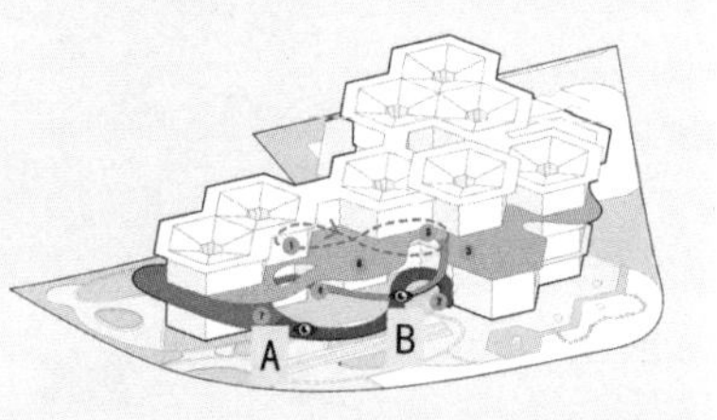

A
B

"Seven Elements"

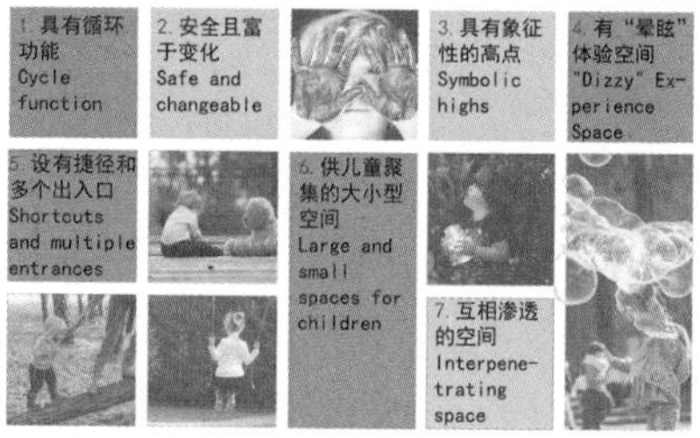

1. 具有循环功能
Cycle function
2. 安全且富于变化
Safe and changeable
3. 具有象征性的高点
Symbolic highs
4. 有"晕眩"体验空间
"Dizzy" Experience Space
5. 设有捷径和多个出入口
Shortcuts and multiple entrances
6. 供儿童聚集的大小型空间
Large and small spaces for children
7. 互相渗透的空间
Interpenetrating space

A

B

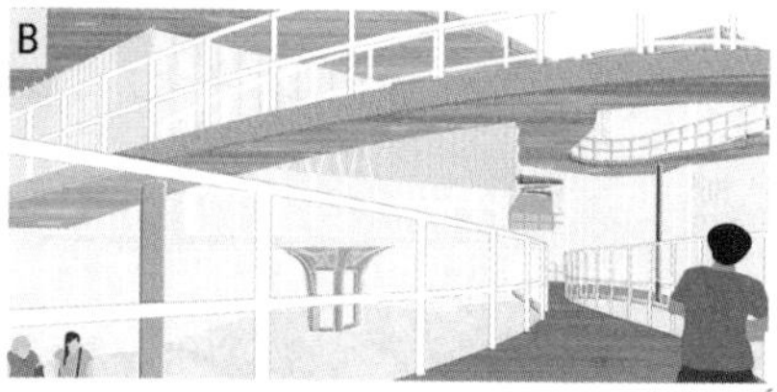

林间风过 Rambling in the Wind

VI

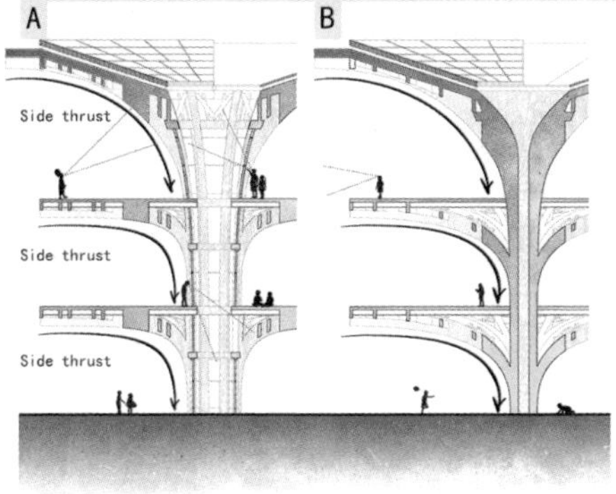
A
B
Side thrust
Side thrust
Side thrust

Column Structure Analysis

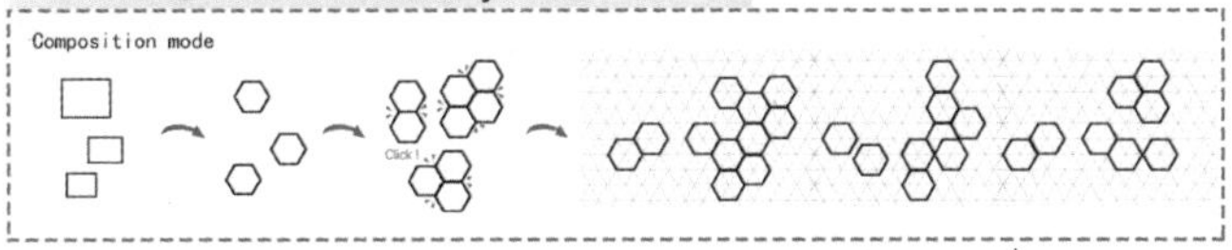
Composition mode
Click!

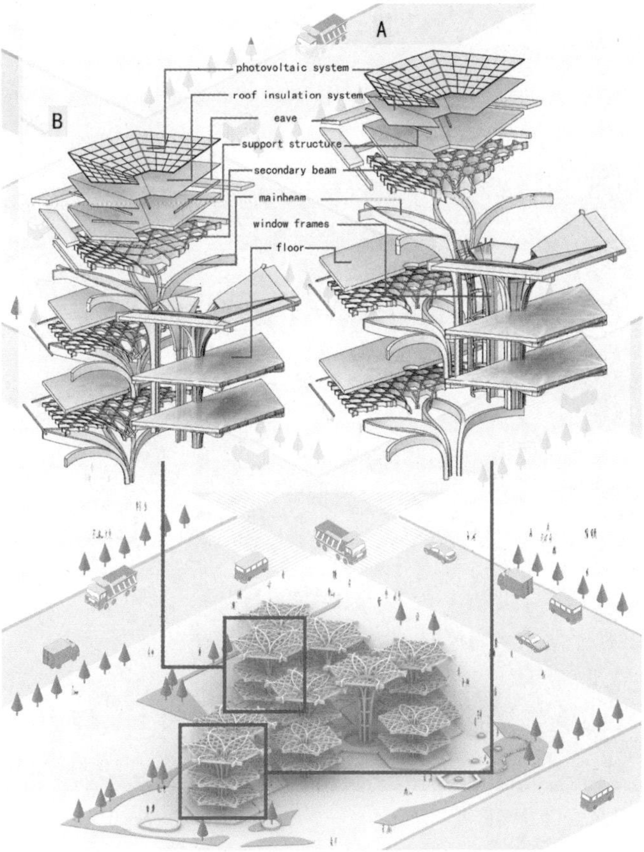
A
B
photovoltaic system
roof insulation system
eave
support structure
secondary beam
mainbeam
window frames
floor

Evaluation of Green Building

安全耐久 Safety and durability

	评分项条文	规范分值	自评得分
		100	77
I 安全	4.2.1 采用基于性能的抗震设计并合理提高建筑的抗震性能安全。	10	6
	4.2.2 采取保障人员安全的防护措施。	15	15
	4.2.3 采用具有安全防护功能的产品或配件。	10	10
	4.2.4 室内外地面或路面设置防滑措施。	10	10
	4.2.5 为行人提供安全步行空间。	8	8
II 耐久	4.2.6 采取提升建筑适变性的措施。	18	14
	4.2.7 采取提升建筑部品部件耐久性的措施。	10	4
	4.2.8 提高建筑结构材料的耐久性。	10	4
	4.2.9 合理采用耐久性好、易维护的装饰装修建筑材料。	9	6

资源节约 Resource conservation

	评分项条文	规范分值	自评得分
		180	114.5
I 节地与土地利用	7.2.1 节约集约利用土地。	12	3
	7.2.2 合理开发利用地下空间。	/	/
	7.2.3 采用机械式停车设施、地下停车库或地面停车楼等方式。	6	0
II 节能与能源利用	7.2.4 建筑围护结构合理采用适宜本地区气候特点的节能技术和产品。	12	9
	7.2.5 优化建筑围护结构的热工性能。	15	10
	7.2.6 采取措施优化建筑供暖空调系统。	10	5
	7.2.7 采取有效措施降低供暖空调系统的末端系统及输配系统的能耗。	5	5
	7.2.8 采用节能型电气设备及节能控制措施。	10	4
	7.2.9 采取措施降低建筑能耗。	10	6
	7.2.10 结合当地气候和自然资源条件合理利用可再生能源。	10	9
	7.2.11 采用风扇调风作为室内热环境改善的辅助措施。	4	4
III 节水与水资源利用	7.2.12 公共区域采用无动力脚踏式水嘴。	2	0
	7.2.13 设置空调冷凝水回收系统。	3	3
	7.2.14 住宅和酒店类建筑中不少于90%的浴室采用淋浴替代浴缸。	/	/
	7.2.15 使用较高用水效率等级的卫生器具。	15	5
	7.2.16 绿化灌溉及空调冷却水系统采用节水设备或技术。	12	12
	7.2.17 结合雨水综合利用设施营造室外景观水体，室外景观水体利用雨水的补水量大于水体蒸发量的60%，且采用保障水体水质的生态水处理技术。	8	8
	7.2.18 使用非传统水源。	10	6.5
IV 节材与绿色建材	7.2.19 建筑所有区域实施土建工程与装修工程一体化设计及施工。	8	6
	7.2.20 合理选用建筑结构材料与构件。	/	/
	7.2.21 建筑装修选用工业化内装部品。	8	3
	7.2.22 选用可再循环材料、可再利用材料及利废建材。	12	12
	7.2.23 选用绿色建材或新工法构件。	8	6

提高与创新 Improvement and innovationlivability

	相关评分项条文摘要	规范分值	自评得分
		137	120
	9.2.1 采取措施进一步降低建筑供暖空调系统的能耗。	30	20
	9.2.2 针对建筑不同使用功能和不同朝向，采用有效的外遮阳措施，并与建筑外观造型融为一体。	5	9
	9.2.3 采用适宜地区特色的建筑风貌设计，因地制宜传承地域建筑文化。	20	16
	9.2.5 在国家规定抗震设防烈度的标准上提高一级及以上烈度设计。	10	10
	9.2.7 采用符合工业化建造要求的结构体系与建筑构件。	10	10
	9.2.8 充分运用循环建筑理念，可再利用建筑材料用量比例达到30%，或可再循环和可再利用建筑材料用量比例达到50%。	10	10
	9.2.10 采取措施降低单位建筑面积碳排放强度。	12	12
	9.2.14 采取节约资源、保护生态环境、保障安全健康、智慧友好运行、传承历史文化等其他创新，并有明显效益。	42	42

健康舒适 Health and comfort

	评分项条文	规范分值	自评得分
		100	83
I 室内空气品质	5.2.1 控制室内主要空气污染物的浓度。	9	9
	5.2.2 选用的装饰装修材料满足国家现行绿色产品评价标准中对有害物质限量的要求。	8	8
II 水质	5.2.3 直饮水、集中生活热水、游泳池水、采暖空调系统用水、景观水体等的水质满足国家现行有关标准。	8	6
	5.2.4 生活饮用水水池、水箱等储水设施采取措施满足卫生要求。	9	7
	5.2.5 所有给水排水管道、设备、设施设置明确、清晰的永久性标识。	8	8
III 声环境与光环境	5.2.6 采取措施优化主要功能房间的室内声环境。	8	4
	5.2.7 主要功能房间的隔声性能良好。	11	9
	5.2.8 充分利用天然光。	14	10
IV 室内热湿环境	5.2.9 具有良好的室内热湿环境。	8	6
	5.2.10 优化建筑空间和平面布局，改善室内通风效果。	8	6
	5.2.11 设置可调节遮阳设施，改善室内热舒适。	9	9

生活便利 Convenience of life

	评分项条文	规范分值	自评得分
		97	52
I 出行与无障碍	6.2.1 场地与公共交通站点联系便捷。	8	0
	6.2.2 建筑室内外公共区域满足全龄化设计要求。	7	4
II 服务设施	6.2.3 提供便利的公共服务。	12	6
	6.2.4 设置电动汽车充电停车位。	5	5
	6.2.5 城市绿地、广场及公共运动场地等开敞空间，步行可达。	5	5
	6.2.6 合理设置健身场地和空间。	7	7
III 智慧运行	6.2.8 设置分类、分级、分项用能实时监测计量系统。	6	2
	6.2.9 设置 CO_2 浓度的空气质量监测系统。	5	0
	6.2.10 设置用水远传计量系统、水质在线监测系统。	6	0
	6.2.11 具有智能化服务系统。	6	4
IV 物业管理	6.2.12 制定完善的节能、节水、节材、绿化的操作规程、应急预案，实施能源资源管理激励机制，且有效实施。	4	0
	6.2.13 建筑平均日用水量满足现行国家标准。	4	4
	6.2.14 定期对建筑运营效果进行评估，并根据结果进行优化。	7	2
	6.2.15 建立绿色教育宣传和实践机制。	6	6
	6.2.16 [illegible]	5	5
	6.2.17 [illegible]	4	2

环境宜居 Environmental livability

	评分项	规范分值	自评得分		评分项	规范分值	自评得分
		54	22			48	30
I 场地生态与景观	8.2.1	5	0	I	8.2.8	5	0
	8.2.2	12	4		8.2.9	9	9
	8.2.3	7	5		8.2.10	6	5
	8.2.4	9	6	II 室外物理环境	8.2.11	6	4
	8.2.5	12	2		8.2.12	6	4
	8.2.6	6	2		8.2.13	8	4
	8.2.7	3	3		8.2.14	8	4

评价结果 Results of evaluation

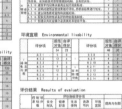

	控制项基础分值	评价指标评分项 安全耐久	健康舒适	生活便利	资源节约	环境宜居	提高与创新
评价分值	400	87	89	74	153.5	60	128
总分 Total score	97.6						
星级 Star rating	★★★						

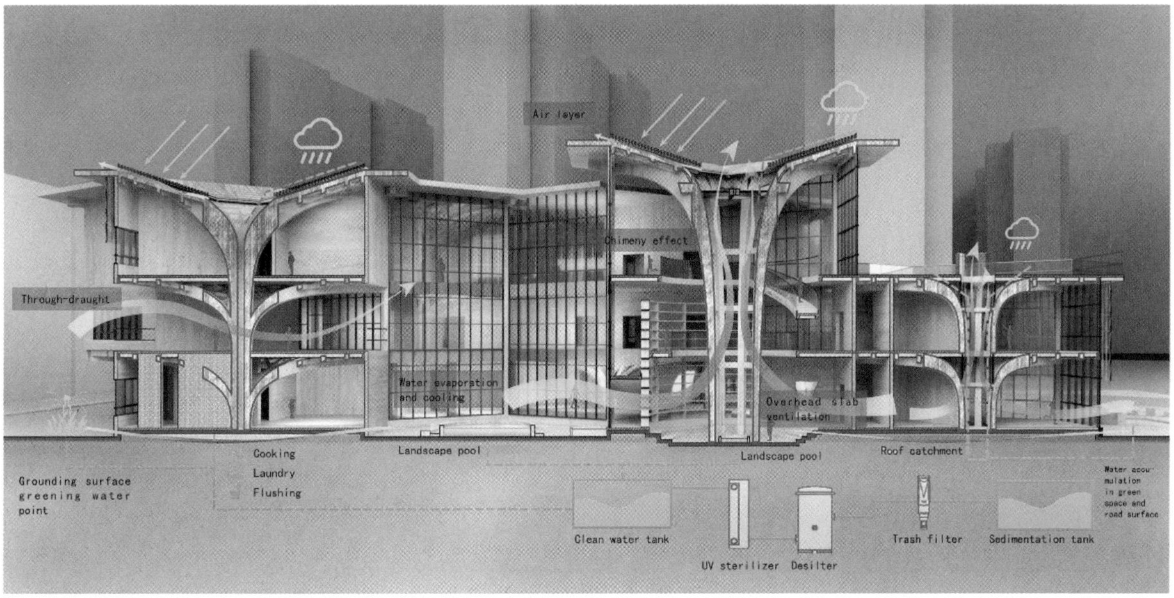
Air layer
Chimeny effect
Through-draught
Water evaporation and cooling
Overhead slab ventilation
Landscape pool
Landscape pool
Roof catchment
Cooking
Laundry
Flushing
Grounding surface greening water point
Clean water tank
UV sterilizer
Desilter
Trash filter
Sedimentation tank
Water accumulation in green space and road surface

综合奖 · 二等奖
General Prize Awarded · Second Prize

注 册 号：7178

项目名称：稚梦 · 星盒（新疆）

Dreamy Star Box（Xinjiang）

作　　者：赵俊逸、许云鹏、史心怡

参赛单位：南京工业大学

指导教师：郗皎如

专家点评：

作品平面布局规整，各功能空间相互独立且联系便捷，交通流线简洁。建筑造型吸纳了当地民居元素，并对夯土墙结构进行创新应用。采用的太阳能烟囱、地道风复合系统等被动式技术和主动式太阳能系统的应用合理且数据分析细致。但部分系统较复杂，可实施性略低。

The work is well-planned in plane layout, independent but conveniently connected in functional spaces, and concise in traffic streamline. It adopts local residential elements in architectural style and makes innovative applications of rammed earth wall structures. It has reasonably applied such passive technologies as the composite system of solar chimneys and air through tunnel and active solar systems, and made meticulous data analysis as well. However, some systems are rather complicated, with less feasibility.

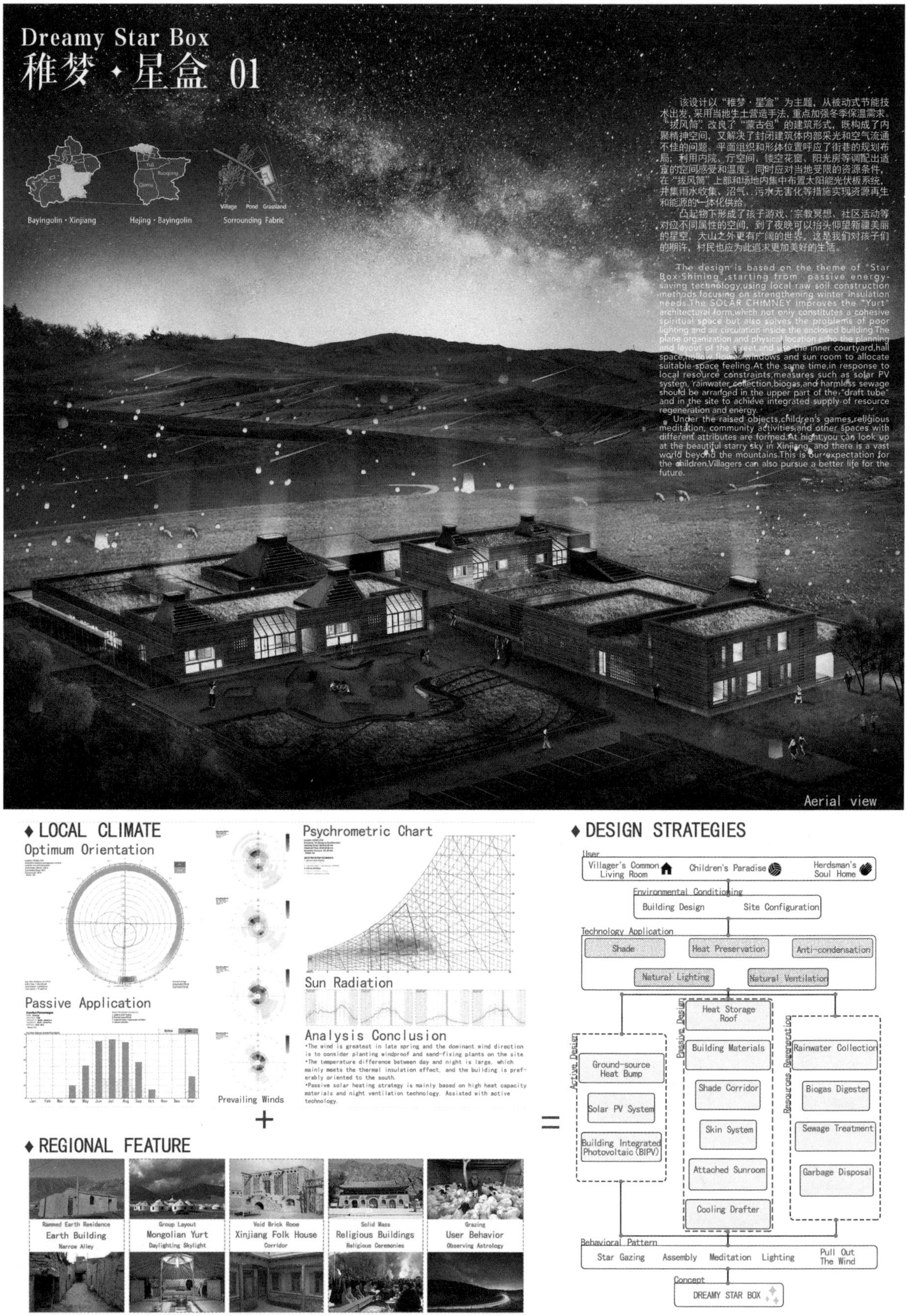

Dreamy Star Box
稚梦 · 星盒 01
Bayingolin · Xinjiang
Hejing · Bayingolin
Village Pond Grassland
Sorrounding Fabric
该设计以“稚梦 · 星盒”为主题，从被动式节能技术出发，采用当地生土营造手法，重点加强冬季保温需求。“拔风筒”改良了“蒙古包”的建筑形式，既构成了内聚精神空间，又解决了封闭建筑体内部采光和空气流通不佳的问题。平面组织和形体位置呼应了街巷的规划布局；利用内院、厅空间、镂空花窗、阳光房等调配出适宜的空间感受和温度。同时应对当地受限的资源条件，在“拔风筒”上部和场地内集中布置太阳能光伏板系统，并集雨水收集、沼气、污水无害化等措施实现资源再生和能源的一体化供给。
凸起物下形成了孩子游戏、宗教冥想、社区活动等对应不同属性的空间，到了夜晚可以抬头仰望新疆美丽的星空，大山之外更有广阔的世界，这是我们对孩子们的期许，村民也应为此追求更加美好的生活。
The design is based on the theme of "Star Box Shining",starting from passive energy-saving technology,using local raw soil construction methods focusing on strengthening winter insulation needs.The SOLAR CHIMNEY improves the "Yurt" architectural form,which not only constitutes a cohesive spiritual space but also solves the problems of poor lighting and air circulation inside the enclosed building.The plane organization and physical location echo the planning and layout of the street,and use the inner courtyard,hall space,hollow flower windows and sun room to allocate suitable space feeling.At the same time,in response to local resource constraints,measures such as solar PV system, rainwater collection,biogas,and harmless sewage should be arranged in the upper part of the "draft tube" and in the site to achieve integrated supply of resource regeneration and energy.
Under the raised objects,children's games,religious meditation, community activities,and other spaces with different attributes are formed.At night,you can look up at the beautiful starry sky in Xinjiang, and there is a vast world beyond the mountains.This is our expectation for the children.Villagers can also pursue a better life for the future.
Aerial view
LOCAL CLIMATE
Optimum Orientation
Passive Application
Prevailing Winds
Psychrometric Chart
Sun Radiation
Analysis Conclusion
REGIONAL FEATURE
Earth Building
Mongolian Yurt
Xinjiang Folk House
Religious Buildings
User Behavior
DESIGN STRATEGIES
User
Villager's Common Living Room
Children's Paradise
Herdsman's Soul Home
Environmental Conditioning
Building Design
Site Configuration
Technology Application
Shade
Heat Preservation
Anti-condensation
Natural Lighting
Natural Ventilation
Active Design
Ground-source Heat Bump
Solar PV System
Building Integrated Photovoltaic (BIPV)
Passive Design
Heat Storage Roof
Building Materials
Shade Corridor
Skin System
Attached Sunroom
Cooling Drafter
Resources Regeneration
Rainwater Collection
Biogas Digester
Sewage Treatment
Garbage Disposal
Behavioral Pattern
Star Gazing
Assembly
Meditation
Lighting
Pull Out The Wind
Concept
DREAMY STAR BOX

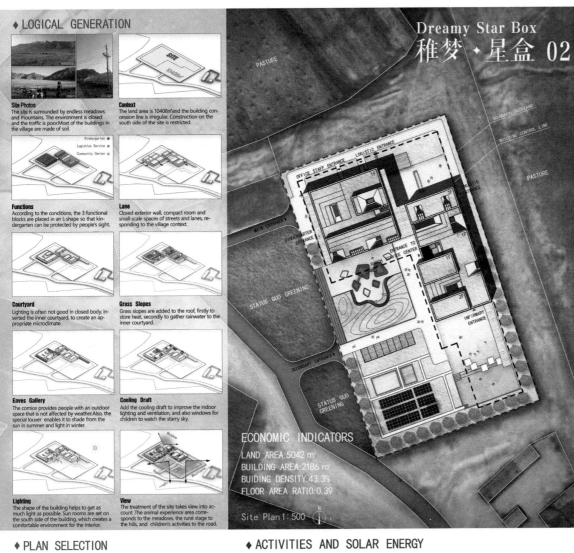

Dreamy Star Box 稚梦·星盒 02

♦ LOGICAL GENERATION

MOUNTAINS
MEADOWS
VILLAGE

Site Photos
The site is surrounded by endless meadows and mountains. The environment is closed and the traffic is poor.Most of the buildings in the village are made of soil.

SITE
10408m²

Context
The land area is 10408m²and the building concession line is irregular. Construction on the south side of the site is restricted.

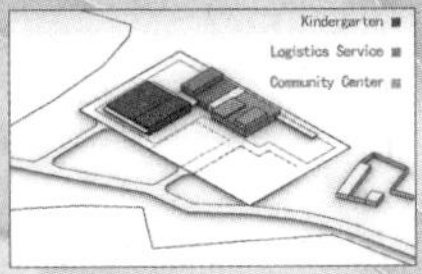
Kindergarten
Logistics Service
Community Center

Functions
According to the conditions, the 3 functional blocks are placed in an L-shape so that kindergarten can be protected by people's sight.

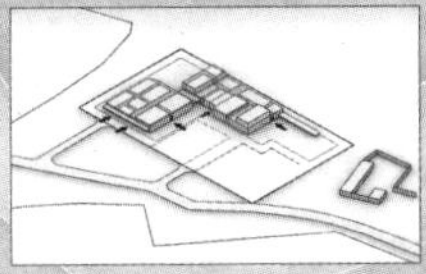

Lane
Closed exterior wall, compact room and small-scale spaces of streets and lanes, responding to the village context.

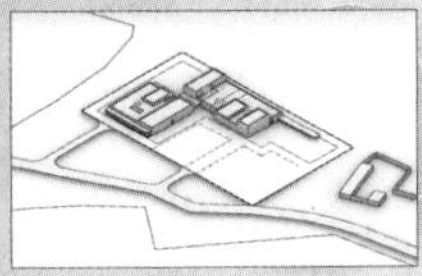

Courtyard
Lighting is often not good in closed body, inserted the inner courtyard, to create an appropriate microclimate.

Grass Slopes
Grass slopes are added to the roof, firstly to store heat, secondly to gather rainwater to the inner courtyard.

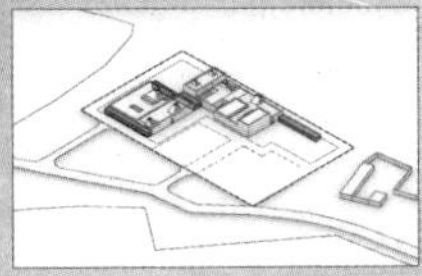

Eaves Gallery
The cornice provides people with an outdoor space that is not affected by weather.Also, the special louver enables it to shade from the sun in summer and light in winter.

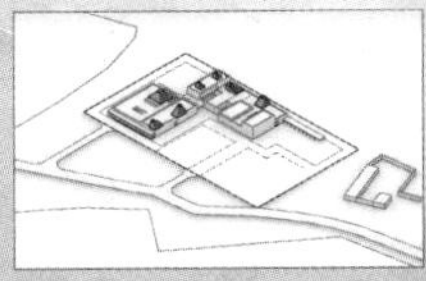

Cooling Draft
Add the cooling draft to improve the indoor lighting and ventilation, and also windows for children to watch the starry sky.

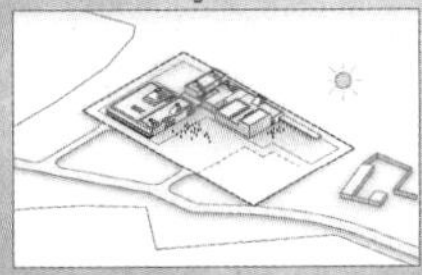

Lighting
The shape of the building helps to get as much light as possible. Sun rooms are set on the south side of the building, which creates a comfortable environment for the interior.

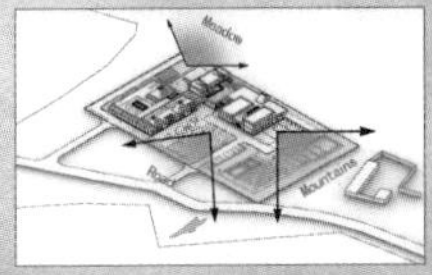
Meadow
Road
Mountains

View
The treatment of the site takes view into account .The animal experience area corresponds to the meadows, the rural stage to the hills, and children's activities to the road.

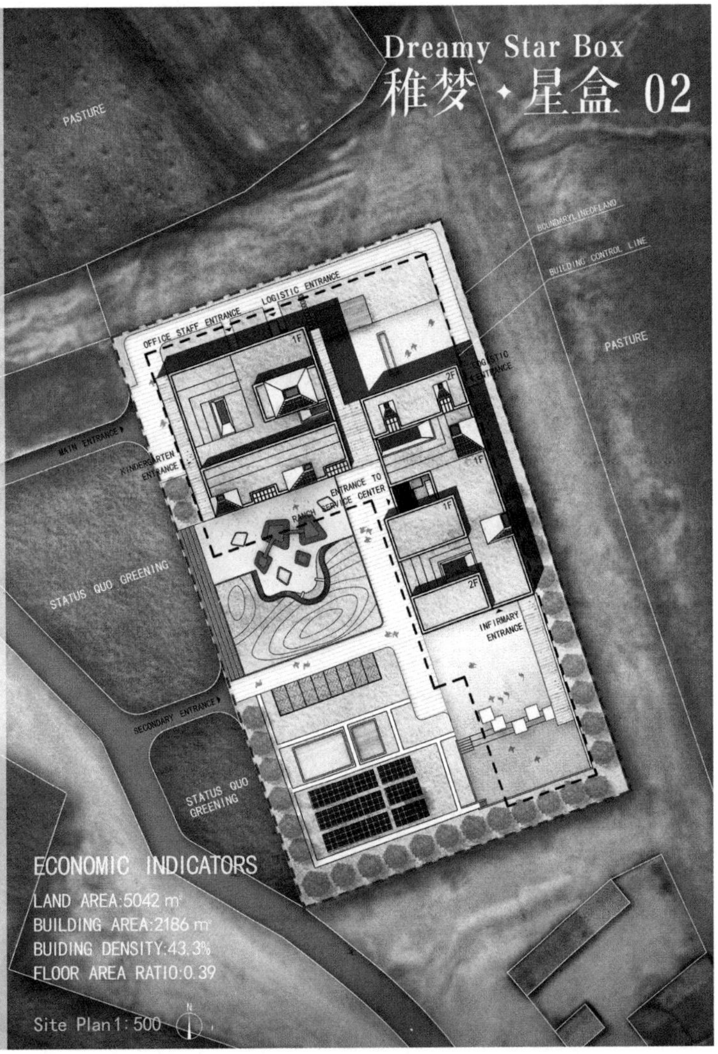
PASTURE
BOUNDARY LINE OF LAND
BUILDING CONTROL LINE
PASTURE
OFFICE STAFF ENTRANCE
LOGISTIC ENTRANCE
LOGISTIC ENTRANCE
MAIN ENTRANCE
KINDERGARTEN ENTRANCE
ENTRANCE TO SERVICE CENTER
RANCH
1F
2F
1F
1F
2F
STATUS QUO GREENING
INFIRMARY ENTRANCE
SECONDARY ENTRANCE
STATUS QUO GREENING
ECONOMIC INDICATORS
LAND AREA:5042 m²
BUILDING AREA:2186 m²
BUIDING DENSITY:43.3%
FLOOR AREA RATIO:0.39
Site Plan1:500

♦ PLAN SELECTION

Plan Layout

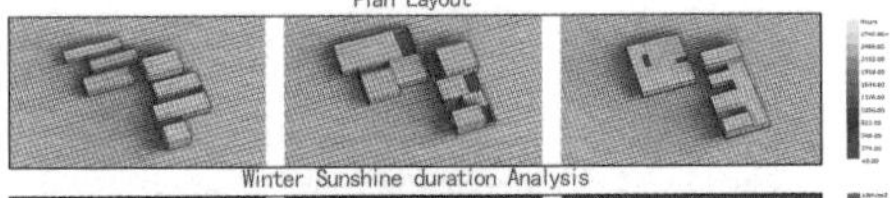

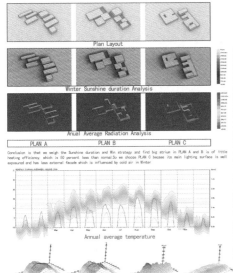

Winter Sunshine duration Analysis

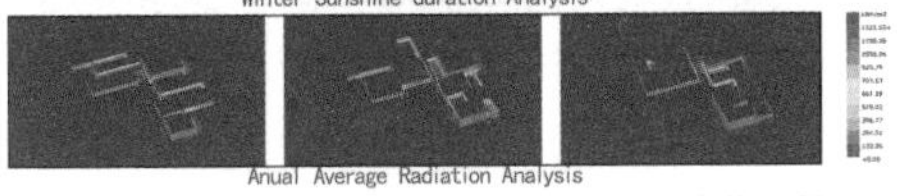

Anual Average Radiation Analysis

PLAN A	PLAN B	PLAN C

Conclusion is that we weigh the Sunshine duration and Win strategy and find big atrium in PLAN A and B is of little heating efficiency. which is 50 percent less than normal.So we choose PLAN C becase its main lighting surface is well exposured and has less external facade which is influenced by cold air in Winter

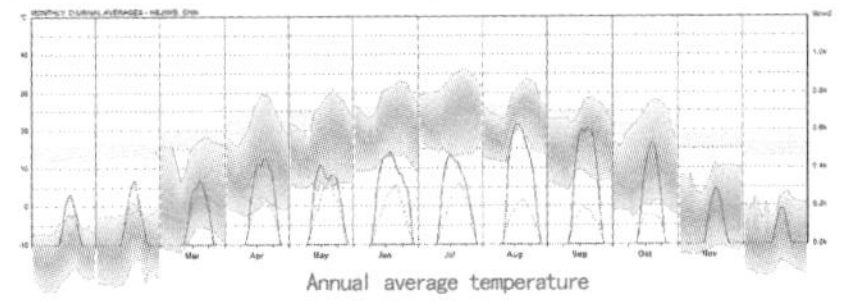

Annual average temperature

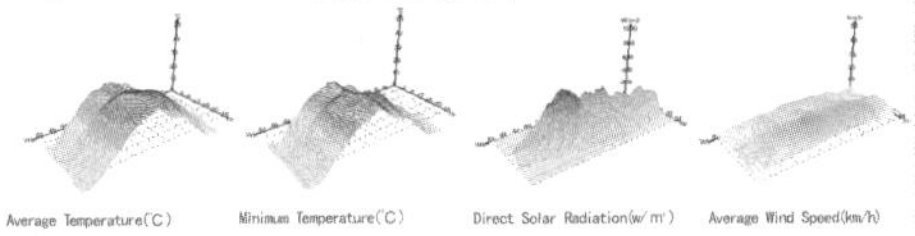
Average Temperature(°C)
Minimum Temperature(°C)
Direct Solar Radiation(w/m²)
Average Wind Speed(km/h)

♦ ACTIVITIES AND SOLAR ENERGY

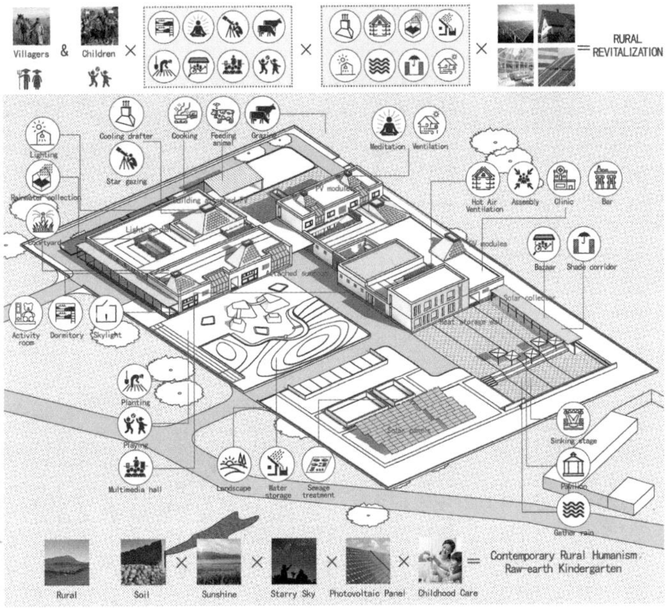
Villagers
&
Children
×
×
×
=
RURAL REVITALIZATION
Cooling drafter
Cooking
Feeding animal
Grazing
Meditation
Ventilation
Lighting
Star gazing
Rainwater collection
Courtyard
Hot Air Ventilation
Assembly
Clinic
Bar
PV modules
Bazaar
Shade corrider
Activity room
Dormitory
Skylight
Planting
Playing
Multimedia hall
Landscape
Water storage
Sewage treatment
Sinking stage
Pavilion
Gether rain
Rural
×
Soil
×
Sunshine
×
Starry Sky
×
Photovoltaic Panel
×
Childhood Care
=
Contemporary Rural Humanism
Raw-earth Kindergarten

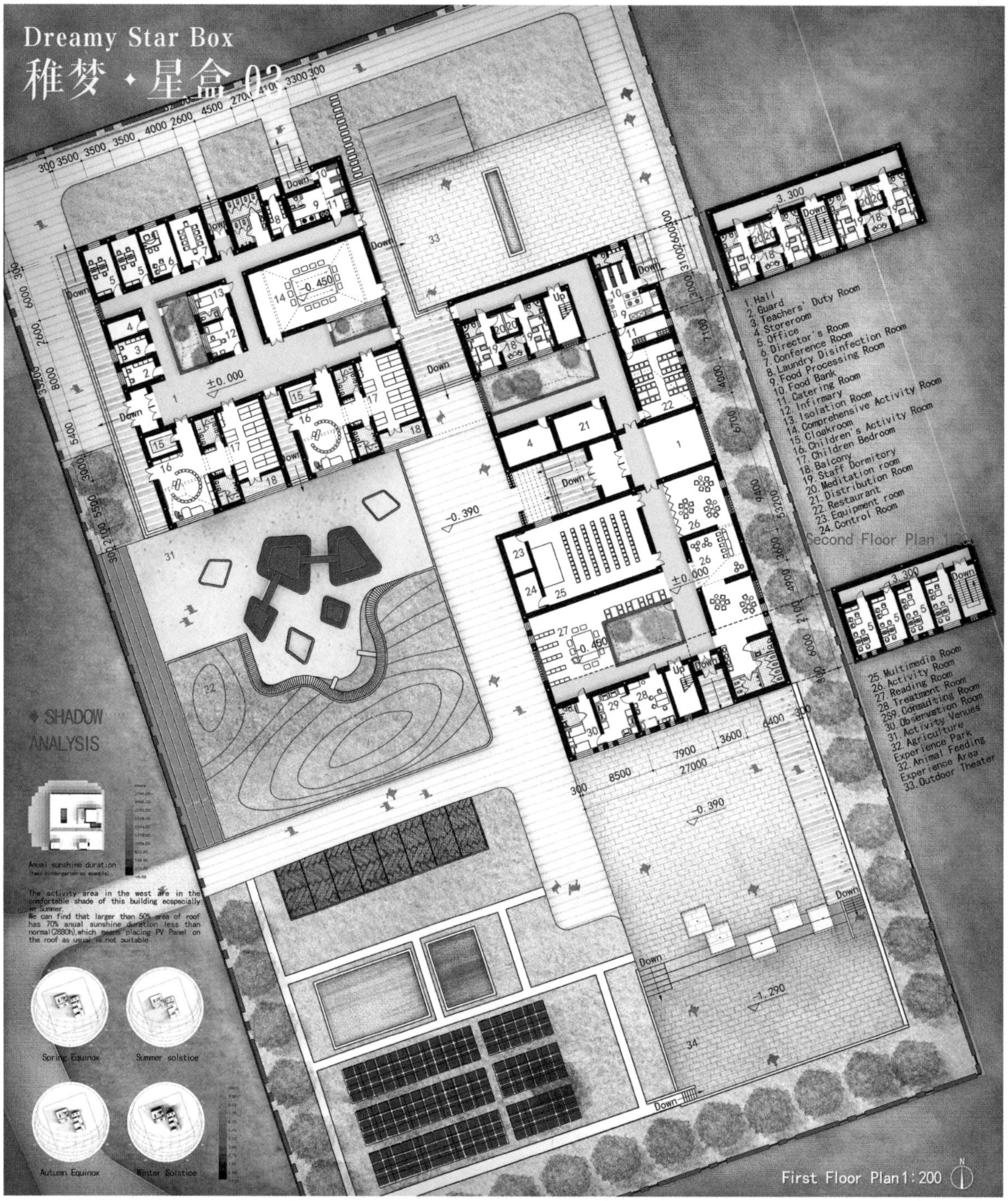
Dreamy Star Box
稚梦 · 星盒 03
1. Hall
2. Guard
3. Teachers' Duty Room
4. Storeroom
5. Office
6. Director's Room
7. Conference Room
8. Laundry Disinfection Room
9. Food Processing Room
10. Food Bank
11. Catering Room
12. Infirmary
13. Isolation Room
14. Comprehensive Activity Room
15. Cloakroom
16. Children's Activity Room
17. Children Bedroom
18. Balcony
19. Staff Dormitory
20. Meditation room
21. Distribution Room
22. Restaurant
23. Equipment room
24. Control Room
Second Floor Plan
25. Multimedia Room
26. Activity Room
27. Reading Room
28. Treatment Room
29. Consulting Room
30. Observation Room
31. Activity Venues
32. Agriculture Experience Park
32. Animal Feeding Experience Area
33. Outdoor Theater
SHADOW ANALYSIS
Anual sunshine duration
The activity area in the west are in the comfortable shade of this building ecspecially in Summer.
We can find that larger than 50% area of roof has 70% anual sunshine duration less than normal(2880h),which means placing PV Panel on the roof as usual is not suitable
Spring Equinox
Summer solstice
Autumn Equinox
Winter Solstice
First Floor Plan1:200

◆ MANUAL MODEL

Dreamy Star Box
稚梦·星盒 04

MEDITATION

ASSEMBLY

STAR GAZING

♦ SOLAR CHIMNEY CONCEPT

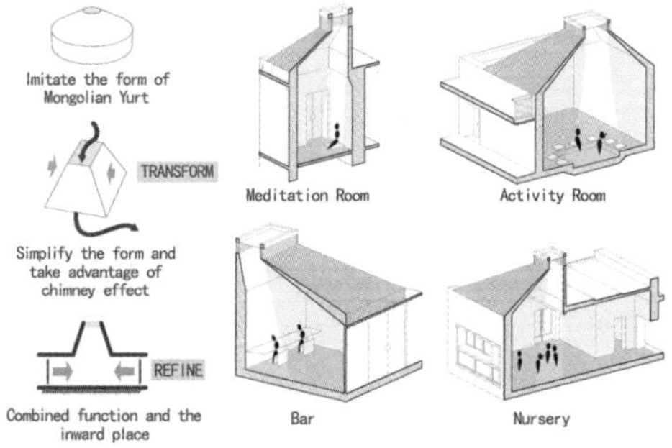
Imitate the form of Mongolian Yurt
TRANSFORM
Simplify the form and take advantage of chimney effect
REFINE
Combined function and the inward place
Meditation Room
Activity Room
Bar
Nursery

♦ SOLAR CHIMNEY ANALYSIS take Nursery as example

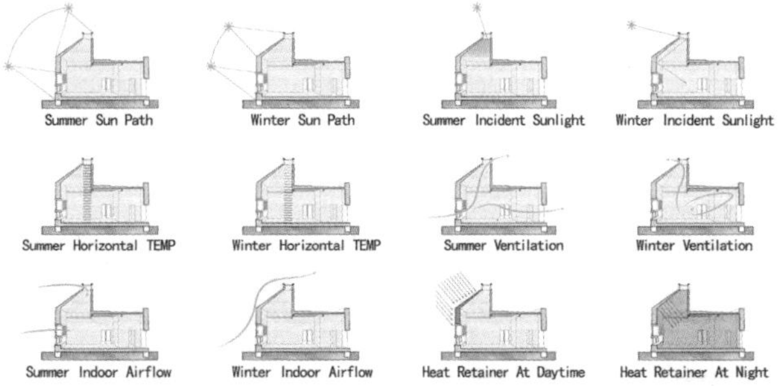
Summer Sun Path
Winter Sun Path
Summer Incident Sunlight
Winter Incident Sunlight
Summer Horizontal TEMP
Winter Horizontal TEMP
Summer Ventilation
Winter Ventilation
Summer Indoor Airflow
Winter Indoor Airflow
Heat Retainer At Daytime
Heat Retainer At Night

♦ SOLAR STRATEGY

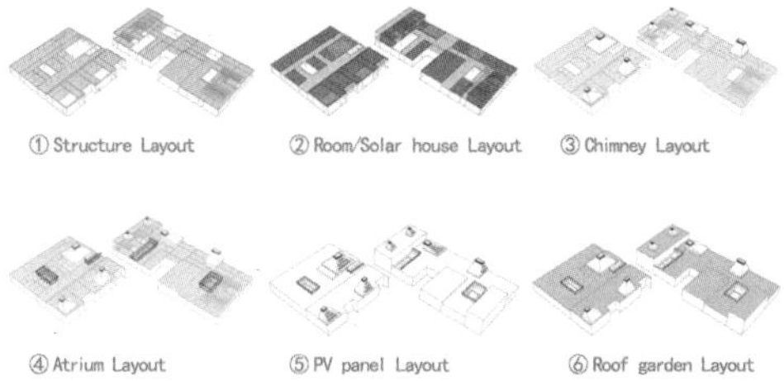
① Structure Layout
② Room/Solar house Layout
③ Chimney Layout
④ Atrium Layout
⑤ PV panel Layout
⑥ Roof garden Layout

First we choose local civil architecture as sample,which has thick wall made of raw soil that has excellent thermal performance.To take advantage of its excellent thermal performance,the external form is closed and we set Climney and atrium to deal with the problem of internal ventliation and light selecting. Besides,PV panel,Solar house,Roof garden are also set in suitable place to make full use of solar energy.

Keywords: Raw soil ,Solar house ,Solar chimney ,Atrium ,PV panel ,Roof garden

♦ UNIT DETAILS

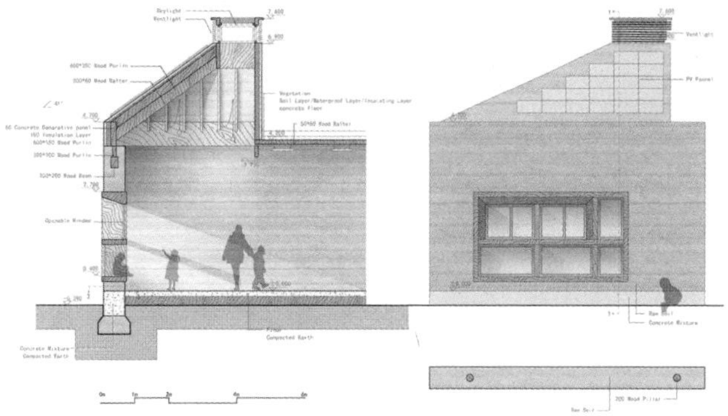

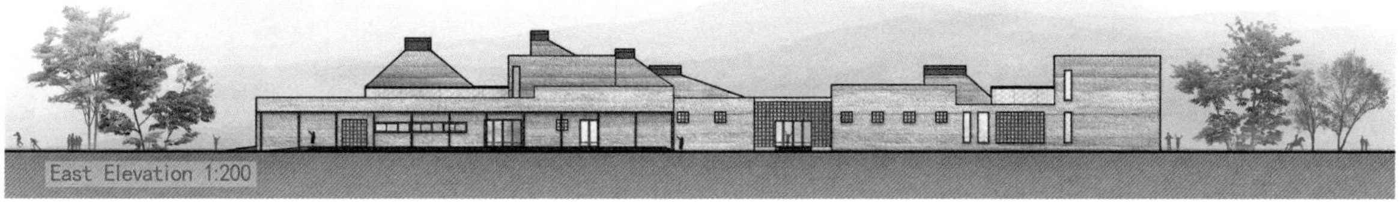
East Elevation 1:200

Dreamy Star Box
稚梦 · 星盒 05

Solar House

The solar house can hold the heat in winter daytime and release in cold winter night to reduce temperarture variation.

Solar Energy

Anual sunshine duration of Hejing is 2963h, which can be a great advantage in solar energy utilization.

Chimney Effect

There is a temperature difference in the large vertical space, which causes the air flow to rise and accelerated temperature adjustment

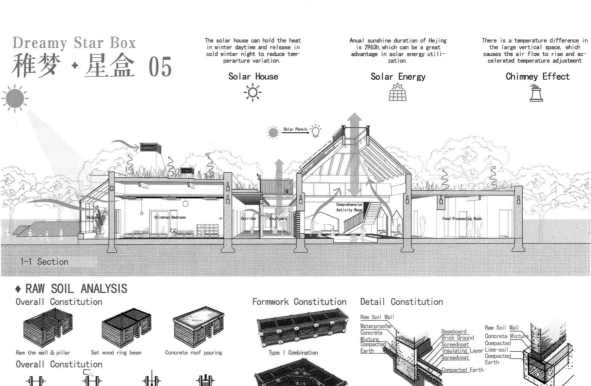

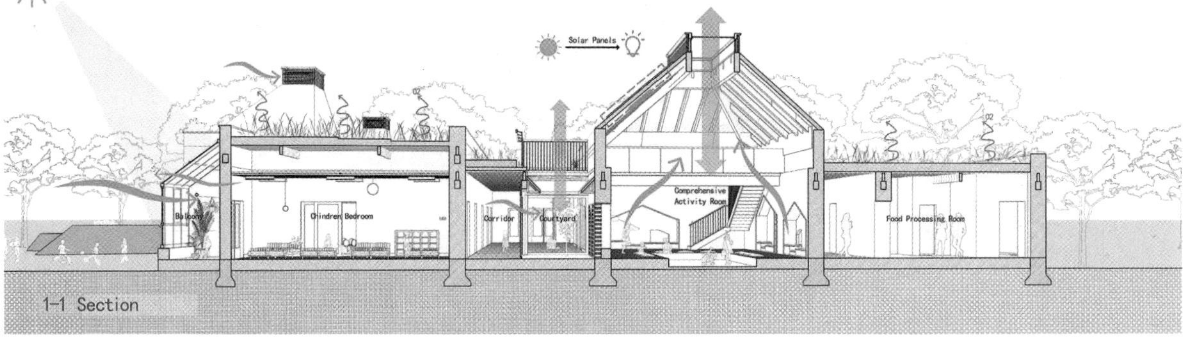
Solar Panels
Balcony
Children Bedroom
Corridor
Courtyard
Comprehensive Activity Room
Food Processing Room
1-1 Section

♦ RAW SOIL ANALYSIS

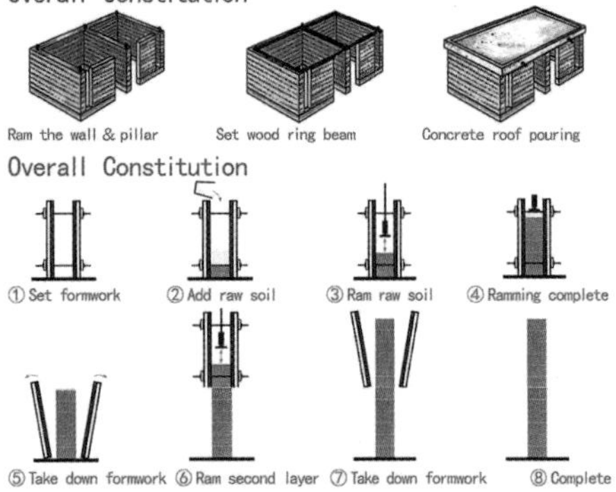
Overall Constitution
Ram the wall & pillar
Set wood ring beam
Concrete roof pouring
Overall Constitution
① Set formwork
② Add raw soil
③ Ram raw soil
④ Ramming complete
⑤ Take down formwork
⑥ Ram second layer
⑦ Take down formwork
⑧ Complete

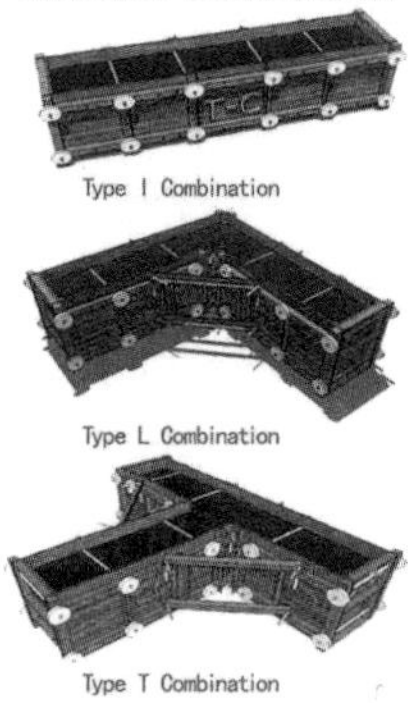
Formwork Constitution
Type I Combination
Type L Combination
Type T Combination

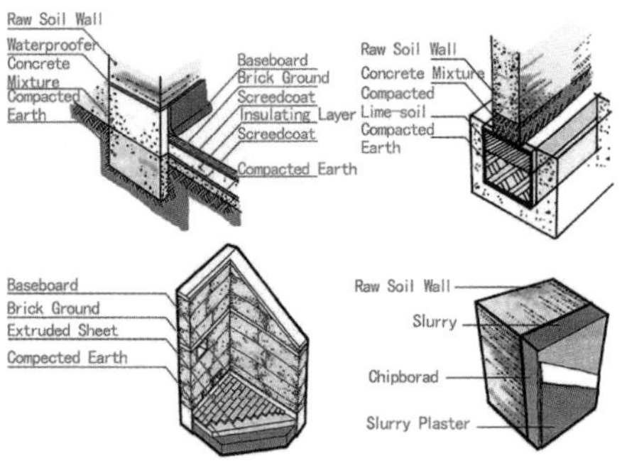
Detail Constitution
Raw Soil Wall
Waterproofer
Concrete Mixture
Compacted Earth
Baseboard
Brick Ground
Screedcoat
Insulating Layer
Screedcoat
Compacted Earth
Raw Soil Wall
Concrete Mixture
Compacted Lime-soil
Compacted Earth
Baseboard
Brick Ground
Extruded Sheet
Compected Earth
Raw Soil Wall
Slurry
Chipborad
Slurry Plaster

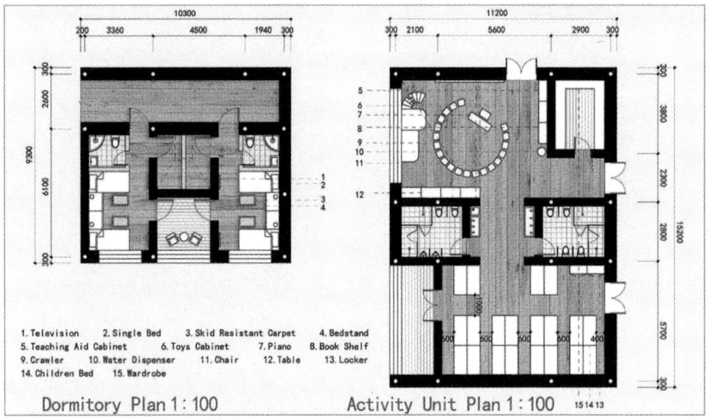
1. Television 2. Single Bed 3. Skid Resistant Carpet 4. Bedstand
5. Teaching Aid Cabinet 6. Toys Cabinet 7. Piano 8. Book Shelf
9. Crawler 10. Water Dispenser 11. Chair 12. Table 13. Locker
14. Children Bed 15. Wardrobe
Dormitory Plan 1 : 100

Activity Unit Plan 1 : 100

♦ CHILDREN SCALE AND UNITS

EVERY SQUARE IS 100mm×100mm

♦ PERFORMANCE ANALYSIS

Cost Analysis

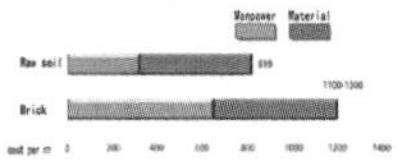
Manpower
Material
Raw soil
Brick

Compared with regular brick buildings, raw soil is almost 30 percent cheaper with extraordinary thermal performance, which makes raw soil a great choice for construction in remote areas and poor areas.

Thermal Performance

	Material	Density (Kg/m³)	Heat Conductivity [W/(m·K)]	Specific Heat [J/(kg·K)]
	Raw soil	2000	1.10	1500
	Concrete Brick	2300	1.63	1000
	Clay Brick	1700	0.7-1.4	800
	Steel	7800	45	480
	Wood	650	0.14	1200
	Stone	2300	1.8	1000

Energy Consumption Analysis

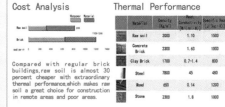

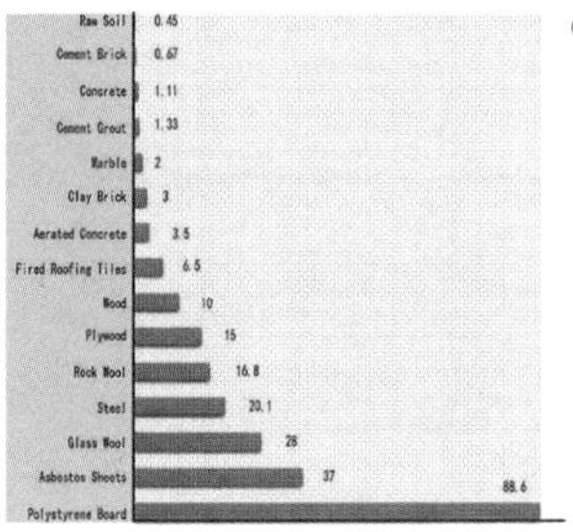
Raw Soil 0.45
Cement Brick 0.67
Concrete 1.11
Cement Grout 1.33
Marble 2
Clay Brick 3
Aerated Concrete 3.5
Fired Roofing Tiles 6.5
Wood 10
Plywood 15
Rock Wool 16.8
Steel 20.1
Glass Wool 28
Asbestos Sheets 37
Polystyrene Board 88.6
Energy consumption for processing (MJ/kg)

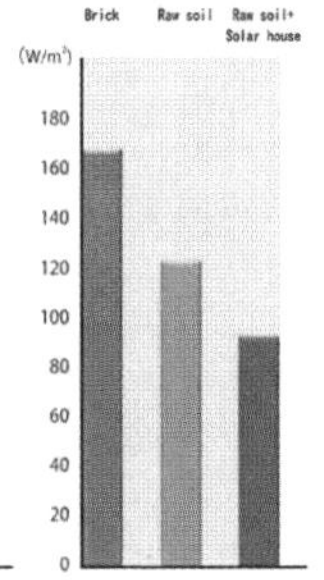
Brick
Raw soil
Raw soil+ Solar house
(W/m²)
180
160
140
120
100
80
60
40
20
0
Heating energy consumption

South Elevation 1:200

Dreamy Star Box
稚梦 · 星盒 06

♦ RAINWATER SYSTEM

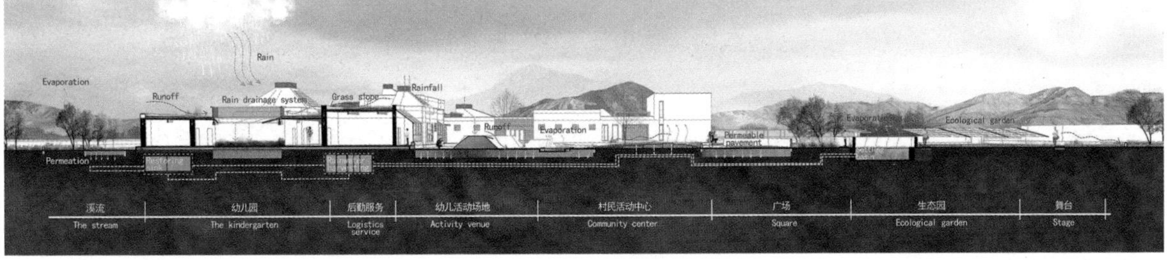

Rain
Evaporation
Runoff
Rain drainage system
Grass slope
Rainfall
Runoff
Evaporation
Permeable pavement
Evaporation
Ecological garden
Permeation
溪流
The stream
幼儿园
The kindergarten
后勤服务
Logistics service
幼儿活动场地
Activity venue
村民活动中心
Community center
广场
Square
生态园
Ecological garden
舞台
Stage

♦ TARGET ANALYSIS

Construction cost

Construction cost						
		raw soil wall	wood pillar&beam&purlin&rafter	roof	roof garden	Total
Building element	area	2186				
	cost/m²	320	20	40	20	
	cost	699520	43720	87440	43720	
Other element		furniture	sanitary ware	pv panel	solar chimney	
	area	2186	102	54	224	
	cost/m²	150	200	600	150	
	cost	327900	20400	32400	33600	1288700

Life Time Cycle Cost
* Reference 《Construction Manual of Rammed earth》 by Mujun

Life time cycle cost					
construction cost		raw soil wall	other building element	energy conservation element	Total
	area	2186			
	cost/m²	320	50	218	
	cost	699520	109300	476548	1285368
use-cost		heating	special maintenance	other energy	Total
	area	2186		40000	
	cost/m²	10	100		
	cost	21860	218600		280460
recovery cost	area	2186	scrap value	construction cost	1285368
	cost/m²	30		rate	0.03
	cost	65580		total	38561.04

The Green Building Assessing Standard Of China GB-T 50378-2019

The result of assessing		
	pre-evaluation	self-evaluation
Control basic score	400	400
Safty and durability	100	83
Healthy and comfortable	100	73
Convenience	100	69
Material saving	200	167
Environment livablity	100	76
Improvement and innovation	100	90
Total	1100	958
Final	100	87
Grade	一星级	

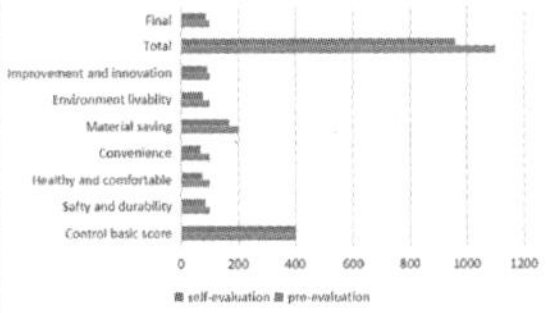

Final
Total
Improvement and innovation
Environment livability
Material saving
Convenience
Healthy and comfortable
Safty and durability
Control basic score
0
200
400
600
800
1000
1200
self-evaluation
pre-evaluation

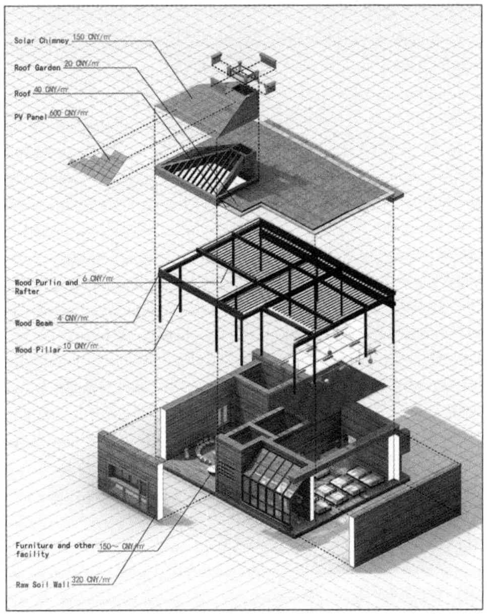

Solar Chimney 150 CNY/m²
Roof Garden 20 CNY/m²
Roof 40 CNY/m²
PV Panel 600 CNY/m²
Wood Purlin and Rafter 6 CNY/m²
Wood Beam 4 CNY/m²
Wood Pillar 10 CNY/m²
Furniture and other facility 150 CNY/m²
Raw Soil Wall 320 CNY/m²

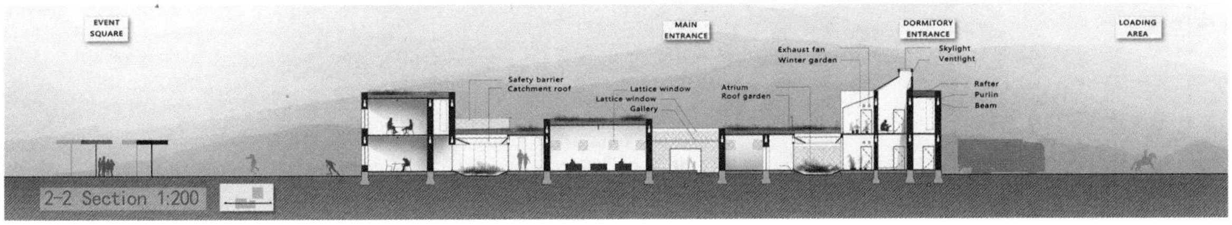

EVENT SQUARE
MAIN ENTRANCE
DORMITORY ENTRANCE
LOADING AREA
Exhaust fan
Winter garden
Skylight
Ventlight
Safety barrier
Catchment roof
Lattice window
Lattice window
Gallery
Atrium
Roof garden
Rafter
Purlin
Beam
2-2 Section 1:200

综合奖 · 二等奖
General Prize Awarded · Second Prize

注 册 号：7441
项目名称：江格尔与日出（新疆）
Janger and the Sunrise（Xinjiang）
作　　者：黄和谷、李崇玮、潘文涛
参赛单位：南京工业大学
指导教师：罗　靖

专家点评：
该作品引入传统地域文化，将相关文化传承中的文字描述以现代的手法进行表现，并营造出丰富而独特的室内外空间，整体设计品质较高。在太阳能集成设计方面，地域式的聚落中冒出太阳能光塔呈现出一种生态技术的识别性，具有创新性。然而，项目在植物利用的方面欠缺考虑。

This work embodies high design quality by introducing traditional regional culture to express the text in related cultural inheritance with modern techniques and create a large and unique indoor and outdoor space. With respect to integrated solar energy design, solar towers emerging from regional clusters present an identification of ecological technologies, which is rather innovative. Nevertheless, the consideration on the utilization of plants is lacking.

1

江格尔与日出
Janger and the Sunrise

建筑技术经济指标
technical and economic index
总建筑面积 Overall floorage：2120 ㎡
容积率 Plot ratio：0.203
建筑高度 Building height：11.3 m

简介 Brief

项目用地位于新疆巴音郭楞州和静县地区。在其众多的文化遗产中，江格尔史诗是不可忽视的内容。
江格尔史诗中有大量关于光、水、山等自然物的描摹。可见其在当地生活和文化中的价值。如今，我们通过现代建筑技术赋予传统价值以新的意义。例如，太阳能、水循环、通风技术等。我们希望儿童能够理解自己的文化和现代技术的关系，并从中找到关于故乡的记忆和认同感。

The site is located in Xinjiang . Among its numerous cultural heritages, Jangel's epic cannot be ignored.
Jangar's epic contains numerous depictions of natural objects such as light, water and mountains. We can see its value in local life and culture. Today, we are giving new meaning to traditional values through modern architectural techniques. Such as solar energy, water circulation, ventilation technology and so on.

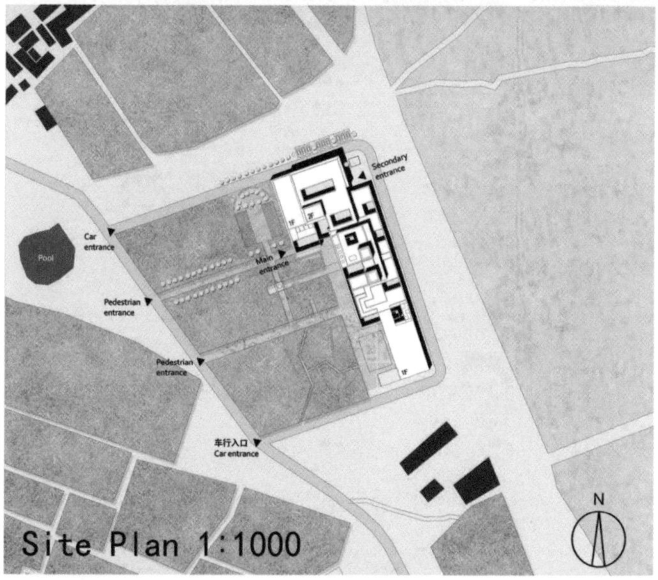
Secondary entrance
Car entrance
Pool
Main entrance
Pedestrian entrance
Pedestrian entrance
车行入口
Car entrance
N

Site Plan 1:1000

Demand Analysis

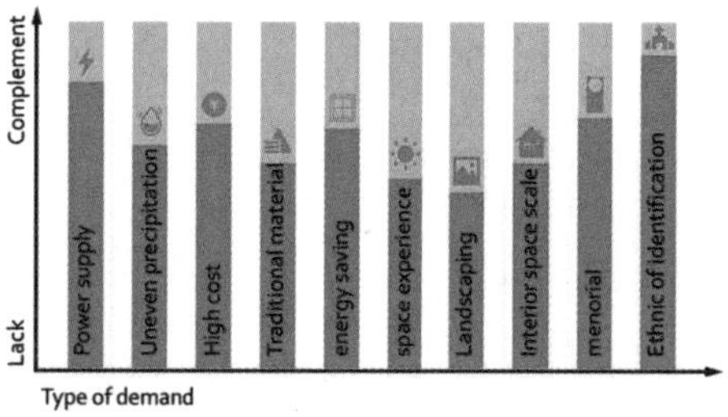
Complement
Lack
Power supply
Uneven precipitation
High cost
Traditional material
energy saving
space experience
Landscaping
Interior space scale
memorial
Ethnic of identification
Type of demand

The project land belongs to the continental climate of warm age, with long cold winter, windy winter and summer and autumn weather.
It is cool and the precipitation is mainly in summer. The surrounding natural landscape is rich, there are Han Great Wall, ancient city ruins.

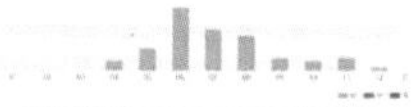

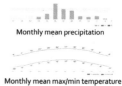

Monthly mean precipitation

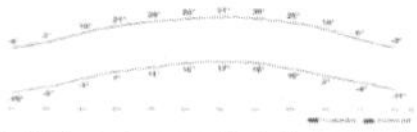

Monthly mean max/min temperature

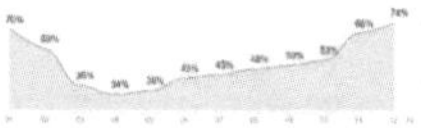

Monthly mean humidity

传统价值 · 现代技术

Traditional Values & Modern Technology

我们从江格尔史诗中截取了七个片段，每个片段对应一种现代技术。这些片段被串联起来，同其他空间一起组成了建筑的整体。

We have taken seven fragments from Jangar's epic, each of which corresponds to a modern technique. These fragments are connected in series to form the whole of the building.

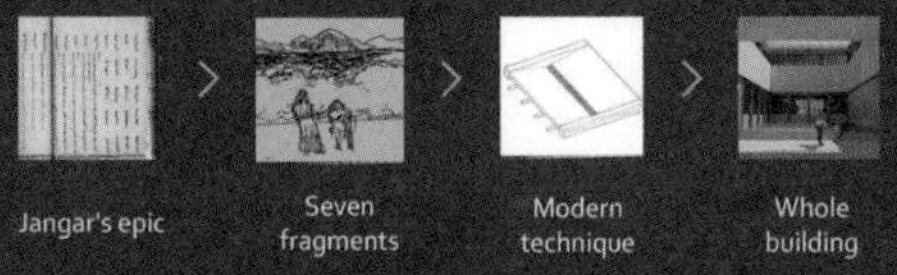

Jangar's epic
Seven fragments
Modern technique
Whole building

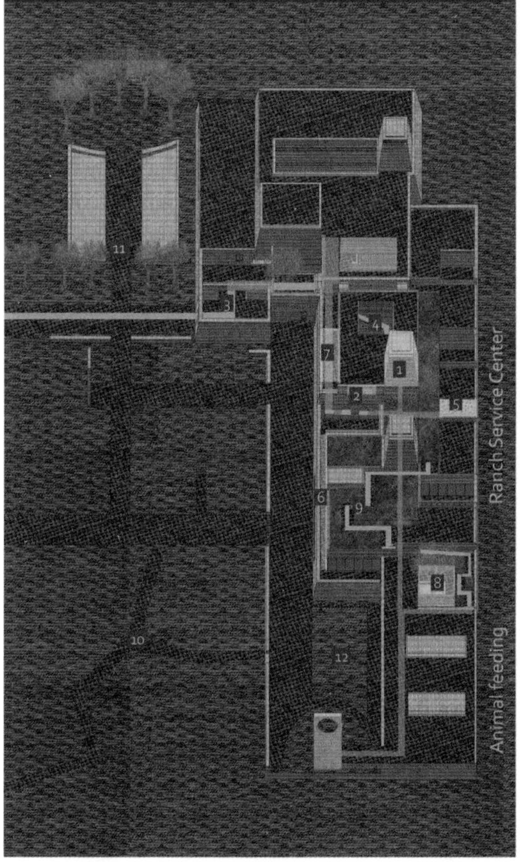

Ranch Service Center
Animal feeding

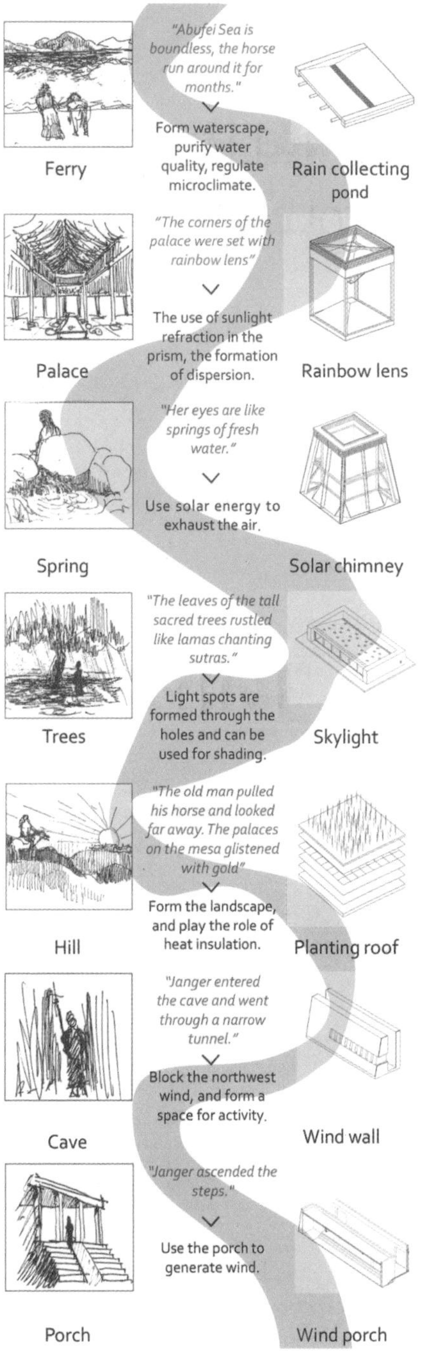

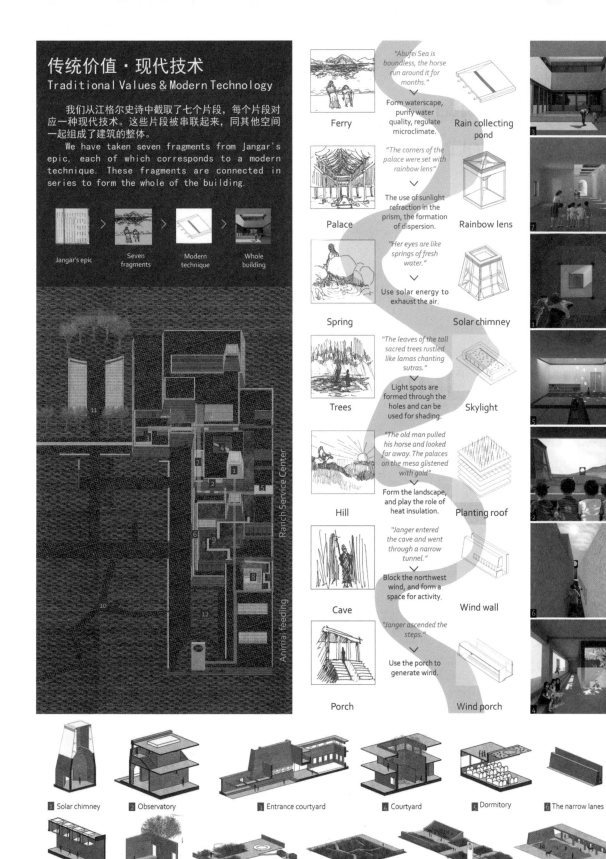

Ferry
"Abufei Sea is boundless, the horse run around it for months."
Form waterscape, purify water quality, regulate microclimate.
Rain collecting pond
Palace
"The corners of the palace were set with rainbow lens"
The use of sunlight refraction in the prism, the formation of dispersion.
Rainbow lens
Spring
"Her eyes are like springs of fresh water."
Use solar energy to exhaust the air.
Solar chimney
Trees
"The leaves of the tall sacred trees rustled like lamas chanting sutras."
Light spots are formed through the holes and can be used for shading.
Skylight
Hill
"The old man pulled his horse and looked far away. The palaces on the mesa glistened with gold"
Form the landscape, and play the role of heat insulation.
Planting roof
Cave
"Janger entered the cave and went through a narrow tunnel."
Block the northwest wind, and form a space for activity.
Wind wall
Porch
"Janger ascended the steps."
Use the porch to generate wind.
Wind porch

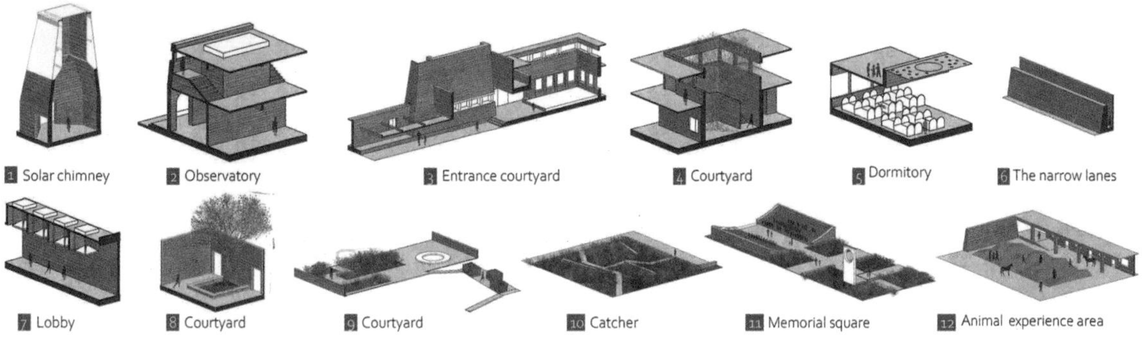

1 Solar chimney
2 Observatory
3 Entrance courtyard
4 Courtyard
5 Dormitory
6 The narrow lanes
7 Lobby
8 Courtyard
9 Courtyard
10 Catcher
11 Memorial square
12 Animal experience area

3

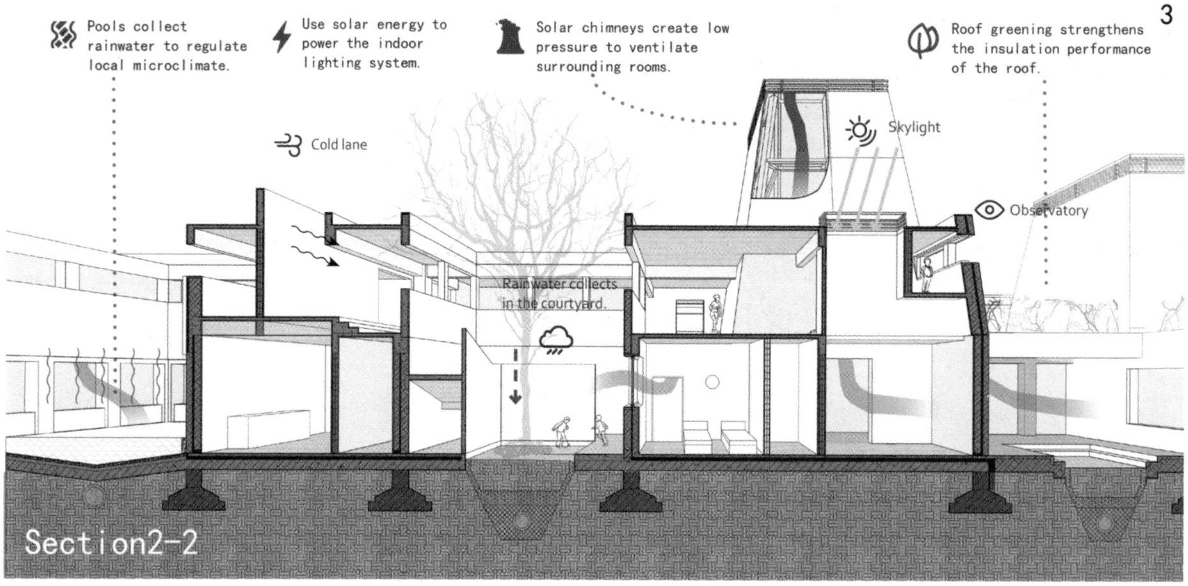
Pools collect rainwater to regulate local microclimate.
Use solar energy to power the indoor lighting system.
Solar chimneys create low pressure to ventilate surrounding rooms.
Roof greening strengthens the insulation performance of the roof.
Cold lane
Skylight
Observatory
Rainwater collects in the courtyard.
Section2-2

Seasonal Activity Range

Spring

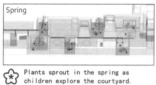

Plants sprout in the spring as children explore the courtyard.

Summer

Open the window sashes for natural ventilation and cool indoor corridors.

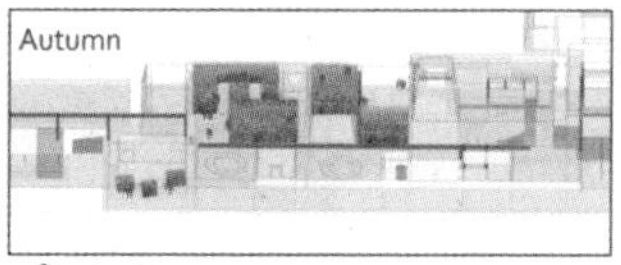
Autumn

Roof crops mature, children field to experience the fun of harvest.

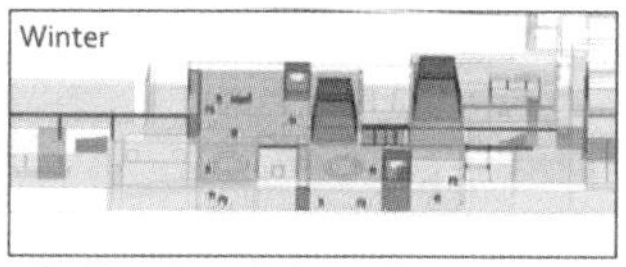
Winter

Use air conditioning for heating, sun room and skylights in winter.

Solar Chimney Structure

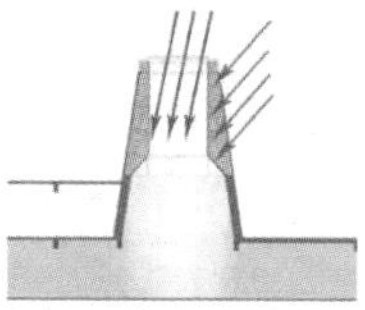

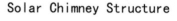

Thermal radiation figure

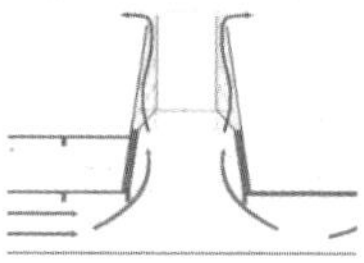

Ventilation figure

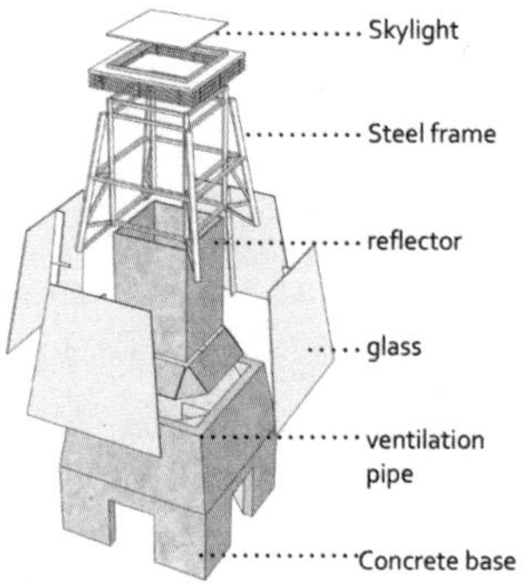
Skylight
Steel frame
reflector
glass
ventilation pipe
Concrete base

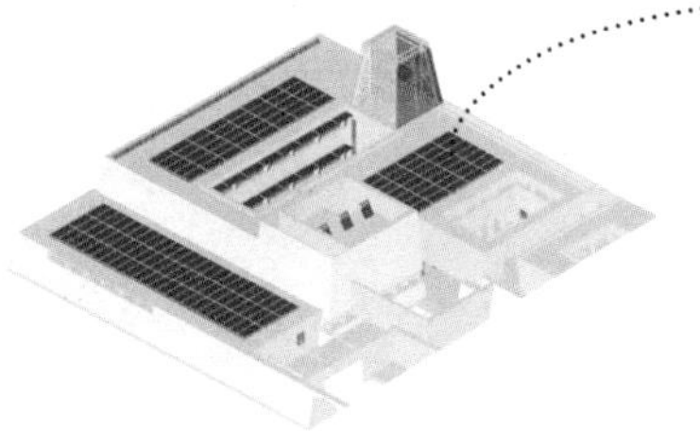

Solar power

Solar photovoltaic panels are installed on the roof ands to supply power for the building.

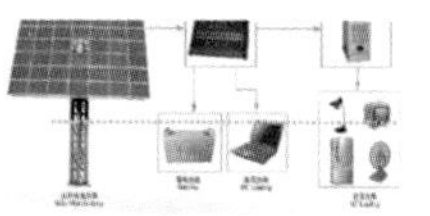

Rammed earth wall

Rammed earth walls can be built from local sources and built by themselves.

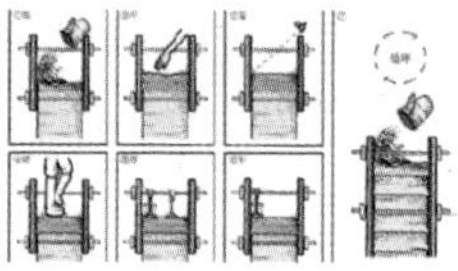

Geothermal heat

Rational use of geothermal energy through equipment.

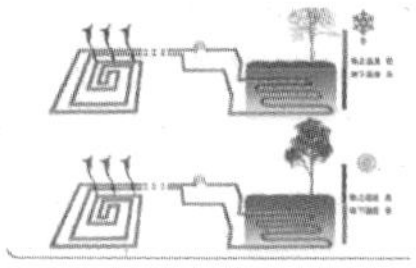

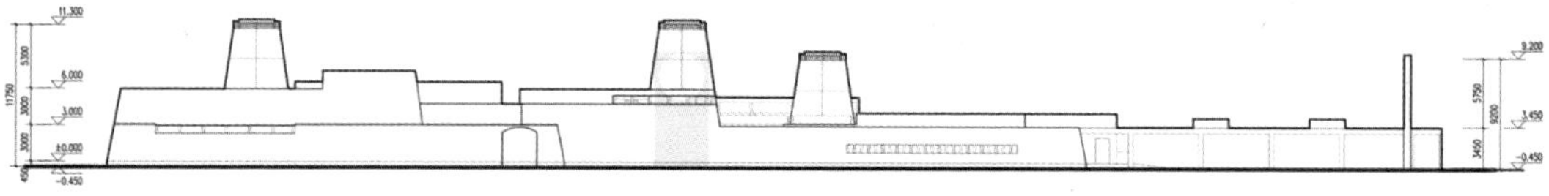
11.300
6.000
3.000
±0.000
-0.450
9.200
3.450

West Elevation 1：250

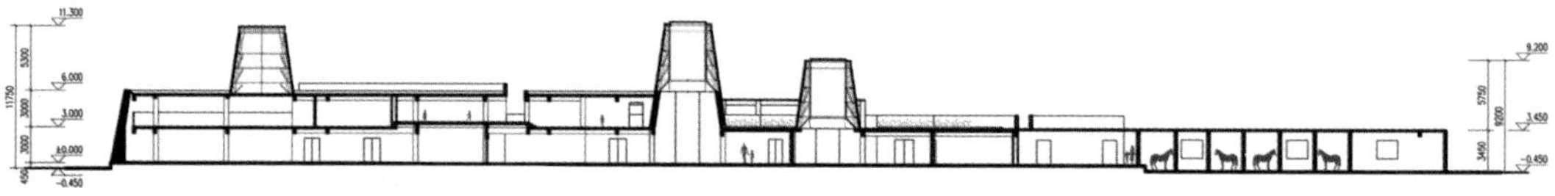
11.300
6.000
3.000
±0.000
-0.450
9.200
3.450

1-1 Section 1：250

Bloke Generation

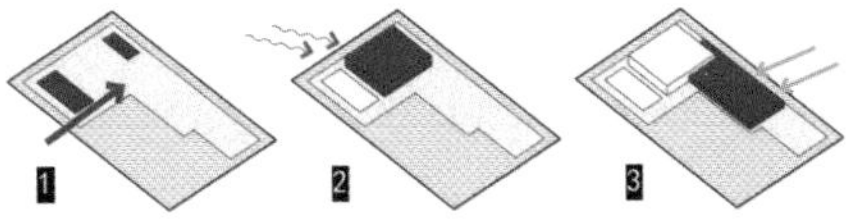

1 In the direction of the people, the buildings give way to form a square.

2 To the northwest of the site, a higher service center is placed to block the monsoon.

3 Kindergartens are placed on the south side, with better lighting.

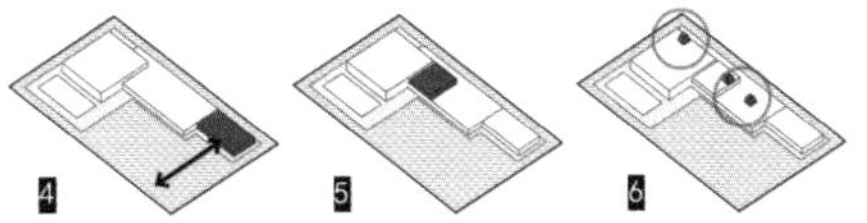

4 The southernmost animal feeding area is well integrated with the field.

5 A semi-outdoor space is added to the roof.

6 Add the chimney to exhaust the air indoors.

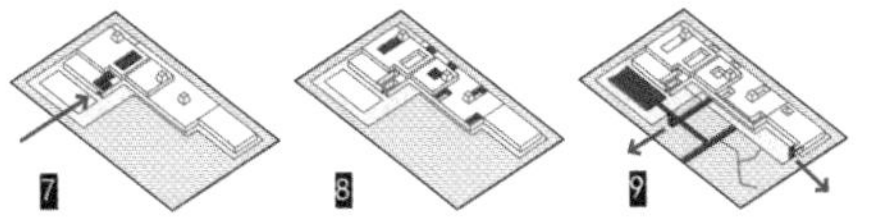

7 Place the foyer at the junction of the two volumes.

8 The courtyard is inserted for light and ventilation.

9 Arrange the site and put the monument in it.

the Wind Environment

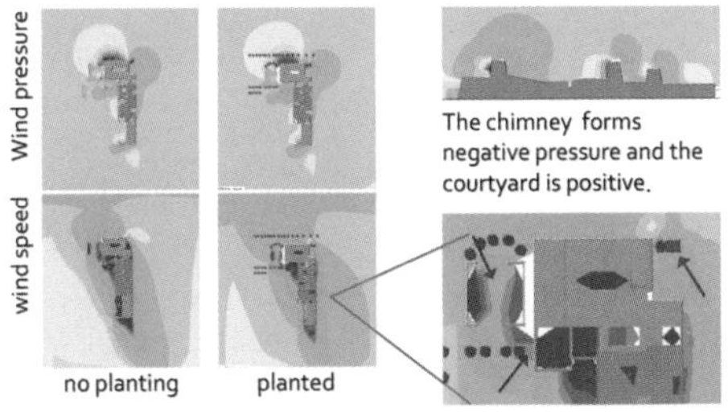
Wind pressure
wind speed
The chimney forms negative pressure and the courtyard is positive.
no planting
planted

Entrance and exit in the shadow of the wind, sand Add trees to improve the environment of squares and corners. At the same time, the garbage disposal room is placed in the area with high wind speed, which is conducive to air drying.

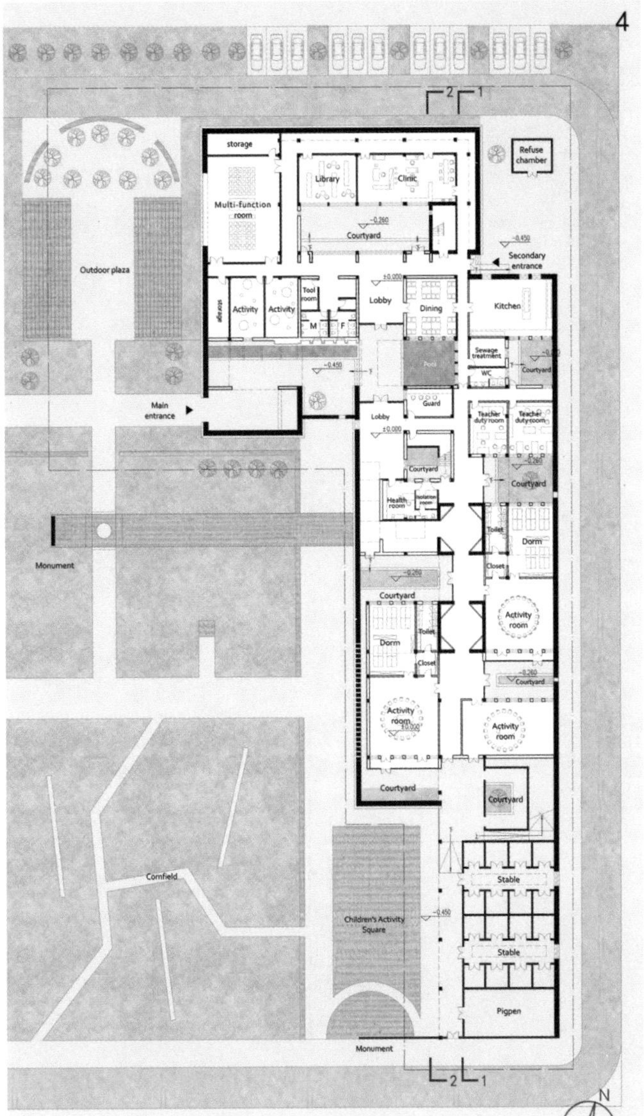
4
storage
Multi-function room
Library
Clinic
Courtyard
Refuse chamber
Secondary entrance
Outdoor plaza
Activity
Activity
Tool room
Lobby
Dining
Kitchen
Sewage treatment
WC
Courtyard
Main entrance
Guard
Lobby
Teacher duty room
Teacher duty room
Courtyard
Courtyard
Health room
Toilet
Dorm
Closet
Monument
Courtyard
Activity room
Dorm
Toilet
Closet
Courtyard
Activity room
Activity room
Courtyard
Courtyard
Cornfield
Stable
Children's Activity Square
Stable
Pigpen
Monument
N

1st Floor Plan 1：250

Explosive View

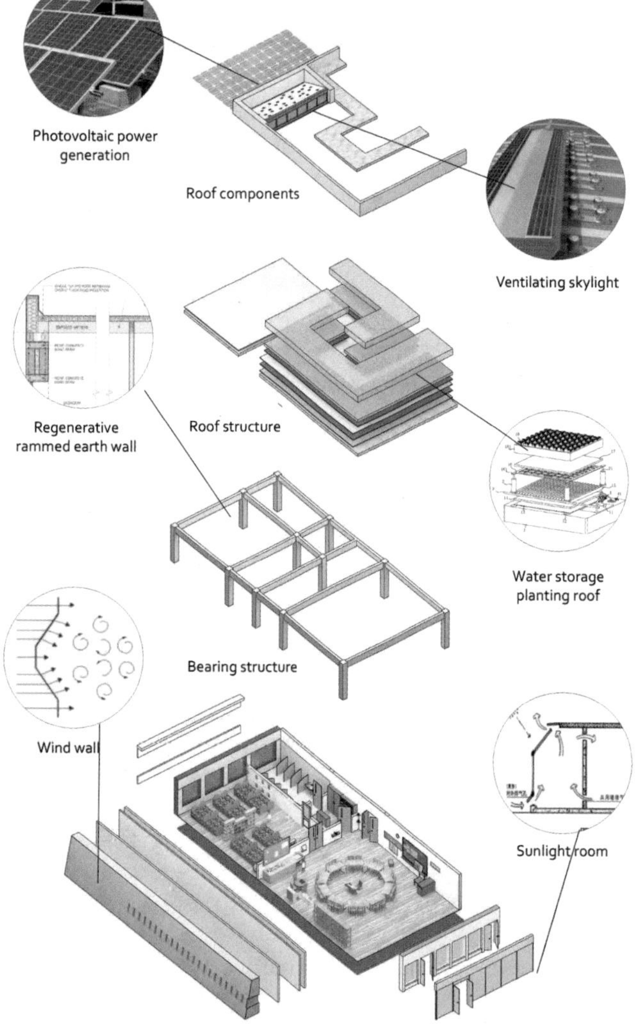
Photovoltaic power generation
Roof components
Ventilating skylight
Regenerative rammed earth wall
Roof structure
Water storage planting roof
Bearing structure
Wind wall
Sunlight room
Maintenance structure

5

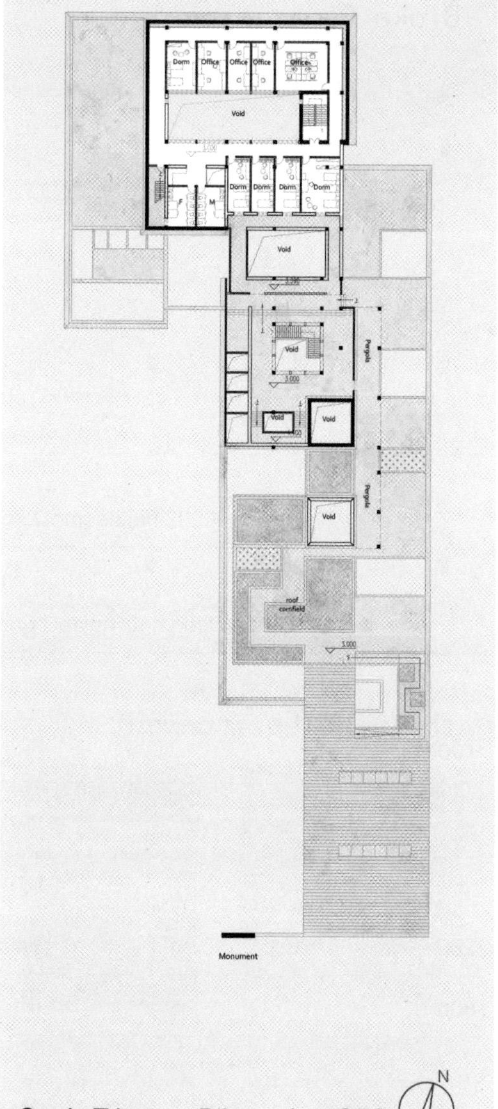
Dorm
Office
Void
Pergola
roof cornfield
Monument

2nd Floor Plan 1 : 250

Unit Thermal Analysis

Insulating glass

Use it for heat insulation, reduce building energy consumption.

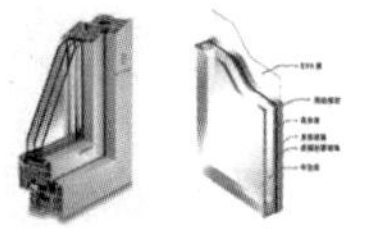

Thermal simulation

Through technologies, a better thermal environment is created for the unit and energy consumption is reduced.

Through analysis, it can be seen that the unit meets the relevant content in the kindergarten design specification.

Insulation roof

Heat insulation roof is used to improve building thermal performance, reduce building energy consumption and isolate solar radiation.

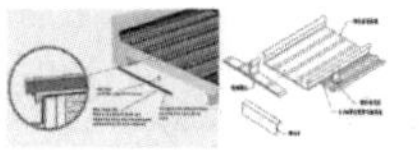

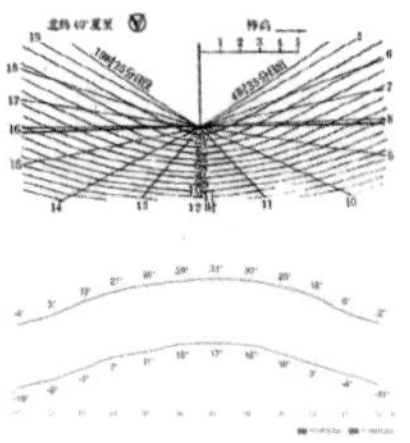

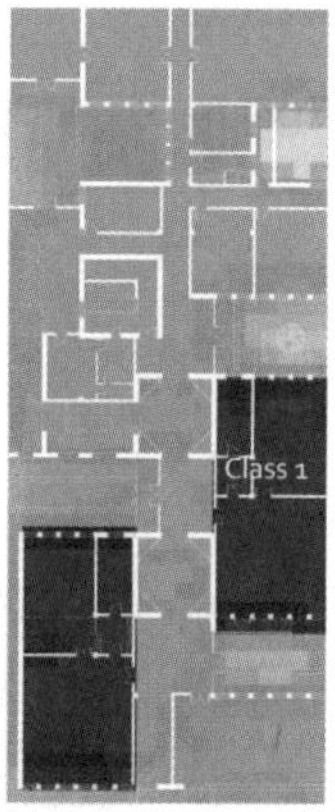
Class 1

Natural light

Reactor thermal analysis of class1

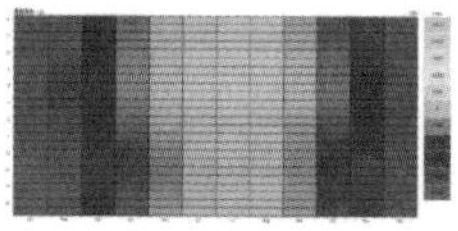

Ventilation heat gain

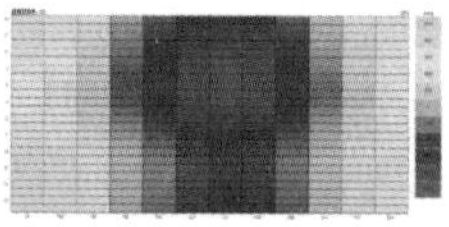

Interregional heat gain

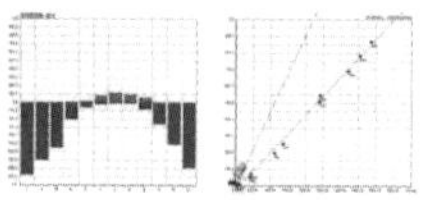

Heat number per month

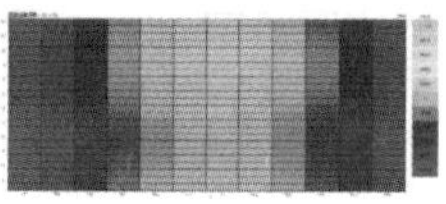

Heat gain of the envelope

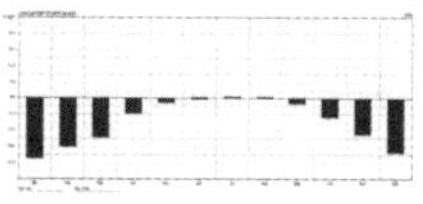

Descomfort degree hours

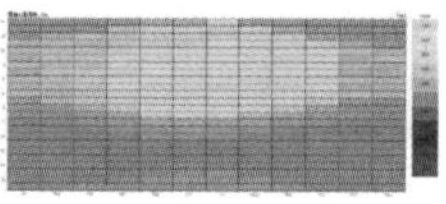

Indirect solar gain

6

Water Circulation

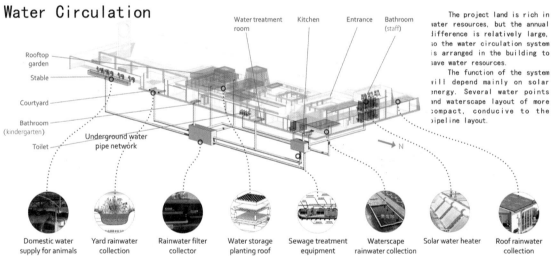

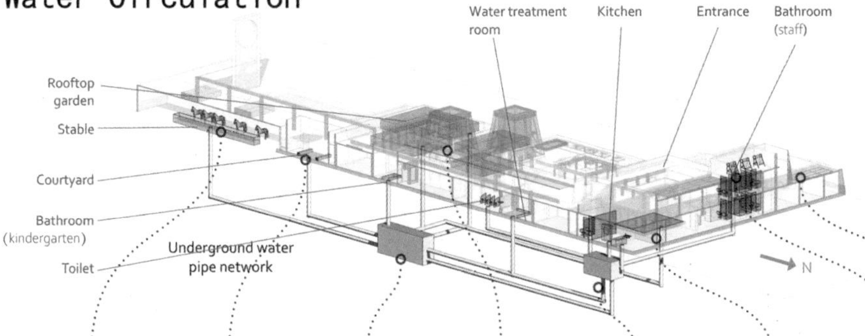
Water treatment room
Kitchen
Entrance
Bathroom (staff)
Rooftop garden
Stable
Courtyard
Bathroom (kindergarten)
Toilet
Underground water pipe network
N

The project land is rich in water resources, but the annual difference is relatively large, so the water circulation system is arranged in the building to save water resources.

The function of the system will depend mainly on solar energy. Several water points and waterscape layout of more compact, conducive to the pipeline layout.

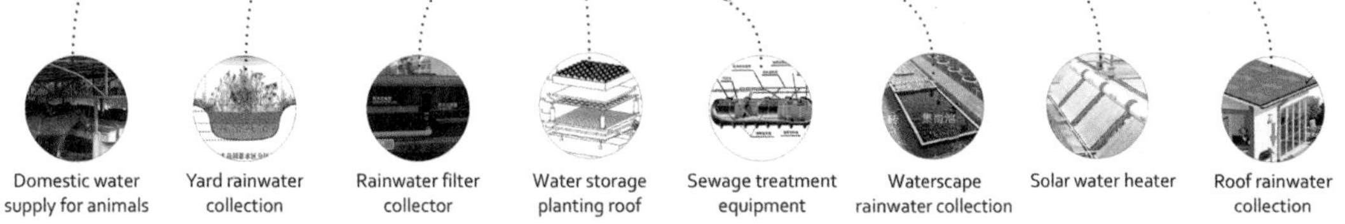
Domestic water supply for animals
Yard rainwater collection
Rainwater filter collector
Water storage planting roof
Sewage treatment equipment
Waterscape rainwater collection
Solar water heater
Roof rainwater collection

Materials and plants

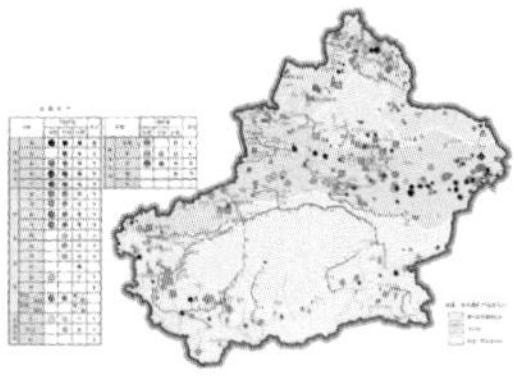

The construction materials use local materials, local materials, reduce transportation costs, and through new processing methods, improve the material related performance.

Local plants from nearby areas are planted on the roofs and courtyards of buildings, which promote children's understanding of the local ecology.

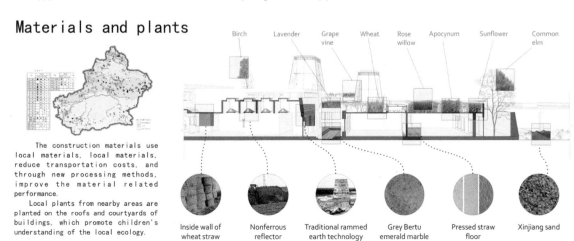

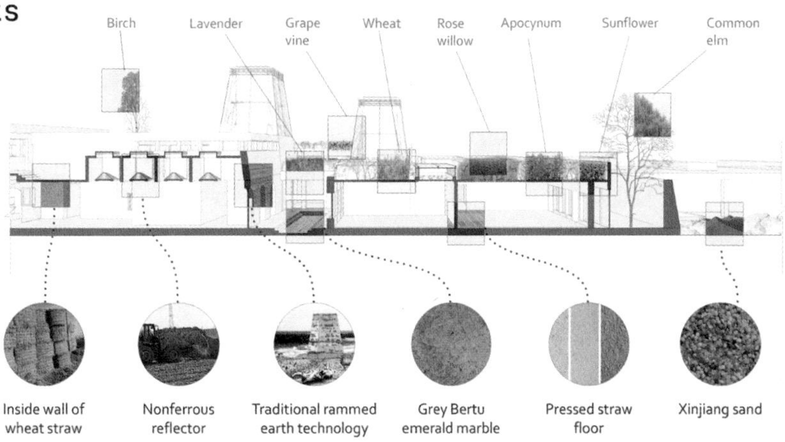
Birch
Lavender
Grape vine
Wheat
Rose willow
Apocynum
Sunflower
Common elm
Inside wall of wheat straw
Nonferrous reflector
Traditional rammed earth technology
Grey Bertu emerald marble
Pressed straw floor
Xinjiang sand

综合奖 · 三等奖
General Prize Awarded · Third Prize

注 册 号：6692

项目名称：风 · 跳跃空间（南平）
Wind · Jumping Space (Nanping)

作　　者：张世超、别　烨

参赛单位：昆明理工大学

指导教师：李莉萍

专家点评：

该作品规划设计合理、功能布局清晰。充分利用南北向的风巷、室内不同高度和室外错位平台，不仅满足了建筑通风、采光和遮阳的需求，还为儿童营造了活泼、悦动的成长与科普空间。主动和被动式技术符合当地气候特征与建筑文化。

The work shows reasonable plans and clear functional layout. It makes full use of the north-south wind ways, different heights indoors and dislocated platforms outdoors to meet the requirements for ventilation, lighting and sun shading, and create a lively growth and popularization of science space for children. Besides, active and passive technologies conform to local climate and architectural culture.

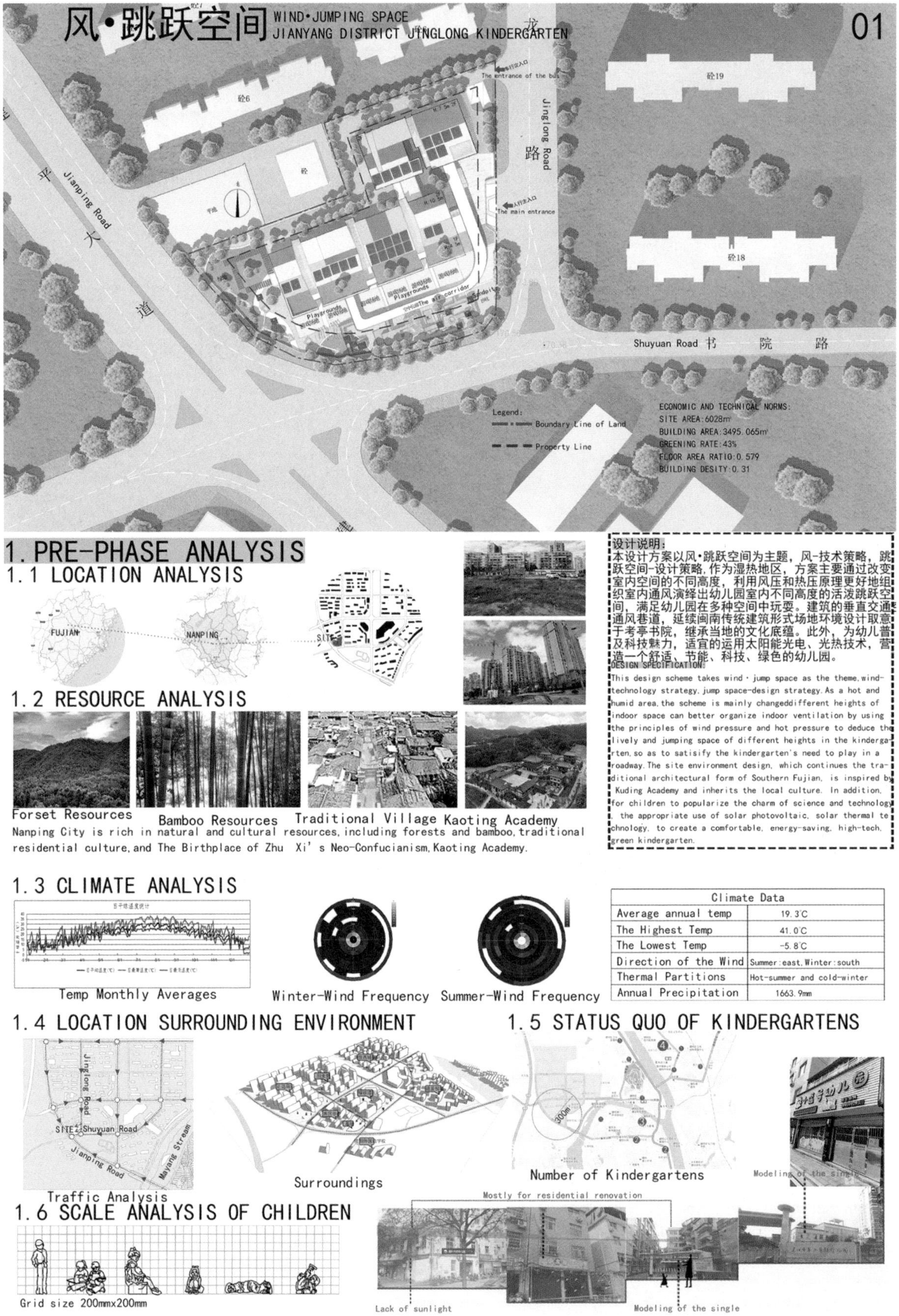

风•跳跃空间
WIND•JUMPING SPACE
JIANYANG DISTRICT JINGLONG KINDERGARTEN
01
Jianping Road
Jinglong Road
Shuyuan Road 书院路
The entrance of the bus
The main entrance
Legend:
Boundary Line of Land
Property Line
ECONOMIC AND TECHNICAL NORMS:
SITE AREA:6028㎡
BUILDING AREA:3495.065㎡
GREENING RATE:43%
FLOOR AREA RATIO:0.579
BUILDING DESITY:0.31
1. PRE-PHASE ANALYSIS
1.1 LOCATION ANALYSIS
FUJIAN
NANPING
SITE
设计说明:
本设计方案以风•跳跃空间为主题，风-技术策略，跳跃空间-设计策略，作为湿热地区，方案主要通过改变室内空间的不同高度，利用风压和热压原理更好地组织室内通风演绎出幼儿园室内不同高度的活泼跳跃空间，满足幼儿园在多种空间中玩耍。建筑的垂直交通通风巷道，延续闽南传统建筑形式场地环境设计取意于考亭书院，继承当地的文化底蕴。此外，为幼儿普及科技魅力，适宜的运用太阳能光电、光热技术，营造一个舒适、节能、科技、绿色的幼儿园。
DESIGN SPECIFICATION:
This design scheme takes wind · jump space as the theme, wind-technology strategy, jump space-design strategy. As a hot and humid area, the scheme is mainly changeddifferent heights of indoor space can better organize indoor ventilation by using the principles of wind pressure and hot pressure to deduce the lively and jumping space of different heights in the kindergarten, so as to satisfy the kindergarten's need to play in a roadway. The site environment design, which continues the traditional architectural form of Southern Fujian, is inspired by Kuding Academy and inherits the local culture. In addition, for children to popularize the charm of science and technology, the appropriate use of solar photovoltaic, solar thermal technology, to create a comfortable, energy-saving, high-tech, green kindergarten.
1.2 RESOURCE ANALYSIS
Forset Resources
Bamboo Resources
Traditional Village
Kaoting Academy
Nanping City is rich in natural and cultural resources, including forests and bamboo, traditional residential culture, and The Birthplace of Zhu Xi's Neo-Confucianism, Kaoting Academy.
1.3 CLIMATE ANALYSIS
Temp Monthly Averages
Winter-Wind Frequency
Summer-Wind Frequency
Climate Data
Average annual temp 19.3℃
The Highest Temp 41.0℃
The Lowest Temp -5.8℃
Direction of the Wind Summer:east, Winter:south
Thermal Partitions Hot-summer and cold-winter
Annual Precipitation 1663.9mm
1.4 LOCATION SURROUNDING ENVIRONMENT
Jinglong Road
SITE
Shuyuan Road
Jianping Road
Mayang Stream
Traffic Analysis
Surroundings
1.5 STATUS QUO OF KINDERGARTENS
300m
Number of Kindergartens
Modeling of the single
1.6 SCALE ANALYSIS OF CHILDREN
Grid size 200mmx200mm
Mostly for residential renovation
Lack of sunlight
Modeling of the single

风·跳跃空间 WIND·JUMPING SPACE JIANYANG DISTRICT JINGLONG KINDERGARTEN

02

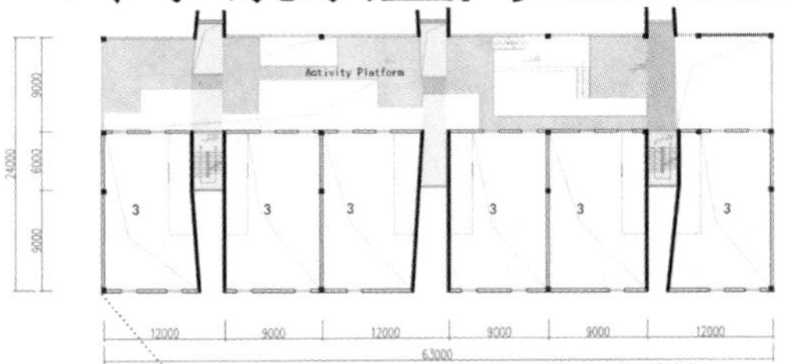
Activity Platform

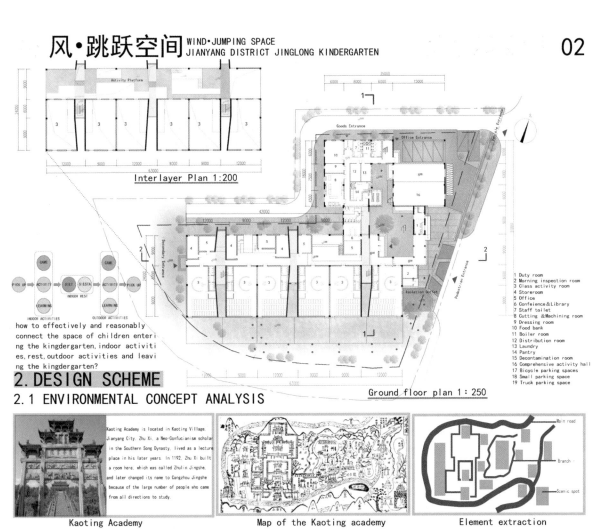

Interlayer Plan 1:200

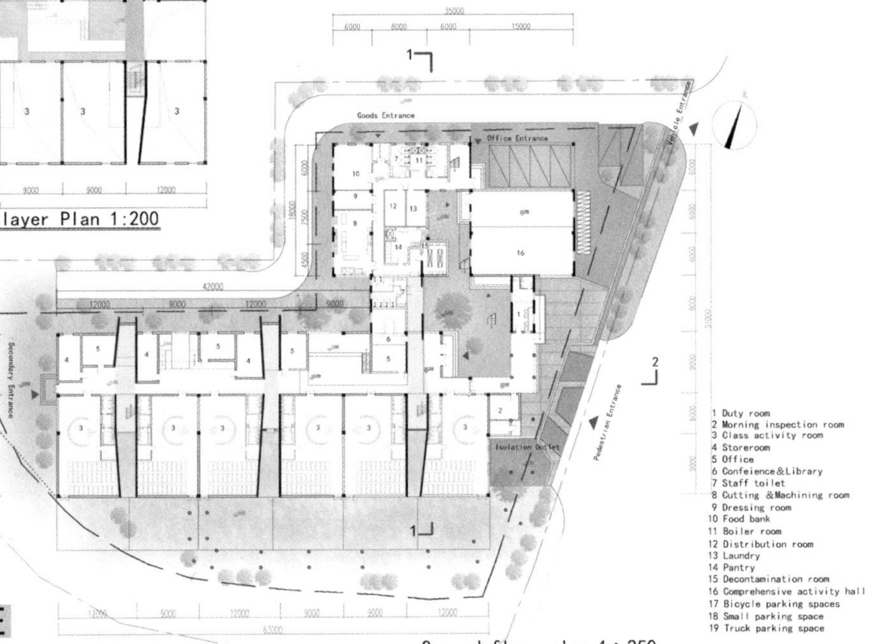
Goods Entrance
Office Entrance
Vehicle Entrance
Secondary Entrance
Pedestrian Entrance
Isolation Outlet
1 Duty room
2 Morning inspection room
3 Class activity room
4 Storeroom
5 Office
6 Confeience&Library
7 Staff toilet
8 Cutting &Machining room
9 Dressing room
10 Food bank
11 Boiler room
12 Distribution room
13 Laundry
14 Pantry
15 Decontamination room
16 Comprehensive activity hall
17 Bicycle parking spaces
18 Small parking space
19 Truck parking space

Ground floor plan 1：250

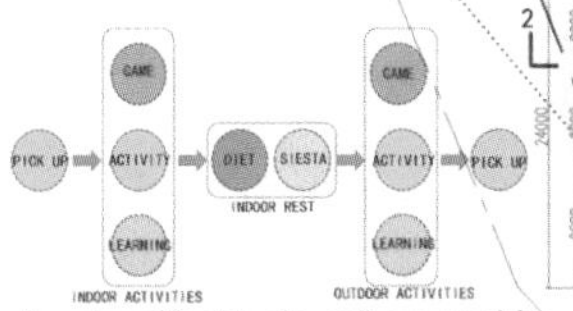
PICK UP
GAME
ACTIVITY
LEARNING
DIET
SIESTA
INDOOR REST
INDOOR ACTIVITIES
OUTDOOR ACTIVITIES

how to effectively and reasonably connect the space of children entering the kingdergarten, indoor activities, rest, outdoor activities and leaving the kingdergarten?

2. DESIGN SCHEME

2.1 ENVIRONMENTAL CONCEPT ANALYSIS

Kaoting Academy is located in Kaoting Village, Jianyang City. Zhu Xi, a Neo-Confucianism scholar in the Southern Song Dynasty, lived as a lecture place in his later years. In 1192, Zhu Xi built a room here, which was called Zhulin Jingshe, and later changed its name to Cangzhou Jingshe because of the large number of people who came from all directions to study.

Kaoting Academy

Map of the Kaoting academy

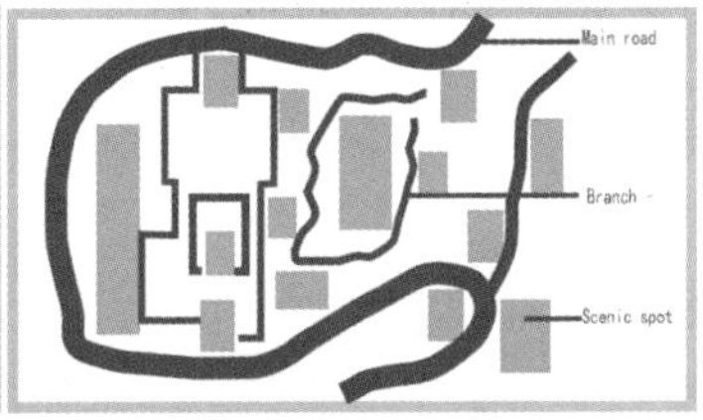
Main road
Branch
Scenic spot

Element extraction

2.3 ARCHITECTURAL DESIGN CONCEPT

Sources of concept

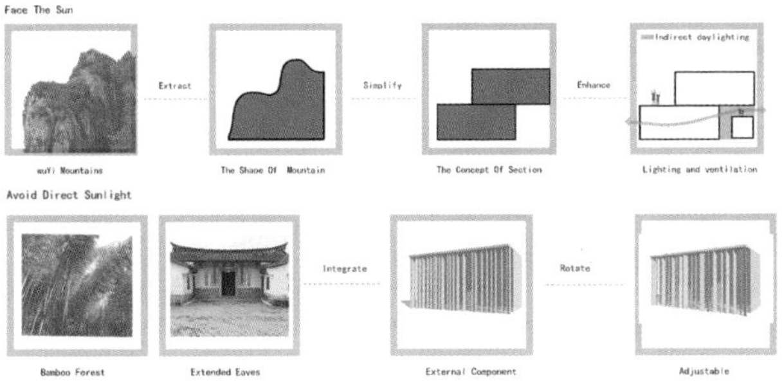
Face The Sun
Extract
Simplify
Enhance
Indirect daylighting
wuYi Mountains
The Shape Of Mountain
The Concept Of Section
Lighting and ventilation
Avoid Direct Sunlight
Integrate
Rotate
Bamboo Forest
Extended Eaves
External Component
Adjustable

Traditional Architecture Analysis

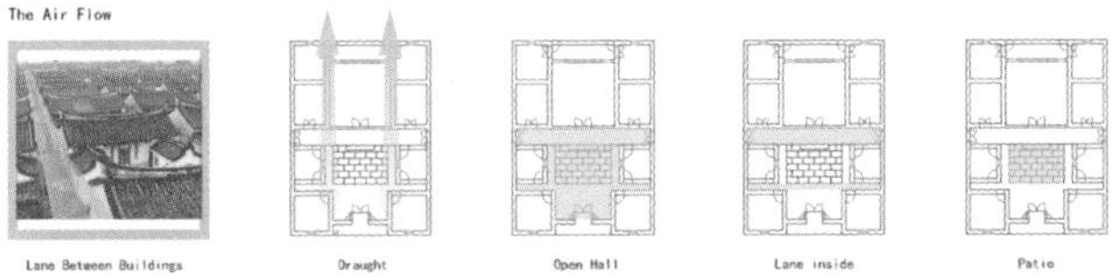
The Air Flow
Lane Between Buildings
Draught
Open Hall
Lane inside
Patio

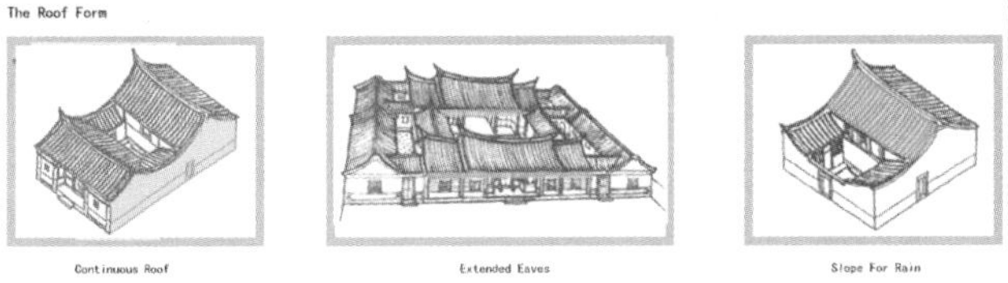
The Roof Form
Continuous Roof
Extended Eaves
Slope For Rain

2.2 SITE FUNCTION ANALYSIS

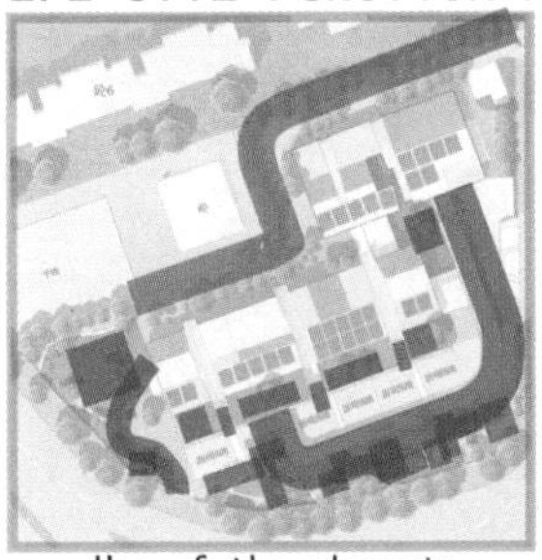

Use of the elements

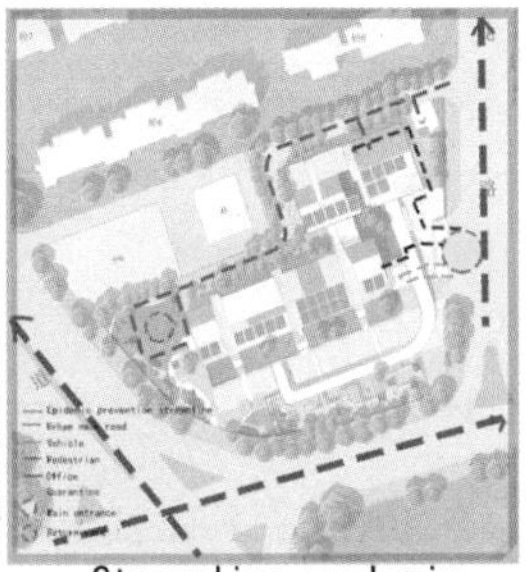

Streamline analysis

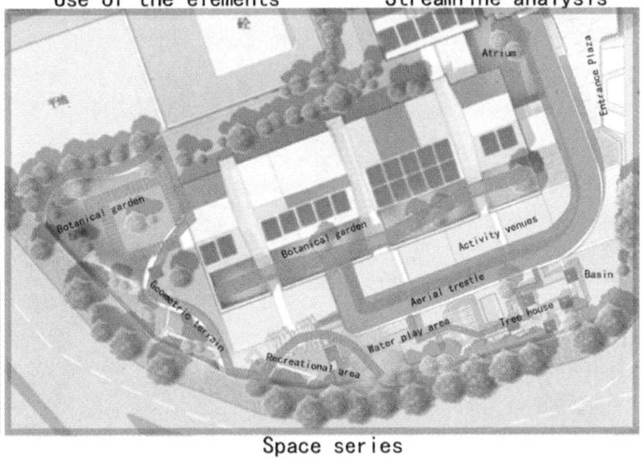
Atrium
Entrance Plaza
Botanical garden
Botanical garden
Activity venues
Basin
Aerial trestle
Tree house
Water play area
Recreational area
Geometric terrain

Space series

风•跳跃空间 WIND•JUMPING SPACE
JIANYANG DISTRICT JINGLONG KINDERGARTEN

03

2. DESIGN SCHEME

2.4 FORMATION OF ARCHITECTURAL FORM

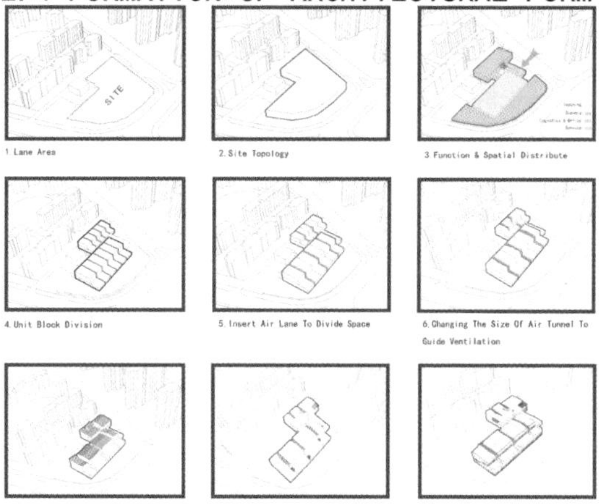
SITE
1 Lane Area
2. Site Topology
3 Function & Spatial Distribute
4 Unit Block Division
5 Insert Air Lane To Divide Space
6 Changing The Size Of Air Tunnel To Guide Ventilation
7 Creating Continuous Sloping Roofs
8 Add Skylight TO Improve Indoor lighting & Ventilation
9 Form architecture

2.5 SECOND FLOOR PLAN 1：250

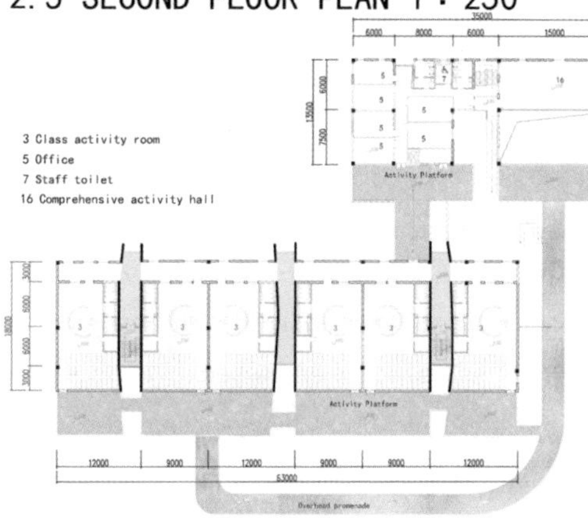
3 Class activity room
5 Office
7 Staff toilet
16 Comprehensive activity hall
35000
6000
8000
6000
15000
13500
6000
7500
Activity Platform
18000
3000
6000
6000
3000
12000
9000
12000
9000
9000
12000
63000
Overhead promenade

2.6 GAMES FACILITIES

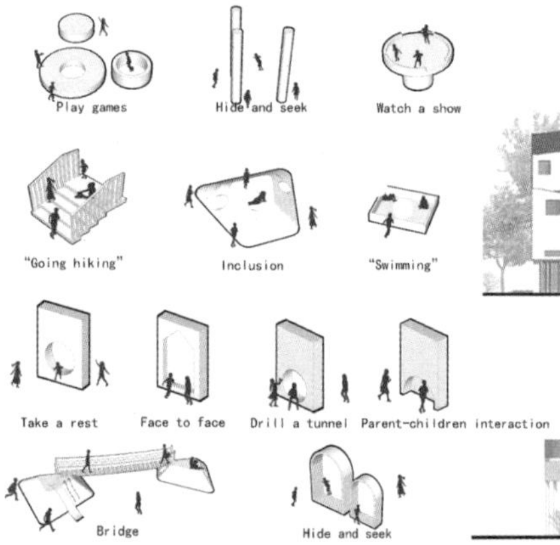
Play games
Hide and seek
Watch a show
"Going hiking"
Inclusion
"Swimming"
Take a rest
Face to face
Drill a tunnel
Parent-children interaction
Bridge
Hide and seek

2.7 SUNSHINE DURATION

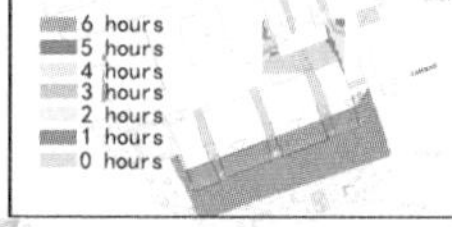
6 hours
5 hours
4 hours
3 hours
2 hours
1 hours
0 hours

Sunshine analysis time:Winter Solstice, 9:00-15:00
Sunshine analysis software:TSun

The kindergarten meets the requirement of three hours of sunshine on the winter solstice

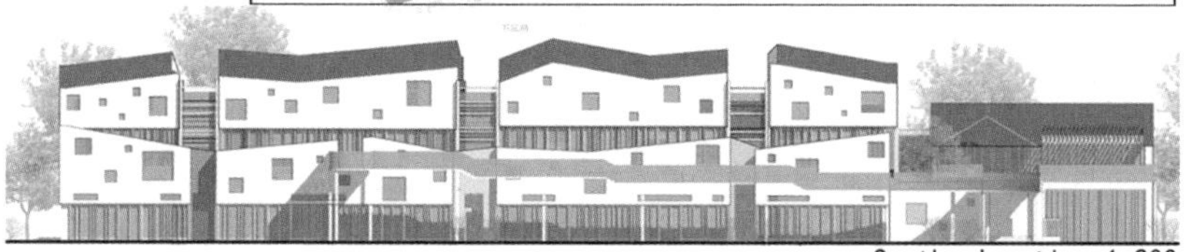

South elevation 1:200

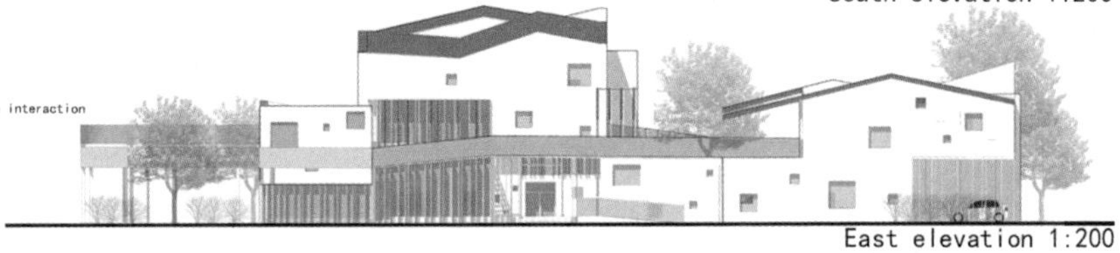

East elevation 1:200

风•跳跃空间 WIND•JUMPING SPACE JIANYANG DISTRICT JINGLONG KINDERGARTEN

04

2. DESIGN SCHEME

2.8 ENLARGED CLASS UNIT

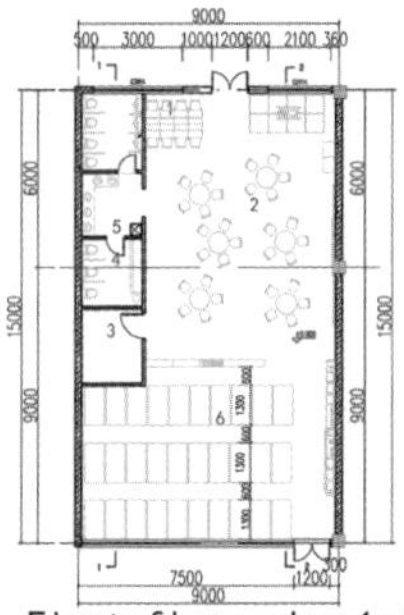
9000
15000
7500

First floor plan 1:50

In order to solve the larger depth of the class unit, bidirectional lighting and roof lighting tube are adopted in the unit to maximize the use of sunlight. Moreover, with the change of time, the light emitted by the tube into the room is also changing, creating different light and shadow spaces for children.

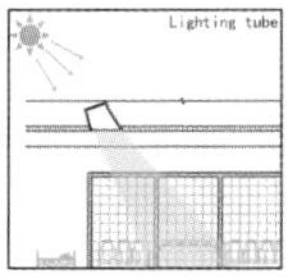
Lighting tube

1 Painting area
2 Activity room
3 Checkroom
4 Female toilet
5 Man toilet
6 Bedroom

Class unit interior renderings

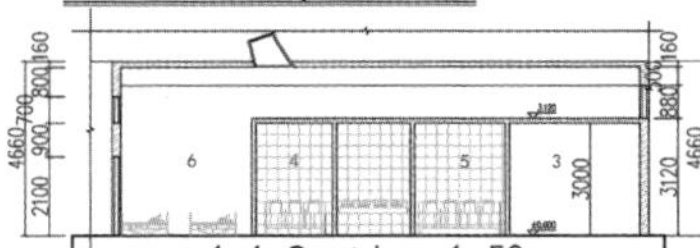

1-1 Section 1:50

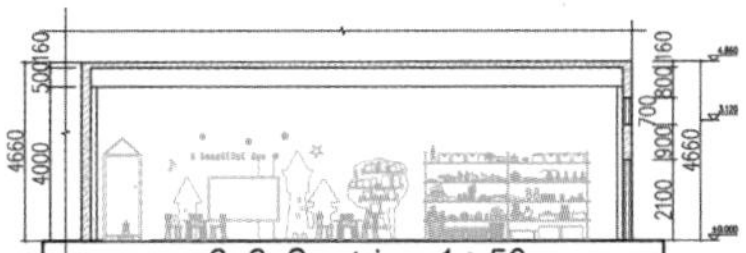

2-2 Section 1:50

2.9 INTERIOR PERSPECTIVE

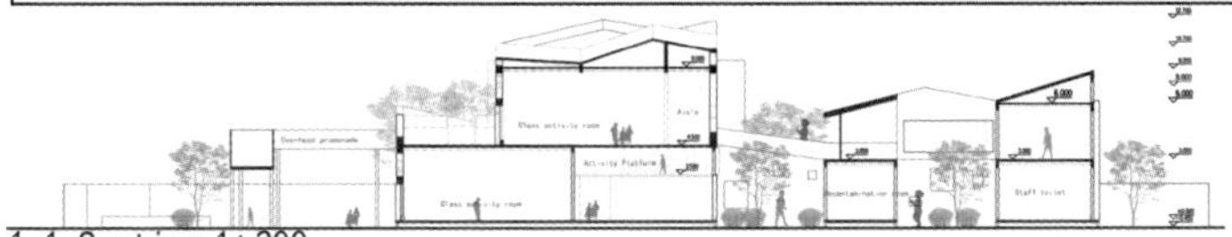

1-1 Section 1:200

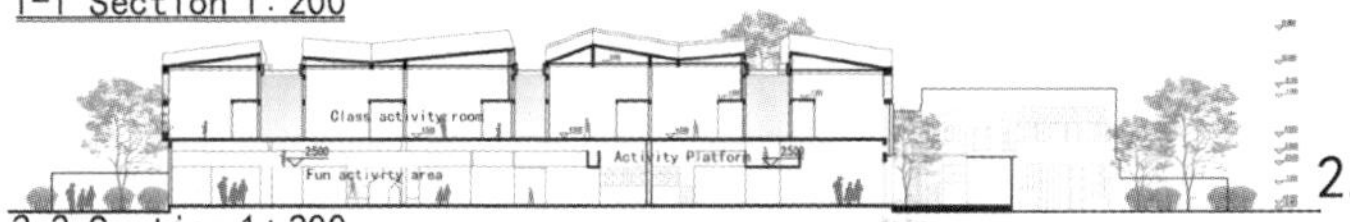
Class activity room
Activity Platform
Fun activity area

2-2 Section 1:200

West elevation 1:200

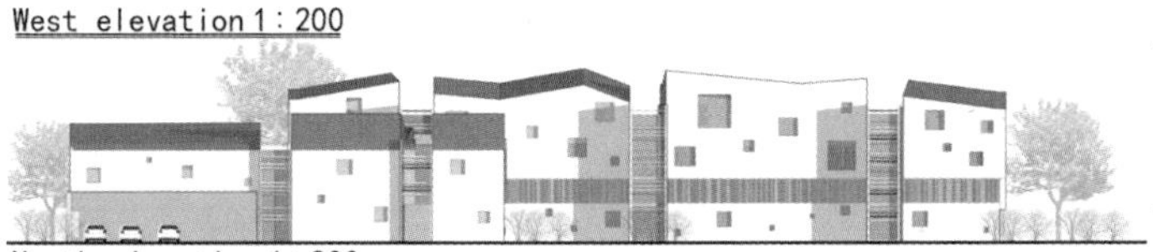

North elevation 1:200

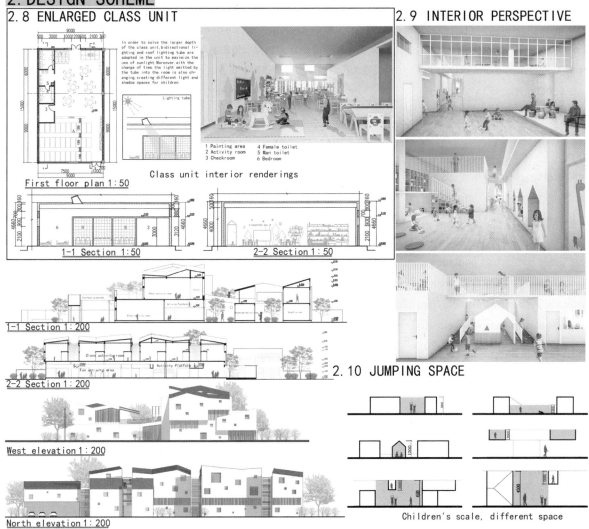

2.10 JUMPING SPACE

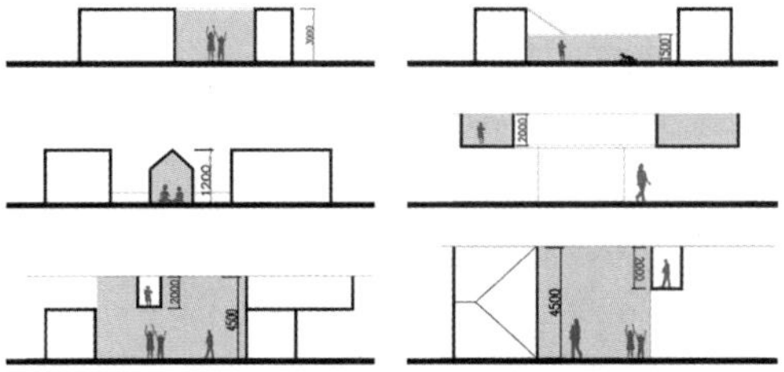

Children's scale, different space

风·跳跃空间 WIND·JUMPING SPACE JIANYANG DISTRICT JINGLONG KINDERGARTEN

05

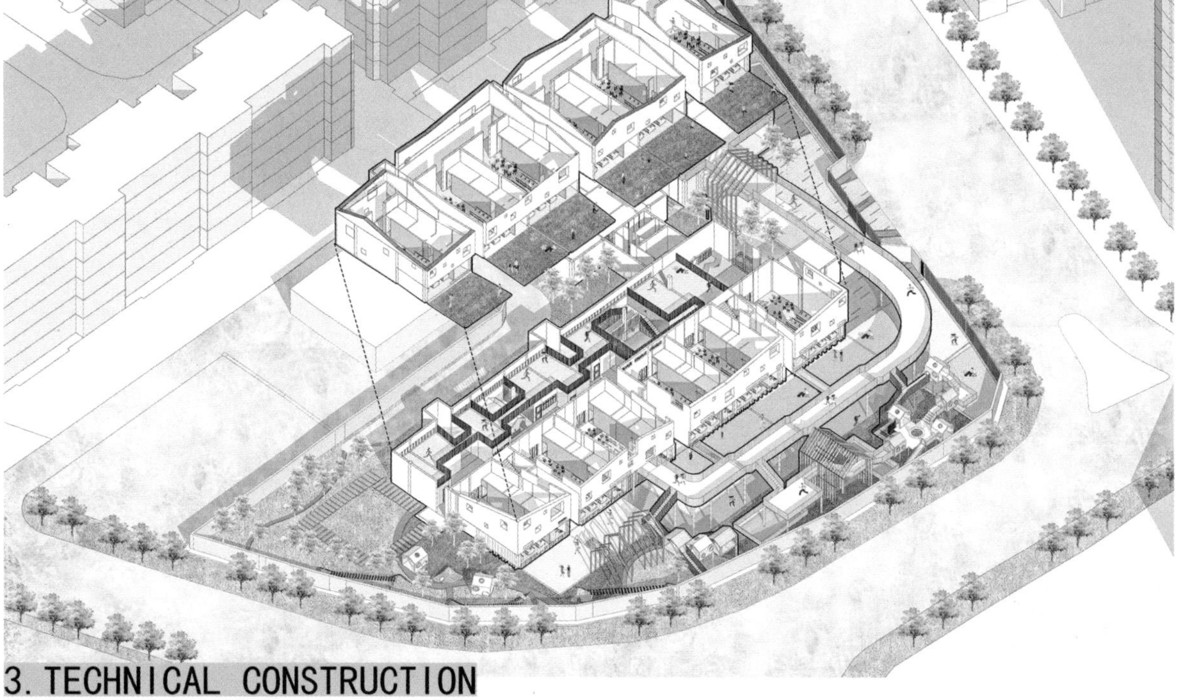

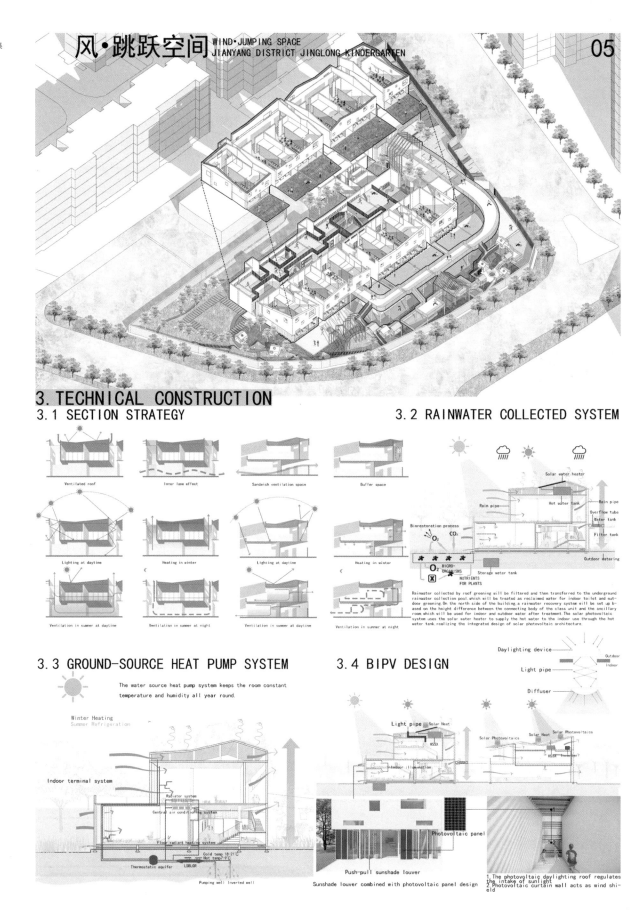

3. TECHNICAL CONSTRUCTION

3.1 SECTION STRATEGY

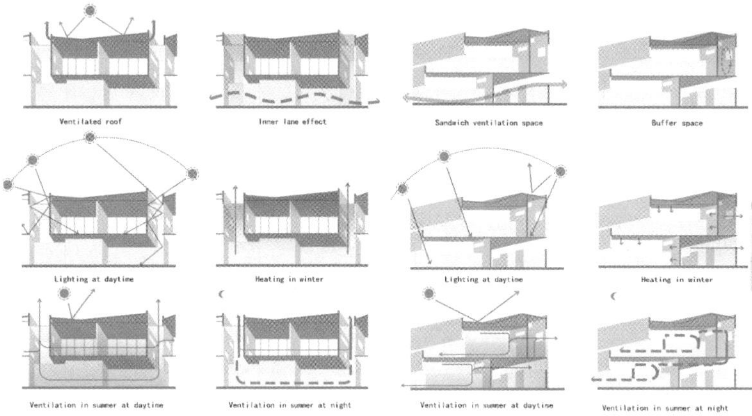

Ventilated roof
Inner lane effect
Sandwich ventilation space
Buffer space
Lighting at daytime
Heating in winter
Lighting at daytime
Heating in winter
Ventilation in summer at daytime
Ventilation in summer at night
Ventilation in summer at daytime
Ventilation in summer at night

3.2 RAINWATER COLLECTED SYSTEM

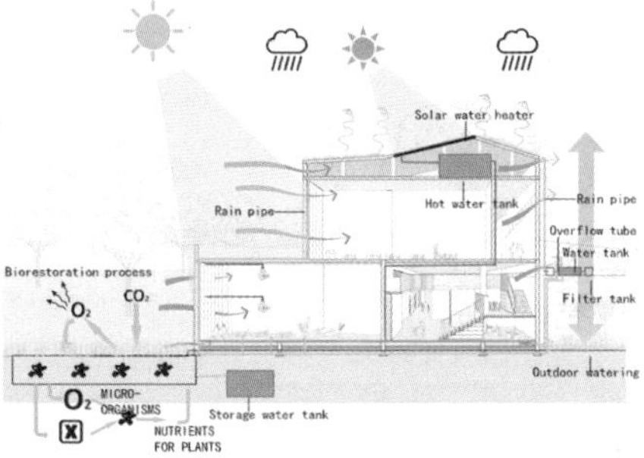

Solar water heater
Hot water tank
Rain pipe
Rain pipe
Overflow tube
Water tank
Filter tank
Biorestoration process
O_2
CO_2
Outdoor watering
O_2
MICRO-ORGANISMS
Storage water tank
NUTRIENTS FOR PLANTS

Rainwater collected by roof greening will be filtered and then transferred to the underground rainwater collection pool,which will be treated as reclaimed water for indoor toilet and outdoor greening.On the north side of the building,a rainwater recovery system will be set up based on the height difference between the connecting body of the class unit and the ancillary room,which will be used for indoor and outdoor water after treatment.The solar photovoltaic system uses the solar water heater to supply the hot water to the indoor use through the hot water tank,realizing the integrated design of solar photovoltaic architecture.

3.3 GROUND-SOURCE HEAT PUMP SYSTEM

The water source heat pump system keeps the room constant temperature and humidity all year round.

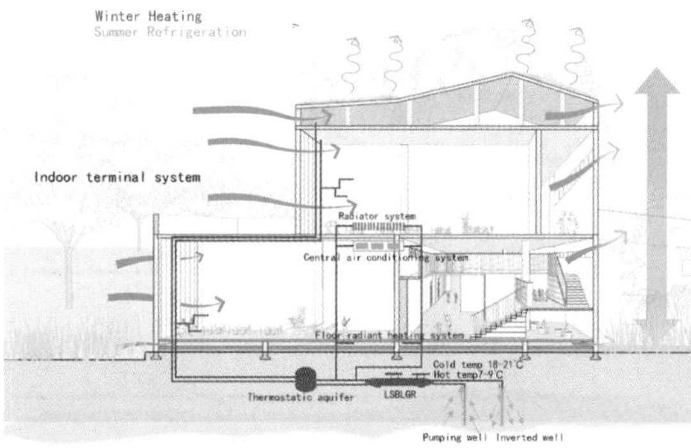

Winter Heating
Summer Refrigeration
Indoor terminal system
Radiator system
Central air conditioning system
Floor radiant heating system
Cold temp 18-21℃
Hot temp7-9℃
Thermostatic aquifer
LSBLGR
Pumping well Inverted well

3.4 BIPV DESIGN

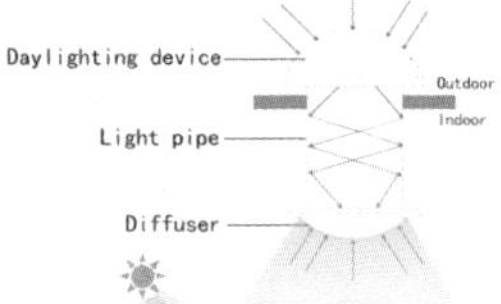

Daylighting device
Outdoor
Indoor
Light pipe
Diffuser

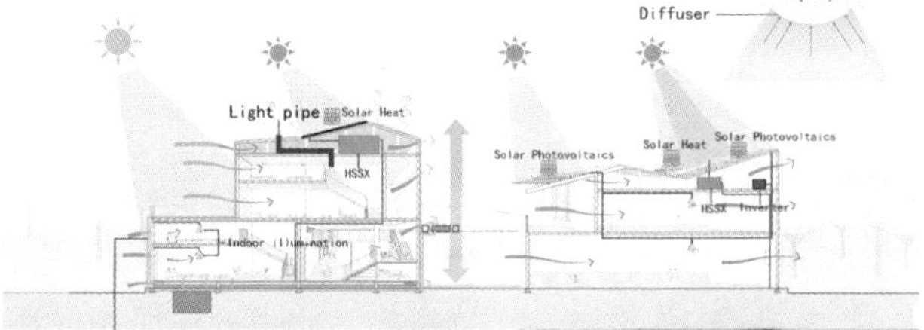

Light pipe
Solar Heat
HSSX
Indoor illumination
Solar Photovoltaics
Solar Heat
Solar Photovoltaics
HSSX

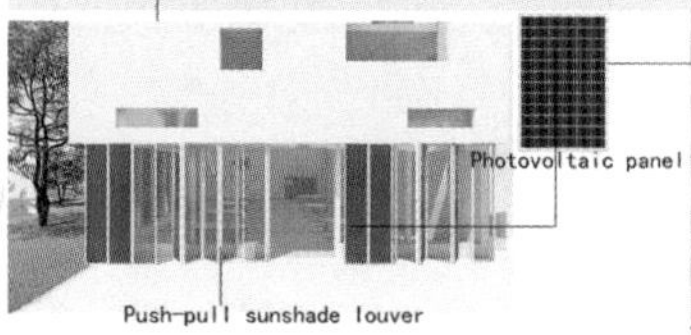

Photovoltaic panel
Push-pull sunshade louver

Sunshade louver combined with photovoltaic panel design

1.The photovoltaic daylighting roof regulates the intake of sunlight
2.Photovoltaic curtain wall acts as wind shield

风·跳跃空间 WIND·JUMPING SPACE JIANYANG DISTRICT JINGLONG KINDERGARTEN

06

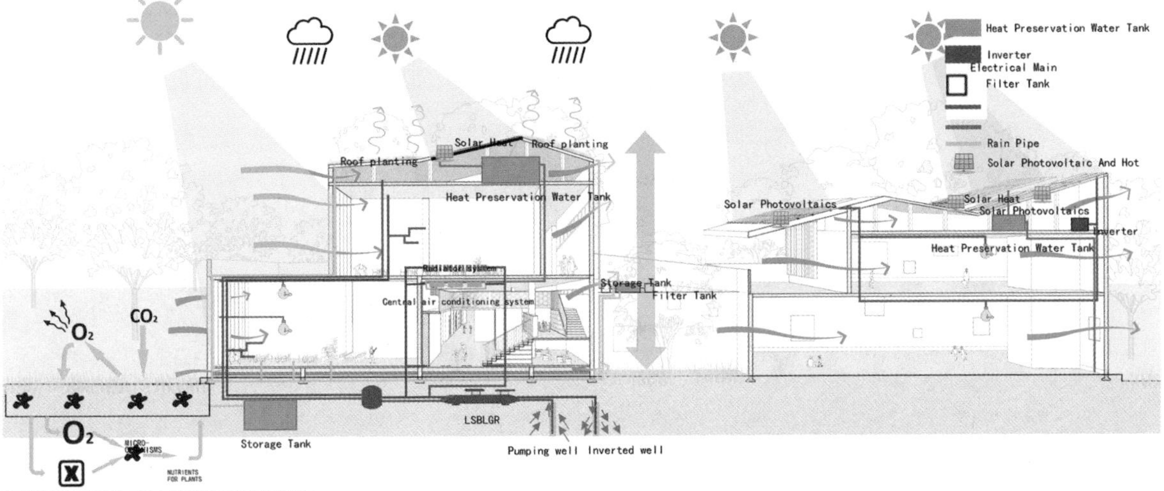

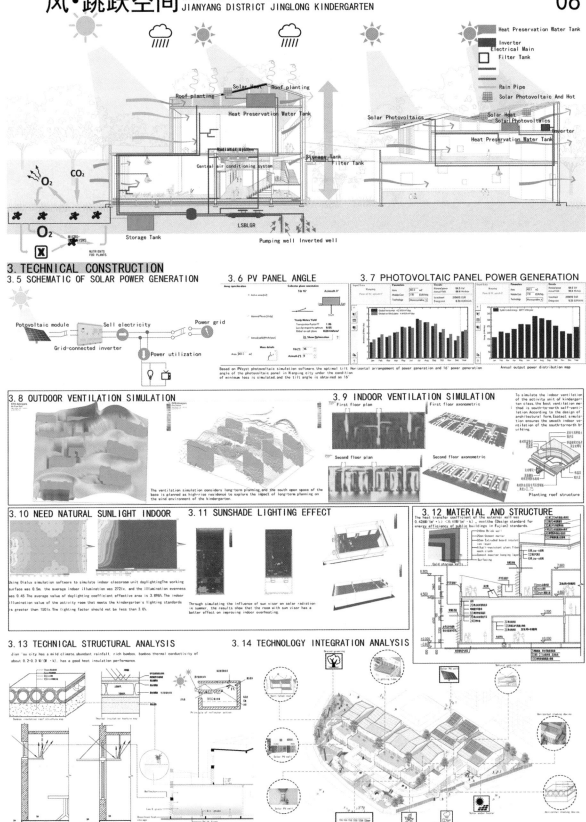

Heat Preservation Water Tank
Inverter
Electrical Main
Filter Tank
Rain Pipe
Solar Photovoltaic And Hot
Roof planting
Solar Heat
Heat Preservation Water Tank
Central air conditioning system
Storage Tank
Filter Tank
Solar Photovoltaics
Solar Heat
Solar Photovoltaics
Inverter
O_2
CO_2
LSBLGR
Storage Tank
Pumping well
Inverted well
NUTRIENTS FOR PLANTS

3. TECHNICAL CONSTRUCTION

3.5 SCHEMATIC OF SOLAR POWER GENERATION

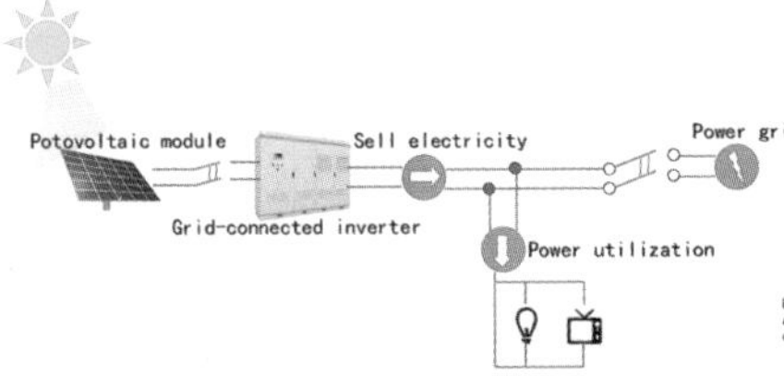

Potovoltaic module
Sell electricity
Power grid
Grid-connected inverter
Power utilization

3.6 PV PANEL ANGLE

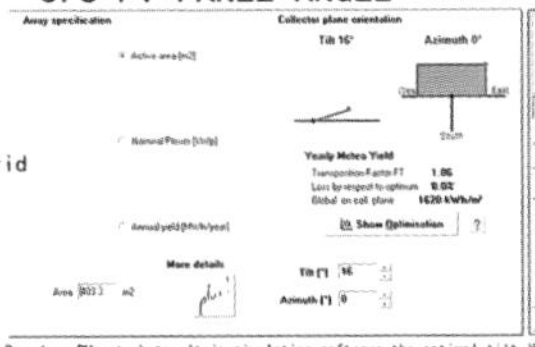

Based on PVsyst photovoltaic simulation software, the optimal tilt angle of the photovoltaic panel in Nanping city under the condition of minimum loss is simulated, and the tilt angle is obtained as 16°

3.7 PHOTOVOLTAIC PANEL POWER GENERATION

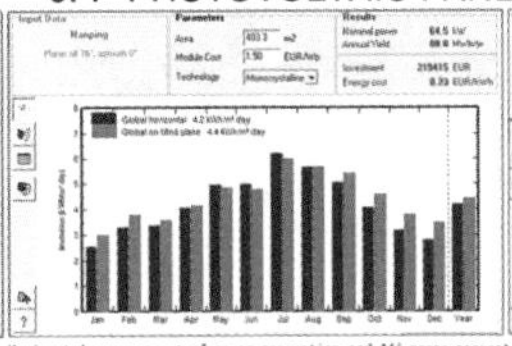

Horizontal arrangement of power generation and 16° power generation

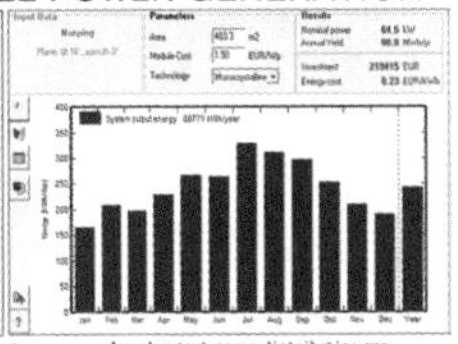

Annual output power distribution map

3.8 OUTDOOR VENTILATION SIMULATION

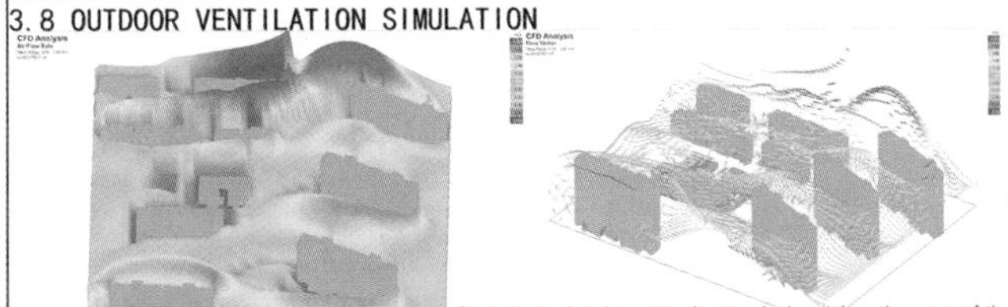

The ventilation simulation considers long-term planning, and the south open space of the base is planned as high-rise residence to explore the impact of long-term planning on the wind environment of the kindergarten.

3.9 INDOOR VENTILATION SIMULATION

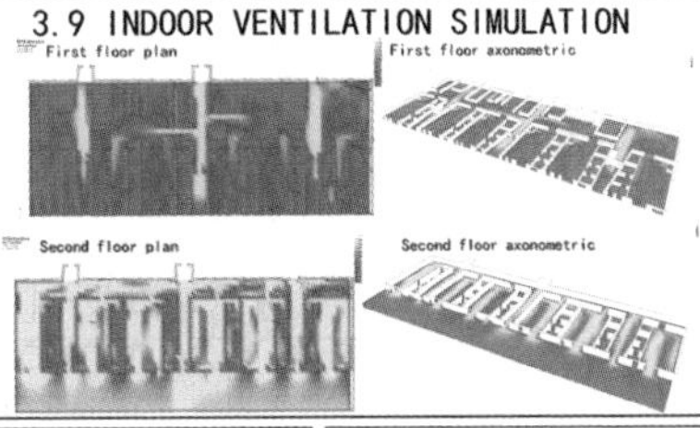

First floor plan
First floor axonometric
Second floor plan
Second floor axonometric

To simulate the indoor ventilation of the activity unit of kindergarten class, the best ventilation method is south-to-north self-ventilation. According to the design of architectural form, Ecotect simulation ensures the smooth indoor ventilation of the south-to-north building.

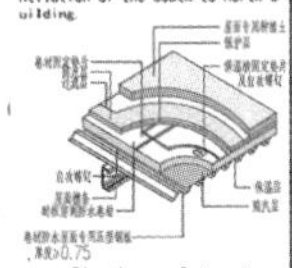

Planting roof structure

3.10 NEED NATURAL SUNLIGHT INDOOR

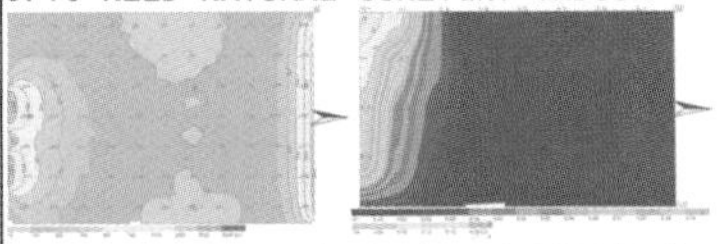

Using Dialux simulation software to simulate indoor classroom unit daylightingThe working surface was 0.5m, the average indoor illumination was 272lx, and the illumination evenness was 0.45 The average value of daylighting coefficient effective area is 3.896%. The indoor illumination value of the activity room that meets the kindergarten's lighting standards is greater than 150lx. The lighting factor should not be less than 3.0%.

3.11 SUNSHADE LIGHTING EFFECT

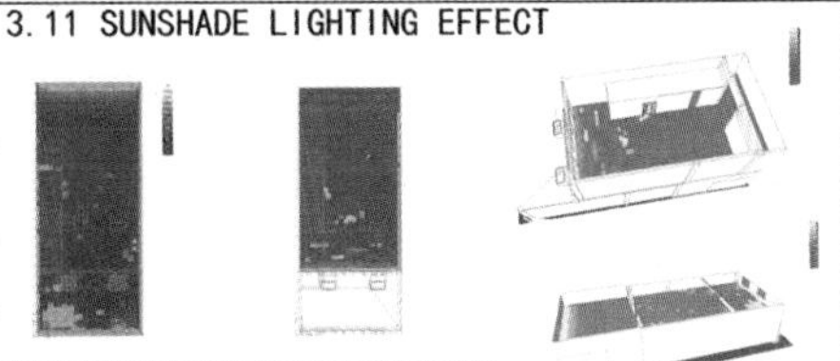

Through simulating the influence of sun visor on solar radiation in summer, the results show that the room with sun visor has a better effect on improving indoor overheating.

3.12 MATERIAL AND STRUCTURE

The heat transfer coefficient of the external wall was 0.426W/(m²·k) <0.60W/(m²·k), meetthe《Design standard for energy efficiency of public buildings in Fujian》standards.

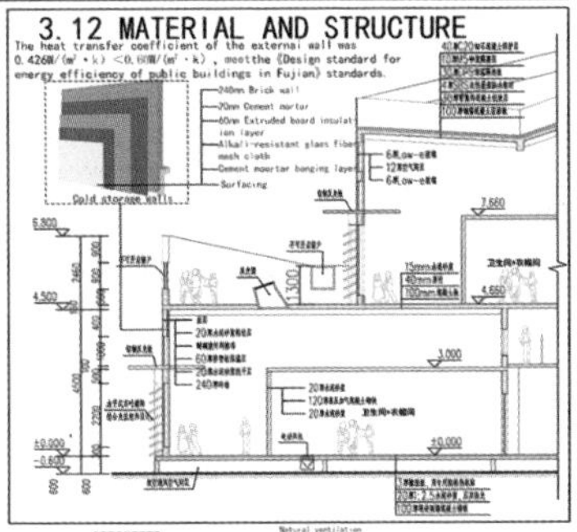

240mm Brick wall
20mm Cement mortar
60mm Extruded board insulation layer
Alkali-resistant glass fiber mesh cloth
Cement moortar banging layer
Surfacing
Cold storage walls

3.13 TECHNICAL STRUCTURAL ANALYSIS

Jian 'ou city has a mild climate, abundant rainfall, rich bamboo, bamboo thermal conductivity of about 0.2-0.3 W/(M·k), has a good heat insulation performance.

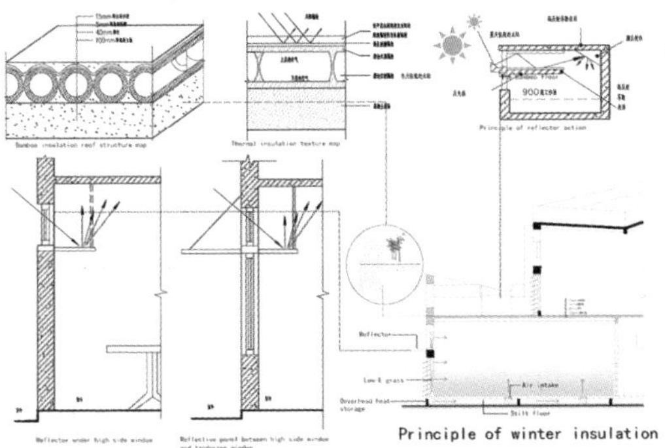

Principle of winter insulation

3.14 TECHNOLOGY INTEGRATION ANALYSIS

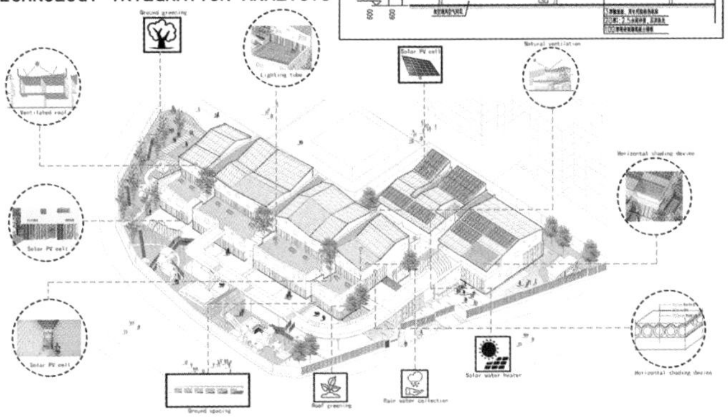

综合奖 · 三等奖
General Prize Awarded · Third Prize

注 册 号：6732
项目名称：漫步园景（南平）
Walking in the Garden (Nanping)
作　　者：李　磊、刘　豪、刘合俊、李丽妹、开　欣
参赛单位：合肥工业大学、苏州大学
指导教师：郑先友、王旭

专家点评：

该作品通过两个相互连接的庭院，合理布局了幼儿园的学习活动空间和生活服务空间。独立的学习活动单元与竖向交通设计，保障了最佳的空气流动、自然采光和室外活动空间。同时，方案还充分吸取了地域传统建筑文化和建筑材料特色。

This work realizes rational layout of the kindergarten's learning space and life space through two interconnected courtyards. The independent learning activity unit and vertical traffic design ensure the best space for air flow, natural lighting and outdoor activities. At the same time, the plan also integrates regional traditional architectural culture and characteristics of building materials.

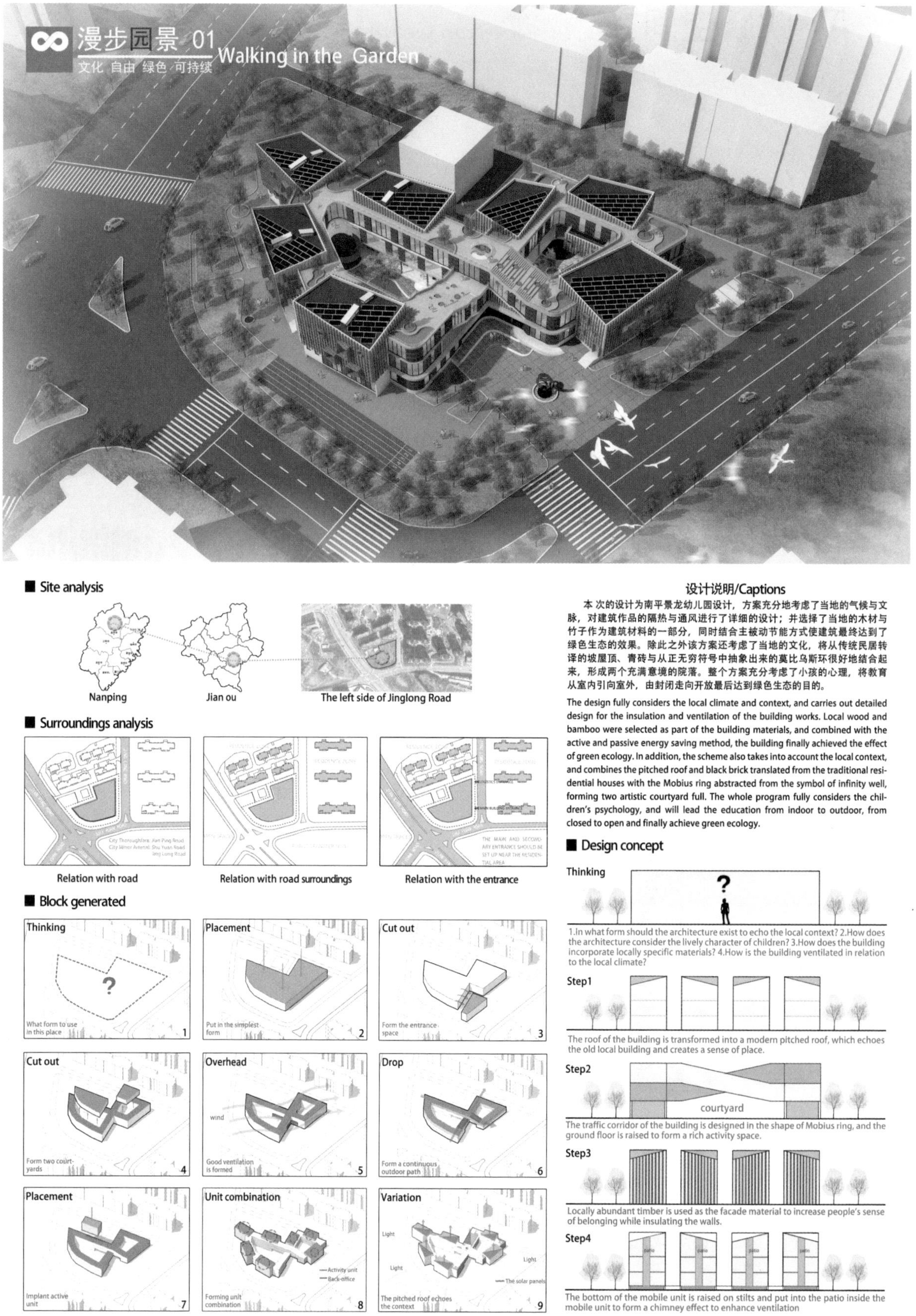
漫步园景 01
文化 自由 绿色 可持续
Walking in the Garden
Site analysis
Nanping
Jian ou
The left side of Jinglong Road
Surroundings analysis
Relation with road
Relation with road surroundings
Relation with the entrance
Block generated
Thinking
What form to use in this place
1
Placement
Put in the simplest form
2
Cut out
Form the entrance space
3
Cut out
Form two courtyards
4
Overhead
wind
Good ventilation is formed
5
Drop
Form a continuous outdoor path
6
Placement
Implant active unit
7
Unit combination
Activity unit
Back-office
Forming unit combination
8
Variation
Light
Light
Light
The solar panels
The pitched roof echoes the context
9
设计说明/Captions
本次的设计为南平景龙幼儿园设计，方案充分地考虑了当地的气候与文脉，对建筑作品的隔热与通风进行了详细的设计；并选择了当地的木材与竹子作为建筑材料的一部分，同时结合主被动节能方式使建筑最终达到了绿色生态的效果。除此之外该方案还考虑了当地的文化，将从传统民居转译的坡屋顶、青砖与从正无穷符号中抽象出来的莫比乌斯环很好地结合起来，形成两个充满意境的院落。整个方案充分考虑了小孩的心理，将教育从室内引向室外，由封闭走向开放最后达到绿色生态的目的。
The design fully considers the local climate and context, and carries out detailed design for the insulation and ventilation of the building works. Local wood and bamboo were selected as part of the building materials, and combined with the active and passive energy saving method, the building finally achieved the effect of green ecology. In addition, the scheme also takes into account the local context, and combines the pitched roof and black brick translated from the traditional residential houses with the Mobius ring abstracted from the symbol of infinity well, forming two artistic courtyard full. The whole program fully considers the children's psychology, and will lead the education from indoor to outdoor, from closed to open and finally achieve green ecology.
Design concept
Thinking
?
1.In what form should the architecture exist to echo the local context? 2.How does the architecture consider the lively character of children? 3.How does the building incorporate locally specific materials? 4.How is the building ventilated in relation to the local climate?
Step1
The roof of the building is transformed into a modern pitched roof, which echoes the old local building and creates a sense of place.
Step2
courtyard
The traffic corridor of the building is designed in the shape of Mobius ring, and the ground floor is raised to form a rich activity space.
Step3
Locally abundant timber is used as the facade material to increase people's sense of belonging while insulating the walls.
Step4
The bottom of the mobile unit is raised on stilts and put into the patio inside the mobile unit to form a chimney effect to enhance ventilation

漫步园景 02 Walking in the Garden

文化 自由 绿色 可持续

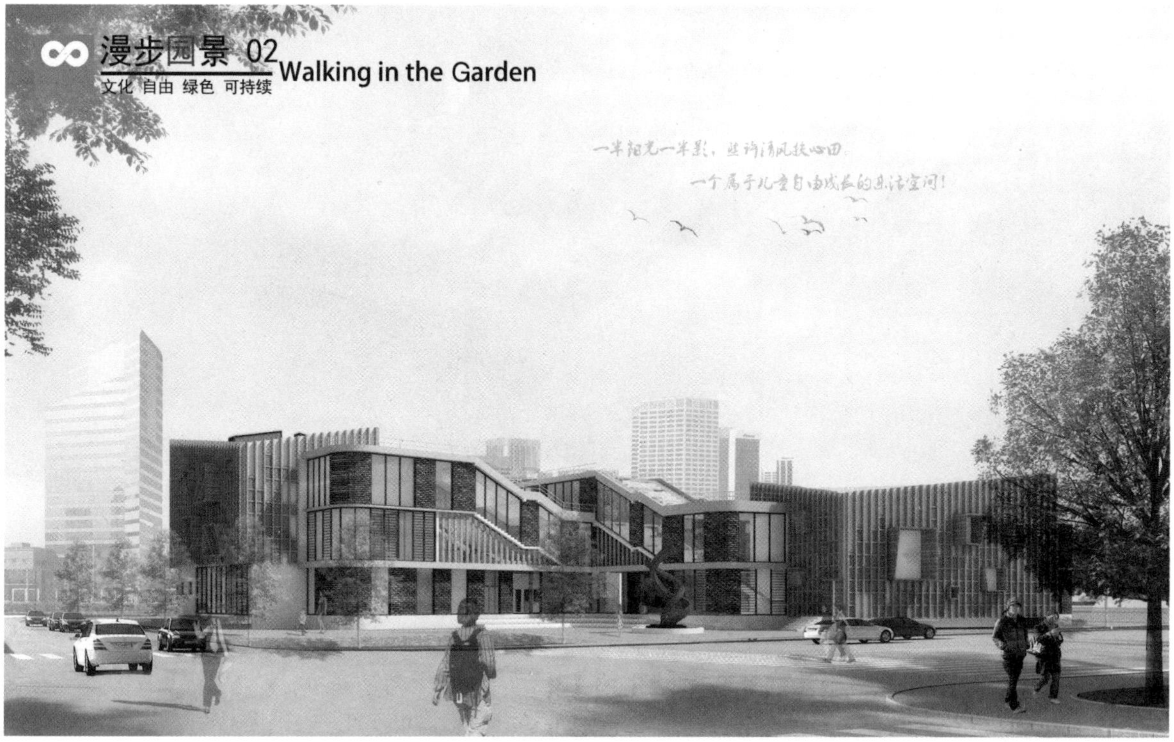

Site plan 1 : 500

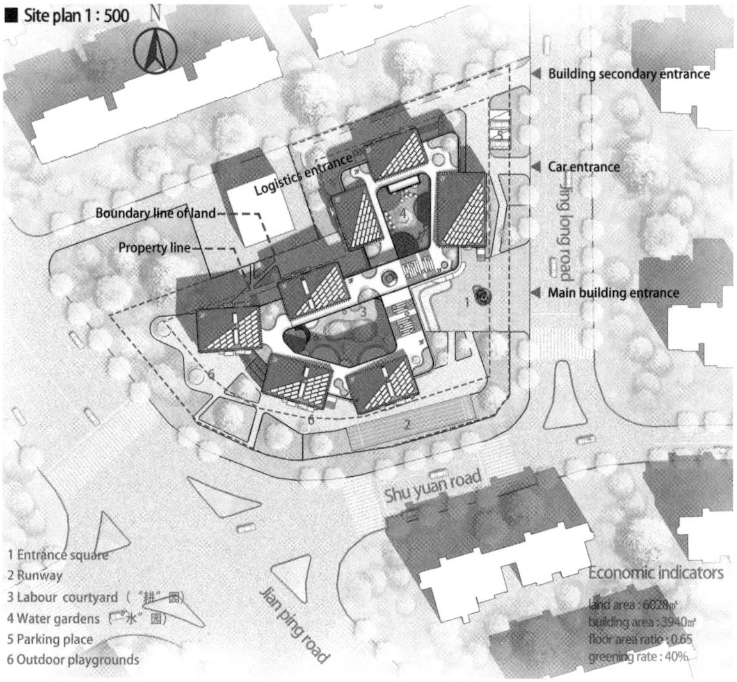
N
Building secondary entrance
Car entrance
Jing long road
Main building entrance
Logistics entrance
Boundary line of land
Property line
Shu yuan road
Jian ping road
1 Entrance square
2 Runway
3 Labour courtyard（"耕"园）
4 Water gardens（"水"园）
5 Parking place
6 Outdoor playgrounds
Economic indicators
land area : 6028㎡
building area : 3940㎡
floor area ratio : 0.65
greening rate : 40%

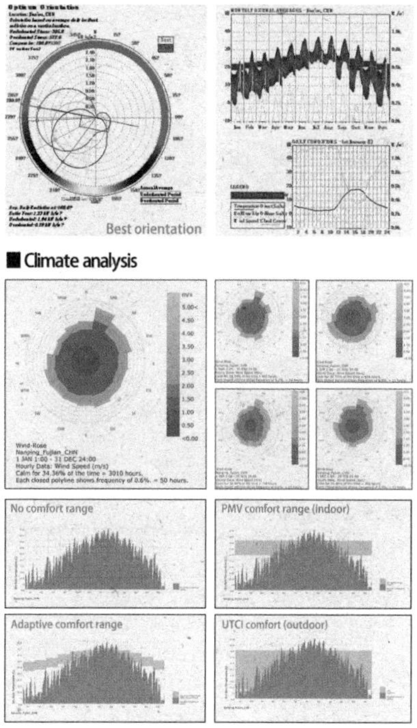
Best orientation
Climate analysis
No comfort range
PMV comfort range (indoor)
Adaptive comfort range
UTCI comfort (outdoor)

East elevation 1 : 300
South elevation 1 : 300

漫步园景 03 Walking in the Garden

文化 自由 绿色 可持续

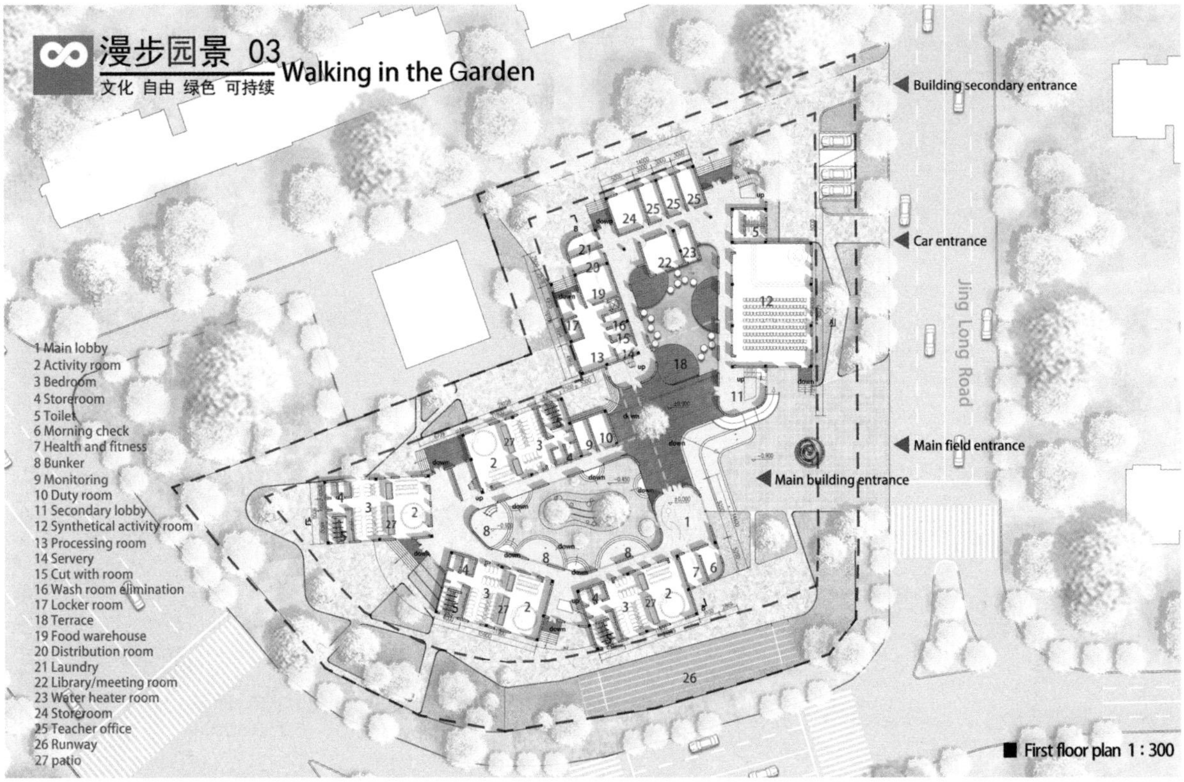
Building secondary entrance
Car entrance
Jing Long Road
Main field entrance
Main building entrance

1 Main lobby
2 Activity room
3 Bedroom
4 Storeroom
5 Toilet
6 Morning check
7 Health and fitness
8 Bunker
9 Monitoring
10 Duty room
11 Secondary lobby
12 Synthetical activity room
13 Processing room
14 Servery
15 Cut with room
16 Wash room elimination
17 Locker room
18 Terrace
19 Food warehouse
20 Distribution room
21 Laundry
22 Library/meeting room
23 Water heater room
24 Storeroom
25 Teacher office
26 Runway
27 patio

■ First floor plan 1：300

■ Unit volume generation

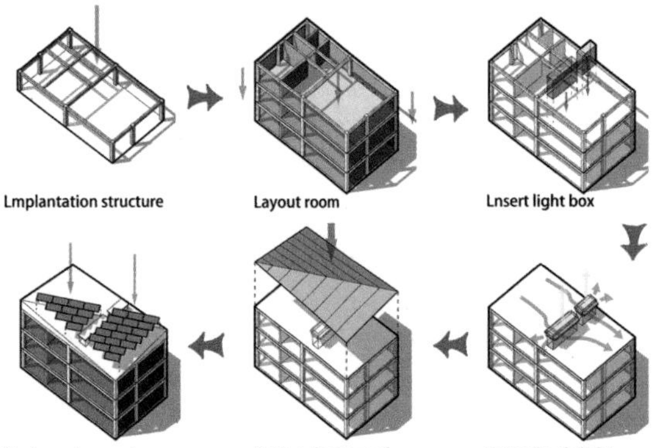
Lmplantation structure
Layout room
Lnsert light box
Placing solar panels
Put in a sloping roof
Put in the skylight

■ Unit ventilation

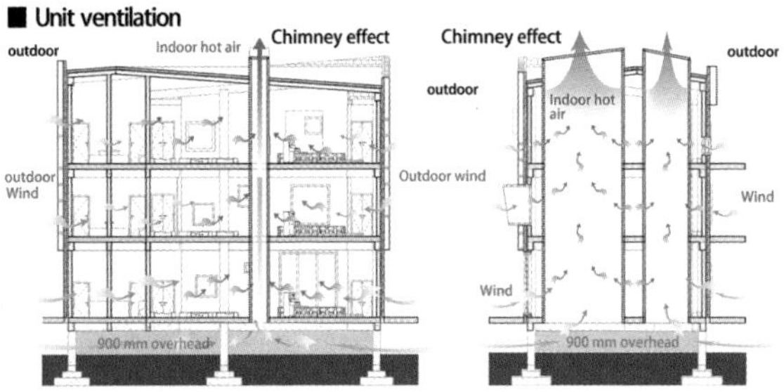
outdoor
Indoor hot air
Chimney effect
outdoor Wind
Outdoor wind
900 mm overhead
Chimney effect
outdoor
outdoor
Indoor hot air
Wind
Wind
900 mm overhead

■ Unit lighting and shading design

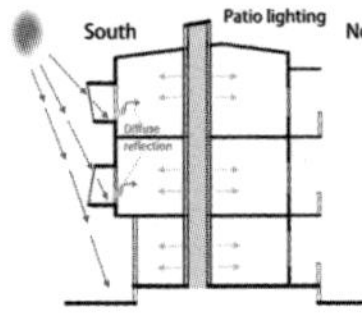
South
Patio lighting
North
Diffuse reflection

A light box is placed in the unit to meet the high lighting requirements of the kindergarten. At the same time, the building facade is prefabricated with window covers to prevent direct sunlight.Prefabricated window covers are made of wooden materials, which diffuse reflex sunlight into the room and form even lighting.

■ Unit moisture-proof design

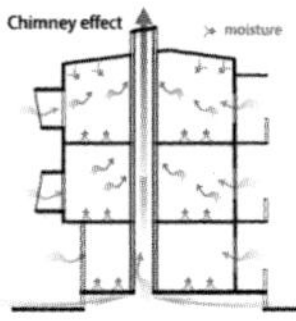
Chimney effect
moisture

The ground floor of the building is built on stilts to prevent moisture erosion. The elevated area will have negative pressure when the wind passes through to take away moisture, and the patio in the unit will have chimney effect, taking away indoor moisture.

1 Activity room
2 Bedroom
3 Storeroom
4 Toilet
5 Lounge hall
6 Mold making room
7 Terrace
8 Display wall
9 Processing room
10 Servery
11 Cut with room
12 Wash room elimination
13 Food warehouse
14 Store
15 Kindergarten teacher office
16 Finance office
17 General office
18 Teacher office
19 patio

■ Second floor plan 1：300

1 Activity room
2 Bedroom
3 Storeroom
4 Toilet
5 Toy storeroom
6 patio

■ Third floor plan 1：300

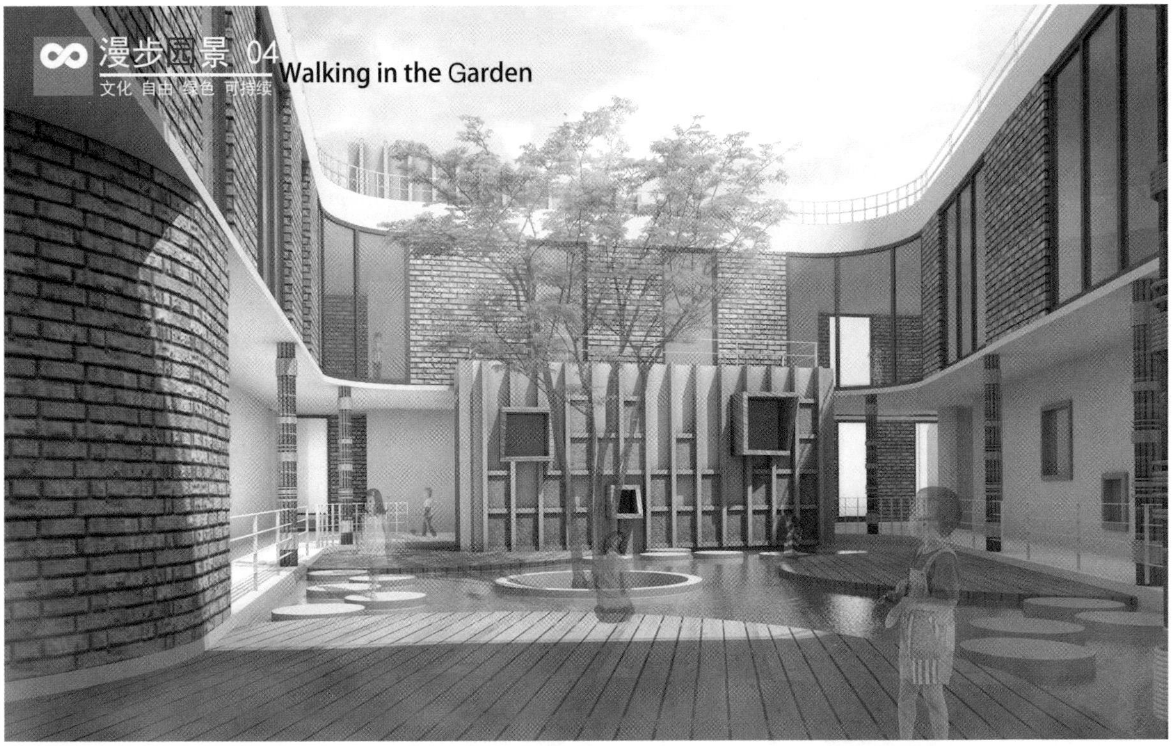
漫步园景 04
Walking in the Garden
文化 自由 绿色 可持续

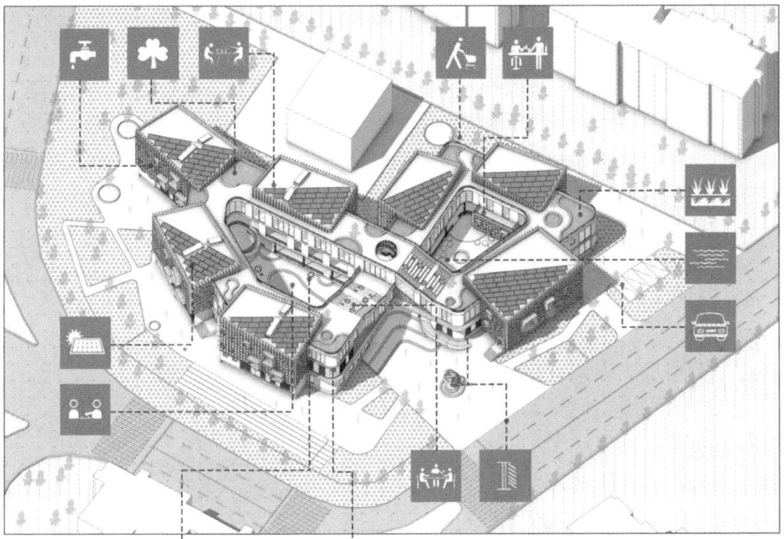

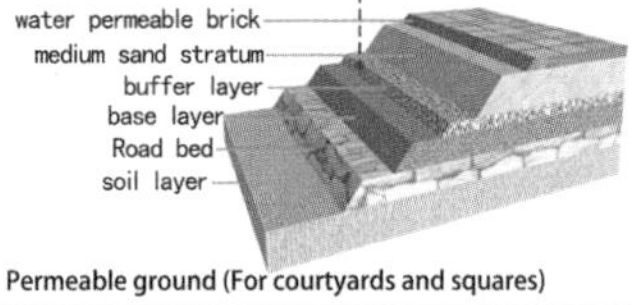
water permeable brick
medium sand stratum
buffer layer
base layer
Road bed
soil layer
Permeable ground (For courtyards and squares)

outdoor window
air layer
heat mirror
inner window
Outdoor
indoor
Outdoor
indoor
Low-E glass (For facade windows)

A-A profile 1 : 300
B-B profile 1 : 300

漫步 景 05
Walking in the Garden
文化 自由 绿色 可持续

Unit decomposition

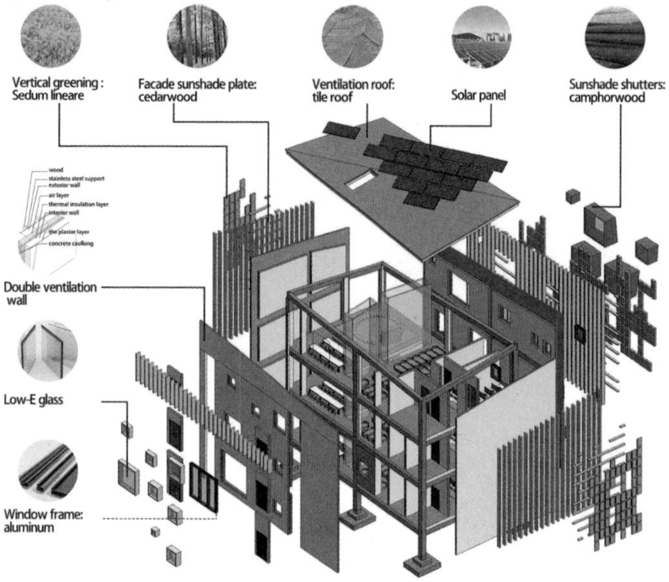
Vertical greening : Sedum lineare
Facade sunshade plate: cedarwood
Ventilation roof: tile roof
Solar panel
Sunshade shutters: camphorwood
wood
stainless steel support
exterior wall
air layer
thermal insulation layer
interior wall
the plaster layer
concrete caulking
Double ventilation wall
Low-E glass
Window frame: aluminum

Epidermal system

The wooden skin and the filled wall form a double-layer wall structure, which reduces the thermal conductivity of the wall. In summer, it can reduce the invasion of outdoor space heat, while in winter, it can reduce the loss of indoor heat

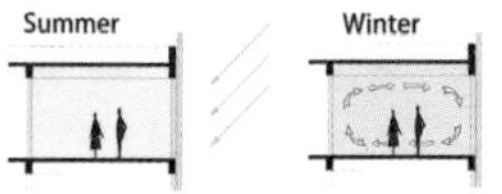
Summer
Winter

Structure split —— Wall —— Steel frame —— Irrigation system —— Local timber —— Green potted

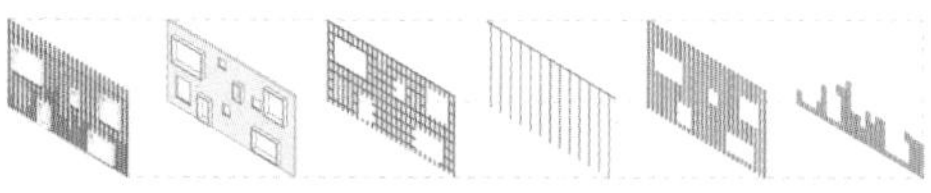

Epidermal optimization

Programme I : grid spacing 750
The average value of the heat radiation of the WALL analyzed by ECO was 54.4kWh, and the value of the heat radiation received by the wall was 72.3% of that without the skin.

Programme 2 : grid spacing 650
The average value of the thermal radiation of the ECO analysis wall was 32.1kWh, which could effectively reduce the indoor impact of solar radiation by reducing the radiation amount by 22.3kWh compared with that of plan 1.

Programme 3 : grid spacing 500
The average value of the thermal radiation of the wall is 23.6kWh, which is 22.3kWh less than that of the Programme 2, Besides, the facade effect is excellent, so this plan is chosen.

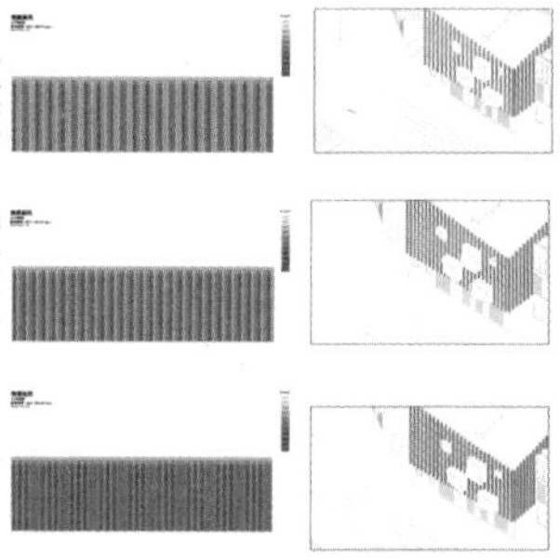

Photovoltaic power generation

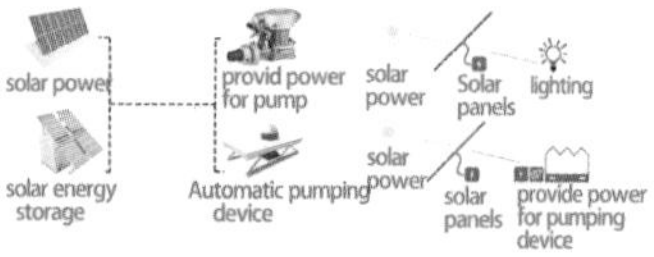
solar power
solar energy storage
provid power for pump
Automatic pumping device
solar power
Solar panels
lighting
solar power
solar panels
provide power for pumping device

The annual average total radiation in Fujian Province is 3800~5300 MJ/㎡, and the total building area of this project is 3940m m. According to the Code for Urban Electric Power Planning GB/50293-014, the power consumption index of this building is 50 W/m. Considering the geographical location of Shenzhen, and for the convenience of bracket design and installation, combined with sloping roof, the annual electricity consumption Q= total building area × annual electricity consumption index = 50 × 0.7 × 8 × 365 × 3940/1000 = 402,668 kWh/year is selected, and the total solar radiation on the horizontal plane is converted to the total solar radiation on the inclined plane, and the slope radiation is adopted According to the recommended value, we take the annual power generation of photovoltaic panel with kop of 0.97 per square meter = 1451.39 × 09 × 0.17 × 0.8 = 17765 (kwh/㎡ year), where 0.17 is the photoelectric conversion rate of photovoltaic panel.

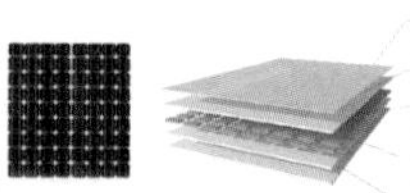

Glass-the main component is silicon dioxide, and the secondary components are soda ash, magnesium oxide, aluminum oxide, carbon nitrate and so on.

EVA- ethylene-vinyl acetate copolymer has water resistance, corrosion resistance and heat preservation.

Battery chip-the core component of battery module, the main components are monocrystalline silicon and polycrystaline silicon.

TPT- backplane protection material is composed of polyvinyl fluoride film -PRT-PVF three-layer film.

Frame-the main material is fully cooked aluminum, which increases the speed of construction.

■ Solar panel structure

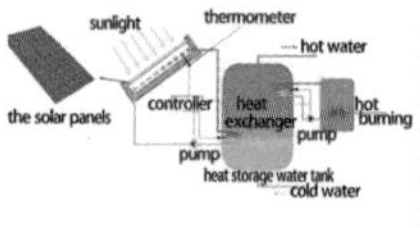
sunlight
thermometer
hot water
the solar panels
controller
heat exchanger
pump
hot burning
pump
heat storage water tank
cold water

■ Anaerobic digestion processor

Anaerobic digestion principle not only play the role of dealing with rubbish.The priciple is through the bacteri-aldecomposition of organic matter into sugar,then transform into all kinds of acid.Material decomposition after fermentation and produce gas.Bacterial action will generate heat and provid quantitfy heat.

Ground source heat pump

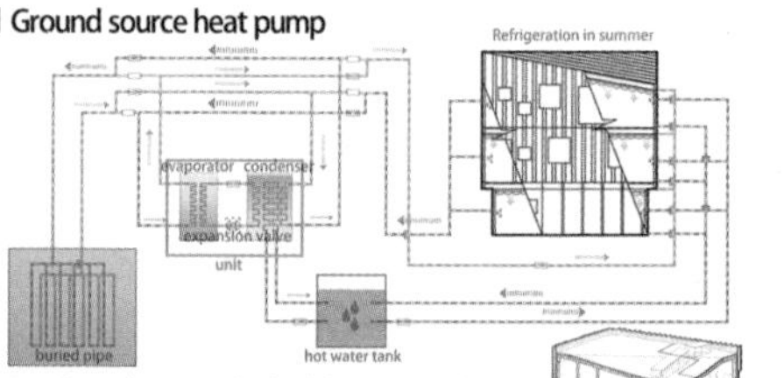
Refrigeration in summer
evaporator condenser
expansion valve
unit
buried pipe
hot water tank

Ground source heat pump technology belongs to renewable energy utilization technology. As the ground source heat pump is a heating and air conditioning system that uses the shallow geothermal resources (usually less than 400 meters deep) on the earth's surface as the cold and heat source for energy conversion. In this design, the energy conversion is realized by this device, and the heat source is delivered to provide bathing indoors through the pipeline device, and the cold air provides centralized cooling indoors.

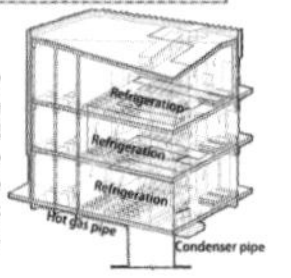
Hot gas pipe
Condenser pipe

漫步园景 06 Walking in the Garden

文化 自由 绿色 可持续

Patio
Activity unit
Patio
Patio
Activity unit
Patio
Patio
Activity unit
Patio
Rest
Rest
Monitoring
The sink
The sink
Fire water
Toilet flushing
Domestic sewage collection
Domestic sewage collection
Green plants to water
Fire pool
Sewage collection tank
A series of filtration processes
Water storage pool
Rainwater collection
Rainwater collection
Irrigate
Road flushing
Wash the car

■ Design of water circulation system

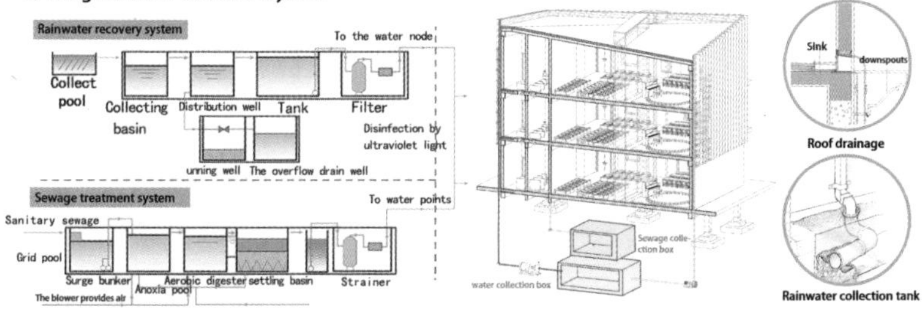
Rainwater recovery system
To the water node
Collect pool
Collecting basin
Distribution well
Tank
Filter
Disinfection by ultraviolet light
The overflow drain well
Sewage treatment system
To water points
Sanitary sewage
Grid pool
Surge bunker
Anoxia pool
Aerobic digester
settling basin
Strainer
The blower provides air
Sewage collection box
water collection box
Sink
downspouts
Roof drainage
Rainwater collection tank

■ Overall green design

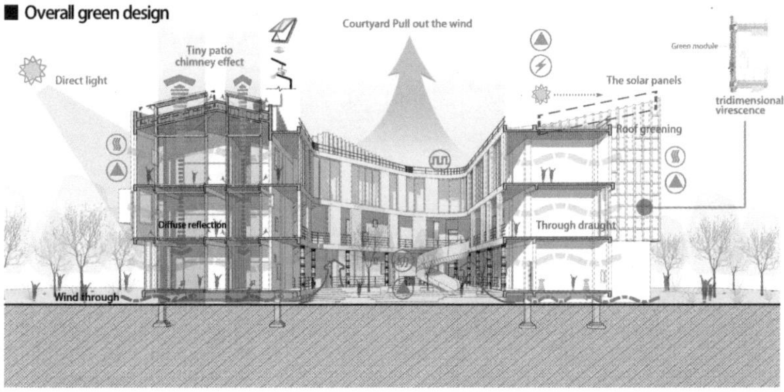
Courtyard Pull out the wind
Tiny patio chimney effect
Direct light
Green module
The solar panels
tridimensional virescence
Roof greening
Diffuse reflection
Through draught
Wind through

■ Wind environment simulation

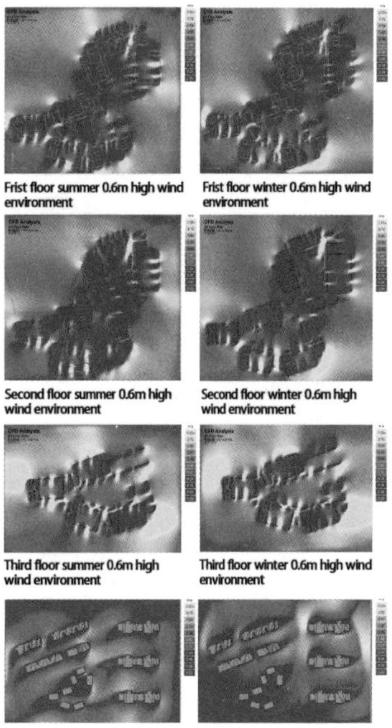

Frist floor summer 0.6m high wind environment

Frist floor winter 0.6m high wind environment

Second floor summer 0.6m high wind environment

Second floor winter 0.6m high wind environment

Third floor summer 0.6m high wind environment

Third floor winter 0.6m high wind environment

Site plan summer 0.6m high wind environment

Site plan winter 0.6m high wind environment

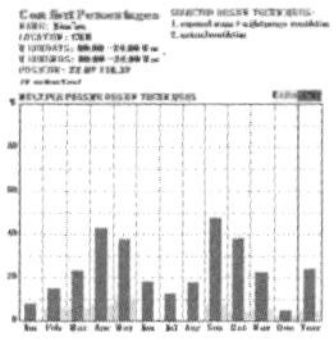

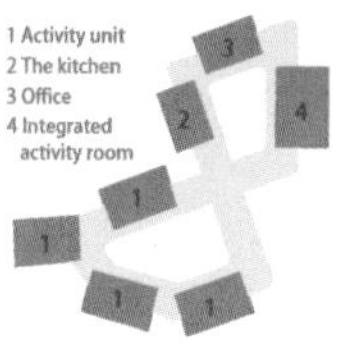
1 Activity unit
2 The kitchen
3 Office
4 Integrated activity room
1
2
3
4

As shown in the figure on the left and the figure above, the effect of natural ventilation and night ventilation in this area will be much better, so the scheme adopts scatter layout, which will facilitate the ventilation of the building. In addition, the chimney effect formed by the patio in the unit will further enhance the ventilation effect of the building.

■ Roof greening

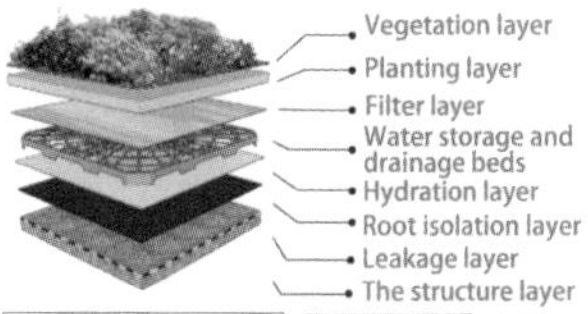
Vegetation layer
Planting layer
Filter layer
Water storage and drainage beds
Hydration layer
Root isolation layer
Leakage layer
The structure layer

Fujia grass

Fujia Grass has the characteristics of shallow root system and easy maintenance

■ shutter shade

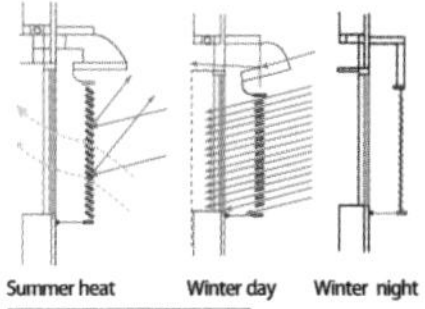
Summer heat
Winter day
Winter night

The bottom unit and local walkway mainly adopt louver shading, louver can adjust itself according to the Angle of sunlight, in order to keep the indoor suitable brightness.

■ Site microclimate

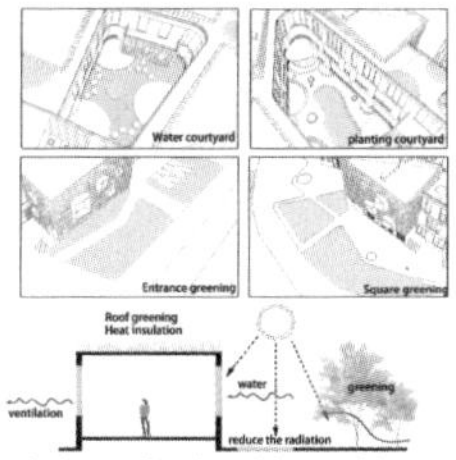
Water courtyard
planting courtyard
Entrance greening
Square greening
Roof greening
Heat insulation
ventilation
water
greening
reduce the radiation

Take advantage of the plants and the nature of the water itself to reduce the sun's radiation

综合奖 · 三等奖
General Prize Awarded · Third Prize

注 册 号：7001
项目名称：林海风语（南平）
Feel the Wind（Nanping）
作　　者：张缤月、袁蜀佳、袁传璞、芦乐平、张瑜珊
参赛单位：山东科技大学
指导教师：冯　巍

专家点评：

该方案采用内庭院环绕式布局，充分借鉴了地方传统院落营造和建筑构造手法。被动式技术应用适当，对于被动式隔热、自然通风、采光遮阳等方面的应用和分析合理，对于主动式太阳能技术以及绿色生态技术的应用可实施性较高。

The plan draws on the techniques of local traditional courtyard and building construction by adopting an inner courtyard wrap-around layout. The application of passive technologies is appropriate, the application and analysis of passive thermal insulation, natural ventilation, lighting and sun shading are reasonable, and the application of active solar technologies and green ecological technologies is highly feasible.

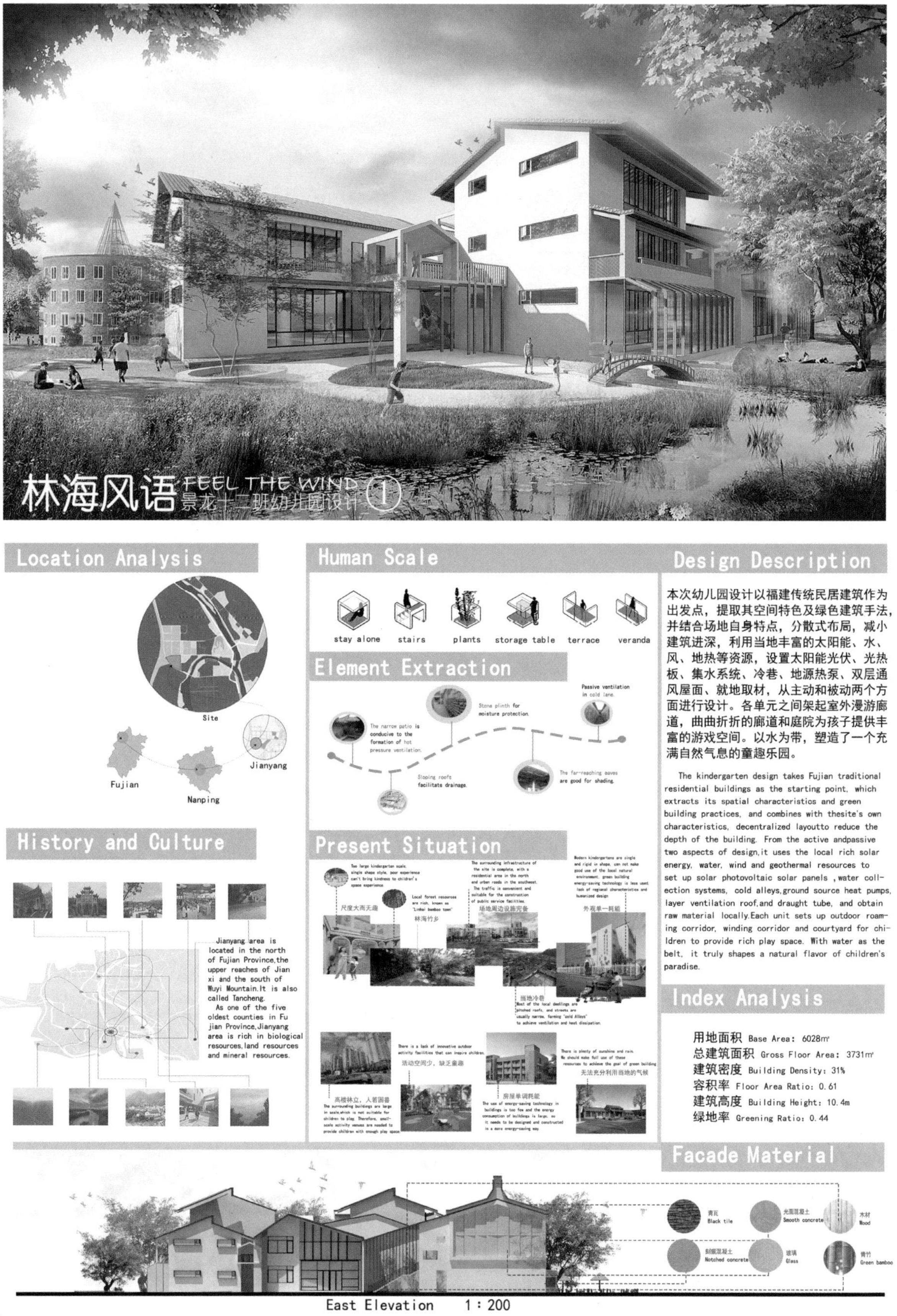
林海风语 FEEL THE WIND
景龙十二班幼儿园设计 ①
Location Analysis
Site
Fujian
Nanping
Jianyang
History and Culture
Jianyang area is located in the north of Fujian Province,the upper reaches of Jian xi and the south of Wuyi Mountain.It is also called Tancheng.
As one of the five oldest counties in Fu jian Province,Jianyang area is rich in biological resources,land resources and mineral resources.
Human Scale
stay alone
stairs
plants
storage table
terrace
veranda
Element Extraction
Present Situation
Design Description
本次幼儿园设计以福建传统民居建筑作为出发点，提取其空间特色及绿色建筑手法，并结合场地自身特点，分散式布局，减小建筑进深，利用当地丰富的太阳能、水、风、地热等资源，设置太阳能光伏、光热板、集水系统、冷巷、地源热泵、双层通风屋面、就地取材，从主动和被动两个方面进行设计。各单元之间架起室外漫游廊道，曲曲折折的廊道和庭院为孩子提供丰富的游戏空间。以水为带，塑造了一个充满自然气息的童趣乐园。
The kindergarten design takes Fujian traditional residential buildings as the starting point, which extracts its spatial characteristics and green building practices, and combines with thesite's own characteristics, decentralized layoutto reduce the depth of the building. From the active andpassive two aspects of design,it uses the local rich solar energy, water, wind and geothermal resources to set up solar photovoltaic solar panels ,water collection systems, cold alleys,ground source heat pumps, layer ventilation roof,and draught tube, and obtain raw material locally.Each unit sets up outdoor roaming corridor, winding corridor and courtyard for children to provide rich play space. With water as the belt, it truly shapes a natural flavor of children's paradise.
Index Analysis
用地面积 Base Area: 6028m²
总建筑面积 Gross Floor Area: 3731m²
建筑密度 Building Density: 31%
容积率 Floor Area Ratio: 0.61
建筑高度 Building Height: 10.4m
绿地率 Greening Ratio: 0.44
Facade Material
East Elevation 1 : 200

Ecotect Analysis

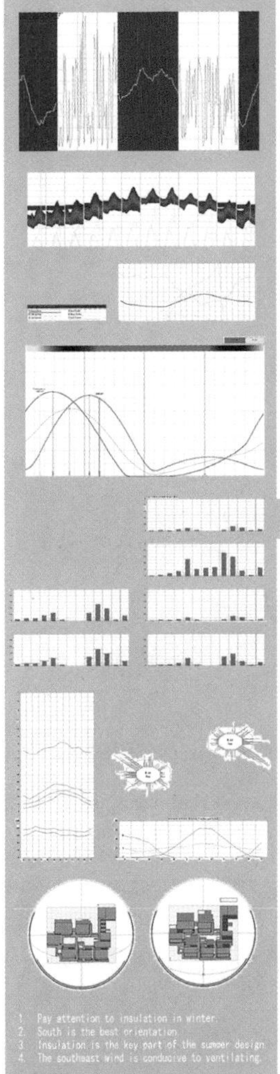
1. Pay attention to insulation in winter.
2. South is the best orientation.
3. Insulation is the key part of the summer design.
4. The southeast wind is conducive to ventilating.

verandas
林海风语 FEEL THE WIND
景龙十二班幼儿园设计 ②

Streamline Analysis

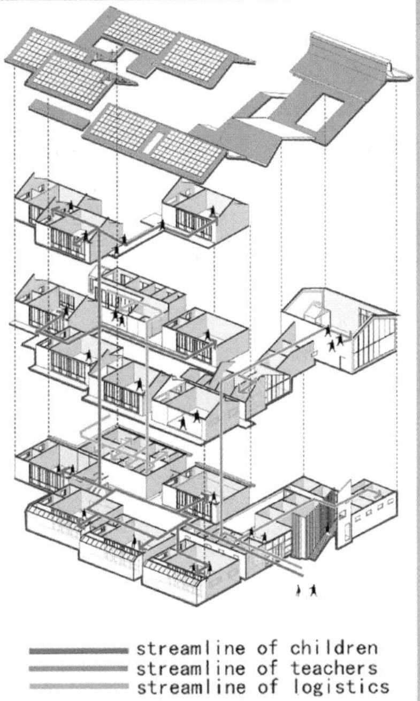
streamline of children
streamline of teachers
streamline of logistics

Kids Of Parts

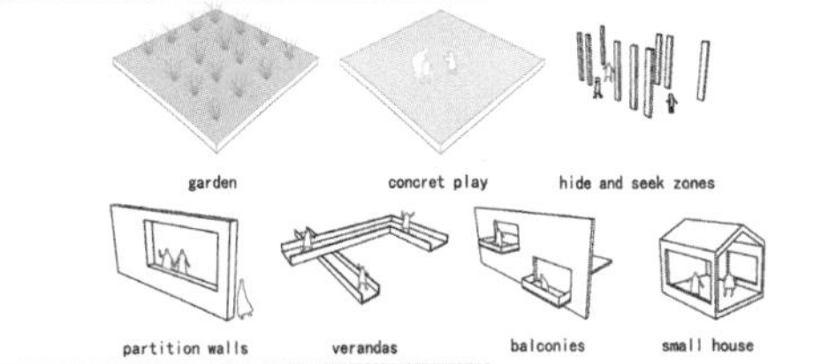
garden
concret play
hide and seek zones
partition walls
verandas
balconies
small house

Action Pattern

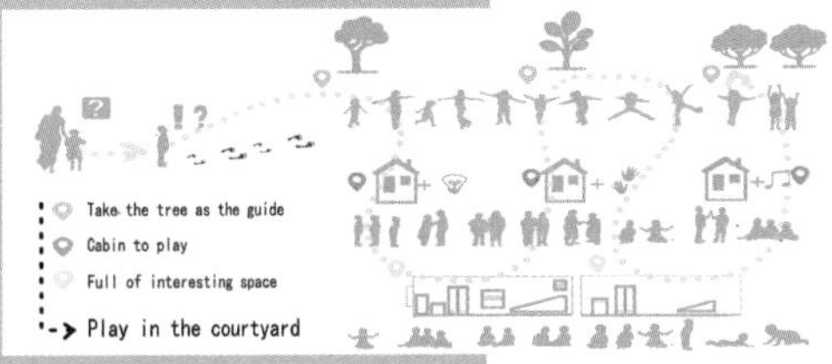
Take the tree as the guide
Cabin to play
Full of interesting space
Play in the courtyard

Block Generrated

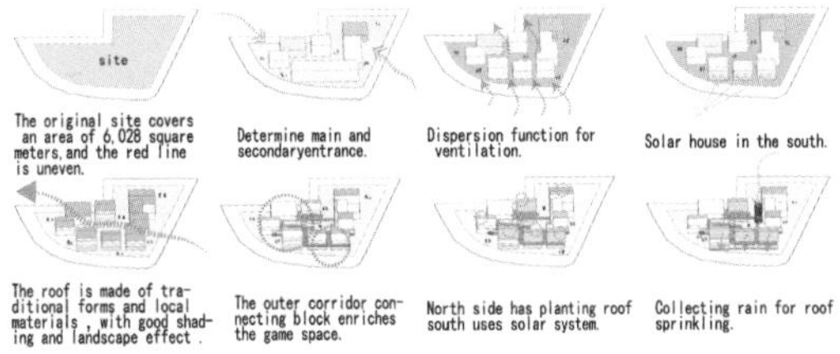
site
The original site covers an area of 6,028 square meters,and the red line is uneven.
Determine main and secondaryentrance.
Dispersion function for ventilation.
Solar house in the south.
The roof is made of traditional forms and local materials , with good shading and landscape effect .
The outer corridor connecting block enriches the game space.
North side has planting roof south uses solar system.
Collecting rain for roof sprinkling.

Having an activity in the multi-function room.

Children are playing in the courtyard.

Southward activity area is full of sunshine.

Solar Energy Technology

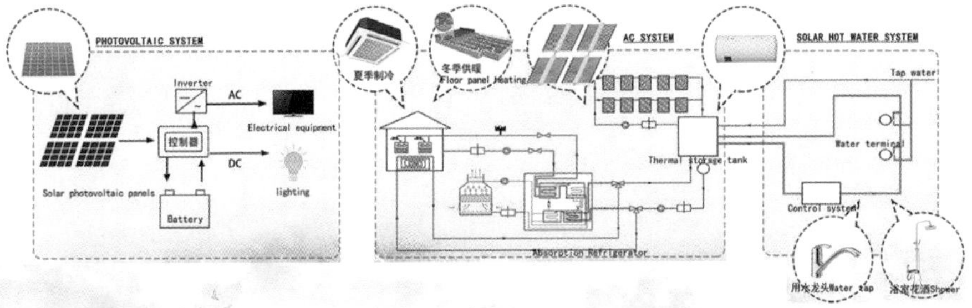
PHOTOVOLTAIC SYSTEM
Inverter
AC
Electrical equipment
控制器
DC
lighting
Solar photovoltaic panels
Battery
夏季制冷
冬季供暖
Floor panel heating
AC SYSTEM
Absorption Refrigerator
SOLAR HOT WATER SYSTEM
Tap water
Water terminal
Thermal storage tank
Control system
用水龙头Water tap
浴室花洒Shower

South Elevation 1：200

林海风语 FEEL THE WIND ③
景龙十二班幼儿园设计

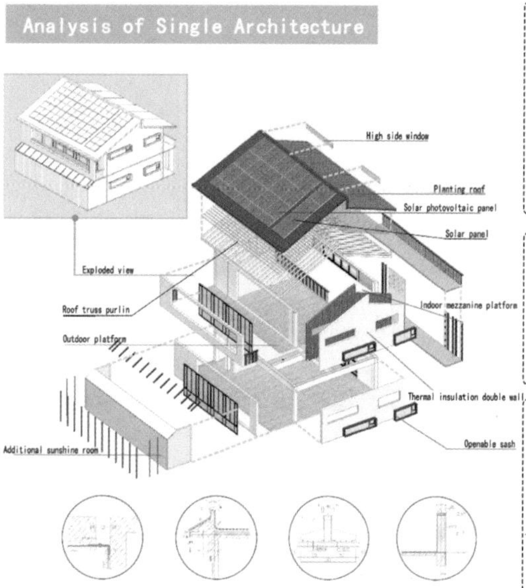
Analysis of Single Architecture
High side window
Planting roof
Solar photovoltaic panel
Solar panel
Exploded view
Roof truss purlin
Indoor mezzanine platform
Outdoor platform
Thermal insulation double wall
Additional sunshine room
Openable sash

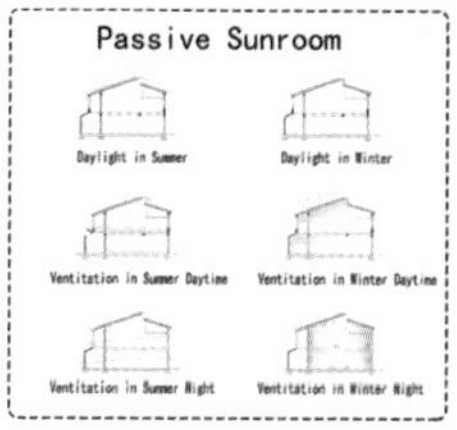
Passive Sunroom
Daylight in Summer
Daylight in Winter
Ventilation in Summer Daytime
Ventilation in Winter Daytime
Ventilation in Summer Night
Ventilation in Winter Night

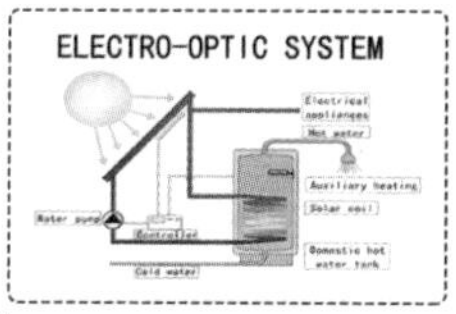
ELECTRO-OPTIC SYSTEM

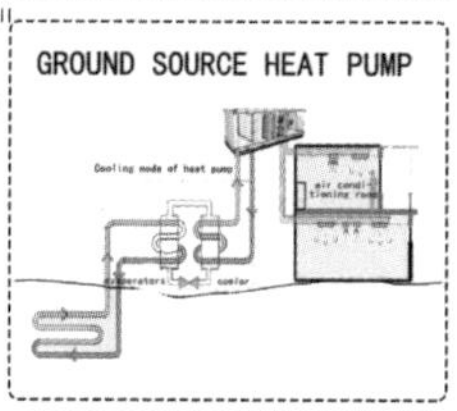
GROUND SOURCE HEAT PUMP

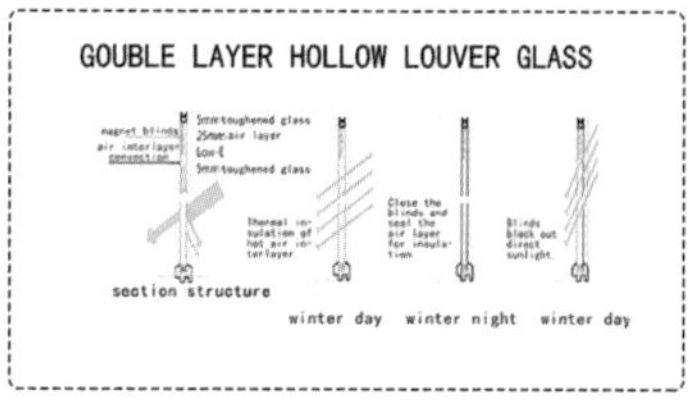
GOUBLE LAYER HOLLOW LOUVER GLASS
section structure
winter day
winter night
winter day

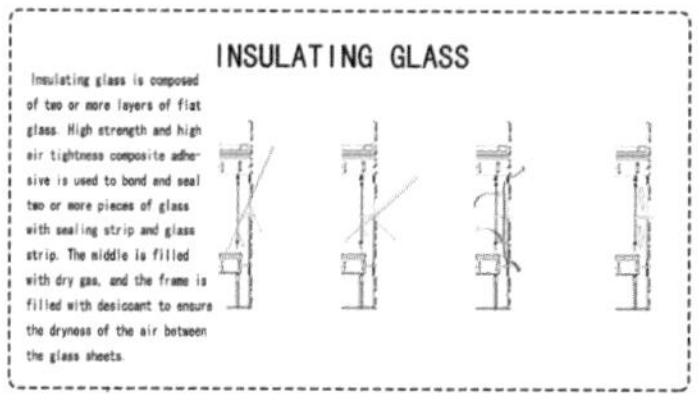
INSULATING GLASS
Insulating glass is composed of two or more layers of flat glass. High strength and high air tightness composite adhesive is used to bond and seal two or more pieces of glass with sealing strip and glass strip. The middle is filled with dry gas, and the frame is filled with desiccant to ensure the dryness of the air between the glass sheets.

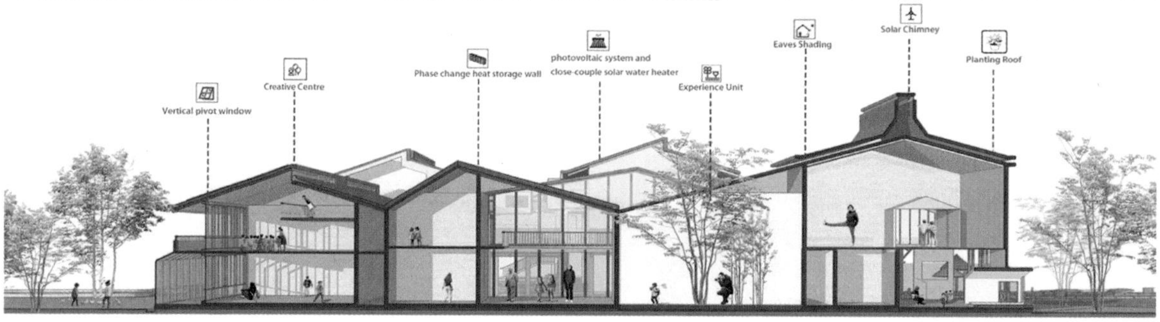
Vertical pivot window
Creative Centre
Phase change heat storage wall
photovoltaic system and
close-couple solar water heater
Experience Unit
Eaves Shading
Solar Chimney
Planting Roof

林海风语 FEEL THE WIND ④ 景龙十二班幼儿园设计

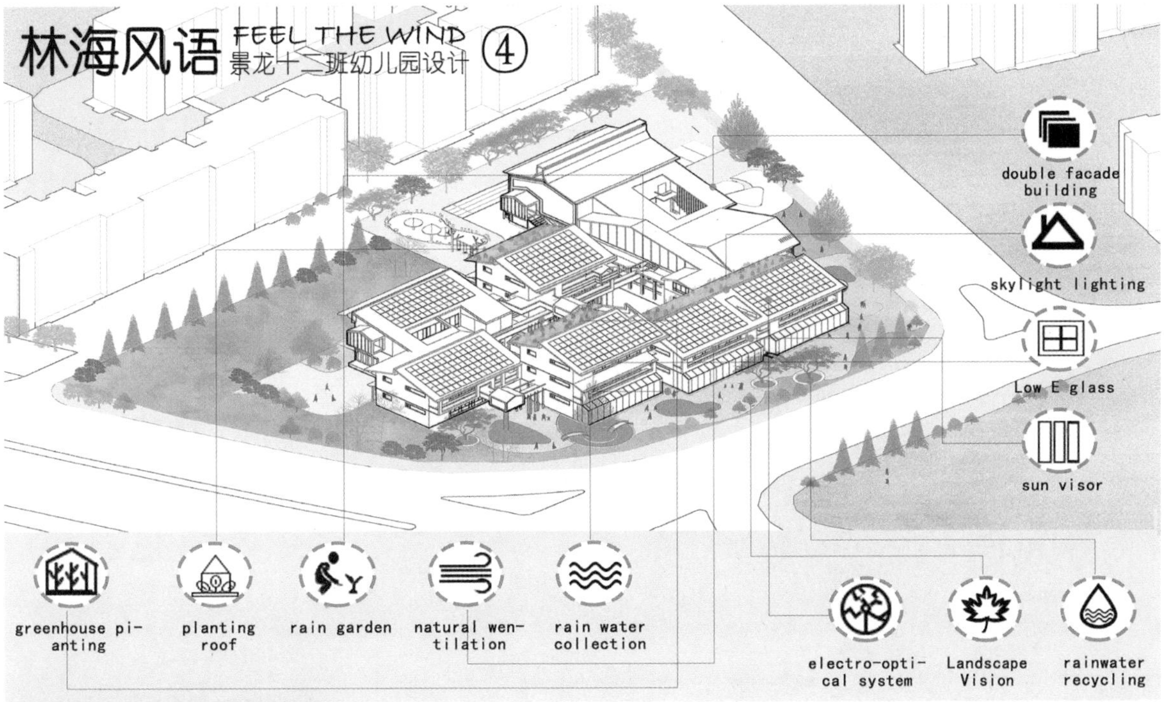
double facade building
skylight lighting
Low E glass
sun visor
greenhouse pi-anting
planting roof
rain garden
natural wen-tilation
rain water collection
electro-optical system
Landscape Vision
rainwater recycling

B-B Section

12.400
8.400
7.200
4.200
3.600
±0.000
-0.450

B-B Section 1：200

First Floor Plan

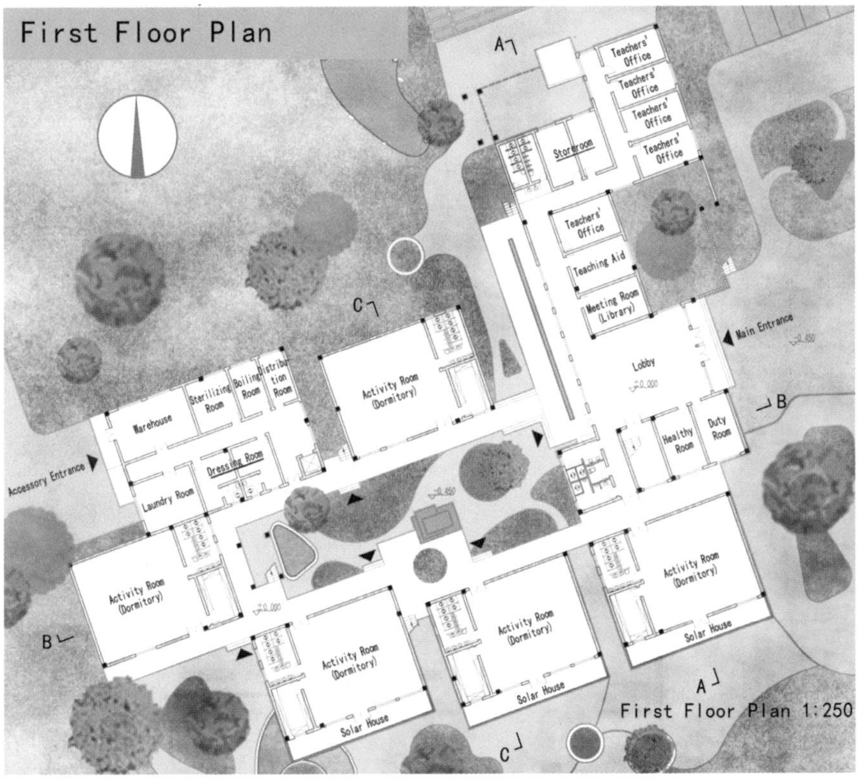
A
B
C
Teachers' Office
Teachers' Office
Teachers' Office
Teachers' Office
Storeroom
Teachers' Office
Teaching Aid
Meeting Room (Library)
Main Entrance
Lobby
Healthy Room
Duty Room
Activity Room (Dormitory)
Warehouse
Sterilizing Room
Boiling Room
Distribution Room
Dressing Room
Laundry Room
Accessory Entrance
Solar House

First Floor Plan 1:250

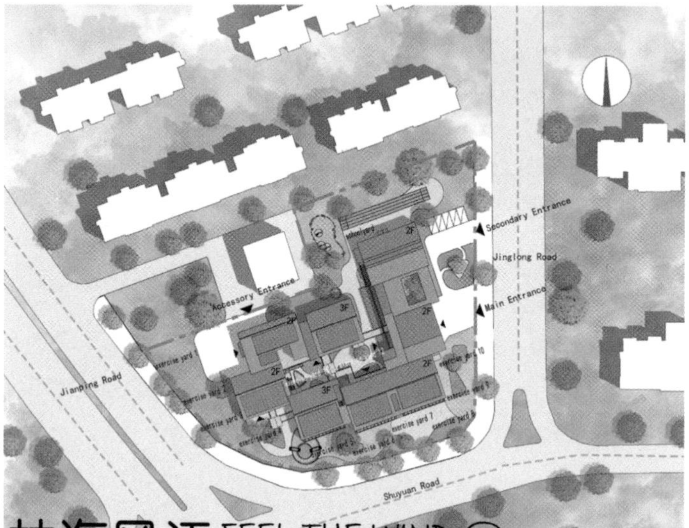
Secondary Entrance
Jinglong Road
Accessory Entrance
Main Entrance
Jianning Road
Shuyuan Road

林海风语 FEEL THE WIND ⑤
景龙十二班幼儿园设计

Eco Technology Analysis

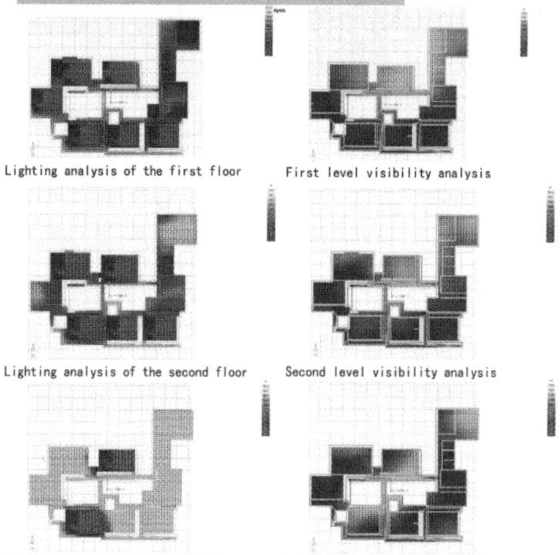
Lighting analysis of the first floor
First level visibility analysis
Lighting analysis of the second floor
Second level visibility analysis
Lighting analysis of the third floor
Third level visibility analysis

Through eco technical analysis, the building lighting and visibility meet the specification requirements, each activity room meets the 3-hour sunshine standard, the building natural lighting meets the requirements, the design meets the human scale, and conforms to the humanized design concept

Unit Enlarged Plan

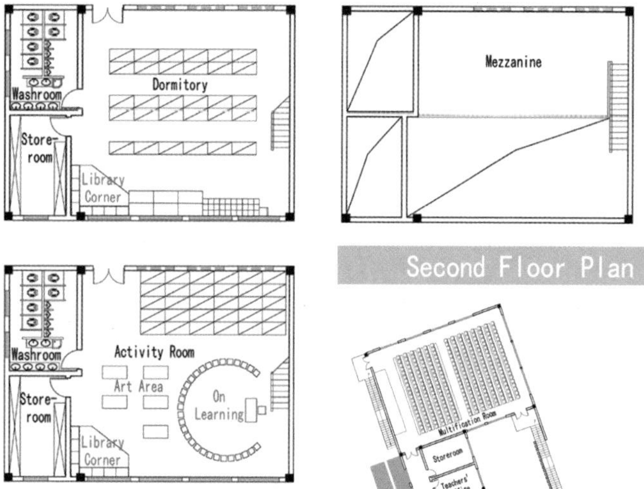
Washroom
Dormitory
Store-room
Library Corner
Mezzanine
Washroom
Activity Room
Art Area
On Learning
Store-room
Library Corner

Second Floor Plan

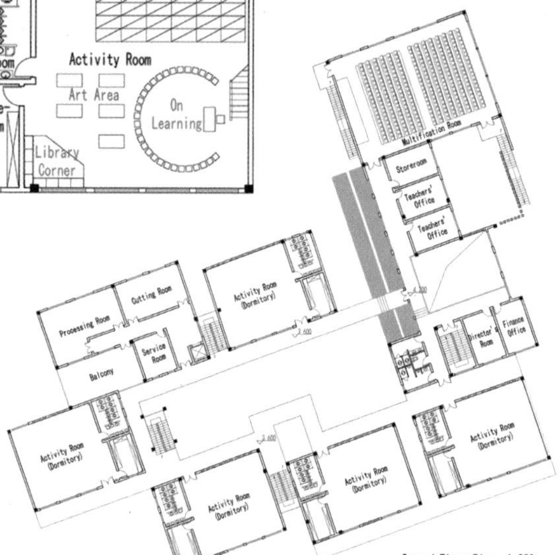
Multifunction Room
Storeroom
Teachers' Office
Teachers' Office
Processing Room
Cutting Room
Activity Room (Dormitory)
Service Room
Balcony
Director's Room
Finance Office
Activity Room (Dormitory)
Activity Room (Dormitory)
Activity Room (Dormitory)
Activity Room (Dormitory)
Activity Room (Dormitory)

Second Floor Plan 1:250

Analysis of Green Building

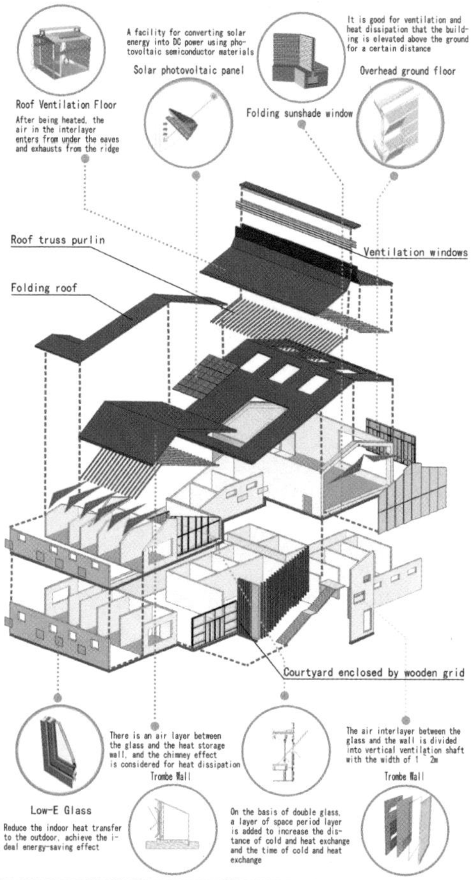
Roof Ventilation Floor
After being heated, the air in the interlayer enters from under the eaves and exhausts from the ridge
A facility for converting solar energy into DC power using photovoltaic semiconductor materials
Solar photovoltaic panel
It is good for ventilation and heat dissipation that the building is elevated above the ground for a certain distance
Overhead ground floor
Folding sunshade window
Roof truss purlin
Ventilation windows
Folding roof
Courtyard enclosed by wooden grid
Low-E Glass
Reduce the indoor heat transfer to the outdoor, achieve the ideal energy-saving effect
There is an air layer between the glass and the heat storage wall, and the chimney effect is considered for heat dissipation
Trombe Wall
On the basis of double glass, a layer of space period layer is added to increase the distance of cold and heat exchange and the time of cold and heat exchange
The air interlayer between the glass and the wall is divided into vertical ventilation shaft with the width of 1 2m
Trombe Wall

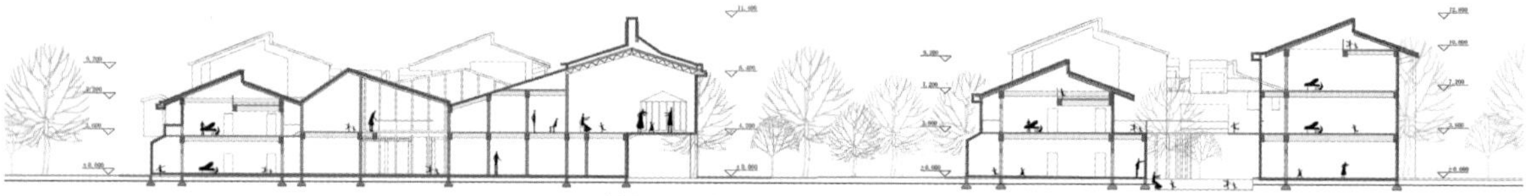

A-A Section 1:250

C-C Section 1:250

林海风语 FEEL THE WIND ⑥
景龙十二班幼儿园设计

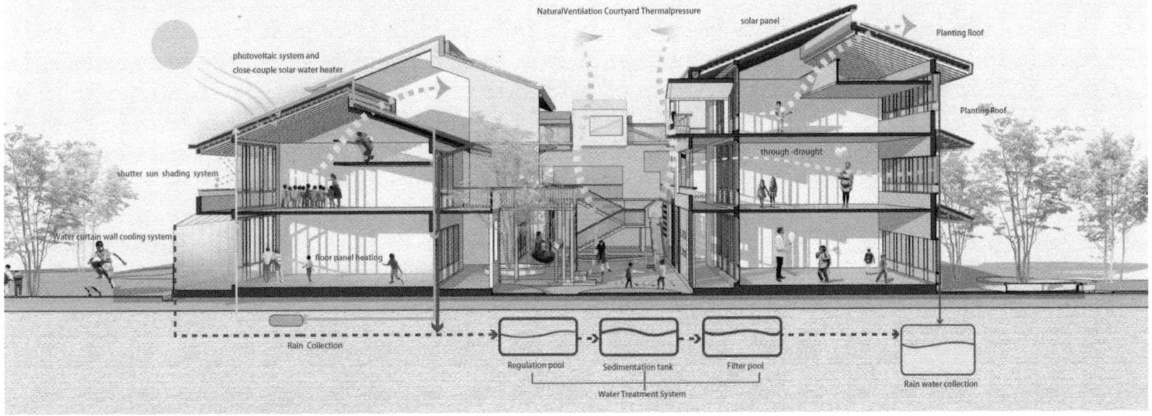

NaturalVentilation Courtyard Thermalpressure
solar panel
Planting Roof
photovoltaic system and
close-couple solar water heater
Planting Roof
through -draught
shutter sun shading system
Water curtain wall cooling system
floor panel heating
Rain Collection
Regulation pool
Sedimentation tank
Filter pool
Water Treatment System
Rain water collection

Comprehensive Technical Analysis

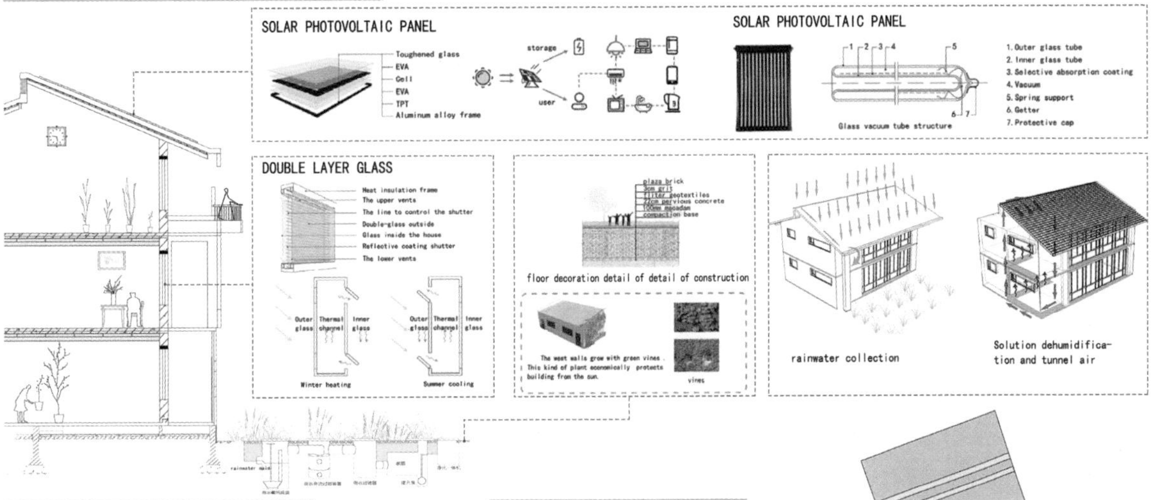

SOLAR PHOTOVOLTAIC PANEL
Toughened glass
EVA
Cell
EVA
TPT
Aluminum alloy frame
storage
user
SOLAR PHOTOVOLTAIC PANEL
1.Outer glass tube
2.Inner glass tube
3.Selective absorption coating
4.Vacuum
5.Spring support
6.Getter
7.Protective cap
Glass vacuum tube structure
DOUBLE LAYER GLASS
Heat insulation frame
The upper vents
The line to control the shutter
Double-glass outside
Glass inside the house
Reflective coating shutter
The lower vents
Outer glass
Thermal channel
Inner glass
Winter heating
Summer cooling
floor decoration detail of detail of construction
The west walls grow with green vines. This kind of plant economically protects building from the sun.
vines
rainwater collection
Solution dehumidification and tunnel air

Ventilation Analysis

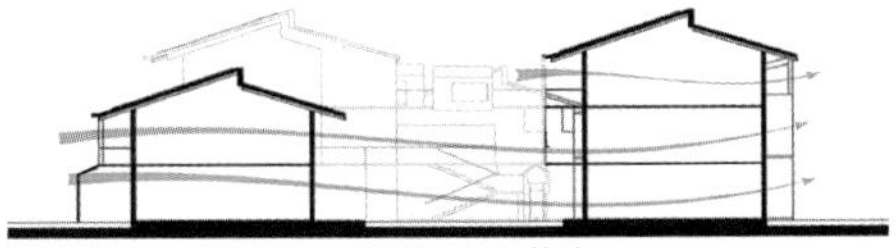
wind pressure ventilation

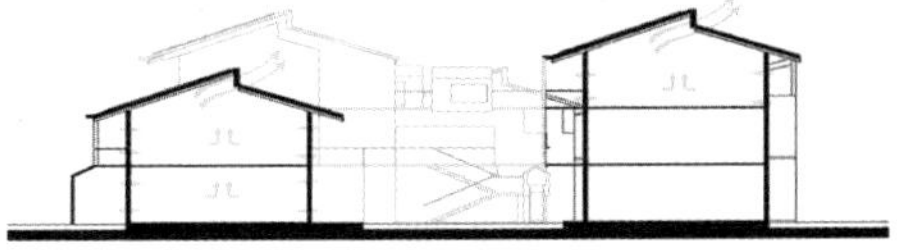
mechanical ventilation

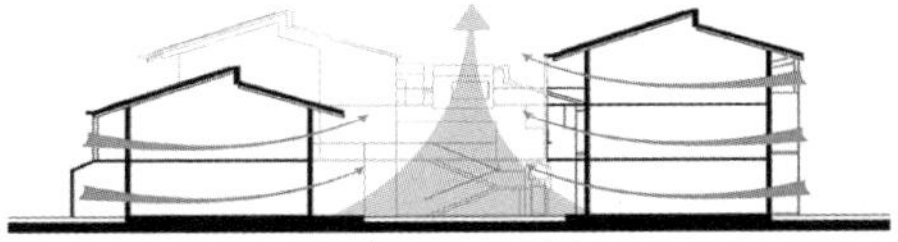
thermal pressure ventilation

Three Floor Plan

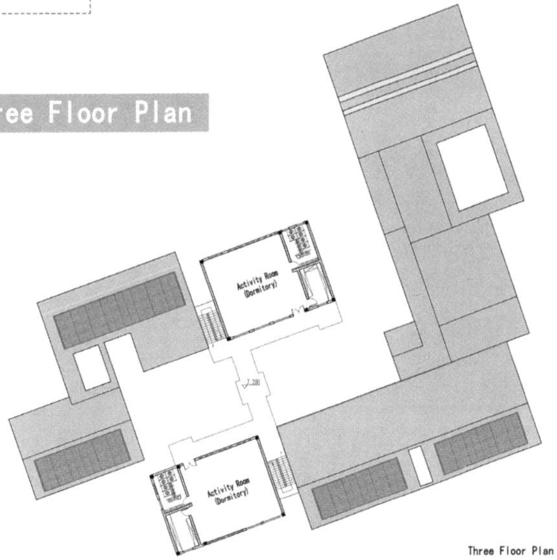

Activity Room (Dormitory)
Activity Room (Dormitory)

Three Floor Plan 1：250

综合奖 · 三等奖
General Prize Awarded · Third Prize

注 册 号：7162

项目名称：层峦 · 沐光行（南平）
Layers of Mountains · Bathe in Light（Nanping）

作　　者：肖婉凝、陈佳怡、郑曼如、刘世恒

参赛单位：福州大学

指导教师：崔育新

专家点评：

该作品规划布局规整。儿童室外活动场地类型丰富，建筑平面及垂直流线简洁，造型活泼。在被动式太阳能技术方面，对架空屋面和太阳能板隔热通风的应用巧妙；在主动式太阳能系统方面，彩色光伏玻璃的应用为室内增加了活跃度。但还要兼顾儿童视觉健康的需求，太阳能地暖系统的设置不够合理。

The layout of the work is well-planned. The work shows many types of outdoor playgrounds for children, concise and lively building plan and vertical streamline. As for the passive solar technology, the application of elevated overhead roofs and solar panels for thermal insulation and ventilation is ingenious. As for the active solar system, the application of colored photovoltaic glass increases indoor vitality, and also considers children's visual health. However, the setting of solar floor heating systems is not reasonable enough.

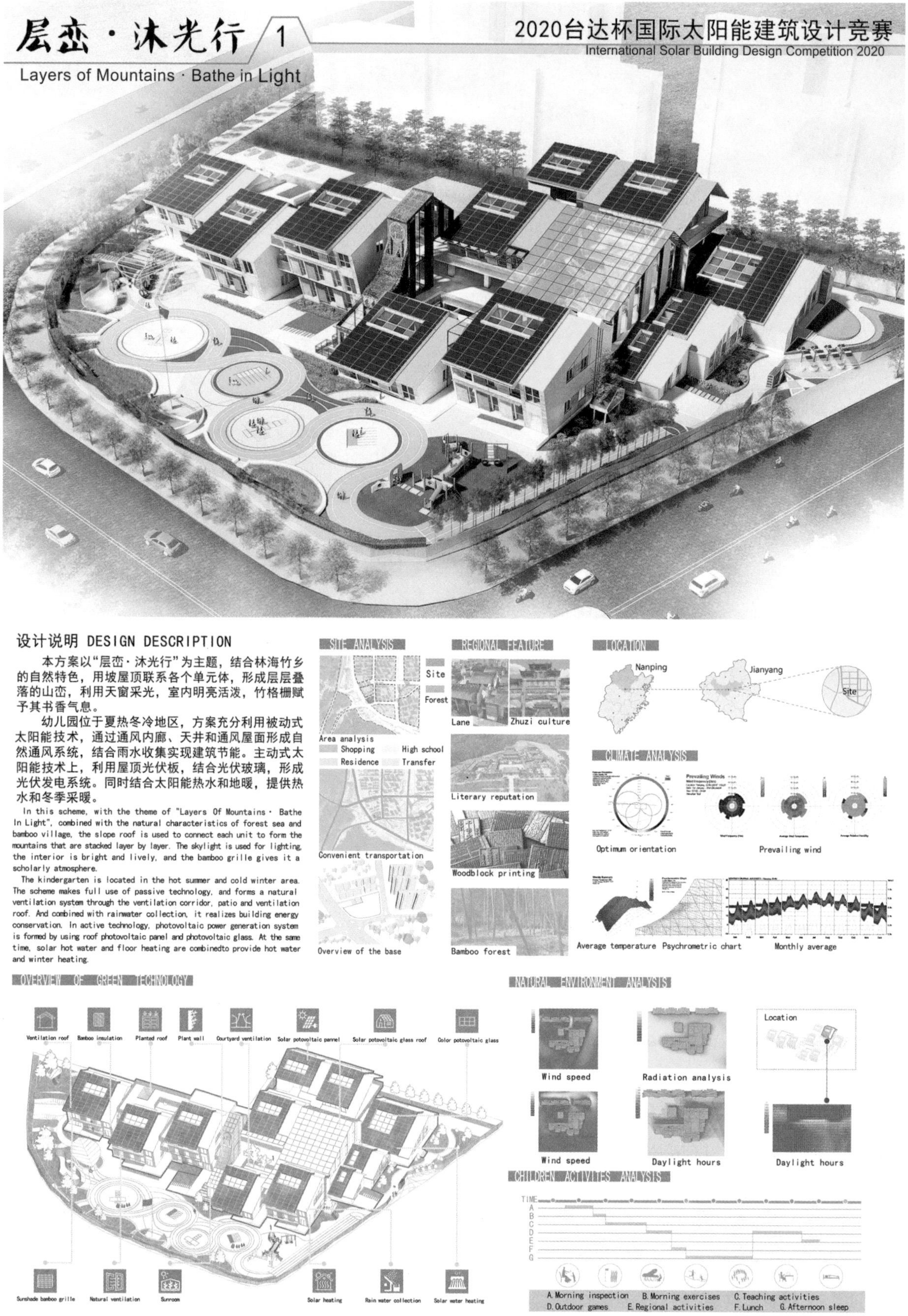
层峦 · 沐光行 1
Layers of Mountains · Bathe in Light
2020台达杯国际太阳能建筑设计竞赛
International Solar Building Design Competition 2020
设计说明 DESIGN DESCRIPTION
本方案以“层峦 · 沐光行”为主题，结合林海竹乡的自然特色，用坡屋顶联系各个单元体，形成层层叠落的山峦，利用天窗采光，室内明亮活泼，竹格栅赋予其书香气息。
幼儿园位于夏热冬冷地区，方案充分利用被动式太阳能技术，通过通风内廊、天井和通风屋面形成自然通风系统，结合雨水收集实现建筑节能。主动式太阳能技术上，利用屋顶光伏板，结合光伏玻璃，形成光伏发电系统。同时结合太阳能热水和地暖，提供热水和冬季采暖。
In this scheme, with the theme of "Layers Of Mountains · Bathe In Light", combined with the natural characteristics of forest sea and bamboo village, the slope roof is used to connect each unit to form the mountains that are stacked layer by layer. The skylight is used for lighting, the interior is bright and lively, and the bamboo grille gives it a scholarly atmosphere.
The kindergarten is located in the hot summer and cold winter area. The scheme makes full use of passive technology, and forms a natural ventilation system through the ventilation corridor, patio and ventilation roof. And combined with rainwater collection, it realizes building energy conservation. In active technology, photovoltaic power generation system is formed by using roof photovoltaic panel and photovoltaic glass. At the same time, solar hot water and floor heating are combinedto provide hot water and winter heating.
SITE ANALYSIS
Site
Forest
Area analysis
Shopping
High school
Residence
Transfer
Convenient transportation
Overview of the base
REGIONAL FEATURE
Lane
Zhuzi culture
Literary reputation
Woodblock printing
Bamboo forest
LOCATION
Nanping
Jianyang
Site
CLIMATE ANALYSIS
Optimum orientation
Prevailing wind
Average temperature
Psychrometric chart
Monthly average
OVERVIEW OF GREEN TECHNOLOGY
Ventilation roof
Bamboo insulation
Planted roof
Plant wall
Courtyard ventilation
Solar potovoltaic pannel
Solar potovoltaic glass roof
Color potovoltaic glass
Sunshade bamboo grille
Natural ventilation
Sunroom
Solar heating
Rain water collection
Solar water heating
NATURAL ENVIRONMENT ANALYSIS
Wind speed
Radiation analysis
Location
Wind speed
Daylight hours
Daylight hours
CHILDREN ACTIVITIES ANALYSIS
TIME
A
B
C
D
E
F
G
A. Morning inspection
B. Morning exercises
C. Teaching activities
D. Outdoor games
E. Regional activities
F. Lunch
G. Afternoon sleep

层峦·沐光行 2

2020台达杯国际太阳能建筑设计竞赛
International Solar Building Design Competition 2020

Layers of Mountains · Bathe in Light

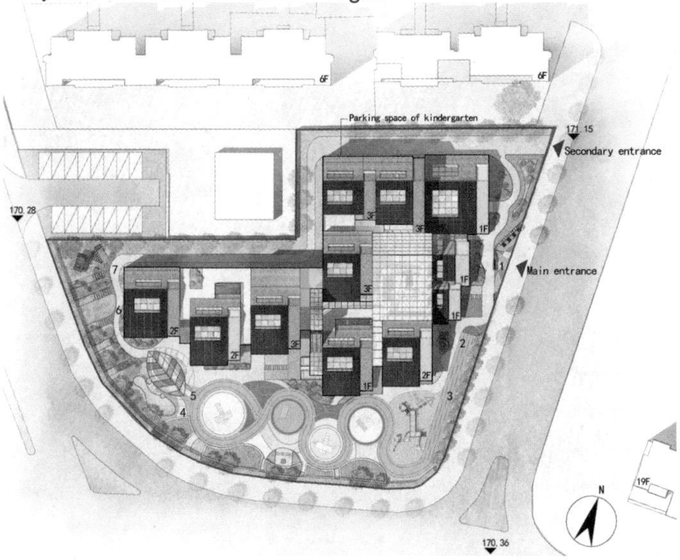

6F
6F
Parking space of kindergarten
171.15
Secondary entrance
170.28
Main entrance
3F
1F
2F
19F
N
170.36

TECHNICAL AND ECONOMIC INDICATORS

SITE PLAN 1:500

Land Area:6028㎡ Total Building Area:4482㎡ Building Occupation Area:2196㎡ Building Density:36.4%
Plot Rratio:0.74 Green space rate:30% Outdoor Playground Area: 1610㎡ Area Of Class Activity Site:760㎡

ANALYSIS OF CHILDREN

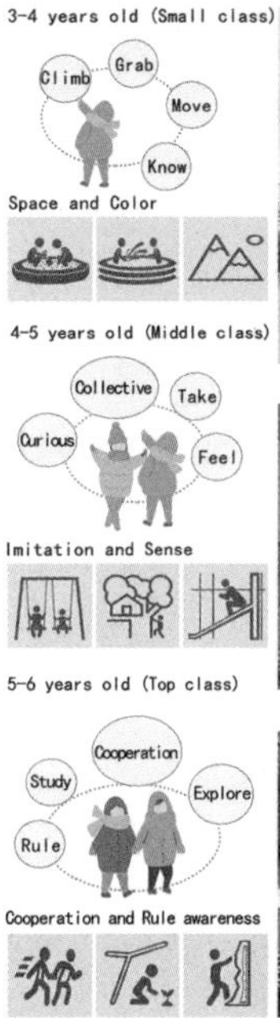

3-4 years old (Small class)
Climb
Grab
Move
Know
Space and Color
4-5 years old (Middle class)
Collective
Take
Curious
Feel
Imitation and Sense
5-6 years old (Top class)
Cooperation
Study
Explore
Rule
Cooperation and Rule awareness

SCENE PERSPECTIVE

1 The main entrance on the east side

2 Treehouse on one side of the runway

3 30 meter track - starting point of the game

4 Play pool with bamboo grille

5 Swing under color photovoltaic shed

6 Tree holes and sand pits on the west side of the slope

7 Tree houses and grass slopes are the end points

SCHEME GENERATION

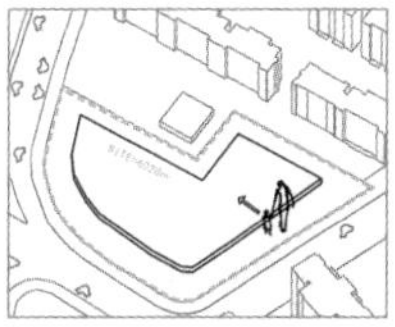
1 Site area:6028㎡, the buildings are arranged according to the site

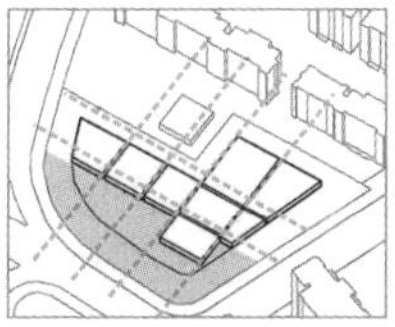
2 Well shaped segmentation, maximization of south facing site

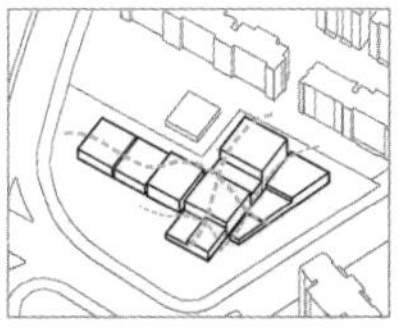
3 Mountain shape appears, four axis control

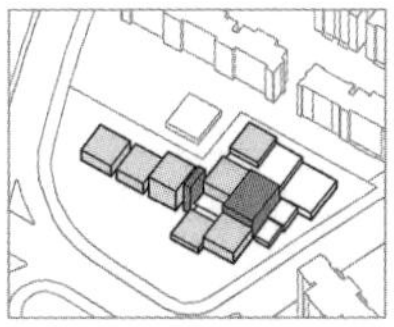
4 Determine the core traffic space, children's class block preliminary determined

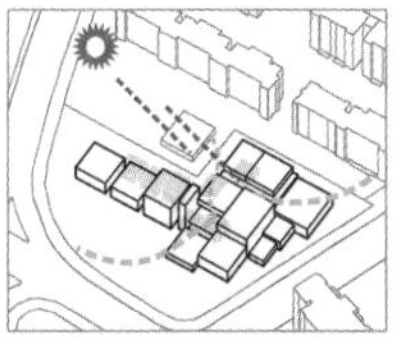
5 Block adjustment based on lighting and passive ventilation

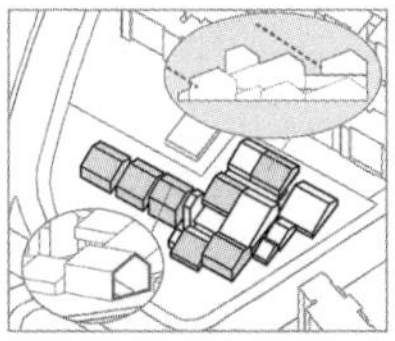
6 The roof is inclined, and the whole is in the shape of a mountain, and provides active solar power generation

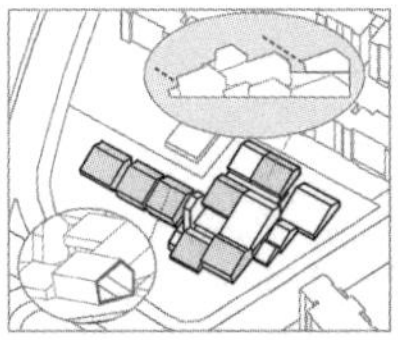
7 The elevation is inclined to improve the mountain shape, providing passive shade for the south facade

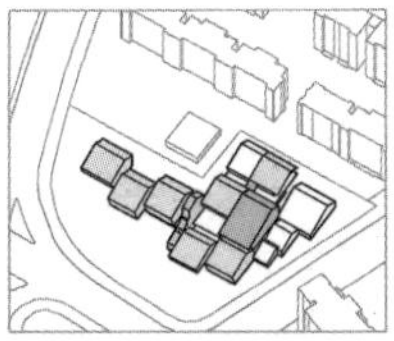
8 The blocks are scattered along the axis to enhance the permeability and weaken the sense of volume

14.300
-0.450

EAST ELEVATION 1:250

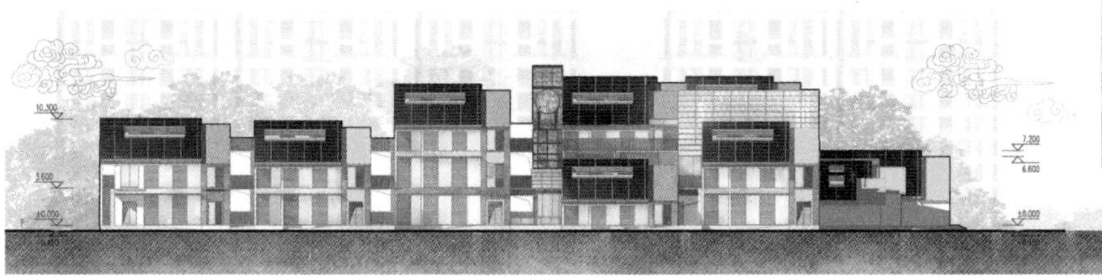

SOUTH ELEVATION 1:250

层峦·沐光行 3

2020台达杯国际太阳能建筑设计竞赛
International Solar Building Design Competition 2020

Layers of Mountains · Bathe in Light

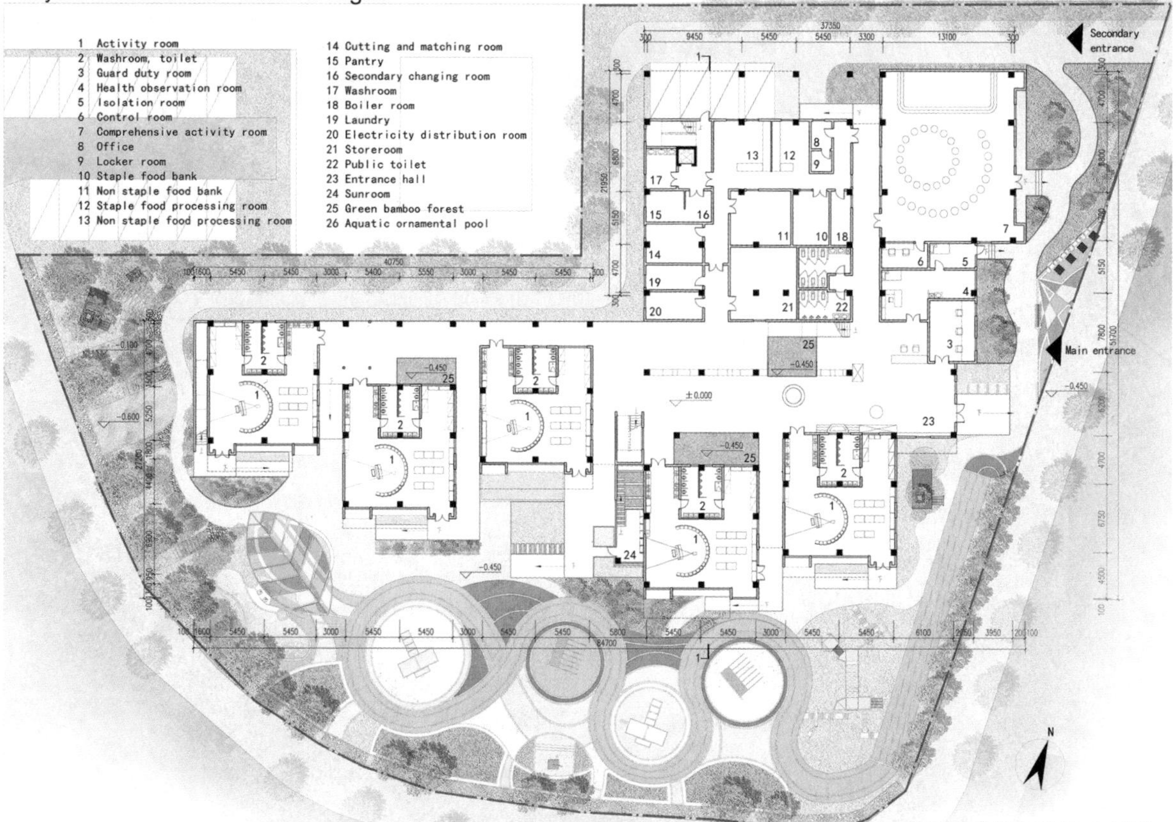
1 Activity room
2 Washroom, toilet
3 Guard duty room
4 Health observation room
5 Isolation room
6 Control room
7 Comprehensive activity room
8 Office
9 Locker room
10 Staple food bank
11 Non staple food bank
12 Staple food processing room
13 Non staple food processing room
14 Cutting and matching room
15 Pantry
16 Secondary changing room
17 Washroom
18 Boiler room
19 Laundry
20 Electricity distribution room
21 Storeroom
22 Public toilet
23 Entrance hall
24 Sunroom
25 Green bamboo forest
26 Aquatic ornamental pool
Secondary entrance
Main entrance
N

FIRST FLOOR PLAN 1：200

INDOOR PERSPECTIVE

A. Entrance Hallway

B. Atrium on the second floor

C. Sunroom

D. Class activity unit

FUNCTION STREAMLINE

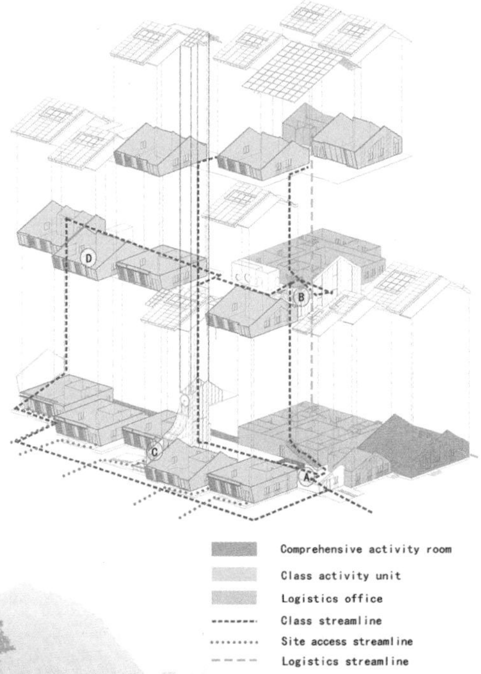
Comprehensive activity room
Class activity unit
Logistics office
Class streamline
Site access streamline
Logistics streamline

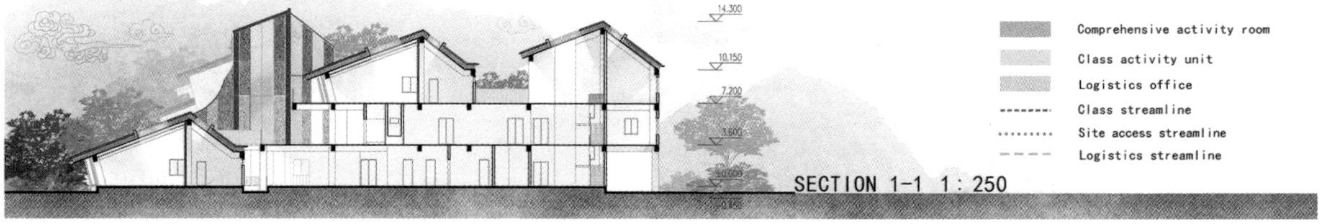

SECTION 1-1 1：250

层峦 · 沐光行 4

Layers of Mountains · Bathe in Light

2020台达杯国际太阳能建筑设计竞赛

International Solar Building Design Competition 2020

1 Activity room
2 Washroom, toilet
3 Public game area
4 Stalls
5 Open construction area
6 Teaching aids workshop
7 Book and conference room
8 Boiler Room
9 Director's Office
10 Finance Office
11 Teachers' Office
12 Leisure platform
13 Tool room
14 Faculty toilet

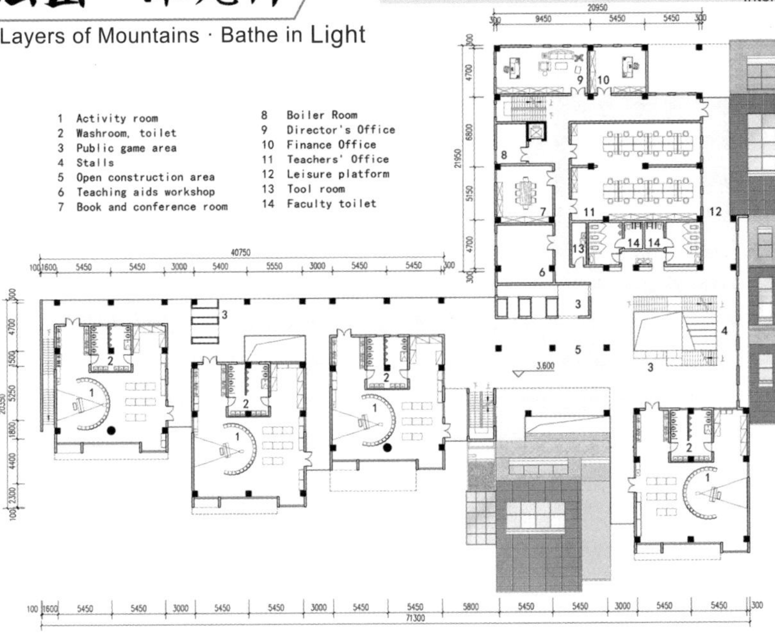

SECOND FLOOR PLAN 1 : 200

PASSIVE TECHNICAL ANALYSIS

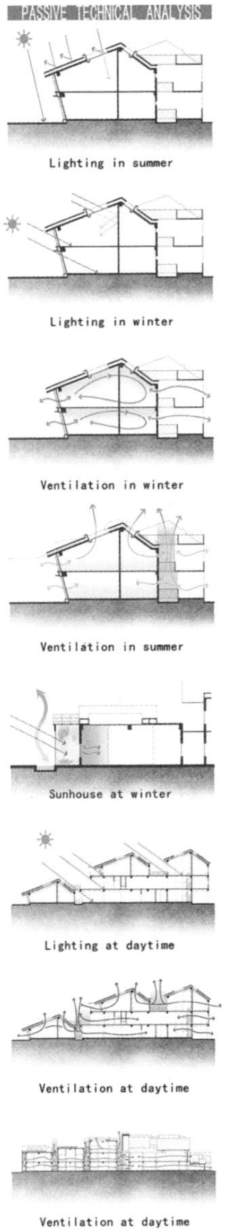
Lighting in summer
Lighting in winter
Ventilation in winter
Ventilation in summer
Sunhouse at winter
Lighting at daytime
Ventilation at daytime
Ventilation at daytime

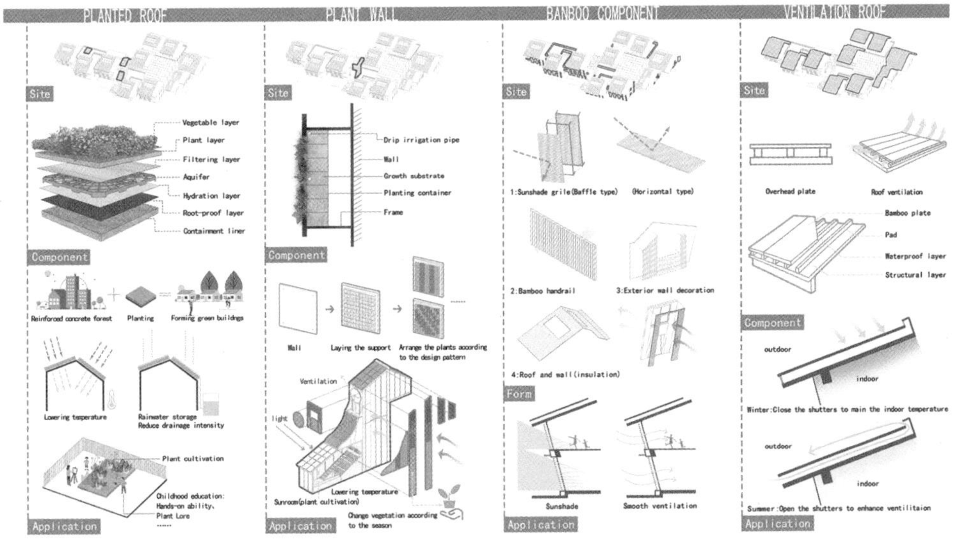
PLANTED ROOF
Site
Vegetable layer
Plant layer
Filtering layer
Aquifer
Hydration layer
Root-proof layer
Containment liner
Component
Reinforced concrete forest
Planting
Forming green buildings
Lowering temperature
Rainwater storage
Reduce drainage intensity
Plant cultivation
Childhood education:
Hands-on ability,
Plant Lore
Application
PLANT WALL
Site
Drip irrigation pipe
Wall
Growth substrate
Planting container
Frame
Component
Wall
Laying the support
Arrange the plants according to the design pattern
Ventilation
light
Lowering temperature
Sunroom(plant cultivation)
Change vegetation according to the season
Application
BANBOO COMPONENT
Site
1:Sunshade grile(Baffle type)
(Horizontal type)
2:Bamboo handrail
3:Exterior wall decoration
4:Roof and wall(insulation)
Form
Sunshade
Smooth ventilation
Application
VENTILATION ROOF
Site
Overhead plate
Roof ventilation
Bamboo plate
Pad
Waterproof layer
Structural layer
Component
outdoor
indoor
Winter:Close the shutters to main the indoor temperature
outdoor
indoor
Summer:Open the shutters to enhance ventilitaion
Application

RAIN GARDEN

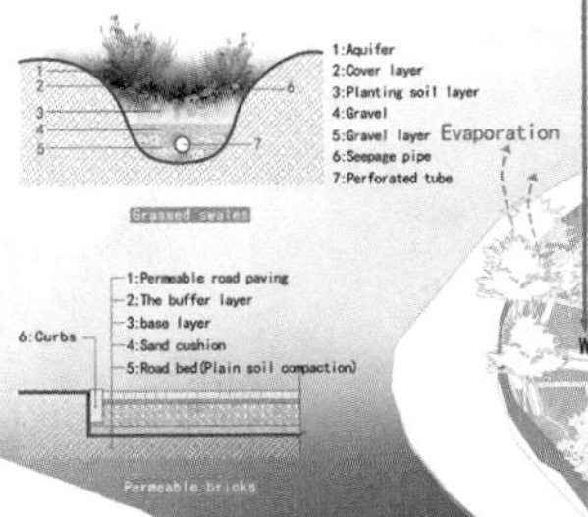
1:Aquifer
2:Cover layer
3:Planting soil layer
4:Gravel
5:Gravel layer
6:Seepage pipe
7:Perforated tube
Evaporation
Grassed swales
1:Permeable road paving
2:The buffer layer
3:base layer
4:Sand cushion
5:Road bed(Plain soil compaction)
6:Curbs
Permeable bricks

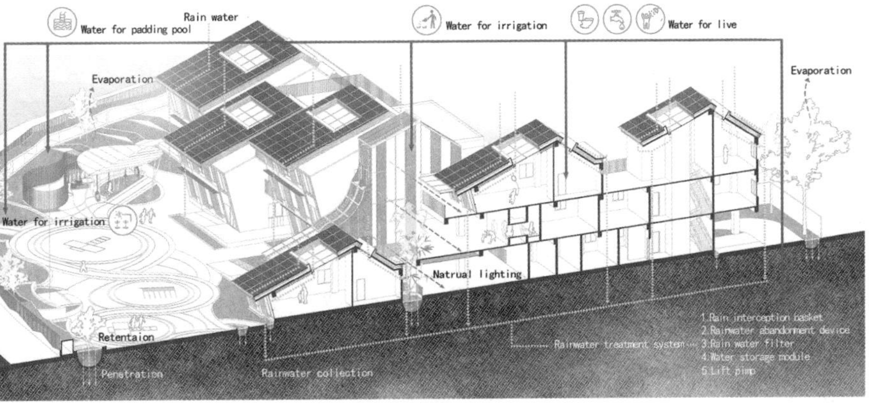
Rain water
Water for padding pool
Water for irrigation
Water for live
Evaporation
Evaporation
Water for irrigation
Natrual lighting
Retentaion
Penetration
Rainwater collection
Rainwater treatment system
1.Rain interception basket
2.Rainwater abandonment device
3.Rain water filter
4.Water storage module
5.Lift pump

层峦·沐光行 / 5

2020台达杯国际太阳能建筑设计竞赛
International Solar Building Design Competition 2020

Layers of Mountains · Bathe in Light

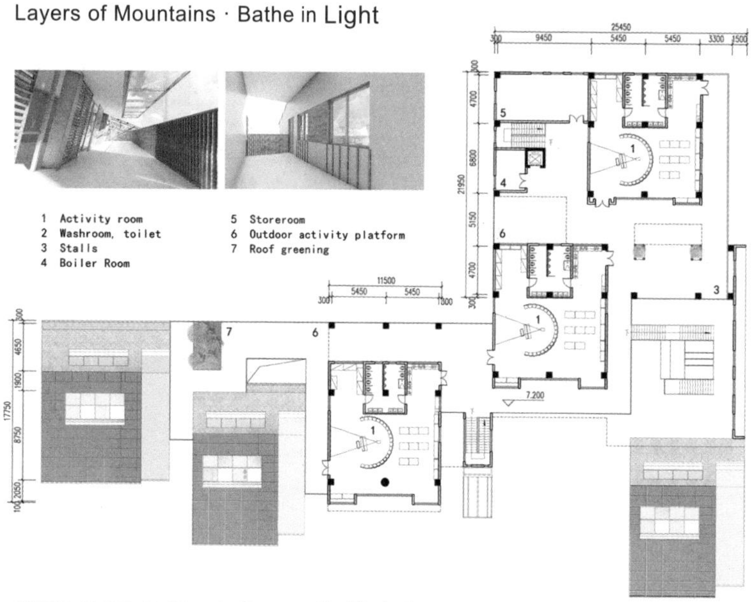

1 Activity room
2 Washroom, toilet
3 Stalls
4 Boiler Room
5 Storeroom
6 Outdoor activity platform
7 Roof greening

THIRD FLOOR PLAN 1 : 200

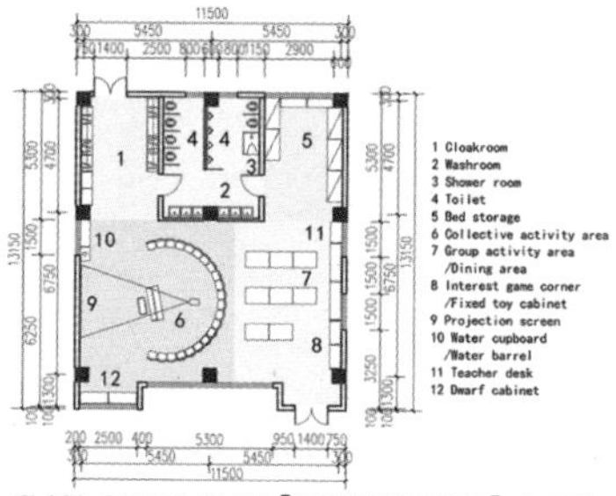

1 Cloakroom
2 Washroom
3 Shower room
4 Toilet
5 Bed storage
6 Collective activity area
7 Group activity area /Dining area
8 Interest game corner /Fixed toy cabinet
9 Projection screen
10 Water cupboard /Water barrel
11 Teacher desk
12 Dwarf cabinet

CLASS SINGLE PLAN [ACTIVITY STATE] 1:150

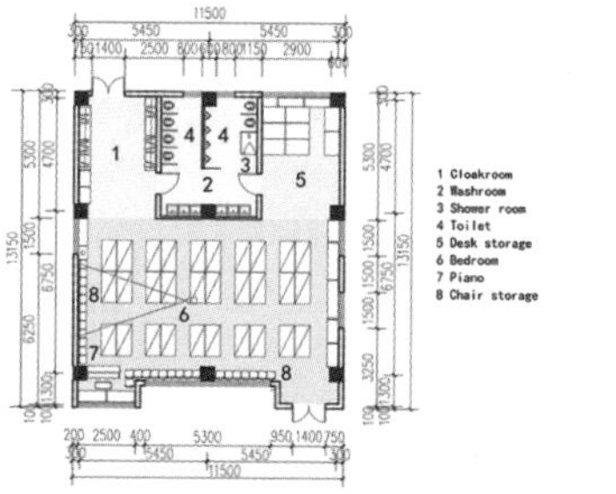

1 Cloakroom
2 Washroom
3 Shower room
4 Toilet
5 Desk storage
6 Bedroom
7 Piano
8 Chair storage

CLASS SINGLE PLAN [LUNCH BREAK] 1:150

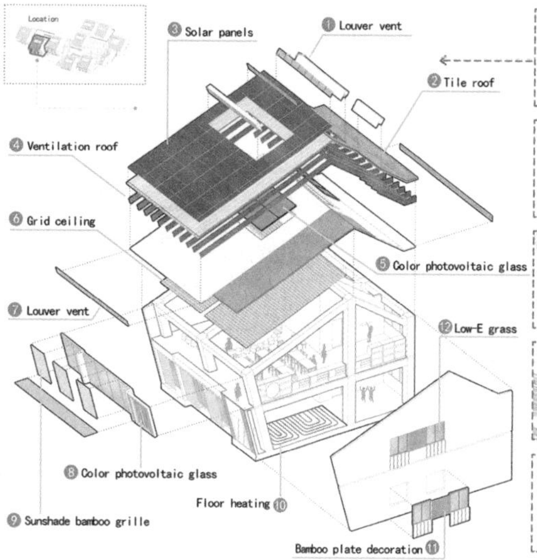

Location
1 Louver vent
2 Tile roof
3 Solar panels
4 Ventilation roof
5 Color photovoltaic glass
6 Grid ceiling
7 Louver vent
8 Color photovoltaic glass
9 Sunshade bamboo grille
10 Floor heating
11 Bamboo plate decoration
12 Low-E grass

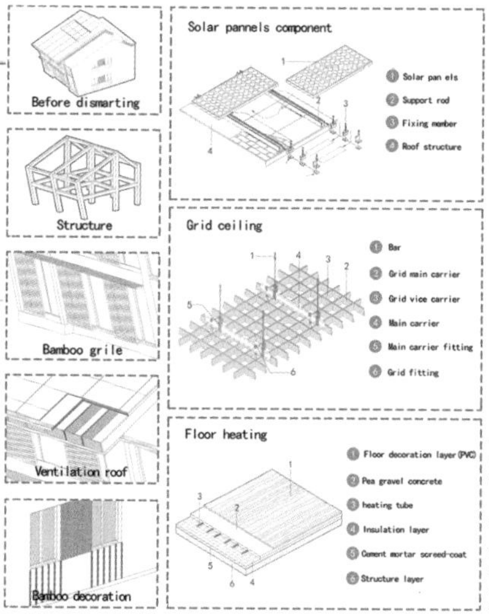

Before dismarting
Structure
Bamboo grile
Ventilation roof
Bamboo decoration
Solar pannels component
1 Solar pan els
2 Support rod
3 Fixing member
4 Roof structure
Grid ceiling
1 Bar
2 Grid main carrier
3 Grid vice carrier
4 Main carrier
5 Main carrier fitting
6 Grid fitting
Floor heating
1 Floor decoration layer (PVC)
2 Pea gravel concrete
3 heating tube
4 Insulation layer
5 Cement mortar screed-coat
6 Structure layer

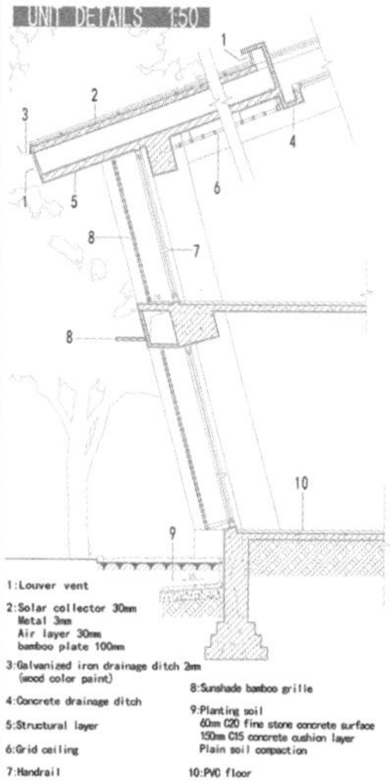

UNIT DETAILS 1:50
1:Louver vent
2:Solar collector 30mm Metal 3mm Air layer 30mm bamboo plate 100mm
3:Galvanized iron drainage ditch 2mm (wood color paint)
4:Concrete drainage ditch
5:Structural layer
6:Grid ceiling
7:Handrail
8:Sunshade bamboo grille
9:Planting soil
10:PVC floor

层峦·沐光行 6

2020台达杯国际太阳能建筑设计竞赛

International Solar Building Design Competition 2020

Layers of Mountains · Bathe in Light

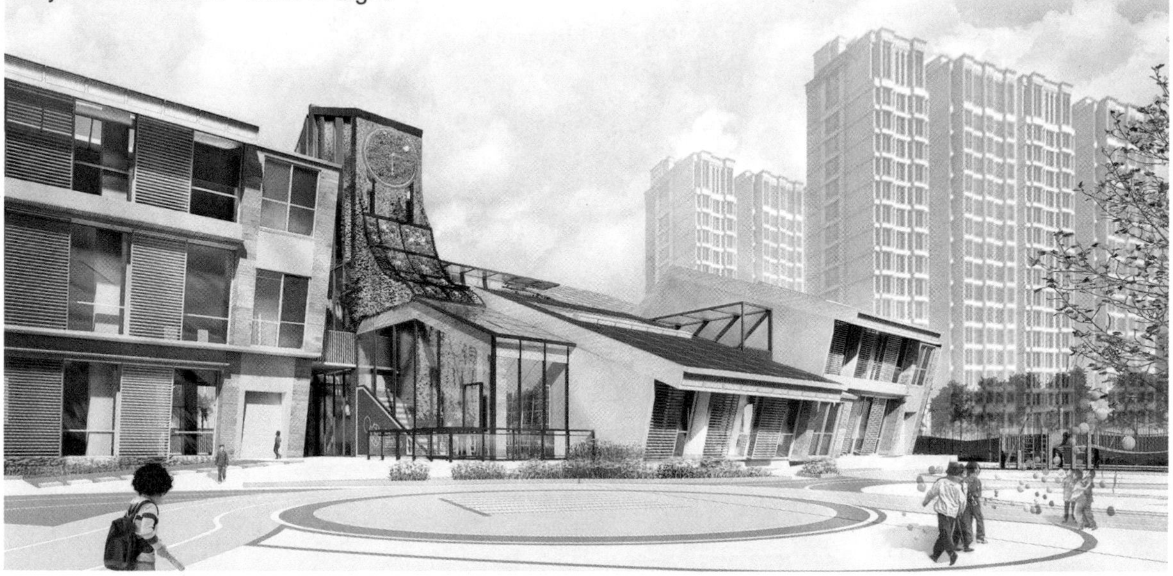

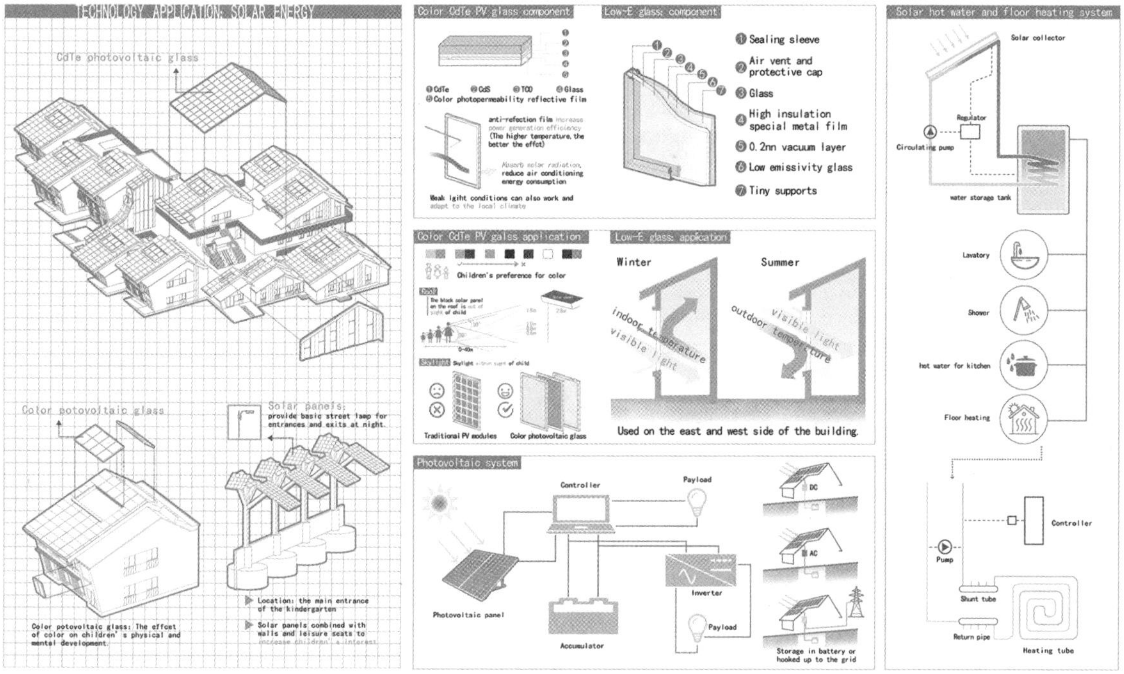
TECHNOLOGY APPLICATION: SOLAR ENERGY
CdTe photovoltaic glass
Color potovoltaic glass
Solar panels: provide basic street lamp for entrances and exits at night.
Color potovoltaic glass: The effect of color on children's physical and mental development.
Location: the main entrance of the kindergarten
Solar panels combined with walls and leisure seats to increase children's interest
Color CdTe PV glass component
CdTe
CdS
TCO
Glass
Color photopermeability reflective film
anti-refection film
(The higher temperature, the better the effct)
Absorb solar radiation, reduce air conditioning energy consumption
Weak light conditions can also work and adapt to the local climate
Low-E glass: component
Sealing sleeve
Air vent and protective cap
Glass
High insulation special metal film
0.2nn vacuum layer
Low emissivity glass
Tiny supports
Color CdTe PV galss application
Children's preference for color
Traditional PV modules
Color photovoltaic glass
Low-E glass: application
Winter
Summer
indoor temperature
visible light
outdoor temperature
Used on the east and west side of the building.
Photovoltaic system
Controller
Payload
Photovoltaic panel
Accumulator
Inverter
DC
AC
Storage in battery or hooked up to the grid
Solar hot water and floor heating system
Solar collector
Regulator
Circulating pump
water storage tank
Lavatory
Shower
hot water for kitchen
Floor heating
Controller
Pump
Shunt tube
Return pipe
Heating tube

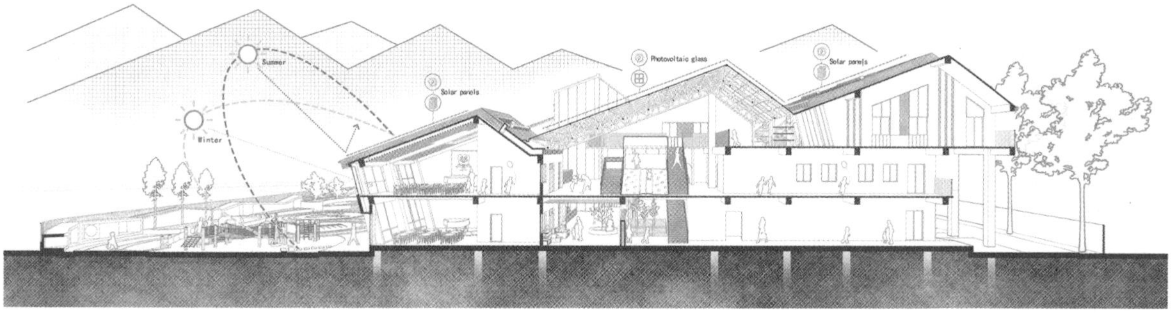
Summer
Winter
Solar panels
Photovoltaic glass
Solar panels

综合奖·三等奖
General Prize Awarded · Third Prize

注 册 号：7167

项目名称：竹巷·风生（南平）
Bamboo Alley · Wind Form (Nanping)

作　　者：陈泉任、欧智凡、许子娴、阮朝锦、魏佳纯

参赛单位：重庆大学、华南理工大学、Royal College of Art

指导教师：周铁军、张海滨

专家点评：

作品规划设计合理，平面布局紧凑，功能合理，注重建筑通风、遮阳等技术与采光有机结合，对“冷巷”的应用较为合理，被动技术合理，简洁，太阳能的主动与被动式利用结合较好，尤其是被动式通风隔热技术利用较好。

The building is reasonable in planning, compact in plane, and reasonable in functions. It focuses on the organic combination of ventilation, sun shading and other technologies with lighting. The work presents reasonable application of “cold lanes”, and reasonable and concise application of passive technologies. In addition, the active and passive solar energy utilization is well combined, and passive technologies of ventilation and thermal insulation are better adopted especially.

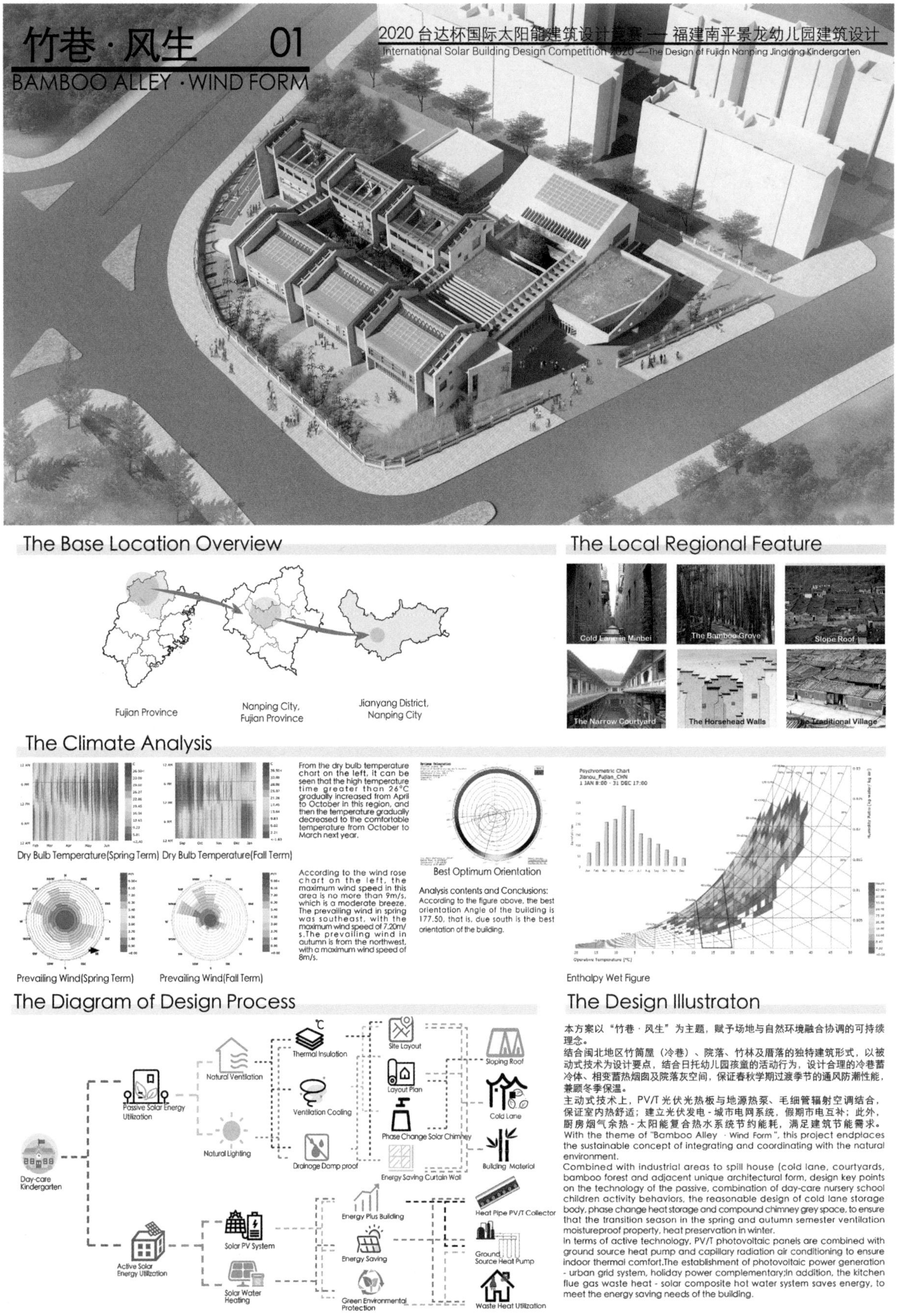
竹巷·风生 01
BAMBOO ALLEY · WIND FORM
2020 台达杯国际太阳能建筑设计竞赛——福建南平景龙幼儿园建筑设计
International Solar Building Design Competition 2020—The Design of Fujian Nanping Jinglong Kindergarten
The Base Location Overview
Fujian Province
Nanping City, Fujian Province
Jianyang District, Nanping City
The Local Regional Feature
Cold Lane in Minbei
The Bamboo Grove
Slope Roof
The Narrow Courtyard
The Horsehead Walls
The Traditional Village
The Climate Analysis
Dry Bulb Temperature(Spring Term)
Dry Bulb Temperature(Fall Term)
From the dry bulb temperature chart on the left, it can be seen that the high temperature time greater than 26°C gradually increased from April to October in this region, and then the temperature gradually decreased to the comfortable temperature from October to March next year.
Prevailing Wind(Spring Term)
Prevailing Wind(Fall Term)
According to the wind rose chart on the left, the maximum wind speed in this area is no more than 9m/s, which is a moderate breeze. The prevailing wind in spring was southeast, with the maximum wind speed of 7.20m/s.The prevailing wind in autumn is from the northwest, with a maximum wind speed of 8m/s.
Best Optimum Orientation
Analysis contents and Conclusions:
According to the figure above, the best orientation Angle of the building is 177.50, that is, due south is the best orientation of the building.
Enthalpy Wet Figure
The Diagram of Design Process
Day-care Kindergarten
Passive Solar Energy Utilization
Natural Ventilation
Natural Lighting
Thermal Insulation
Ventilation Cooling
Drainage Damp proof
Site Layout
Layout Plan
Phase Change Solar Chimney
Energy Saving Curtain Wall
Sloping Roof
Cold Lane
Building Material
Active Solar Energy Utilization
Solar PV System
Solar Water Heating
Energy Plus Building
Energy Saving
Green Environmental Protection
Heat Pipe PV/T Collector
Ground Source Heat Pump
Waste Heat Utilization
The Design Illustraton
本方案以“竹巷·风生”为主题，赋予场地与自然环境融合协调的可持续理念。
结合闽北地区竹筒屋（冷巷）、院落、竹林及厝落的独特建筑形式，以被动式技术为设计要点，结合日托幼儿园孩童的活动行为，设计合理的冷巷蓄冷体、相变蓄热烟囱及院落灰空间，保证春秋学期过渡季节的通风防潮性能，兼顾冬季保温。
主动式技术上，PV/T 光伏光热板与地源热泵、毛细管辐射空调结合，保证室内热舒适；建立光伏发电 - 城市电网系统，假期市电互补；此外，厨房烟气余热 - 太阳能复合热水系统节约能耗，满足建筑节能需求。
With the theme of "Bamboo Alley · Wind Form", this project endplaces the sustainable concept of integrating and coordinating with the natural environment.
Combined with industrial areas to spill house (cold lane, courtyards, bamboo forest and adjacent unique architectural form, design key points on the technology of the passive, combination of day-care nursery school children activity behaviors, the reasonable design of cold lane storage body, phase change heat storage and compound chimney grey space, to ensure that the transition season in the spring and autumn semester ventilation moistureproof property, heat preservation in winter.
In terms of active technology, PV/T photovoltaic panels are combined with ground source heat pump and capillary radiation air conditioning to ensure indoor thermal comfort.The establishment of photovoltaic power generation - urban grid system, holiday power complementary;In addition, the kitchen flue gas waste heat - solar composite hot water system saves energy, to meet the energy saving needs of the building.

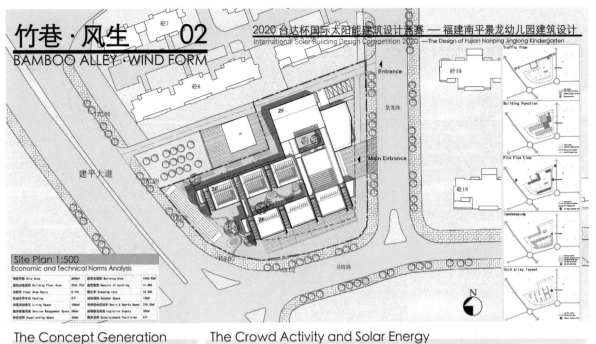

竹巷·风生 02
BAMBOO ALLEY·WIND FORM

2020 台达杯国际太阳能建筑设计竞赛 — 福建南平景龙幼儿园建筑设计
International Solar Building Design Competition 2020 —The Design of Fujian Nanping Jinglong Kindergarten

Entrance
Main Entrance
景龙路
建平大道
书院路
2F
砼7
砼6
砼19
砼18
N

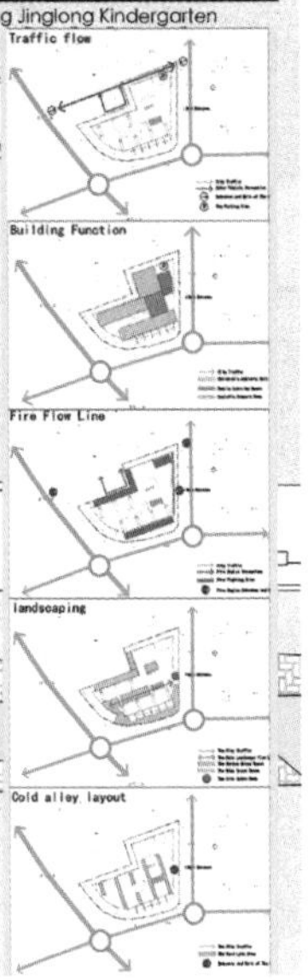
Traffic flow
Building Function
Fire Flow Line
landscaping
Cold alley layout

Site Plan 1:500
Economic and Technical Norms Analysis

项目用地 Site Area	6028m²	建筑总面积 Building Area	4465.92m²
建筑占地面积 Building Floor Area	2524.72m²	建筑密度 Density of buidling	41.88%
容积率 Floor Area Ratio	0.74%	绿化率 Greening rate	32.86%
机动车停车位 Parking	5个	活动场地 Outdoor Space	720m²
班级活动单元 Living Space	1586m²	音体综合活动室 Music & Sports Space	270.45m²
服务管理用房 Service Management Space	306m²	后勤服务用房 Logistics Supply	302m²
体验空间 Experiencing Space	326m²	娱乐设施 Entertainment Facilities	5个

The Concept Generation

Regional Technology: Cold Lane

Regional Material: Bamboo

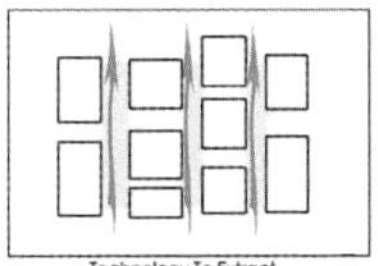
Technology To Extract

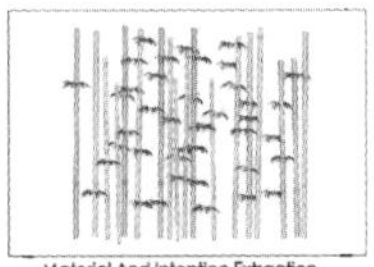
Material And Intention Extraction

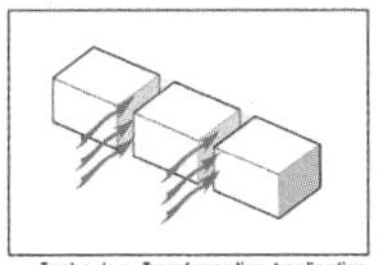
Technology Transformation Application

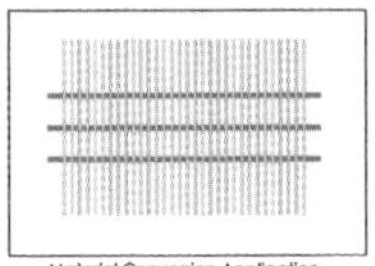
Material Conversion Application

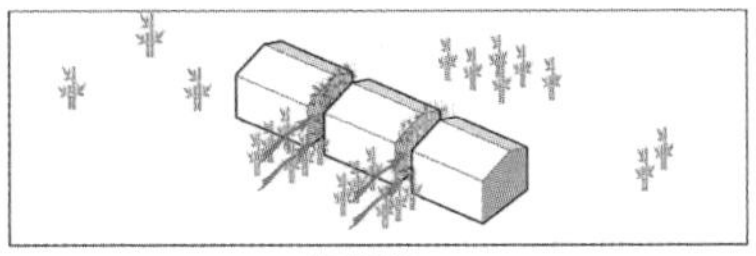
Bamboo lane

The Crowd Activity and Solar Energy

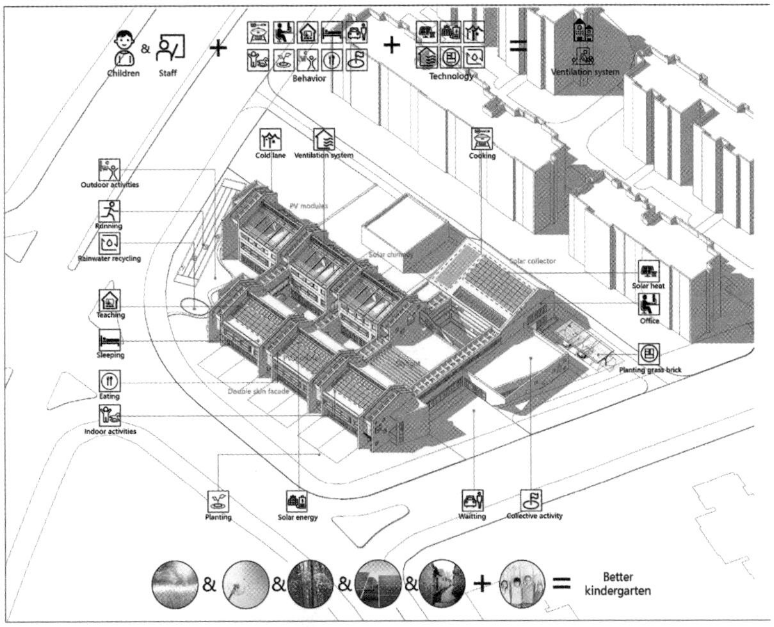
Children
&
Staff
Behavior
Technology
Ventilation system
Cold lane
Ventilation system
Cooking
Outdoor activities
Running
Rainwater recycling
PV modules
Solar chimney
Solar collector
Solar heat
Office
Teaching
Sleeping
Eating
Indoor activities
Double skin facade
Planting grass brick
Planting
Solar energy
Waiting
Collective activity
Better kindergarten

The Solar Analysis

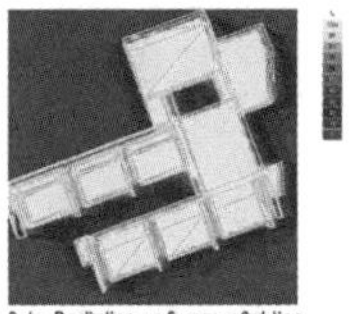
Solar Radiation on Summer Solstice

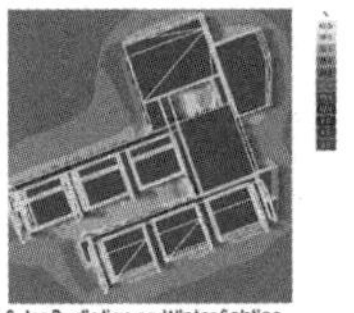
Solar Radiation on Winter Solstice

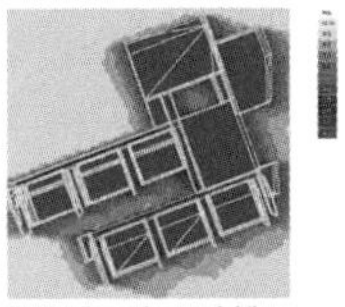
Sun Shading on Summer Solstice

Direct Solar Radiation on Winter Solstice

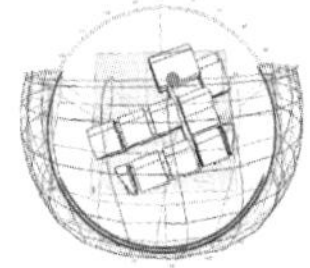
Shadow Map on Summer Solstice

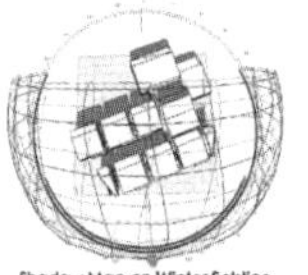
Shadow Map on Winter Solstice

The Block Analysis

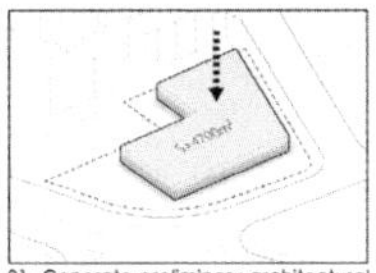
01--Generate preliminary architectural form according to the form of the site.

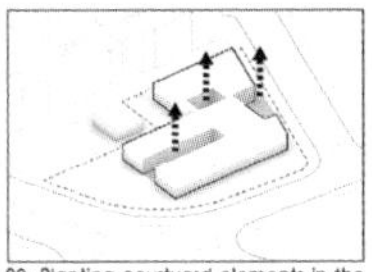
02--Planting courtyard elements in the original volume.

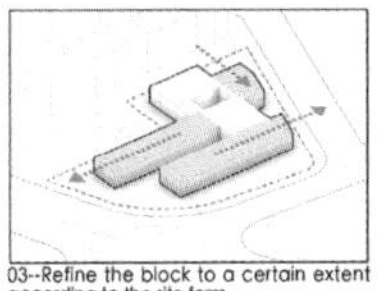
03--Refine the block to a certain extent according to the site form.

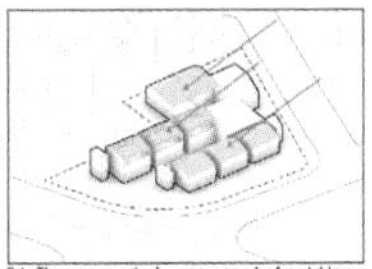
04--The green design concept of cold lane is implanted in the building to achieve the goal of energy conservation.

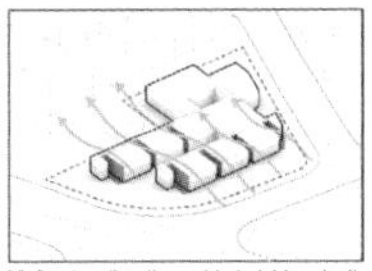
05--By elevating the gable height on both sides of the movable unit, the function of cold lane is further strengthened to achieve effective energy saving effect.

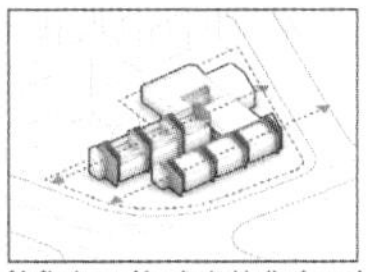
06--Sloping roof is adopted in the form of building roof, which not only keeps the characteristics of traditional folk houses, but also enhances the energy-saving effect of cold lane.

竹巷 · 风生 03

BAMBOO ALLEY · WIND FORM

2020 台达杯国际太阳能建筑设计竞赛 — 福建南平景龙幼儿园建筑设计

International Solar Building Design Competition 2020 — The Design of Fujian Nanping Jinglong Kindergarten

Entrance
Main Entrance
景龙路
大道
1 Hall
2 Duty room
3 Isolation room
4 Morning inspection room
5 Health room
6 Storage Room
7 Parent Reception Room
8 Classroom
9 Cold lane
10 Secondary foyer
11 Multi-funtional room
12 Kitchen
13 Storage room
14 laundry room
15 The lounge
16 Water room
17 Switch room
18 Bathroom

The First Floor Plan 1 : 250

The Diagram of Mobile Unit Technology

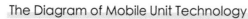

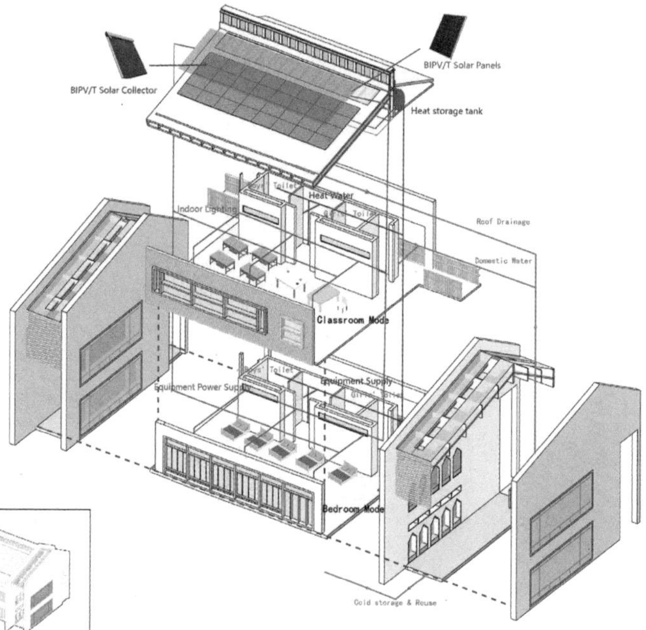
BIPV/T Solar Collector
BIPV/T Solar Panels
Heat storage tank
Heat Water
Indoor Lighting
Roof Drainage
Domestic Water
Classroom Mode
Equipment Power Supply
Equipment Supply
Bedroom Mode
Cold storage & Reuse

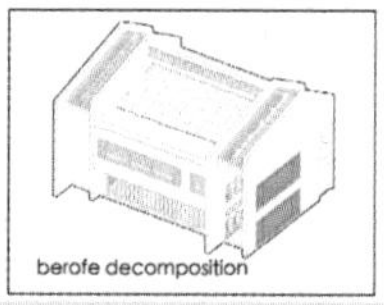

berofe decomposition

The Room in Active State

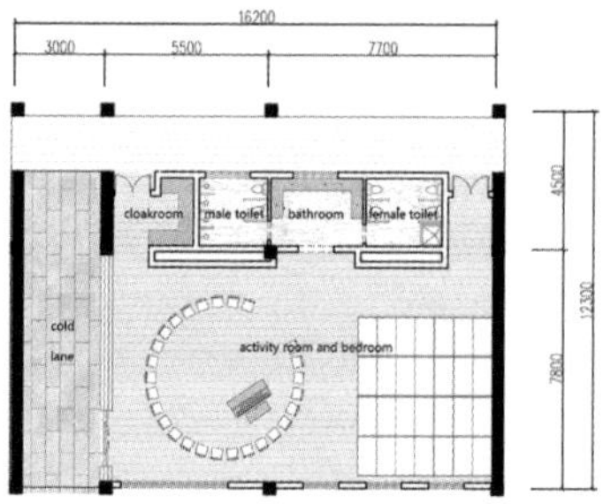
16200
3000
5500
7700
cloakroom
male toilet
bathroom
female toilet
cold lane
activity room and bedroom
4500
7800
12300

The Room in Reading State

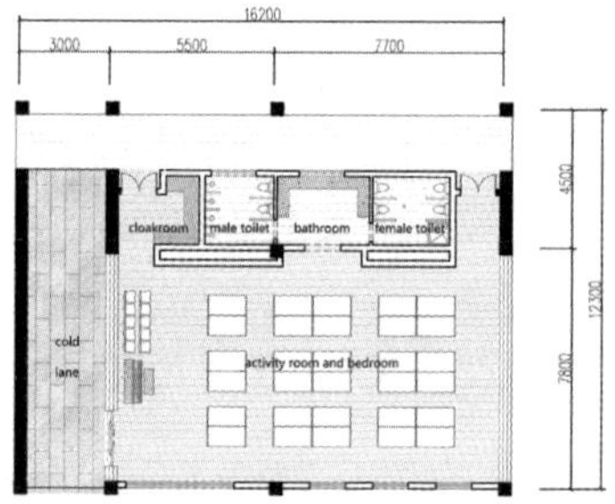
16200
3000
5500
7700
cloakroom
male toilet
bathroom
female toilet
cold lane
activity room and bedroom
4500
7800
12300

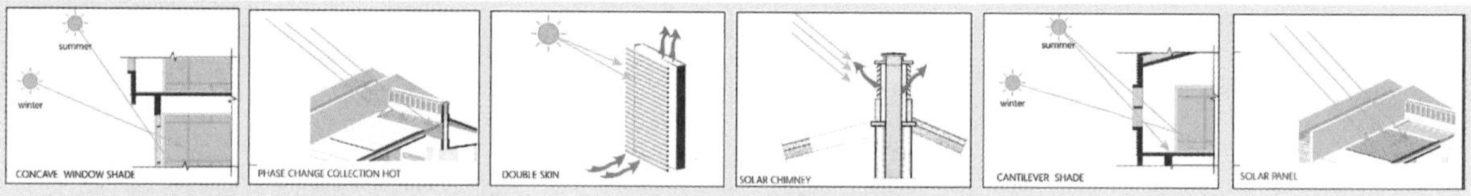
summer
winter
CONCAVE WINDOW SHADE
PHASE CHANGE COLLECTION HOT
DOUBLE SKIN
SOLAR CHIMNEY
CANTILEVER SHADE
SOLAR PANEL

竹巷 · 风生 04

BAMBOO ALLEY · WIND FORM

2020 台达杯国际太阳能建筑设计竞赛 — 福建南平景龙幼儿园建筑设计

International Solar Building Design Competition 2020—The Design of Fujian Nanping Jinglong Kindergarten

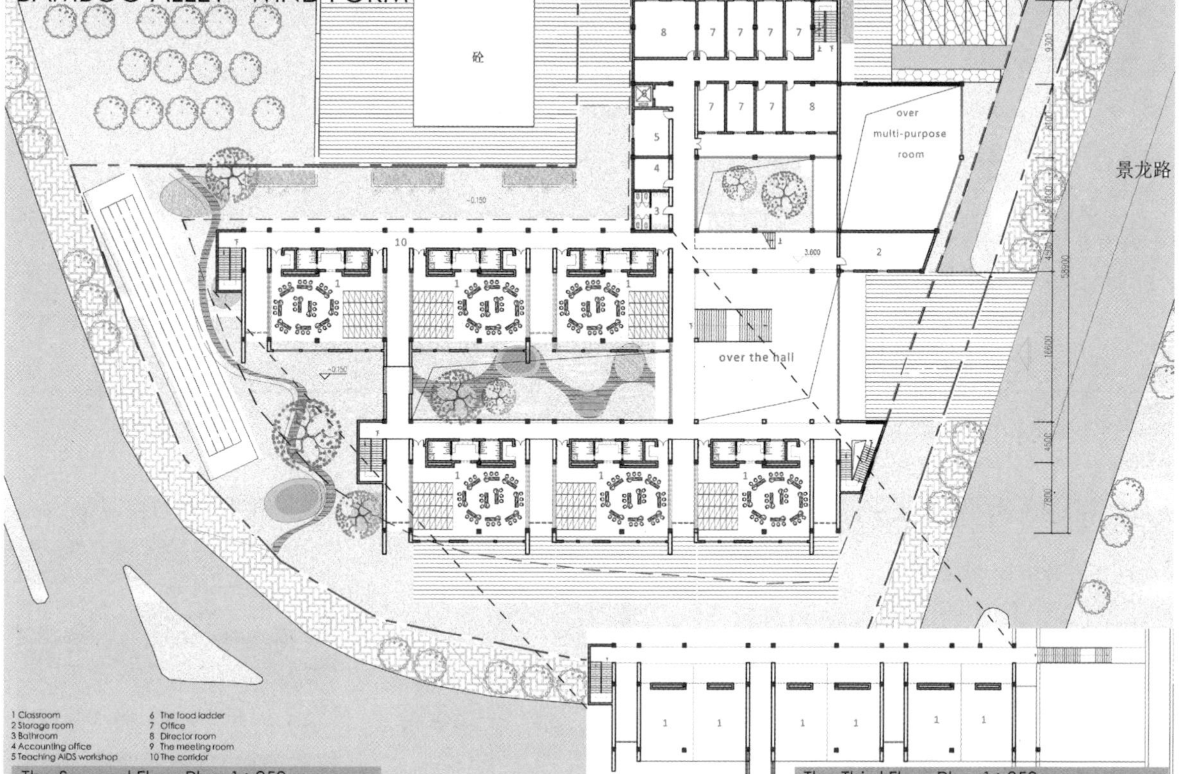
over
multi-purpose
room
over the hall
景龙路
1 Classroom
2 Storage room
3 Bathroom
4 Accounting office
5 Teaching AIDS workshop
6 The food ladder
7 Office
8 Director room
9 The meeting room
10 The corridor

The Second Floor Plan 1 : 250

The Third Floor Plan 1 : 250

The Active Unit Profile Analysis

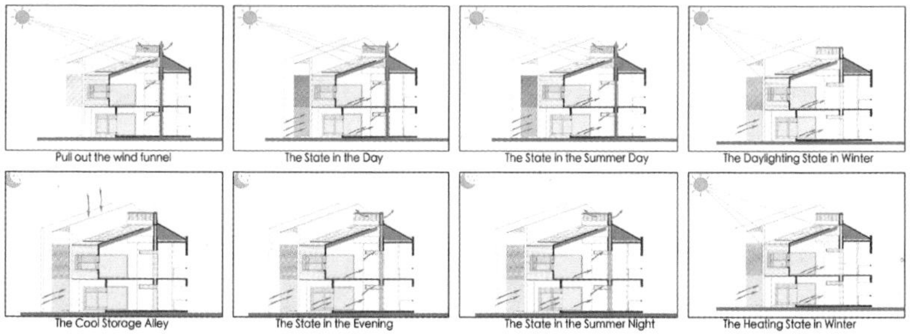
Pull out the wind funnel
The State in the Day
The State in the Summer Day
The Daylighting State in Winter
The Cool Storage Alley
The State in the Evening
The State in the Summer Night
The Heating State in Winter

The Wind Simulation

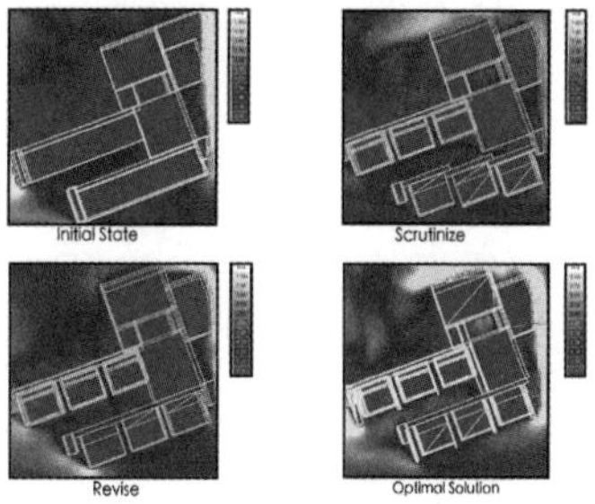
Initial State
Scrutinize
Revise
Optimal Solution

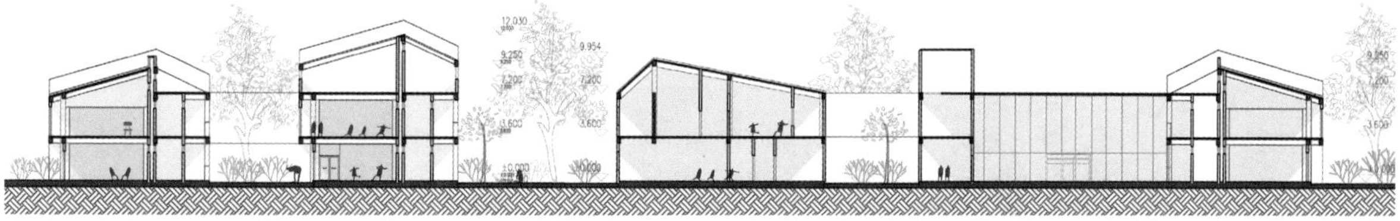

1-1 Section 1 : 200

2-2 Section 1 : 200

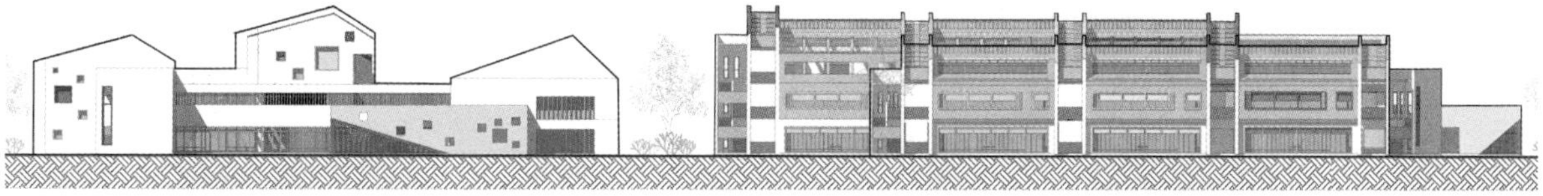

The South Elevation 1 : 200

The East Elevation 1 : 200

竹巷 · 风生 05
BAMBOO ALLEY · WIND FORM

2020 台达杯国际太阳能建筑设计竞赛 ---- 福建南平景龙幼儿园建筑设计
International Solar Building Design Competition 2020 ---The Design of Fujian Nanping Jinglong Kindergarten

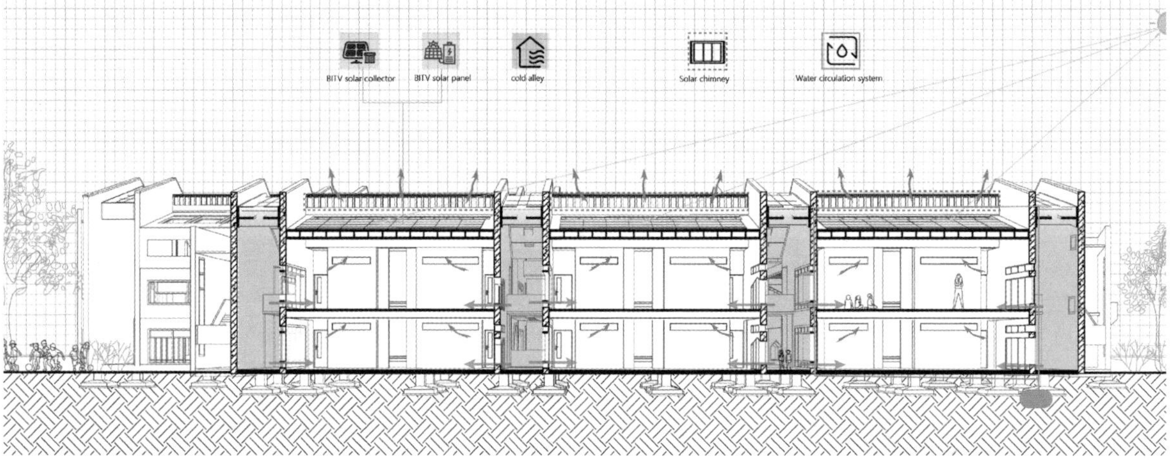

BITV solar collector
BITV solar panel
cold alley
Solar chimney
Water circulation system

Active Solar Energy Utilization

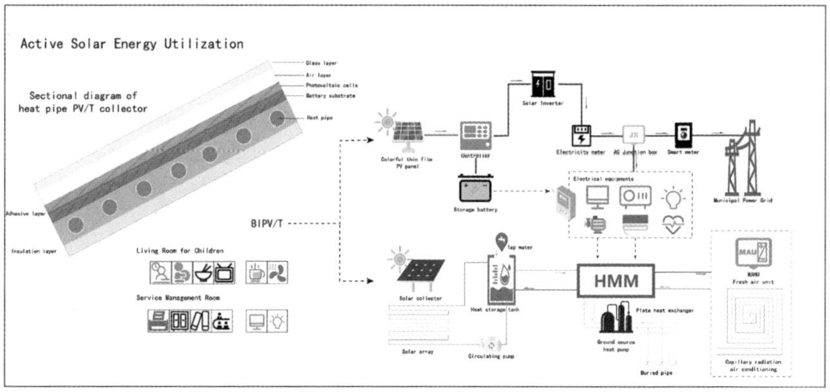

Active Solar Energy Utilization
Sectional diagram of heat pipe PV/T collector
BIPV/T
Living Room for Children
Service Management Room
HMM

The Partial Perspective

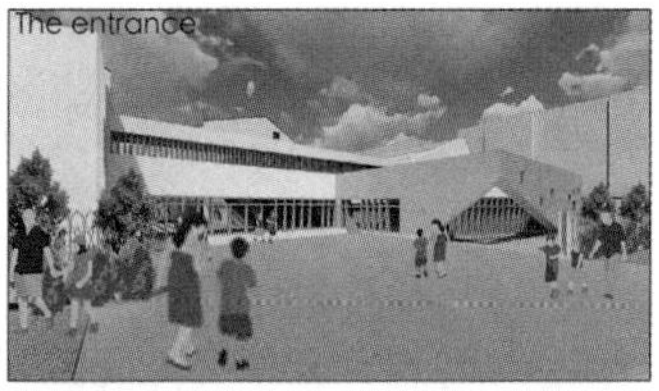
The entrance

The hall

Over the hallway

Courtyard

Bamboo Alley

Basic Motor Skill Setting for Children

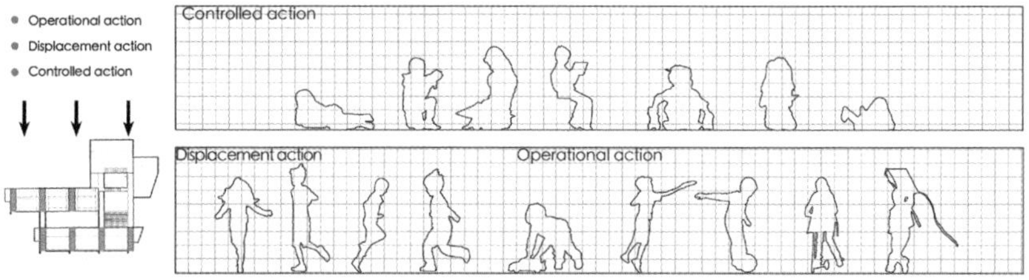

Operational action
Displacement action
Controlled action
Controlled action
Displacement action
Operational action

竹巷·风生 06

BAMBOO ALLEY · WIND FORM

2020 台达杯国际太阳能建筑设计竞赛 — 福建南平景龙幼儿园建筑设计

International Solar Building Design Competition 2020 —The Design of Fujian Nanping Jinglong Kindergarten

Energy Consumption Index of Building Envelope (EEV⩾0.20)

Index Classification	Energy-saving Index		Physical Quantity	Unit	EV	EV_c	EV_{min}	EEV
Basic indicator	Roof		K	W/(m²·K)	0.58	<0.8	<0.4	0.55
	Roof light		HWs	——	0.27	<0.15~0.35	——	0.23
			r	%	0.03	<0.2	——	0.85
Sectional Specification	Exterior window	East	K	W/(m²·K)	1.85	3.00	——	0.38
		West			1.85	3.00	——	0.38
		South			1.85	3.50	——	0.47
		North			1.85	3.50	——	0.47
	Exterior window SHGC	East	SHGC	——	0.21	0.30	0.15	0.60
		West			0.21	0.30	0.15	0.60
		South			0.21	0.40	0.20	0.95
		North			0.21	0.40	0.20	0.95
	Exterior wall		K	W/(m²·K)	1.24	2.00	——	0.38

Notes:Reference to Assessment standard for green building between both sides of Taiwan Strait.
The energy-saving efficiency of building envelope should not be less than 0.2.

Energy-saving Efficiency of Indoor Lighting System (EL<1.0) EL 0.33

β₂	8.0;Preferential coefficient of renewable energy;Solar photovoltaic system of power selling type		
δ₁	0.05;Energy system management	δ₂	0.00↑;Light Pipe
ni	4~8;Number of unit lamps	wi	11-20W;Lamp Power
Ci	0.75;Automatic Dimming Control	Di	0.90;Anti-Glare & Reflective coating
Aj	Floor area of single function room	LPDcj	9.0;Classroom & Office

Calculation sheet of solar energy utilization

Parameter		Unit	Designed Kindergarten
Annual solar radiation in Nanping	H_A	MJ/m²	4609
Solar panel area (PV/T)	A	m²	300

Energy Plus building parameter	Unit	Designed Kindergarten	Reference Standard
Photovoltaic power generation	kWh/d	111.9	180
Storage battery per element	kWh	10.6	——
Solar water heating production based on 400 persons	L/d·p	15.2	20
Recovery of flue gas waste heat based on 1500 m³/h exhaust	kJ/s	9.54	——

Notes:This sheet is calculated by theoretical value.
The reference standard is based on the design of Urban Municipal Kindergarten's electricity and water consumption.

Comparison chart of Solar System Performance

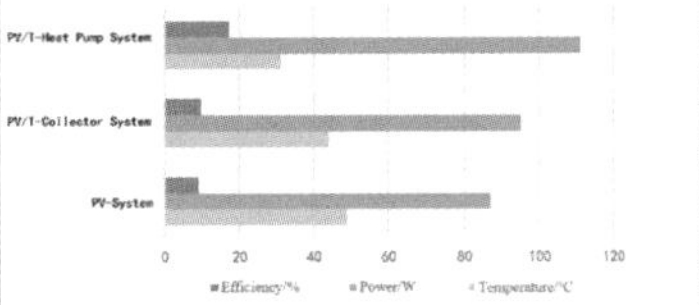
PV/T-Heat Pump System
PV/T-Collector System
PV-System
0
20
40
60
80
100
120
Efficiency/%
Power/W
Temperature/°C

Indoor Natural Ventilation Simulation

Solor Chimney

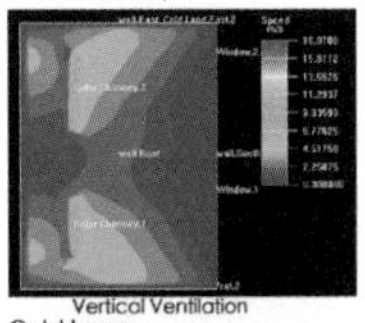
Vertical Ventilation

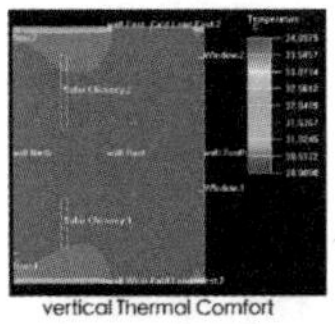
vertical Thermal Comfort

Cold Lane

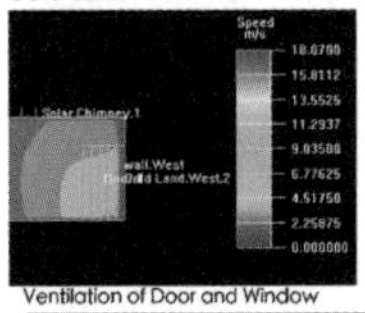
Ventilation of Door and Window

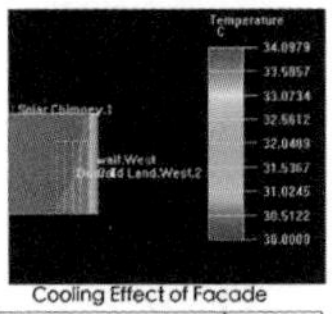
Cooling Effect of Facade

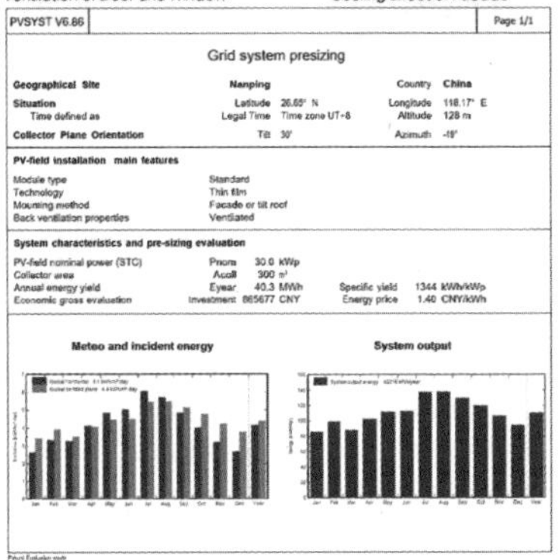
PVSYST V6.86
Page 1/1
Grid system presizing
Geographical Site
Nanping
Country
China
Situation
Latitude 26.65° N
Longitude 118.17° E
Time defined as
Legal Time Time zone UT+8
Altitude 128 m
Collector Plane Orientation
Tilt 30°
Azimuth -19°
PV-field installation main features
Module type
Standard
Technology
Thin film
Mounting method
Facade or tilt roof
Back ventilation properties
Ventilated
System characteristics and pre-sizing evaluation
PV-field nominal power (STC)
Pnom 30.0 kWp
Collector area
Acoll 300 m²
Annual energy yield
Eyear 40.3 MWh
Specific yield 1344 kWh/kWp
Economic gross evaluation
Investment 865677 CNY
Energy price 1.40 CNY/kWh
Meteo and incident energy
System output

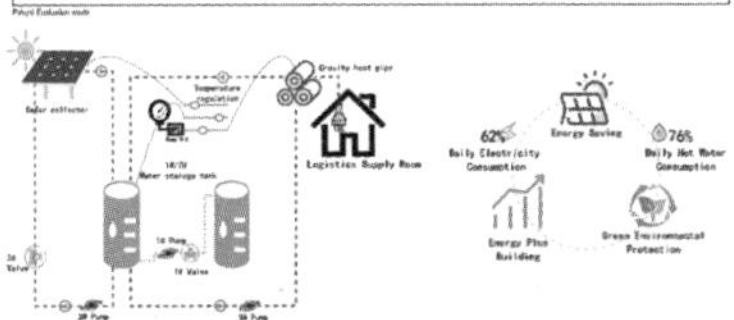
62%
Daily Electricity Consumption
Energy Saving
76%
Daily Hot Water Consumption
Energy Plus Building
Green Environmental Protection
Logistics Supply Room

Assessment standard for green building between both sides of Taiwan Strait

SAFETY & DURABILITY
1. Safety Protection Measures for Children
2. Durable Materials

HEALTH & COMFORT
1. Indoor Air Quality
2. Sound Insulation
3. Daylighting and Shading
4. Natural Ventilation

OCCUPANT CONVENIENCE
1. Public Transportation Accessibility
2. Charging Parking Space
3. Intelligent Monitoring

RESOURCES SAVING
1. Enclosure Performance
2. Unit Energy Efficiency
3. Renewable Energy

ENVIRONMENT LIVABILITY
1. Ecological Landscape
2. Educational Logo
3. Reasonable Physical Environment

PROMOTION & INNOVATION
1. Regional Architectural Culture
2. Passive Energy Saving Design
3. Energy Cycle System

Chart Analysis of Green Building

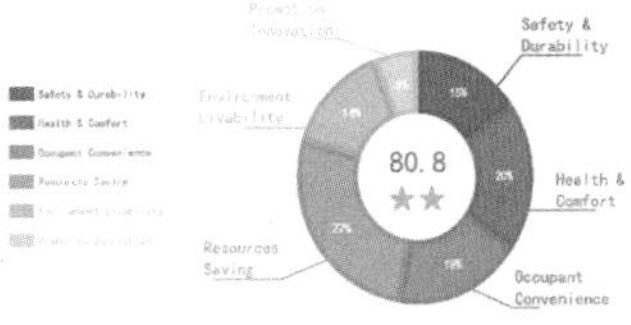
Safety & Durability
Health & Comfort
Occupant Convenience
Resources Saving
Environment Livability
80.8
★★

综合奖 · 三等奖
General Prize Awarded · Third Prize

注 册 号：7116

项目名称：太阳升起的地方（新疆）

Follow the Sun（Xinjiang）

作　　者：梁　瑞、郑　涛、陈兆哲

参赛单位：西安建筑科技大学

指导教师：李　涛、张　斌

专家点评：该作品布局紧凑、功能分区与交通流线合理，将当地传统的夯土式构造与现代的结构体系结合，采用方台无窗的设计，除了给室内带来阳光和散热通风的功效外，也成了荒漠中可识别的风景，对屋顶活动空间的利用较为巧妙。建筑的逻辑显现出“被动优先”的原则。

The work embodies compact layout, reasonable functional zoning and reasonable traffic streamline. It combines local traditional rammed earth structure with modern structural system, and adopts the design of square platforms without windows, thus bringing sunlight to the interior, realizing the effect of heat dissipation and ventilation, and also becoming a recognizable landscape in the desert. The application of roof space is more ingenious. In addition, the logic of the building shows the principle of “passive priority”.

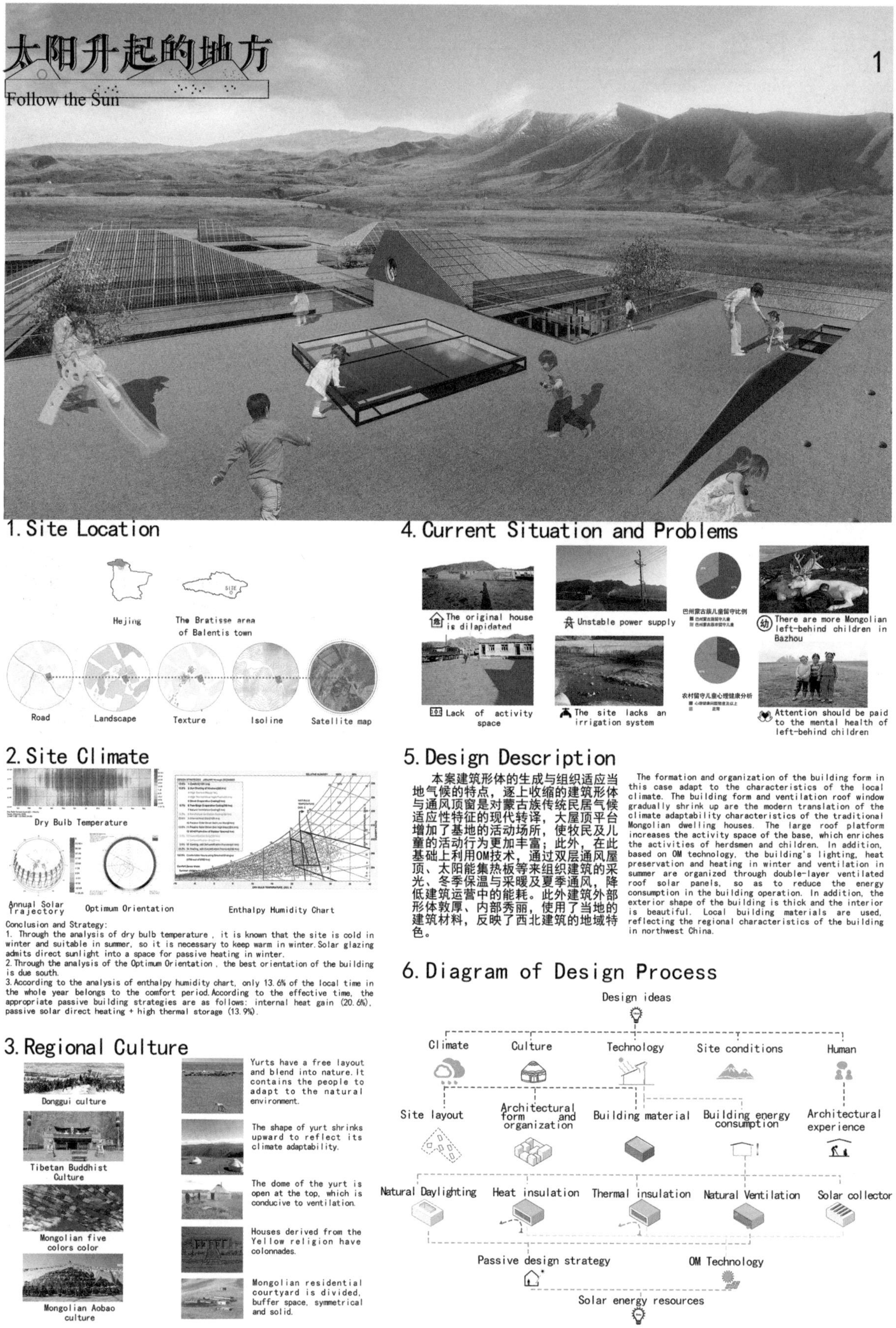

太阳升起的地方
Follow the Sun
1
1. Site Location
Hejing
The Bratisse area of Balentis town
Road
Landscape
Texture
Isoline
Satellite map
4. Current Situation and Problems
The original house is dilapidated
Unstable power supply
There are more Mongolian left-behind children in Bazhou
Lack of activity space
The site lacks an irrigation system
Attention should be paid to the mental health of left-behind children
2. Site Climate
Dry Bulb Temperature
Annual Solar Trajectory
Optimum Orientation
Enthalpy Humidity Chart
Conclusion and Strategy:
1. Through the analysis of dry bulb temperature , it is known that the site is cold in winter and suitable in summer, so it is necessary to keep warm in winter. Solar glazing admits direct sunlight into a space for passive heating in winter.
2. Through the analysis of the Optimum Orientation , the best orientation of the building is due south.
3. According to the analysis of enthalpy humidity chart, only 13.6% of the local time in the whole year belongs to the comfort period. According to the effective time, the appropriate passive building strategies are as follows: internal heat gain (20.6%), passive solar direct heating + high thermal storage (13.9%).
5. Design Description
本案建筑形体的生成与组织适应当地气候的特点，逐上收缩的建筑形体与通风顶窗是对蒙古族传统民居气候适应性特征的现代转译，大屋顶平台增加了基地的活动场所，使牧民及儿童的活动行为更加丰富；此外，在此基础上利用OM技术，通过双层通风屋顶、太阳能集热板等来组织建筑的采光、冬季保温与采暖及夏季通风，降低建筑运营中的能耗。此外建筑外部形体敦厚、内部秀丽，使用了当地的建筑材料，反映了西北建筑的地域特色。
The formation and organization of the building form in this case adapt to the characteristics of the local climate. The building form and ventilation roof window gradually shrink up are the modern translation of the climate adaptability characteristics of the traditional Mongolian dwelling houses. The large roof platform increases the activity space of the base, which enriches the activities of herdsmen and children. In addition, based on OM technology, the building's lighting, heat preservation and heating in winter and ventilation in summer are organized through double-layer ventilated roof solar panels, so as to reduce the energy consumption in the building operation. In addition, the exterior shape of the building is thick and the interior is beautiful. Local building materials are used, reflecting the regional characteristics of the building in northwest China.
3. Regional Culture
Donggui culture
Tibetan Buddhist Culture
Mongolian five colors color
Mongolian Aobao culture
Yurts have a free layout and blend into nature. It contains the people to adapt to the natural environment.
The shape of yurt shrinks upward to reflect its climate adaptability.
The dome of the yurt is open at the top, which is conducive to ventilation.
Houses derived from the Yellow religion have colonnades.
Mongolian residential courtyard is divided, buffer space, symmetrical and solid.
6. Diagram of Design Process
Design ideas
Climate
Culture
Technology
Site conditions
Human
Site layout
Architectural form and organization
Building material
Building energy consumption
Architectural experience
Natural Daylighting
Heat insulation
Thermal insulation
Natural Ventilation
Solar collector
Passive design strategy
OM Technology
Solar energy resources

太阳升起的地方

Follow the Sun

2

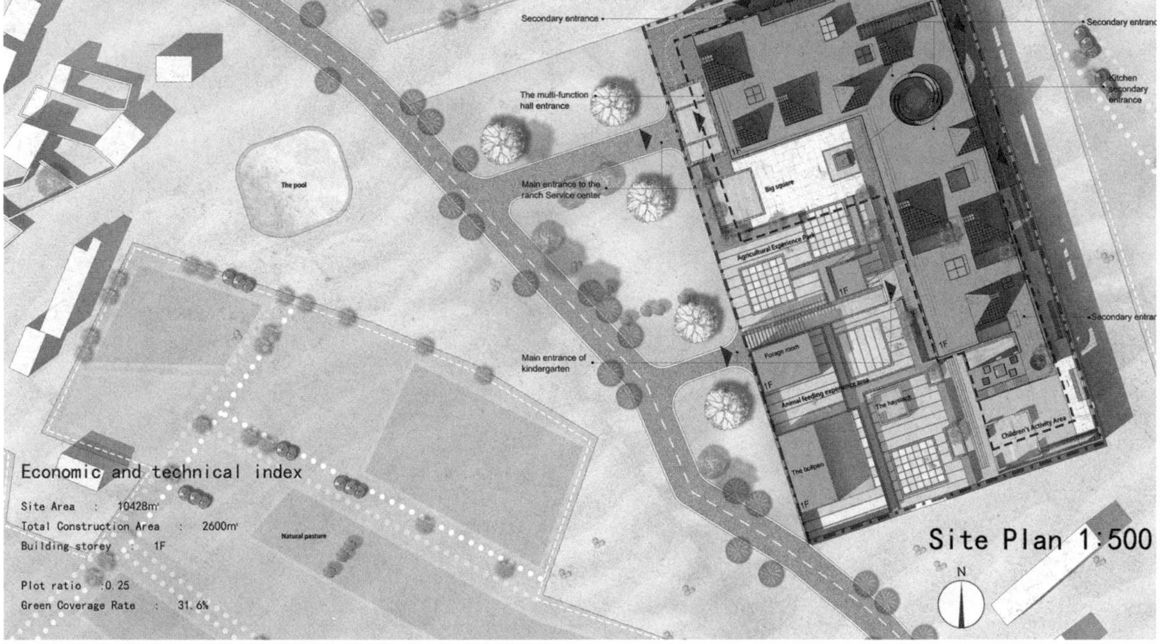

Dormitory Secondary entrance
Secondary entrance
Secondary entrance
Kitchen secondary entrance
The multi-function hall entrance
Main entrance to the ranch Service center
Main entrance of kindergarten
Secondary entrance
The pool
Natural pasture
Big square
Agricultural Experience Area
Forage room
Animal feeding experience area
The haystack
The bullpen
Children's Activity Area
1F
Economic and technical index
Site Area : 10428m²
Total Construction Area : 2600m²
Building storey : 1F
Plot ratio : 0.25
Green Coverage Rate : 31.6%
Site Plan 1:500
N

Wind Simulation

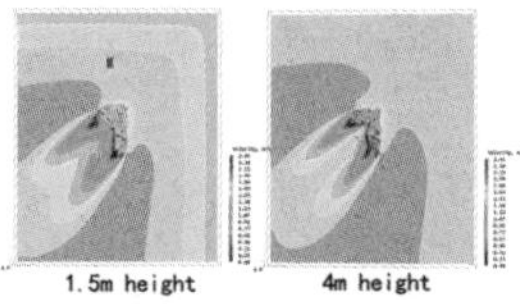

1.5m height
4m height

Site Solar Radiation

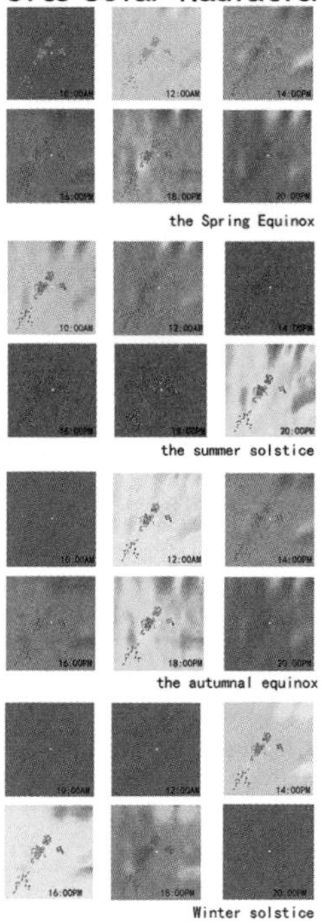

the Spring Equinox
the summer solstice
the autumnal equinox
Winter solstice

Architectural form Organization

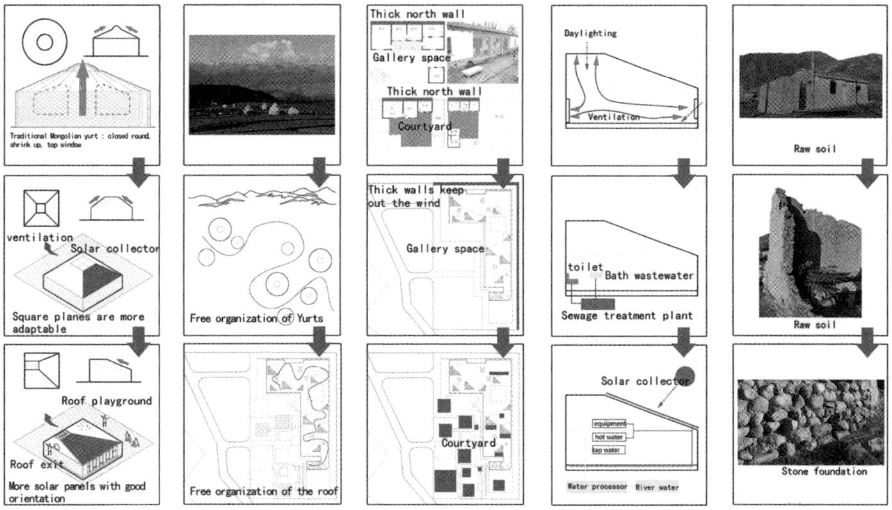

Traditional Mongolian yurt : closed round, shrink up, top window
Thick north wall
Gallery space
Thick north wall
Courtyard
Daylighting
Ventilation
Raw soil
ventilation
Solar collector
Square planes are more adaptable
Free organization of Yurts
Thick walls keep out the wind
Gallery space
toilet
Bath wastewater
Sewage treatment plant
Raw soil
Roof playground
Roof exit
More solar panels with good orientation
Free organization of the roof
Courtyard
Solar collector
equipment
hot water
tap water
Water processor
River water
Stone foundation

Logical Generation

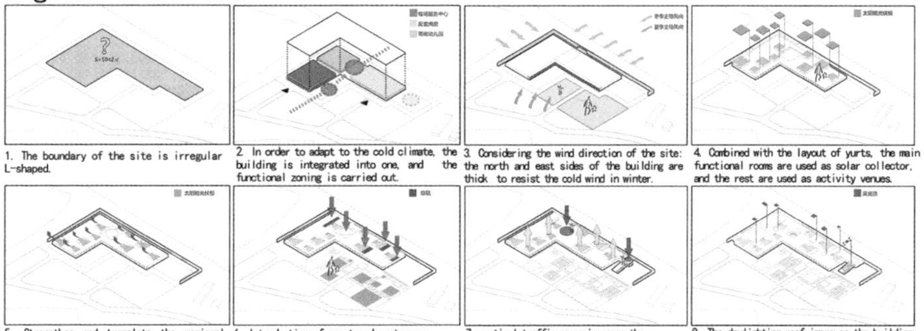

S=10412㎡

1. The boundary of the site is irregular L-shaped.

2. In order to adapt to the cold climate, the building is integrated into one, and the functional zoning is carried out.

3. Considering the wind direction of the site: the north and east sides of the building are thick to resist the cold wind in winter.

4. Combined with the layout of yurts, the main functional rooms are used as solar collector, and the rest are used as activity venues.

5. Strengthen and translate the regional characteristics of yurts.

6. Introduction of courtyard: get more lighting and ventilation.

7. vertical traffic space increase the possibility of contact between ground and roof.

8. The daylighting roof improves the building light environment.

太阳升起的地方

Follow the Sun

3

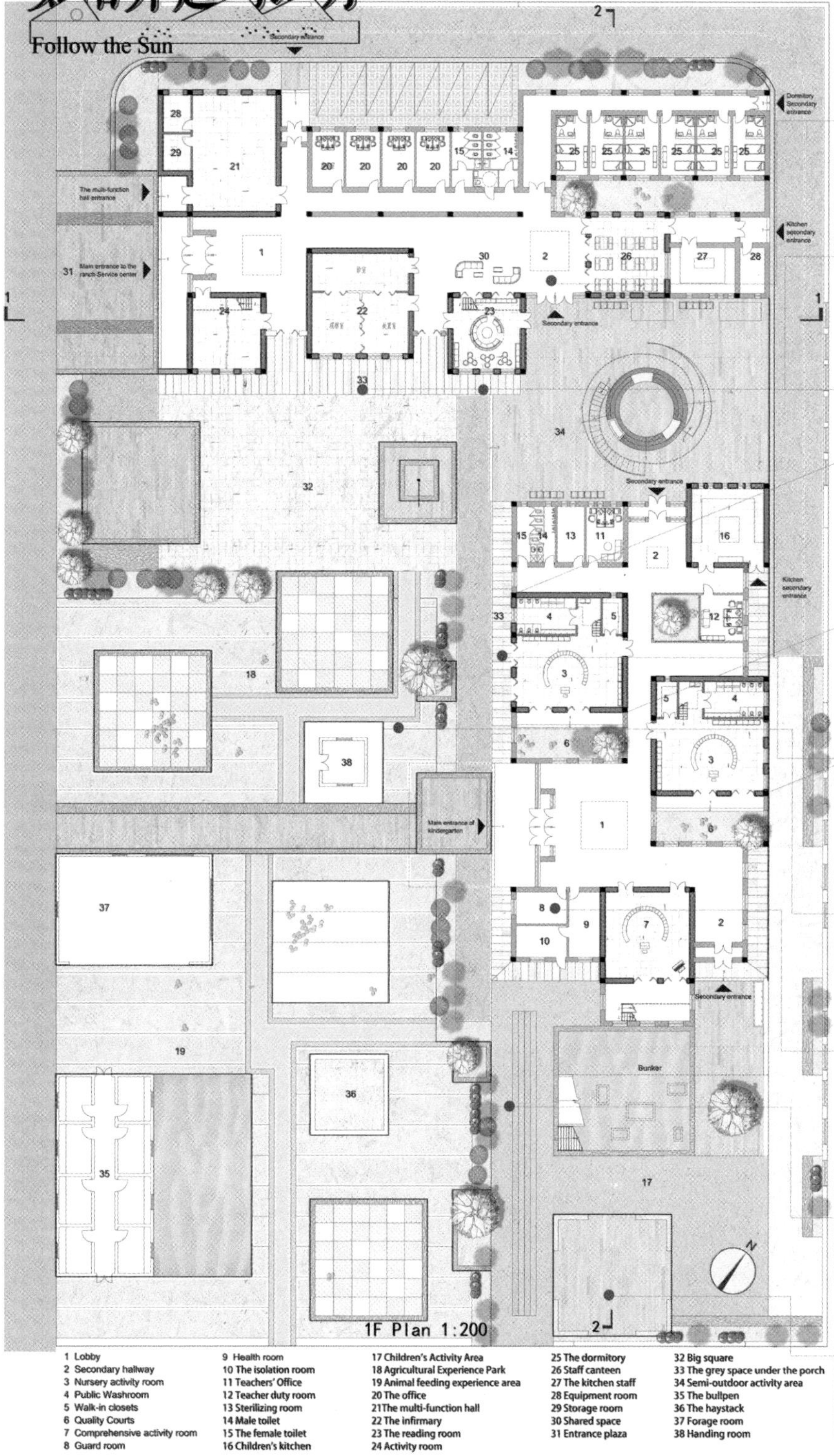
Secondary entrance
Dormitory Secondary entrance
The multi-function hall entrance
Main entrance to the ranch Service center
Kitchen secondary entrance
Secondary entrance
Main entrance of kindergarten
Bunker

1F Plan 1:200

Roof playground

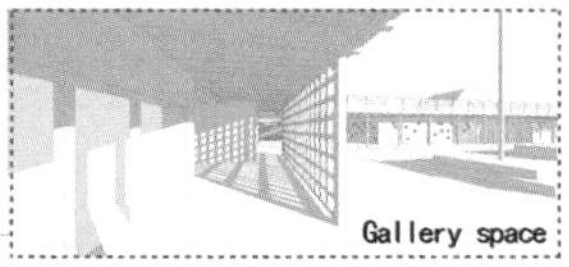

Gallery space

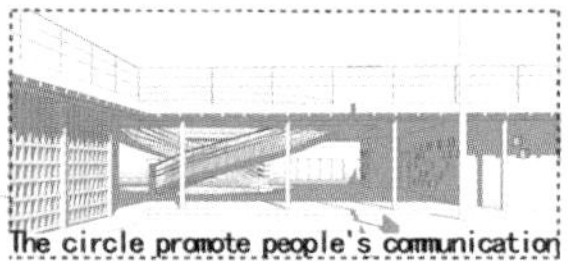

The circle promote people's communication

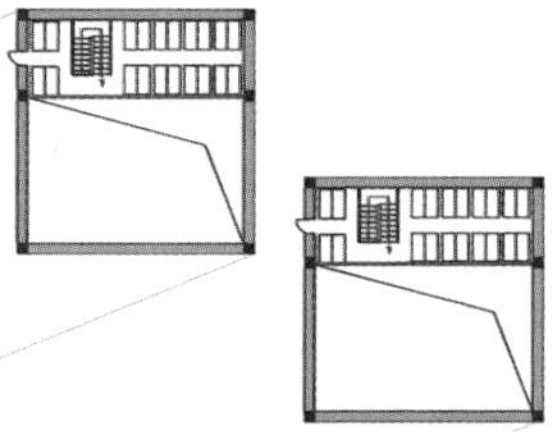

Interlayer Plan 1:200

Gallery space

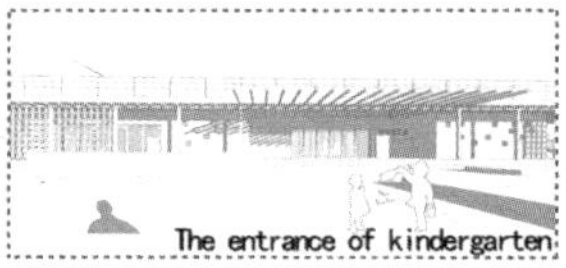

The entrance of kindergarten

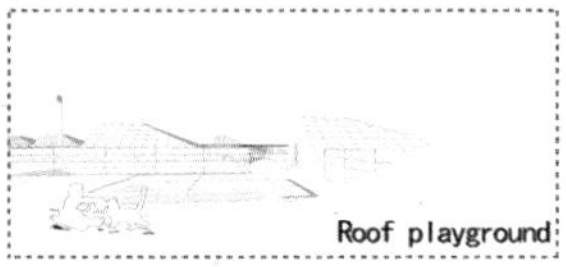

Roof playground

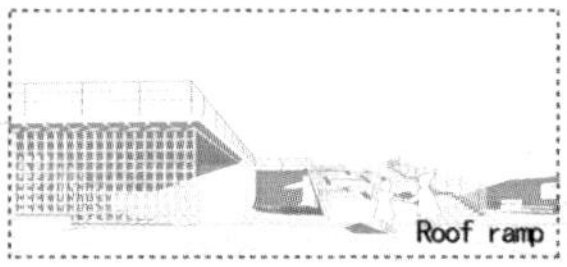

Roof ramp

Kindergarten activity site

1 Lobby
2 Secondary hallway
3 Nursery activity room
4 Public Washroom
5 Walk-in closets
6 Quality Courts
7 Comprehensive activity room
8 Guard room
9 Health room
10 The isolation room
11 Teachers' Office
12 Teacher duty room
13 Sterilizing room
14 Male toilet
15 The female toilet
16 Children's kitchen
17 Children's Activity Area
18 Agricultural Experience Park
19 Animal feeding experience area
20 The office
21The multi-function hall
22 The infirmary
23 The reading room
24 Activity room
25 The dormitory
26 Staff canteen
27 The kitchen staff
28 Equipment room
29 Storage room
30 Shared space
31 Entrance plaza
32 Big square
33 The grey space under the porch
34 Semi-outdoor activity area
35 The bullpen
36 The haystack
37 Forage room
38 Handing room

West Elevation 1:200

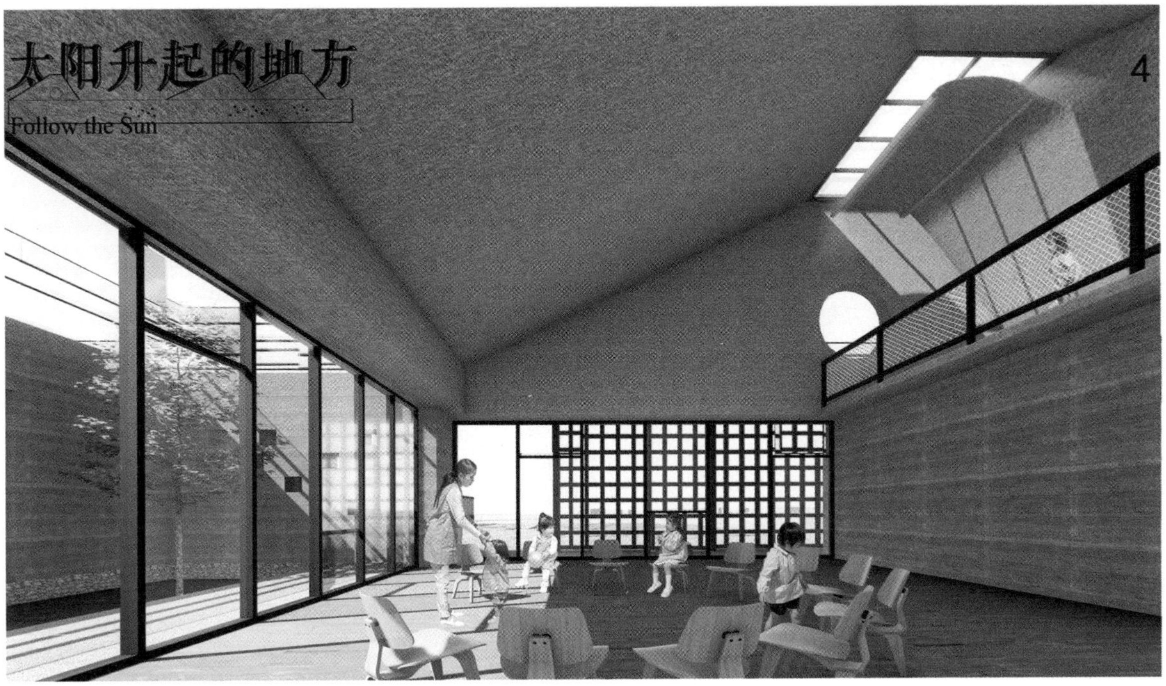
太阳升起的地方
Follow the Sun
4

Wind Simulation of Monomer

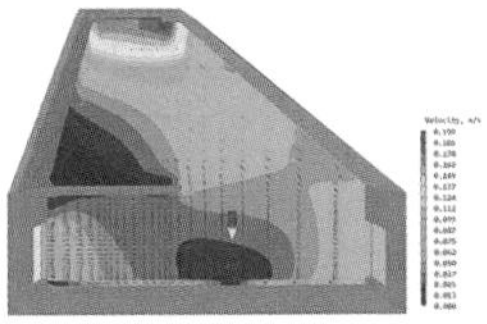

The height of the building makes the ventilation effect of the building remarkable.

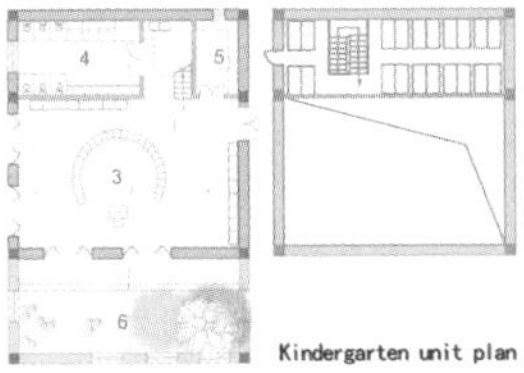
Kindergarten unit plan

Building Section Strategy

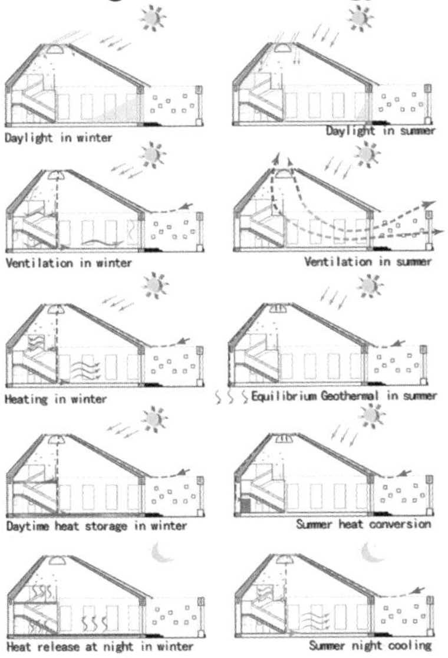
Daylight in winter
Daylight in summer
Ventilation in winter
Ventilation in summer
Heating in winter
Equilibrium Geothermal in summer
Daytime heat storage in winter
Summer heat conversion
Heat release at night in winter
Summer night cooling

Natural Lighting Analysis

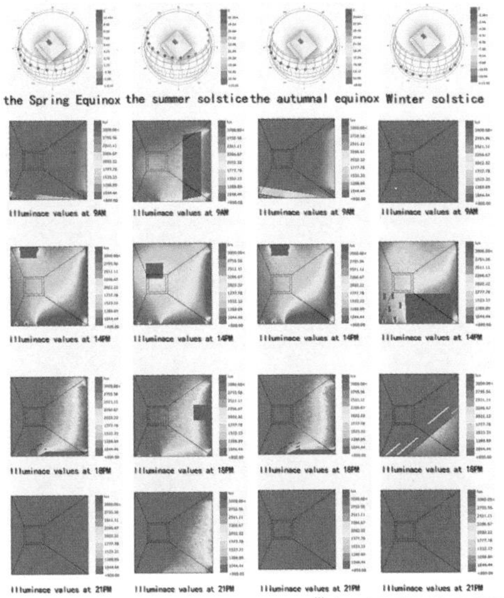
the Spring Equinox
the summer solstice
the autumnal equinox
Winter solstice
Illuminace values at 9AM
Illuminace values at 14PM
Illuminace values at 18PM
Illuminace values at 21PM

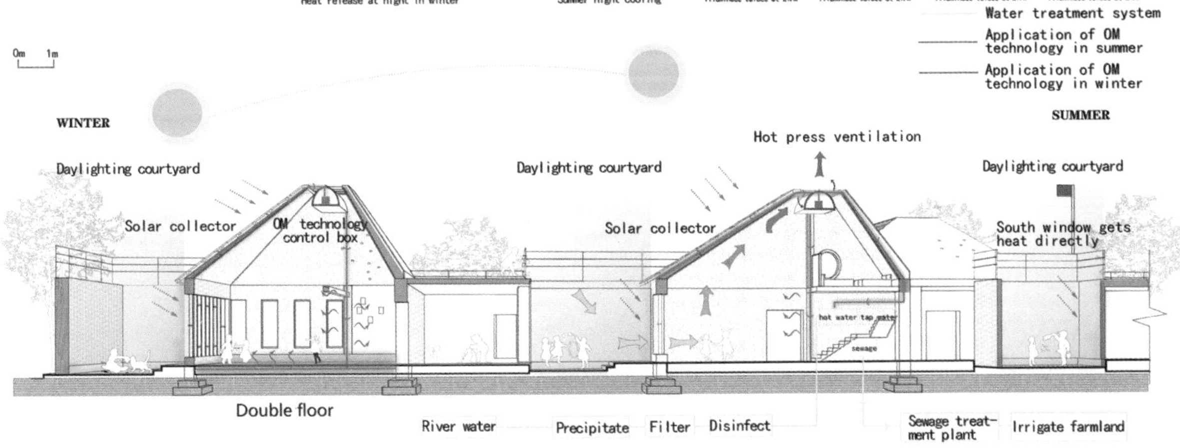
Water treatment system
Application of OM technology in summer
Application of OM technology in winter
0m 1m
WINTER
SUMMER
Hot press ventilation
Daylighting courtyard
Daylighting courtyard
Daylighting courtyard
Solar collector
OM technology control box
Solar collector
South window gets heat directly
hot water tap water
sewage
Double floor
River water
Precipitate
Filter
Disinfect
Sewage treatment plant
Irrigate farmland

太阳升起的地方

Follow the Sun

5

OM Technology Diagram

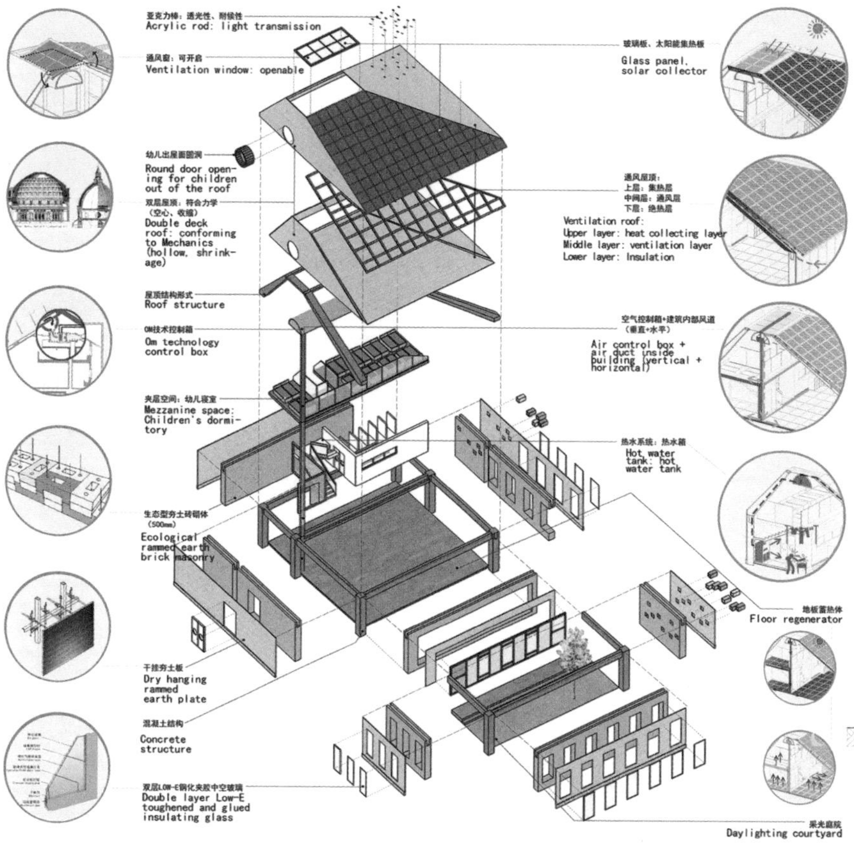

亚克力棒：透光性、耐候性
Acrylic rod: light transmission
通风窗：可开启
Ventilation window: openable
幼儿出屋面圆洞
Round door opening for children out of the roof
双层屋顶：符合力学（空心、收缩）
Double deck roof: conforming to Mechanics (hollow, shrinkage)
屋顶结构形式
Roof structure
OM技术控制箱
Om technology control box
夹层空间：幼儿寝室
Mezzanine space: Children's dormitory
生态型夯土砖砌体（500mm）
Ecological rammed earth brick masonry
干挂夯土板
Dry hanging rammed earth plate
混凝土结构
Concrete structure
双层LOW-E钢化夹胶中空玻璃
Double layer Low-E toughened and glued insulating glass
玻璃板、太阳能集热板
Glass panel, solar collector
通风屋顶：
上层：集热层
中间层：通风层
下层：绝热层
Ventilation roof:
Upper layer: heat collecting layer
Middle layer: ventilation layer
Lower layer: Insulation
空气控制箱+建筑内部风道（垂直+水平）
Air control box + air duct inside building (vertical + horizontal)
热水系统：热水箱
Hot water tank: hot water tank
地板蓄热体
Floor regenerator
采光庭院
Daylighting courtyard

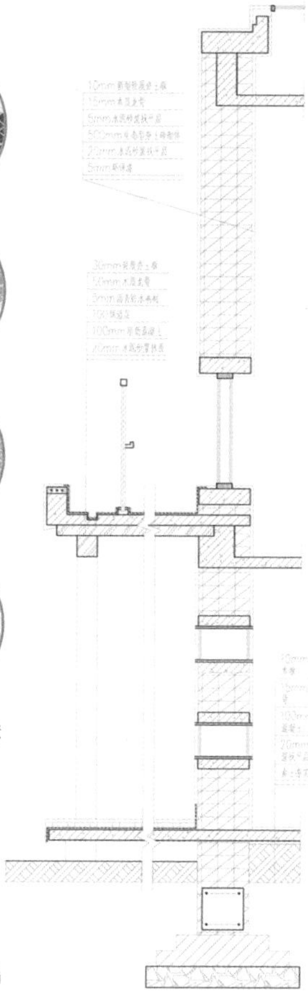

Wall detail 1:20

South Elevation 1:200

太阳升起的地方
Follow the Sun
6

Building Operation Diagram

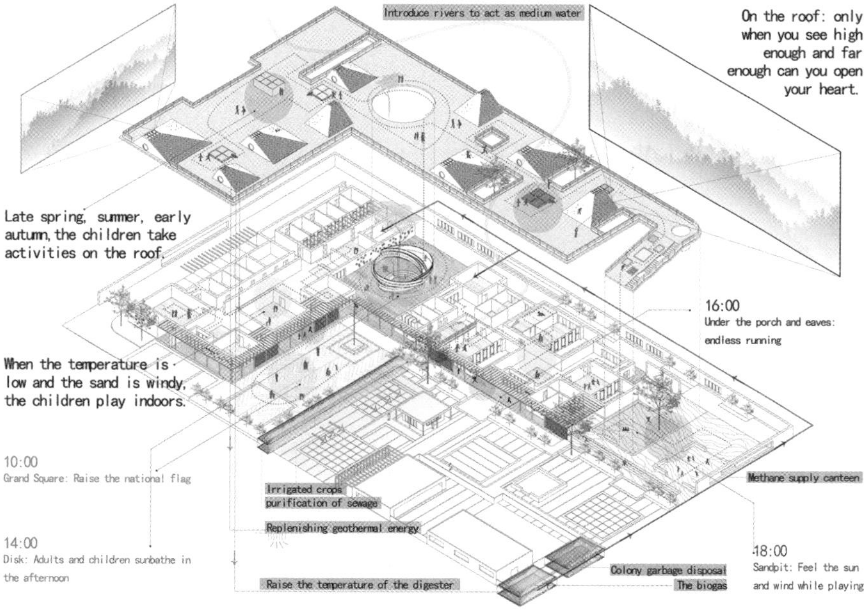
Introduce rivers to act as medium water
On the roof: only when you see high enough and far enough can you open your heart.
Late spring, summer, early autumn, the children take activities on the roof.
16:00
Under the porch and eaves: endless running
When the temperature is low and the sand is windy, the children play indoors.
10:00
Grand Square: Raise the national flag
Irrigated crops
purification of sewage
Replenishing geothermal energy
Methane supply canteen
14:00
Disk: Adults and children sunbathe in the afternoon
18:00
Sandpit: Feel the sun and wind while playing
Colony garbage disposal
The biogas
Raise the temperature of the digester

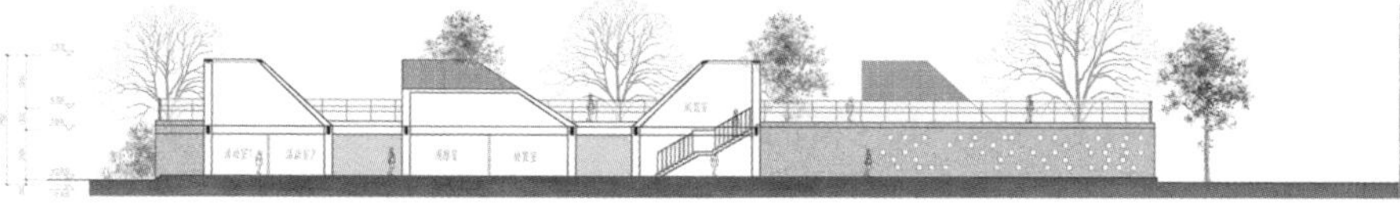

1-1 Section 1:200

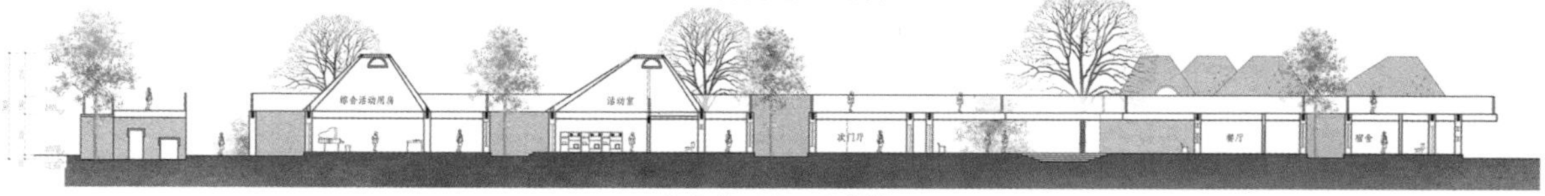

2-2 Section 1:200

Architectural Model

综合奖 · 优秀奖
General Prize Awarded · Honorable Mention Prize

注 册 号：7017

项目名称：光之谷（南平）

The Valley of Light（Nanping）

作　　者：张明远、程咏梅、崔帅杰、王诗雨、岳雅琦、李尚谦

参赛单位：北京工业大学、大连理工大学

指导教师：戴　俭

光之谷 | The Valley of Light

SITE PLAN 1：1000

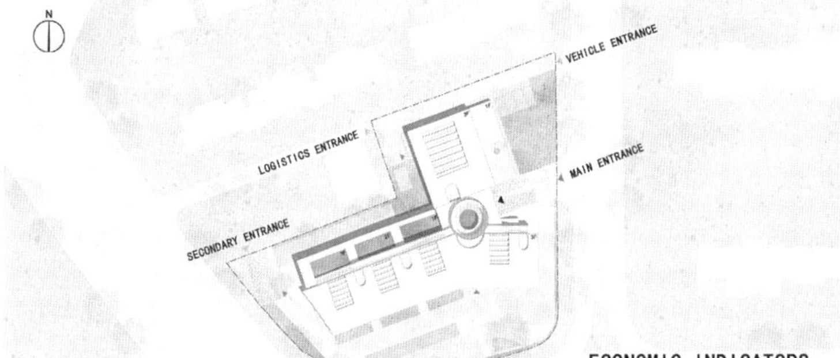

N
VEHICLE ENTRANCE
LOGISTICS ENTRANCE
MAIN ENTRANCE
SECONDARY ENTRANCE

ECONOMIC INDICATORS

LAND AREA: 6028㎡

BUILDING AREA: 3919 ㎡

BUILDING DENSITY: 39.4%

FLOOR AREA RATIO: 0.65

DESIGN DESCRIPTION

本方案以“自然共生”为设计理念，通过趣味“山洞”的空间形态，将室外的自然环境引入建筑内部。节能方面，将福建传统民居中的“采光天井、拔风烟囱、冷巷”等被动式节能方式与建筑空间进行有机融合，形成以采光中庭的拔风烟囱为核心驱动，两翼冷巷为动脉，每个活动房间的送风竖井为血管的脉络式节能系统，利用风压和热压原理带动室内空气流动，除湿降温。同时结合太阳能光热、光电等主动式节能技术，形成主被动一体的节能体系。

The project takes "natural symbiosis" as the design concept, through the interest spatial form of "cave", introduces the outdoor natural environment to the interior of the building. In terms of energy conservation, fujian traditional residential houses will be passive energy saving, such as "light patio, ventilation chimney and cold lane". The way and the building space are organically integrated to form the daylighting. The ventilation chimney of the courtyard is driven by the core, and the cold passages of the two wings are arteries. The air supply shaft in each movable room is a vein type energy saving system, using the principle of wind pressure and hot pressure to drive indoor air flow, Dehumidify and cool down. At the same time combined with solar thermal, photovoltaic and other active type energy saving technology, the formation of a passive energy saving system.

CHAPTER 1 : PRE-PROGRA

1.1 BACKGROUND ANALYSIS

• NANPING IMAGE ANALYSIS

Typical region of highland

Kaoting institute

Water lane

• SITE ANALYSIS

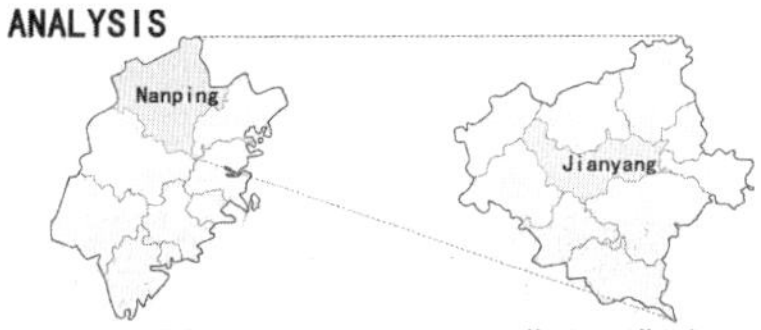

Nanping
Jianyang

Nanping · Fujian　　Jianyang · Nanping

The project site is located in the upper part of Fujian Province, middle of Nanping. And it is located in the center of the city with convenienttransportation, slightly higher than the territory lying North-East, south-west gradually lower. Nanping is located in the Hot Summer & Warm Winter area.

the Monthly Mean Temperature and Precipitation of Nanping

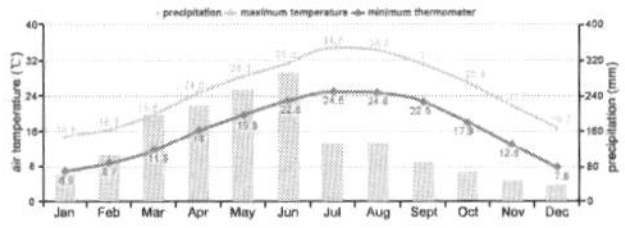

precipitation
maximum temperature
minimum thermometer
air temperature (℃)
precipitation (mm)
Jan
Feb
Mar
Apr
May
Jun
Jul
Aug
Sept
Oct
Nov
Dec

Subtropical Monsoon Climate Characteristics

SUMMER: High Temperature - More Rain

WINTER: Mild , Less Rain

RAIN HOT DURING THE SAME PERIOD

The kindergarden should be designed primarily in line with summer weather conditions.

1.2 TRADITIONAL ARCHITECTERAL ANALYSIS

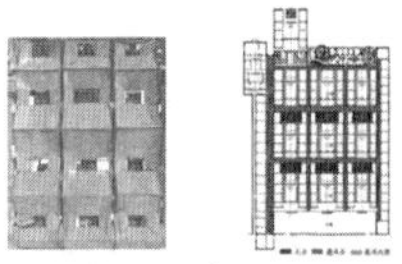

Wuzhai, Guitan Lane, Quanzhou

translate

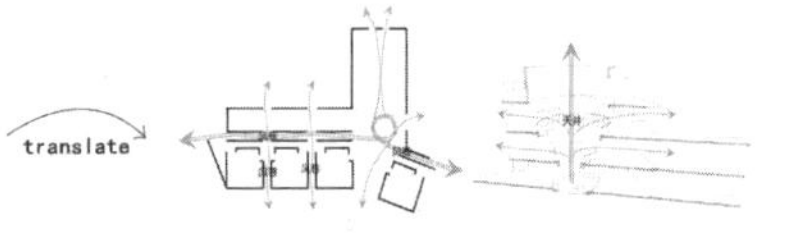

· wind lane: to reduce the temperature in summer; strengthen the ventilation

· courtyard: enhance natural lighting and ventilation; provide public space

'ancuo' roofs in south Fujian

translate

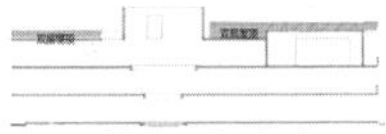

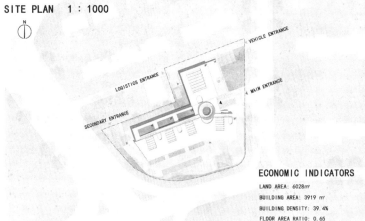

· double roof: provide heat insulation; provide sun protection

光之谷 II The Valley of Light

1.3 URBAN PHYSICAL ENVIRONMENT ANALYSIS

CHAPTER 2 : CONCEPT

· CLIMATE SUMMARY

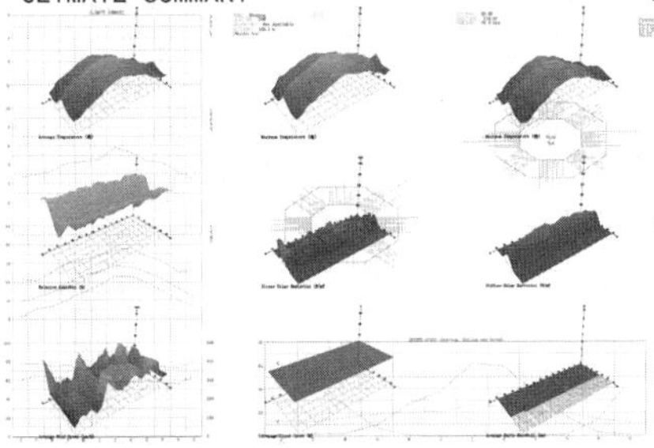

· PREVAILING WINDS

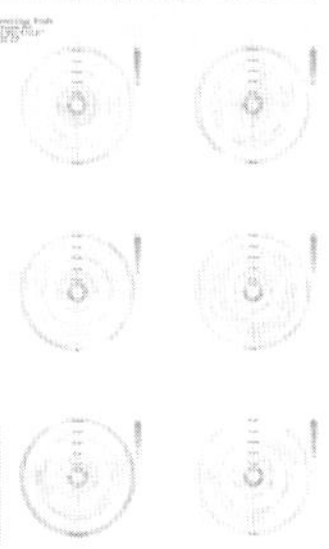

· GENERATION OF ARCHITECTURAL FORM

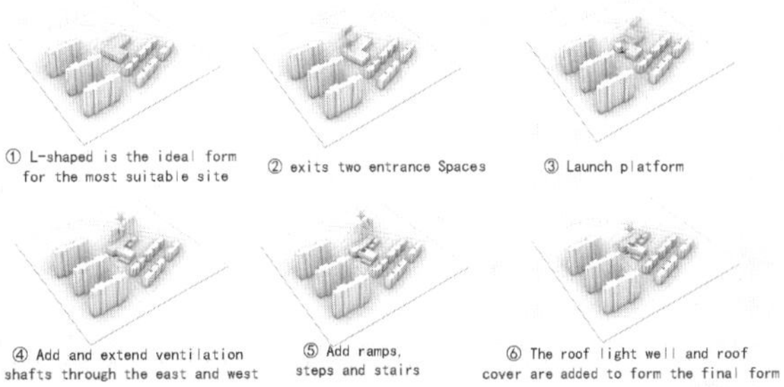
① L-shaped is the ideal form for the most suitable site
② exits two entrance Spaces
③ Launch platform
④ Add and extend ventilation shafts through the east and west
⑤ Add ramps, steps and stairs
⑥ The roof light well and roof cover are added to form the final form

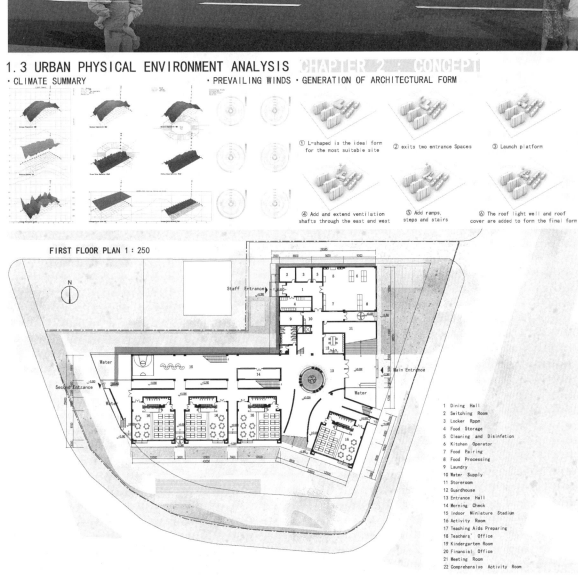

FIRST FLOOR PLAN 1：250

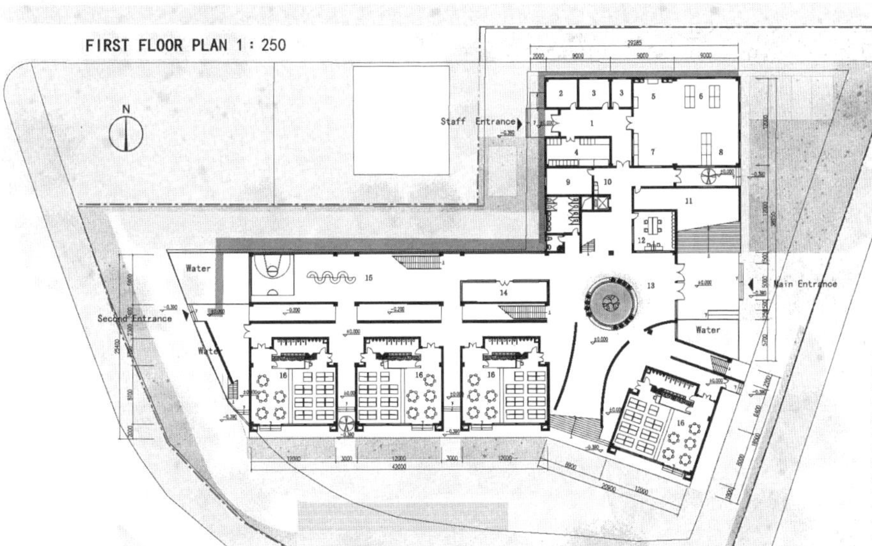
N
Staff Entrance
Main Entrance
Second Entrance
Water
Water
Water

1 Dining Hall
2 Switching Room
3 Locker Rppm
4 Food Storage
5 Cleaning and Disinfetion
6 Kitchen Operator
7 Food Pairing
8 Food Processing
9 Laundry
10 Water Supply
11 Storeroom
12 Guardhouse
13 Entrance Hall
14 Morning Check
15 Indoor Miniature Stadium
16 Activity Room
17 Teaching Aids Preparing
18 Teachers` Office
19 Kindergarten Room
20 Financial Office
21 Meeting Room
22 Comprehensive Activity Room

光之谷III The Valley of Light

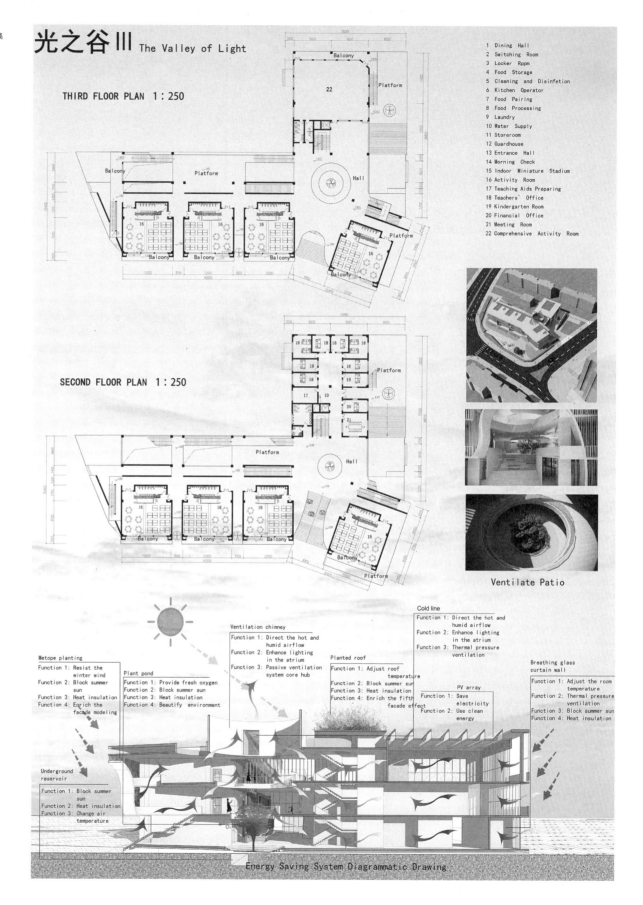

THIRD FLOOR PLAN 1：250

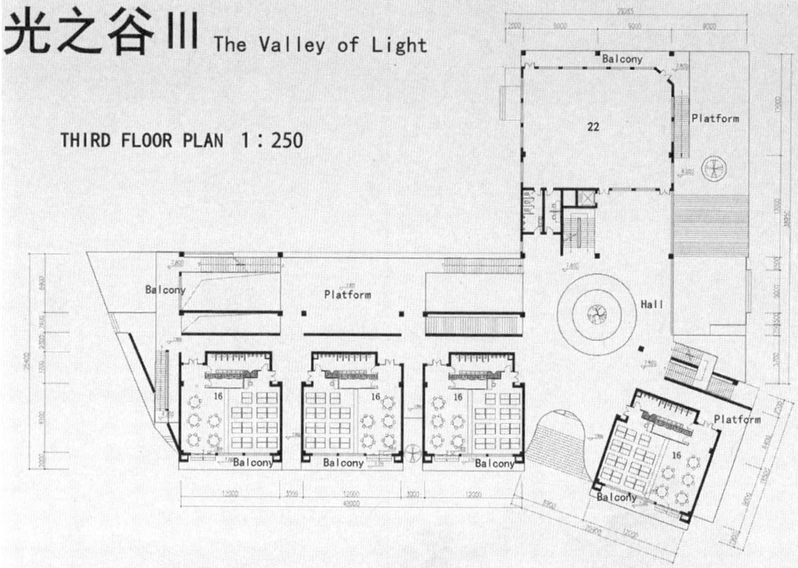
Balcony
Platform
22
Hall
16
Balcony
Platform
Balcony
Balcony
Balcony
Platform
Balcony

1 Dining Hall
2 Switching Room
3 Locker Rppm
4 Food Storage
5 Cleaning and Disinfetion
6 Kitchen Operator
7 Food Pairing
8 Food Processing
9 Laundry
10 Water Supply
11 Storeroom
12 Guardhouse
13 Entrance Hall
14 Morning Check
15 Indoor Miniature Stadium
16 Activity Room
17 Teaching Aids Preparing
18 Teachers` Office
19 Kindergarten Room
20 Financial Office
21 Meeting Room
22 Comprehensive Activity Room

SECOND FLOOR PLAN 1：250

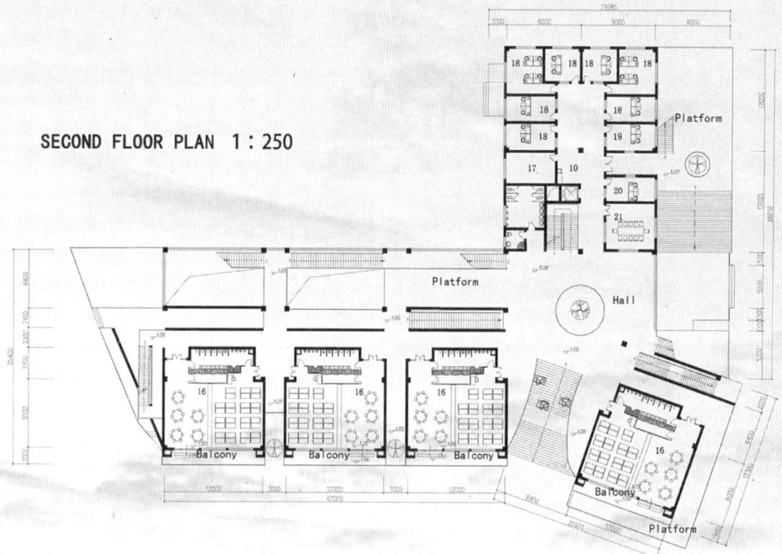
18
Platform
17
10
19
20
21
Platform
Hall
16
Balcony
Balcony
Balcony
Balcony
Platform

Ventilate Patio

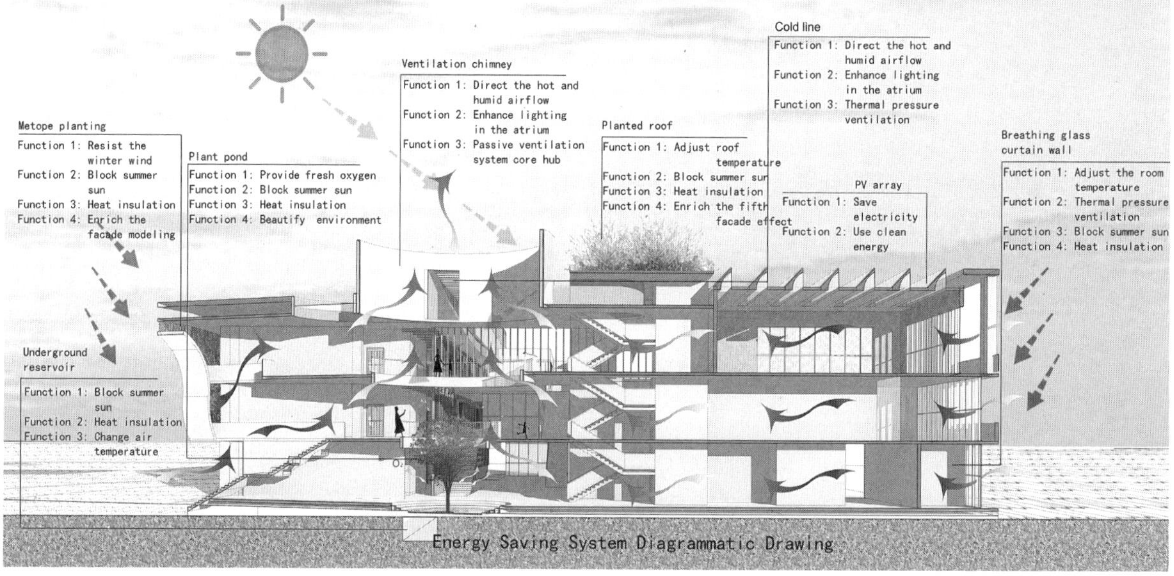
Cold line
Function 1: Direct the hot and humid airflow
Function 2: Enhance lighting in the atrium
Function 3: Thermal pressure ventilation
Ventilation chimney
Function 1: Direct the hot and humid airflow
Function 2: Enhance lighting in the atrium
Function 3: Passive ventilation system core hub
Metope planting
Function 1: Resist the winter wind
Function 2: Block summer sun
Function 3: Heat insulation
Function 4: Enrich the facade modeling
Plant pond
Function 1: Provide fresh oxygen
Function 2: Block summer sun
Function 3: Heat insulation
Function 4: Beautify environment
Planted roof
Function 1: Adjust roof temperature
Function 2: Block summer sun
Function 3: Heat insulation
Function 4: Enrich the fifth facade effect
PV array
Function 1: Save electricity
Function 2: Use clean energy
Breathing glass curtain wall
Function 1: Adjust the room temperature
Function 2: Thermal pressure ventilation
Function 3: Block summer sun
Function 4: Heat insulation
Underground reservoir
Function 1: Block summer sun
Function 2: Heat insulation
Function 3: Change air temperature

Energy Saving System Diagrammatic Drawing

光之谷Ⅳ The Valley of Light

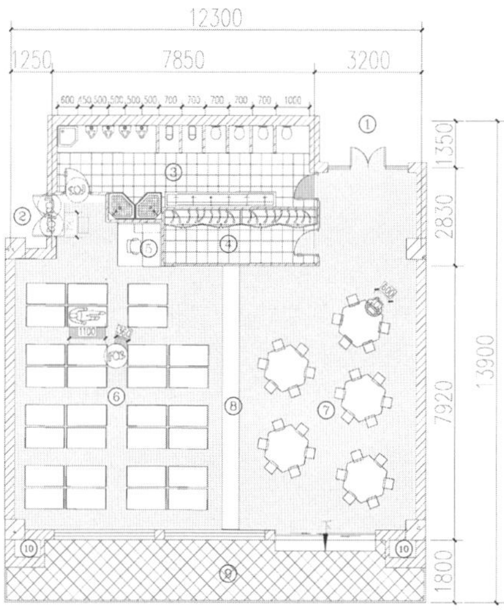
12300
1250
7850
3200
1350
2830
7920
13900
1800

1. 入口1
The First Entrance
2. 入口2
The Second Entrance
3. 卫生间
Toilet
4. 储藏间
Storeroom
5. 值班处
Duty room
6. 寝室
Bedroom
7. 活动室
Activity room
8. 趣味矮墙
Funny Parapet
9. 阳台
Terrace
10. 通风井
Vent Shaft

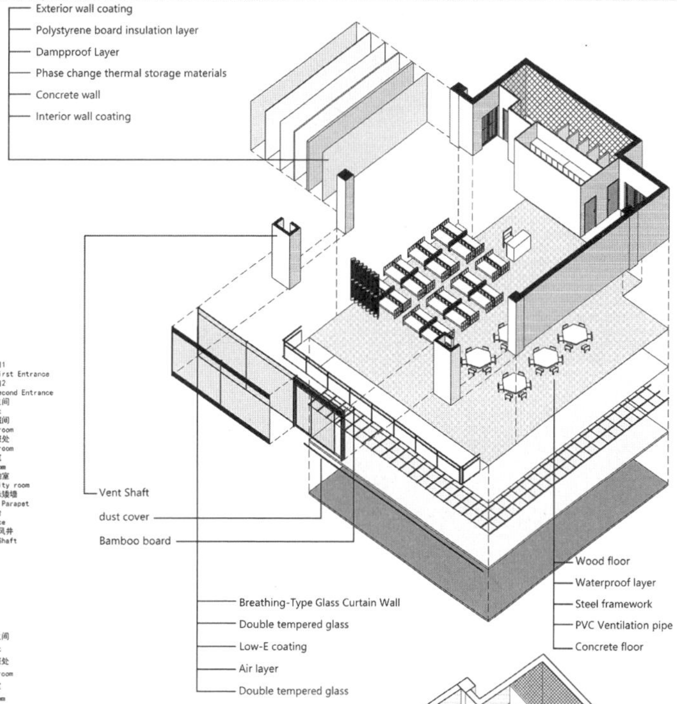
Exterior wall coating
Polystyrene board insulation layer
Dampproof Layer
Phase change thermal storage materials
Concrete wall
Interior wall coating
Vent Shaft
dust cover
Bamboo board
Wood floor
Waterproof layer
Steel framework
PVC Ventilation pipe
Concrete floor
Breathing-Type Glass Curtain Wall
Double tempered glass
Low-E coating
Air layer
Double tempered glass

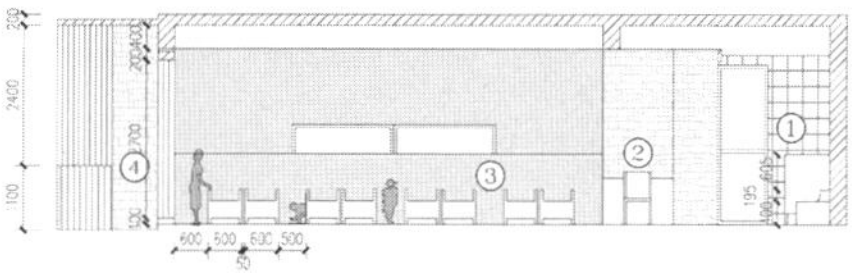

1. 卫生间
Toilet
2. 值班处
Duty room
3. 寝室
Bedroom
4. 阳台
Terrace

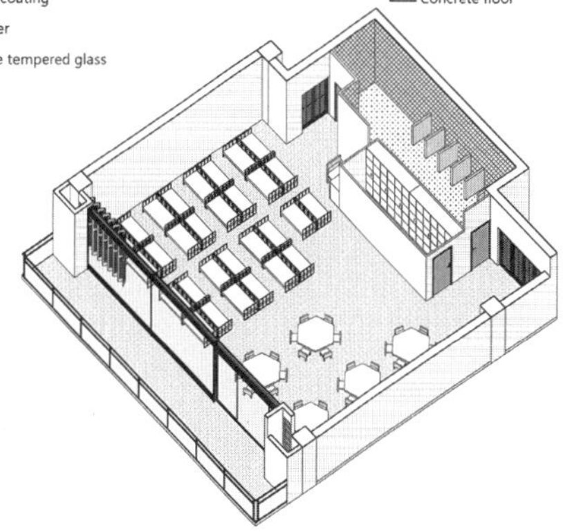

幼儿的身高：
The hight of childern;

3-4岁：94~103cm;
3-4 years old:94~103cm;
4-5岁：102~109cm;
4-5 years old:102~109cm;
5-6岁：108~109cm;
5-6 years old:10~109cm;

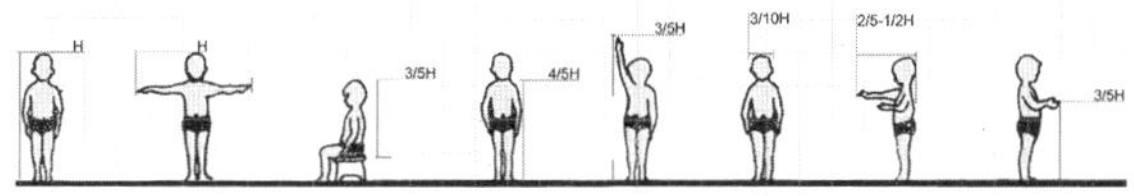
H
H
3/5H
4/5H
3/5H
3/10H
2/5-1/2H
3/5H

幼儿的活动尺度数据：
The scale date of child's activity;

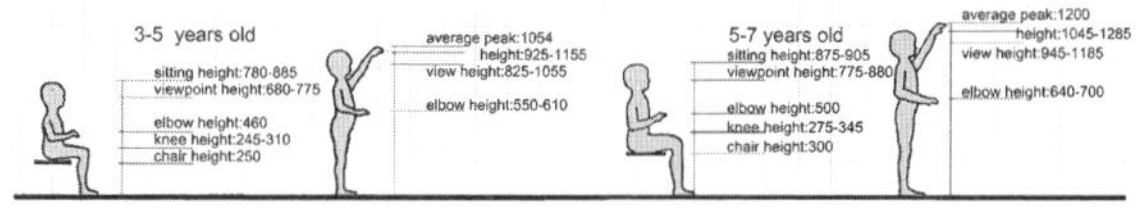
3-5 years old
sitting height:780-885
viewpoint height:680-775
elbow height:460
knee height:245-310
chair height:250
average peak:1054
height:925-1155
view height:825-1055
elbow height:550-610
5-7 years old
sitting height:875-905
viewpoint height:775-880
elbow height:500
knee height:275-345
chair height:300
average peak:1200
height:1045-1285
view height:945-1185
elbow height:640-700

光之谷V The Valley of Light

太阳能发电系统

Solar power system

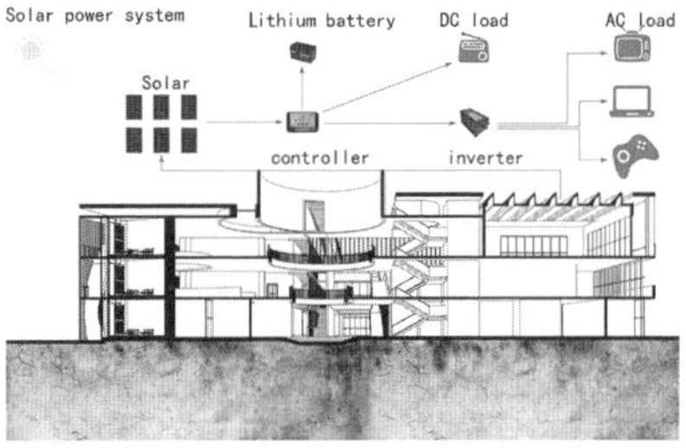
Lithium battery
DC load
AC load
Solar
controller
inverter

太阳能发电不会产生任何废弃物，没有污染、噪声等公害，对环境无不良影响，是理想的清洁能源。

Solar power generation will not produce any waste, pollution, noise and other public hazards, no adverse impact on the environment, is an ideal clean energy.

呼吸式玻璃幕墙

Breathing-Type Glass Curtain Wall

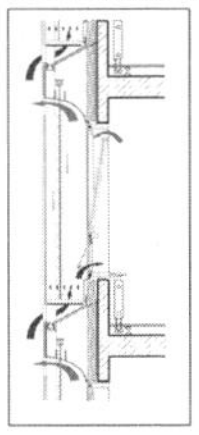

呼吸式幕墙内外两层幕墙形成通风换气层。夏季时，打开换气层的进排风口，在阳光的照射下换气层形成自下而上的空气流，降低内层玻璃表面的温度。

The inner and outer two-layer curtain wall forms ventilation layer. In summer, the air inlet and outlet of the ventilation layer are opened, and the air flow from the bottom to the top is formed in the air exchange layer under the sunlight, which reduces the temperature of the inner glass surface.

雨水回收系统

Rainwater recovery system

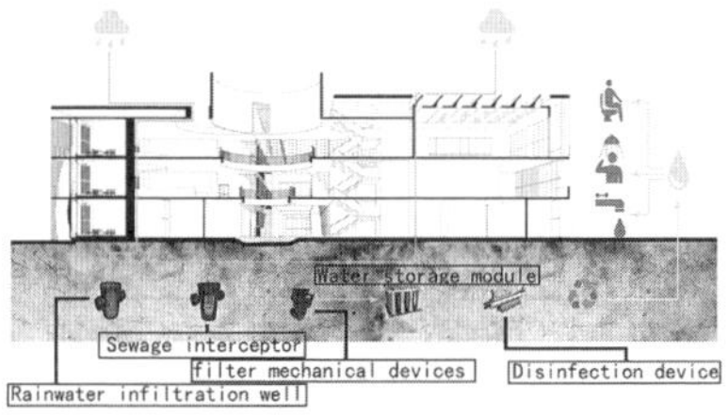
Water storage module
Sewage interceptor
filter mechanical devices
Disinfection device
Rainwater infiltration well

雨水通过过滤系统，满足处理要求的雨水进入蓄水系统，蓄水系统中的水由消毒装置进行净化后进入供水系统。

The rainwater passes through the filtration system, and the rainwater meeting the treatment requirements enters into the water storage system. The water in the storage system is purified by the disinfection device and then enters the water supply system.

烟囱效应

Chimney effect

烟囱效应是在从底部到顶部具有通畅的流通空间的建筑物，空气靠密度差的作用，沿着通道很快进行扩散或排出建筑物的现象，即为烟囱效应。

Chimney effect refers to the phenomenon that air diffuses or exhausts the building rapidly along the passage due to the effect of density difference in the building with unobstructed circulation space from the bottom to the top.

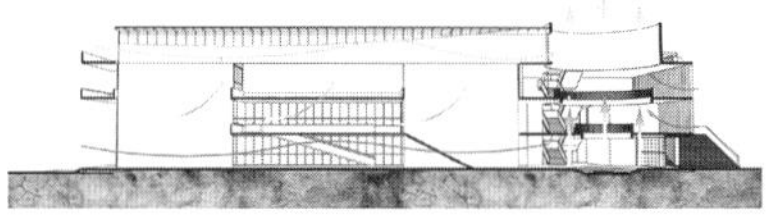

季节策略

Seasonal strategy

日照 Sunshine | 烟囱效应 Chimney effect | 通风 Ventilation

夏季 Summer

Sunshine in summer

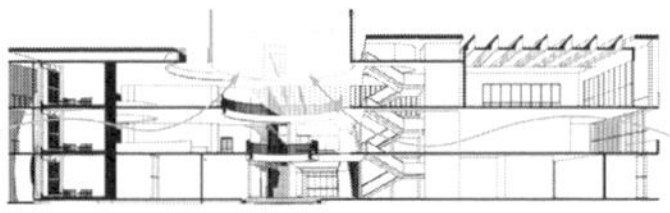

Chimney effect in summer

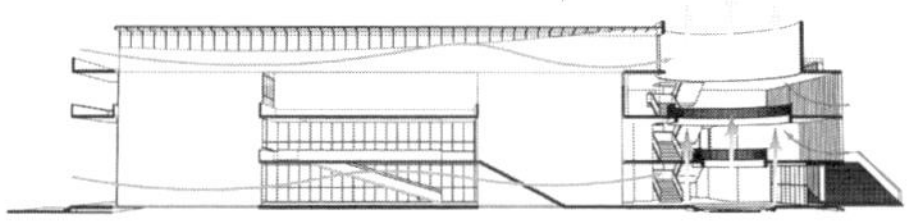

Ventilation in summer

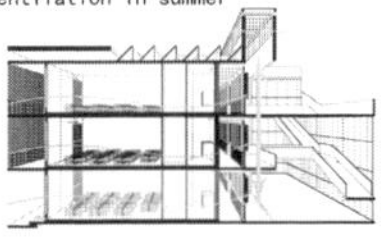

冬季 winter

Sunshine in winter

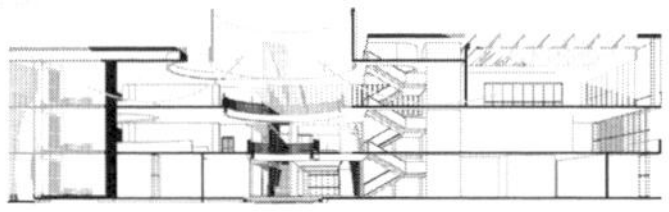

Chimney effect in winter

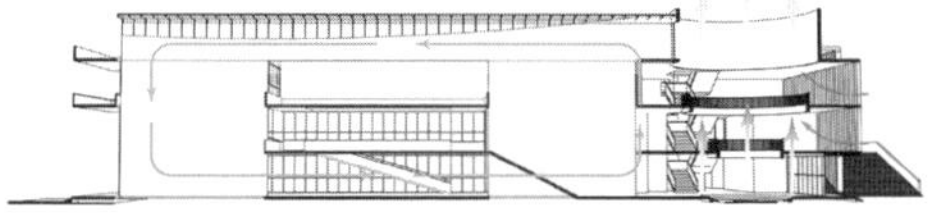

Ventilation in winter

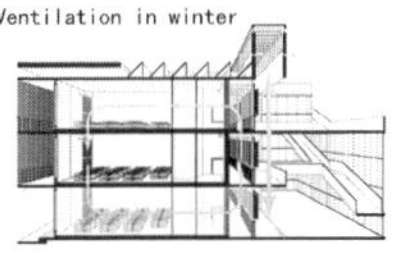

光之谷VI The Valley of Light

Look out from the atrium

· ENERGY SAVING ANALYSIS

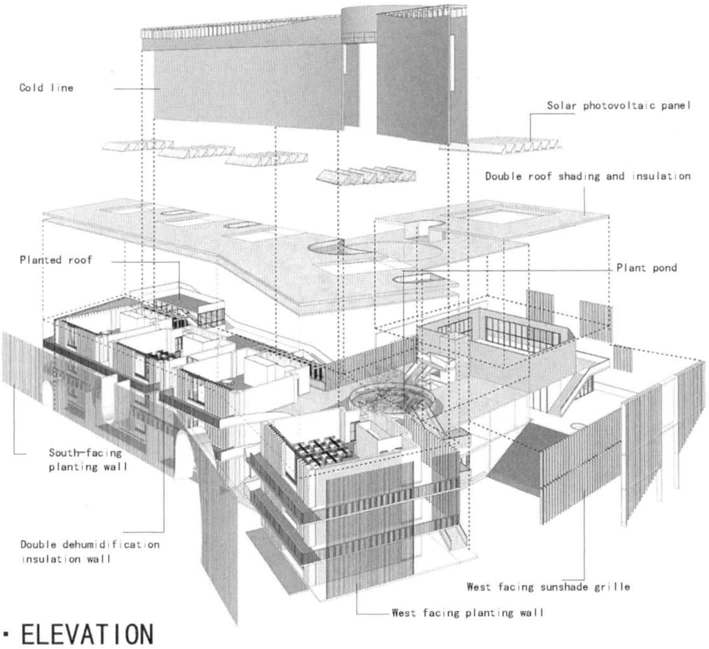
Cold line
Solar photovoltaic panel
Double roof shading and insulation
Planted roof
Plant pond
South-facing planting wall
Double dehumidification insulation wall
West facing sunshade grille
West facing planting wall

· INTERIOR RENDERINGS

The slope

The light well at the entrance to the classroom

· ELEVATION

NORTH EVELATION 1:250

WEST EVELATION 1:250

SOUTH EVELATION 1:250

EAST EVELATION 1:250

综合奖 · 优秀奖
General Prize Awarded · Honorable Mention Prize

注 册 号：7034

项目名称：律动 · 魔方（南平）
Rotating Cubes（Nanping）

作　　者：李　航、陈峙宇、Yang Yang

参赛单位：南京工业大学、
Syracuse University

指导教师：杨亦陵

Rotating Cubes
律动·魔方 1

Design Notes

本次幼儿园设计，我们试图回答下述问题:幼儿园的**城市尺度**，幼儿园自身的**建筑特色**，**绿色技术**的融入。

我们认为，用简洁有力、富于韵律的立方体，通过旋转和变化，将能够解决幼儿园的城市尺度问题。位于快速道路上的移动视线，将被这些简洁律动的方块所吸引。儿童有着自己的世界，他们的思想边界也许超出我们的想象。我们试图在方案中通过抽象的立体主义的风格，以及丰富多样的室内空间，来呼应他们的这种特质。由此，我们确定了**“律动魔方”**这一主题来展开我们的设计。

对于绿色技术，主要通过被动式的太阳能技术来解决幼儿园夏季的隔热通风以及冬季的保温和防湿，也运用多种其他的绿色技术。详见太阳能及绿色技术措施。

In this project, we attempted to explore the following areas: the **urban scale** of a kindergarten besides city expressway; the **unique characteristics** of kindergarten; using **green technologies** in kindergarten.

We believe this conflict of the urban scale will be solved by designing a series simple, powerful, rhythmic rolling cubes. The moving eye view on the expressway will be drawn to the simple and rhythmic cubes. Children have a world of their own, and the boundaries of their minds may be beyond our imagination. We try to respond to this quality in our design through an abstract cubist style and a variety of interior spaces. Thus, we focus the theme "**Rotating Cubes**" to develop our design.

The green technologies in the design include mainly passive solar technology for the thermal insulation and ventilation of the kindergarten in summer and the insulation and dehumidification in winter, but also other green technologies are used. Detail in Solar and green technology strategies.

Location Analysis

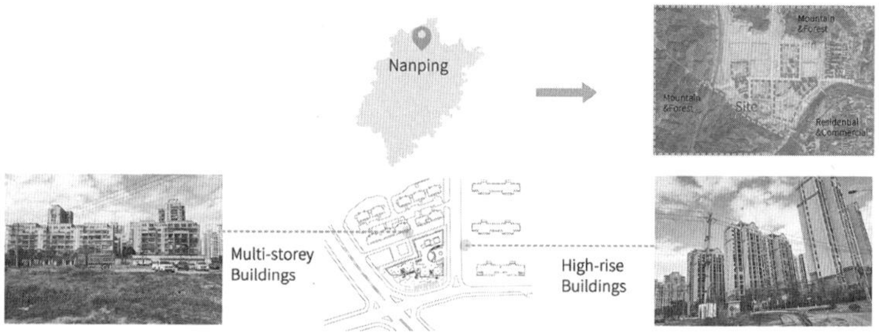
Nanping
Multi-storey Buildings
High-rise Buildings

Design Concept

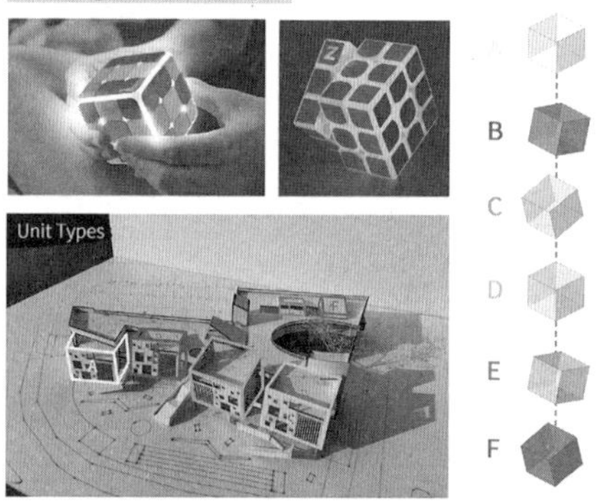
Unit Types
B
C
D
E
F

Behavior per day

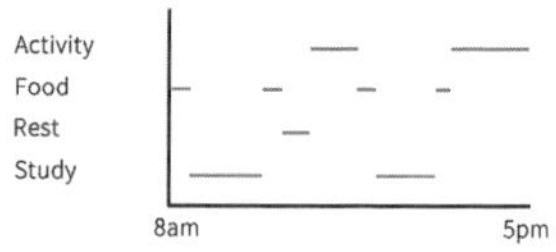
Activity
Food
Rest
Study
8am
5pm

Behavior in kindergarten

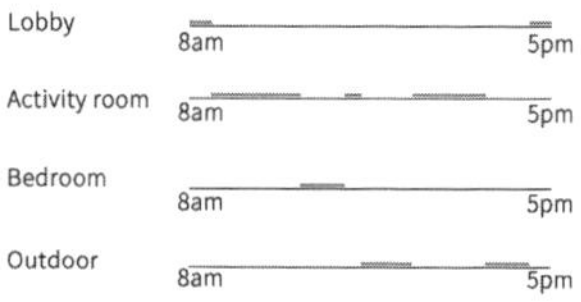
Lobby
Activity room
Bedroom
Outdoor
8am
5pm

Rotating Cubes
律动·魔方 2

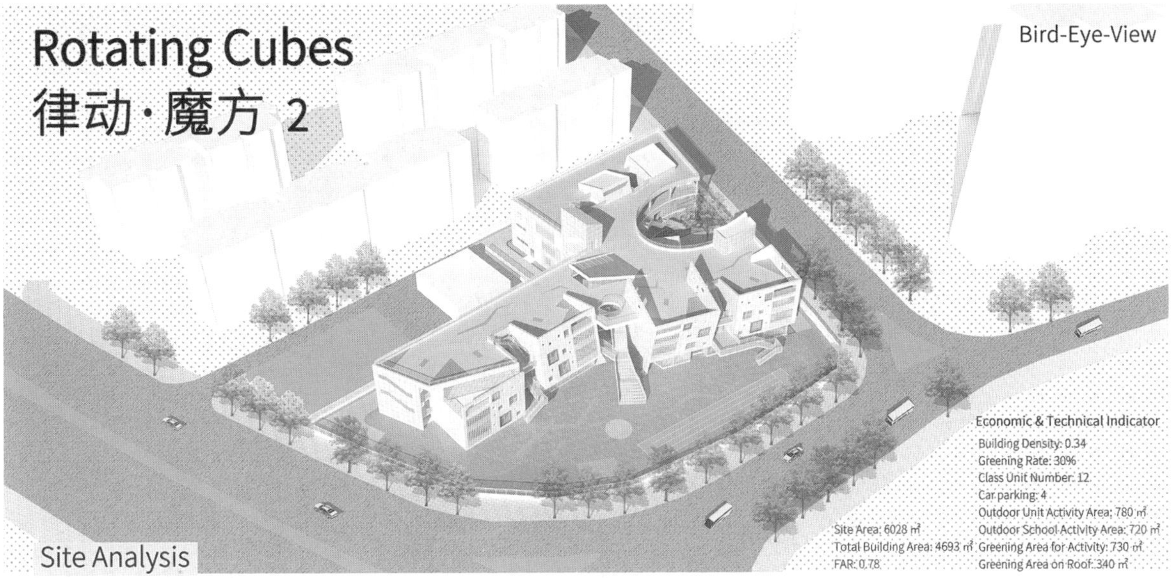
Bird-Eye-View
Economic & Technical Indicator
Building Density: 0.34
Greening Rate: 30%
Class Unit Number: 12
Car parking: 4
Outdoor Unit Activity Area: 780 ㎡
Outdoor School Activity Area: 720 ㎡
Greening Area for Activity: 730 ㎡
Greening Area on Roof: 340 ㎡
Site Area: 6028 ㎡
Total Building Area: 4693 ㎡
FAR: 0.78

Site Analysis

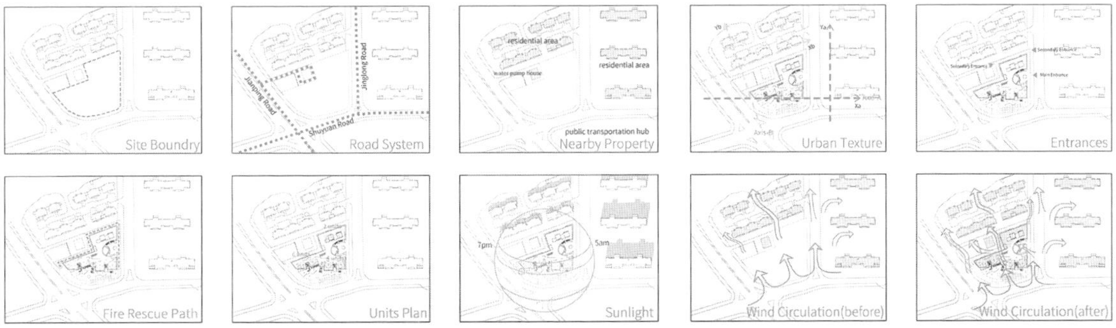
Site Boundry
Road System
Jinglong Road
Jianping Road
Shuyuan Road
residential area
residential area
public transportation hub
Nearby Property
Urban Texture
Entrances
Fire Rescue Path
Units Plan
7pm
5am
Sunlight
Wind Circulation(before)
Wind Circulation(after)

N
7F
7F
6F
6F
Residential Community
19F
19F
18F
3F
1F
Jinglong Road
Jianping Road
Shuyuan Road

Site Plan 1:500

Climate Analysis

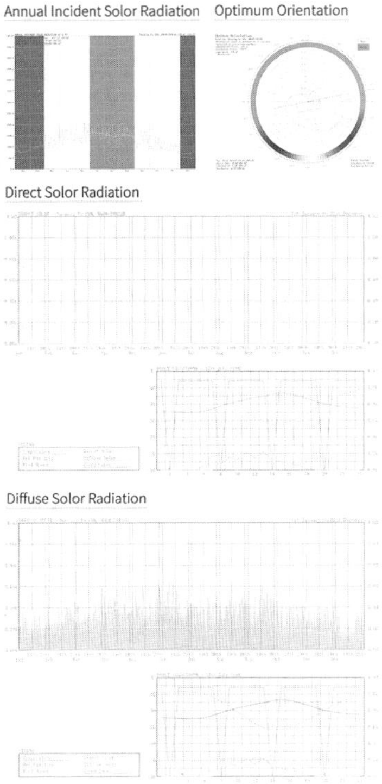
Annual Incident Solor Radiation
Optimum Orientation
Direct Solor Radiation
Diffuse Solor Radiation

Rotating Cubes
律动·魔方 3

1. Activity room
2. Bedroom
3. Cloak room
4. Multi Function
5. Reading room
6. Lobby
7. Medical check
8. Medical care
9. Gatehouse
10. Monitor room
11. Storage
12. Laundry
13. Electric system
14. Kitchen
15. Dish washing room
16. Pantry(备餐间)
17. Cold storage
18. Male staff changing room
19. Female staff changing room
20. Basic processing
21. Main food storage
22. Fresh food storage
23. Male toilet
24. Female toilet
25. Disabled toilet
26. Corridor

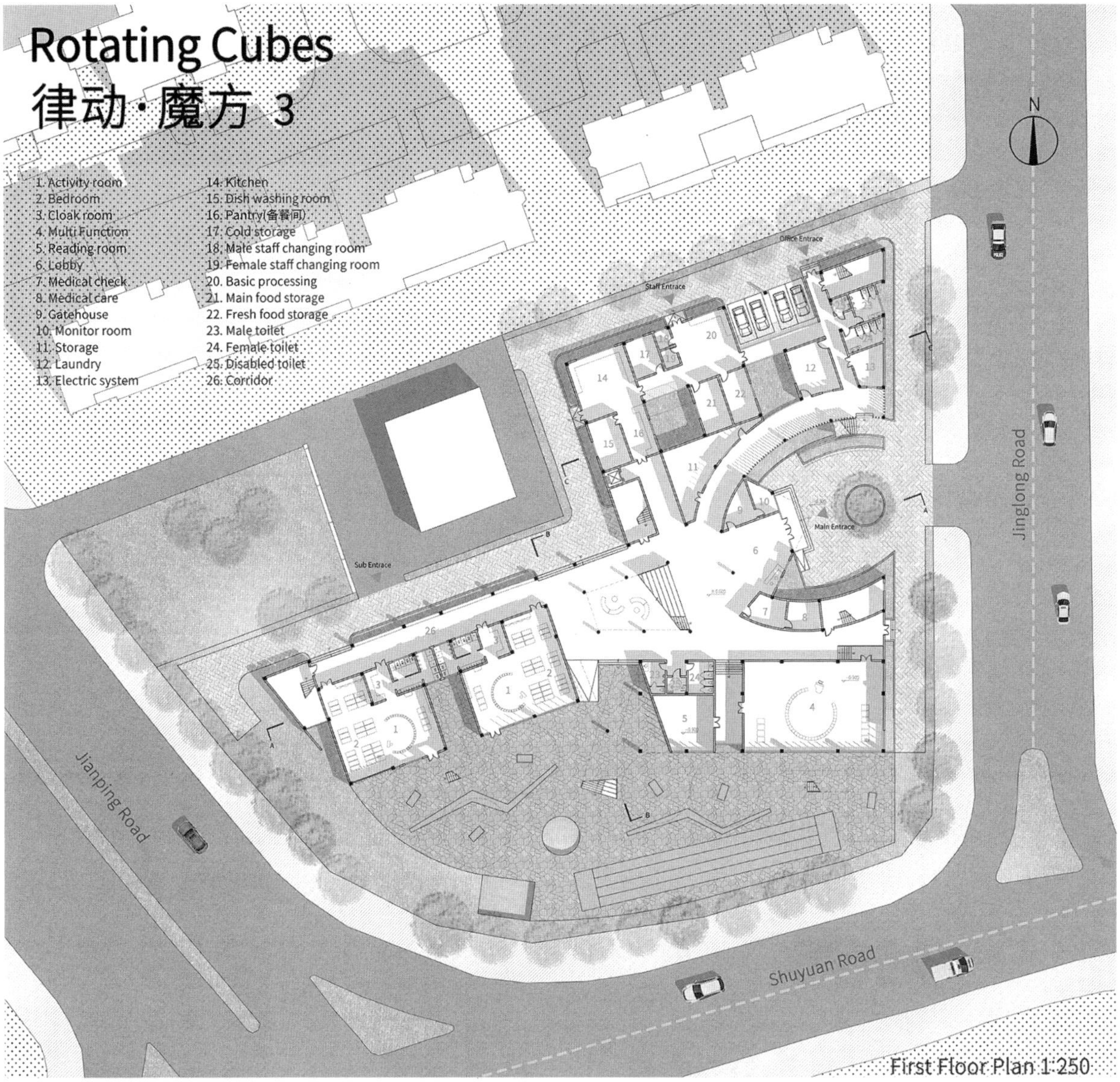
N
Office Entrance
Staff Entrance
Main Entrace
Sub Entrace
Jinglong Road
Jianping Road
Shuyuan Road

First Floor Plan 1:250

A-A Perspective Section

Rotating Cubes
律动·魔方 4

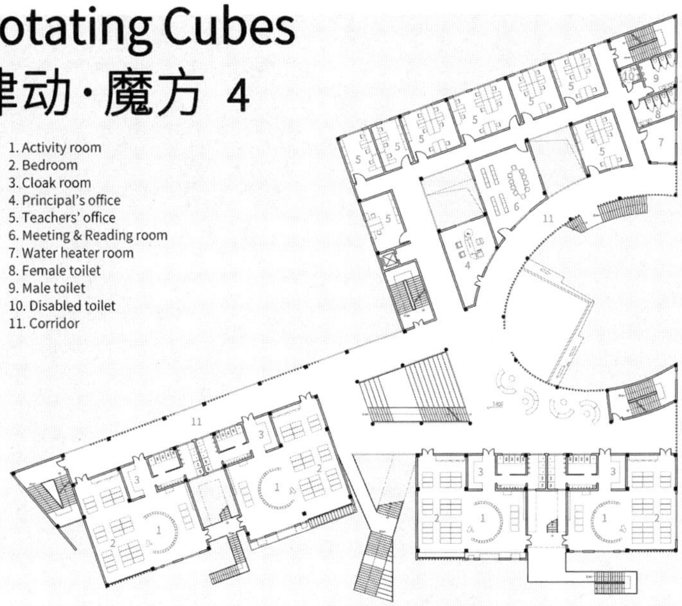
1. Activity room
2. Bedroom
3. Cloak room
4. Principal's office
5. Teachers' office
6. Meeting & Reading room
7. Water heater room
8. Female toilet
9. Male toilet
10. Disabled toilet
11. Corridor

Second Floor Plan 1:250

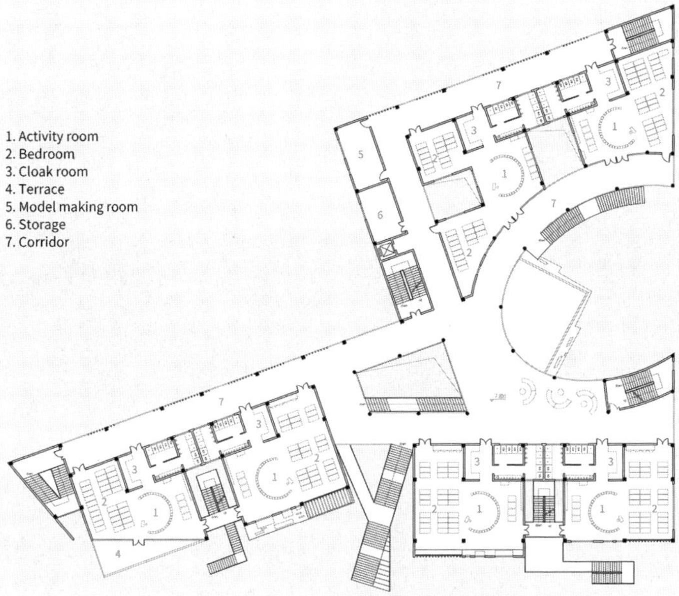
1. Activity room
2. Bedroom
3. Cloak room
4. Terrace
5. Model making room
6. Storage
7. Corridor

Third Floor Plan 1:250

East Facade 1:250

South Facade 1:250

Program Analysis

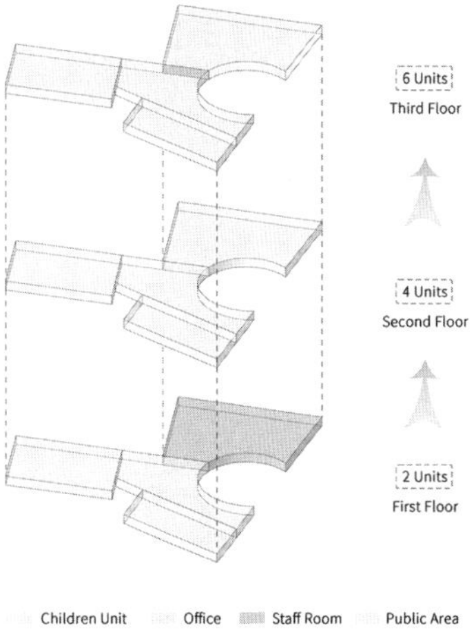
6 Units
Third Floor
4 Units
Second Floor
2 Units
First Floor
Children Unit
Office
Staff Room
Public Area

Circulation Analysis

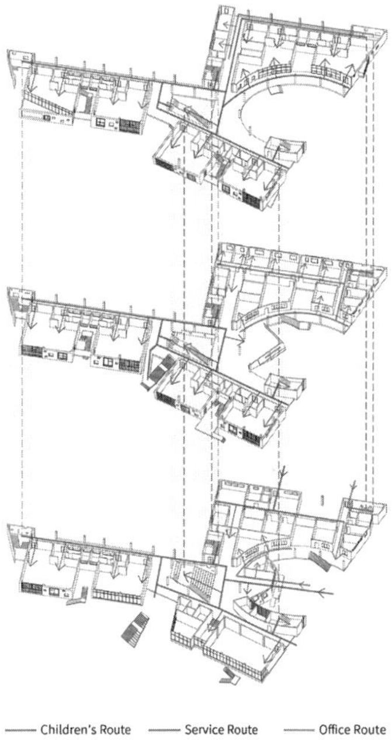
Children's Route
Service Route
Office Route

Playgrounds Analysis

Children's Outdoor Activity Space
Solar Photovoltaic Array

Rotating Cubes
律动·魔方 5

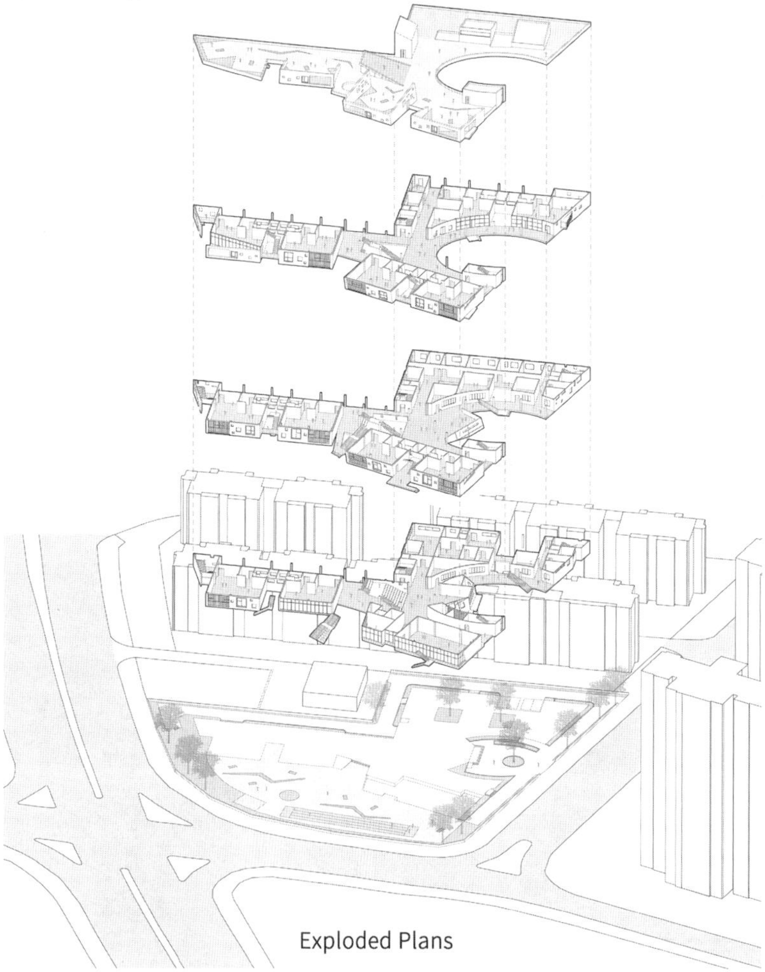

Exploded Plans

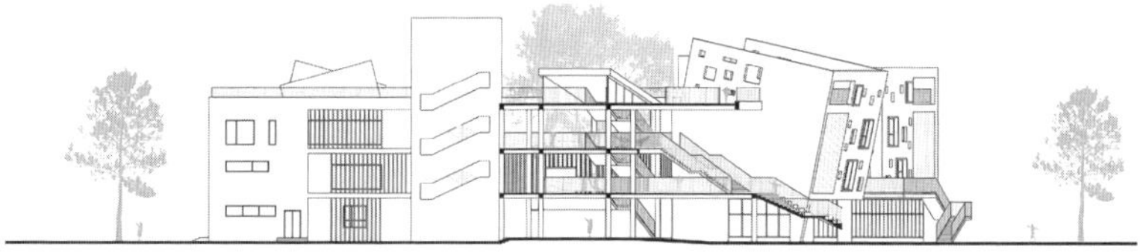

B-B Section 1：250

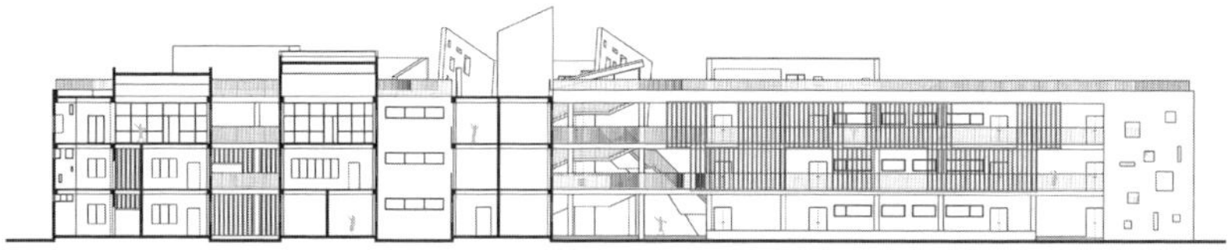

C-C Section 1：250

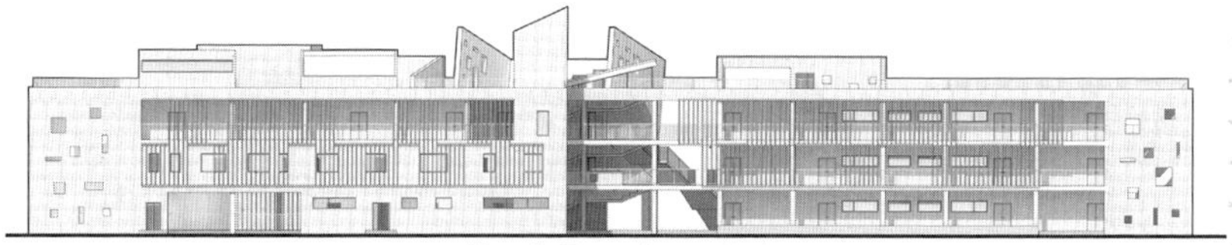

North Facade 1：250

Climate Analysis

Prevailing Winds

Psychrometric Chart

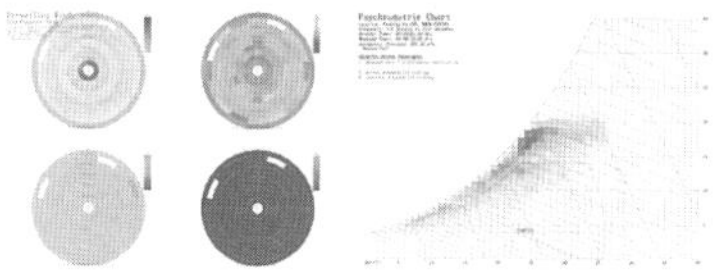

Dry Bulb Temperature

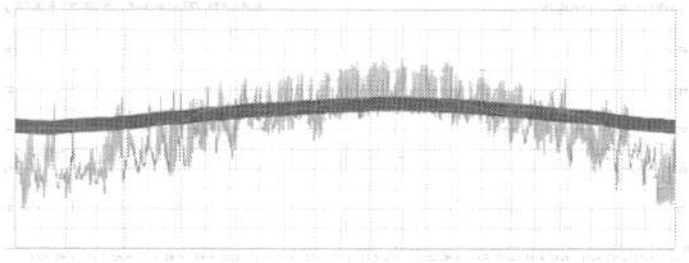

Wet Bulb Temperature

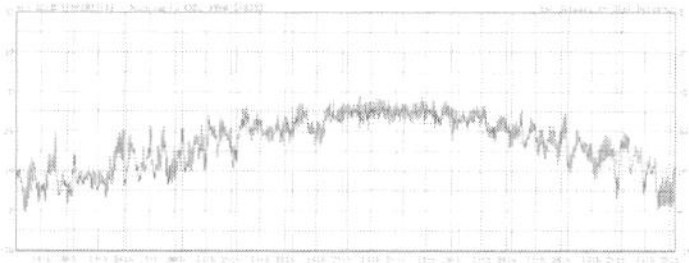

Dry Bulb X Relative Humidity

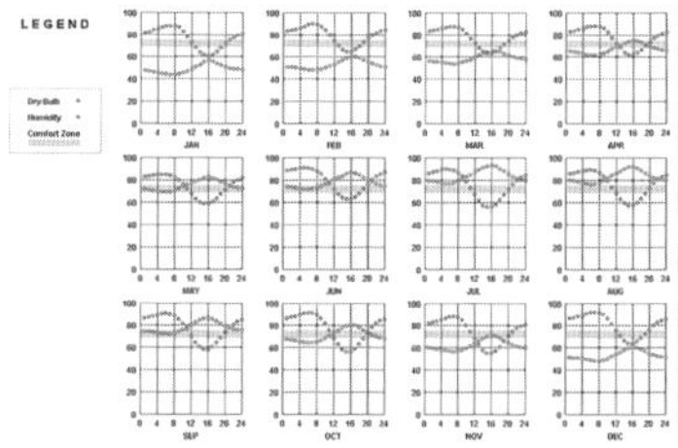

Monthly Diurnal Average

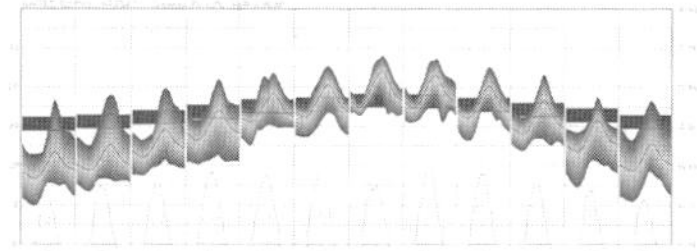

The Coolest Day 1.5

The Hottest Day 8.3

Weekly Summary

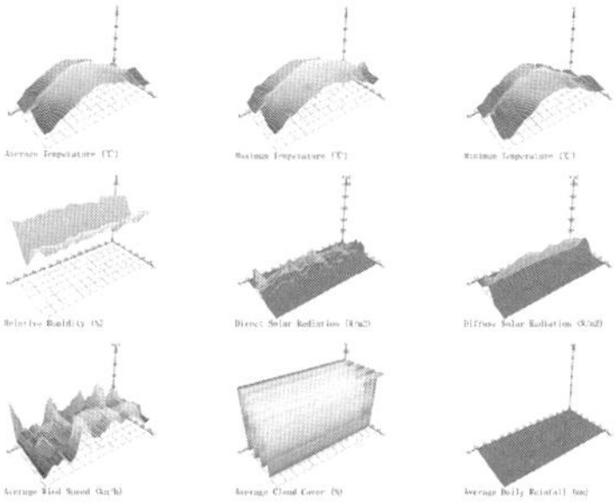

Heating & Cooling Potential of Passive Architecture

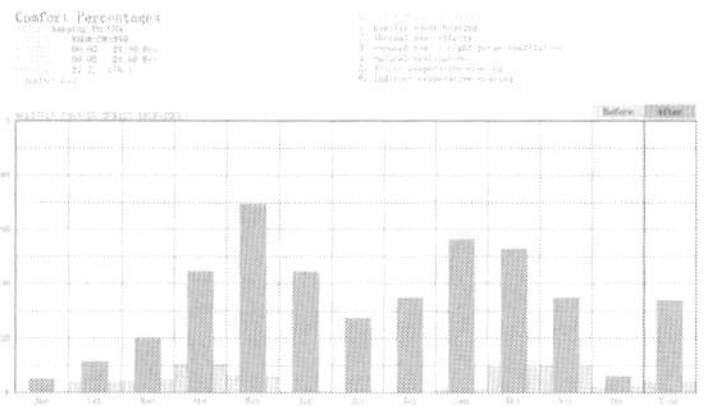

Rotating Cubes

律动·魔方 6

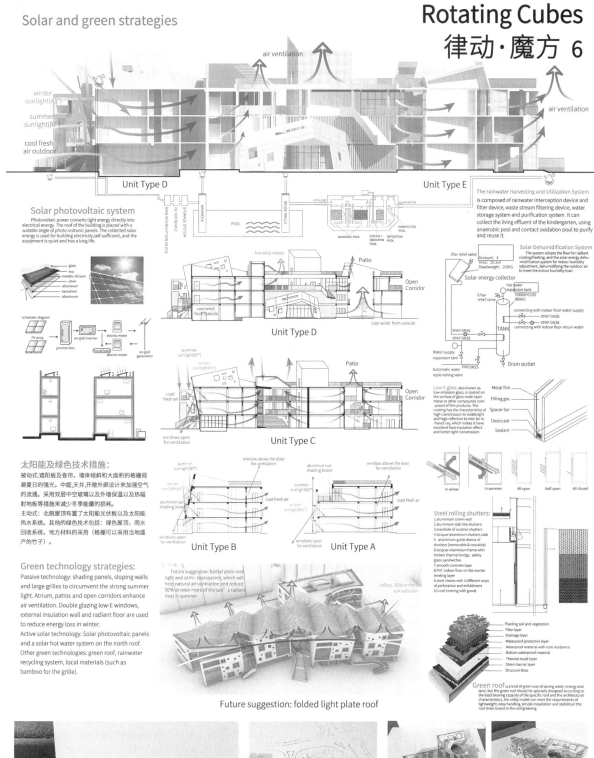

Solar and green strategies

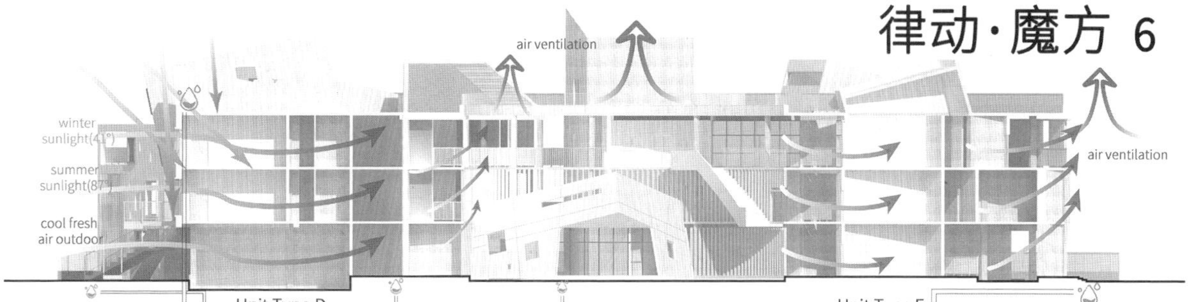
air ventilation
winter sunlight(41°)
summer sunlight(87°)
cool fresh air outdoor
air ventilation
Unit Type D
Unit Type E

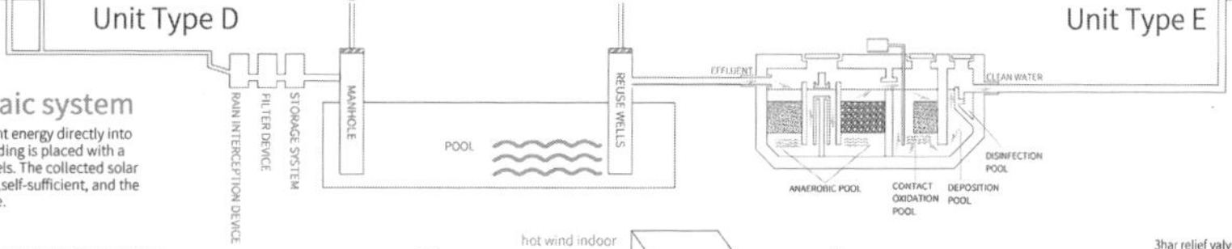
RAIN INTERCEPTION DEVICE
FILTER DEVICE
STORAGE SYSTEM
MANHOLE
POOL
REUSE WELLS
EFFLUENT
CLEAN WATER
ANAEROBIC POOL
CONTACT OXIDATION POOL
DEPOSITION POOL
DISINFECTION POOL

The rainwater Harvesting and Utilization System is composed of rainwater interception device and filter device, waste stream filtering device, water storage system and purification system. It can collect the living effluent of the kindergarten, using anaerobic pool and contact oxidation pool to purify and reuse it.

Solar photovoltaic system

Photovoltaic power converts light energy directly into electrical energy. The roof of the building is placed with a suitable angle of photo-voltanic panels. The collected solar energy is used for building electricity,self-sufficient, and the equipment is quiet and has a long life.

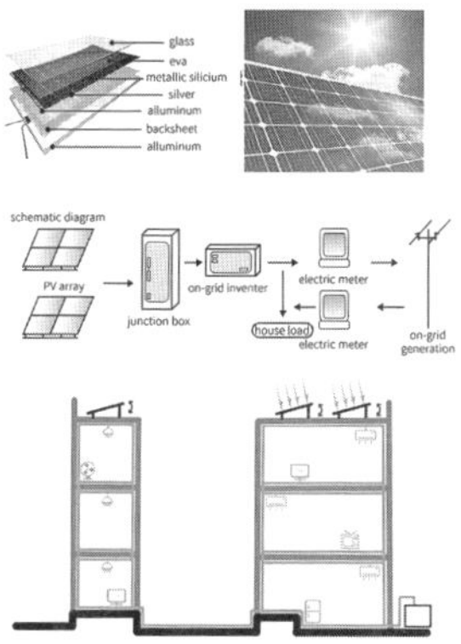
glass
eva
metallic silicium
silver
alluminum
backsheet
alluminum
schematic diagram
PV array
junction box
on-grid inverter
electric meter
house load
electric meter
on-grid generation

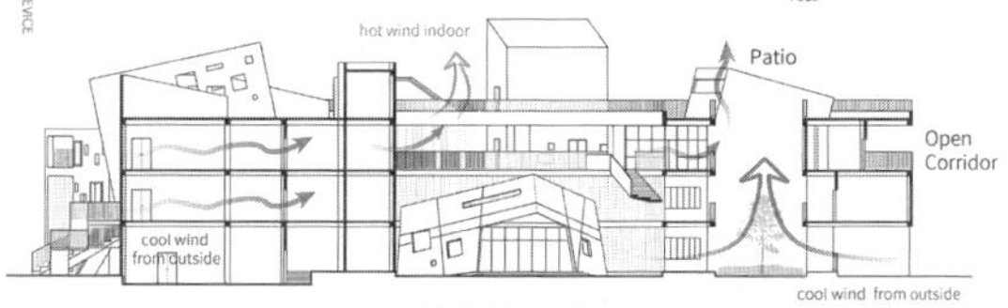
hot wind indoor
Patio
Open Corridor
cool wind from outside
cool wind from outside

Unit Type D

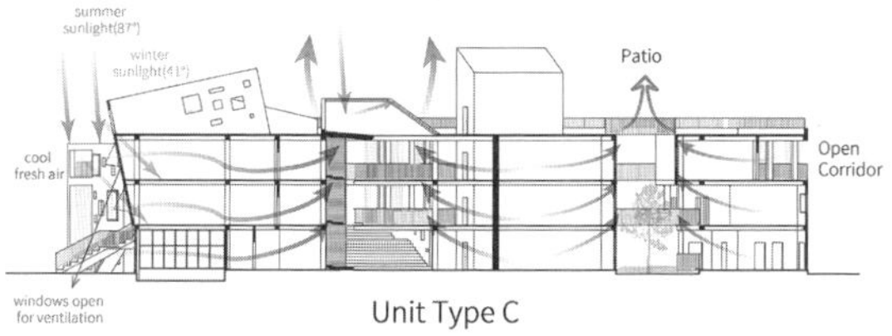
summer sunlight(87°)
winter sunlight(41°)
cool fresh air
Patio
Open Corridor
windows open for ventilation

Unit Type C

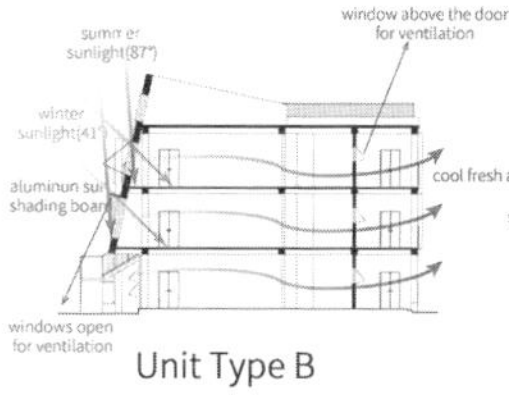
window above the door for ventilation
cool fresh air
aluminum sun shading board
windows open for ventilation

Unit Type B

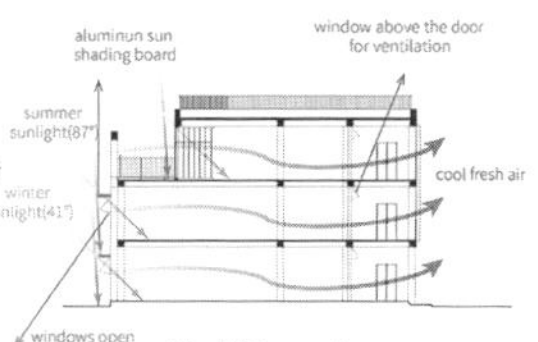
aluminun sun shading board
window above the door for ventilation
summer sunlight(87°)
winter sunlight(41°)
cool fresh air
windows open for ventilation

Unit Type A

太阳能及绿色技术措施:

被动式:遮阳板及卷帘，墙体倾斜和大面积的格栅规避夏日的强光。中庭,天井,开敞外廊设计来加强空气的流通。采用双层中空玻璃以及外墙保温以及热辐射地板等措施来减少冬季能量的损耗。

主动式：北侧屋顶布置了太阳能光伏板以及太阳能热水系统。其他的绿色技术包括：绿色屋顶，雨水回收系统，地方材料的采用（格栅可以采用当地盛产的竹子）。

Green technology strategies:

Passive technology: shading panels, sloping walls and large grilles to circumvent the strong summer light. Atrium, patios and open corridors enhance air ventilation. Double glazing low-E windows, external insulation wall and radiant floor are used to reduce energy loss in winter.

Active solar technology: Solar photovoltaic panels and a solar hot water system on the north roof.

Other green technologies: green roof, rainwater recycling system, local materials (such as bamboo for the grille).

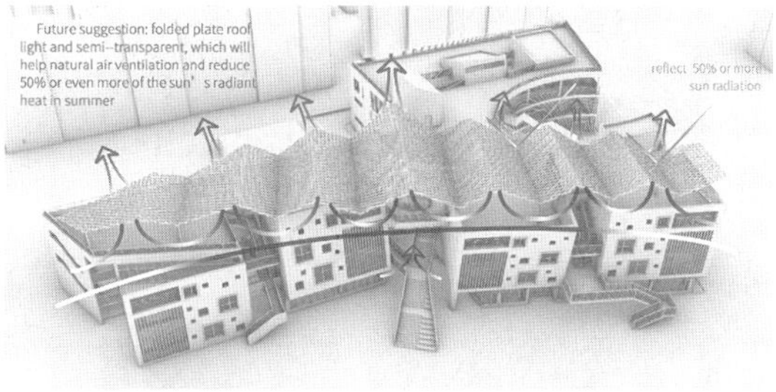
Future suggestion: folded plate roof light and semi--transparent, which will help natural air ventilation and reduce 50% or even more of the sun' s radiant heat in summer
reflect 50% or more sun radiation

Future suggestion: folded light plate roof

Solar Dehumidification System

The system adopts the floor for radiant cooling/heating, and the solar energy dehu--midification system for indoor humidity adjustment, dehumidifying the outdoor air to meet the indoor humidity load.

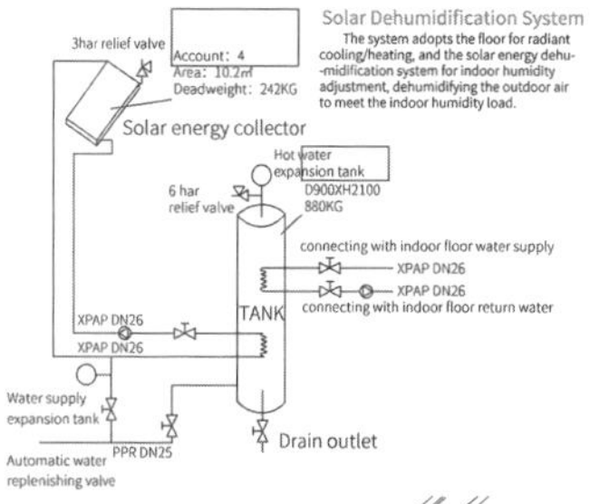
3har relief valve
Account: 4
Area: 10.2㎡
Deadweight: 242KG
Solar energy collector
Hot water expansion tank
6 har relief valve
D900XH2100
880KG
connecting with indoor floor water supply
XPAP DN26
XPAP DN26
connecting with indoor floor return water
TANK
XPAP DN26
XPAP DN26
Water supply expansion tank
PPR DN25
Automatic water replenishing valve
Drain outlet

Low-E glass, also known as low-emission glass, is coated on the surface of glass multi-layer metal or other compounds com--posed of film products. The coating has the characteristics of high transmission to visiblelight and high reflection to mid-far in--frared ray, which makes it have excellent heat insulation effect and better light transmission.

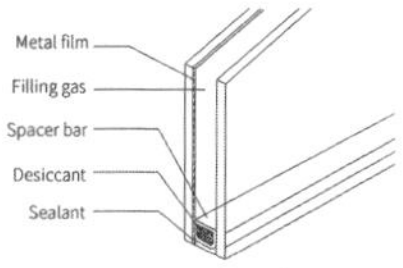
Metal film
Filling gas
Spacer bar
Desiccant
Sealant

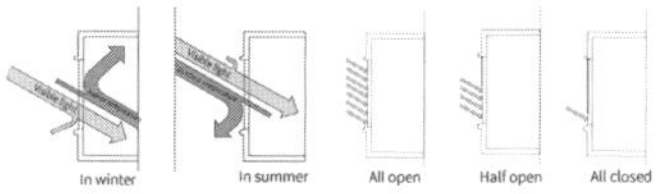
In winter
In summer
All open
Half open
All closed

Steel rolling shutters:

1.aluminium crown wall
2.aluminium slab like shutters
3.manhole of ourdoor shutters
4.lacquer aluminium shutters slab
5. aluminium guide device of shutters (immovable & movable)
6.lacquer aluminium frame with broken thermal bridge, safety glass sandwiches
7.smooth concrete layer
8.PVC indoor floor on the mortar leveling layer
9.steel sheets with 3 different ways of perforation and enfoldment
10.roof civering with gravel

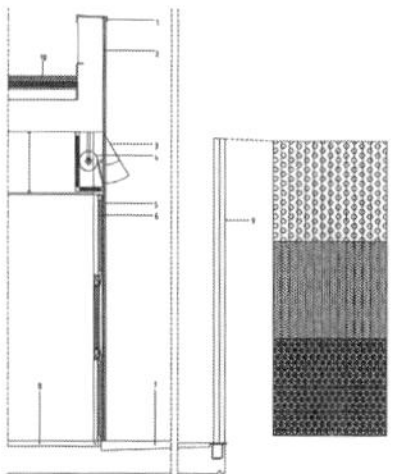

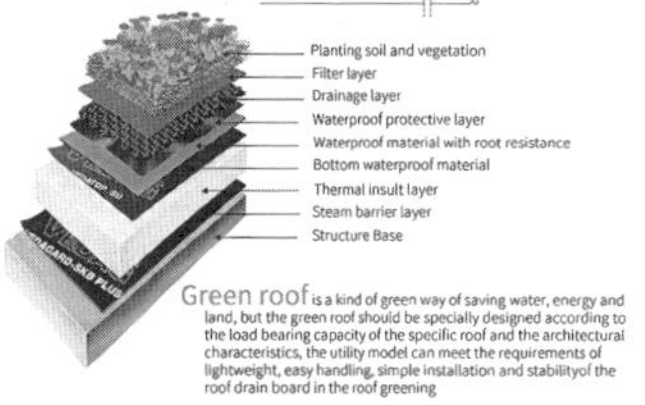
Planting soil and vegetation
Filter layer
Drainage layer
Waterproof protective layer
Waterproof material with root resistance
Bottom waterproof material
Thermal insult layer
Steam barrier layer
Structure Base

Green roof is a kind of green way of saving water, energy and land, but the green roof should be specially designed according to the load bearing capacity of the specific roof and the architectural characteristics, the utility model can meet the requirements of lightweight, easy handling, simple installation and stabilityof the roof drain board in the roof greening

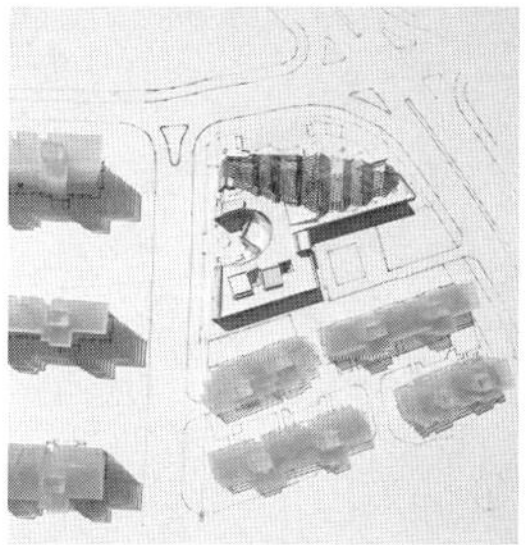

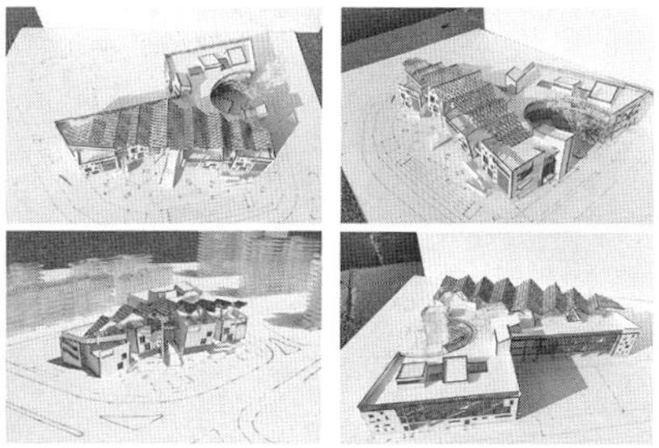

综合奖·优秀奖
General Prize Awarded · Honorable Mention Prize

注 册 号：7038

项目名称：阳光之下·田园之上（南平）
Sunshine Pouring · Children Geting（Nanping）

作　　者：李起航、高雨凯、李　思

参赛单位：中国矿业大学（北京）

指导教师：贺丽洁、郑利军、曹　颖

阳光之下·田园之上
SUNSHINE POURING · CHILDREN GETING
01

PROJECT BACKGROUND

DIAOJIAO BUILDING
A WIND-GUIDING BUILDING FORM
PRODUCTIVE LANDSCAPE
SURROUNDING ENVIRONMENT
SURROUNDING ENVIRONMENT
SURROUNDING ENVIRONMENT

PROJECT LOCATION

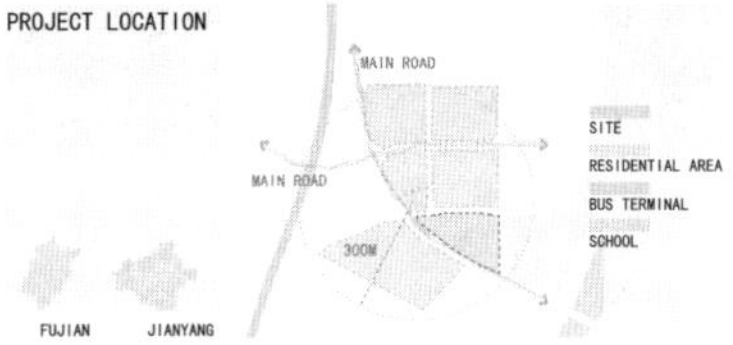

MAIN ROAD
MAIN ROAD
300M
SITE
RESIDENTIAL AREA
BUS TERMINAL
SCHOOL
FUJIAN
JIANYANG

TODDLER ACTIVITY SCHEDULE

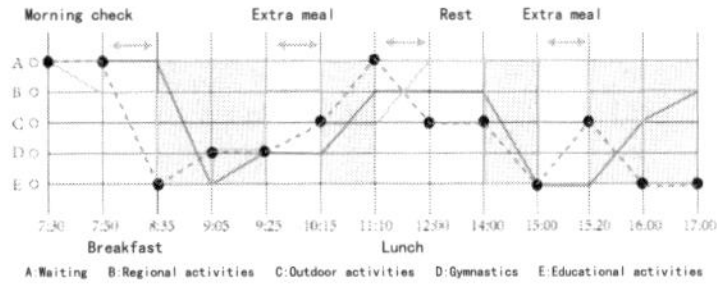

Morning check
Extra meal
Rest
Extra meal
A
B
C
D
E
7:30
7:50
8:35
9:05
9:25
10:15
11:10
12:00
14:00
15:00
15:20
16:00
17:00
Breakfast
Lunch
A:Waiting
B:Regional activities
C:Outdoor activities
D:Gymnastics
E:Educational activities

WEATHER ANALYSIS

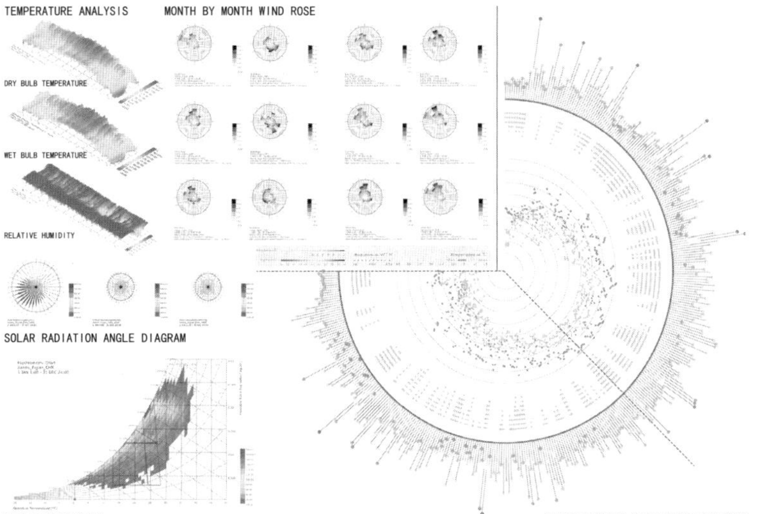

TEMPERATURE ANALYSIS
MONTH BY MONTH WIND ROSE
DRY BULB TEMPERATURE
WET BULB TEMPERATURE
RELATIVE HUMIDITY
SOLAR RADIATION ANGLE DIAGRAM
ENTHALPY DIAGRAM
DAILY WEATHER CONDITIONS DIAGRAM

设计说明 Design description

闽北地区属典型的中亚热带湿润季风气候，雨量充沛，光照丰富，气候湿热，因此隔热通风策略在建筑中的应用是我们这次幼儿园设计的重点。

该方案主要以冬夏主导风向为根据控制建筑形体，通过合理的场地布局达到夏季导风，冬季挡风的效果。具体建筑策略方面结合主动式与被动式节能技术，并且利用太阳能通风塔与特朗勃墙结合，采用底层架空、通风吊顶、北向天窗等方式来降低场地内部的湿热感，同时考虑为建筑提供良好的南向采光，为儿童创造舒适愉悦的环境。

The northern part of Fujian has a typical mid-subtropical humid monsoon climate, with abundant rainfall, abundant sunlight, and humid and hot climate. Therefore, the application of insulation and ventilation strategies in the building is the focus of our kindergarten design.

This plan mainly controls the building shape based on the dominant wind direction in winter and summer, and achieves the purpose of guiding the wind in summer and blocking the wind in winter through the layout of the building and site. In terms of specific building strategies, it combines active and passive energy-saving technologies, and uses the principles of solar ventilation towers and Trumb walls to solve the dampness and heat inside the site by using overhead ground, ventilated ceilings, and north-facing skylights. At the same time, it is considered to provide a good building South-facing daylighting creates a comfortable and pleasant environment for children.

Obtained from meteorological analysis, the design points are as follows:

1. Use solar panels to improve comfort during low temperature periods.
2. Use desiccant dehumidification technology to improve comfort during high humidity.
3. Use ventilated roof technology to increase air flow to reduce temperature and improve comfort
4. Adopt heat storage roof for thermal insulation

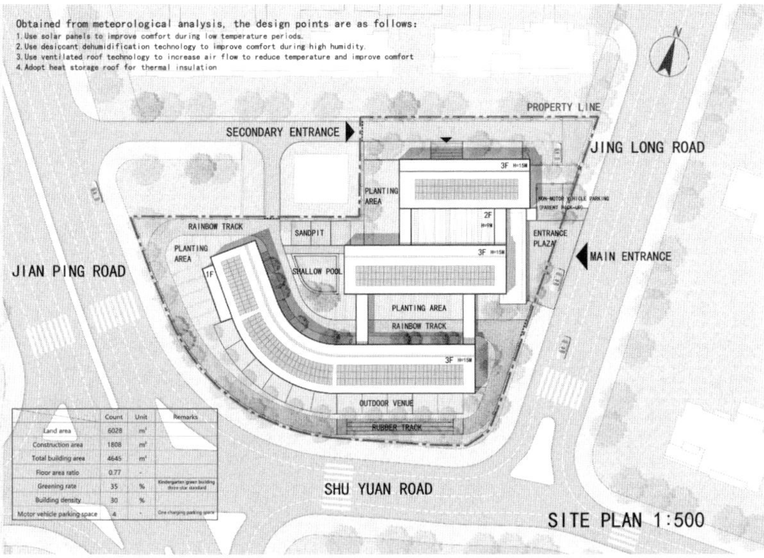

PROPERTY LINE
SECONDARY ENTRANCE
JING LONG ROAD
PLANTING AREA
3F H=15M
NON-MOTOR VEHICLE PARKING
2F H=9M
RAINBOW TRACK
SANDPIT
ENTRANCE PLAZA
PLANTING AREA
MAIN ENTRANCE
JIAN PING ROAD
1F
SHALLOW POOL
3F H=15M
PLANTING AREA
RAINBOW TRACK
3F H=15M
OUTDOOR VENUE
RUBBER TRACK
SHU YUAN ROAD

	Count	Unit	Remarks
Land area	6028	m²	
Construction area	1808	m²	
Total building area	4645	m²	
Floor area ratio	0.77	-	
Greening rate	35	%	Kindergarten green building three-star standard
Building density	30	%	
Motor vehicle parking space	4	-	One charging parking space

SITE PLAN 1:500

阳光之下·田园之上

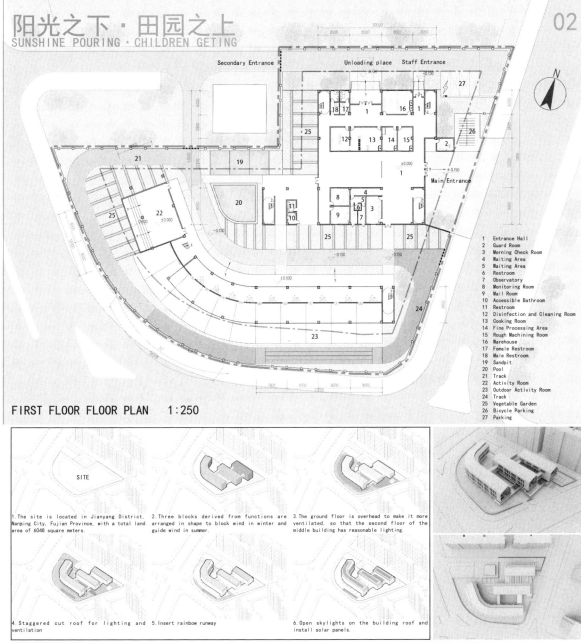

SUNSHINE POURING · CHILDREN GETING

02

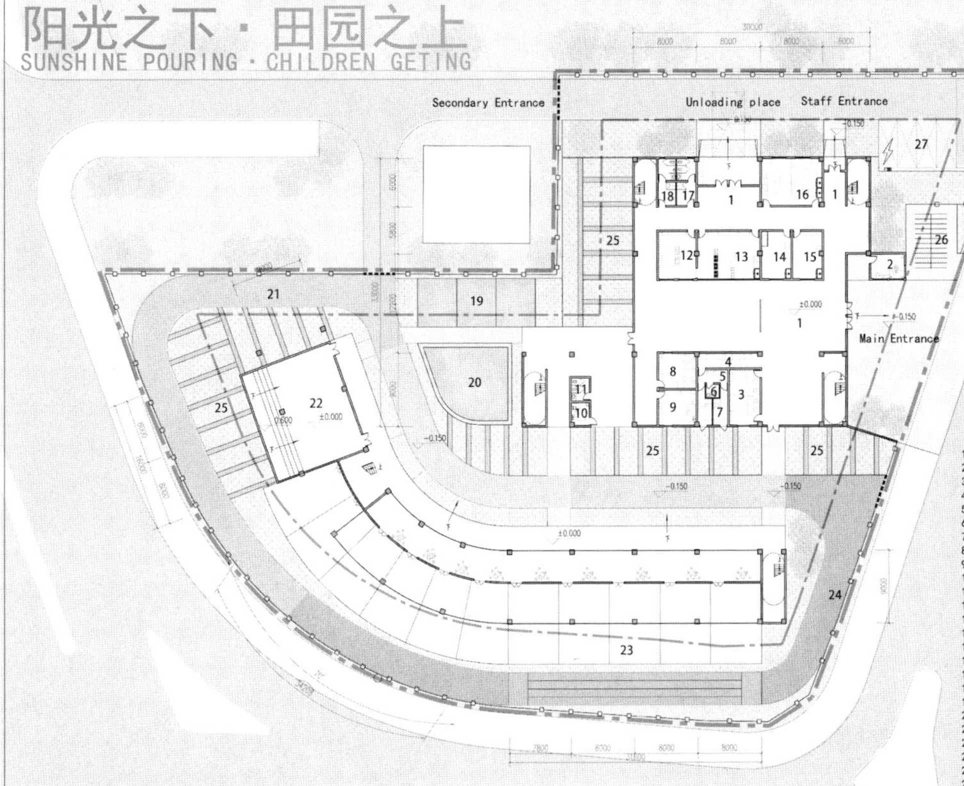
Secondary Entrance
Unloading place
Staff Entrance
Main Entrance

FIRST FLOOR FLOOR PLAN 1:250

1 Entrance Hall
2 Guard Room
3 Morning Check Room
4 Waiting Area
5 Waiting Area
6 Restroom
7 Observatory
8 Monitoring Room
9 Mail Room
10 Accessible Bathroom
11 Restroom
12 Disinfection and Cleaning Room
13 Cooking Room
14 Fine Processing Area
15 Rough Machining Room
16 Warehouse
17 Female Restroom
18 Male Restroom
19 Sandpit
20 Pool
21 Track
22 Activity Room
23 Outdoor Activity Room
24 Track
25 Vegetable Garden
26 Bicycle Parking
27 Parking

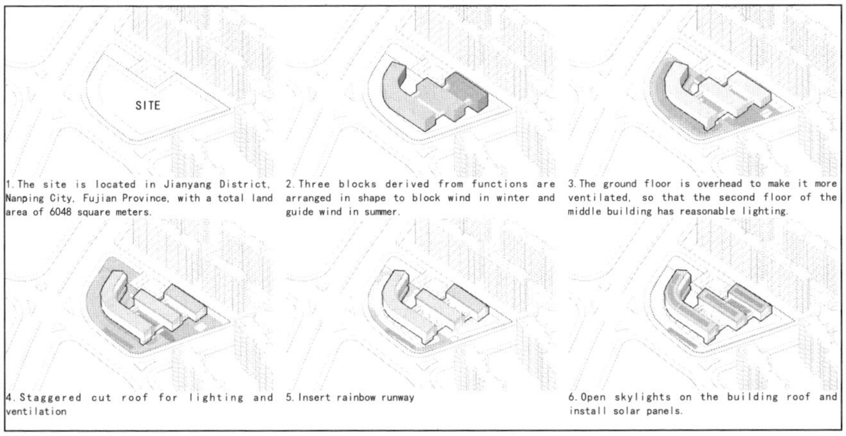
SITE
1.The site is located in Jianyang District, Nanping City, Fujian Province, with a total land area of 6048 square meters.
2.Three blocks derived from functions are arranged in shape to block wind in winter and guide wind in summer.
3.The ground floor is overhead to make it more ventilated, so that the second floor of the middle building has reasonable lighting.
4.Staggered cut roof for lighting and ventilation
5.Insert rainbow runway
6.Open skylights on the building roof and install solar panels.

BODY ANALYSIS CHART

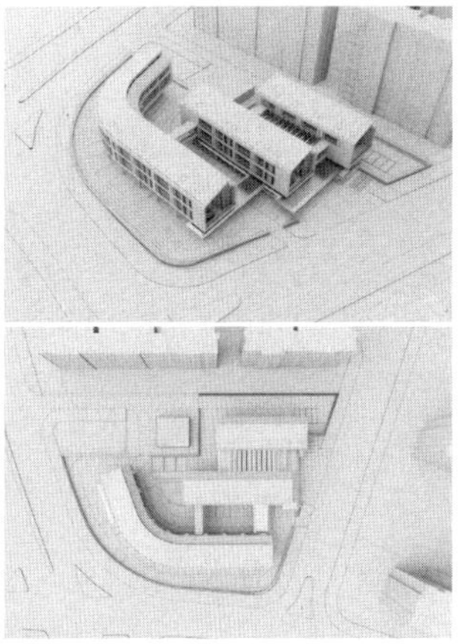

MODEL PHOTOS

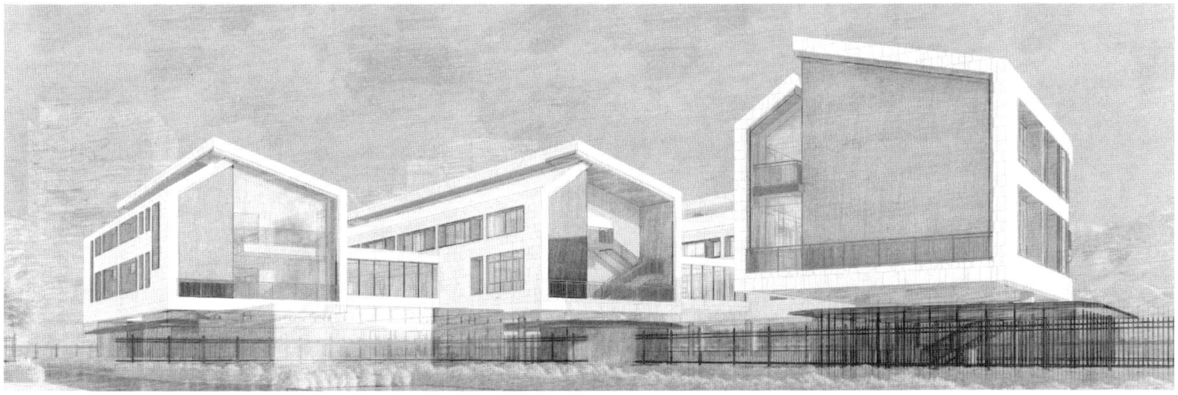

阳光之下 · 田园之上
SUNSHINE POURING · CHILDREN GETING

03

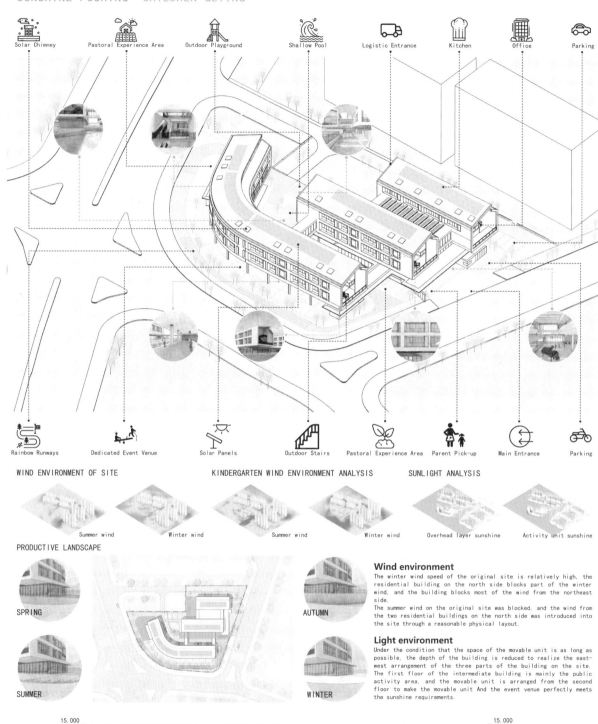

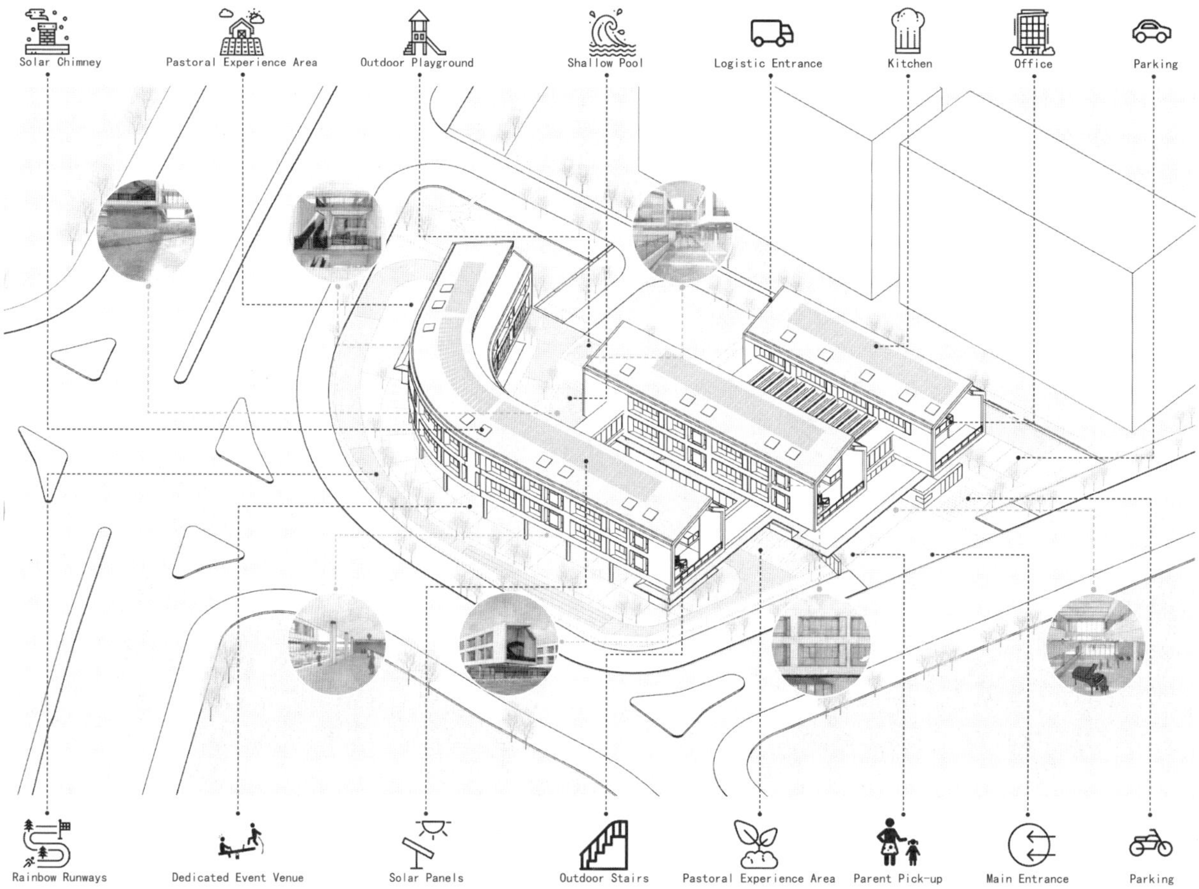
Solar Chimney
Pastoral Experience Area
Outdoor Playground
Shallow Pool
Logistic Entrance
Kitchen
Office
Parking
Rainbow Runways
Dedicated Event Venue
Solar Panels
Outdoor Stairs
Pastoral Experience Area
Parent Pick-up
Main Entrance
Parking

WIND ENVIRONMENT OF SITE　　KINDERGARTEN WIND ENVIRONMENT ANALYSIS　　SUNLIGHT ANALYSIS

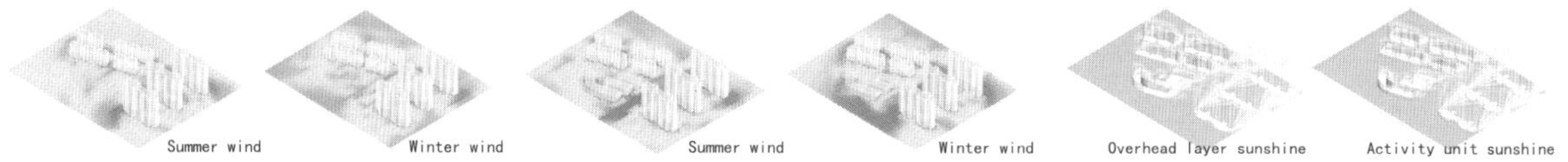
Summer wind
Winter wind
Summer wind
Winter wind
Overhead layer sunshine
Activity unit sunshine

PRODUCTIVE LANDSCAPE

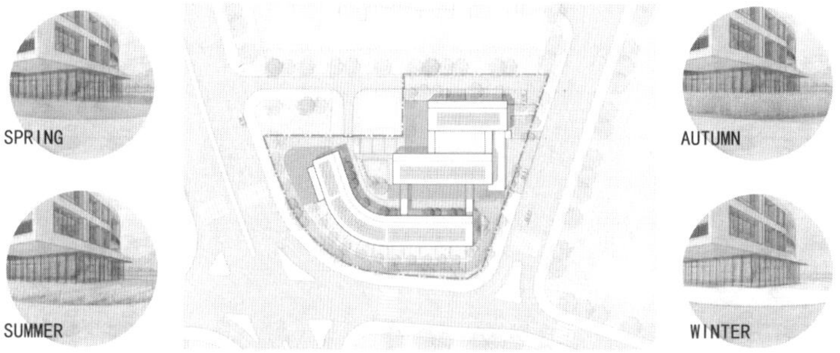
SPRING
AUTUMN
SUMMER
WINTER

Wind environment

The winter wind speed of the original site is relatively high, the residential building on the north side blocks part of the winter wind, and the building blocks most of the wind from the northeast side.

The summer wind on the original site was blocked, and the wind from the two residential buildings on the north side was introduced into the site through a reasonable physical layout.

Light environment

Under the condition that the space of the movable unit is as long as possible, the depth of the building is reduced to realize the east-west arrangement of the three parts of the building on the site. The first floor of the intermediate building is mainly the public activity area, and the movable unit is arranged from the second floor to make the movable unit And the event venue perfectly meets the sunshine requirements.

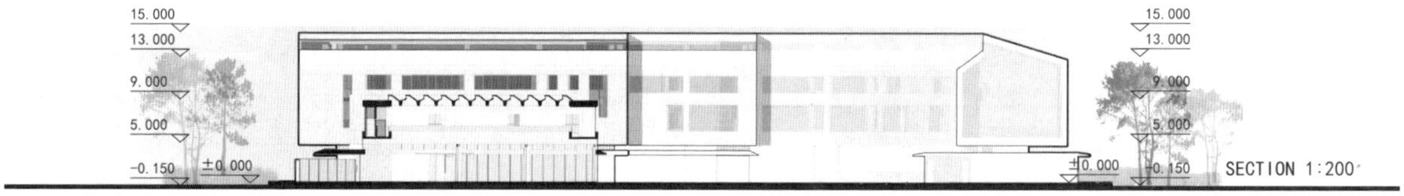
15.000
13.000
9.000
5.000
-0.150
±0.000
SECTION 1:200

阳光之下·田园之上

SUNSHINE POURING·CHILDREN GETING

04

PROFILE STRATEGY

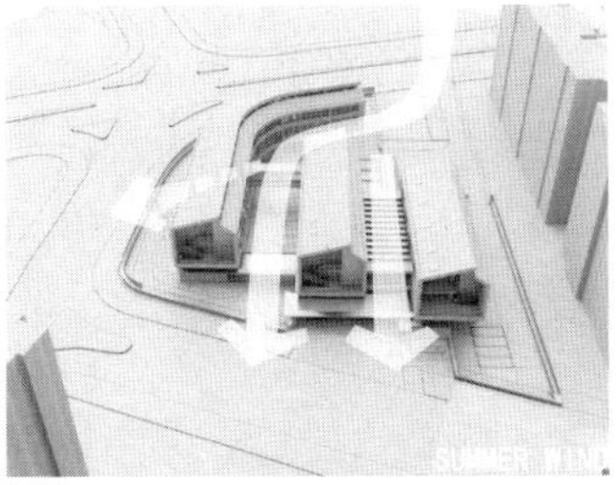

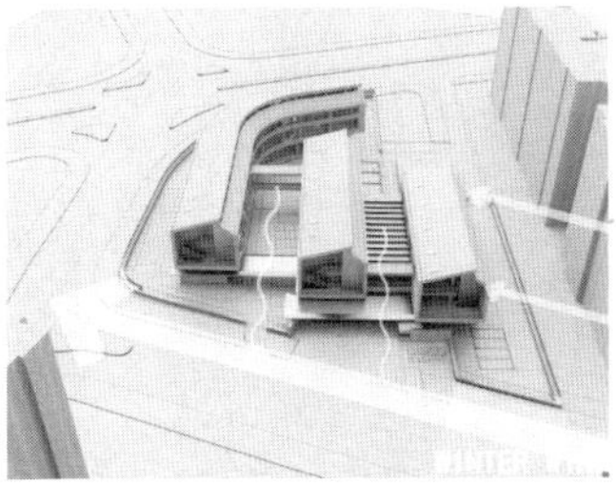

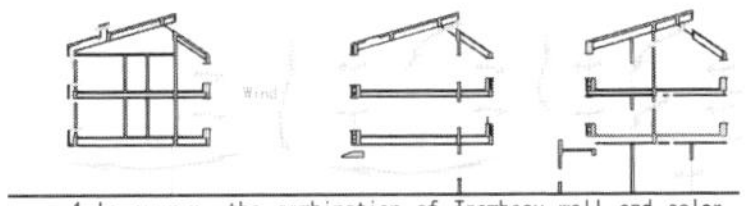

1. In summer, the combination of Trombeau wall and solar wind tower, aerial bottom, ventilated skylight, ventilated ceiling, etc. are used to enhance the ventilation in the site to remove heat.

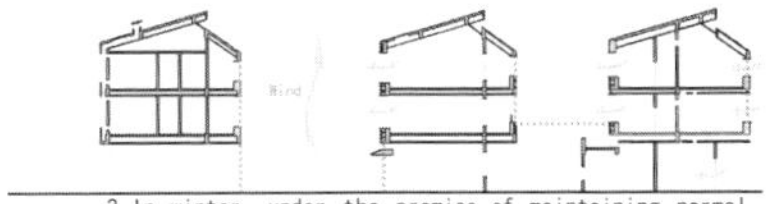

2. In winter, under the premise of maintaining normal indoor ventilation, close the building weather boundary to reduce the wind speed in the site.

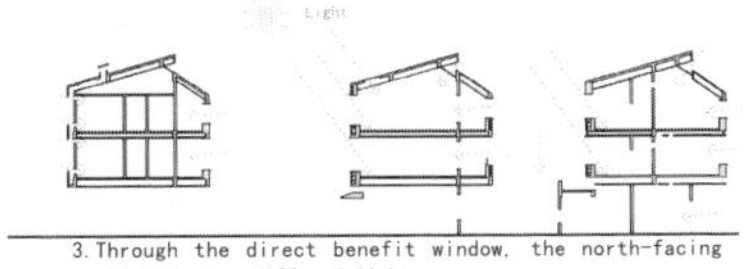

3. Through the direct benefit window, the north-facing skylight brings diffused light.

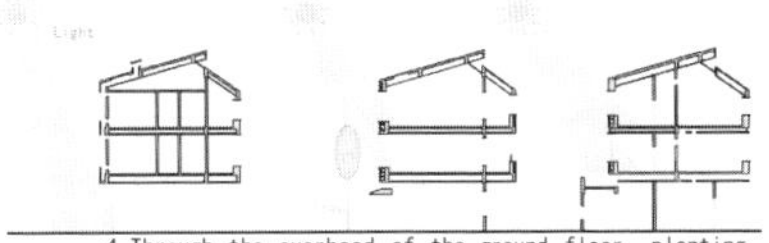

4. Through the overhead of the ground floor, planting trees and adopting recessed windows to block the positive part of the sun.

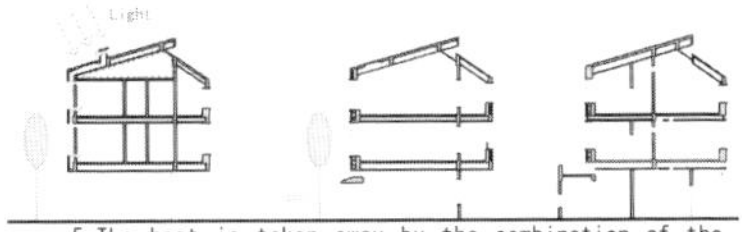

5. The heat is taken away by the combination of the Trombone wall and the solar tower, the heat storage roof collects part of the heat, and the transpiration of the plants takes away a certain amount of heat.

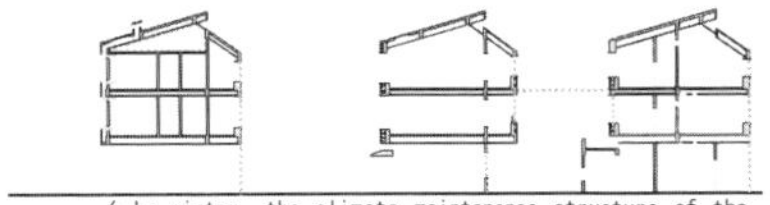

6. In winter, the climate maintenance structure of the building is closed for heat preservation, the heat storage roof will release a certain amount of heat, and the Trombeau wall collects a lot of radiant heat.

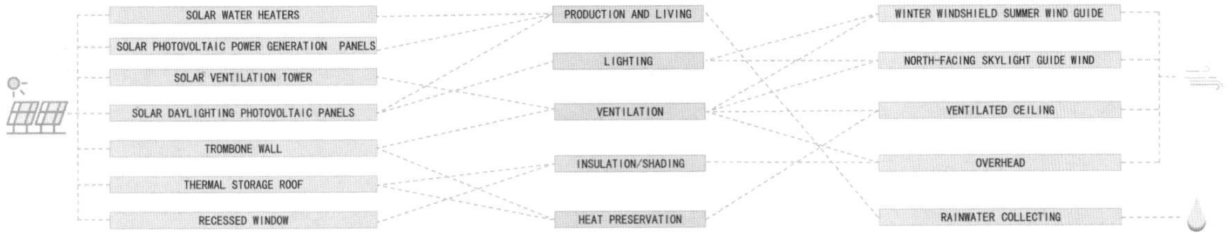

SOLAR WATER HEATERS
SOLAR PHOTOVOLTAIC POWER GENERATION PANELS
SOLAR VENTILATION TOWER
SOLAR DAYLIGHTING PHOTOVOLTAIC PANELS
TROMBONE WALL
THERMAL STORAGE ROOF
RECESSED WINDOW
PRODUCTION AND LIVING
LIGHTING
VENTILATION
INSULATION/SHADING
HEAT PRESERVATION
WINTER WINDSHIELD SUMMER WIND GUIDE
NORTH-FACING SKYLIGHT GUIDE WIND
VENTILATED CEILING
OVERHEAD
RAINWATER COLLECTING

PERSPECTIVE ANALYSIS

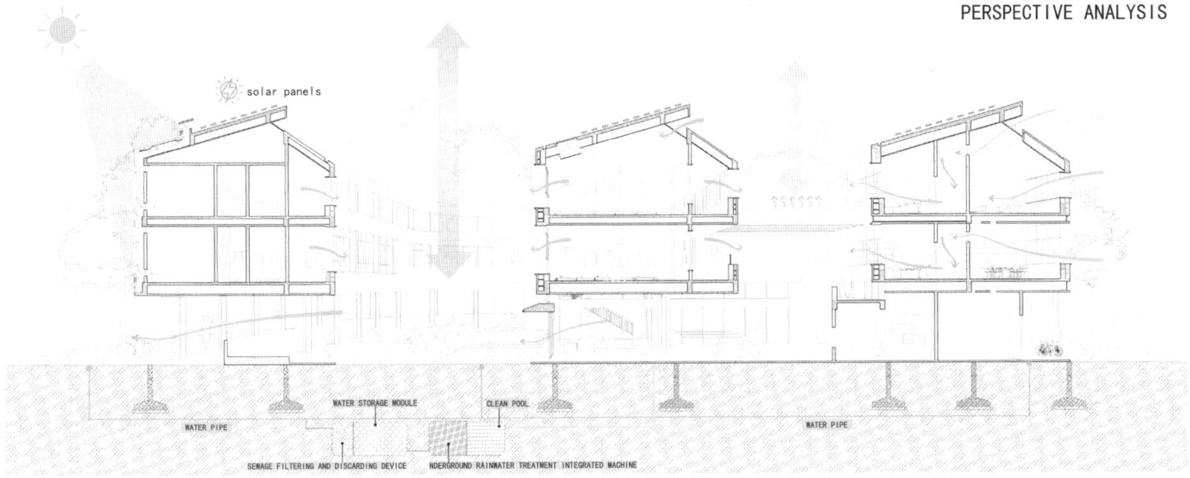

solar panels
WATER STORAGE MODULE
CLEAN POOL
WATER PIPE
WATER PIPE
SEWAGE FILTERING AND DISCARDING DEVICE
NDERGROUND RAINWATER TREATMENT INTEGRATED MACHINE

CONSTRUCTION DETAILS

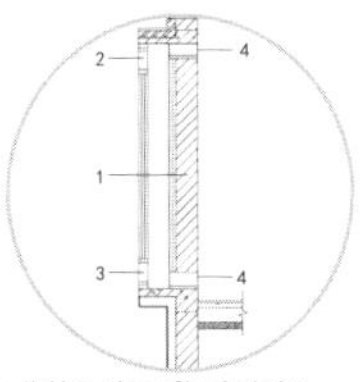

1: Hollow glass fixed window
Air cavity
Brush with matt black paint
Paint 25 thick M5 cement mortar
Outer wall
2: Outdoor air outlet
3: Outdoor air return duct
4: Finished shutters

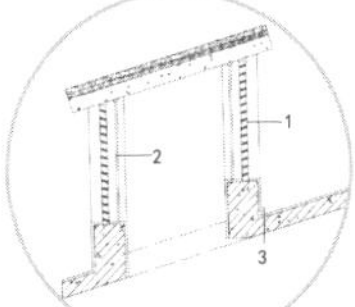

1: Movable blinds
2: Metal insect net
3: Waterproof mortar

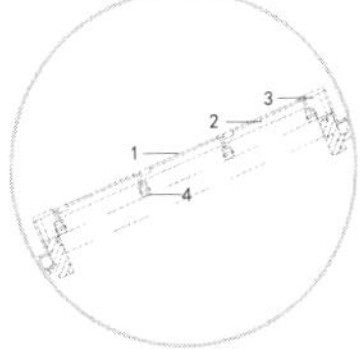

1: Photovoltaic module
2: Main structure
3: Aluminum main beam
4: Aluminum secondary beam

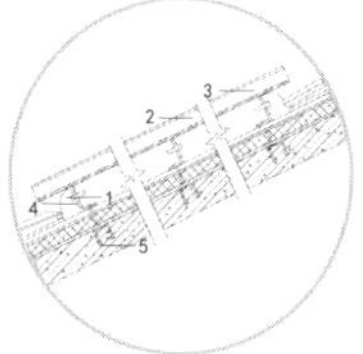

1: 1:3 edge sealing with cement mortar
2: Photovoltaic module
3: Junction Box
4: Steel frame
5: Embedded parts120*120*8

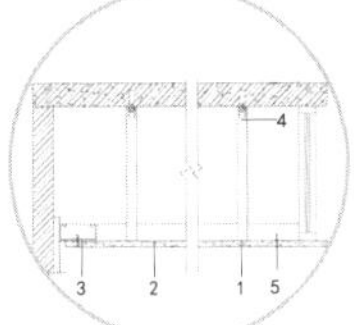

1: Gypsum board
2: Self-tapping screws
3: Cross brace
4: U-shaped clip
5: Secondary keel

阳光之下 · 田园之上

SUNSHINE POURING · CHILDREN GETING

05

A

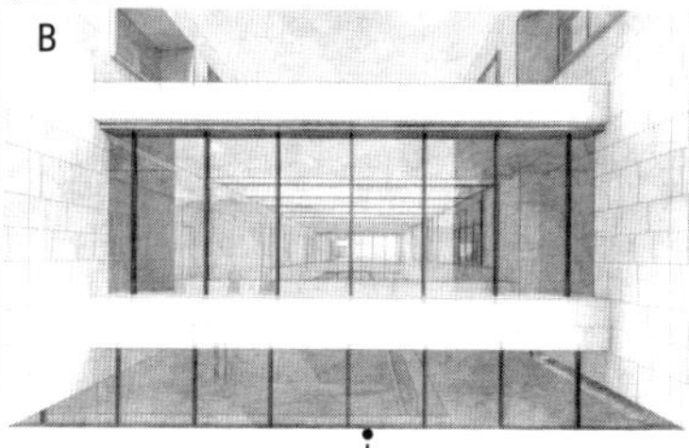
B

C

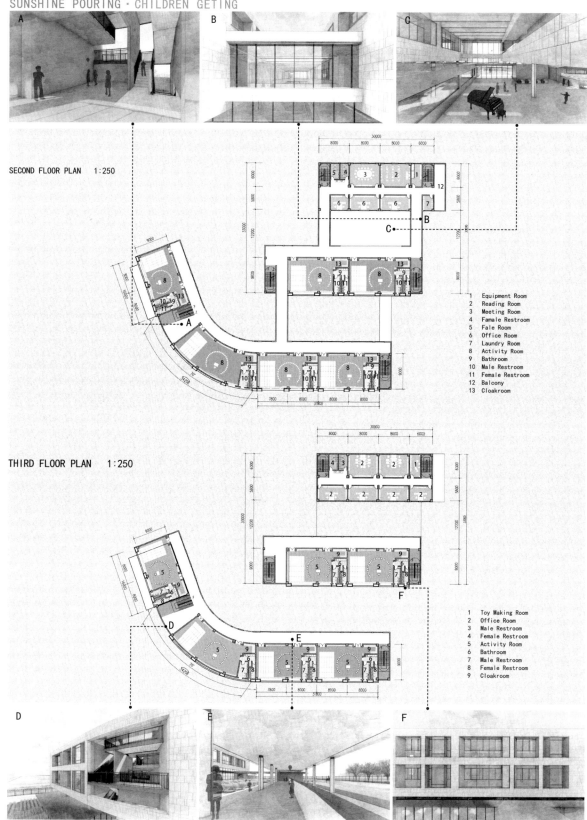

SECOND FLOOR PLAN 1:250

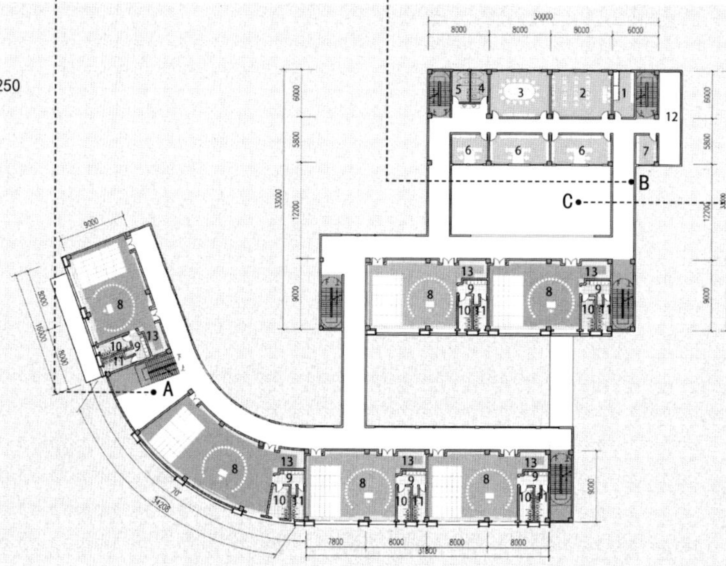

1 Equipment Room
2 Reading Room
3 Meeting Room
4 Famale Restroom
5 Fale Room
6 Office Room
7 Laundry Room
8 Activity Room
9 Bathroom
10 Male Restroom
11 Female Restroom
12 Balcony
13 Cloakroom

THIRD FLOOR PLAN 1:250

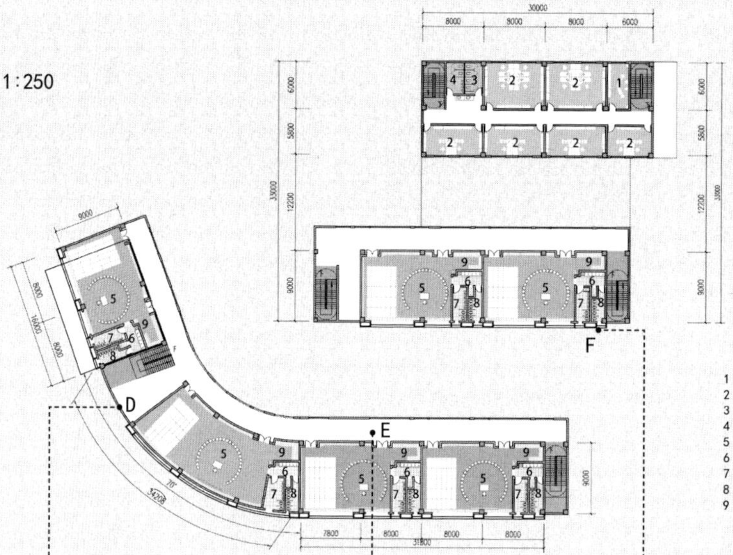

1 Toy Making Room
2 Office Room
3 Male Restroom
4 Female Restroom
5 Activity Room
6 Bathroom
7 Male Restroom
8 Female Restroom
9 Cloakroom

D

E

F

阳光之下 · 田园之上
SUNSHINE POURING · CHILDREN GETING

06

SPATIAL LOCATION ANALYSIS

Space A+B　　SCENE1: TODDLER SPORTS

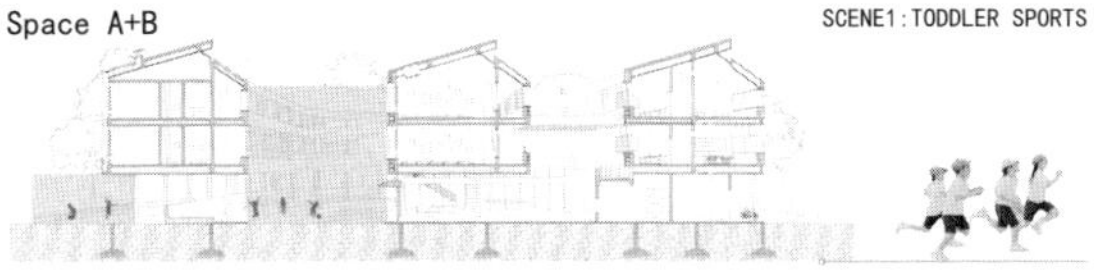

Space B　　SCENE2: PASTORAL EXPERIENCE

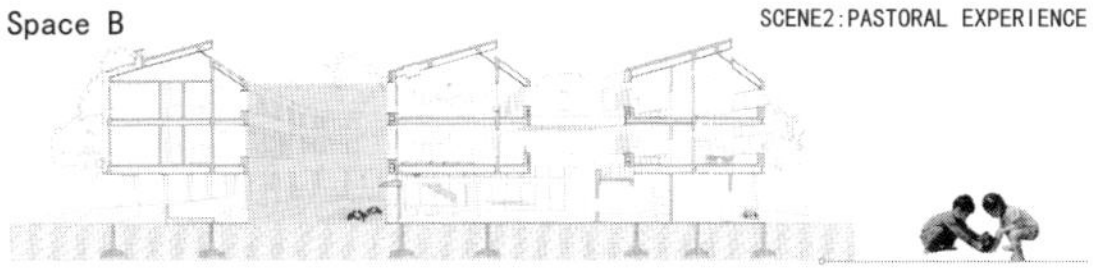

Space C　　SCENE3: INDOOR ACTIVITIES

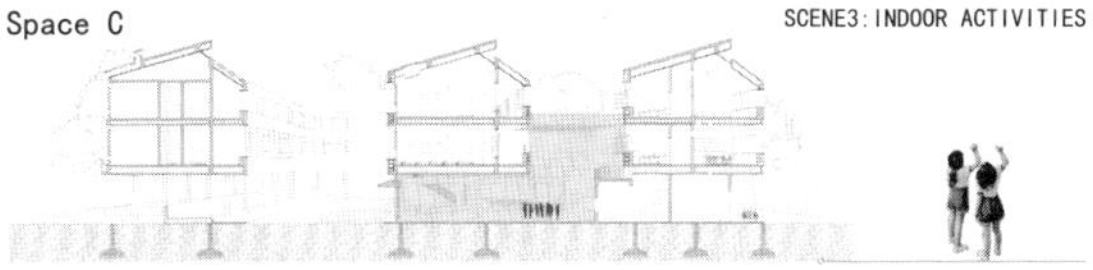

Space C　　SCENE4: SHOW AND VISIT

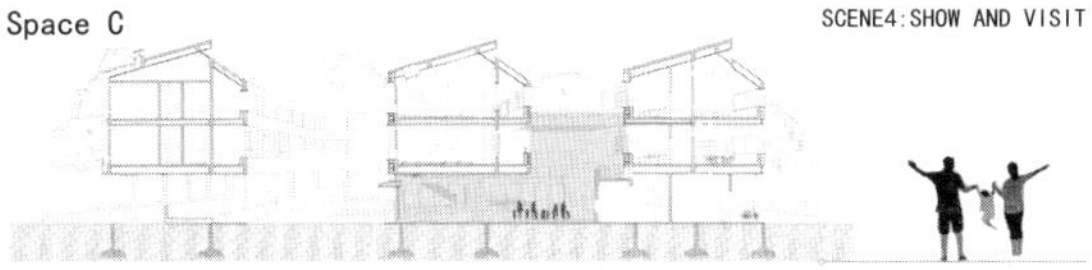

Space A+B+C　　SCENE5: FREE TIME

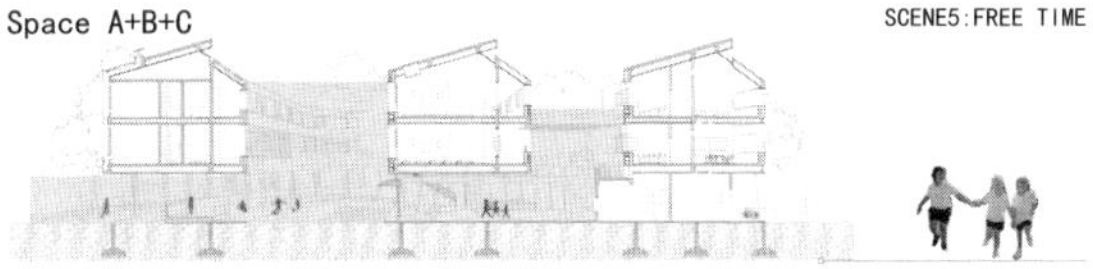

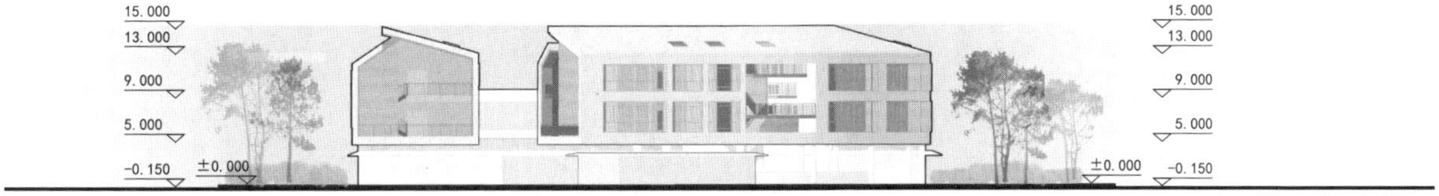
15.000
13.000
9.000
5.000
-0.150
±0.000

West Elevation 1:200

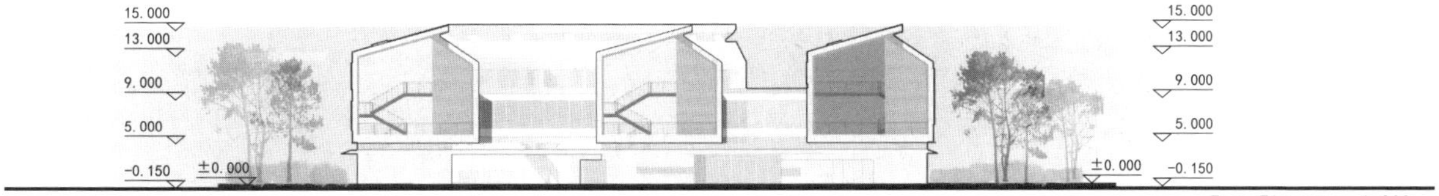
15.000
13.000
9.000
5.000
-0.150
±0.000

East Elevation 1:200

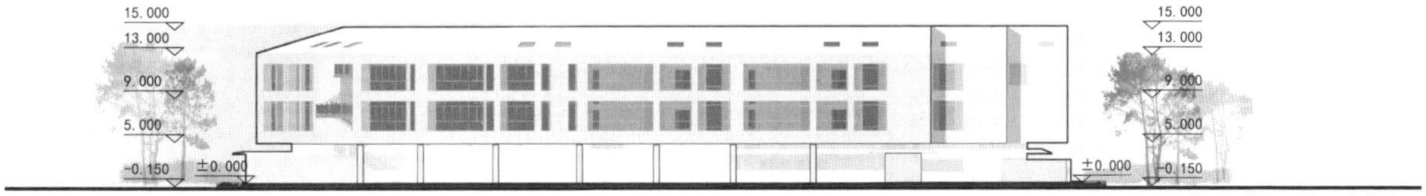
15.000
13.000
9.000
5.000
-0.150
±0.000

South Elevation 1:200

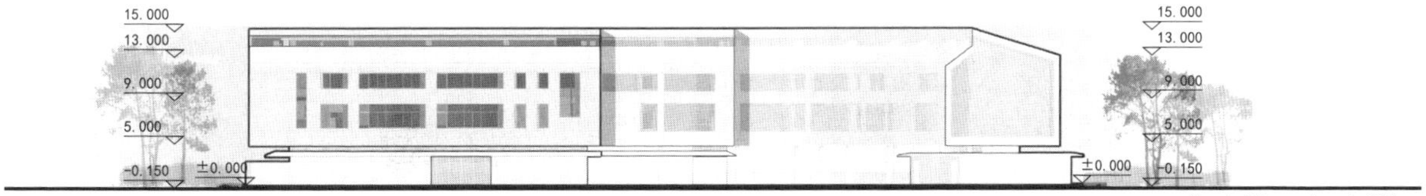
15.000
13.000
9.000
5.000
-0.150
±0.000

North Elevation 1:200

综合奖 · 优秀奖
General Prize Awarded · Honorable Mention Prize

注 册 号：7061

项目名称：晨 · 星（南平）

Sun & Star（Nanping）

作　　者：邹道圳、陈树强、黄宇敏、严淑园

参赛单位：惠州学院

指导教师：黄汇雯

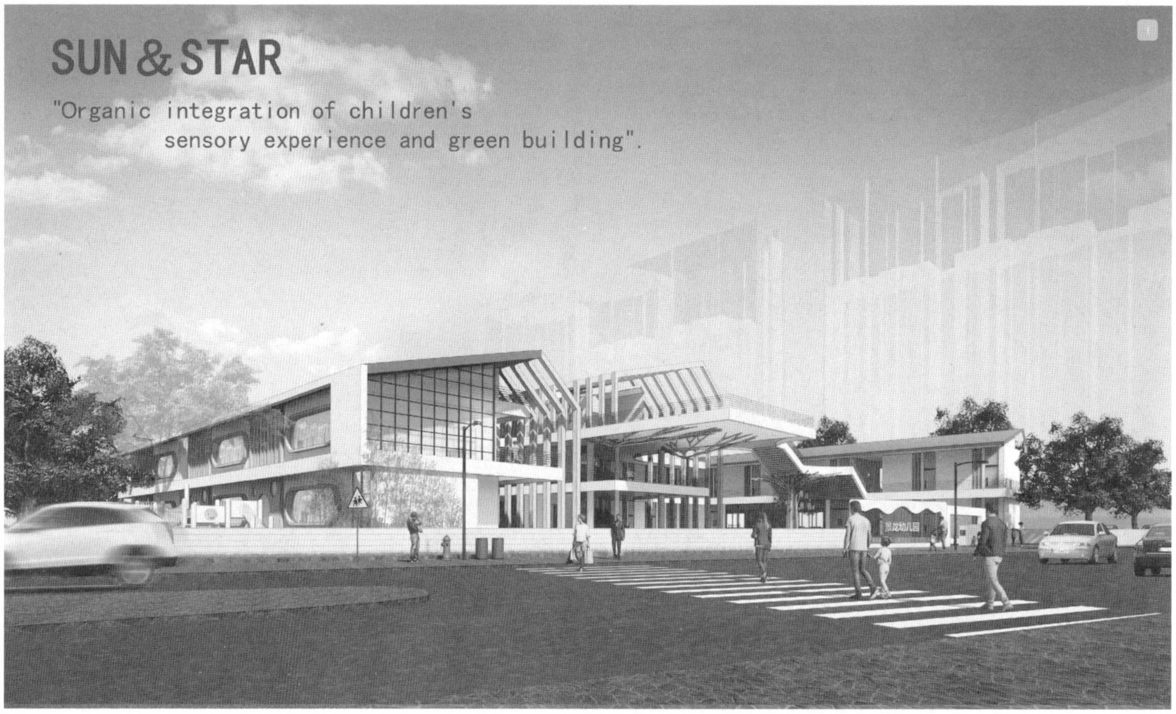
SUN & STAR
"Organic integration of children's
sensory experience and green building".

Preliminary Analysis

■ Location

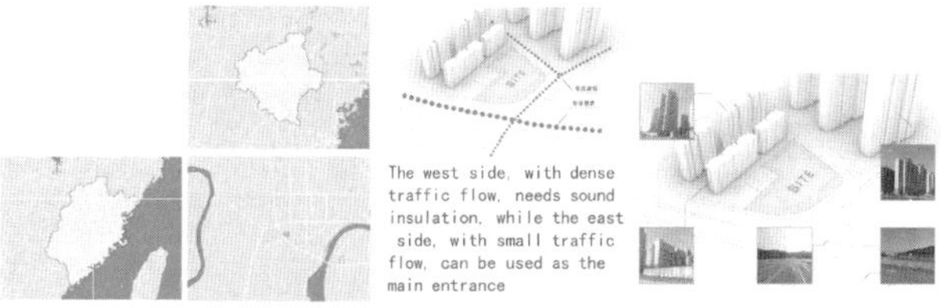

The west side, with dense traffic flow, needs sound insulation, while the east side, with small traffic flow, can be used as the main entrance

The project is located in the new west city of Jianyang District, Nanping City, Fujian Province. There are three roads around the base, north is residential area, east has high-rise residential area, west wide view

■ Extraction Element

Hakkas round house	Earth-buildings	Gargoyle	Phyllostachys edulis
⇩	⇩	⇩	⇩
roof form	The atrium intention	The atrium drainage	Corridor decoration

Starting from the extraction of local elements, the architectural design organically integrates cultural elements into the detailed design of the corridor facade, in which the public activity spacein the atrium combines the gargoyles with the intention of the enclosure, creating the effect of rain-curtain enclosure

■ Form-Creation

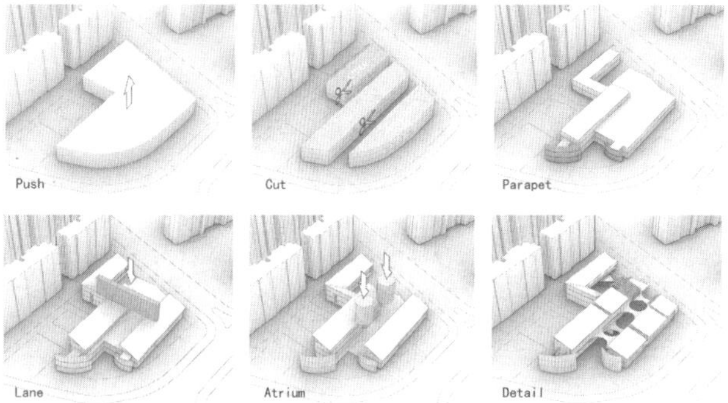
Push
Cut
Parapet
Lane
Atrium
Detail

■ Design Concept

The plan is based on the concept of "organic integration of children's sensory experience and green building".

Based on the traditional architectural culture of Nanping, a kindergarten designed to stimulate children's five senses experience is designed.

Through wind environment simulation and sunshine analysis to assist in the design of buildings, and use solar energy, solar chimneys, venturi effect, cold alleys and other green building techniques to create a green kindergarten with a comfortable environment.

Combining the traditional architectural cultural elements of Fujian, "enclosed house" and "dripping beast", to create.A fun public activity space with multi-sensory experience-"Rain Screen Wall".

方案以“ 幼儿感官体验与绿色建筑有机融合”为理念。基于南平传统建筑文化，设计一个能激发儿童五感体验的幼儿园。

通过风环境模拟及日照分析辅助设计建筑，并运用太阳能、太阳能烟囱、文丘里效应、冷巷等绿色建筑手法打造具有舒适环境的绿色幼儿园。

结合福建传统建筑文化要素“围屋”与“滴水兽”，营造具有多感官体验的趣味公共活动空间——“雨幕围屋”。

■ Economic and Technical Norms

Land area: 6028 m²	Area covered: 1402 m²
Construction area: 4515 m²	Plot ratio: 0.75
Building height: 14.7 m	Green area: 2290 m²
Building layers: 3 floors	Greening rate: 38.0%

■ Strategy

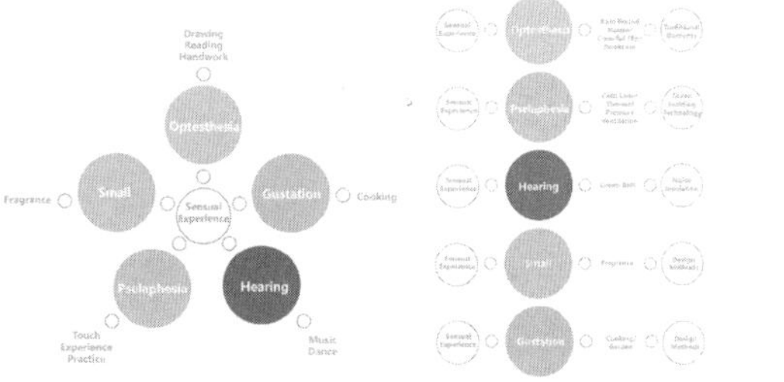
Drawing
Reading
Handwork
Optesthesia
Smell
Gustation
Cooking
Fragrance
Sensual Experience
Hearing
Touch
Experience
Practice
Music
Dance

Streamline Analysis

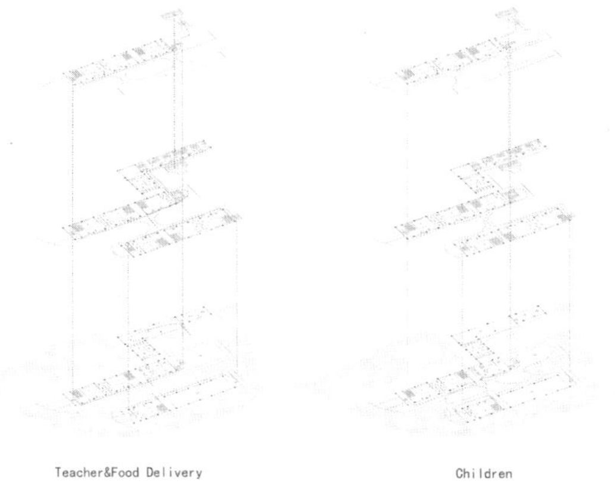
Teacher&Food Delivery
Children

Site Plan 1：500

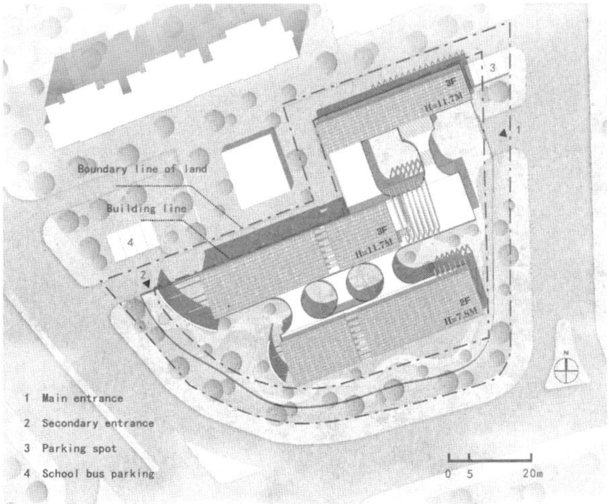
Boundary line of land
Building line
3F H=11.7M
3F H=11.7M
2F H=7.8M
1 Main entrance
2 Secondary entrance
3 Parking spot
4 School bus parking
0 5 20m

Explosive View

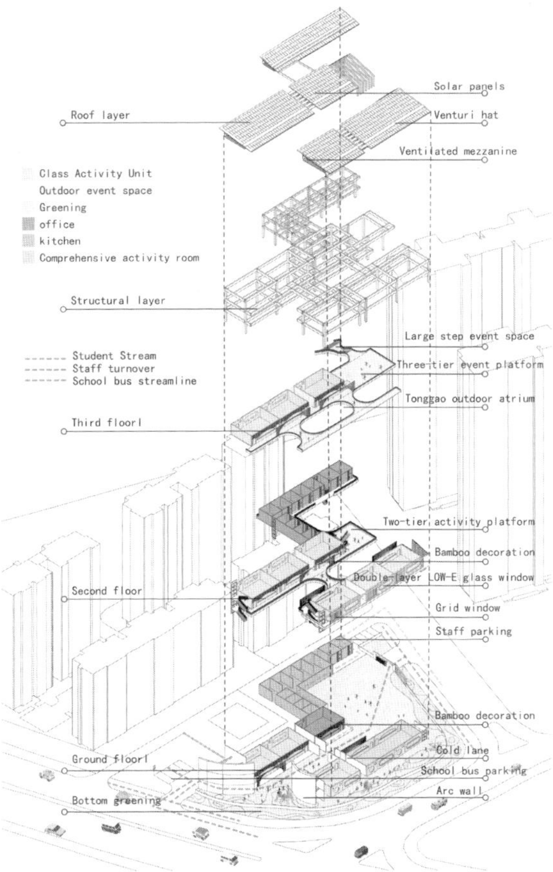
Solar panels
Roof layer
Venturi hat
Ventilated mezzanine
Class Activity Unit
Outdoor event space
Greening
office
kitchen
Comprehensive activity room
Structural layer
Large step event space
Student Stream
Staff turnover
School bus streamline
Three-tier event platform
Tonggao outdoor atrium
Third floorl
Two-tier activity platform
Bamboo decoration
Double-layer LOW-E glass window
Second floor
Grid window
Staff parking
Bamboo decoration
Cold lane
Ground floorl
School bus parking
Arc wall
Bottom greening

01 Active Region
02 Washroom
03 Lavatory
04 Locker Room
05 Activity Corner
06 Teachers' Office
07 Women's Room
08 Men's Room
09 Accessible Toilet
10 Water Heater Room
11 Storeroom
12 Washhouse
13 Kitchen
14 Health Room
15 Concierge
16 Open Lobby
17 Activity Room

1st Floor Plan 1:200

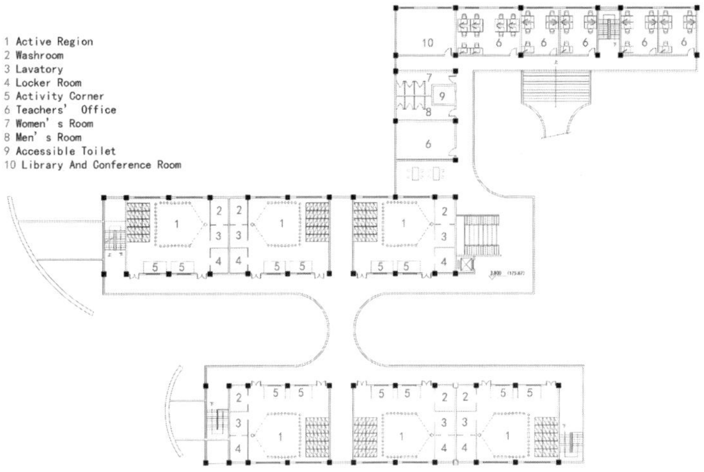
1 Active Region
2 Washroom
3 Lavatory
4 Locker Room
5 Activity Corner
6 Teachers' Office
7 Women's Room
8 Men's Room
9 Accessible Toilet
10 Library And Conference Room

2nd Floor Plan 1:200

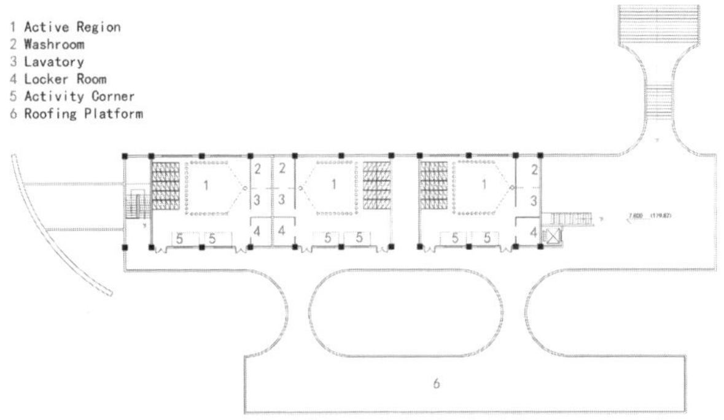
1 Active Region
2 Washroom
3 Lavatory
4 Locker Room
5 Activity Corner
6 Roofing Platform

3rd Floor Plan 1:200

Extraction Element

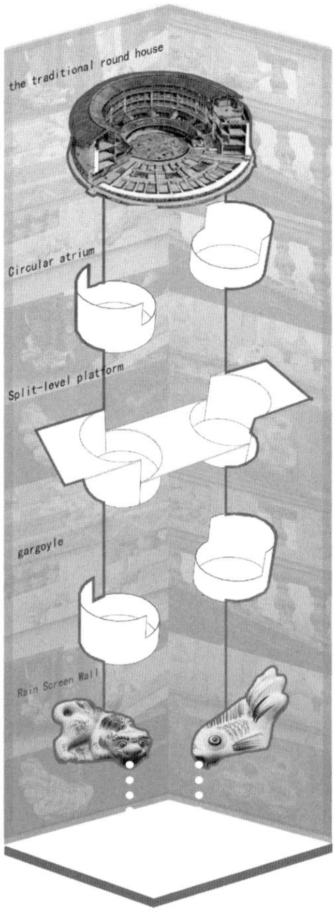
the traditional round house
Circular atrium
Split-level platform
gargoyle
Rain Screen Wall

Sensory Experience Analysis

What are you going to eat today? ! I smell the scent! ! Let me take a look.
KITCHEN

Transparent kitchen arouses children's curiosity and stimulates appetite.

Every time pass by after school I want to put it Neat

The rainbow-colored carousel attracts children to have fun and stimulate their creativity.

I like to be with my sister Watching Ultraman in the sun !

Classroom color scheme adopts a combination of white, wood and color.

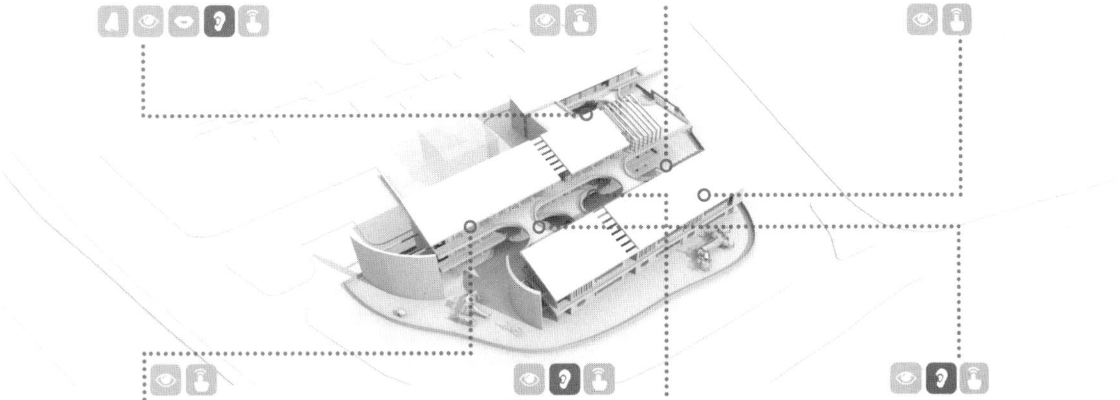

I heard this flying thing behind it's made of bamboo grown by grandpa!

The exterior walls and corridors are decorated with local bamboo, creating a rich visual and spatial experience.

Why didn't it rain today?! still want to watch fish together Spit water, but also want to be under the rainbow play with Xiaohong.

The circular atrium created by the misaligned platform is surrounded by dripping beasts. When it rains, it creates a unique atmosphere of the enclosure.

Which monster is Altman going tofight today?

Along the corridor, a colorful cabinet with storage and rest functions is designed for children.

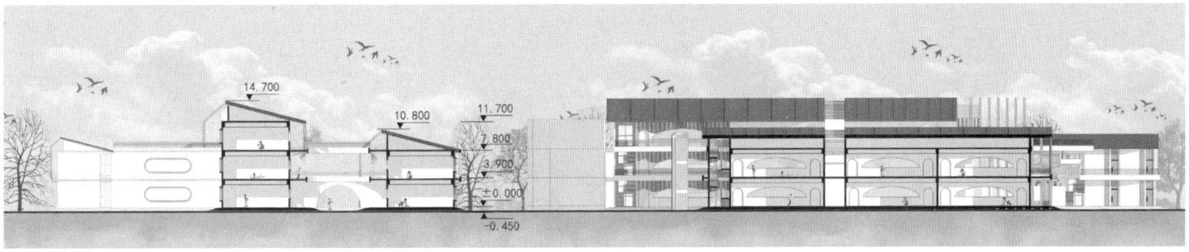
14.700
10.800
11.700
7.800
3.900
±0.000
-0.450

1-1 Section View 1：200

2-2 Section View 1：200

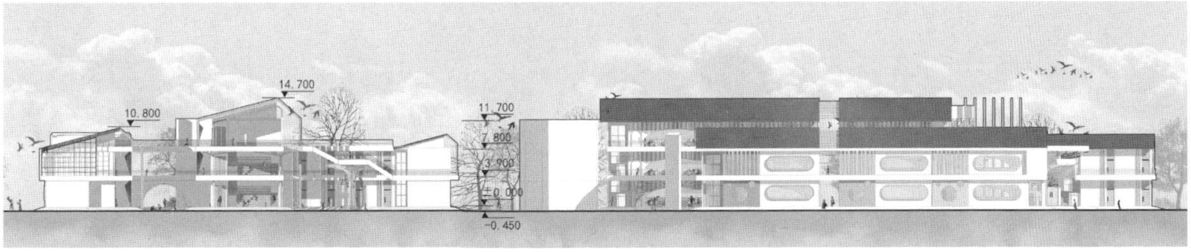
14.700
10.800
11.700
7.800
3.900
±0.000
-0.450

East Elevation 1：200

South Elevation 1：200

Green Construction Analysis and Strategy

Base Analysis

Year-round wind rose illustration

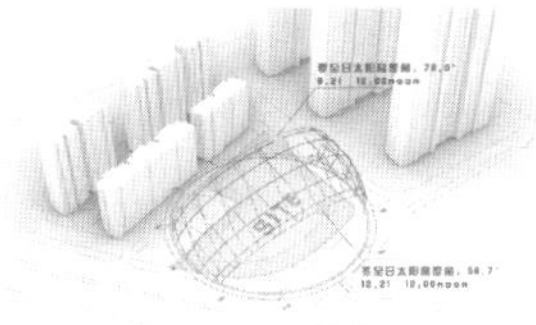

Solar altitude angle

On the summer solstice, the solar altitude is 78° ,and on the winter solstice, the solar altitude is 58.6° .The site is covered by tall buildings to the east and north, while the south and west are wider.

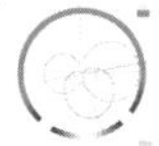

Combined with the site's sunshine, ventilation and other factors, the best orientation of the building is obtained: 10° south-east.

There are winds from all directions in Nanping all year round, and the winds are widely distributed, mainly from northeast to southwest. Southeastof summer. The wind to the southwest is the dominant direction, and the northeast wind is the winter lead the way. Buildings on the north and east will block the wind.

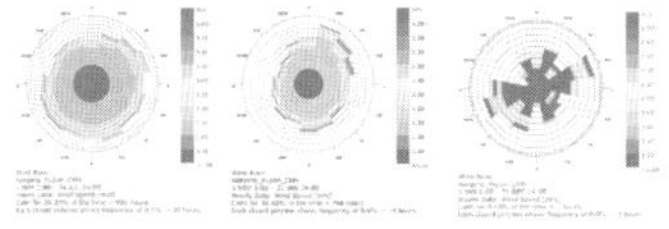

Summer Winter Able to meet natural ventilation

By setting the conditions, the situation that satisfies the temperature>24°C wind speed>2m/s humidity>70% is located to meet the situation of natural ventilation. The directions that can meet the natural ventilation conditions are mainly northeast and southwest.

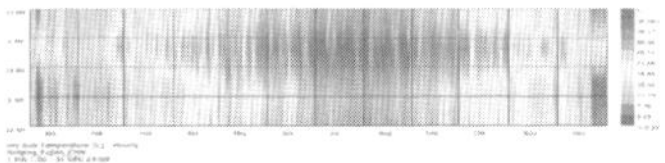

The summer is extremely hot, sultry, and cloudy; the winter is cold and partly cloudy; and it is wet all year round.

Passive Design Strategy

This design is mainly passive design, through the use of arc walls, cold alleys, venturi hats, overhead ventilation layers, eaves, atrium and other design methods to reduce energy consumption for architectural lighting, heating and air-conditioning.

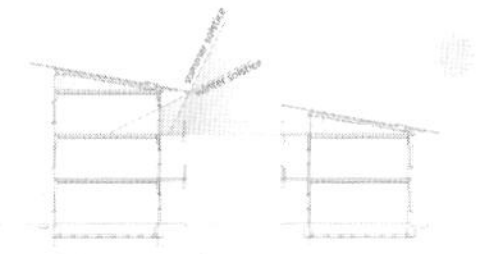

Solar altitude angle

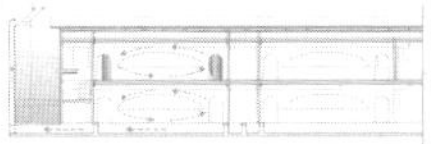

Ventilated passive wall (Daytime)

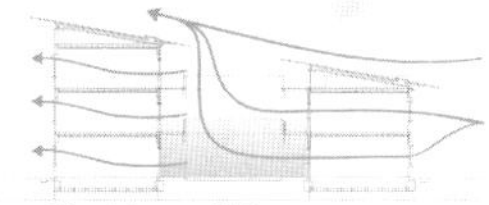

The atrium ventilation

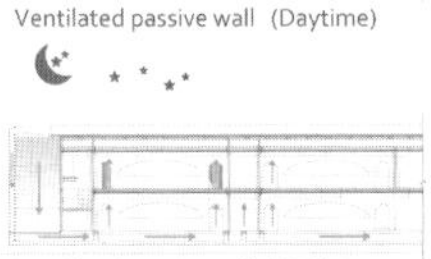

Ventilated passive wall (Night)

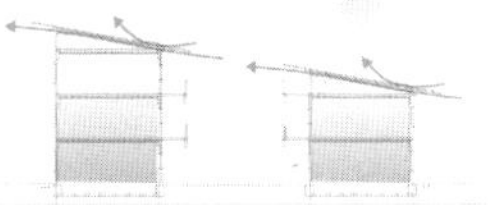

Venturi cap

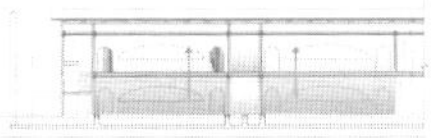

Lane

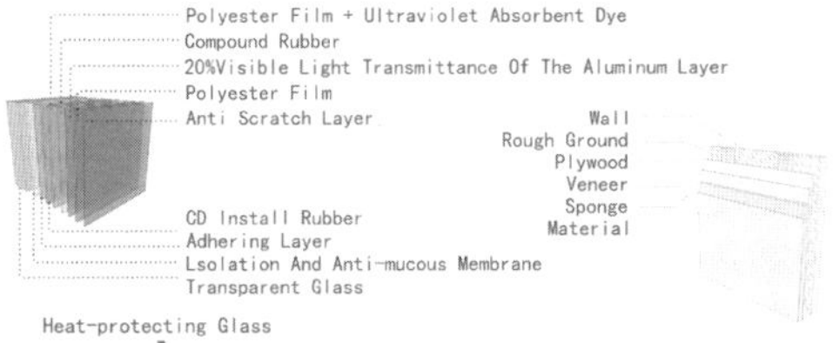
Polyester Film + Ultraviolet Absorbent Dye
Compound Rubber
20%Visible Light Transmittance Of The Aluminum Layer
Polyester Film
Anti Scratch Layer
CD Install Rubber
Adhering Layer
Lsolation And Anti-mucous Membrane
Transparent Glass
Wall
Rough Ground
Plywood
Veneer
Sponge
Material

Heat-protecting Glass

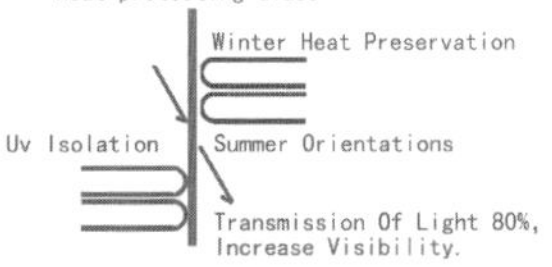
Winter Heat Preservation
Uv Isolation
Summer Orientations
Transmission Of Light 80%,
Increase Visibility.

In terms of materials, we have adopted a heat-insulating glass system to improve the building's thermal insulation capabilities and use sound insulation walls to block the noise of the west side roads, so as to improve the building's comfort.

Proactive Design Strategy

The building uses active solar technology on the roof and solar energy conversion and control devices are placed in the mezzanine of the sloped roof. In addition, the ground source heat pump technology is used, which is cooled by fan coils in summer and heated by floor heating pipes in winter, so that the indoor temperature of the kindergarten is maintained at 20-26°C all year round.

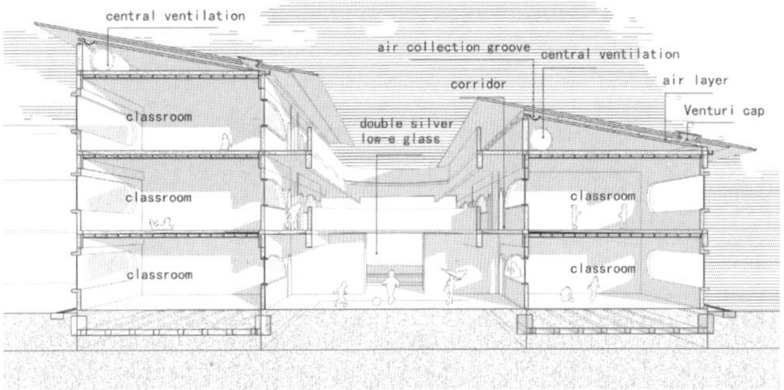
central ventilation
air collection groove
central ventilation
corridor
air layer
Venturi cap
double silver
low-e glass
classroom
classroom
classroom
classroom
classroom

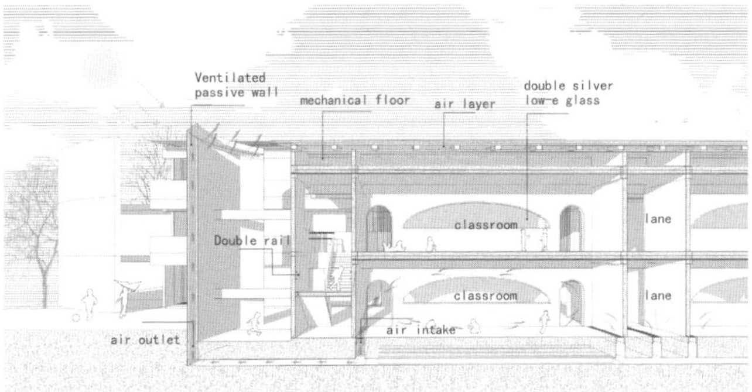
Ventilated
passive wall
mechanical floor
air layer
double silver
low-e glass
Double rail
classroom
classroom
lane
lane
air outlet
air intake

Ventilation Simulation

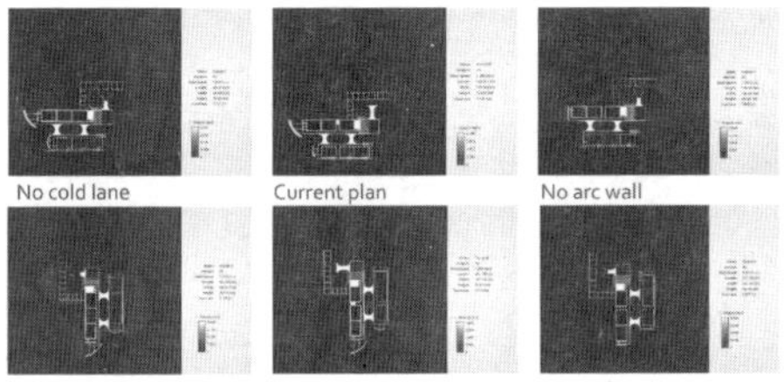
No cold lane
Current plan
No arc wall

By using the control variable method, the actual effects of cold lanes and arc walls in the plan are compared.Using the data obtained from the previous survey, the summer average wind speed of 1.3m/s was used to simulate the wind in the southeast and southwest directions. It was concluded that the cold lane and arc wall played a key role in increasing the wind speed of the atrium. The embodiment of passive design.

SunlightHours Analysis

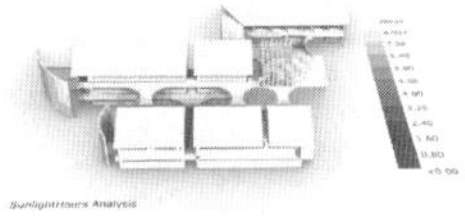

Ladybug software is used to simulate the sunshine of the building, and it can be concluded that the classroom of the kindergarten meets the requirements of sunshine regulations and the cold lane and arc wall can block the sunshine and play a cooling role.

Green Construction Node Renderings

It's blowing!
Let's go
kite flying!

The Venturi effect is used in the atrium of the building to enhance the ventilation effect of the atrium.

Why is it so
cool here?And Xiaohong
and Xiaolan what to
play outside! ?

The cold alley space is not only an important activity space but also an important green building element.

It's super
wide here !The sun
cannot reach us again.
We like to play here
the most!

Create a gray space in the entrance space to allowchildren to play here without having to be exposed to the sun.

■ Reduce Thermal Bridges

The separation of the arc wall and the main body of the building can greatly reduce the direct heating of the building by heat, and reduce the indoor and outdoor temperature difference between late autumn and early winter due to poor indoor ventilation, frequent contact with cold and hot air, and uneven heat conduction of the wall insulation layer. The thermal bridge effect causes condensation, mildew and even dripping on the interior walls of the house.

■ Venturi Hat

The venturi effect is used to enhance the ventilation effect of the double-sloped roof, which can quickly take away the heat. The Venturi effect means that a low pressure will be generated near the high-speed flowing fluid, thereby generating adsorption.

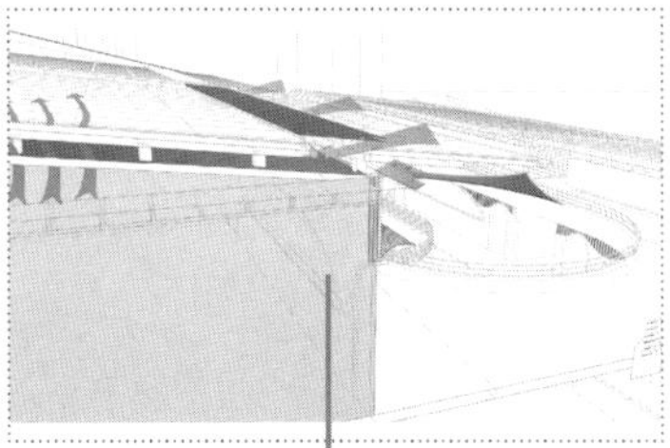

■ Atrium Pull

The air flows from a place with a high density to a place with a low density. Due to the blockage of the building, the air temperature in the lower part of the patio is low and the density is high, and the temperature in the upper part is high and the density is small.

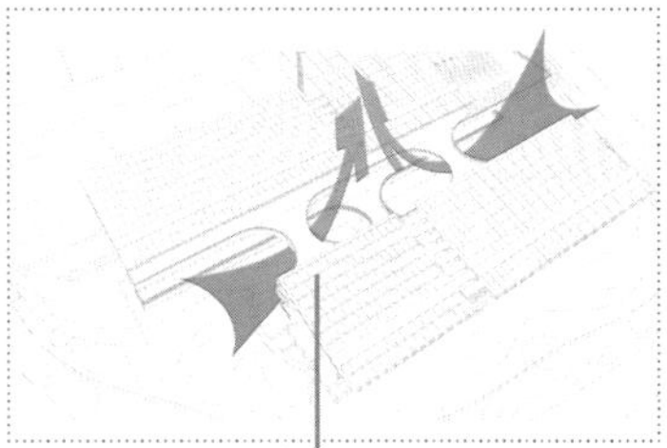

■ Photovoltaic Solar

Facility that converts solar energy into DC power by using photovoltaic effect of photovoltaic semiconductor material (photovoltaic effect refers to the phenomenon that semiconductor generates electromotive force when exposed to light). The core of a photovoltaic facility is a solar panel. The site of this kindergarten is located in Nanping, Fujian. The solar energy resources belong to the third type of area. Assuming that the average is full of 4 hours per day, about 1200 hours per year, it takes about 5 years to recover costs.

■ Node Analysis

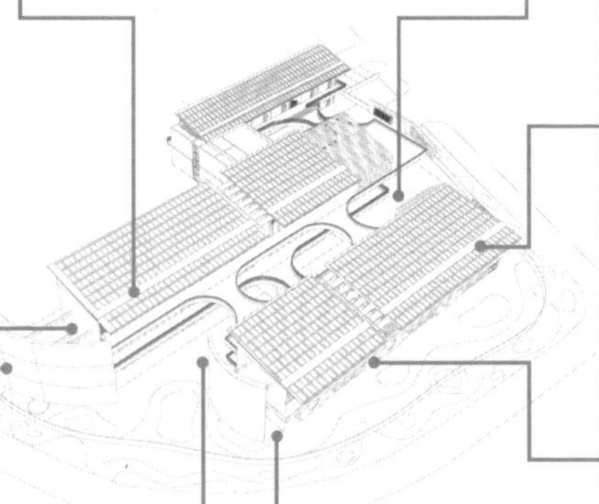

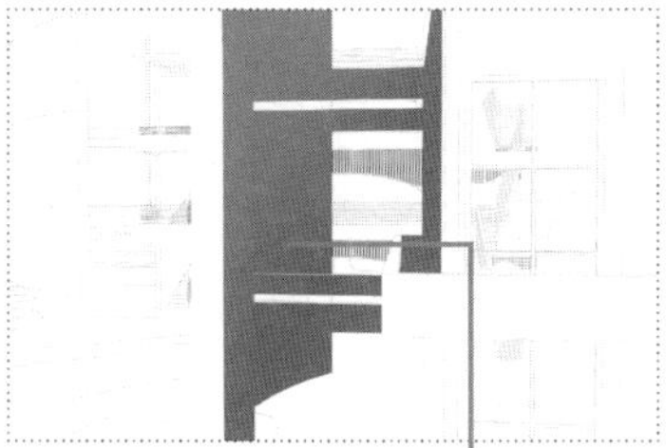

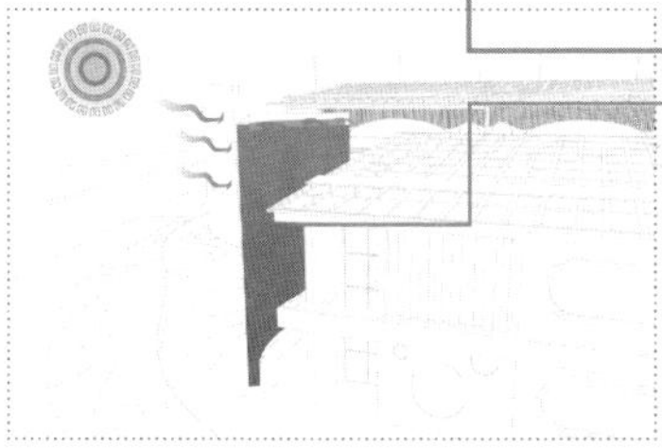

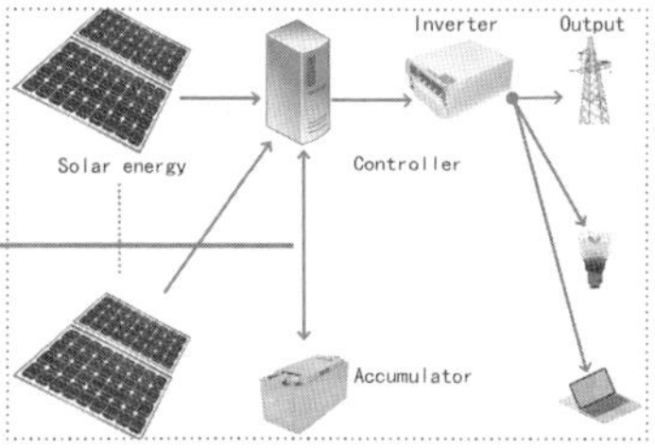
Inverter
Output
Solar energy
Controller
Accumulator

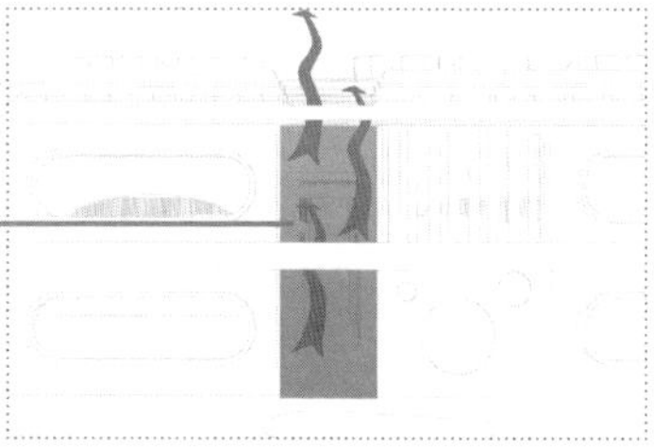

■ Curved Wall

The design of the arc wall is mainly aimed at strengthening the indoor air circulation and preventing the west sun and the noise of the west side road. Among them, the effect of solar chimney is used, and the hollow wall is heated by the sun to generate internal lift, and finally the indoor turbid air is discharged.

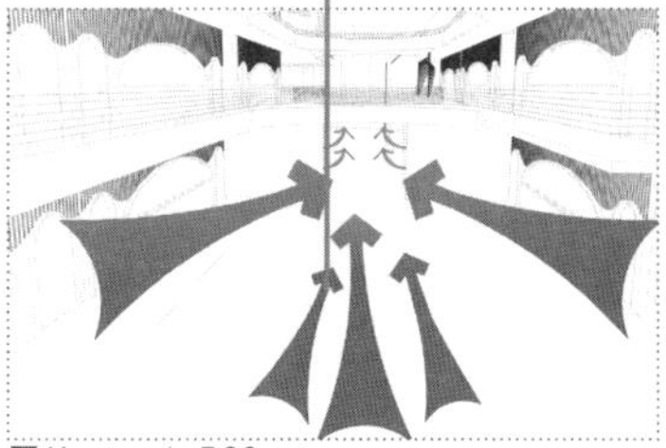

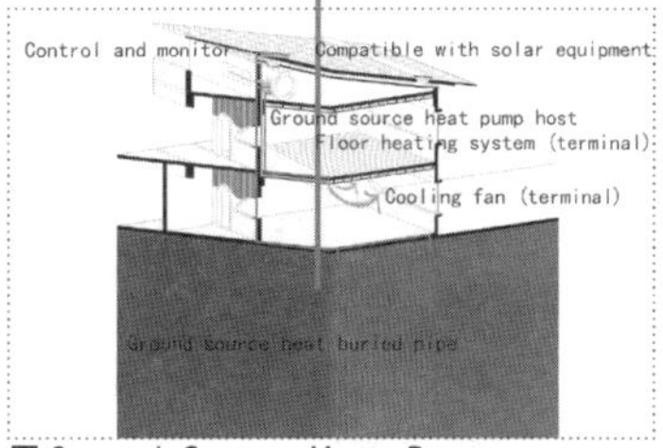
Control and monitor
Compatible with solar equipment
Ground source heat pump host
Floor heating system (terminal)
Cooling fan (terminal)
Ground source heat buried pipe

■ Cold Lane

The cold alley of the building is produced by the activities of children's venues combined with the local traditional architectural wisdom. Through the narrow cold alley, a cool activity space is created between the building blocks. At the same time, the combination of hot-pressed ventilation and the venturi effect of the atrium can make the cold alley produce more physical cooling effect.

■ Venturi Effect

The atrium space cooperates with the arc wall and the cold lane to produce the Venturi effect. In layman's terms, this effect means that a low pressure will be generated near the high-speed flowing fluid, thereby generating adsorption.

■ Ground Source Heat Pump

The ground source heat pump is an air conditioning system that uses shallow underground geothermal resources for cooling and heating. The operating conditions of the ground source heat pump are very stable. In summer, it is cooled by fan coils, and in winter, it is heated by ground heating pipes. The cooling and heating effect is good. The indoor temperature is maintained at 20-26° C all year round.

■ Rainwater Recycling System

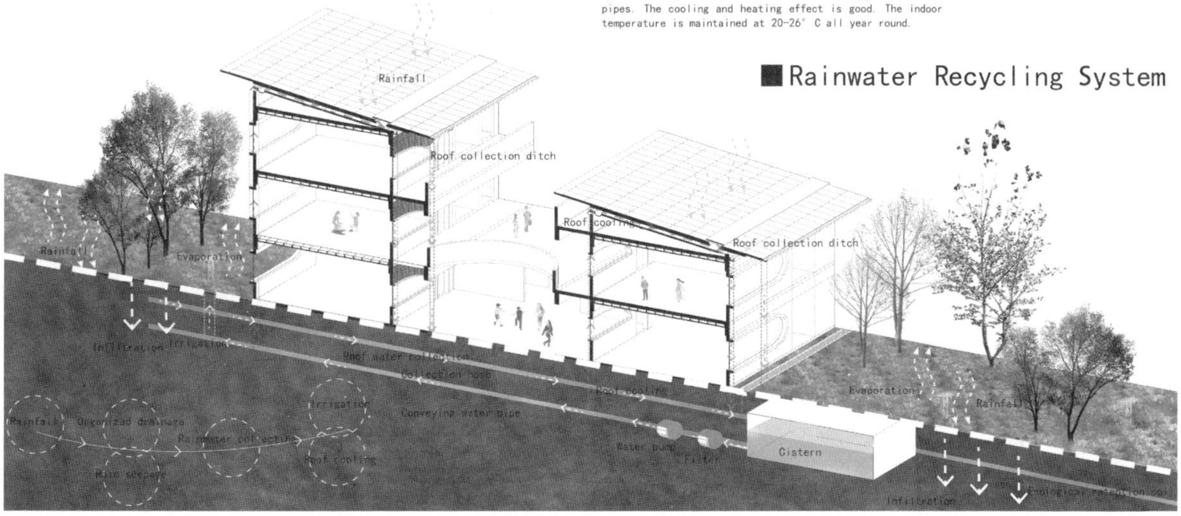
Rainfall
Roof collection ditch
Roof cooling
Roof collection ditch
Rainfall
Evaporation
Infiltration
Irrigation
Roof water collection
Collection hose
Conveying water pipe
Irrigation
Rainfall
Organized drainage
Rainwater collection
Roof cooling
Roof cooling
Water pump
Filter
Cistern
Evaporation
Rainfall
Infiltration

阳光·稚梦
童之梦·光之巢 壹
The Dream of Children · the · Nest of Light
GEOGRAPHICAL SITE
Nanyang City
Jianyang District
City Location
Base Location
CLIMATE ANALYSIS
Average daily minimum temperature(℃)
15 17 21 26 30 32 36 35 33 28 22 16
Average daily minimum temperature(℃)
7 8 12 17 21 25 26 26 24 18 14 8
Historical maximum temperature(℃)
26 28 31 37 37 38 40 38 37 36 30 27
Historical minimum temperature(℃)
-4 -2 4 6 14 18 22 20 16 0 1 -1
average rainfall (mm)
60 104 183 213 282 283 136 137 97 65 46 41
Mean wind temperature
DESIGN DESCRIPTION
设计说明：
幼儿园的设计针对当地气候、地形条件，规划建筑群体布局、建筑设计上，以六边形为单元模块进行排列组合。为了获得最佳日照采用凹形建筑形式，以幼儿活动特点设计室内空间，被动式设计上，南向设置的由六边形两条边组成的大开角“V”形窗可以在冬季获得足够的太阳热能，利用挑檐等遮阳，避免夏季室内温度过高，同时在建筑的中间部位设置了廊院以便通风降温，主动式设计上，采用屋面太阳能集热系统，与雨水收集系统。产业化上，主要采用了装配式框架大阪结构和模数化建造，用盛产的材料作为主要材料。
According to the local climate and topographical conditions, the kindergarten is arranged and combined with hexagon as unit module in the planning of architectural group layout and architectural design. In order to obtain the best sunshine, concave building form is adopted to design the indoor space according to the characteristics of children's activities. In passive design, the large open angle "V" window composed of hexagon and two sides is set in the south direction, which can obtain enough solar heat energy in winter. The overhanging eaves and other sunshade are used to avoid the indoor temperature being too high in summer. Meanwhile, a corridor is set in the middle of the building for ventilation and cooling. In active design, rooftop solar collector system and rainwater collection system are adopted. In terms of industrialization, it mainly adopts prefabricated frame Osaka structure and modular construction, with abundant materials as the main materials.
ARCHITECTURAL CONCEPT
Average temprature
Relative humidity
Direct solar radiation
Scattered radiation
BASE WEEKLY DATA
Average temperature(℃)
Maximum temperature (℃)
Minimum temperature (℃)
Relative humidity (%)
Direct solar radiation(W/㎡)
Diffuse solar radiation (W/㎡)
Average wind speed (km/h)
Average cloud cover(%)
Average daily rainfall(mm)
ARCHITECTURAL CONCEPT
Fujian Tulou
prototype
circle
tangency
hexagon
transfrom
result
Variety of topological directions
Good topological property
Good topological relationship

综合奖·优秀奖
General Prize Awarded · Honorable Mention Prize

注 册 号：7082

项目名称：童之梦·光之巢（南平）
The Dream of Children · the Nest of Light（Nanping）

作　　者：王　宇、李　婧、何玉龙、赵嘉诚

参赛单位：中国矿业大学（北京）

指导教师：贺丽洁

阳光 · 稚梦

童之梦 · 光之巢 贰

The Dream of Children · the · Nest of Light

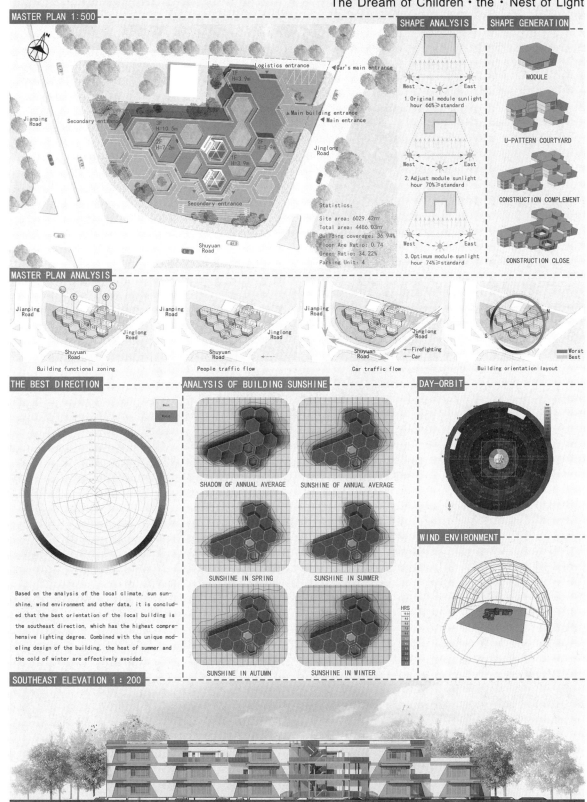

MASTER PLAN 1:500

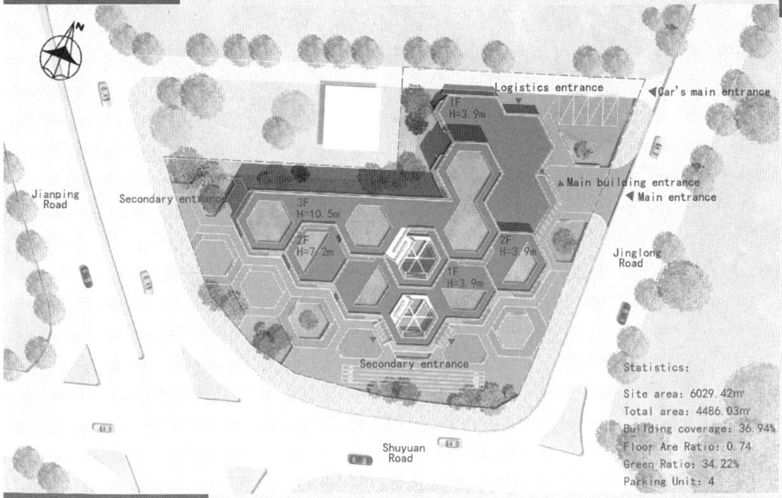

Logistics entrance
Car's main entrance
1F
H=3.9m
Jianping Road
Secondary entrance
Main building entrance
Main entrance
3F
H=10.5m
2F
H=7.2m
2F
H=3.9m
1F
H=3.9m
Jinglong Road
Secondary entrance
Shuyuan Road
Statistics:
Site area: 6029.42㎡
Total area: 4486.03㎡
Building coverage: 36.94%
Floor Are Ratio: 0.74
Green Ratio: 34.22%
Parking Unit: 4

SHAPE ANALYSIS

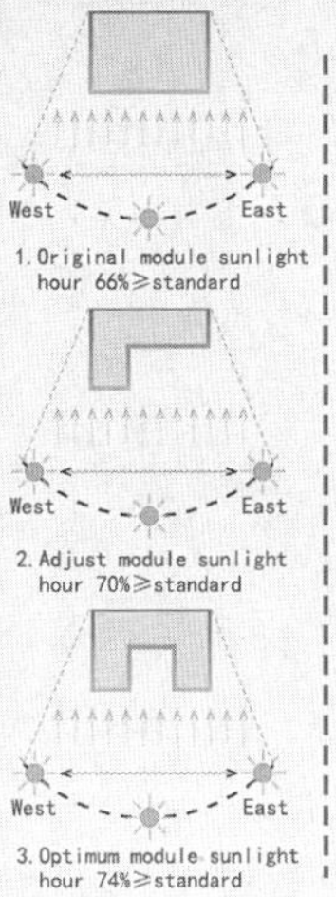

West
East
1. Original module sunlight hour 66%≥standard
West
East
2. Adjust module sunlight hour 70%≥standard
West
East
3. Optimum module sunlight hour 74%≥standard

SHAPE GENERATION

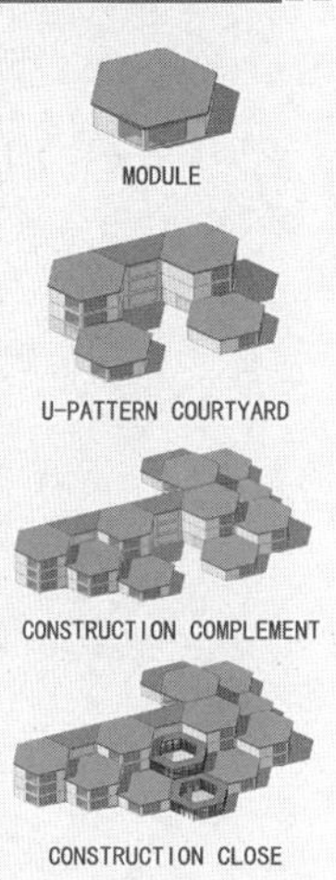

MODULE
U-PATTERN COURTYARD
CONSTRUCTION COMPLEMENT
CONSTRUCTION CLOSE

MASTER PLAN ANALYSIS

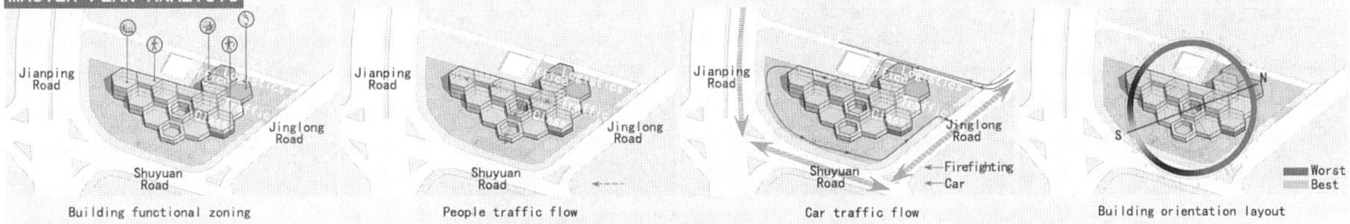

Jianping Road
Jinglong Road
Shuyuan Road
Building functional zoning
Jianping Road
Jinglong Road
Shuyuan Road
People traffic flow
Jianping Road
Jinglong Road
Shuyuan Road
Firefighting
Car
Car traffic flow
N
S
Worst
Best
Building orientation layout

THE BEST DIRECTION

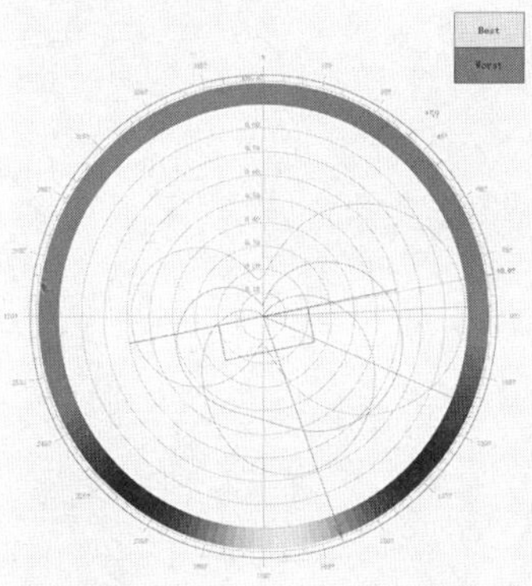

Best
Worst

Based on the analysis of the local climate, sun sunshine, wind environment and other data, it is concluded that the best orientation of the local building is the southeast direction, which has the highest comprehensive lighting degree. Combined with the unique modeling design of the building, the heat of summer and the cold of winter are effectively avoided.

ANALYSIS OF BUILDING SUNSHINE

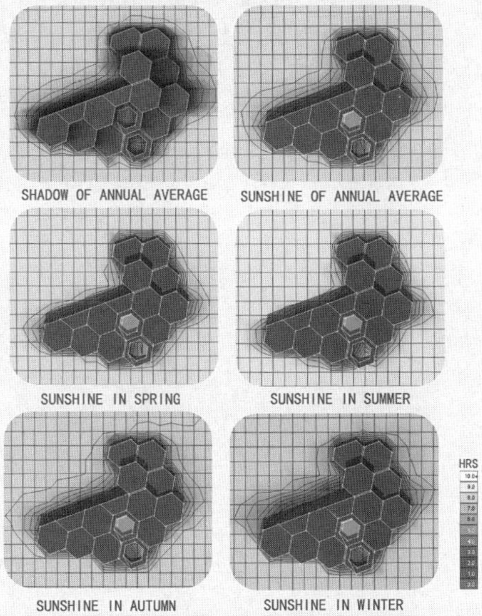

SHADOW OF ANNUAL AVERAGE
SUNSHINE OF ANNUAL AVERAGE
SUNSHINE IN SPRING
SUNSHINE IN SUMMER
SUNSHINE IN AUTUMN
SUNSHINE IN WINTER
HRS

DAY-ORBIT

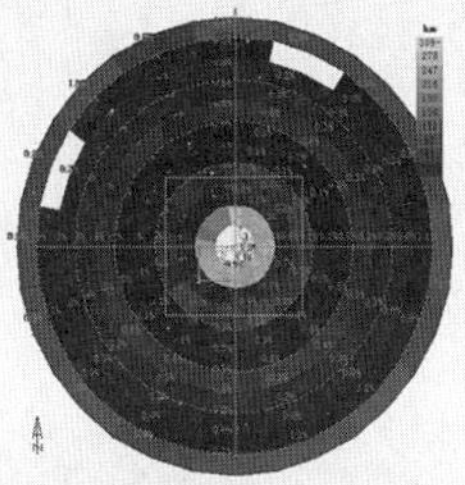

WIND ENVIRONMENT

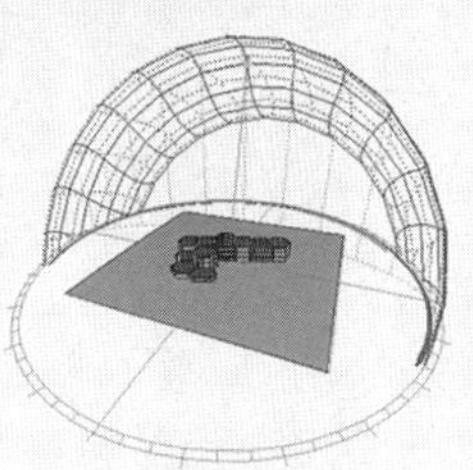

SOUTHEAST ELEVATION 1 : 200

阳光 · 稚梦

童之梦 · 光之巢 叁

The Dream of Children · the · Nest of Light

GROUND PLOOR PLAN 1 : 200

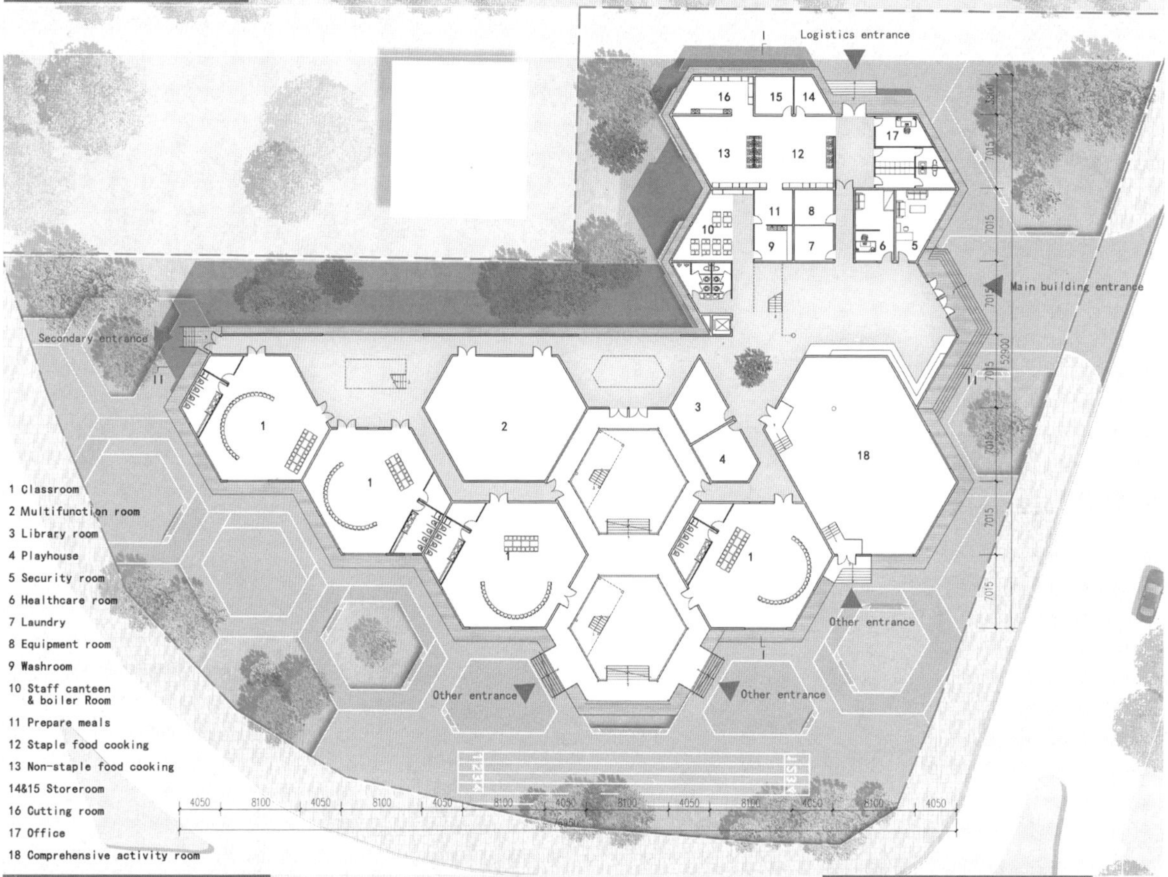

Logistics entrance
Main building entrance
Secondary entrance
Other entrance
Other entrance
Other entrance
1 Classroom
2 Multifunction room
3 Library room
4 Playhouse
5 Security room
6 Healthcare room
7 Laundry
8 Equipment room
9 Washroom
10 Staff canteen & boiler Room
11 Prepare meals
12 Staple food cooking
13 Non-staple food cooking
14&15 Storeroom
16 Cutting room
17 Office
18 Comprehensive activity room

PLANE WENTILATION ANALYSIS

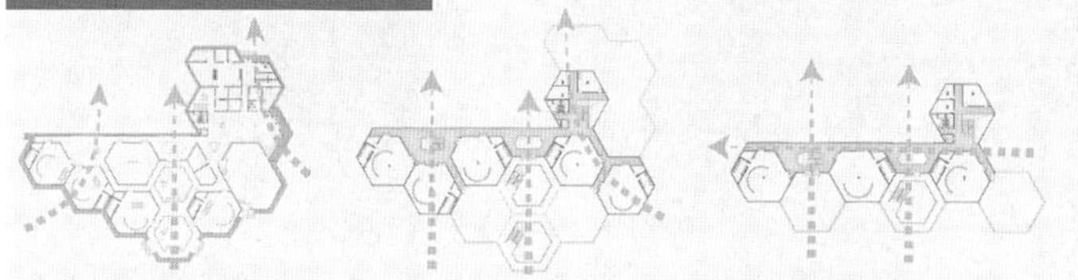

Airflow lineson on first floor　Airflow lineson on first floor　Airflow lineson on thrid floor

Through the overall breathability and step-like design of the building, the wind is guided from multiple directions into each indoor building layer, reducing indoor temperature and promoting air renewal

PLANE WENTILATION ANALYSIS

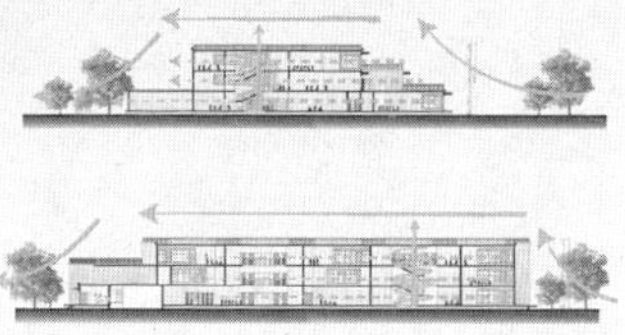

Airflow lines on vertical section

Building special joined the likely to lead to temperature difference in the design of the courtyard, because the courtyard of the grille and the reasons for the shutter to reduce direct sunlight, passive add different building area and building the difference in temperature between inside and outside air, promote the air flow and drive the indoor air renewal, reduce indoor temperature, the introduction of fresh air, promote microcirculation.

I-I SECTION 1 : 200

II-II SECTION 1 : 200

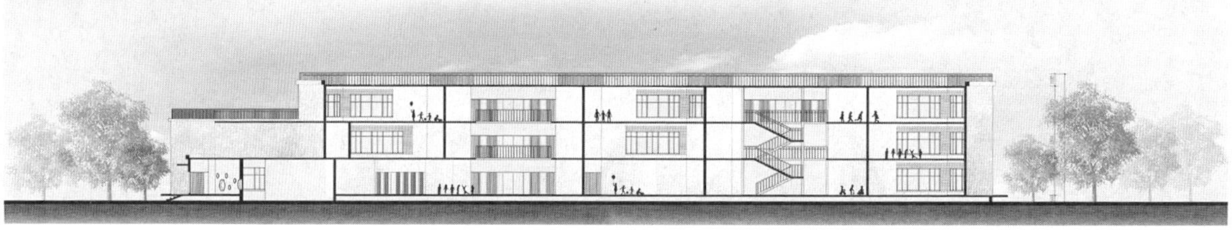

阳光·稚梦

童之梦·光之巢 肆

The Dream of Children · the · Nest of Light

CELL DESIGN

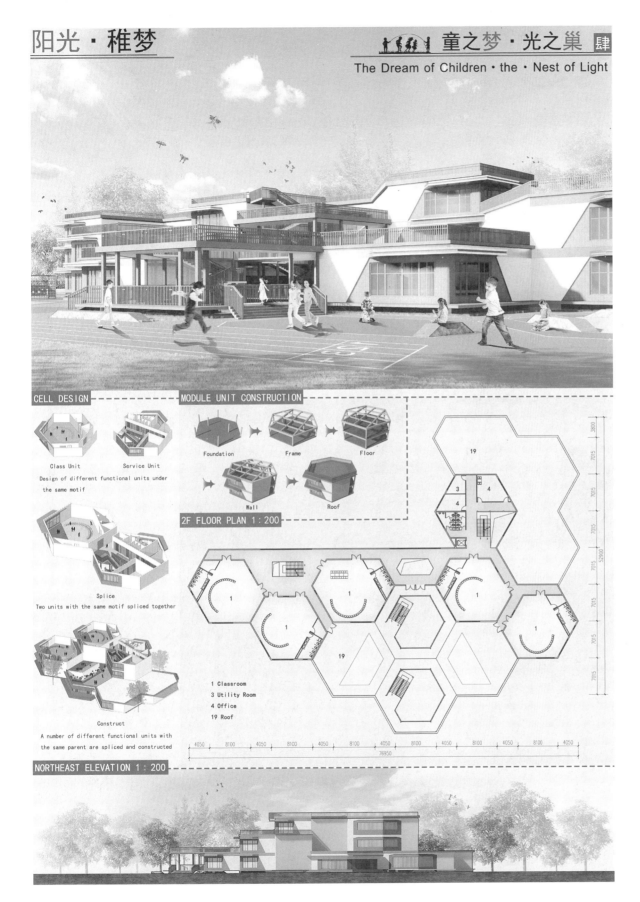

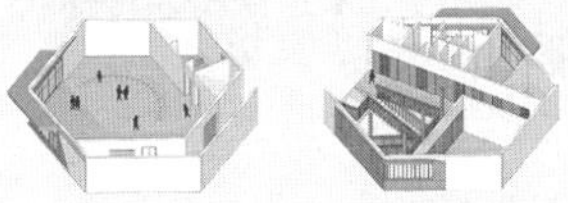
Class Unit
Service Unit

Design of different functional units under the same motif

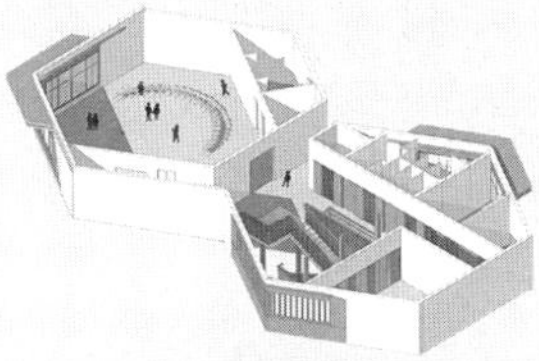

Splice

Two units with the same motif spliced together

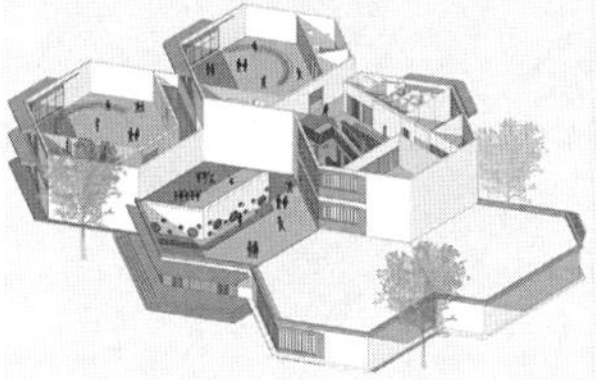

Construct

A number of different functional units with the same parent are spliced and constructed

MODULE UNIT CONSTRUCTION

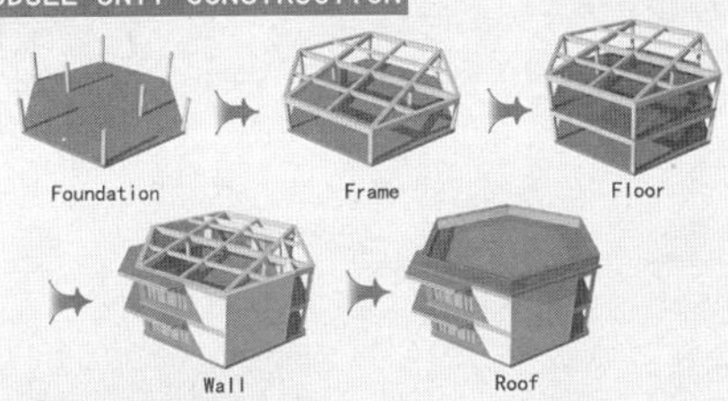
Foundation
Frame
Floor
Wall
Roof

2F FLOOR PLAN 1 : 200

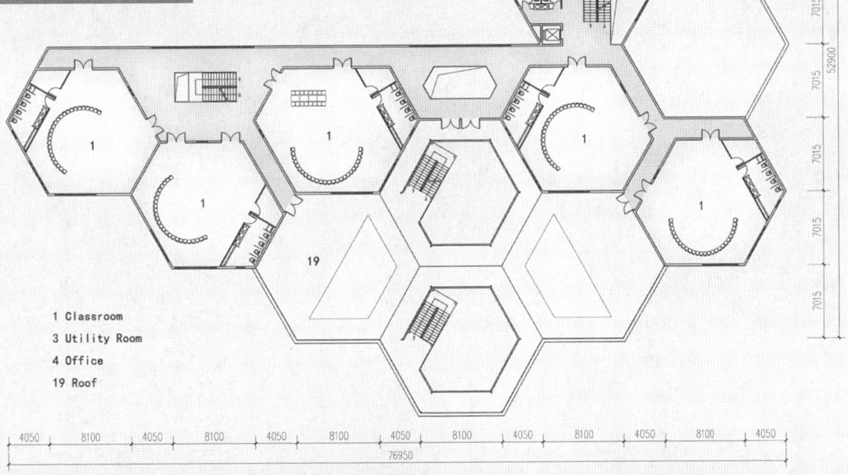
19
3
4
1
1
1
1
1
19
3800
7015
52900
4050
8100
76950

1 Classroom
3 Utility Room
4 Office
19 Roof

NORTHEAST ELEVATION 1 : 200

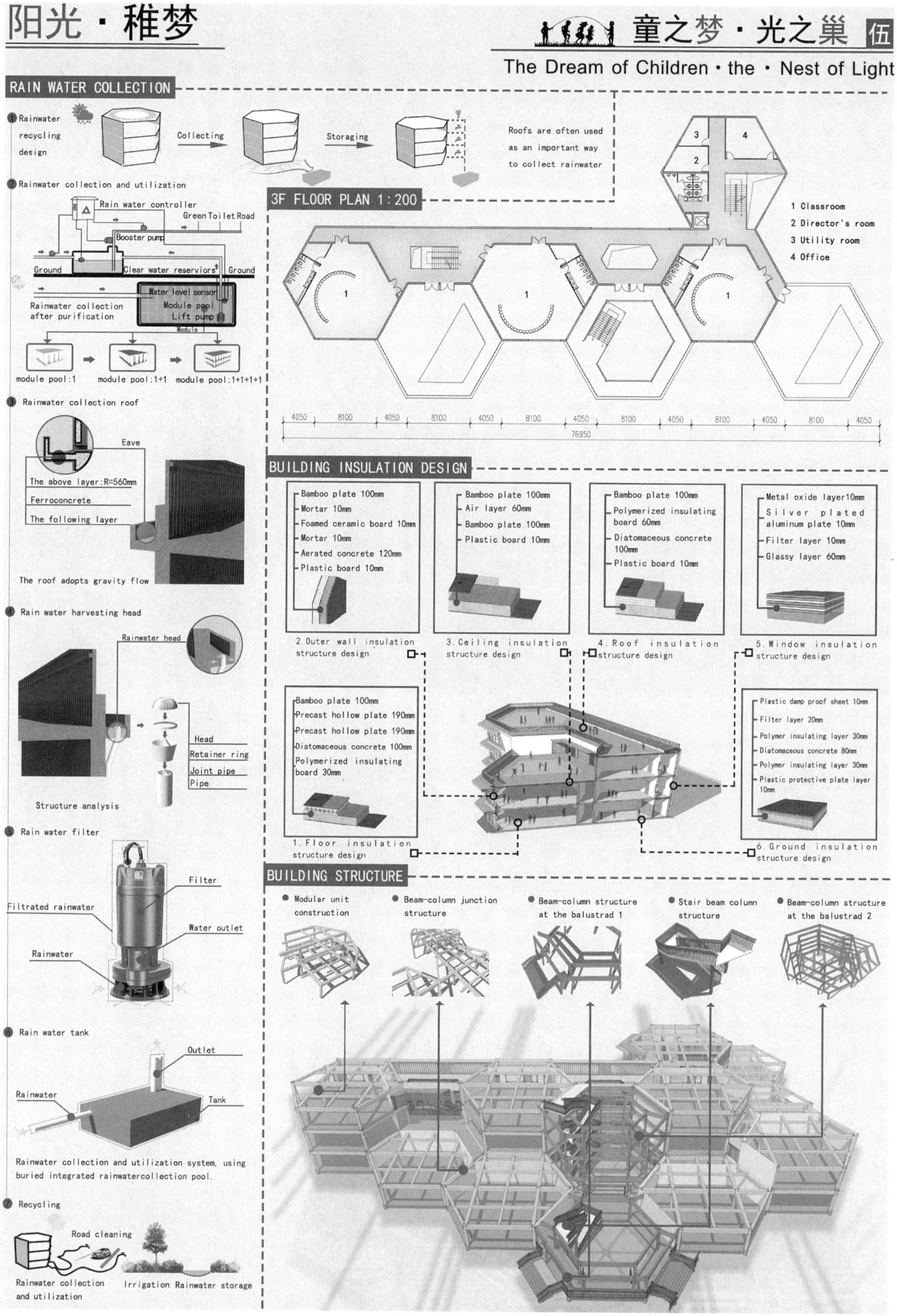
阳光・稚梦
童之梦・光之巢 伍
The Dream of Children · the · Nest of Light
RAIN WATER COLLECTION
1 Rainwater recycling design
Collecting
Storaging
Roofs are often used as an important way to collect rainwater
2 Rainwater collection and utilization
Rain water controller
Green Toilet Road
Booster pump
Ground
Clear water reserviors
Ground
Water level sensor
Module pool
Lift pump
Rainwater collection after purification
Module
module pool:1
module pool:1+1
module pool:1+1+1+1
3 Rainwater collection roof
Eave
The above layer:R=560mm
Ferroconcrete
The following layer
The roof adopts gravity flow
4 Rain water harvesting head
Rainwater head
Head
Retainer ring
Joint pipe
Pipe
Structure analysis
5 Rain water filter
Filter
Filtrated rainwater
Water outlet
Rainwater
6 Rain water tank
Outlet
Rainwater
Tank
Rainwater collection and utilization system, using buried integrated rainwatercollection pool.
7 Recycling
Road cleaning
Rainwater collection and utilization
Irrigation Rainwater storage
3F FLOOR PLAN 1:200
1 Classroom
2 Director's room
3 Utility room
4 Office
4050 8100 4050 8100 4050 8100 4050 8100 4050 8100 4050 8100 4050
76950
BUILDING INSULATION DESIGN
Bamboo plate 100mm
Mortar 10mm
Foamed ceramic board 10mm
Mortar 10mm
Aerated concrete 120mm
Plastic board 10mm
2.Outer wall insulation structure design
Bamboo plate 100mm
Air layer 60mm
Bamboo plate 100mm
Plastic board 10mm
3.Ceiling insulation structure design
Bamboo plate 100mm
Polymerized insulating board 60mm
Diatomaceous concrete 100mm
Plastic board 10mm
4.Roof insulation structure design
Metal oxide layer10mm
Silver plated aluminum plate 10mm
Filter layer 10mm
Glassy layer 60mm
5.Window insulation structure design
Bamboo plate 100mm
Precast hollow plate 190mm
Precast hollow plate 190mm
Diatomaceous concrete 100mm
Polymerized insulating board 30mm
1.Floor insulation structure design
Plastic damp proof sheet 10mm
Filter layer 20mm
Polymer insulating layer 30mm
Diatomaceous concrete 80mm
Polymer insulating layer 30mm
Plastic protective plate layer 10mm
6.Ground insulation structure design
BUILDING STRUCTURE
Modular unit construction
Beam-column junction structure
Beam-column structure at the balustrad 1
Stair beam column structure
Beam-column structure at the balustrad 2

阳光 · 稚梦

童之梦 · 光之巢 陆

The Dream of Children · the · Nest of Light

PERCENTAGE OF EACH MATERIAL

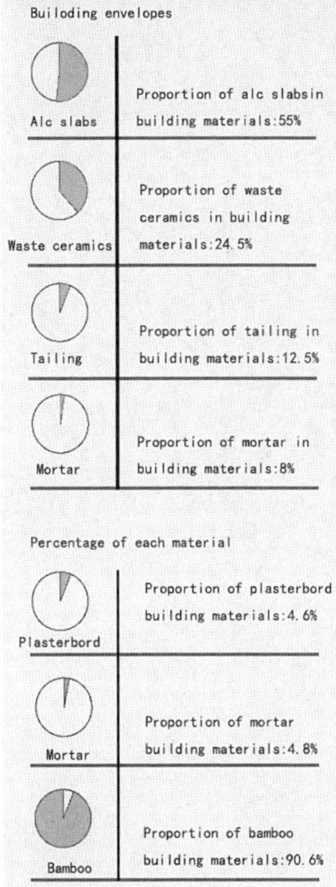

Builoding envelopes
Alc slabs
Proportion of alc slabsin building materials:55%
Waste ceramics
Proportion of waste ceramics in building materials:24.5%
Tailing
Proportion of tailing in building materials:12.5%
Mortar
Proportion of mortar in building materials:8%
Percentage of each material
Plasterbord
Proportion of plasterbord building materials:4.6%
Mortar
Proportion of mortar building materials:4.8%
Bamboo
Proportion of bamboo building materials:90.6%

The design specially customized the mortar and materials in a specific proportion, and carefully matched the materials to meet the good environmental friendliness and sustainability.

SUMMARY OF DESIGN STRATEGIES

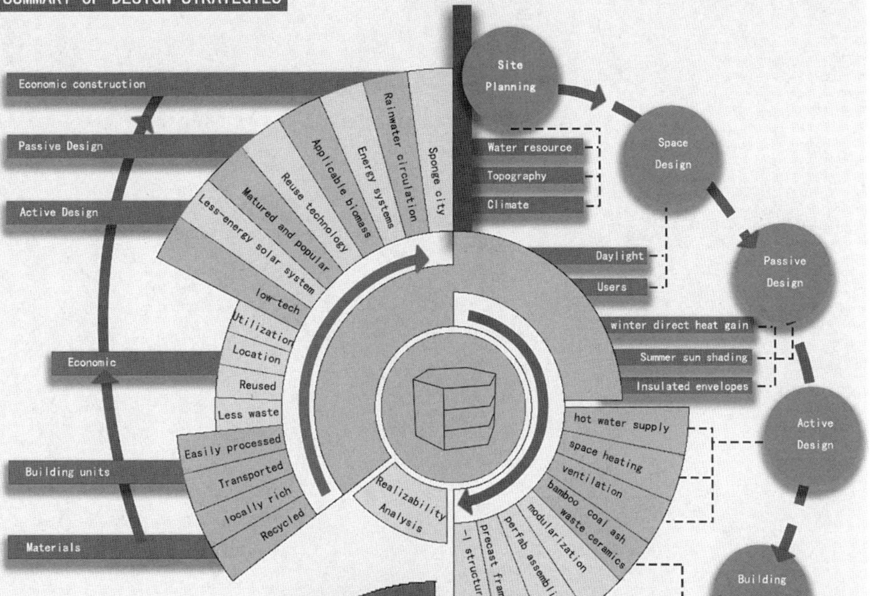

Economic construction
Passive Design
Active Design
Economic
Building units
Materials
Sponge city
Rainwater circulation
Energy systems
Applicable biomass
Reuse technology
Matured and popular
Less-energy solar system
low-tech
Utilization
Location
Reused
Less waste
Easily processed
Transported
locally rich
Recycled
Realizability Analysis
Water resource
Topography
Climate
Daylight
Users
winter direct heat gain
Summer sun shading
Insulated envelopes
hot water supply
space heating
ventilation
bamboo coal ash
waste ceramics
modularization
perfab assembling
precast frame pane
-l structure
Site Planning
Space Design
Passive Design
Active Design
Building Materials
Industria -lization
Happy Healthy

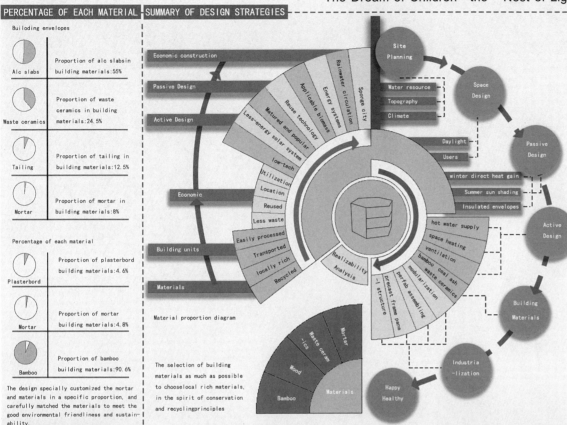

Material proportion diagram

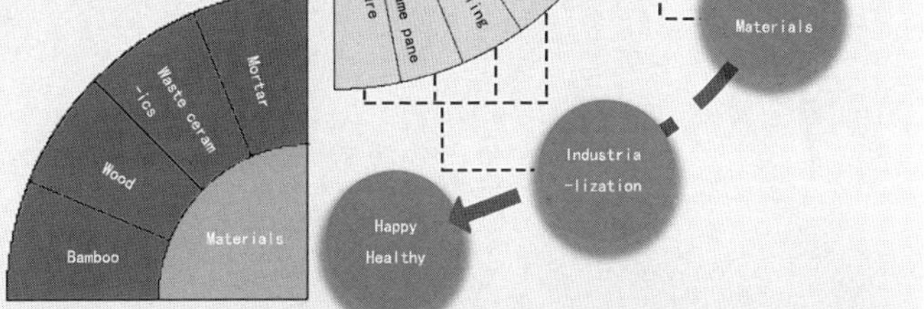

Mortar
Waste ceram -ics
Wood
Bamboo
Materials

The selection of building materials as much as possible to chooselocal rich materials, in the spirit of conservation and recyclingprinciples

GREEN BUILDING DESIGN

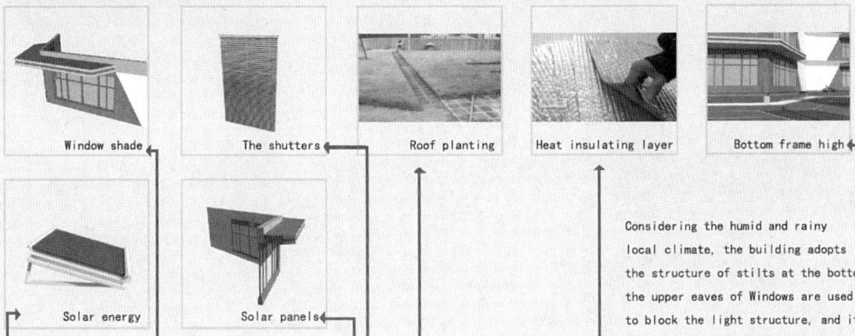

Window shade
The shutters
Roof planting
Heat insulating layer
Bottom frame high
Solar energy
Solar panels

Considering the humid and rainy local climate, the building adopts the structure of stilts at the bottom, the upper eaves of Windows are used to block the light structure, and it is combined with passive ventilation, and the green plants on the roof are combined with solar energy.

BUILDING INDUSTRIALIZATION

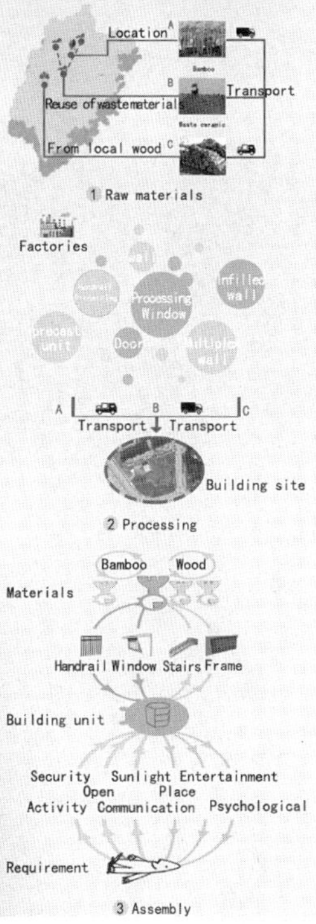

Location
Transport
Reuse of wastematerial
From local wood
1 Raw materials
Factories
Processing Window
Door
A Transport B Transport C
Building site
2 Processing
Bamboo
Wood
Materials
Handrail Window Stairs Frame
Building unit
Security Sunlight Entertainment
Open Place
Activity Communication Psychological
Requirement
3 Assembly

综合奖·优秀奖
General Prize Awarded · Honorable Mention Prize

注 册 号：7124
项目名称：光·遇（南平）
Meet the Light in the Corner (Nanping)
作　　者：贝莎莎、岳　乐
参赛单位：长安大学
指导教师：任　娟

光·遇 1
Meet the Light in the Corner

Location

Nanping City　Jianyang Area

Regional Feature

Folk House

Sloping Roof

Narrow Lane

Tea Culture

Bamboo Forest

Abundant Rainfall

Climate

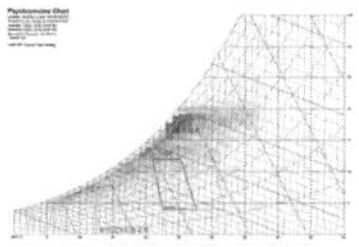
Psychrometric Chart

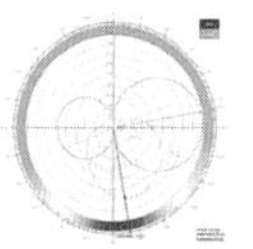
Optimum Orietation

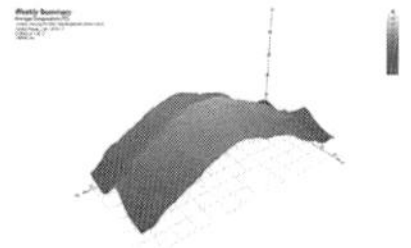
The Average Wendy

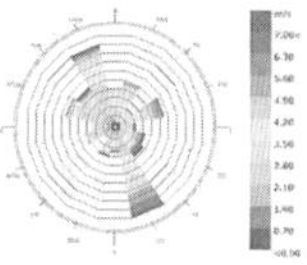
Prevailing Winds

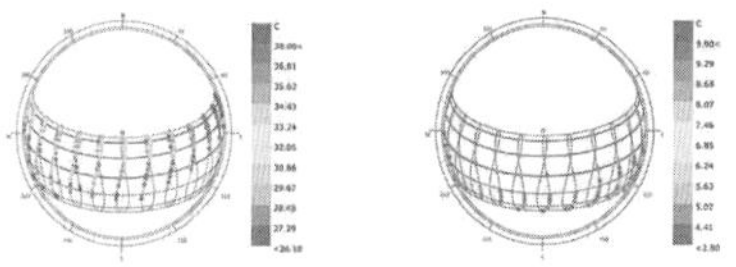
The Path of the Sun

Design Description

本次设计位于福建省南平县，方案以“光·遇”为主题，利用建筑大平台和中庭结合的设计，打造出独具一格的光影效果，赋予建筑更多趣味性。

在建筑取材上，利用当地竹资源丰富的特点，就地取材，在建筑外墙和屋顶上采用竹子作为外饰，并利用竹子中空的特性，形成独特的外墙雨水收集系统及屋顶排水系统。在被动式太阳能利用技术上，应用建筑朝向、集热蓄热墙、建筑材料、“冷巷”通风等方式，并采用太阳能集热板、太阳能空调系统、光伏建筑系统、雨水收集系统、海绵城市等主动式设计，实现了对光、风、水等可再生资源的利用。

This design is located in Nanping County, Fujian Province. The plan takes "light encounter" as the theme and USES the building. The combination of the large platform and the atrium creates a unique light and shadow effect, giving the building more Lots of fun.

In terms of architectural materials, the rich characteristics of local bamboo resources are utilized, and local materials are used in architecturen Bamboo is used as exterior decoration on the exterior wall and roof, and its hollow character is utilized to form a unique feature external wall rainwater collection system and roof drainage system. In passive solar technologies, Application of building orientation, heat collection and storage wall, building materials, "cold alley" ventilation and other ways, and adopted Solar panels, solar air conditioning systems, photovoltaic building systems, rainwater harvesting systems, Sponge city and other active designs realize the utilization of renewable resources such as light, wind and water.

Site Analysis

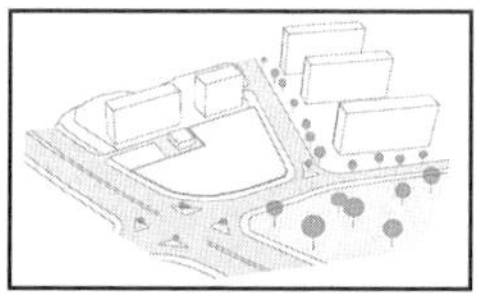
The site is open to the south.

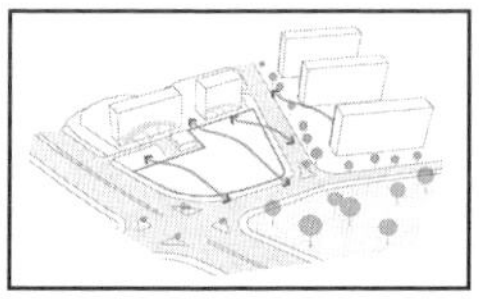
Prevailing southeasterly.

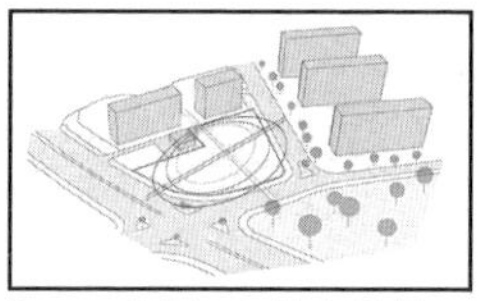
Surrounded by tall buildings.

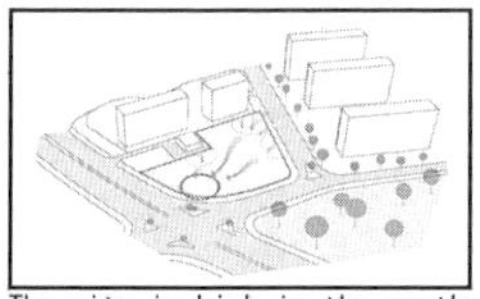
The site is high in the northeast and low in the southwest.

光·遇 2

Meet the Light in the Corner

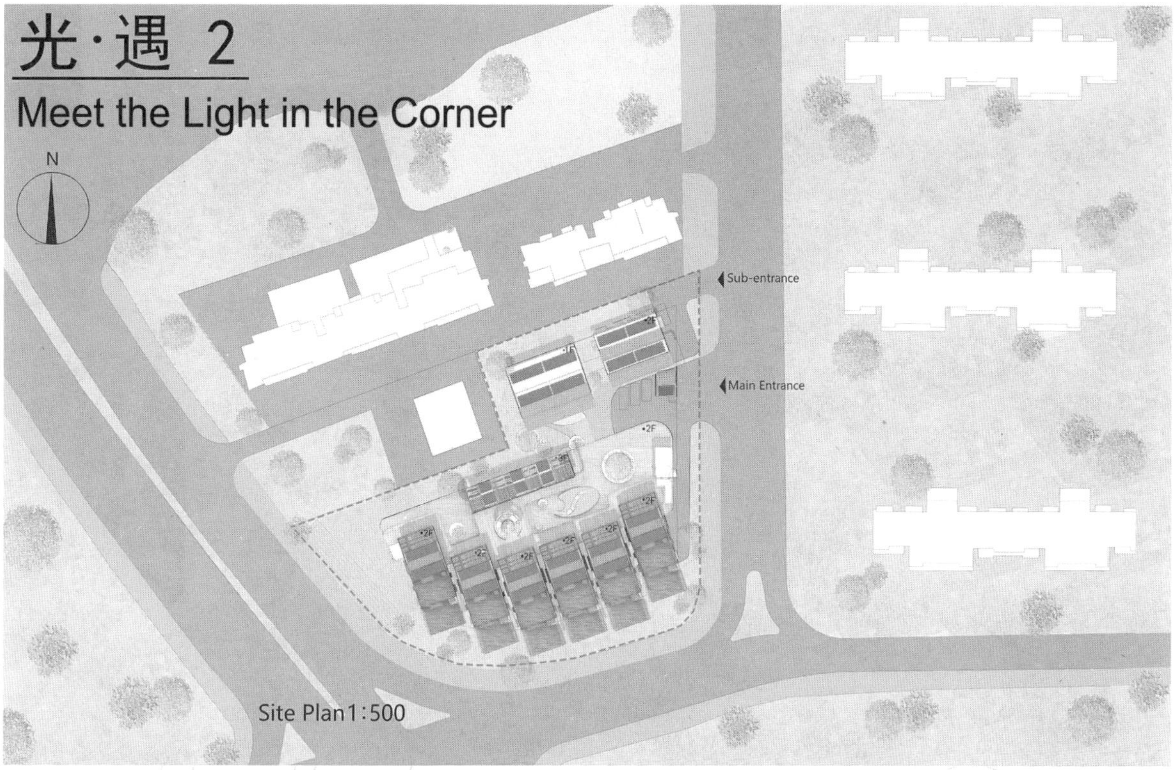
N
Sub-entrance
Main Entrance
Site Plan1:500

Sunlight Analusisx

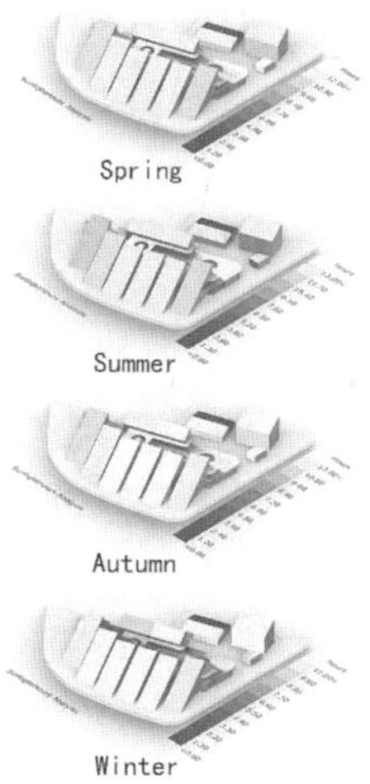
Spring
Summer
Autumn
Winter

Human Behavior and Building systems

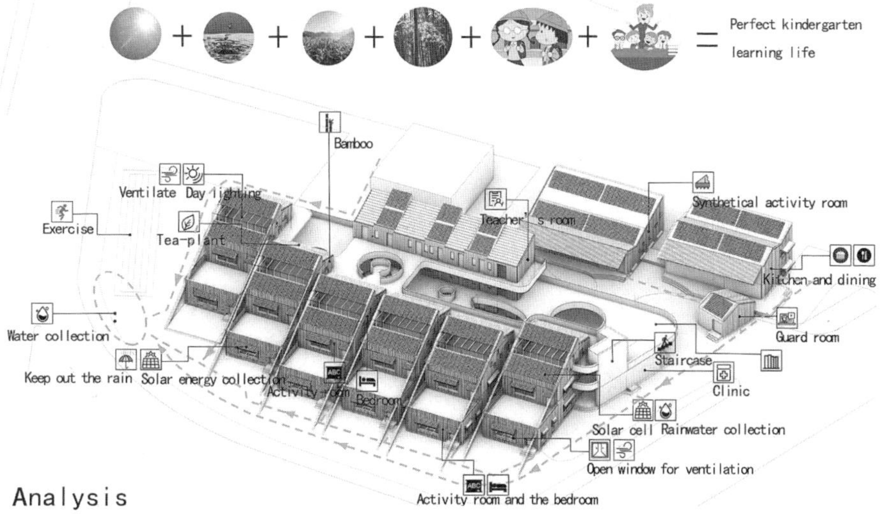
Perfect kindergarten learning life
Bamboo
Ventilate
Day lighting
Exercise
Tea-plant
Teacher's room
Synthetical activity room
Kitchen and dining
Water collection
Guard room
Staircase
Keep out the rain
Solar energy collection
Activity room
Bedroom
Clinic
Solar cell
Rainwater collection
Open window for ventilation
Activity room and the bedroom

Blcok Generation Analysis

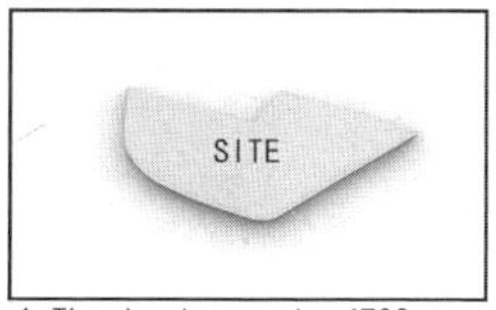
SITE

1. The land area is 4700㎡

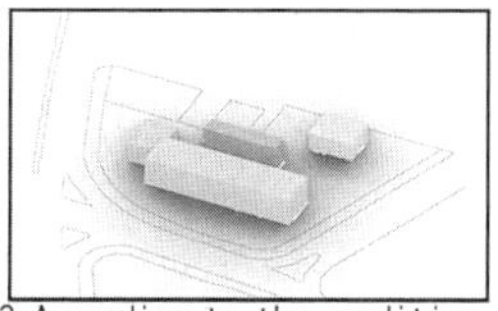

2. According to the condition, we divide the functional partition

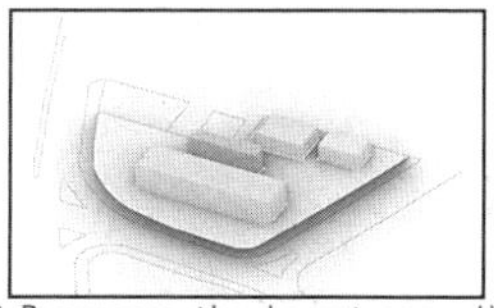

3. Rearrange the layout according to the wind direction

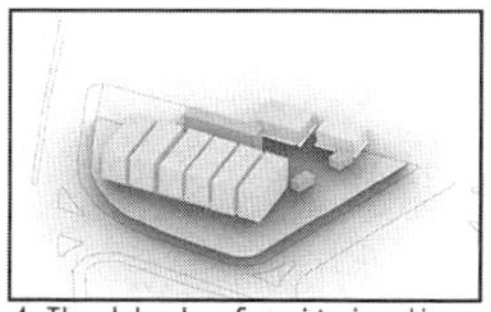

4. The block of unit is divided according to the principle of cold tunnel

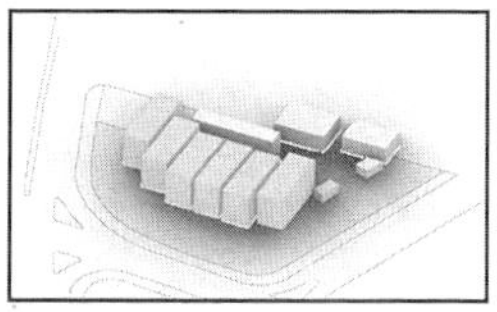

5. The unit volume is rearranged

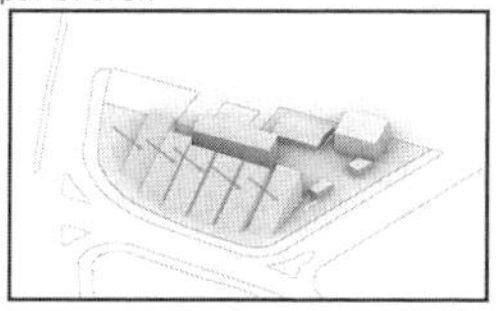

6. Use the best angle of direct sunlight to generate the volume

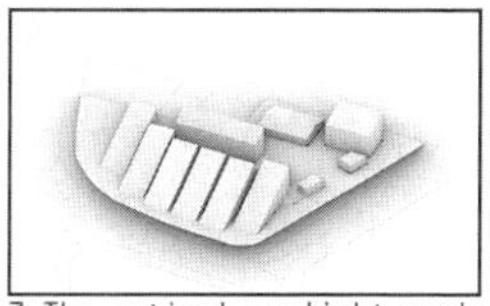

7. The optimal sunlight angle of all volumes was selected for cutting

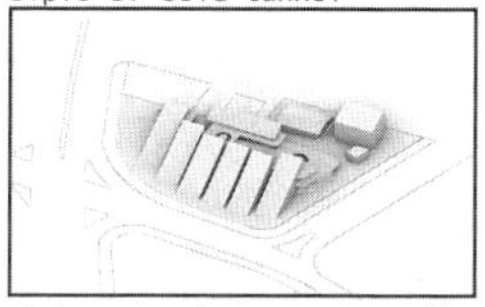

8. Organized the laminar flow lines of the roof by enclosing the courtyard

光·遇 3

Meet the Light in the Corner

1 Bedroom
2 Activity Room
3 Class Activity Area
4 Medical Room
5 Guard Room
6 Toilet
7 Storage Room
8 Teacher's Office
9 Switching Room
10 Boiling Room
11 Laundry
12 Reading And Meeting Room
13 Teaching Tool Making Room
14 Synthetical Activity Room
15 Lounge
16 Locker Room
17 Kitchen
18 Chopping Area
19 Washing Room
20 Servery
21 Food Library

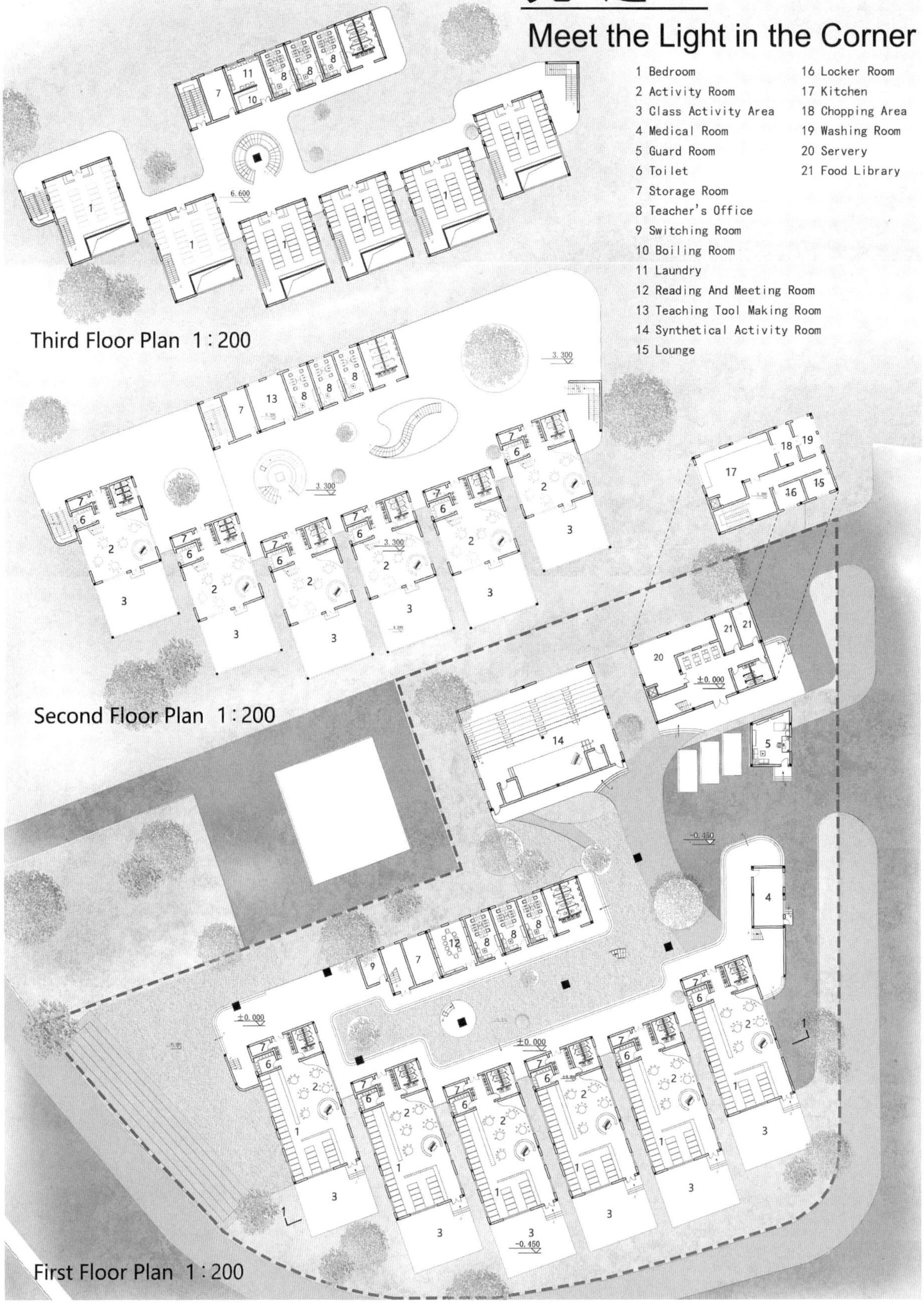
Third Floor Plan 1:200
6.600
3.300
Second Floor Plan 1:200
±0.000
-0.450
First Floor Plan 1:200

光·遇 4

Meet the Light in the Corner

Dormitory Unit

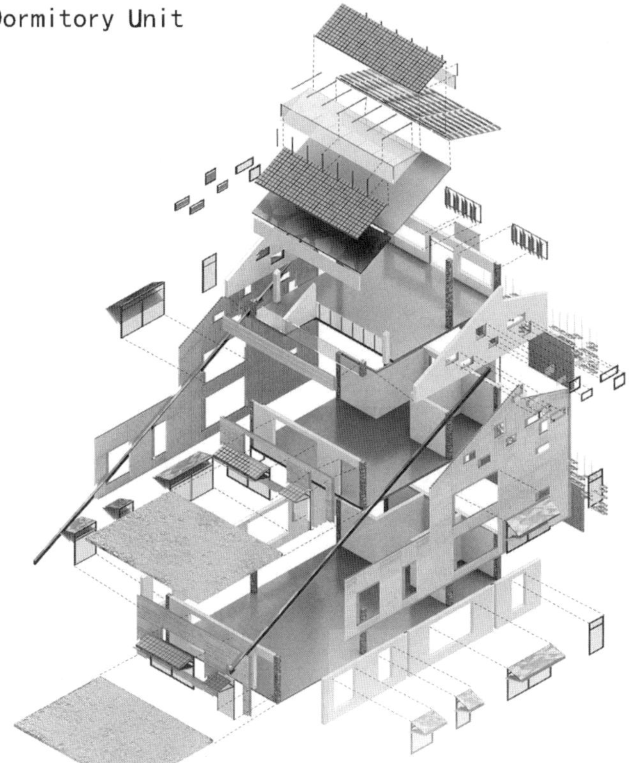

Children's Plant Adoption Wall

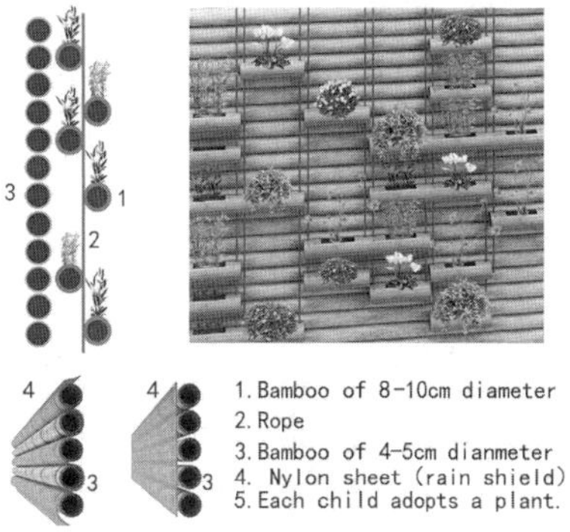
1. Bamboo of 8-10cm diameter
2. Rope
3. Bamboo of 4-5cm dianmeter
4. Nylon sheet (rain shield)
5. Each child adopts a plant.

Removable Folding Window

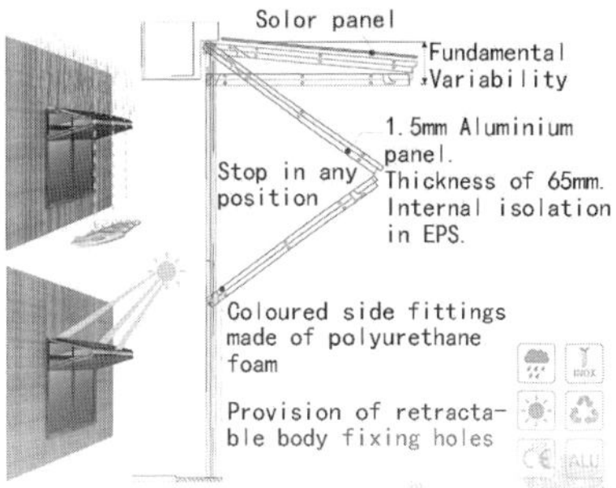
Solor panel
Fundamental Variability
1.5mm Aluminium panel.
Thickness of 65mm.
Internal isolation in EPS.
Stop in any position
Coloured side fittings made of polyurethane foam
Provision of retractable body fixing holes

11.600
9.300
6.600
3.300
±0.000
-0.450

South Elevation 1：200

光·遇 5

Meet the Light in the Corner

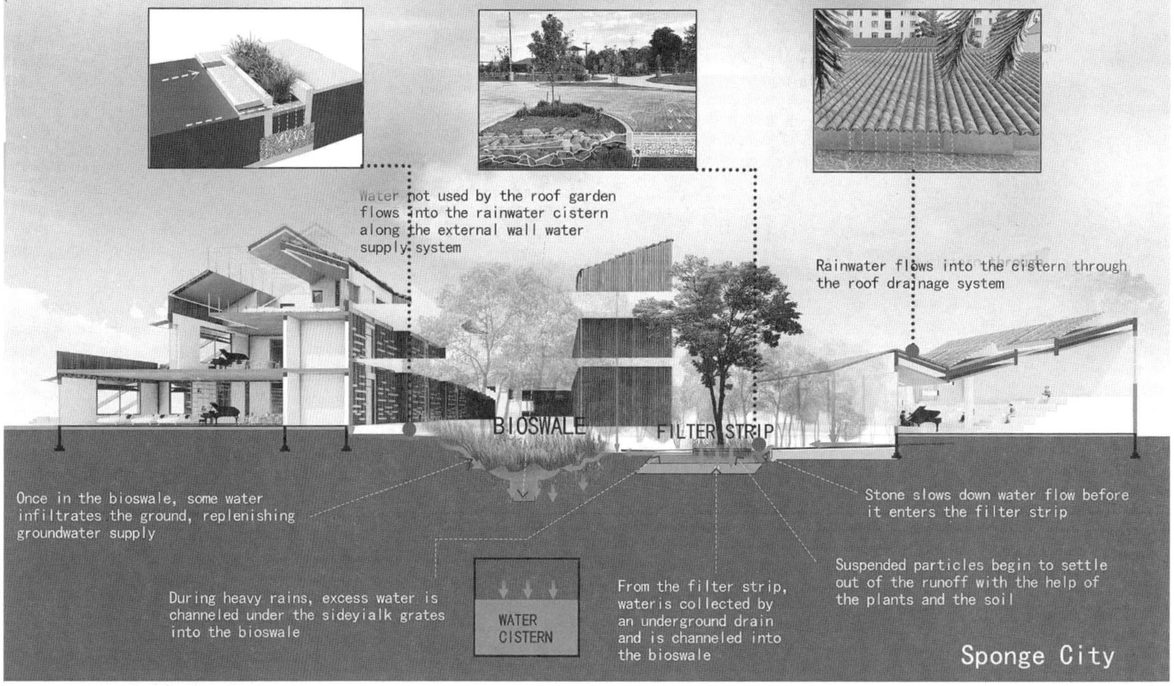
Water not used by the roof garden
flows into the rainwater cistern
along the external wall water
supply system
Rainwater flows into the cistern through
the roof drainage system
BIOSWALE
FILTER STRIP
Once in the bioswale, some water
infiltrates the ground, replenishing
groundwater supply
Stone slows down water flow before
it enters the filter strip
Suspended particles begin to settle
out of the runoff with the help of
the plants and the soil
During heavy rains, excess water is
channeled under the sideyialk grates
into the bioswale
WATER
CISTERN
From the filter strip,
wateris collected by
an underground drain
and is channeled into
the bioswale
Sponge City

Bioswale Cross Section

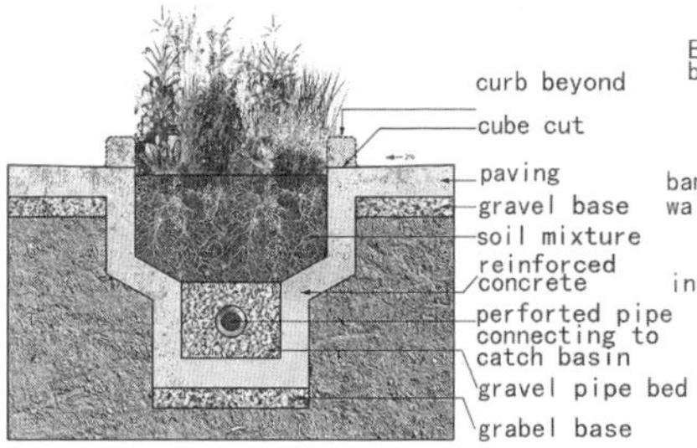
curb beyond
cube cut
paving
gravel base
soil mixture
reinforced
concrete
perforted pipe
connecting to
catch basin
gravel pipe bed
grabel base

Exterior Wall

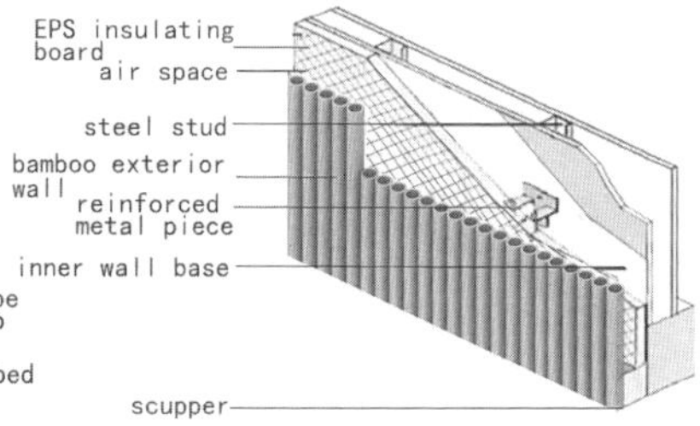
EPS insulating
board
air space
steel stud
bamboo exterior
wall
reinforced
metal piece
inner wall base
scupper

Roof Garden

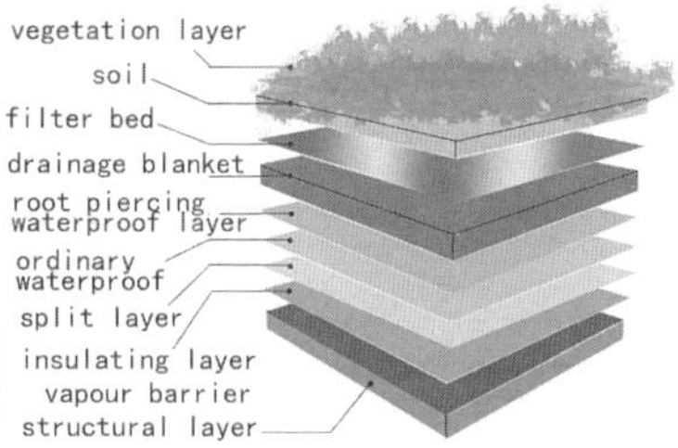
vegetation layer
soil
filter bed
drainage blanket
root piercing
waterproof layer
ordinary
waterproof
split layer
insulating layer
vapour barrier
structural layer

Water System Circulation

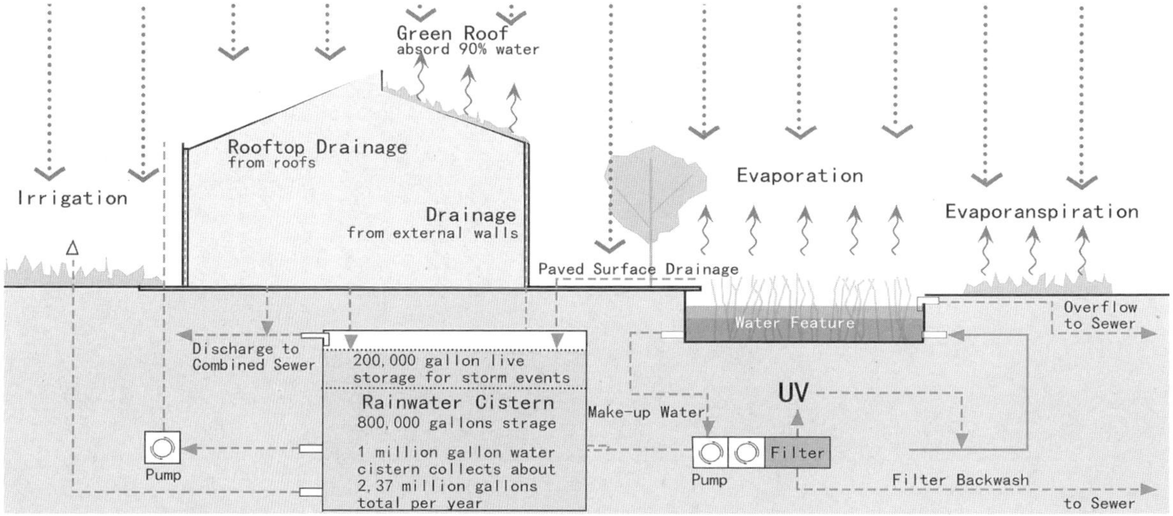
Green Roof
absord 90% water
Rooftop Drainage
from roofs
Irrigation
Drainage
from external walls
Evaporation
Evaporanspiration
Paved Surface Drainage
Water Feature
Overflow
to Sewer
Discharge to
Combined Sewer
200,000 gallon live
storage for storm events
Rainwater Cistern
800,000 gallons strage
1 million gallon water
cistern collects about
2,37 million gallons
total per year
Make-up Water
UV
Pump
Pump
Filter
Filter Backwash
to Sewer

光·遇 6

Meet the Light in the Corner

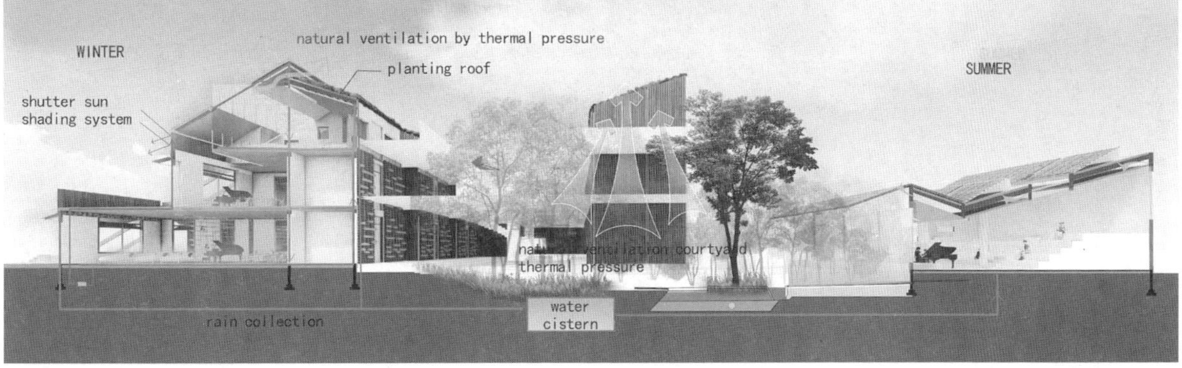
WINTER
natural ventilation by thermal pressure
planting roof
shutter sun shading system
SUMMER
thermal pressure
rain collection
water cistern

Sunshine Analysis

In Summer　　In Winter

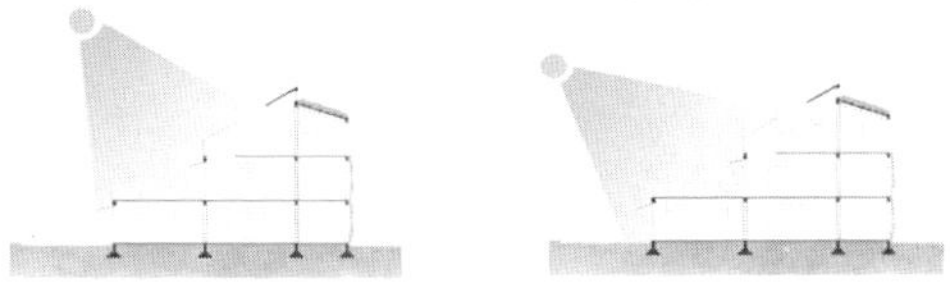

The sun is higher in summer than in winter.

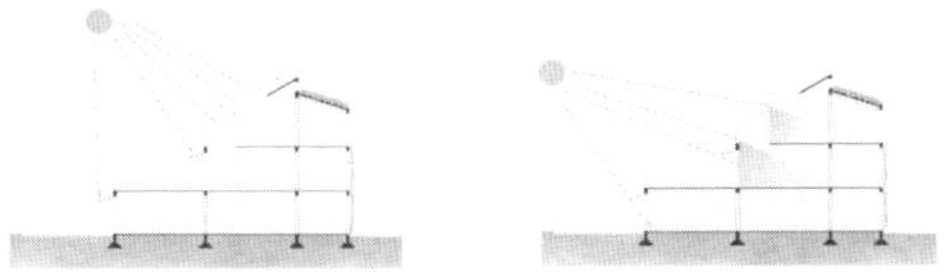

Control weather sunlight enters the room through awnings, windows, etc.

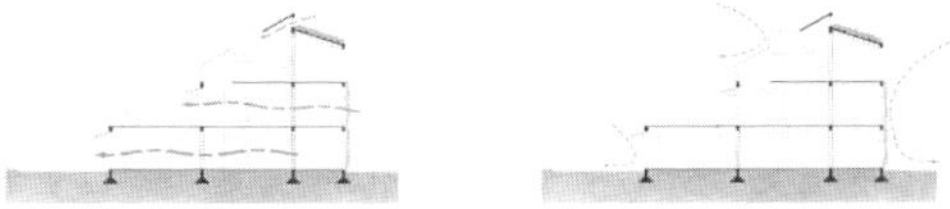

Use windoes to channel cold air in the summer and block out in the winter.

Use cool air to cool in summer to discharge heat, roof garden can insulate partial heat.

Summer Sunshine in Cold Alley

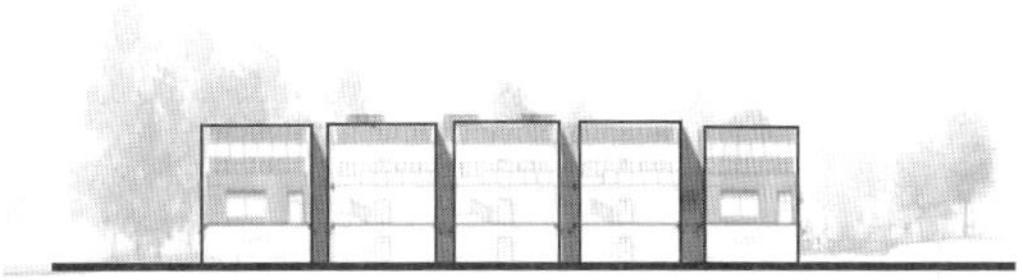

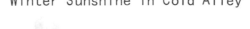

Winter Sunshine in Cold Alley

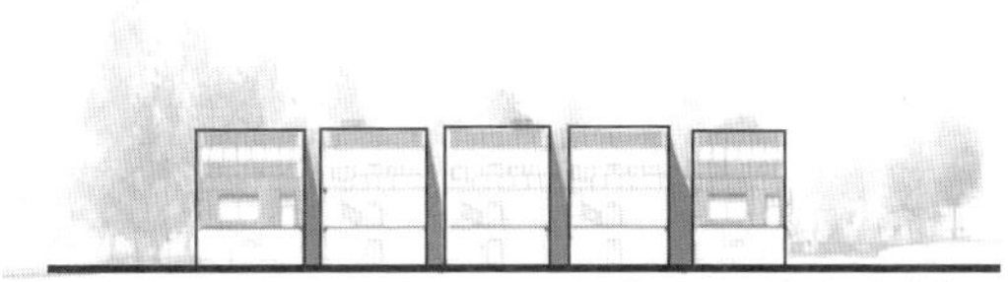

Summer Temperature in Cold Alley

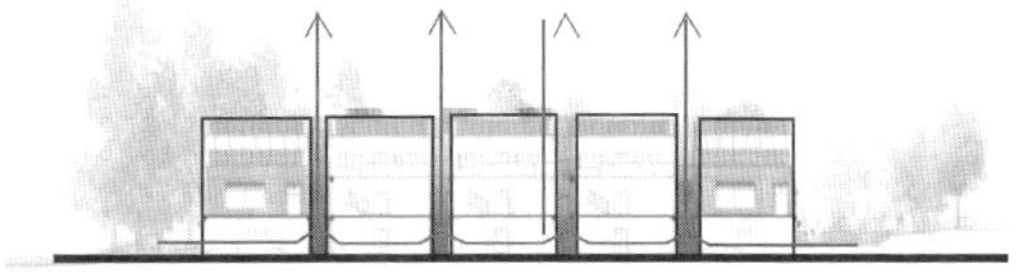

Winter Temperature in Cold Alley

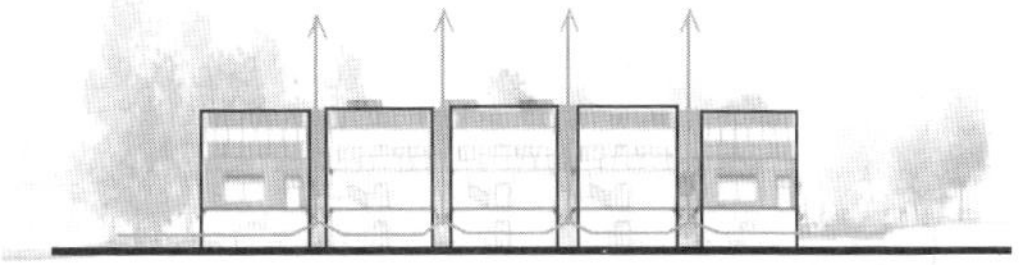

Local Perspective

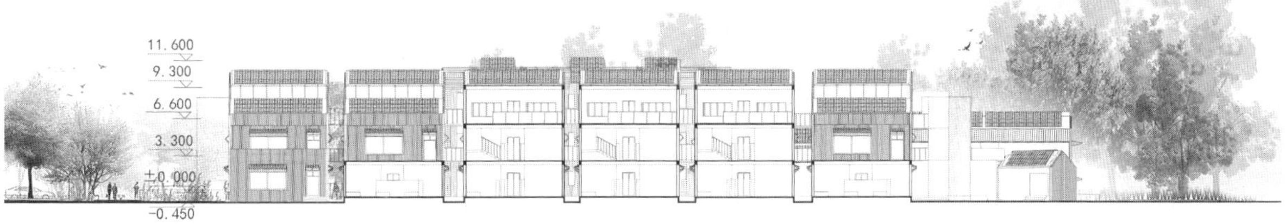
11.600
9.300
6.600
3.300
±0.000
-0.450

1-1 Section 1∶200

综合奖 · 优秀奖
General Prize Awarded · Honorable Mention Prize

注 册 号：7161

项目名称：山行 · 竹荫里（南平）
Mountain Trip · In the Shade of Bamboo（Nanping）

作　　者：郑曼如、刘世恒、陈佳怡、肖婉凝

参赛单位：福州大学

指导教师：崔育新

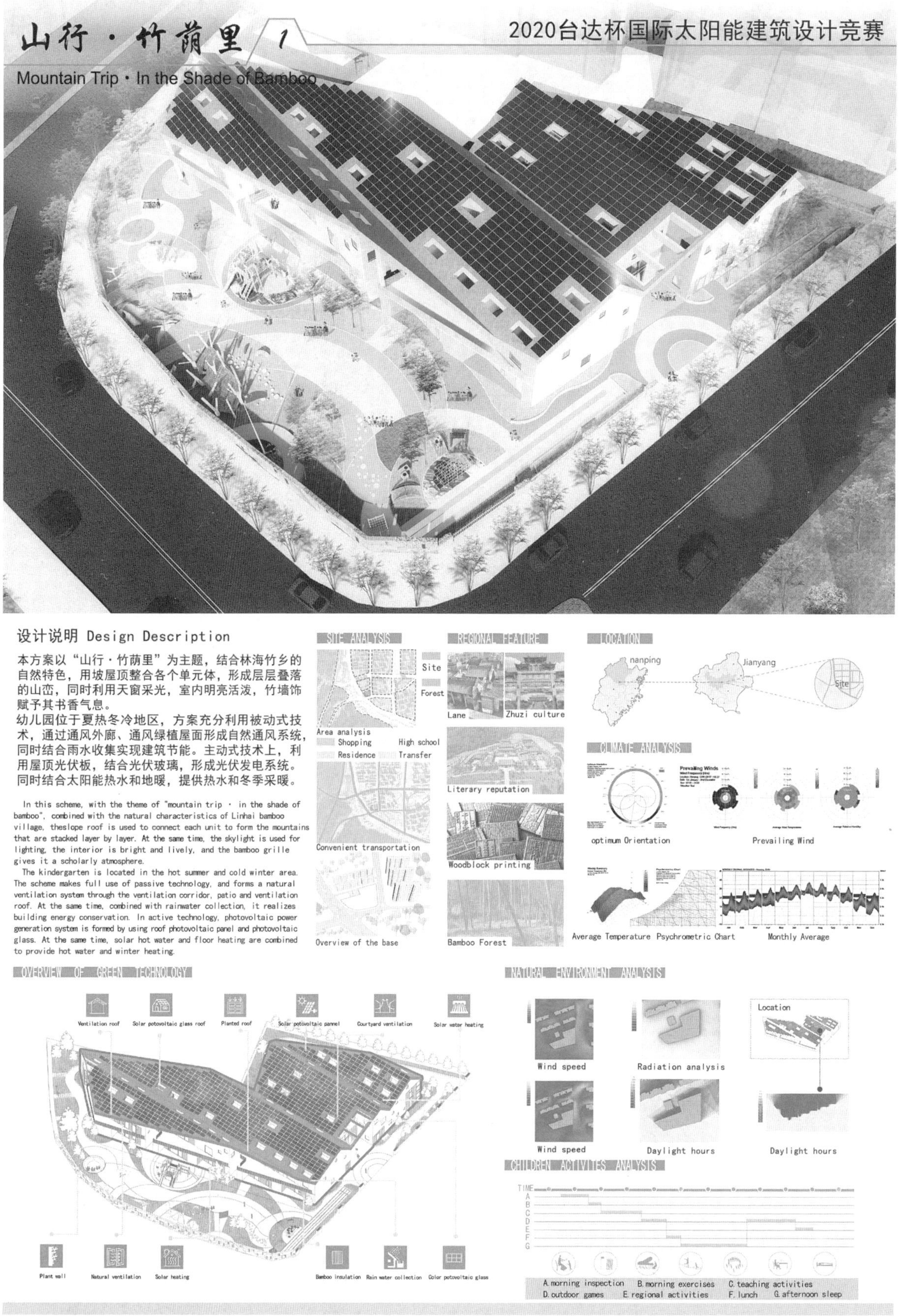
山行 · 竹荫里 1
2020台达杯国际太阳能建筑设计竞赛
Mountain Trip · In the Shade of Bamboo
设计说明 Design Description
本方案以“山行 · 竹荫里”为主题，结合林海竹乡的自然特色，用坡屋顶整合各个单元体，形成层层叠落的山峦，同时利用天窗采光，室内明亮活泼，竹墙饰赋予其书香气息。
幼儿园位于夏热冬冷地区，方案充分利用被动式技术，通过通风外廊、通风绿植屋面形成自然通风系统，同时结合雨水收集实现建筑节能。主动式技术上，利用屋顶光伏板，结合光伏玻璃，形成光伏发电系统。同时结合太阳能热水和地暖，提供热水和冬季采暖。
In this scheme, with the theme of "mountain trip · in the shade of bamboo", combined with the natural characteristics of Linhai bamboo village, theslope roof is used to connect each unit to form the mountains that are stacked layer by layer. At the same time, the skylight is used for lighting, the interior is bright and lively, and the bamboo grille gives it a scholarly atmosphere.
The kindergarten is located in the hot summer and cold winter area. The scheme makes full use of passive technology, and forms a natural ventilation system through the ventilation corridor, patio and ventilation roof. At the same time, combined with rainwater collection, it realizes building energy conservation. In active technology, photovoltaic power generation system is formed by using roof photovoltaic panel and photovoltaic glass. At the same time, solar hot water and floor heating are combined to provide hot water and winter heating.
SITE ANALYSIS
Site
Forest
Area analysis
Shopping
High school
Residence
Transfer
Convenient transportation
Overview of the base
REGIONAL FEATURE
Lane
Zhuzi culture
Literary reputation
Woodblock printing
Bamboo Forest
LOCATION
nanping
Jianyang
Site
CLIMATE ANALYSIS
optimum Orientation
Prevailing Wind
Average Temperature
Psychrometric Chart
Monthly Average
OVERVIEW OF GREEN TECHNOLOGY
Ventilation roof
Solar potovoltaic glass roof
Planted roof
Solar potovoltaic pannel
Courtyard ventilation
Solar water heating
Plant wall
Natural ventilation
Solar heating
Bamboo insulation
Rain water collection
Color potovoltaic glass
NATURAL ENVIRONMENT ANALYSIS
Wind speed
Radiation analysis
Location
Wind speed
Daylight hours
Daylight hours
CHILDREN ACTIVITIES ANALYSIS
TIME
A
B
C
D
E
F
G
A. morning inspection
B. morning exercises
C. teaching activities
D. outdoor games
E. regional activities
F. lunch
G. afternoon sleep

山行 · 竹荫里 2

2020台达杯国际太阳能建筑设计竞赛

Mountain Trip · In the Shade of Bamboo

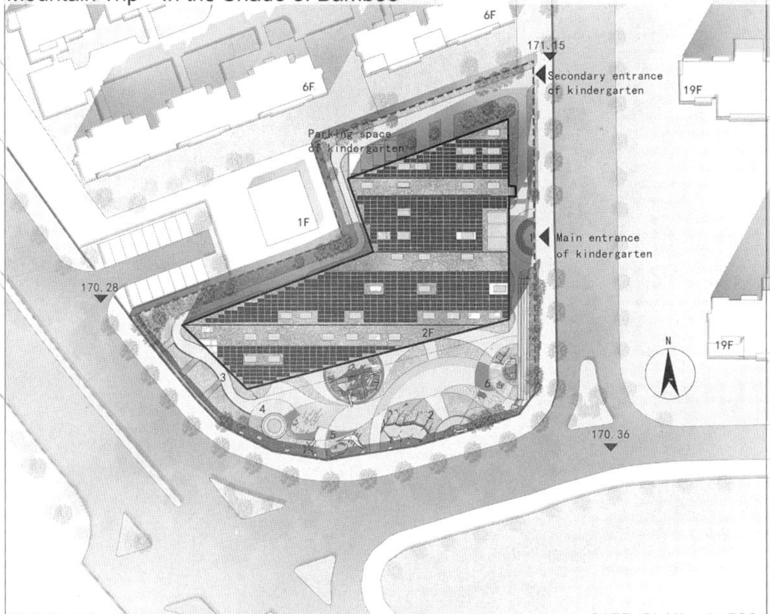

6F
6F
171.15
Secondary entrance of kindergarten
19F
Parking space of kindergarten
1F
Main entrance of kindergarten
170.28
2F
N
19F
170.36

TECHNICAL AND ECONOMIC INDICATORS

SITE PLAN 1:500

Land Area:6028㎡ Total Building Area:4685㎡ Building Occupation Area:2354㎡ Building Density:39.0%
Plot Rratio:0.78 Green space rate:30% Outdoor Playground Area: 1452㎡ Area Of Class Activity Site:760㎡

SCHEME GENERATION

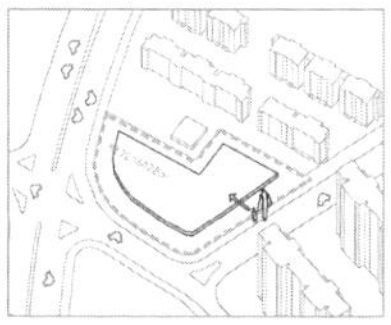

1 Site area:6028㎡, the buildings are arranged according to the site

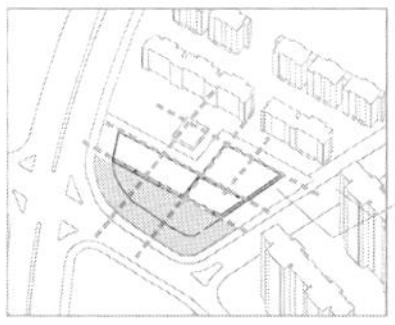

2 Well shaped segmentation, maximization of south facing site

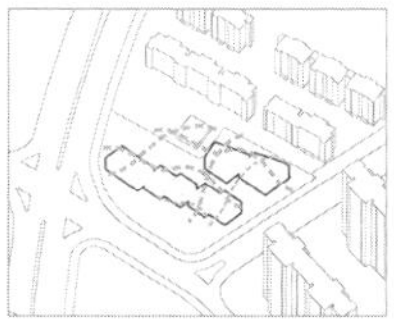

3 Mountain shape appears, two axis control

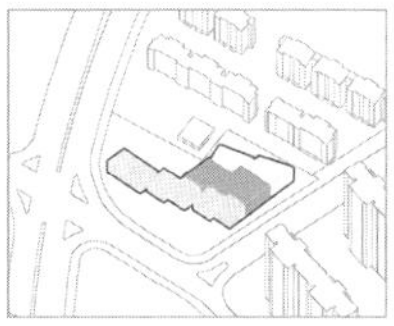

4 Determine the core traffic space, children's class block preliminary determined

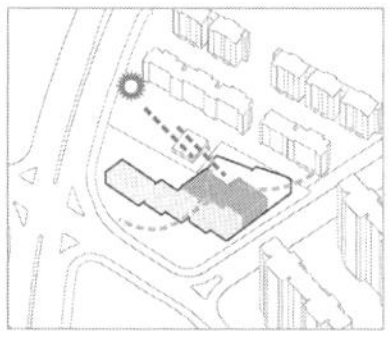

5 Block adjustment based on lighting and passive ventilation

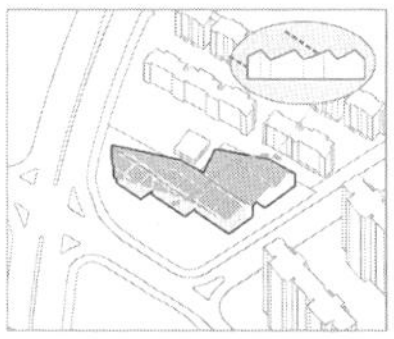

6 The roof is inclined, and the whole is in the shape of a mountain, and provides active solar power generation

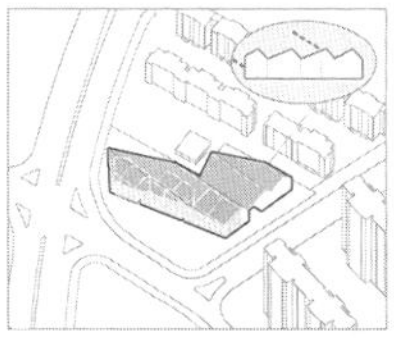

7 The lower layer forms a retreat platform to enrich the vertical space activities

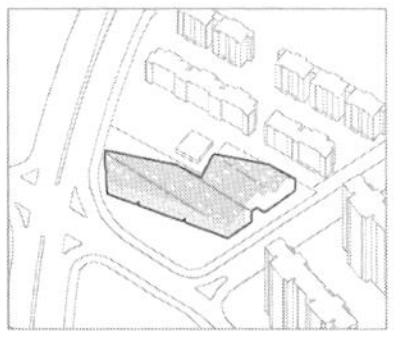

8 Planting roof is set to form the feeling of hillside. And open the skylight to increase daylighting

WEST ELEVATION 1:250

SOUTH ELEVATION 1:250

SCENE PERSPECTIVE

1 The main entrance on the east side

2 Green space beside the national flag

3 The secondary entrance on the west side

4 Play pool on the square

5 Play area on the square

6 Facilities beside 30m runway

7 Slide in front of the classroom

山行·竹荫里 3

2020台达杯国际太阳能建筑设计竞赛

Mountain Trip · In the Shade of Bamboo

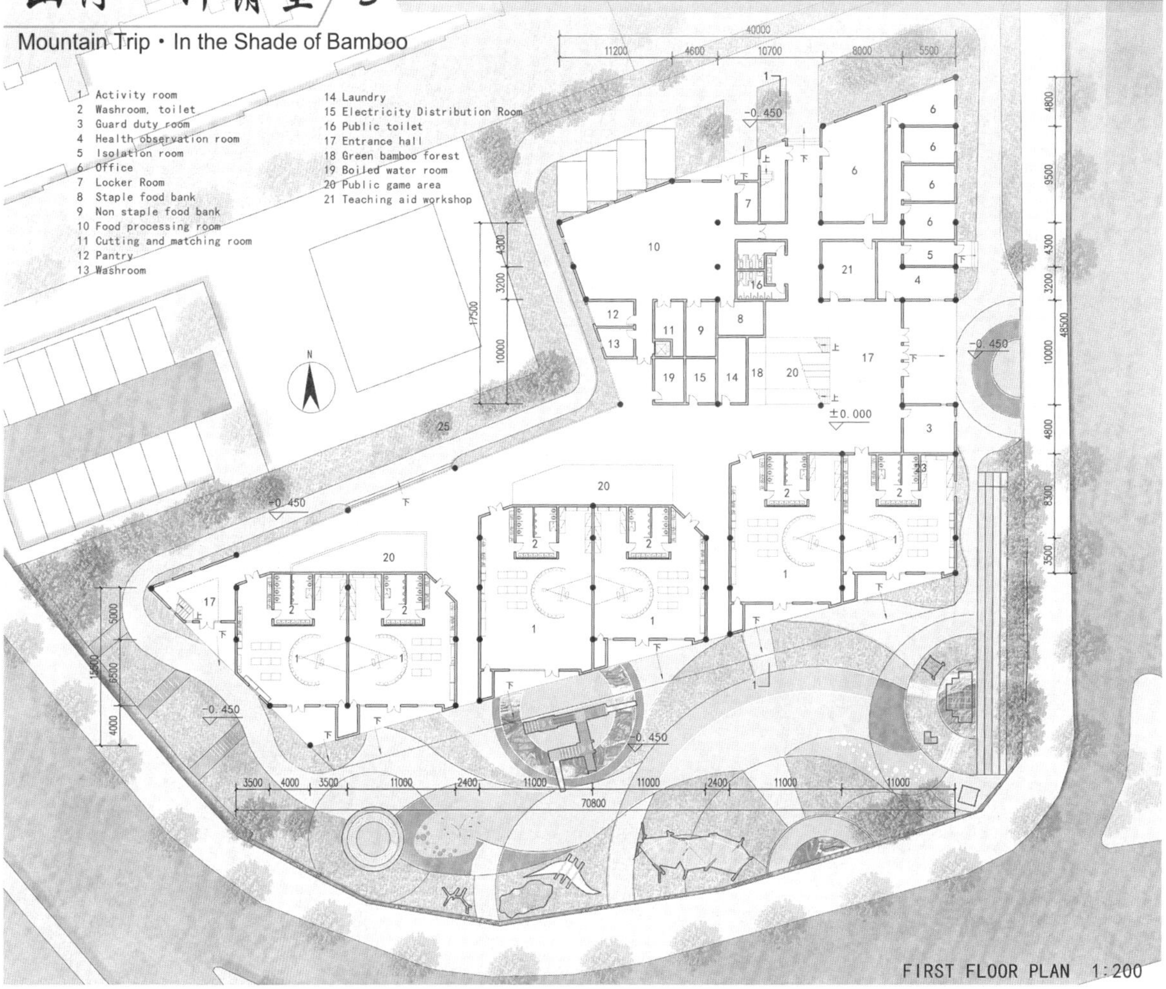
1 Activity room
2 Washroom, toilet
3 Guard duty room
4 Health observation room
5 Isolation room
6 Office
7 Locker Room
8 Staple food bank
9 Non staple food bank
10 Food processing room
11 Cutting and matching room
12 Pantry
13 Washroom
14 Laundry
15 Electricity Distribution Room
16 Public toilet
17 Entrance hall
18 Green bamboo forest
19 Boiled water room
20 Public game area
21 Teaching aid workshop
N
-0.450
±0.000
70800

FIRST FLOOR PLAN 1:200

INDOOR PERSPECTIVE

A. Climbing space

B. Creeping tunnel and ocean balls

C. Stalls

D. Children's reading area

FUNCTION STREAM

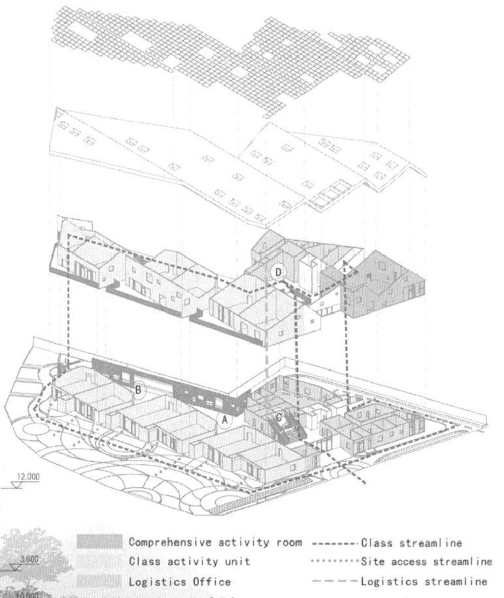
Comprehensive activity room
Class activity unit
Logistics Office
Class streamline
Site access streamline
Logistics streamline

12.000
3.600
±0.000

SECTION 1-1 1:250

山行·竹荫里 4

2020台达杯国际太阳能建筑设计竞赛

Mountain Trip · In the Shade of Bamboo

1 Activity room
2 Washroom, toilet
3 Public game area
4 Stalls
5 Reading area
6 Office
7 Conference room
8 Boiled water room
9 Director's office
10 Finance office
11 Storage room
12 Leisure platform
13 Tool room
14 Toilet
15 Control room
16 Comprehensive activity room

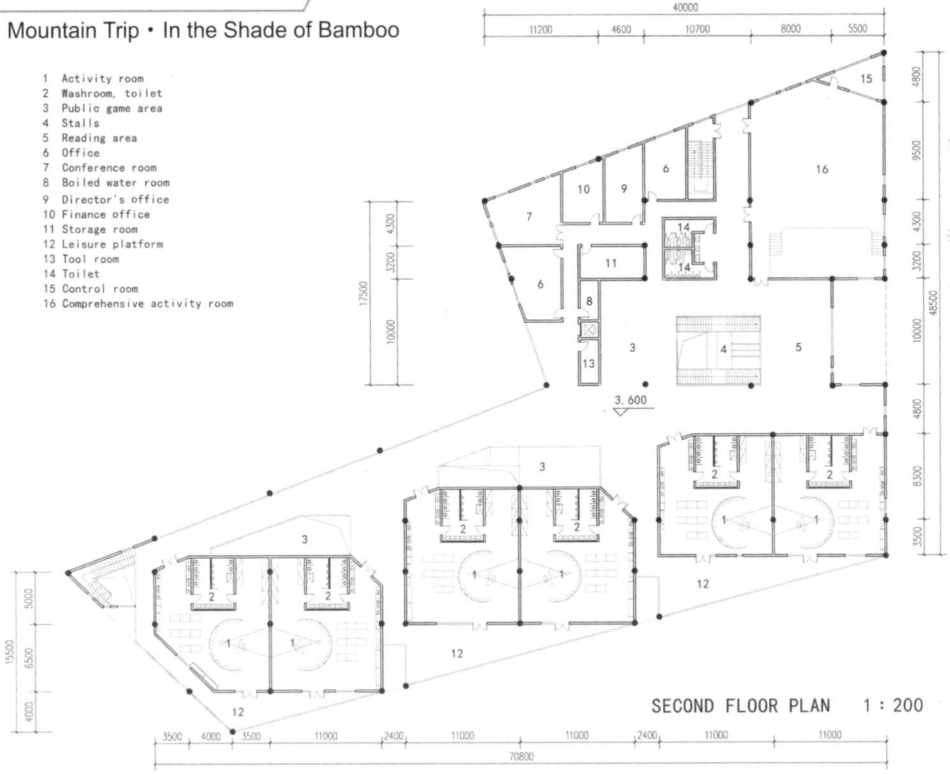

40000
11200
4600
10700
8000
5500
48500
70800
17500
15500
3.600
SECOND FLOOR PLAN 1 : 200

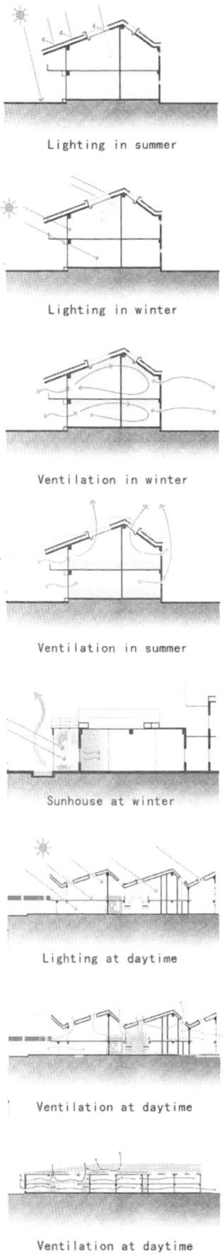

Lighting in summer
Lighting in winter
Ventilation in winter
Ventilation in summer
Sunhouse at winter
Lighting at daytime
Ventilation at daytime
Ventilation at daytime

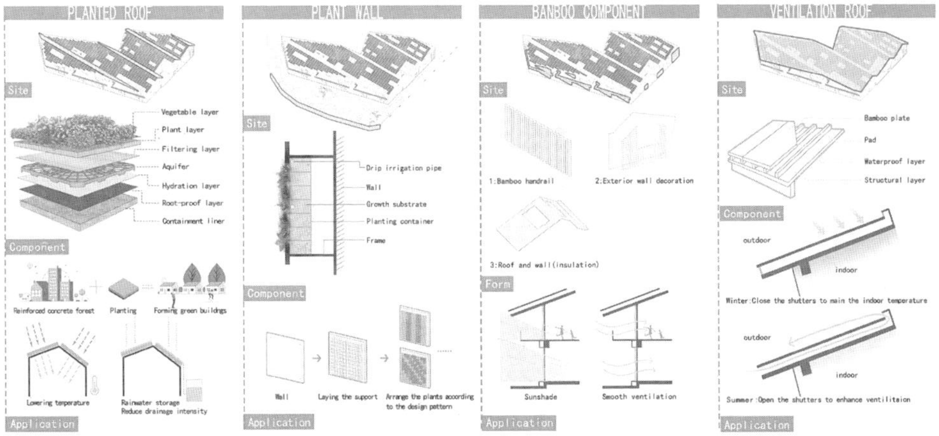

PLANTED ROOF
Site
Vegetable layer
Plant layer
Filtering layer
Aquifer
Hydration layer
Root-proof layer
Containment liner
Component
Reinforced concrete forest
Planting
Forming green buildings
Lowering temperature
Rainwater storage
Reduce drainage intensity
Application
PLANT WALL
Site
Drip irrigation pipe
Wall
Growth substrate
Planting container
Frame
Component
Wall
Laying the support
Arrange the plants according to the design pattern
Application
BAMBOO COMPONENT
Site
1:Bamboo handrail
2:Exterior wall decoration
3:Roof and wall(insulation)
Form
Sunshade
Smooth ventilation
Application
VENTILATION ROOF
Site
Bamboo plate
Pad
Waterproof layer
Structural layer
Component
outdoor
indoor
Winter:Close the shutters to main the indoor temperature
outdoor
indoor
Summer:Open the shutters to enhance ventilitaion
Application

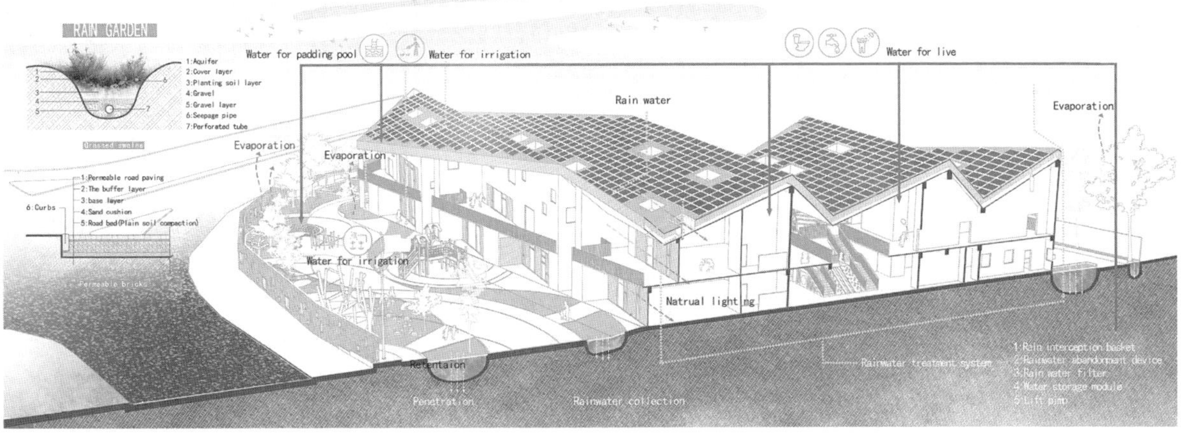

RAIN GARDEN
1:Aquifer
2:Cover layer
3:Planting soil layer
4:Gravel
5:Gravel layer
6:Seepage pipe
7:Perforated tube
1:Permeable road paving
2:The buffer layer
3:base layer
4:Sand cushion
5:Road bed(Plain soil compaction)
6:Curbs
Permeable bricks
Water for padding pool
Water for irrigation
Water for live
Rain water
Evaporation
Evaporation
Evaporation
Water for irrigation
Natrual lighting
Retentaion
Penetration
Rainwater collection
Rainwater treatment system
1:Rain interception basket
2:Rainwater abandonment device
3:Rain water filter
4:Water storage module
5:Lift pump

山行·竹荫里 / 5

2020台达杯国际太阳能建筑设计竞赛

Mountain Trip · In the Shade of Bamboo

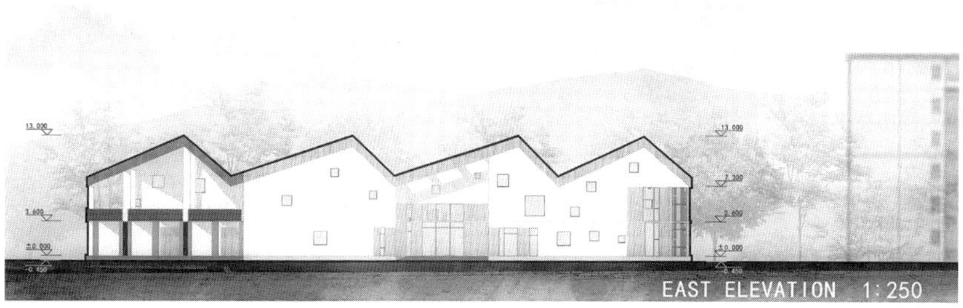
13.000
3.600
±0.000
EAST ELEVATION 1:250

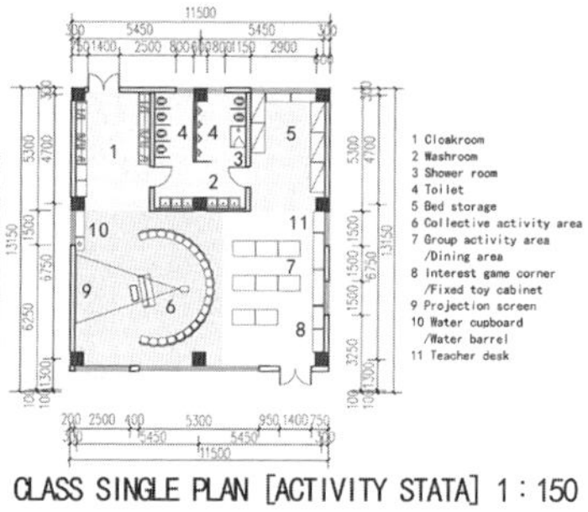
1 Cloakroom
2 Washroom
3 Shower room
4 Toilet
5 Bed storage
6 Collective activity area
7 Group activity area /Dining area
8 Interest game corner /Fixed toy cabinet
9 Projection screen
10 Water cupboard /Water barrel
11 Teacher desk
CLASS SINGLE PLAN [ACTIVITY STATA] 1：150

13.000
3.600
±0.000
NORTH ELEVATION 1:250

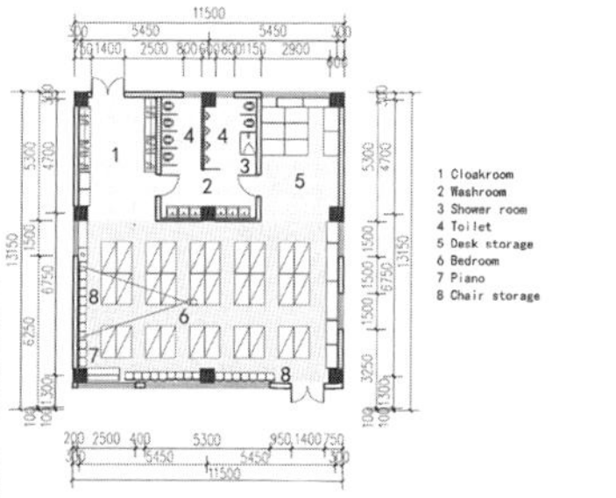
1 Cloakroom
2 Washroom
3 Shower room
4 Toilet
5 Desk storage
6 Bedroom
7 Piano
8 Chair storage
CLASS SINGLE PLAN [LUNCH BREAK] 1：150

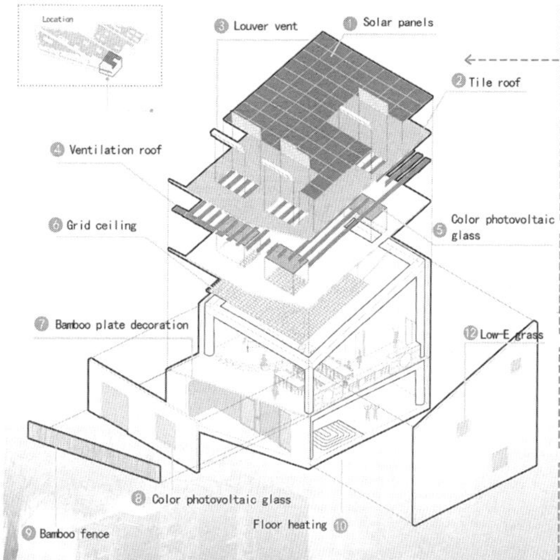
Location
① Solar panels
③ Louver vent
② Tile roof
④ Ventilation roof
⑥ Grid ceiling
⑤ Color photovoltaic glass
⑦ Bamboo plate decoration
⑫ Low-E grass
⑧ Color photovoltaic glass
⑨ Bamboo fence
Floor heating ⑪

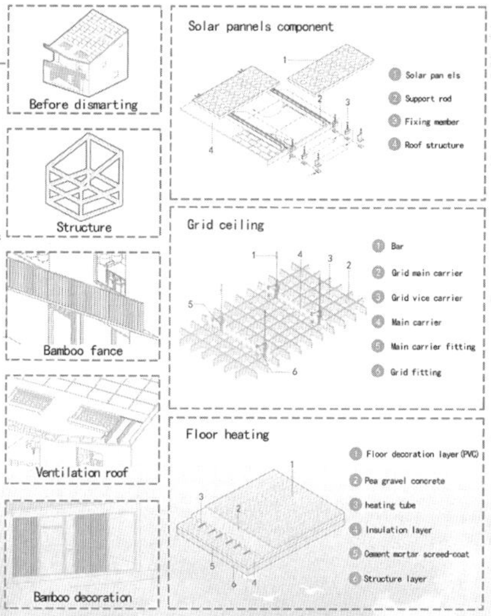
Before dismarting
Structure
Bamboo fance
Ventilation roof
Bamboo decoration
Solar pannels component
① Solar panels
② Support rod
③ Fixing member
④ Roof structure
Grid ceiling
① Bar
② Grid main carrier
③ Grid vice carrier
④ Main carrier
⑤ Main carrier fitting
⑥ Grid fitting
Floor heating
① Floor decoration layer (PVC)
② Pea gravel concrete
③ heating tube
④ Insulation layer
⑤ Cement mortar screed-coat
⑥ Structure layer

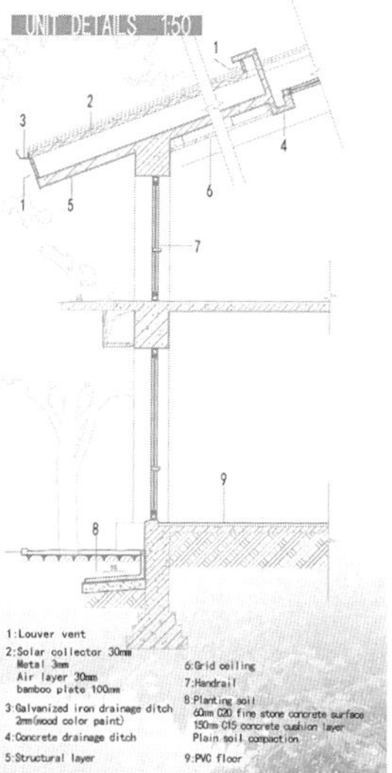
UNIT DETAILS 1:50
1:Louver vent
2:Solar collector 30mm
Metal 3mm
Air layer 30mm
bamboo plate 100mm
3:Galvanized iron drainage ditch 2mm(wood color paint)
4:Concrete drainage ditch
5:Structural layer
6:Grid ceiling
7:Handrail
8:Planting soil
9:PVC floor

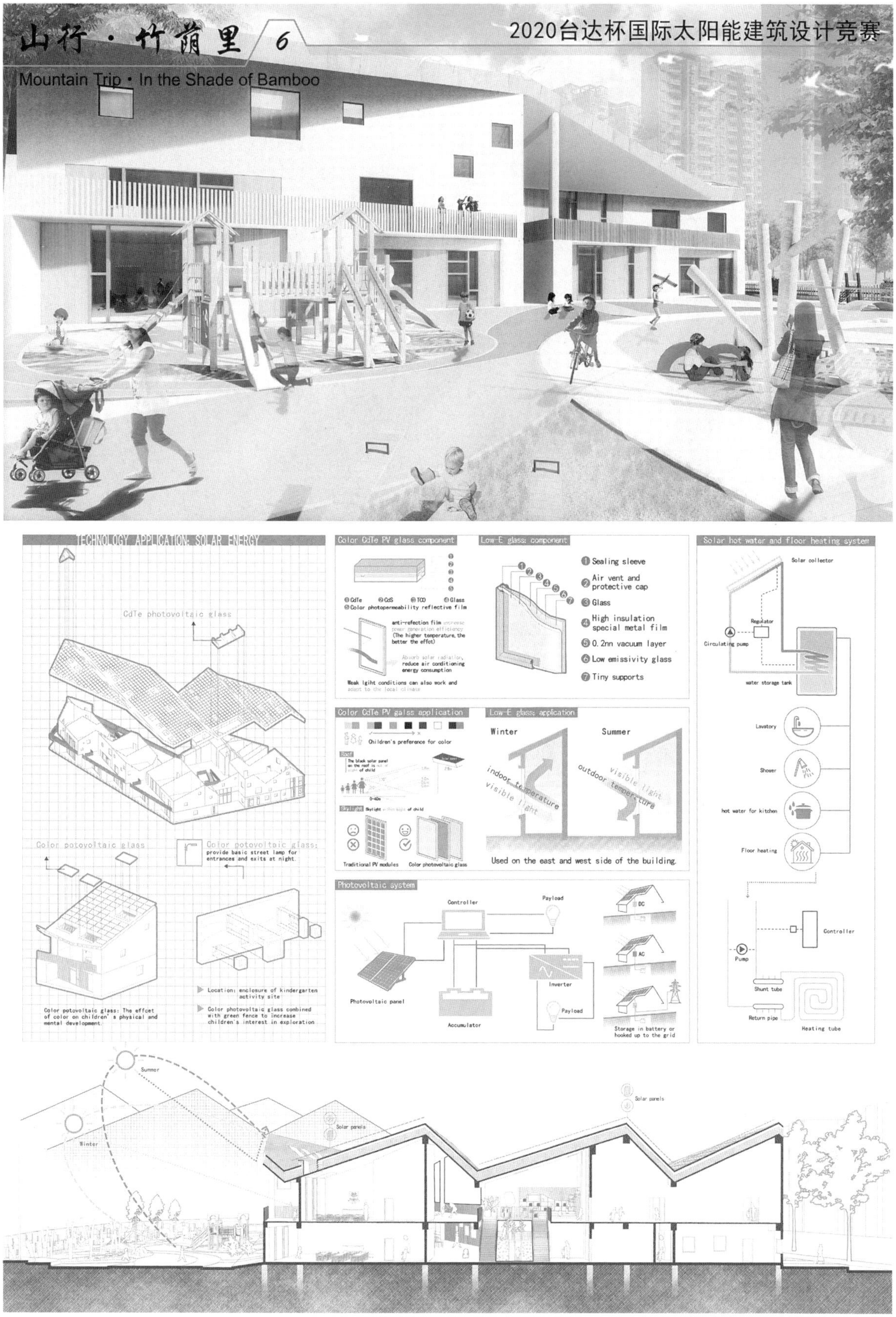

山行·竹荫里 6
2020台达杯国际太阳能建筑设计竞赛
Mountain Trip · In the Shade of Bamboo
TECHNOLOGY APPLICATION: SOLAR ENERGY
CdTe photovoltaic glass
Color potovoltaic glass
Color potovoltaic glass: provide basic street lamp for entrances and exits at night.
Color potovoltaic glass: The effcet of color on children' s physical and mental development.
Location: enclosure of kindergarten activity site
Color photovoltaic glass combined with green fence to increase children's interest in exploration
Color CdTe PV glass component
Low-E glass component
Sealing sleeve
Air vent and protective cap
Glass
High insulation special metal film
0.2nn vacuum layer
Low emissivity glass
Tiny supports
Weak light conditions can also work and adapt to the local climate
Color CdTe PV galss application
Children's preference for color
Traditional PV modules
Color photovoltaic glass
Low-E glass application
Winter
Summer
indoor temperature
visible light
outdoor temperature
Used on the east and west side of the building.
Photovoltaic system
Controller
Payload
Photovoltaic panel
Inverter
Accumulator
Storage in battery or hooked up to the grid
Solar hot water and floor heating system
Solar collector
Regulator
Circulating pump
water storage tank
Lavatory
Shower
hot water for kitchen
Floor heating
Controller
Pump
Shunt tube
Return pipe
Heating tube
Summer
Winter
Solar panels

综合奖 · 优秀奖
General Prize Awarded · Honorable Mention Prize

注 册 号：7165

项目名称：光合积木（南平）
Building Blocks of Sunlight (Nanping)

作　　者：明　健、张蔚杰

参赛单位：华南理工大学

指导教师：王　静

2020年台达杯国际太阳能建筑设计竞赛
2020 Taida Cup International solar Building Design competition

光合积木 I
Building Blocks of Sunlight

本方案以“光合积木”为主题，基于场地临山靠江地域特色、湿热多雨的气候特点，结合以太阳能利用为代表的一系列绿色建筑技术，贯彻“积木”主题概念与幼儿教育建筑特点，根据“积木缝隙”（空间）至“积木实体”（实体）这一系列虚实层次变化的绿色建筑设计策略，打造本地化绿色太阳能幼儿教育建筑。

主动式技术结合建筑屋顶，设有太阳能光伏与光热系统利用太阳能，同时利用屋面角度实现雨水收集与雨水花园设计；在被动式技术方面，设计利用自然采光降低能耗、太阳热辐射结合双层屋面实现热压通风、建筑形体缝隙实现自然通风、微气候庭院与灰空间达到建筑节能效果。

Description of design

This scheme takes "photosynthetic building blocks" as the theme. It combines the natural characteristics of the site to build a localized green solar energy early childhood education building.Based on the regional characteristics of the site near the mountain and river, the humid, hot and rainy climate, the green building technology is used to realize the green building design strategy from "building block gap" to "building block entity", which is a virtual and real level change, combined with the concept of "building block" and the demand of preschool education building.

In terms of passive technology, the design uses natural lighting to reduce energy consumption, solar thermal radiation combined with double-layer roof to achieve thermal pressure ventilation, building body gap to achieve natural ventilation, microclimate courtyard and gray space to achieve energy saving effect; active technology combined with building roof, with solar energy photovoltaic and photothermal system to use solar energy, while using roof angle to achieve rainwater collection and rainwater Garden design.

Diagram of process

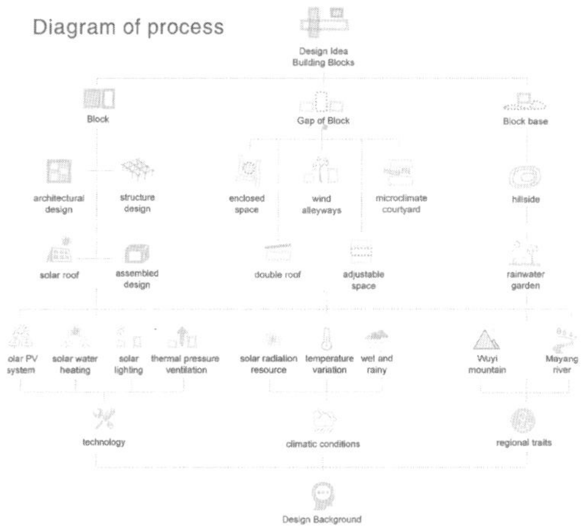
Design Idea
Building Blocks
Block
Gap of Block
Block base
architectural design
structure design
enclosed space
wind alleyways
microclimate courtyard
hillside
solar roof
assembled design
double roof
adjustable space
rainwater garden
olar PV system
solar water heating
solar lighting
thermal pressure ventilation
solar radiation resource
temperature variation
wet and rainy
Wuyi mountain
Mayang river
technology
climatic conditions
regional traits
Design Background

Regional feature

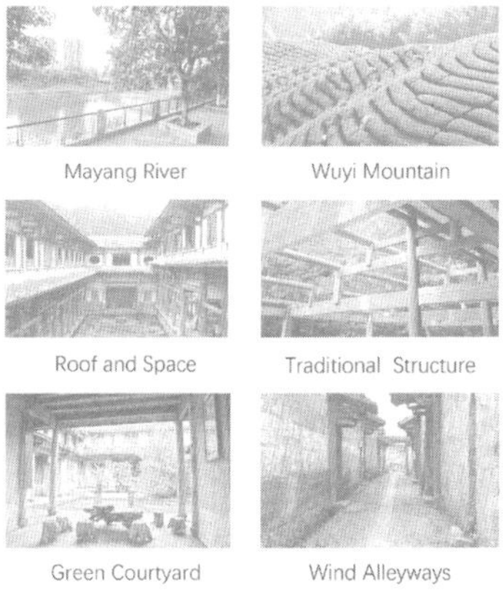
Mayang River
Wuyi Mountain
Roof and Space
Traditional Structure
Green Courtyard
Wind Alleyways

Prevailing Winds

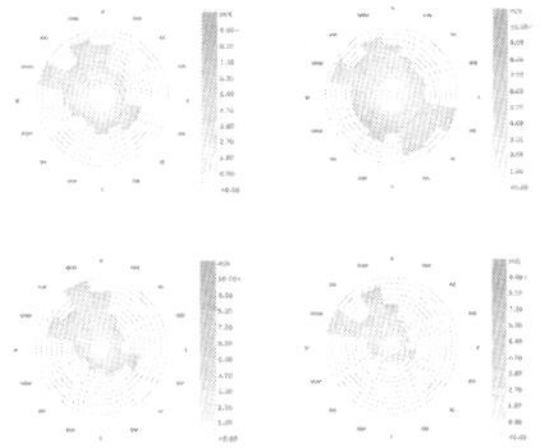

Psychrometric Chart

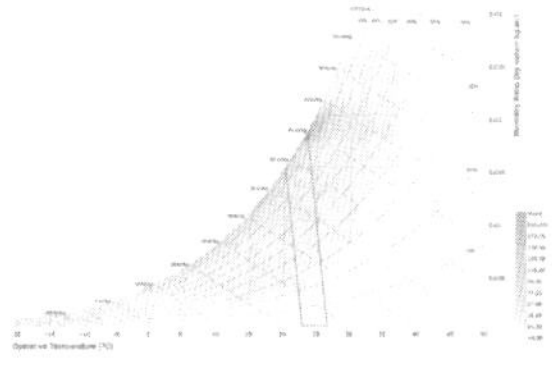

Location

Monthly Average

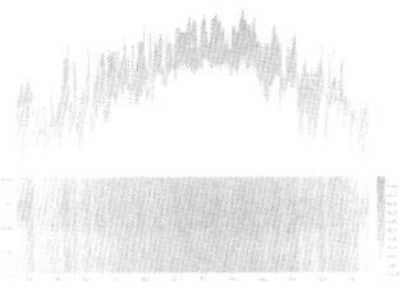

Optimum Orietation

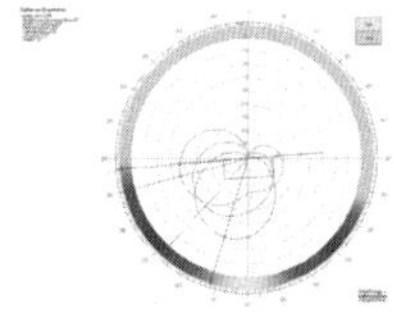

Site Analvsis

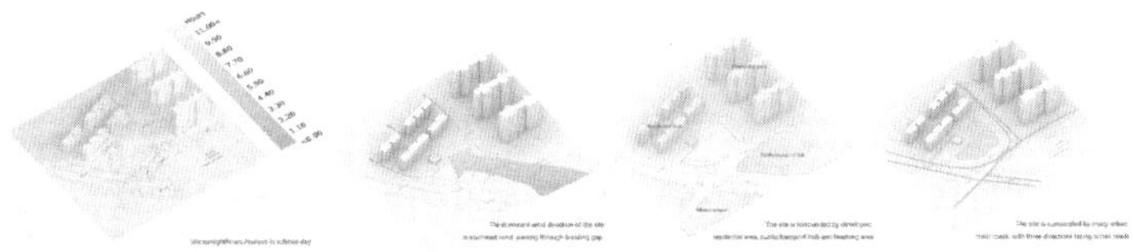

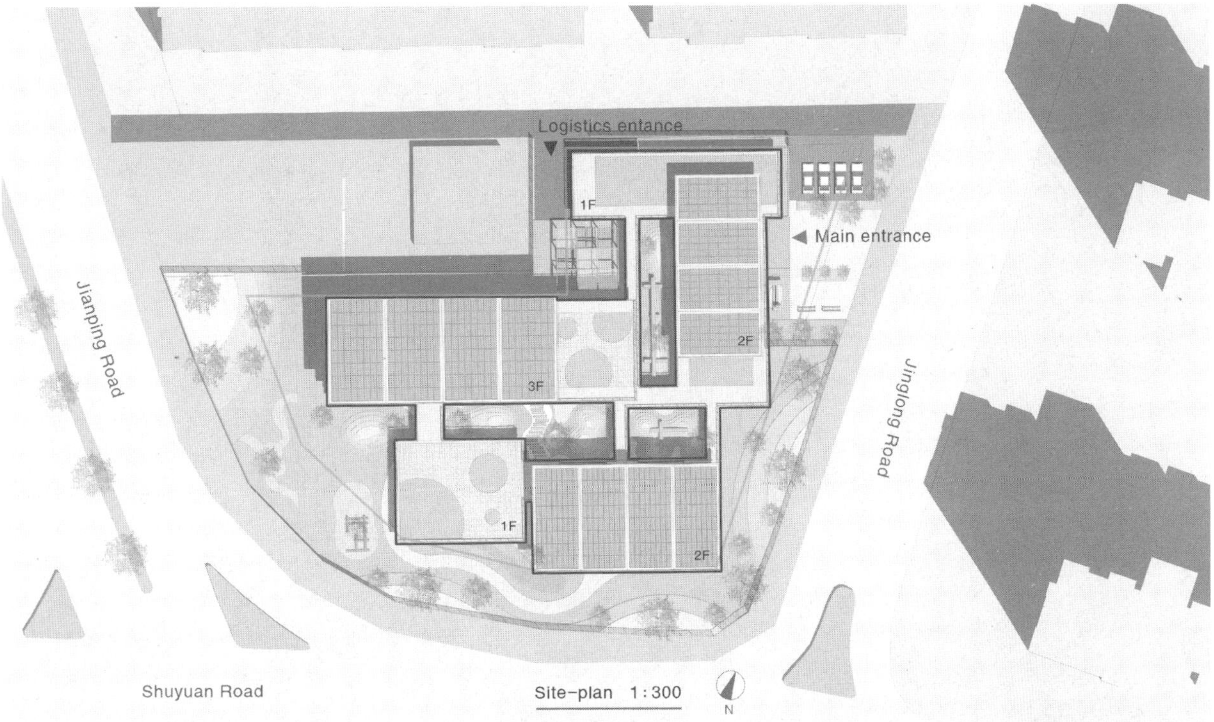
Logistics entrance
Main entrance
1F
2F
3F
1F
2F
Jianping Road
Jinglong Road
Shuyuan Road
N

Site-plan 1:300

Exploded view

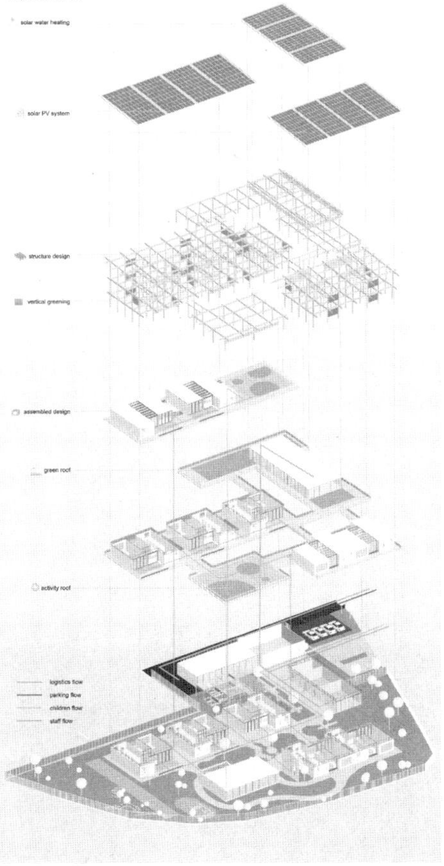
solar water heating
solar PV system
structure design
vertical greening
assembled design
green roof
activity roof
logistics flow
parking flow
children flow
staff flow

Wind simulation

Wind simulation at a height of 0.5m from the surface | Wind simulation at a height of 6m from the surface | Wind simulation at a height of 9m from the surface

Sunshine simulation

Winter solstice without building | Winter solstice | Summer solstice

Logical Generation

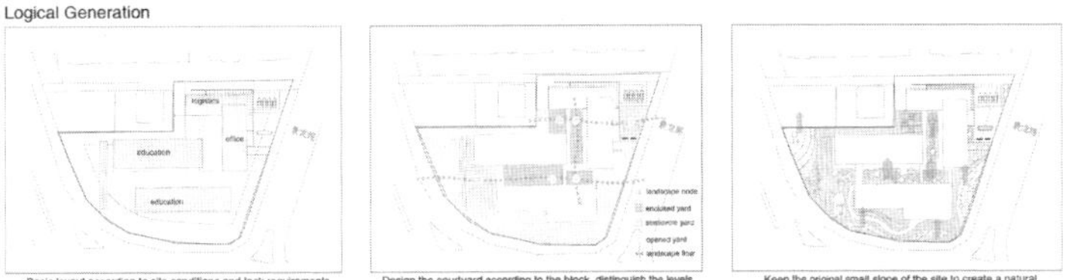

Basic layout according to site conditions and task requirements | Design the courtyard according to the block, distinguish the levels of the courtyard, and form the landscape flow | Keep the original small slope of the site to create a natural landscape for children

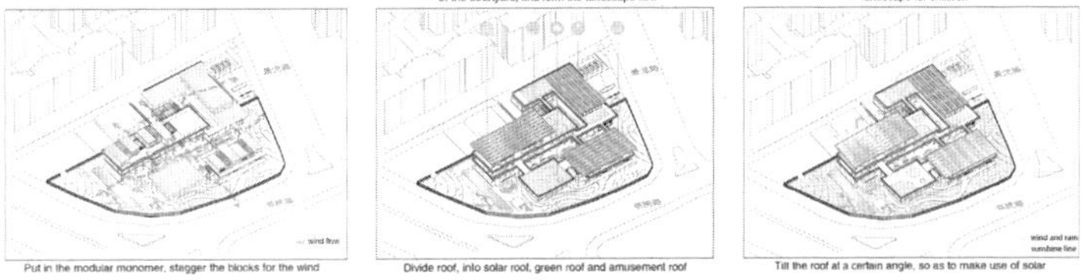

Put in the modular monomer, stagger the blocks for the wind | Divide roof, into solar roof, green roof and amusement roof | Tilt the roof at a certain angle, so as to make use of solar energy, ventilate and collect rainwater

Urban scale green strategy

As the urban daylight station, the surplus power is provided to the surrounding residents | As the green island of the city, clean the surrounding air | As the wind valley of the city, a wind road is reserved for the surrounding buildings

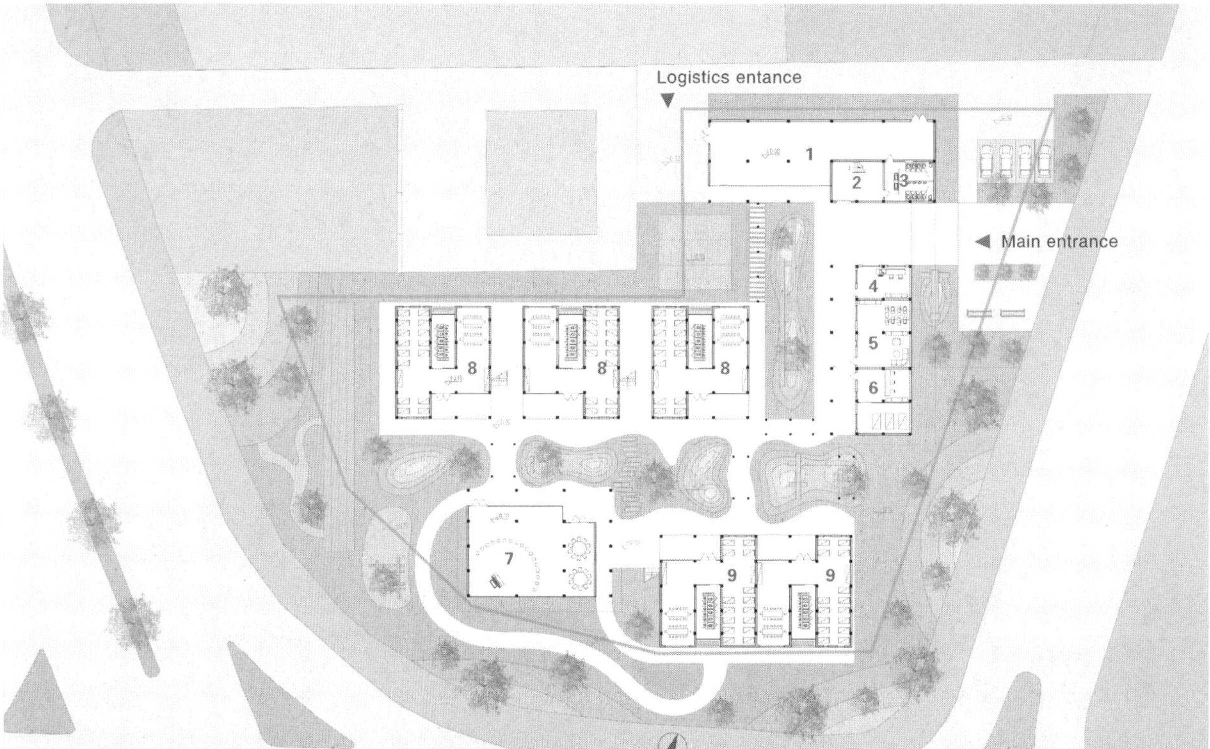
Logistics entance
Main entrance
N

First Floor Plan 1 : 250

1.Kitchen/ logistics
2.Utility room
3.Staff barhroom
4.Guard room
5.Health care room
6.Storeroom
7.Activity room
8.Middle class
9.Bottom class
10.Top class
11.Meetingroom
12.Teacher's office

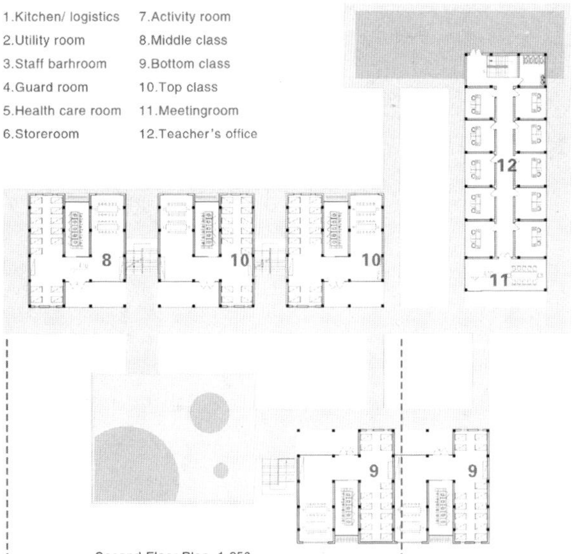

Second Floor Plan 1:250

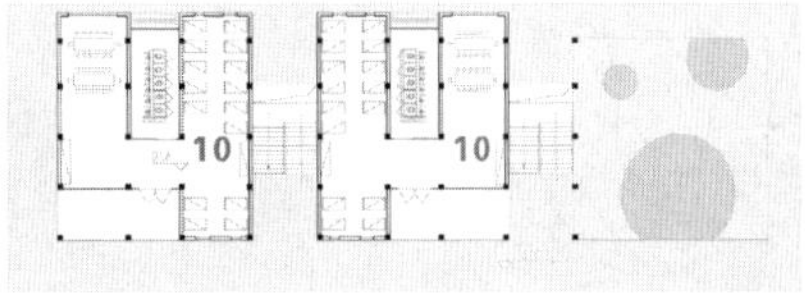

Third Floor Plan 1 : 250

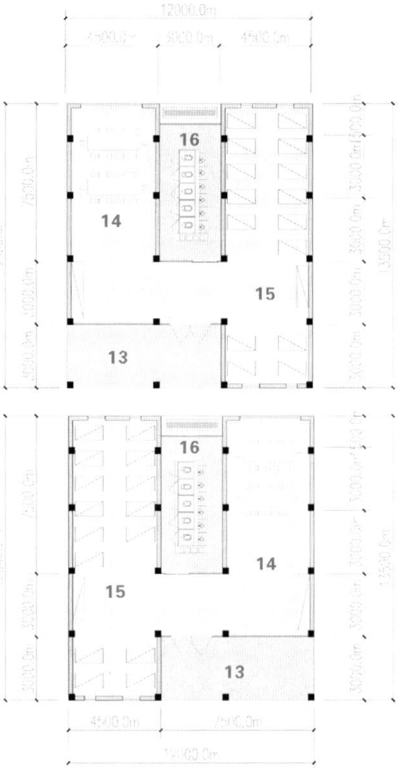

13.Outdoor activity space
14.Indoor activity space
15.Bedroom/cloakroom
16.Bathroom

Class Unit Plan 1 : 100

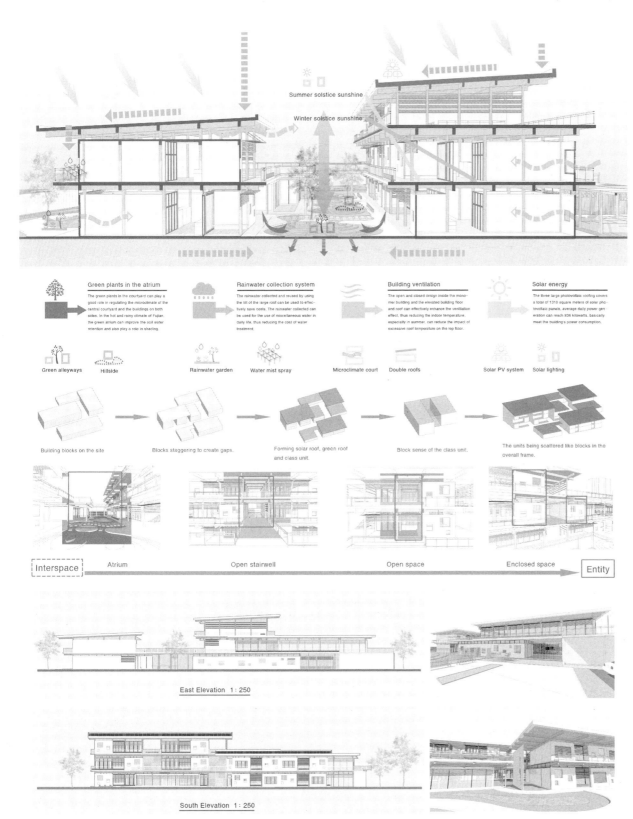

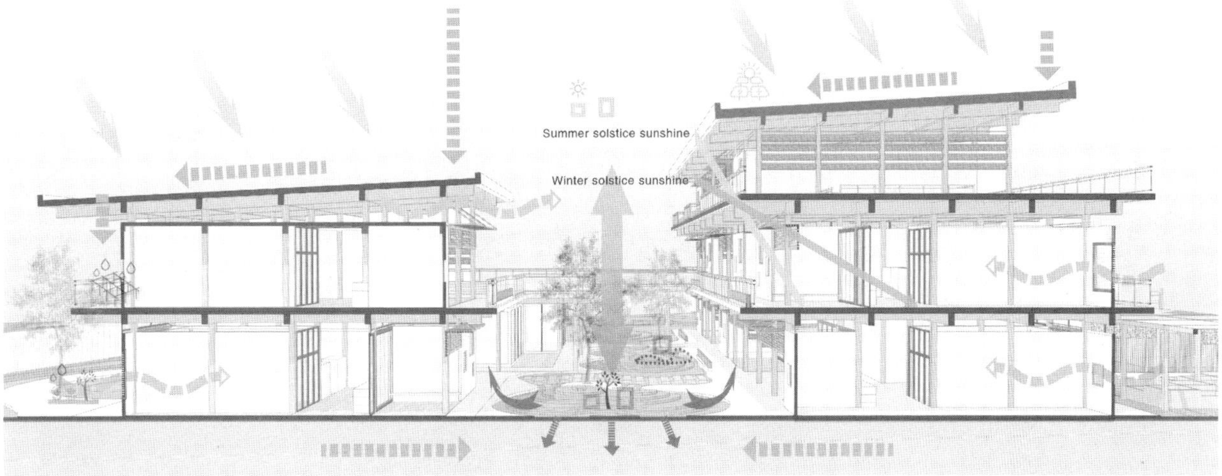
Summer solstice sunshine
Winter solstice sunshine

Green plants in the atrium

The green plants in the courtyard can play a good role in regulating the microclimate of the central courtyard and the buildings on both sides. In the hot and rainy climate of Fujian, the green atrium can improve the soil water retention and also play a role in shading.

Rainwater collection system

The rainwater collected and reused by using the tilt of the large roof can be used to effectively save costs. The rainwater collected can be used for the use of miscellaneous water in daily life, thus reducing the cost of water treatment.

Building ventilation

The open and closed design inside the monomer building and the elevated building floor and roof can effectively enhance the ventilation effect, thus reducing the indoor temperature, especially in summer, can reduce the impact of excessive roof temperature on the top floor.

Solar energy

The three large photovoltaic roofing covers a total of 1310 square meters of solar photovoltaic panels, average daily power generation can reach 836 kilowatts, basically meet the building's power consumption.

Green alleyways　Hillside　Rainwater garden　Water mist spray　Microclimate court　Double roofs　Solar PV system　Solar lighting

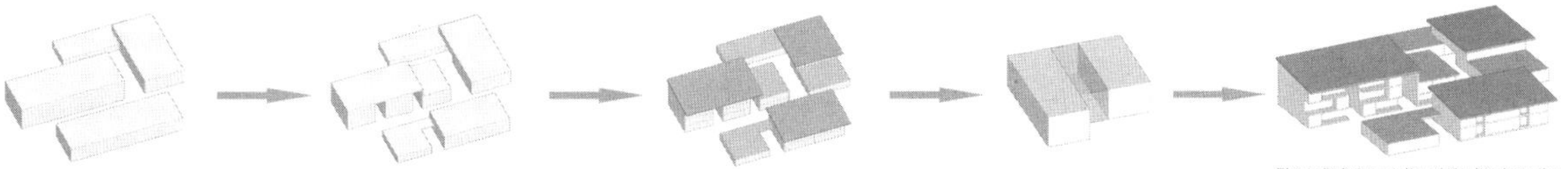

Building blocks on the site　Blocks staggering to create gaps.　Forming solar roof, green roof and class unit.　Block sense of the class unit.　The units being scattered like blocks in the overall frame.

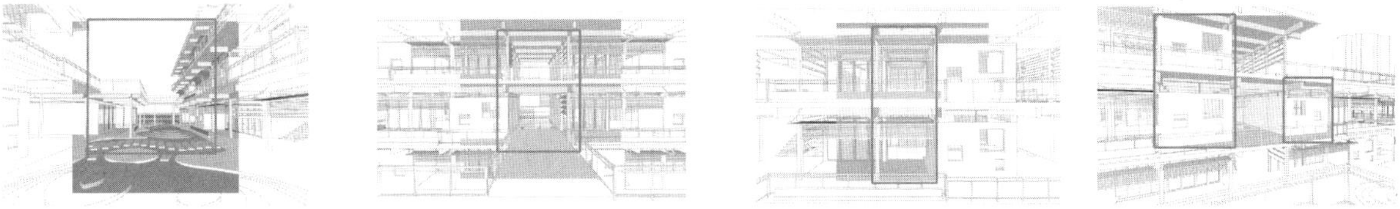

Interspace　Atrium　Open stairwell　Open space　Enclosed space　Entity

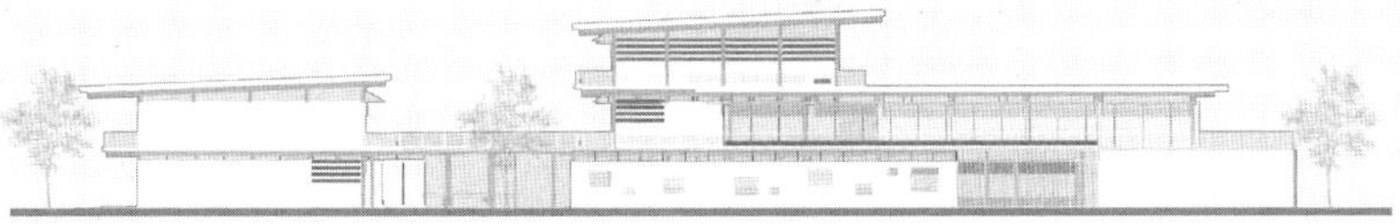

East Elevation 1：250

South Elevation 1：250

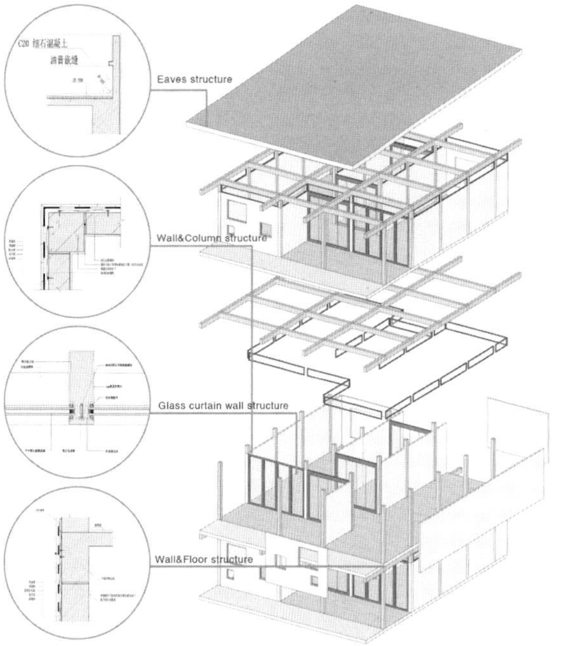
Eaves structure
Wall&Column structure
Glass curtain wall structure
Wall&Floor structure

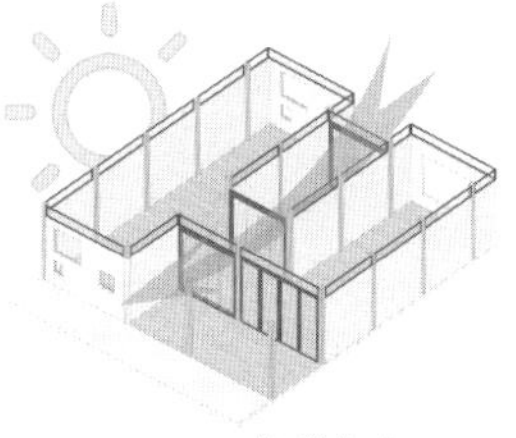

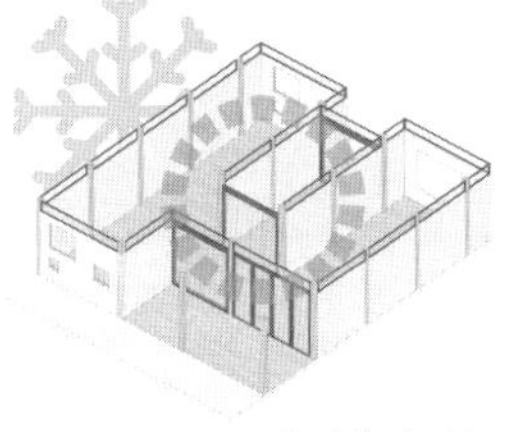

Ventilation in summer

In hot and humid summer, the middle of the glass partition can be opened, is the central space north-south transparent, indoor ventilation dehumidification, also can reduce the summer indoor temperature, save air conditioning cost especially after the opening of the toilet, can prevent damp breeding bacteria, can also avoid odor.

Insulation in winter

When it is cold in winter, the glass partition in the middle can be closed. When the sunlight is shining indoors in winter, this sunshine greenhouse can be formed to raise the indoor temperature and provide a more comfortable indoor environment. Meanwhile, large area of glass also enables more sunlight to enter the room in winter and create a bright environment.

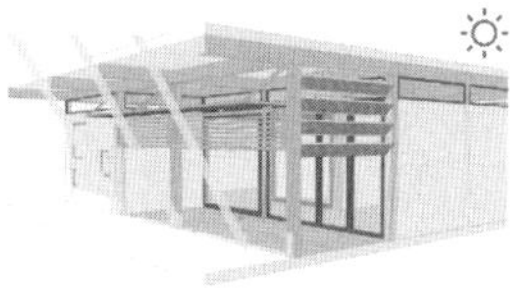

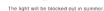

The light will be blocked out in summer.

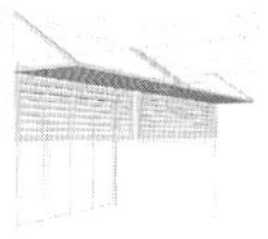

sunshade component

The wooden shading components can block the sun when children play in the outdoor areas, and reduce the spatial scale, giving children more sense of closeness and interaction.

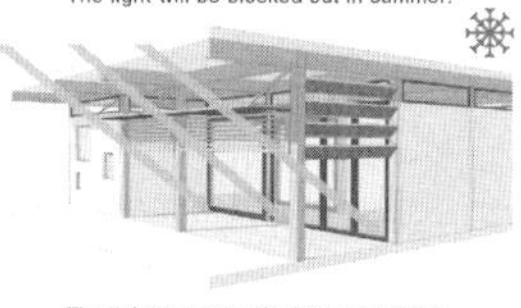

The light can enter the house in winter.

Plant accessories

The plant accessories located in the east-west direction can cover the problem of east and west sun in summer, while the rainwater collection of the roof can also be used for watering the plants.

Land saving and outdoor environment

节地与室外环境				
	标准条文	规范分值	权重	得分
控制项	4.1.1 项目选址应符合所在地城乡规划，且应符合各类保护区、文物古迹保护的建设控制要求	应达到	-	达到
	4.1.2 场地应无洪涝、滑坡、泥石流等自然灾害的威胁，无危险化学品、易燃易爆危险源的威胁，无电磁辐射、含氡土壤等危害	应达到	-	达到
	4.1.3 场地内不应有排放超标的污染源	应达到	-	达到
	4.1.4 建筑规划布局应满足日照标准，且不得降低周边建筑的日照标准	应达到	-	达到
评分项	4.2.1 节约集约利用土地	8	0.19	1.52
	4.2.2 场地内合理设置绿化用地	5	0.19	0.95
	4.2.3 合理开发利用地下空间	4	0.19	0
	4.2.4 建筑及照明设计避免产生光污染	4	0.19	0.57
	4.2.5 场地内环境噪声符合现行国家标准《声环境质量标准》的有关规定	4	0.19	0.57
	4.2.6 场地内风环境有利于室外行走、活动舒适和建筑的自然通风	4	0.19	0.76
	4.2.7 采取措施降低热岛强度	2	0.19	0.38
	4.2.8 场地与公共交通设施具有便捷的联系	5	0.19	0.57
	4.2.9 场地内人行通道采用无障碍设计	2	0.19	0.38
	4.2.10 合理设置停车场所	3	0.19	0.57
	4.2.11 提供便利的公共服务	5	0.19	0.76
	4.2.12 结合现状地形地貌进行场地设计与建筑布局，保护场地内原有的自然水域、湿地和植被，采取表层土利用等生态补偿措施	3	0.19	0.57
	4.2.13 充分利用场地空间合理设置绿色雨水基础设施，对大于 10hm2 的场地进行雨水专项规划设计	4	0.19	0.76
	4.2.14 合理规划地表与屋面雨水径流，对场地雨水实施外排总量控制，其场地年径流总量控制率达到 55%	2	0.19	0.19
	4.2.15 合理选择绿化方式，科学配置绿化植物	3	0.19	0.57

Indoor environmental quality

室内环境质量				
	标准条文	规范分值	权重	得分
控制项	8.1.1 主要功能房间的室内噪声级应满足现行国家标准《民用建筑隔声设计规范》GB 50118 中的低限要求	应达到	-	达到
	8.1.2 主要功能房间的外墙、隔墙、楼板和门窗的隔声性能应满足现行国家标准《民用建筑隔声设计规范》GB 50118 中的低限要求	应达到	-	达到
	8.1.3 建筑照明数量和质量应符合现行国家标准《建筑照明设计标准》GB 50034 的规定	应达到	-	达到
		应达到	-	达到
	8.1.4 采用集中供暖空调系统的建筑，房间内的温度、湿度、新风量等设计参数应符合现行国家标准《民用建筑供暖通风与空气调节设计规范》GB 50736 的规定	应达到	-	达到
	8.1.5 在室内设计温、湿度条件下，建筑围护结构内表面不得结露	应达到	-	达到
	8.1.6 屋顶和东西外墙隔热性能应满足现行国家标准《民用建筑热工设计规范》GB 50176 的要求	应达到	-	达到
	8.1.7 室内空气中的氨、甲醛、苯、总挥发性有机物、氡等污染物浓度应符合现行国家标准《室内空气质量标准》GB/T 18883 的有关规定	应达到	-	达到
评分项	8.2.1 主要功能房间室内噪声级，评价总分值为 6 分。噪声级达到现行国家标准《民用建筑隔声设计规范》GB 50118 中的低限标准限值和高要求标准限值的平均值	2	0.19	0.38
	8.2.2 主要功能房间的隔声性能良好	6	0.19	1.14
	8.2.3 采取减少噪声干扰的措施	4	0.19	0.76
	8.2.4 公共建筑中的多功能厅、接待大厅、大型会议室和其他有声学要求的重要房间进行专项声学设计，满足相应功能要求	2	0.19	0.19
	8.2.5 建筑主要功能房间具有良好的户外视野，评价分值为 3 分。对居住建筑，其与相邻建筑的直接间距超过 18m；对公共建筑，其主要功能房间能通过外窗看到室外自然景观，无明显视线干扰	4	0.19	0.76
	8.2.6 主要功能房间的采光系数满足现行国家标准《建筑采光设计标准》GB 50033 的要求	4	0.19	0.76
	8.2.7 改善建筑室内天然采光效果	7	0.19	1.33
	8.2.8 采取可调节遮阳措施，降低夏季太阳辐射得热	8	0.19	1.52
	8.2.9 供暖空调系统末端现场可独立调节	6	0.19	0.38
	8.2.10 优化建筑空间、平面布局和构造设计，改善自然通风效果	7	0.19	1.33
	8.2.11 气流组织合理	6	0.19	1.14
	8.2.12 主要功能房间中人员密度较高且随时间变化大的区域设置室内空气质量监控系统	4	0.19	0.38
	8.2.13 地下车库设置与排风设备联动的一氧化碳浓度监测装置	5	0.19	0

Energy saving and using

节能与能源利用				
	标准条文	规范分值	权重	得分
控制项	5.1.1 建筑设计应符合国家现行有关建筑节能设计标准中强制性条文的规定	应达到	-	达到
	5.1.2 不应采用电直接加热设备作为供暖空调系统的供暖热源和空气加湿热源	应达到	-	达到
	5.1.3 冷热源、输配系统和照明等各部分能耗应进行独立分项计量	应达到	-	达到
	5.1.4 各房间或场所的照明功率密度值不得高于现行国家标准《建筑照明设计标准》GB 50034 中规定的现行值	应达到	-	达到
评分项	5.2.1 结合场地自然条件，对建筑的体形、朝向、楼距、窗墙比等进行优化设计	3	0.24	0.72
	5.2.2 外窗、玻璃幕墙的可开启部分能使建筑获得良好的通风	3	0.24	0.72
	5.2.3 围护结构热工性能指标优于国家现行有关建筑节能设计标准的规定	6	0.24	1.2
	5.2.4 供暖空调系统的冷、热源机组能效均优于现行国家标准《公共建筑节能设计标准》GB 50189 的规定以及现行有关国家标准能效限定值的要求	4	0.24	0.96
	5.2.5 集中供暖系统热水循环泵的耗电输热比和通风空调系统风机的单位风量耗功率符合现行国家标准《公共建筑节能设计标准》GB 50189 等的有关规定	4	0.24	0.72
	5.2.6 合理选择和优化供暖、通风与空调系统	4	0.24	0.72
	5.2.7 采取措施降低过渡季节供暖、通风与空调系统能耗	3	0.24	0.72
	5.2.8 采取措施降低部分负荷、部分空间使用下的供暖、通风与空调系统能耗	6	0.24	1.44
	5.2.9 走廊、楼梯间、门厅、大堂、大空间、地下停车场等场所的照明系统采取分区、定时、感应等节能控制措施	3	0.24	0.72
	5.2.10 照明功率密度值达到现行国家标准《建筑照明设计标准》GB 50034 中的目标值规定	6	0.24	1.44
	5.2.11 合理选用电梯和自动扶梯，并采取电梯群控、扶梯自动启停等节能控制措施	2	0.24	0
	5.2.12 合理选用节能型电气设备	2	0.24	0.48
	5.2.13 排风能量回收系统设计合理并运行可靠	2	0.24	0.72
	5.2.14 合理采用蓄冷蓄热系统	0	0.24	0
	5.2.15 合理利用余热废热解决建筑的蒸汽、供暖或生活热水需求	0	0.24	0
	5.2.16 根据当地气候和自然资源条件，合理利用可再生能源	6	0.24	1.44

Water saving and using

节水与水资源利用				
	标准条文	规范分值	权重	得分
控制项	6.1.1 应制定水资源利用方案，统筹利用各种水资源	应达到	-	达到
	6.1.2 给排水系统设置应合理、完善、安全	应达到	-	达到
	6.1.3 应采用节水器具	应达到	-	达到
评分项	6.2.1 建筑平均日用水量满足现行国家标准《民用建筑节水设计标准》GB 50555 中的节水用水定额的要求	6	0.19	1.14
	6.2.2 采取有效措施避免管网漏损	5	0.19	0.38
	6.2.3 给水系统无超压出流现象	4	0.19	0.76
	6.2.4 设置用水计量装置	4	0.19	0.76
	6.2.5 公用浴室采取节水措施	0	0.19	0
	6.2.6 使用较高用水效率等级的卫生器具	8	0.19	1.14
	6.2.7 绿化灌溉采用节水灌溉方式	6	0.19	1.14
	6.2.8 空调设备或系统采用节水冷却技术	8	0.19	1.52
	6.2.9 除卫生器具、绿化灌溉和冷却塔外的其他用水采用了节水技术或措施	3	0.19	0.57
	6.2.10 合理使用非传统水源	10	0.19	1.33
	6.2.11 冷却水补水使用非传统水源	0	0.19	0
	6.2.12 结合雨水利用设施进行景观水体设计，景观水体利用水的补水量大于其水体蒸发量的 60%，且采用生态水处理技术保障水体质	6	0.19	0.76

Overview

	节地与室外环境	节能与能源利用	节水与水资源利用	节材与材料资源利用	室内环境质量
控制项	√	√	√	√	√
评分项	48	50	50	56	53
权重	0.19	0.24	0.19	0.19	0.19
得分	9.12	12	9.5	10.64	10.07
总分	51.33				
星级	★ ★				

综合奖 · 优秀奖
General Prize Awarded · Honorable Mention Prize

注 册 号：7200

项目名称：童年 · 里巷 · 巢下（南平）
Childhood · Cold Lane · Under the Nest（Nanping）

作　　者：谭若晨、付浩然、高　力、王胜娟、刘宸溪、牛静仁

参赛单位：石家庄铁道大学、中铁建安工程设计有限公司

指导教师：樊海彬、高力强

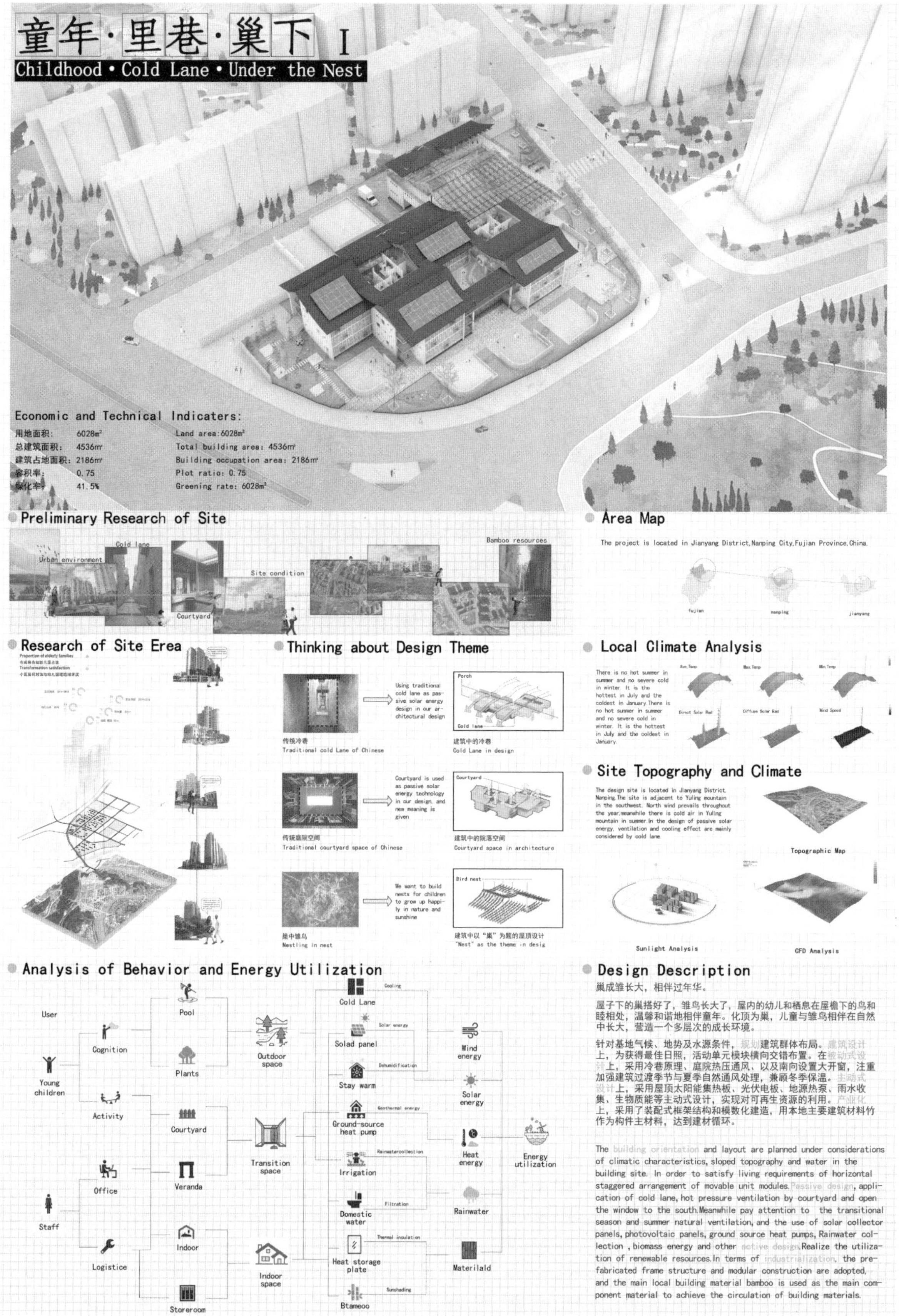
童年 · 里巷 · 巢下 I
Childhood • Cold Lane • Under the Nest
Economic and Technical Indicaters:
用地面积：6028m²
总建筑面积：4536m²
建筑占地面积：2186m²
容积率：0.75
绿化率：41.5%
Land area:6028m²
Total building area: 4536m²
Building occupation area: 2186m²
Plot ratio: 0.75
Greening rate: 6028m²
Preliminary Research of Site
Urban environment
Cold lane
Courtyard
Site condition
Bamboo resources
Area Map
The project is located in Jianyang District,Nanping City,Fujian Province,China.
Research of Site Erea
Thinking about Design Theme
传统冷巷
Traditional cold Lane of Chinese
建筑中的冷巷
Cold Lane in design
传统庭院空间
Traditional courtyard space of Chinese
建筑中的院落空间
Courtyard space in architecture
巢中雏鸟
Nestling in nest
Local Climate Analysis
Site Topography and Climate
Topographic Map
Sunlight Analysis
CFD Analysis
Analysis of Behavior and Energy Utilization
User
Young children
Staff
Cognition
Activity
Office
Logistice
Pool
Plants
Courtyard
Veranda
Indoor
Storeroom
Outdoor space
Transition space
Indoor space
Cold Lane
Solad panel
Stay warm
Ground-source heat pump
Irrigation
Domestic water
Heat storage plate
Btameoo
Wind energy
Solar energy
Heat energy
Rainwater
Materilald
Energy utilization
Design Description
巢成雏长大，相伴过年华。
屋子下的巢搭好了，雏鸟长大了，屋内的幼儿和栖息在屋檐下的鸟和睦相处，温馨和谐地相伴童年。化顶为巢，儿童与雏鸟相伴在自然中长大，营造一个多层次的成长环境。
针对基地气候、地势及水源条件，规划建筑群体布局。建筑设计上，为获得最佳日照，活动单元模块横向交错布置。在被动式设计上，采用冷巷原理、庭院热压通风、以及南向设置大开窗，注重加强建筑过渡季节与夏季自然通风处理，兼顾冬季保温。主动式设计上，采用屋顶太阳能集热板、光伏电板、地源热泵、雨水收集、生物质能等主动式设计，实现对可再生资源的利用。产业化上，采用了装配式框架结构和模数化建造，用本地主要建筑材料竹作为构件主材料，达到建材循环。
The building orientation and layout are planned under considerations of climatic characteristics, sloped topography and water in the building site. In order to satisfy living requirements of horizontal staggered arrangement of movable unit modules. Passive design, application of cold lane, hot pressure ventilation by courtyard and open the window to the south. Meanwhile pay attention to the transitional season and summer natural ventilation, and the use of solar collector panels, photovoltaic panels, ground source heat pumps, Rainwater collection , biomass energy and other active design. Realize the utilization of renewable resources. In terms of industrialization, the prefabricated frame structure and modular construction are adopted, and the main local building material bamboo is used as the main component material to achieve the circulation of building materials.

童年·里巷·巢下 II
Childhood • Cold Lane • Under the Nest

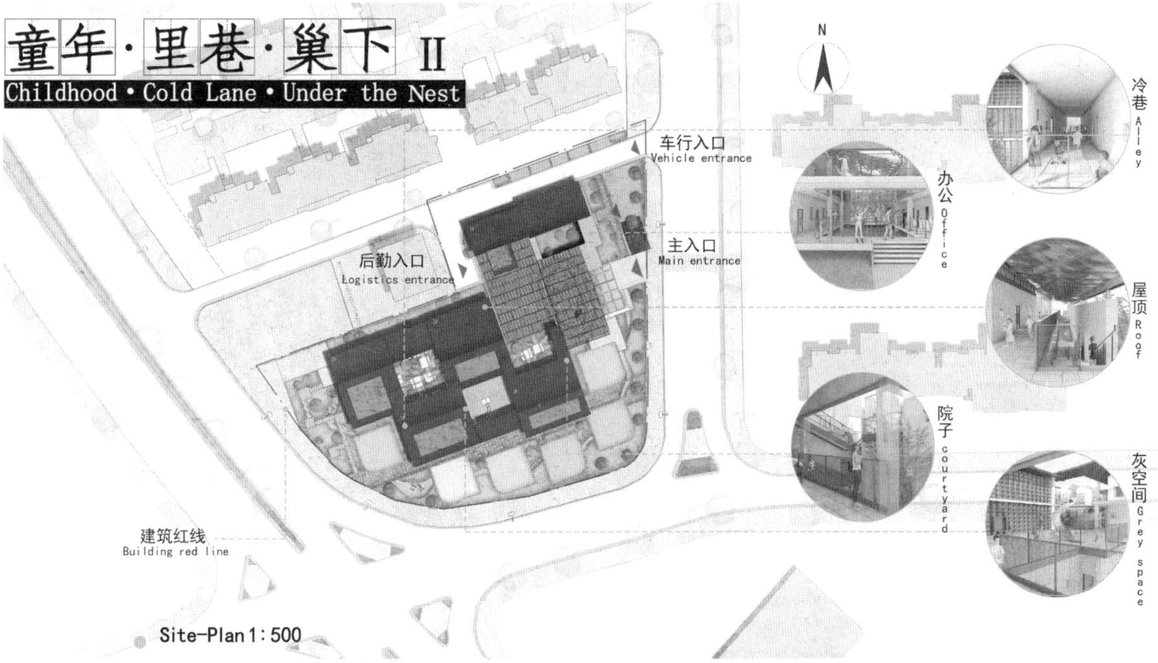
N
车行入口
Vehicle entrance
主入口
Main entrance
后勤入口
Logistics entrance
建筑红线
Building red line
冷巷
Alley
办公
Office
屋顶
Roof
院子
courtyard
灰空间
Grey space

Site-Plan 1∶500

Logical Generation

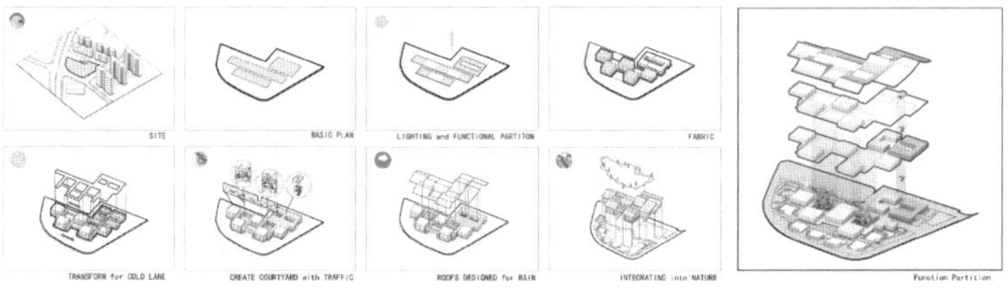
SITE
BASIC PLAN
LIGHTING and FUNCTIONAL PARTITION
FABRIC
TRANSFORM for COLD LANE
CREATE COURTYARD with TRAFFIC
ROOFS DESIGNED for RAIN
INTEGRATING into NATURE
Function Partition

Analysis of Functions and Streamline

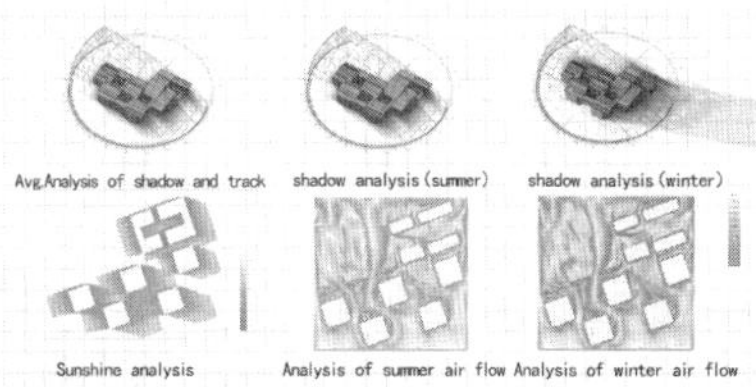
Avg.Analysis of shadow and track
shadow analysis (summer)
shadow analysis (winter)
Sunshine analysis
Analysis of summer air flow
Analysis of winter air flow

Green building Analysis

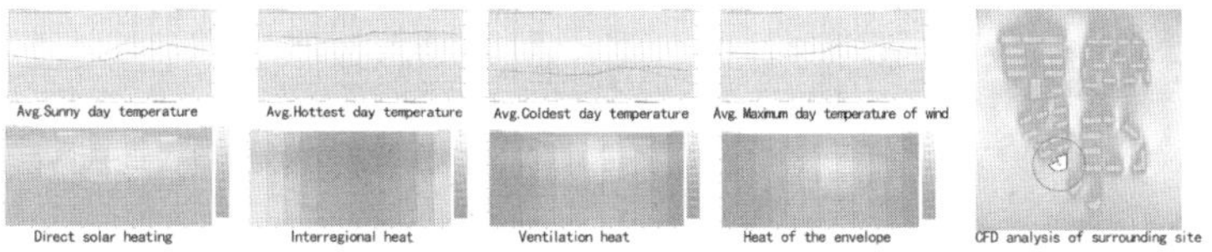
Avg.Sunny day temperature
Avg.Hottest day temperature
Avg.Coldest day temperature
Avg. Maximum day temperature of wind
Direct solar heating
Interregional heat
Ventilation heat
Heat of the envelope
CFD analysis of surrounding site

Climate Analysis

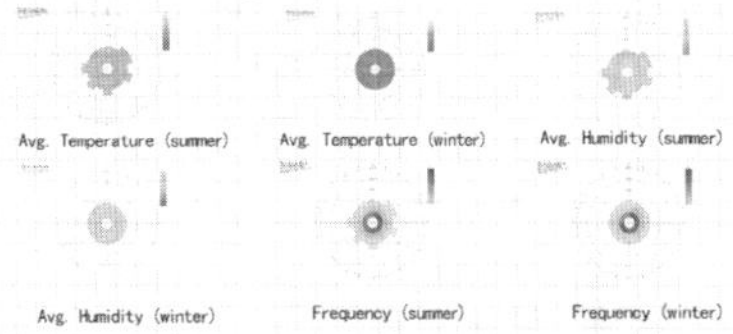
Avg. Temperature (summer)
Avg. Temperature (winter)
Avg. Humidity (summer)
Avg. Humidity (winter)
Frequency (summer)
Frequency (winter)

Crowd Activity and Solar Eneragy

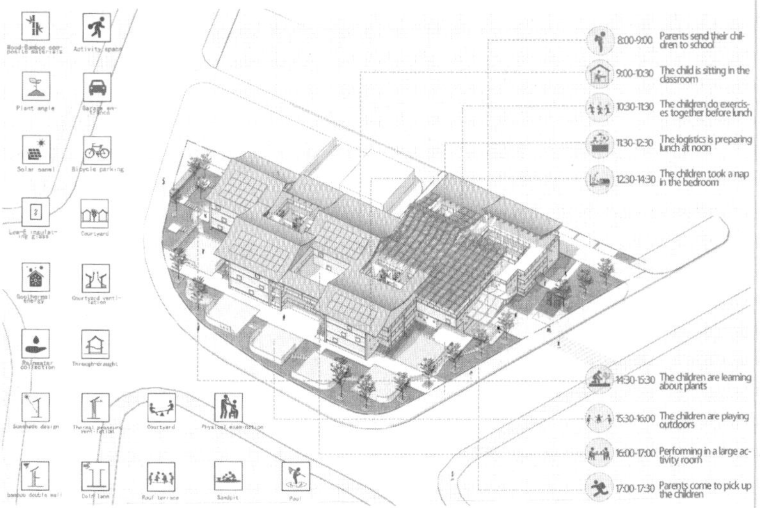
Activity space
Plant angle
Solar panel
Bicycle parking
Courtyard
Rainwater collection
Throughdraught
Courtyard
Sandpit
Pool
8:00-9:00 Parents send their children to school
9:00-10:30 The child is sitting in the classroom
10:30-11:30 The children do exercises together before lunch
11:30-12:30 The logistics is preparing lunch at noon
12:30-14:30 The children took a nap in the bedroom
14:30-15:30 The children are learning about plants
15:30-16:00 The children are playing outdoors
16:00-17:00 Performing in a large activity room
17:00-17:30 Parents come to pick up the children

Energy and Material Saving Effects

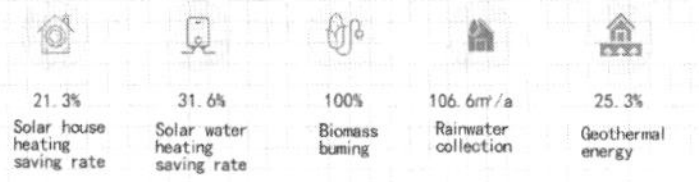
21.3%
Solar house heating saving rate
31.6%
Solar water heating saving rate
100%
Biomass burning
106.6m³/a
Rainwater collection
25.3%
Geothermal energy

Material Cycles of Ceramics

The building materials are rich in the areas around the project site.

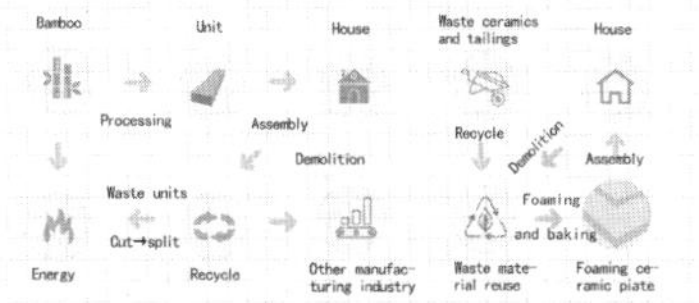
Bamboo
Unit
House
Waste ceramics and tailings
House
Processing
Assembly
Demolition
Recycle
Demolition
Assembly
Waste units
Cut→split
Foaming and baking
Energy
Recycle
Other manufacturing industry
Waste material reuse
Foaming ceramic plate

Percentage of Each Material

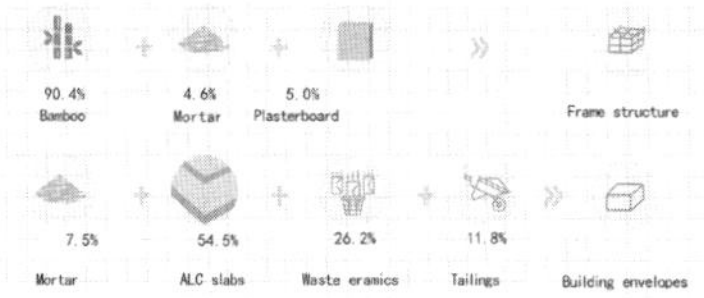
90.4%
Bamboo
4.6%
Mortar
5.0%
Plasterboard
Frame structure
7.5%
Mortar
54.5%
ALC slabs
26.2%
Waste ceramics
11.8%
Tailings
Building envelopes

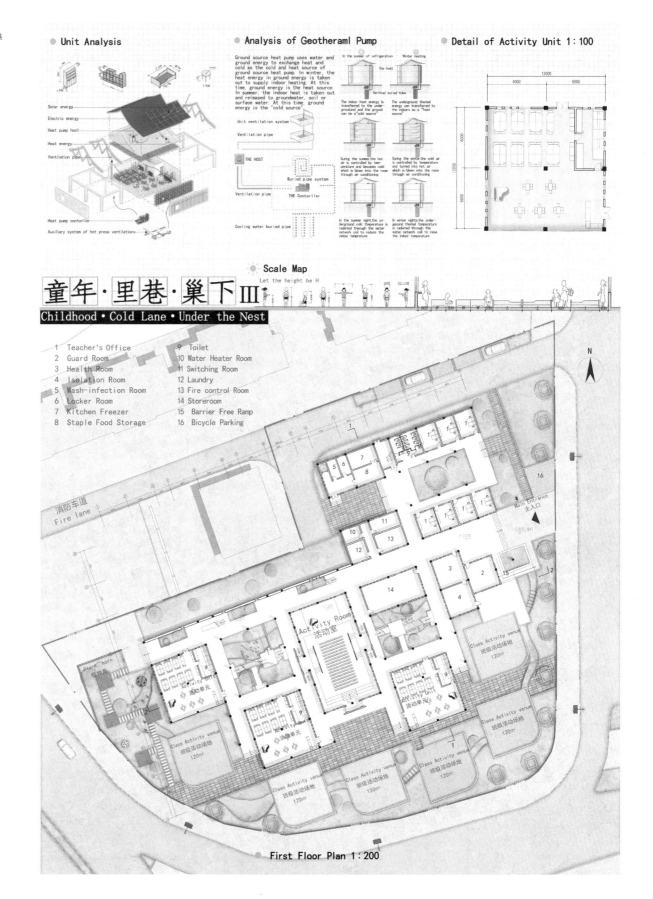

Unit Analysis

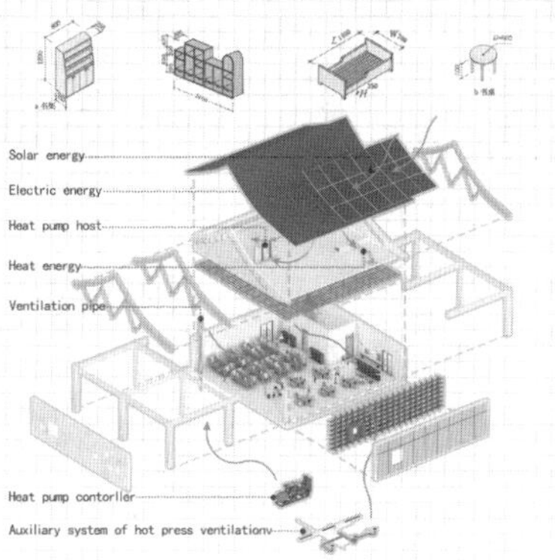
Solar energy
Electric energy
Heat pump host
Heat energy
Ventilation pipe
Heat pump contorller
Auxiliary system of hot press ventilation

Analysis of Geotheraml Pump

Ground source heat pump uses water and ground energy to exchange heat and cold as the cold and heat source of ground source heat pump. In winter, the heat energy in ground energy is taken out to supply indoor heating. At this time, ground energy is the heat source. In summer, the indoor heat is taken out and released to groundwater, soil or surface water. At this time, ground energy is the "cold source".

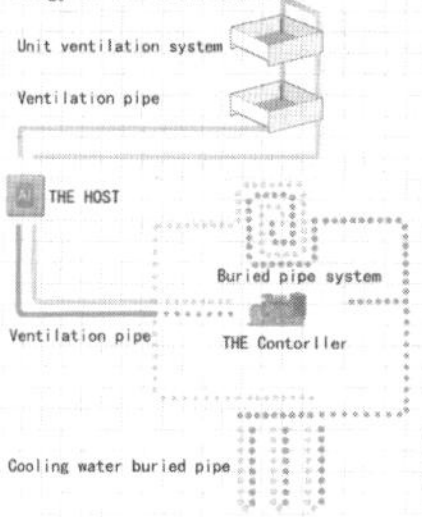
Unit ventilation system
Ventilation pipe
THE HOST
Buried pipe system
Ventilation pipe
THE Contorller
Cooling water buried pipe

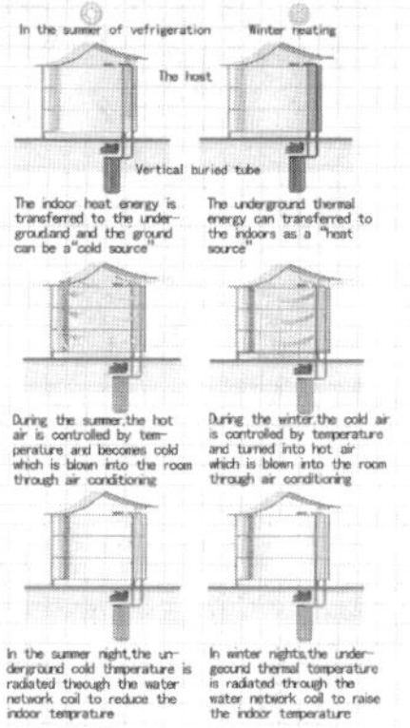
In the summer of vefrigeration
Winter heating
The host
Vertical buried tube
The indoor heat energy is transferred to the undergroud.and and the ground can be a"cold source"
The underground thermal energy can transferred to the indoors as a "heat source"
During the summer,the hot air is controlled by temperature and becomes cold which is blown into the room through air conditioning
During the winter,the cold air is controlled by temperature and turned into hot air which is blown into the room through air conditioning
In the summer night,the underground cold thmperature is radiated theough the water network coil to reduce the indoor temprature
In winter nights,the undergeound thermal temperature is radiated through the water network coil to raise the indoor temperature

Detail of Activity Unit 1 : 100

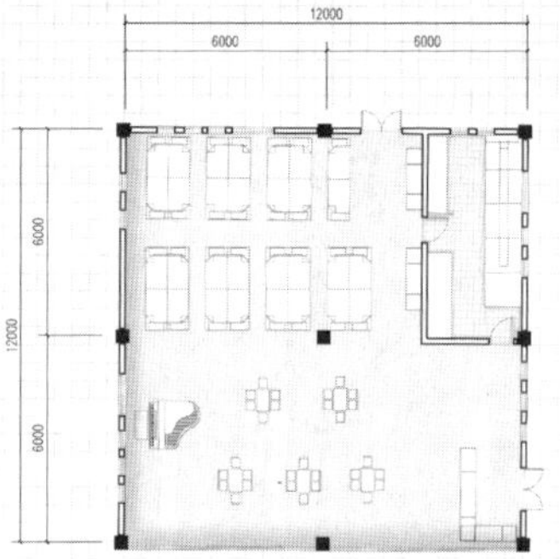
12000
6000
6000
6000
12000
6000

Scale Map

Let the height be H

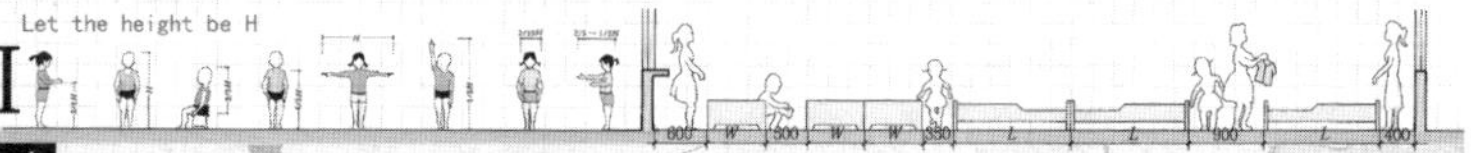

童年·里巷·巢下III

Childhood • Cold Lane • Under the Nest

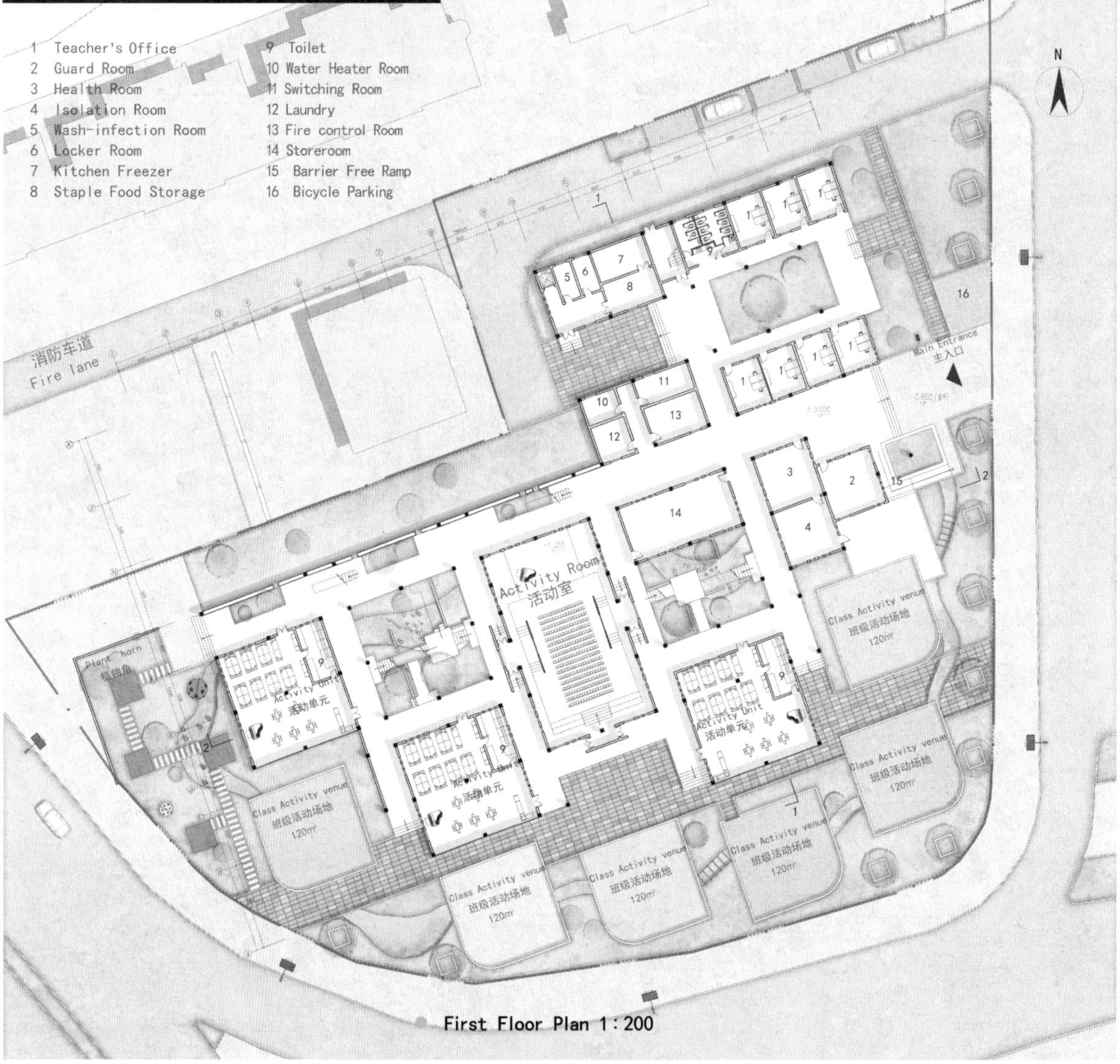
1 Teacher's Office
2 Guard Room
3 Health Room
4 Isolation Room
5 Wash-infection Room
6 Locker Room
7 Kitchen Freezer
8 Staple Food Storage
9 Toilet
10 Water Heater Room
11 Switching Room
12 Laundry
13 Fire control Room
14 Storeroom
15 Barrier Free Ramp
16 Bicycle Parking
N
消防车道
Fire lane
Main Entrance
主入口
Activity Room
活动室
Activity Unit
活动单元
Class Activity venue
班级活动场地
120㎡
Plant horn
植物角

First Floor Plan 1 : 200

童年·里巷·巢下 IV

Childhood • Cold Lane • Under the Nest

1 Teacher's Office
2 Model Studio
3 Book and Conference Room
4 Toilet
5 Non-staple food processing area
6 Cuttting Room
7 Staple food processing area
8 Snack area
9 Dressing Rooom
10 Pantry
11 Stalls

Second Floor Plan 1:300

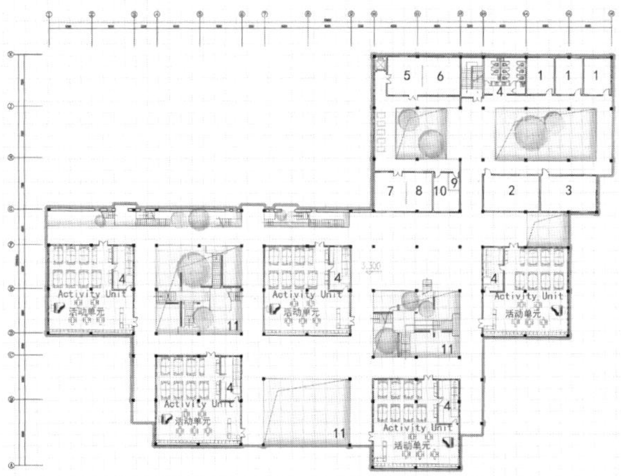

Activity Unit
活动单元

Third Floor Plan 1:300

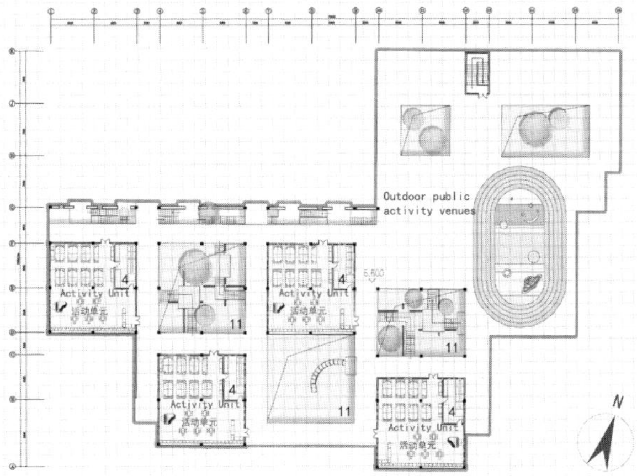

Outdoor public activity venues
Activity Unit
活动单元
N

Explosion Analysis

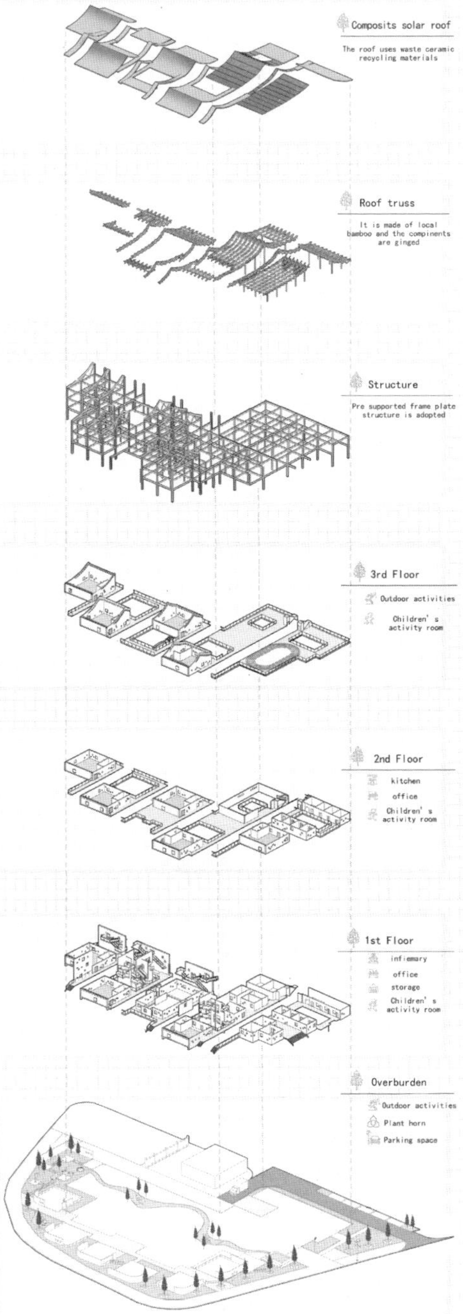

Composits solar roof
The roof uses waste ceramic recycling materials
Roof truss
It is made of local bamboo and the compinents are ginged
Structure
Pre supported frame plate structure is adopted
3rd Floor
Outdoor activities
Children's activity room
2nd Floor
kitchen
office
Children's activity room
1st Floor
infirmary
office
storage
Children's activity room
Overburden
Outdoor activities
Plant horn
Parking space

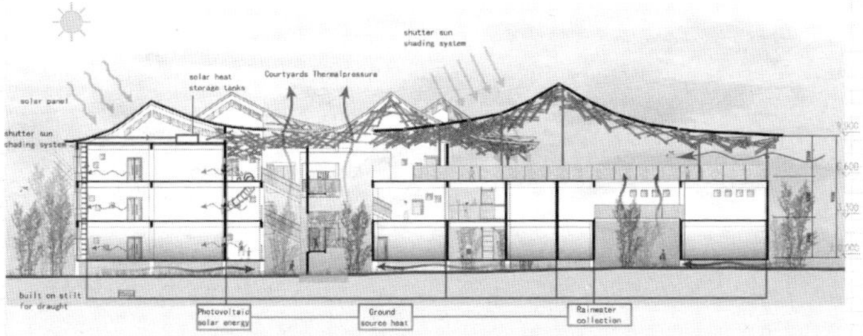

shutter sun shading system
solar heat storage tanks
Courtyards Thermalpressure
solar panel
shutter sun shading system
built on stilt for draught
Photovoltaic solar energy
Ground source heat
Rainwater collection

1-1 Section 1:200

童年·里巷·巢下 V
Childhood · Cold Lane · Under the Nest

Chinese characters
Pictograph
Roof truss
Birds
Nesting
Children with the bird
Live in harmony with nature
Herring gull
Larus saundersi
Halcyon coromanda
Eurystomus orientalis
Megalaima zeylanica
Oriolus chinensis

Transformation of Traditional Ventilation

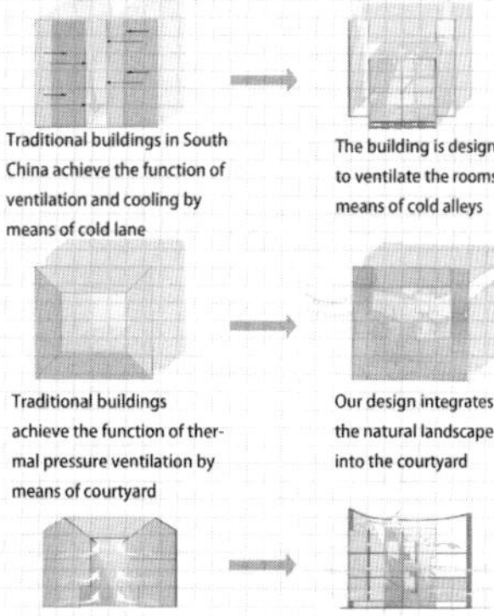

Traditional buildings in South China achieve the function of ventilation and cooling by means of cold lane

The building is designed to ventilate the rooms by means of cold alleys

Traditional buildings achieve the function of thermal pressure ventilation by means of courtyard

Our design integrates the natural landscape into the courtyard

Traditional buildings achieve the function of Pull the wind and Drainage of rainwater by the sloping

In addition to the basic function, a lot of children's activities are integrated into the courtyard

Analysis of Solar Energy Technology

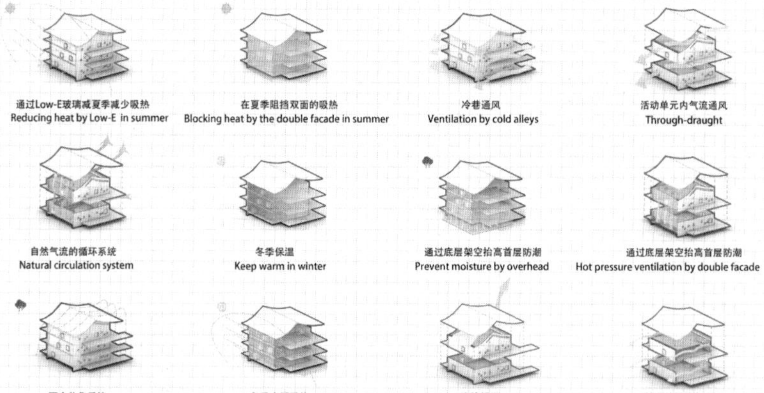

通过Low-E玻璃减夏季减少吸热
Reducing heat by Low-E in summer

在夏季阻挡双面的吸热
Blocking heat by the double facade in summer

冷巷通风
Ventilation by cold alleys

活动单元内气流通风
Through-draught

自然气流的循环系统
Natural circulation system

冬季保温
Keep warm in winter

通过底层架空抬高首层防潮
Prevent moisture by overhead

通过底层架空抬高首层防潮
Hot pressure ventilation by double facade

雨水收集系统
Rain water collection

冬季光照吸热
Warm sunshine in winter

庭院拔风
Pull the wind by the courtyard

地源热泵换气通风
Air exchange of ground source heat pump

The exposure range of sunlight to the room on the ground floor

The representation of airflow in building section

Representation of rainwater collection in building section

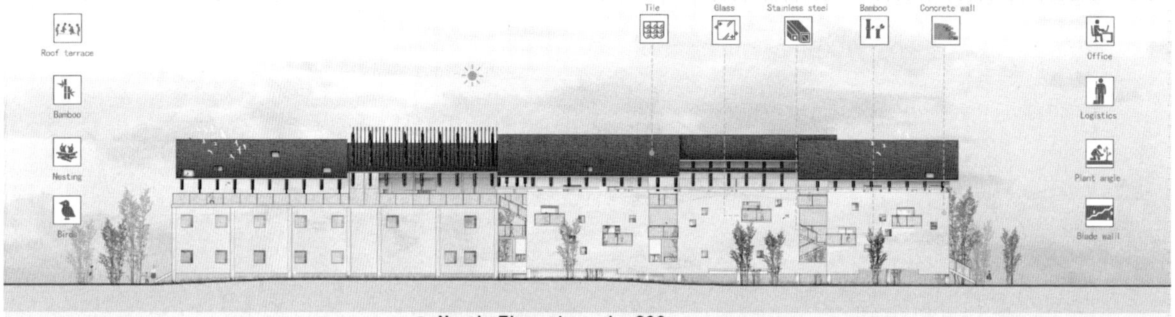
Roof terrace
Bamboo
Nesting
Birds
Tile
Glass
Stainless steel
Bamboo
Concrete wall
Office
Logistics
Plant angle
Blade wall

North Elevation 1 : 200

童年·里巷·巢下 VI

Childhood • Cold Lane • Under the Nest

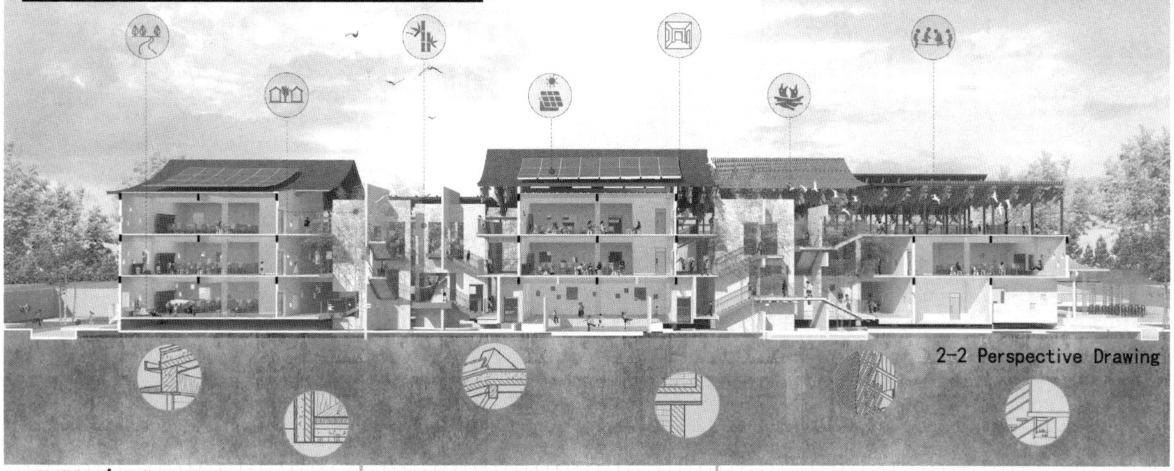

2-2 Perspective Drawing

Children's Activities

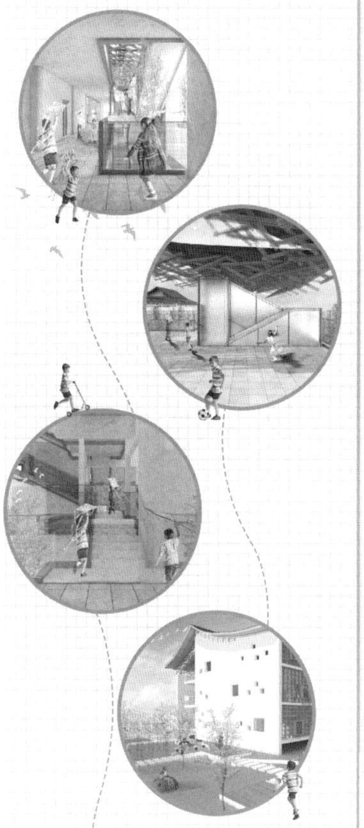

Active Solar Energy Utilizaton

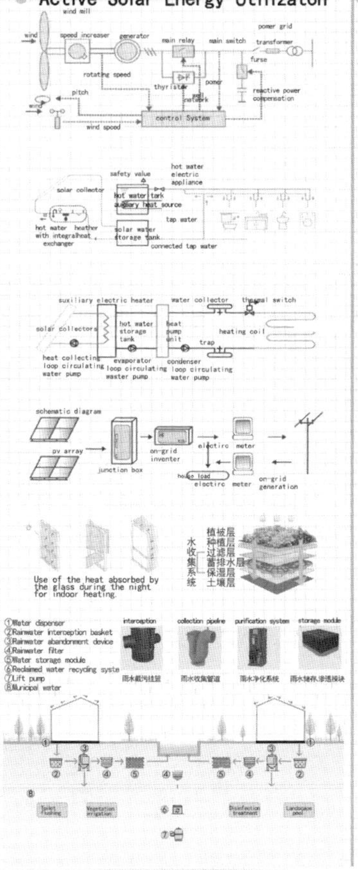

Summary of Design Strategies

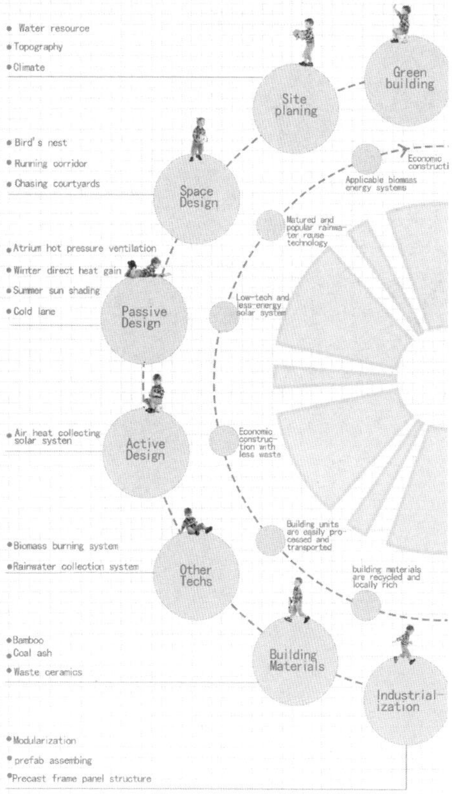

Water resource
Topography
Climate
Site planing
Green building
Bird's nest
Running corridor
Chasing courtyards
Space Design
Atrium hot pressure ventilation
Winter direct heat gain
Summer sun shading
Cold lane
Passive Design
Air heat collecting solar system
Active Design
Biomass burning system
Rainwater collection system
Other Techs
Bamboo
Coal ash
Waste ceramics
Building Materials
Industrialization
Modularization
prefab assembing
Precast frame panel structure

West Elevation 1:200

Courtyard Rendering 1:200

综合奖·优秀奖
General Prize Awarded · Honorable Mention Prize

注 册 号：7292

项目名称：稚梦·风光（南平）

Childlike Dream · Wind and Light（Nanping）

作　　者：邓冠中、费若雯、刘　敏

参赛单位：石家庄铁道大学

指导教师：高力强、樊海彬

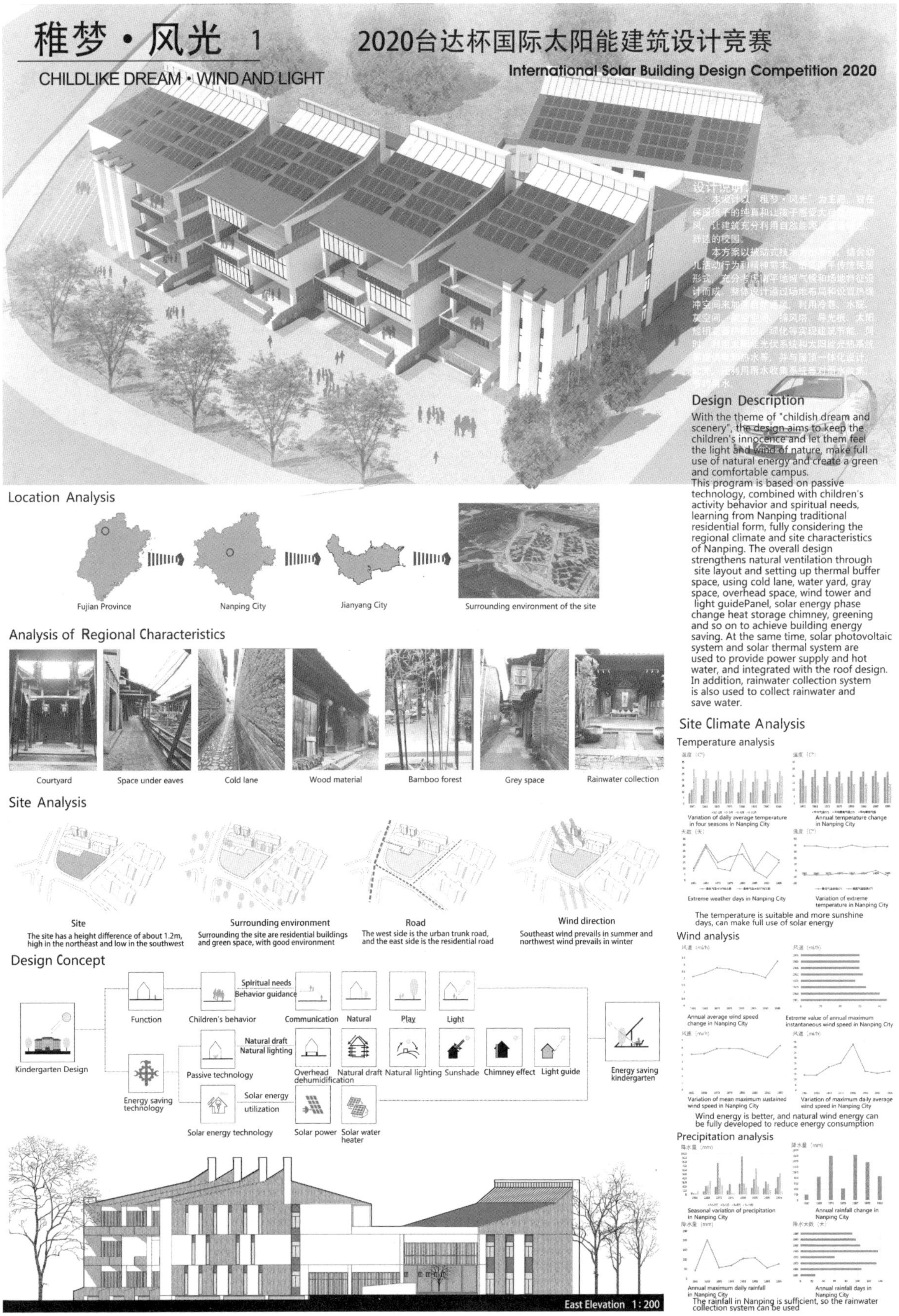
稚梦·风光 1
CHILDLIKE DREAM · WIND AND LIGHT
2020台达杯国际太阳能建筑设计竞赛
International Solar Building Design Competition 2020
设计说明
Design Description
With the theme of "childish dream and scenery", the design aims to keep the children's innocence and let them feel the light and wind of nature, make full use of natural energy and create a green and comfortable campus.
This program is based on passive technology, combined with children's activity behavior and spiritual needs, learning from Nanping traditional residential form, fully considering the regional climate and site characteristics of Nanping. The overall design strengthens natural ventilation through site layout and setting up thermal buffer space, using cold lane, water yard, gray space, overhead space, wind tower and light guidePanel, solar energy phase change heat storage chimney, greening and so on to achieve building energy saving. At the same time, solar photovoltaic system and solar thermal system are used to provide power supply and hot water, and integrated with the roof design. In addition, rainwater collection system is also used to collect rainwater and save water.
Location Analysis
Fujian Province
Nanping City
Jianyang City
Surrounding environment of the site
Analysis of Regional Characteristics
Courtyard
Space under eaves
Cold lane
Wood material
Bamboo forest
Grey space
Rainwater collection
Site Analysis
Site
The site has a height difference of about 1.2m, high in the northeast and low in the southwest
Surrounding environment
Surrounding the site are residential buildings and green space, with good environment
Road
The west side is the urban trunk road, and the east side is the residential road
Wind direction
Southeast wind prevails in summer and northwest wind prevails in winter
Design Concept
Kindergarten Design
Function
Children's behavior
Spiritual needs
Behavior guidance
Communication
Natural
Play
Light
Energy saving technology
Passive technology
Natural draft
Natural lighting
Overhead dehumidification
Natural draft
Natural lighting
Sunshade
Chimney effect
Light guide
Solar energy utilization
Solar energy technology
Solar power
Solar water heater
Energy saving kindergarten
East Elevation 1:200
Site Climate Analysis
Temperature analysis
Variation of daily average temperature in four seasons in Nanping City
Annual temperature change in Nanping City
Extreme weather days in Nanping City
Variation of extreme temperature in Nanping City
The temperature is suitable and more sunshine days, can make full use of solar energy
Wind analysis
Annual average wind speed change in Nanping City
Extreme value of annual maximum instantaneous wind speed in Nanping City
Variation of mean maximum sustained wind speed in Nanping City
Variation of maximum daily average wind speed in Nanping City
Wind energy is better, and natural wind energy can be fully developed to reduce energy consumption
Precipitation analysis
Seasonal variation of precipitation in Nanping City
Annual rainfall change in Nanping City
Annual maximum daily rainfall in Nanping City
Annual rainfall days in Nanping City
The rainfall in Nanping is sufficient, so the rainwater collection system can be used

稚梦·风光 2

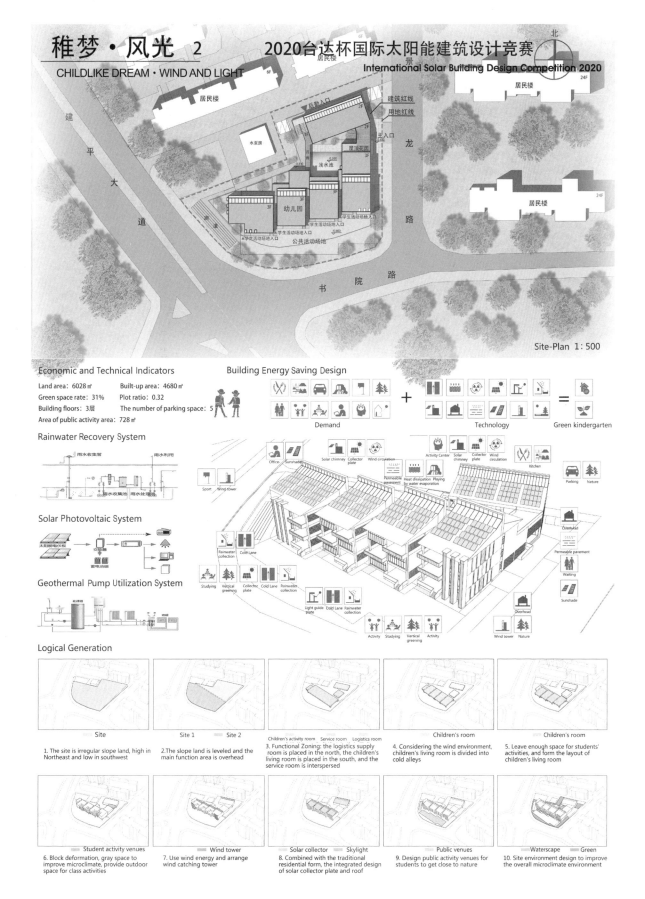

CHILDLIKE DREAM · WIND AND LIGHT

2020台达杯国际太阳能建筑设计竞赛

International Solar Building Design Competition 2020

北
居民楼
6F
居民楼
24F
居民楼
24F
居民楼
建筑红线
用地红线
后勤入口
主入口
水泵房
屋顶花园
浅水池
幼儿园
学生活动场地入口
公共活动场地
建平大道
龙景路
书院路

Site-Plan 1:500

Economic and Technical Indicators

Land area: 6028㎡ Built-up area: 4680㎡

Green space rate: 31% Plot ratio: 0.32

Building floors: 3层 The number of parking space: 5

Area of public activity area: 728㎡

Building Energy Saving Design

Demand
Technology
Green kindergarten

Rainwater Recovery System

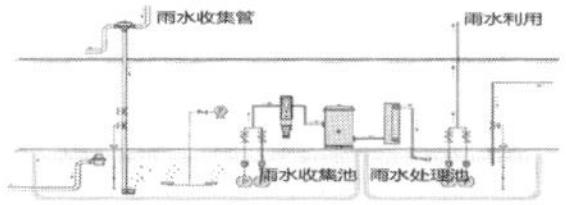
雨水收集管
雨水利用
雨水收集池
雨水处理池

Solar Photovoltaic System

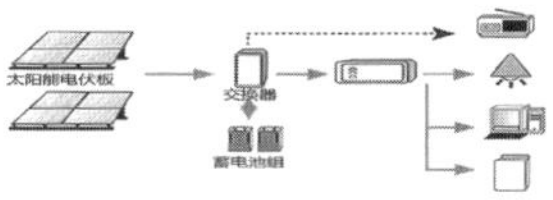
太阳能电伏板
蓄电池组

Geothermal Pump Utilization System

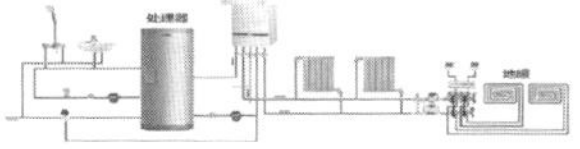

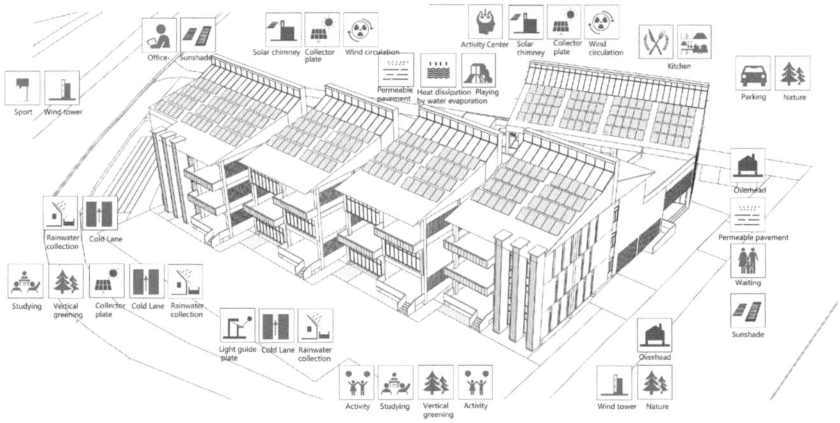
Office
Sunshade
Solar chimney
Collector plate
Wind circulation
Permeable pavement
Heat dissipation by water evaporation
Playing
Activity Center
Solar chimney
Collector plate
Wind circulation
Kitchen
Parking
Nature
Sport
Wind tower
Overhead
Permeable pavement
Waiting
Sunshade
Rainwater collection
Cold Lane
Studying
Vertical greening
Collector plate
Cold Lane
Rainwater collection
Light guide plate
Cold Lane
Rainwater collection
Overhead
Activity
Studying
Vertical greening
Activity
Wind tower
Nature

Logical Generation

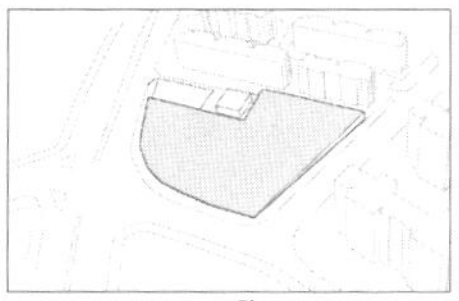

Site

1. The site is irregular slope land, high in Northeast and low in southwest

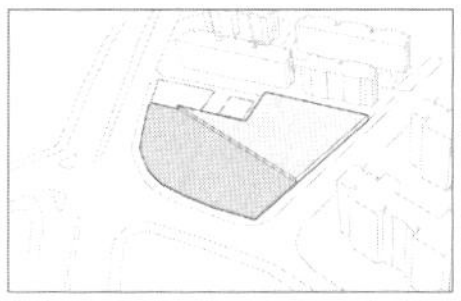

Site 1 Site 2

2.The slope land is leveled and the main function area is overhead

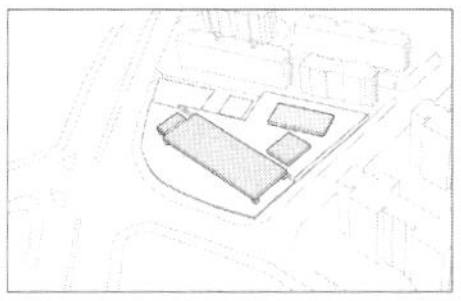

Children's activity room Service room Logistics room

3. Functional Zoning: the logistics supply room is placed in the north, the children's living room is placed in the south, and the service room is interspersed

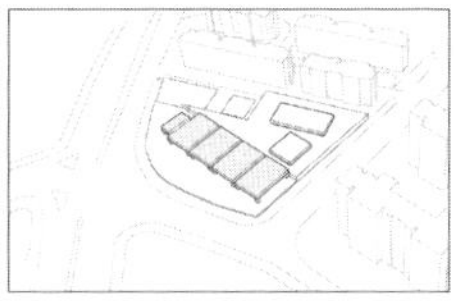

Children's room

4. Considering the wind environment, children's living room is divided into cold alleys

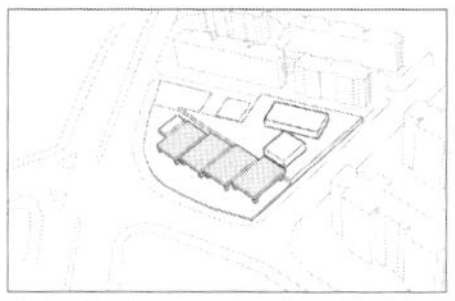

Children's room

5. Leave enough space for students' activities, and form the layout of children's living room

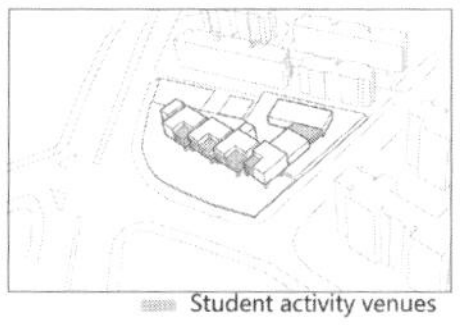

Student activity venues

6. Block deformation, gray space to improve microclimate, provide outdoor space for class activities

Wind tower

7. Use wind energy and arrange wind catching tower

Solar collector Skylight

8. Combined with the traditional residential form, the integrated design of solar collector plate and roof

Public venues

9. Design public activity venues for students to get close to nature

Waterscape Green

10. Site environment design to improve the overall microclimate environment

稚梦·风光 3

CHILDLIKE DREAM · WIND AND LIGHT

2020台达杯国际太阳能建筑设计竞赛

International Solar Building Design Competition 2020

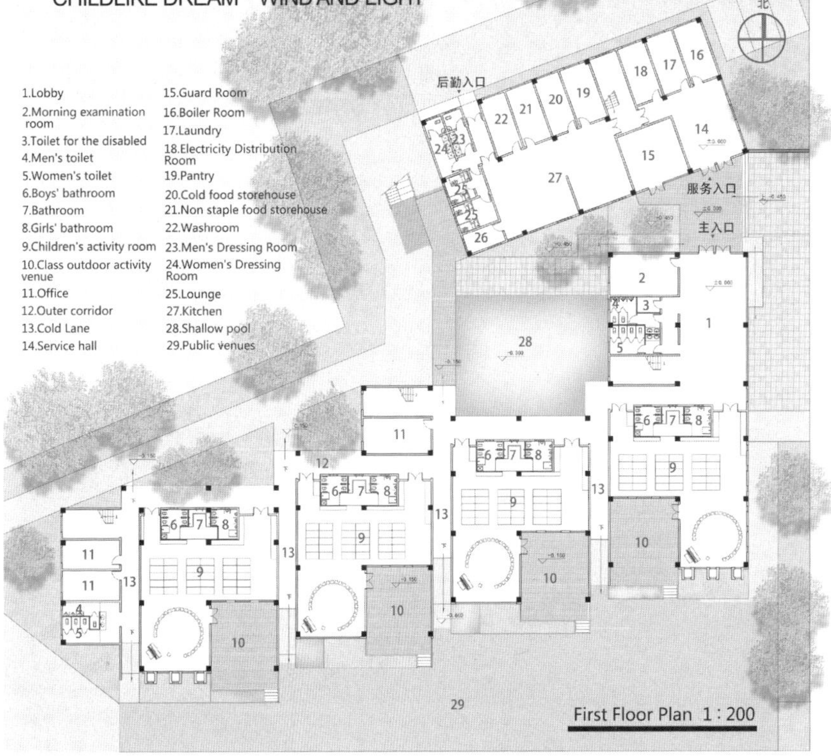

北
后勤入口
服务入口
主入口

1.Lobby
2.Morning examination room
3.Toilet for the disabled
4.Men's toilet
5.Women's toilet
6.Boys' bathroom
7.Bathroom
8.Girls' bathroom
9.Children's activity room
10.Class outdoor activity venue
11.Office
12.Outer corridor
13.Cold Lane
14.Service hall
15.Guard Room
16.Boiler Room
17.Laundry
18.Electricity Distribution Room
19.Pantry
20.Cold food storehouse
21.Non staple food storehouse
22.Washroom
23.Men's Dressing Room
24.Women's Dressing Room
25.Lounge
27.Kitchen
28.Shallow pool
29.Public venues

First Floor Plan 1：200

Cold Lane

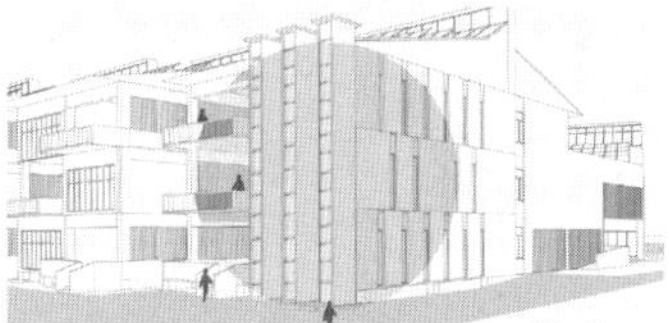

Wind tower

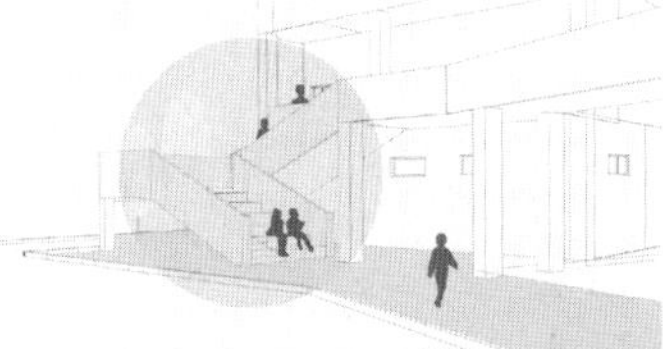

Children's playground

Entrance grey space

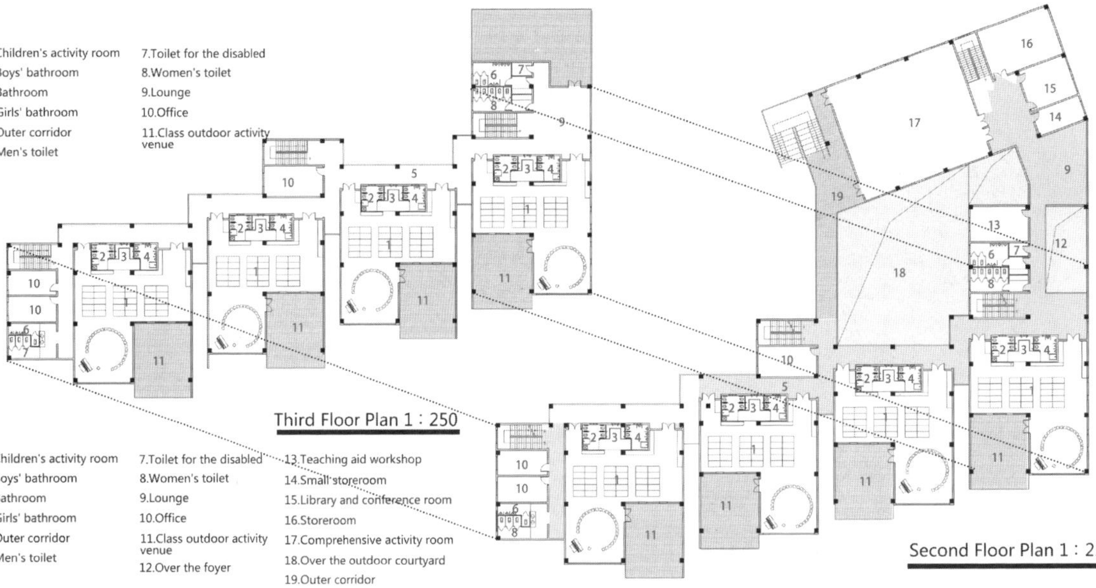

1.Children's activity room
2.Boys' bathroom
3.Bathroom
4.Girls' bathroom
5.Outer corridor
6.Men's toilet
7.Toilet for the disabled
8.Women's toilet
9.Lounge
10.Office
11.Class outdoor activity venue

Third Floor Plan 1：250

1.Children's activity room
2.Boys' bathroom
3.Bathroom
4.Girls' bathroom
5.Outer corridor
6.Men's toilet
7.Toilet for the disabled
8.Women's toilet
9.Lounge
10.Office
11.Class outdoor activity venue
12.Over the foyer
13.Teaching aid workshop
14.Small storeroom
15.Library and conference room
16.Storeroom
17.Comprehensive activity room
18.Over the outdoor courtyard
19.Outer corridor

Second Floor Plan 1：250

West Elevation 1：200

稚梦 · 风光 4

CHILDLIKE DREAM · WIND AND LIGHT

2020台达杯国际太阳能建筑设计竞赛

International Solar Building Design Competition 2020

Energy Saving Strategy

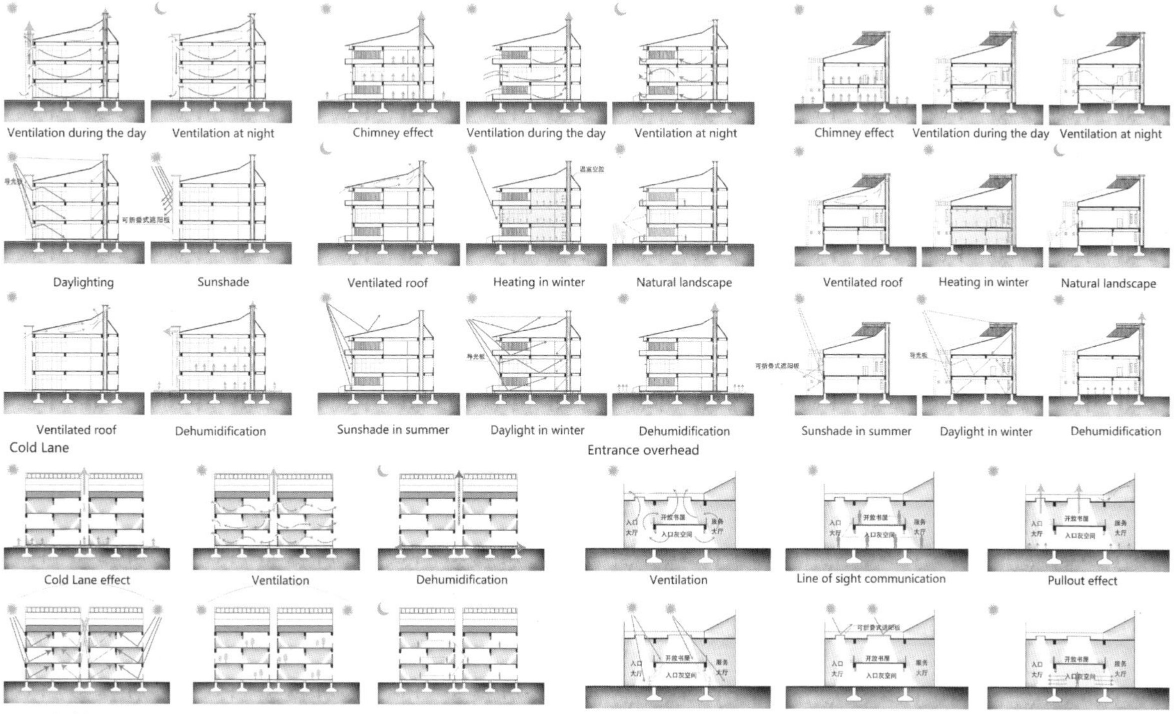
Wind tower
Ventilation during the day
Ventilation at night
Daylighting
Sunshade
Ventilated roof
Dehumidification
Children's activity room
Chimney effect
Ventilation during the day
Ventilation at night
Ventilated roof
Heating in winter
Natural landscape
Sunshade in summer
Daylight in winter
Dehumidification
Service room
Chimney effect
Ventilation during the day
Ventilation at night
Ventilated roof
Heating in winter
Natural landscape
Sunshade in summer
Daylight in winter
Dehumidification
Cold Lane
Cold Lane effect
Ventilation
Dehumidification
Daylight
Natural landscape
Line of sight communication
Entrance overhead
Ventilation
Line of sight communication
Pullout effect
Daylight
Sunshade in summer
Flow guidance

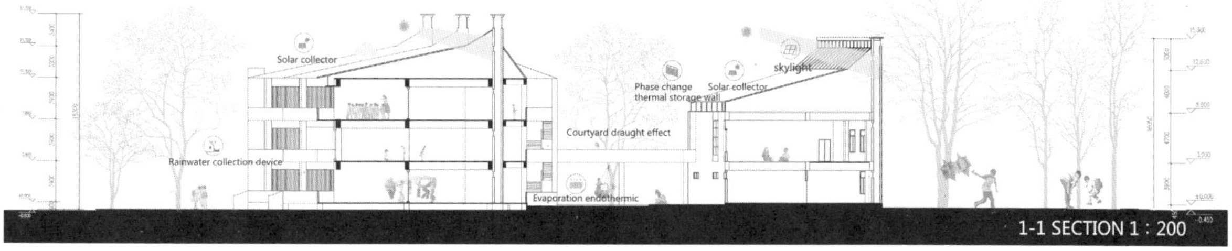
Solar collector
Rainwater collection device
Courtyard draught effect
Evaporation endothermic
Phase change thermal storage wall
Solar collector
skylight

1-1 SECTION 1 : 200

综合奖 · 优秀奖
General Prize Awarded · Honorable Mention Prize

注册号：7307
项目名称：树下幼儿园（南平）
Under the Tree（Nanping）
作　　者：黄承佳、江逸妍
参赛单位：华南理工大学
指导教师：王　静、肖毅强

树　下

2020台达杯福建南平阳光幼儿园设计
UNDER THE TREE

DESIGN DESCRIPTION This program aims to create a kindergarten under the tree for children, and to provide a small world where children in the city can coexist with nature and roam freely. It achieves educational significance while maintaing the goal of nearly zero-energy consumption within the whole life span. From the perspective of caring for the healthy growth of young children, the design draws on the cultural image of Nanping in Fujian Province -- a forest city, based on the characteristics of the mountain and forests around the site, and finally introduces the TREE into the architecture, which is a natural element inseparable from the healthy childhood of the young generation.

设计说明 此方案旨在为孩子们创造一个树下的幼儿园，为城市的孩子打造一处能与自然和谐共处、无拘束漫游的小世界；同时，在满足近零能耗的绿色建筑要求之上，实现对幼儿的太阳能教育意义。本设计从关爱幼儿健康成长的角度出发，汲取了福建南宁作为森林城市的文化背景，基于场地周边独特的山林特征，并最终将“树”——这一与儿童的健康成长息息相关、不可分割的自然元素引入建筑内部。

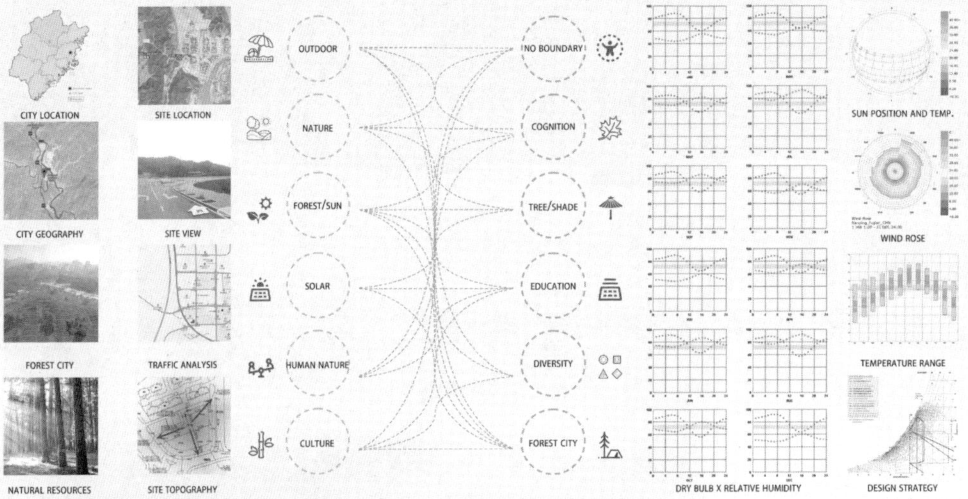
CITY LOCATION
SITE LOCATION
CITY GEOGRAPHY
SITE VIEW
FOREST CITY
TRAFFIC ANALYSIS
NATURAL RESOURCES
SITE TOPOGRAPHY
OUTDOOR
NATURE
FOREST/SUN
SOLAR
HUMAN NATURE
CULTURE
NO BOUNDARY
COGNITION
TREE/SHADE
EDUCATION
DIVERSITY
FOREST CITY
SUN POSITION AND TEMP.
WIND ROSE
TEMPERATURE RANGE
DRY BULB X RELATIVE HUMIDITY
DESIGN STRATEGY

BACKGROUND ANALYSIS　　DESIGN CONCEPT　　CLIMATEBACKGROUND

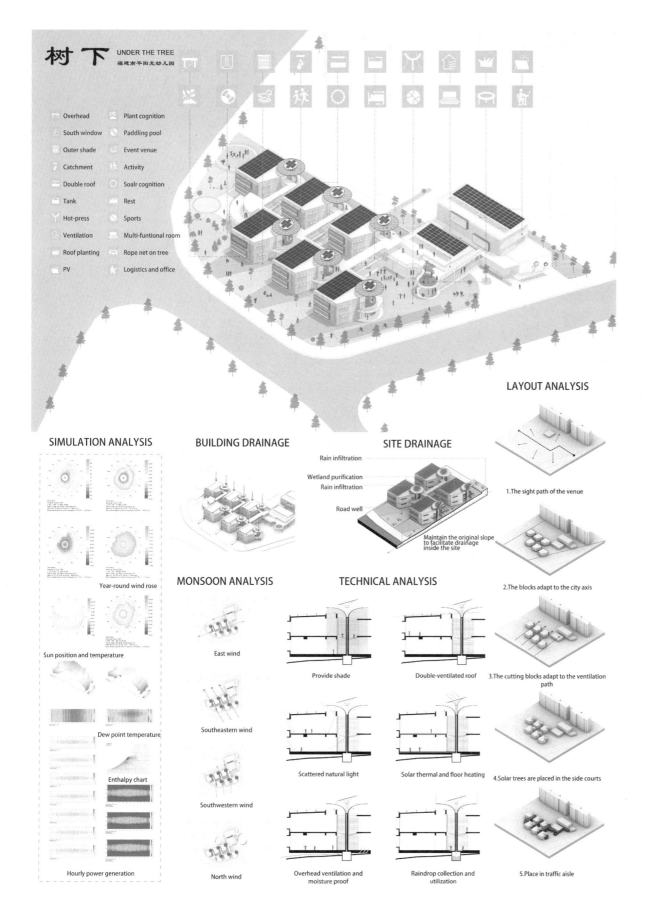

树下 UNDER THE TREE

福建南平阳光幼儿园

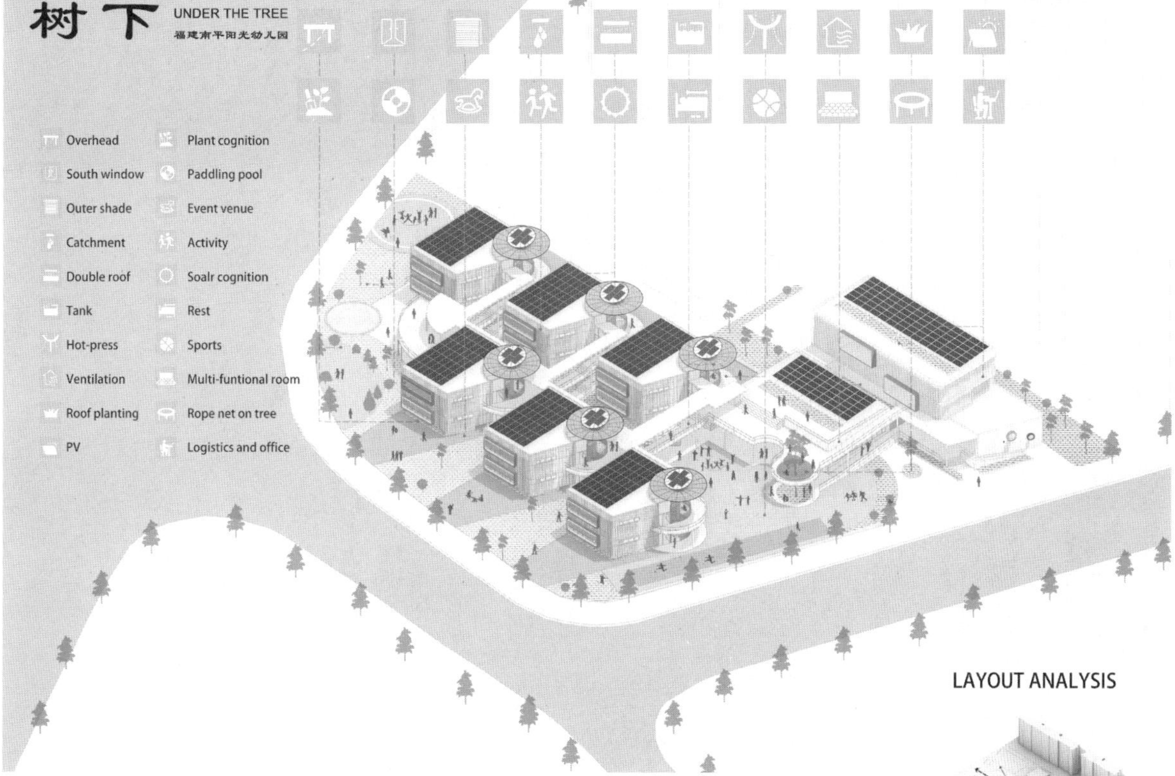
Overhead
South window
Outer shade
Catchment
Double roof
Tank
Hot-press
Ventilation
Roof planting
PV
Plant cognition
Padding pool
Event venue
Activity
Soalr cognition
Rest
Sports
Multi-funtional room
Rope net on tree
Logistics and office

SIMULATION ANALYSIS

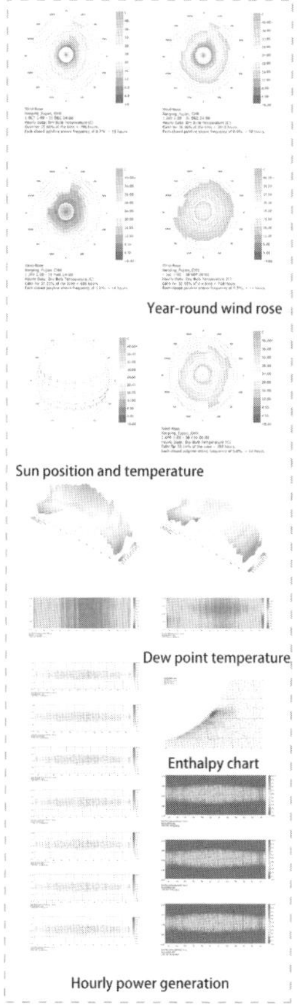
Year-round wind rose
Sun position and temperature
Dew point temperature
Enthalpy chart
Hourly power generation

BUILDING DRAINAGE

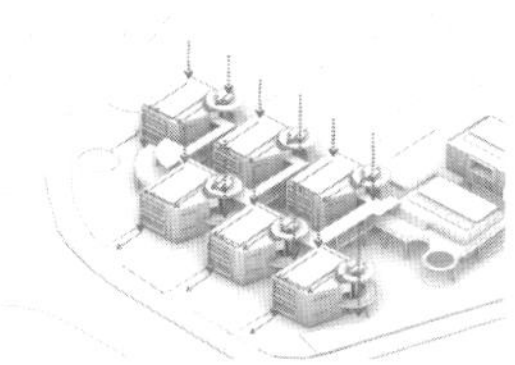

SITE DRAINAGE

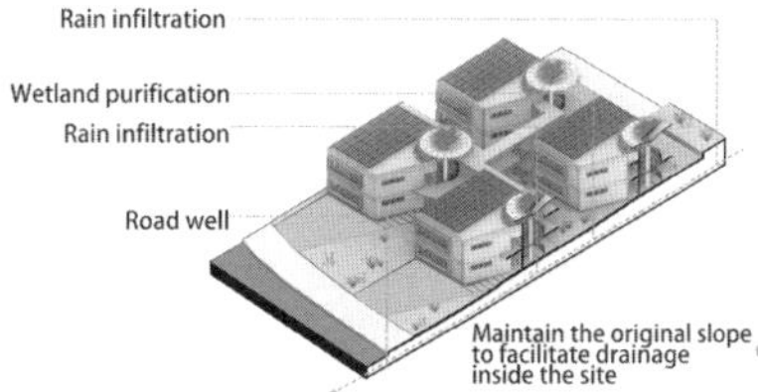
Rain infiltration
Wetland purification
Rain infiltration
Road well
Maintain the original slope to facilitate drainage inside the site

LAYOUT ANALYSIS

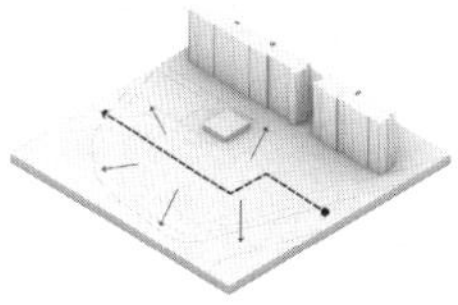

1.The sight path of the venue

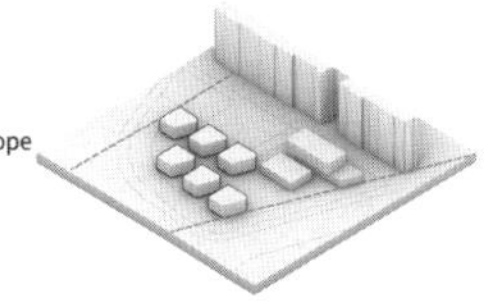

2.The blocks adapt to the city axis

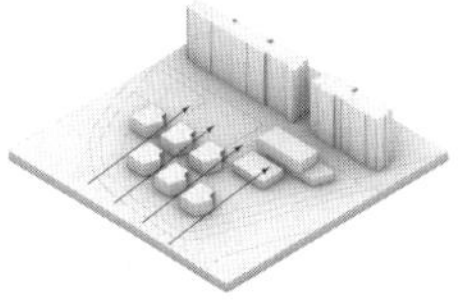

3.The cutting blocks adapt to the ventilation path

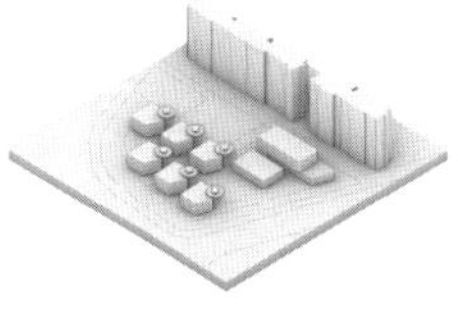

4.Solar trees are placed in the side courts

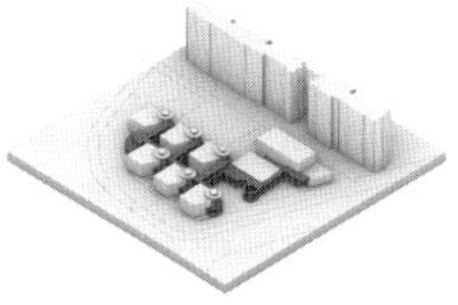

5.Place in traffic aisle

MONSOON ANALYSIS

East wind

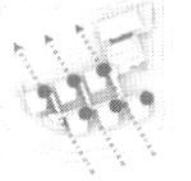

Southeastern wind

Southwestern wind

North wind

TECHNICAL ANALYSIS

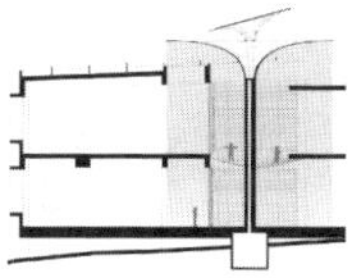

Provide shade

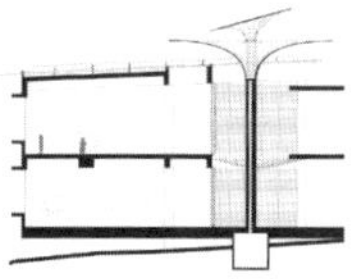

Double-ventilated roof

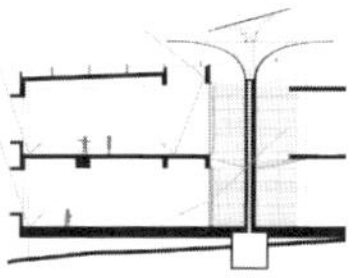

Scattered natural light

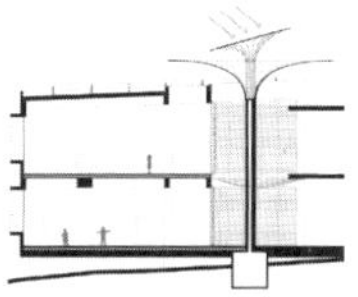

Solar thermal and floor heating

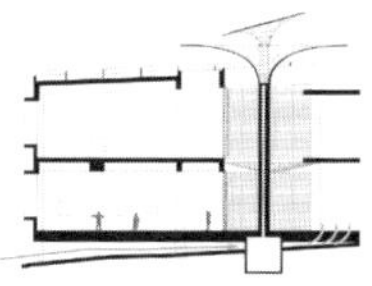

Overhead ventilation and moisture proof

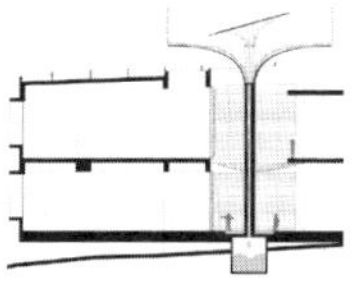

Raindrop collection and utilization

树 下
UNDER THE TREE
福建南平阳光幼儿园

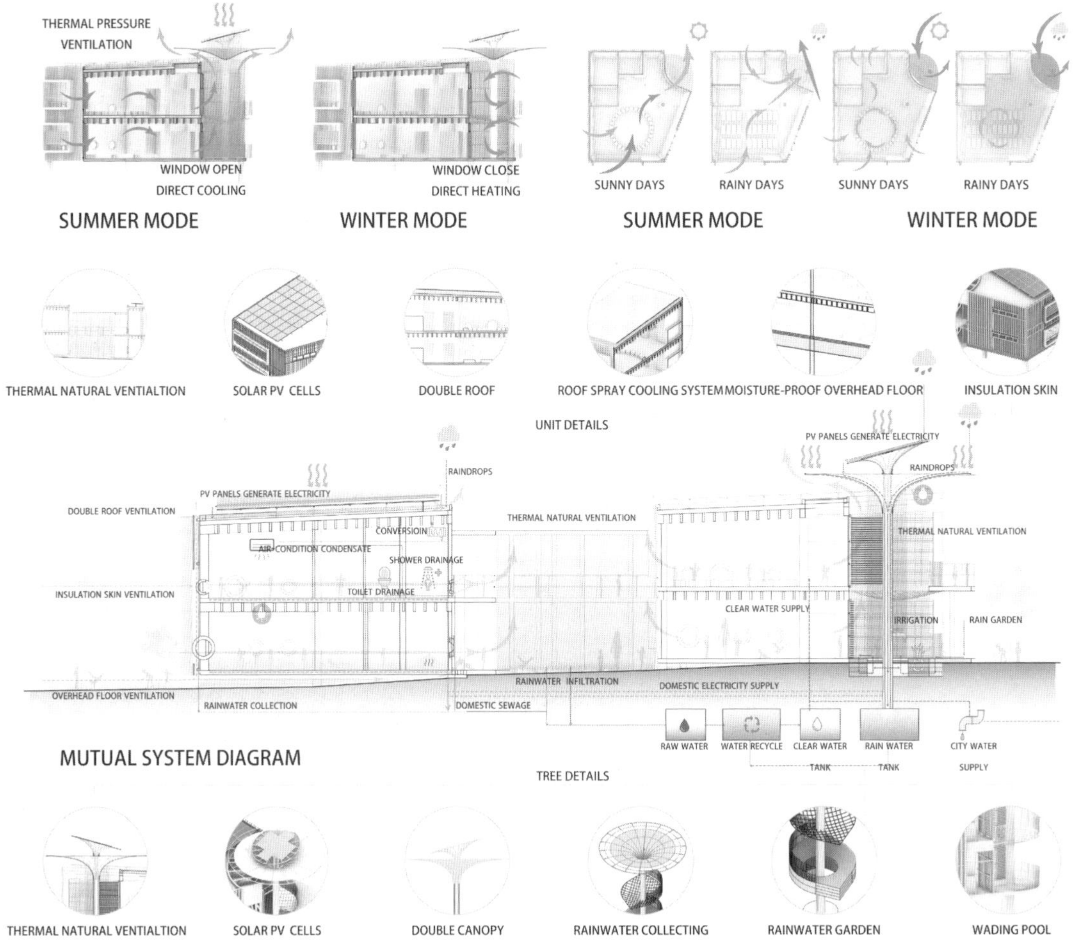
THERMAL PRESSURE
VENTILATION
WINDOW OPEN
DIRECT COOLING
WINDOW CLOSE
DIRECT HEATING
SUNNY DAYS
RAINY DAYS
SUNNY DAYS
RAINY DAYS
SUMMER MODE
WINTER MODE
SUMMER MODE
WINTER MODE
THERMAL NATURAL VENTIALTION
SOLAR PV CELLS
DOUBLE ROOF
ROOF SPRAY COOLING SYSTEM
MOISTURE-PROOF OVERHEAD FLOOR
INSULATION SKIN
UNIT DETAILS
PV PANELS GENERATE ELECTRICITY
RAINDROPS
DOUBLE ROOF VENTILATION
PV PANELS GENERATE ELECTRICITY
RAINDROPS
THERMAL NATURAL VENTILATION
CONVERSION
AIR-CONDITION CONDENSATE
SHOWER DRAINAGE
TOILET DRAINAGE
THERMAL NATURAL VENTILATION
INSULATION SKIN VENTILATION
CLEAR WATER SUPPLY
IRRIGATION
RAIN GARDEN
RAINWATER INFILTRATION
DOMESTIC ELECTRICITY SUPPLY
OVERHEAD FLOOR VENTILATION
RAINWATER COLLECTION
DOMESTIC SEWAGE
RAW WATER
WATER RECYCLE
CLEAR WATER TANK
RAIN WATER TANK
CITY WATER SUPPLY
MUTUAL SYSTEM DIAGRAM
TREE DETAILS
THERMAL NATURAL VENTIALTION
SOLAR PV CELLS
DOUBLE CANOPY
RAINWATER COLLECTING
RAINWATER GARDEN
WADING POOL

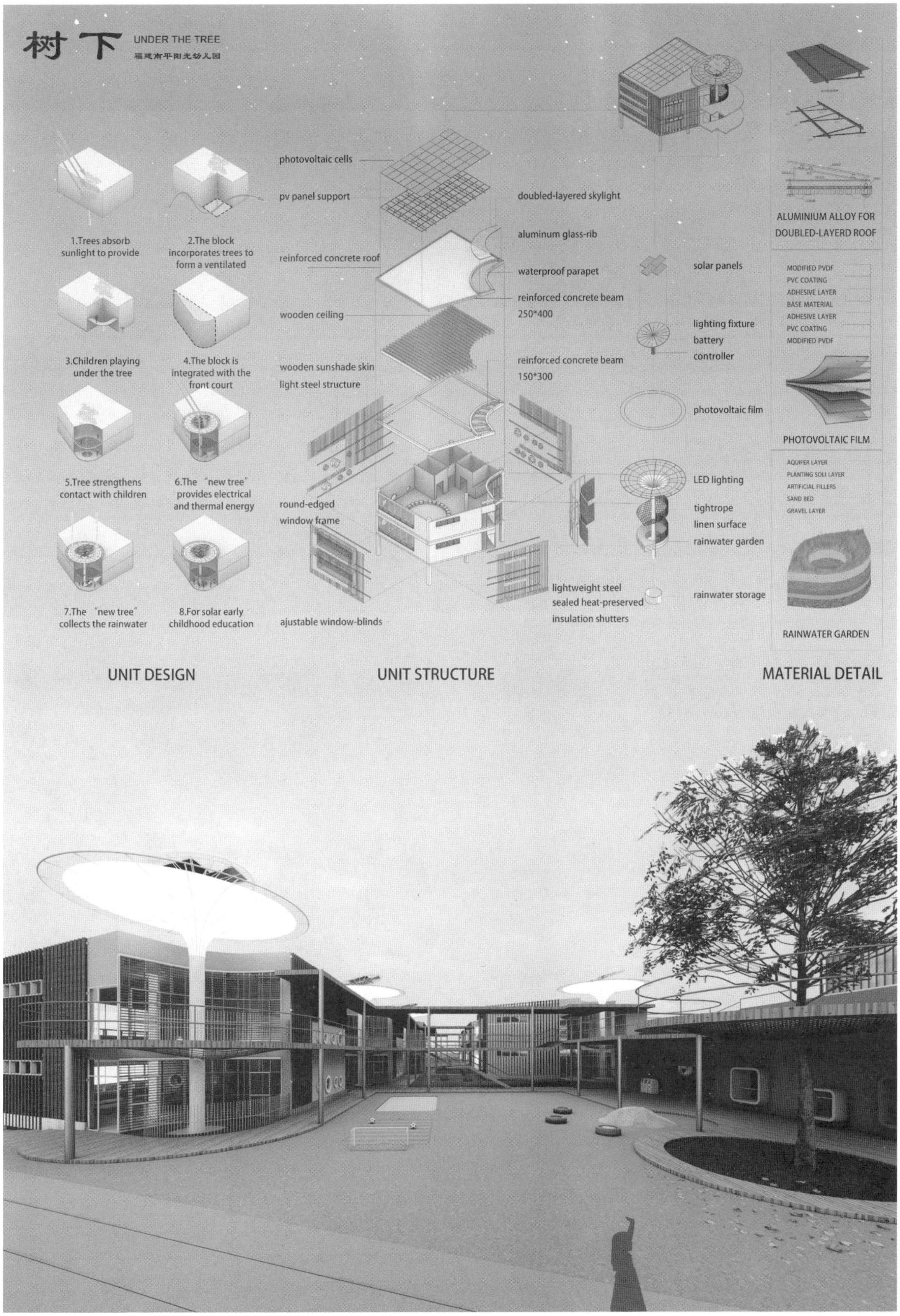
树 下
UNDER THE TREE
福建南平阳光幼儿园
1.Trees absorb sunlight to provide
2.The block incorporates trees to form a ventilated
3.Children playing under the tree
4.The block is integrated with the front court
5.Tree strengthens contact with children
6.The "new tree" provides electrical and thermal energy
7.The "new tree" collects the rainwater
8.For solar early childhood education
UNIT DESIGN
photovoltaic cells
pv panel support
reinforced concrete roof
wooden ceiling
wooden sunshade skin
light steel structure
round-edged window frame
ajustable window-blinds
doubled-layered skylight
aluminum glass-rib
waterproof parapet
reinforced concrete beam 250*400
reinforced concrete beam 150*300
lightweight steel
sealed heat-preserved
insulation shutters
UNIT STRUCTURE
solar panels
lighting fixture
battery
controller
photovoltaic film
LED lighting
tightrope
linen surface
rainwater garden
rainwater storage
ALUMINIUM ALLOY FOR DOUBLED-LAYERD ROOF
MODIFIED PVDF
PVC COATING
ADHESIVE LAYER
BASE MATERIAL
ADHESIVE LAYER
PVC COATING
MODIFIED PVDF
PHOTOVOLTAIC FILM
AQUIFER LAYER
PLANTING SOLI LAYER
ARTIFICIAL FILLERS
SAND BED
GRAVEL LAYER
RAINWATER GARDEN
MATERIAL DETAIL

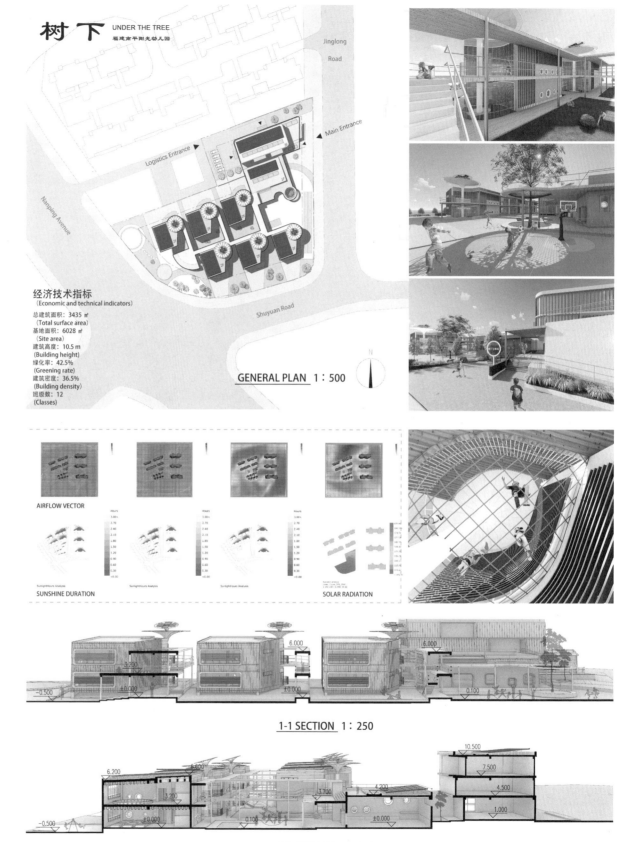

树下 UNDER THE TREE

福建南平阳光幼儿园

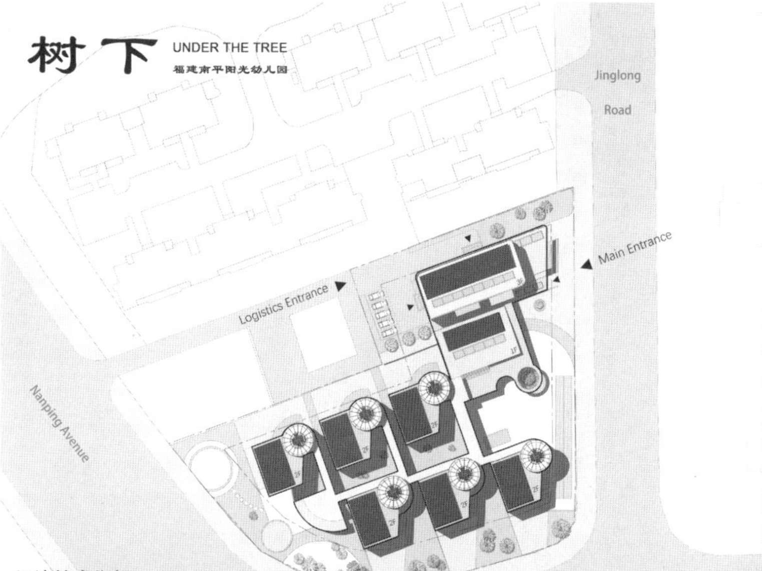
Jinglong
Road
Main Entrance
Logistics Entrance
Nanping Avenue
Shuyuan Road
N

经济技术指标
(Economic and technical indicators)

总建筑面积：3435 ㎡
(Total surface area)
基地面积：6028 ㎡
(Site area)
建筑高度：10.5 m
(Building height)
绿化率：42.5%
(Greening rate)
建筑密度：36.5%
(Building density)
班级数：12
(Classes)

GENERAL PLAN 1：500

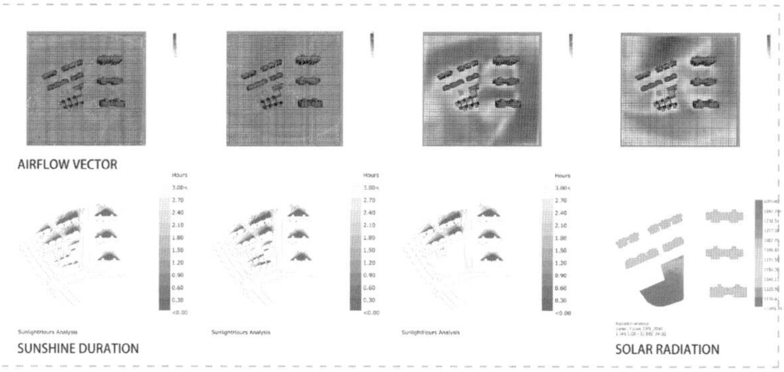
AIRFLOW VECTOR
SUNSHINE DURATION
SOLAR RADIATION

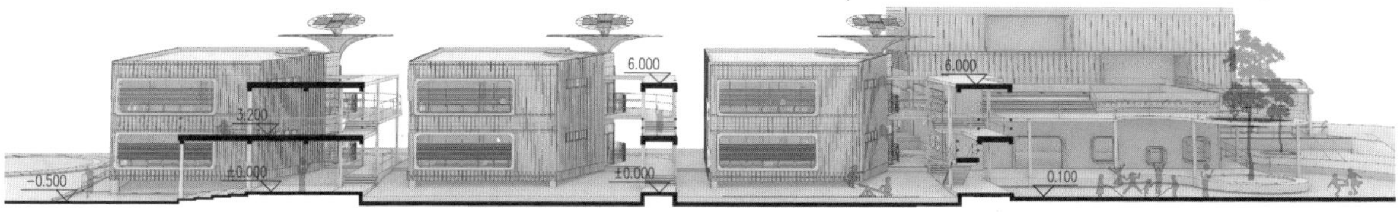
6.000
6.000
3.200
-0.500
±0.000
±0.000
0.100

1-1 SECTION 1：250

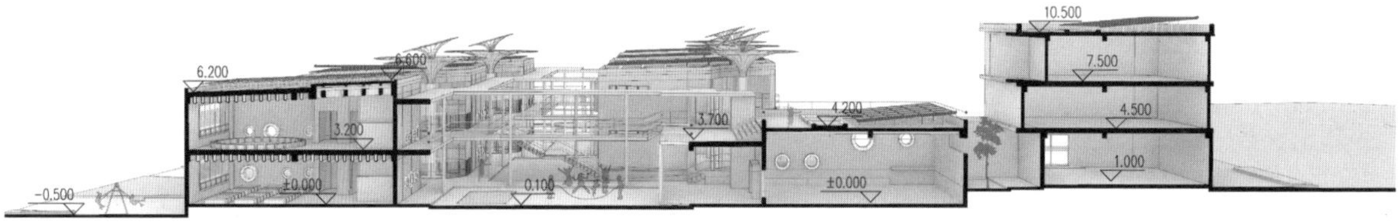
10.500
6.200
6.600
7.500
4.200
3.700
4.500
3.200
1.000
-0.500
±0.000
0.100
±0.000

2-2 SECTION 1：250

树下 UNDER THE TREE

福建南平阳光幼儿园

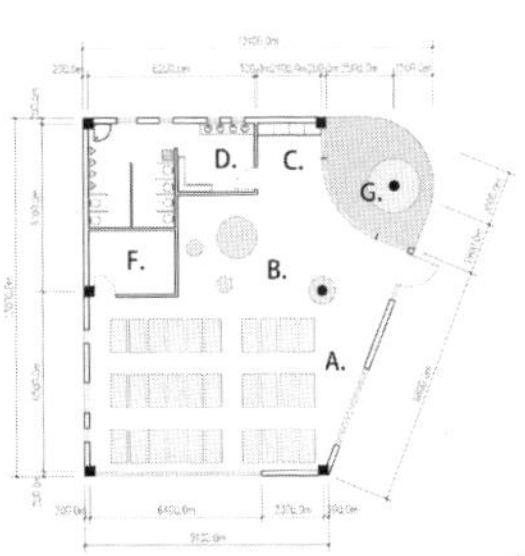

D.
C.
G.
F.
B.
A.

Unit Plan I 1：150

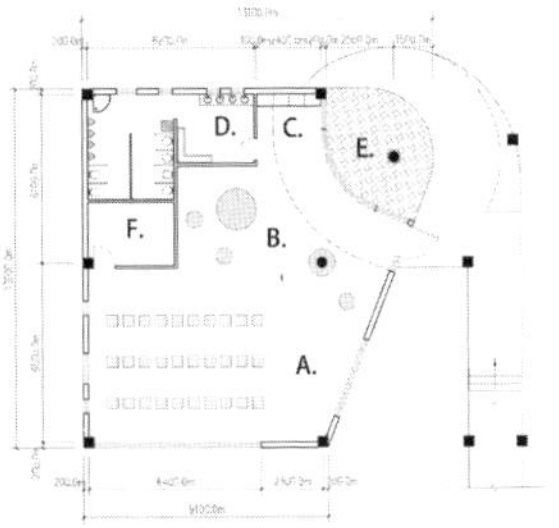

D.
C.
E.
F.
B.
A.

Unit Plan II 1：150

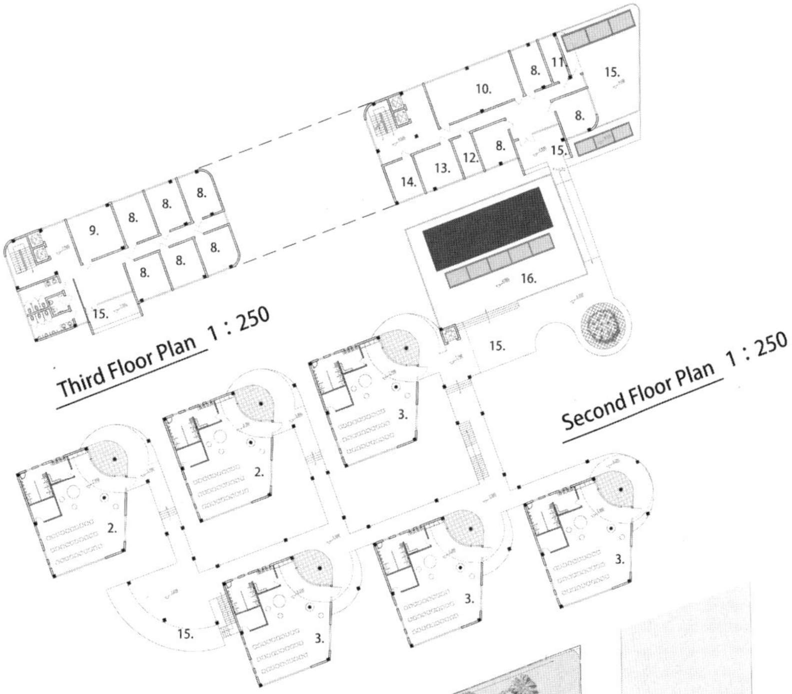

Third Floor Plan 1：250

Second Floor Plan 1：250

A.main activity area
B.rest area
C.locker
D.toilet
E.rope net
F.storage
G.small garden

Logistics entrance
Main entrance
N

1.Small class
2.Big class
3.Middle class
4.Comprehensive activity room
5.Kitchen
6.Health room
7.Guard room
8.Office
9.Books and conference room
10.Storage
11.Water room
12.Distribution room
13.Teaching aids production
14.Laundry
15.Terrace
16.Planting cognitive area
17.Class venue
18.Paddling pool
19.Equipment venue
20.Comprehensive venue
21.Main entrance plaza
22.Logistics plaza

Ground Floor Plan 1：250

综合奖·优秀奖
General Prize Awarded · Honorable Mention Prize

注 册 号：7323

项目名称：向阳·梦园（南平）

Sunny · Dream Garden (Nanping)

作　　者：张淑娴、邓盛炜、曾小婷

参赛单位：福州大学

指导教师：吴木生、林志森

向阳·梦园 Sunny · Dream Garden 01

RESOURCE CONDITION

BAMBOO
SUN
WATER
HEAT
WIND

SITE LOCATION

NAN PING
JIAN YANG
SITE

DESIGN DESCRIPTION

本方案以“向阳·梦园”为主题，将主动式太阳能技术与屋顶设计、儿童趣味空间相结合，在不同层次的形体变化中创造丰富的幼儿室内外空间；在被动式太阳能技术应用上，设置了适应冬季和夏季的节能模式，重点考虑雨水收集、滴灌系统的应用，同时设置合适的地源热泵等能源系统；在材料利用上，使用当地盛产的竹材，打造垂直绿化竹墙等特殊构筑物，进一步丰富幼儿园的内部环境，创造既满足节能要求又为幼儿提供良好成长环境的“梦园”。

Taking "Sunny·Dream Garden" as the theme, this scheme combines active solar energy technology with roof design and children's interesting space to create rich indoor and outdoor space for children in different levels of body changes; in the application of passive solar energy technology, energy-saving mode suitable for winter and summer seasons is set, with the application of rainwater collection and drip irrigation system considered, and appropriate settings are set In terms of material utilization, the local abundant bamboo materials are used to create vertical green bamboo walls and other special structures, so as to further enrich the internal environment of kindergartens and create a "dream garden" that can meet the energy-saving requirements and provide a good growth environment for children.

REGIONAL FEATURE

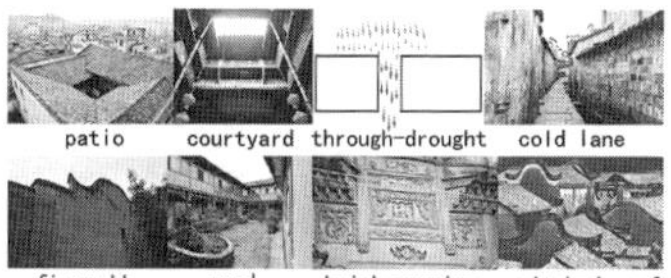
patio
courtyard
through-drought
cold lane
firewall
wood
brick curving
pitched roof

JIANYANG CULTURE

kiln and lamp
Zhu Xi
academy
ancient street

SITE ANALYSIS

site
site
site
surrounding roads -convenient but have noise problem
sunshine-south good but west needs sunscreen
surrounding buildings-residential area mainly

CLIMATE ANALYSIS

Jianyang belongs to the subtropical monsoon climate, with short winter and long summer, annual average temperature of 18.1 ℃, and annual average sunshine of 1802 hours

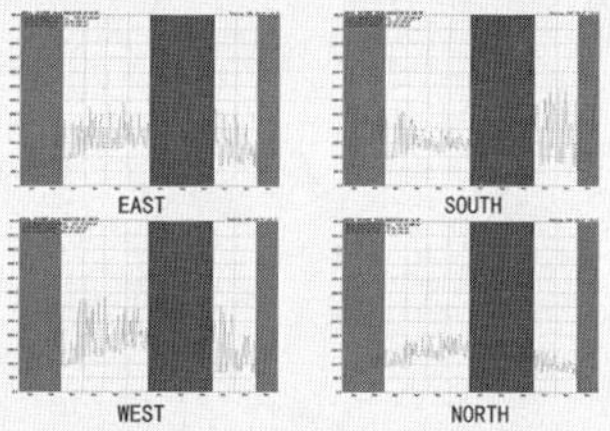
EAST
SOUTH
WEST
NORTH

the East and West solar radiation is relatively large in spring and summer, and sunscreen should be paid attention to. The fluctuation of solar radiation i n winter in the south is relatively large, and the overall solar radiation in the north is relatively small.

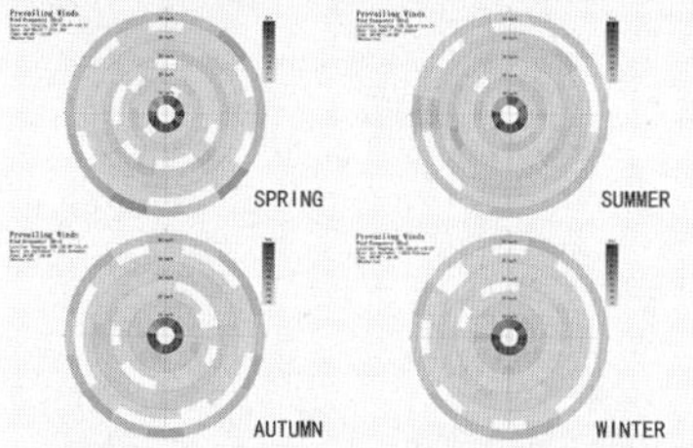
SPRING
SUMMER
AUTUMN
WINTER

There are more calm winds in Nanping, southeast wind prevails in summer and northerly wind prevails in winter.

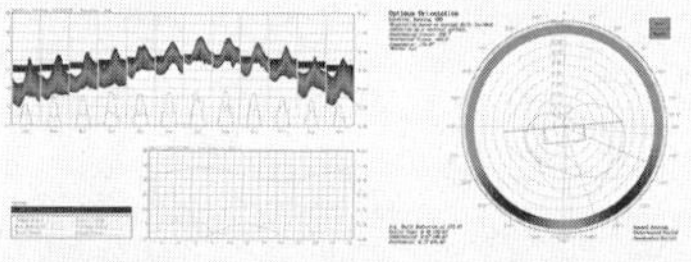

the climate in May and September is humid and hot, and the annual solar scattering is relatively large. From July to November, the direct radiation is relatively large, and the best direction is about 5 ° south by East

Abundant rainwater resources- Drainage and rainwater collection should be considered in the building

MIND MAPPING

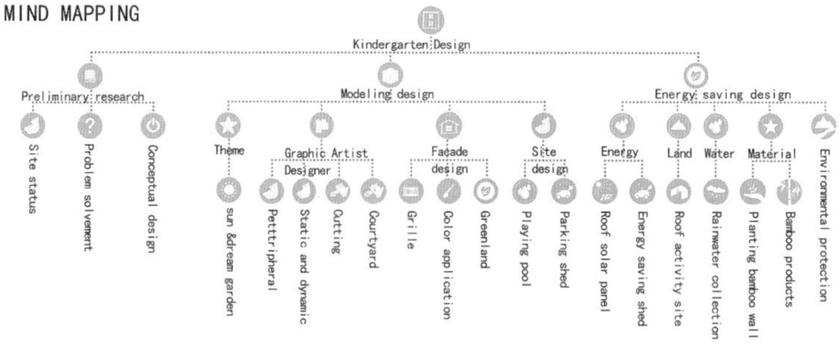
Kindergarten Design
Preliminary research
Site status
Problem solvement
Conceptual design
Modeling design
Theme
sun &dream garden
Graphic Artist Designer
Petttripheral
Static and dynamic
Cutting
Courtyard
Facade design
Grille
Color application
Greenland
Site design
Playing pool
Parking shed
Energy saving design
Energy
Roof solar panel
Energy saving shed
Land
Roof activity site
Water
Rainwater collection
Material
Planting bamboo wall
Bamboo products
Environmental protection

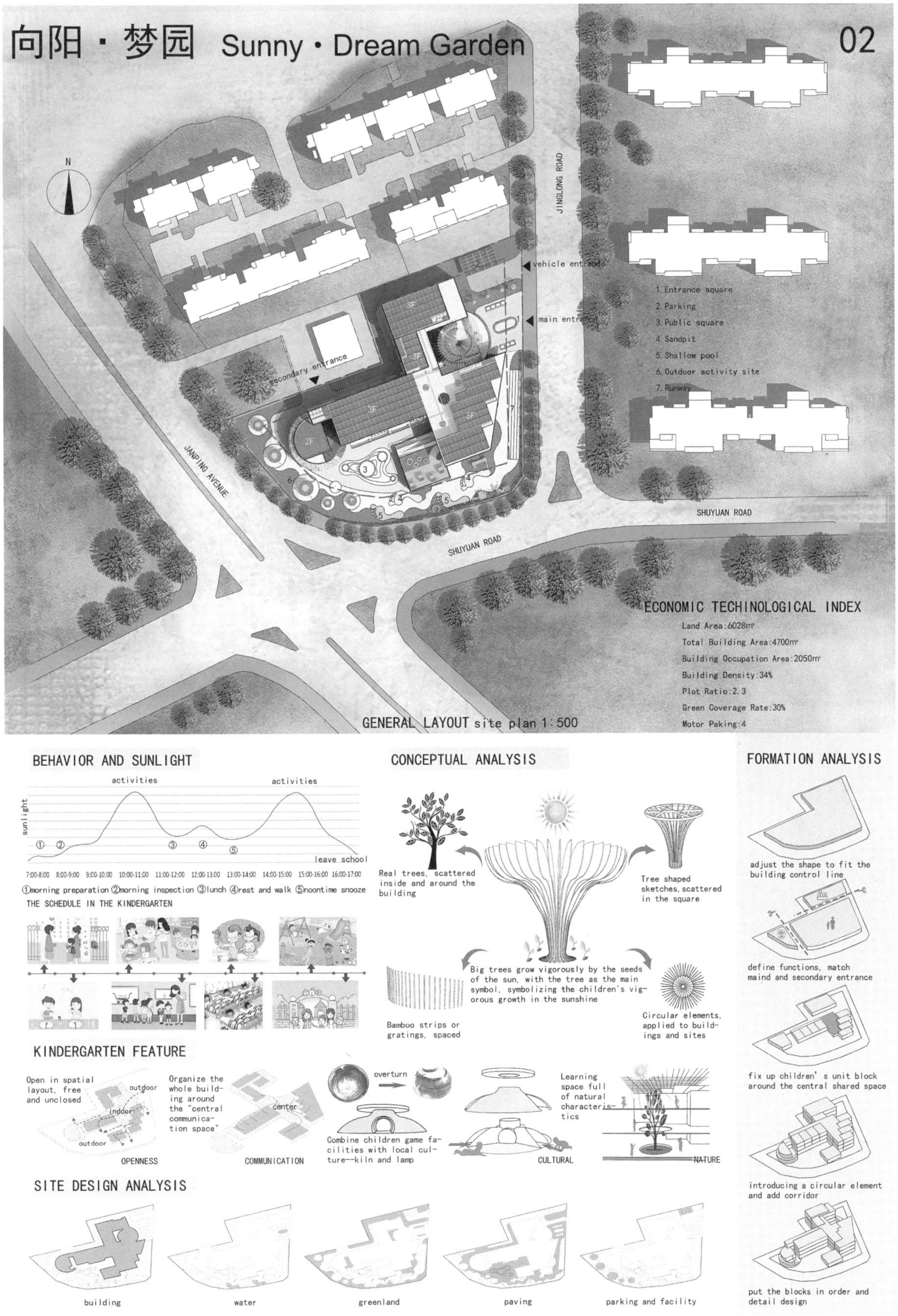

向阳 · 梦园 Sunny · Dream Garden
02
N
JINGLONG ROAD
vehicle entrance
main entrance
secondary entrance
JANPING AVENUE
SHUYUAN ROAD
SHUYUAN ROAD
1. Entrance square
2. Parking
3. Public square
4. Sandpit
5. Shallow pool
6. Outdoor activity site
7. Runway
ECONOMIC TECHINOLOGICAL INDEX
Land Area: 6028m²
Total Building Area: 4700m²
Building Occupation Area: 2050m²
Building Density: 34%
Plot Ratio: 2.3
Green Coverage Rate: 30%
Motor Paking: 4
GENERAL LAYOUT site plan 1:500
BEHAVIOR AND SUNLIGHT
activities
activities
sunlight
leave school
7:00-8:00 8:00-9:00 9:00-10:00 10:00-11:00 11:00-12:00 12:00-13:00 13:00-14:00 14:00-15:00 15:00-16:00 16:00-17:00
①morning preparation ②morning inspection ③lunch ④rest and walk ⑤noontime snooze
THE SCHEDULE IN THE KINDERGARTEN
CONCEPTUAL ANALYSIS
Real trees, scattered inside and around the building
Tree shaped sketches, scattered in the square
Big trees grow vigorously by the seeds of the sun, with the tree as the main symbol, symbolizing the children's vigorous growth in the sunshine
Bamboo strips or gratings, spaced
Circular elements, applied to buildings and sites
FORMATION ANALYSIS
adjust the shape to fit the building control line
define functions, match maind and secondary entrance
fix up children's unit block around the central shared space
introducing a circular element and add corridor
put the blocks in order and detail design
KINDERGARTEN FEATURE
Open in spatial layout, free and unclosed
outdoor
indoor
outdoor
OPENNESS
Organize the whole building around the "central communication space"
center
COMMUNICATION
overturn
Combine children game facilities with local culture—kiln and lamp
CULTURAL
Learning space full of natural characteristics
NATURE
SITE DESIGN ANALYSIS
building
water
greenland
paving
parking and facility

向阳 · 梦园 Sunny · Dream Garden

03

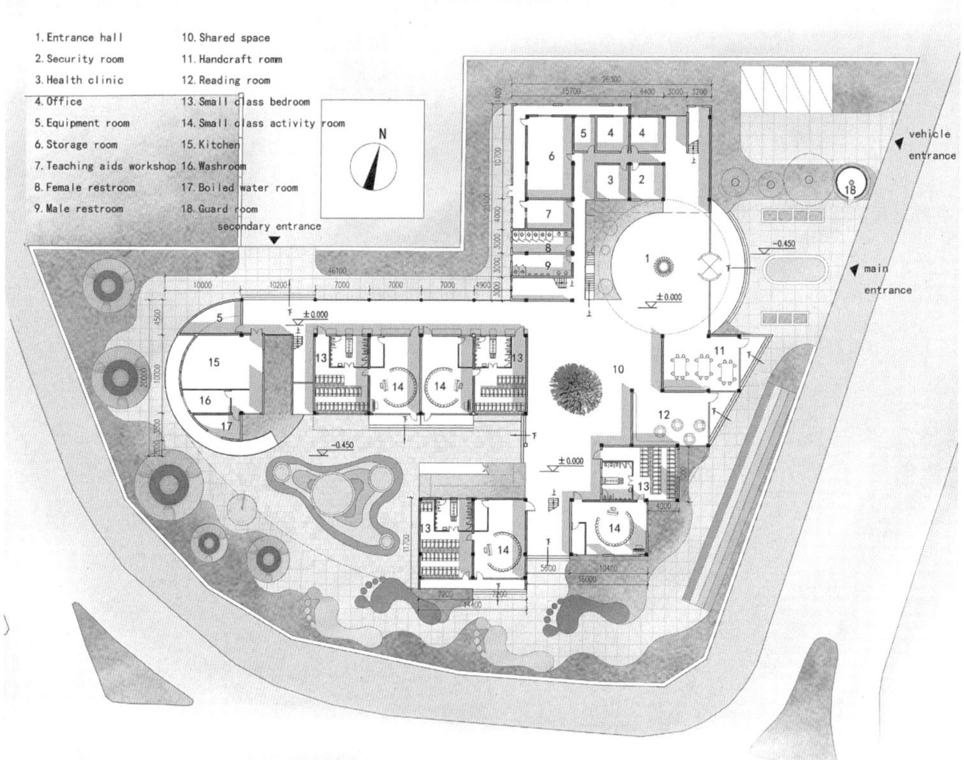
1. Entrance hall
2. Security room
3. Health clinic
4. Office
5. Equipment room
6. Storage room
7. Teaching aids workshop
8. Female restroom
9. Male restroom
10. Shared space
11. Handcraft romm
12. Reading room
13. Small class bedroom
14. Small class activity room
15. Kitchen
16. Washroom
17. Boiled water room
18. Guard room
N
secondary entrance
vehicle entrance
main entrance
±0.000
-0.450

FIRST FLOOR PLAN site plan 1:250

FUNCTION ANALYSIS

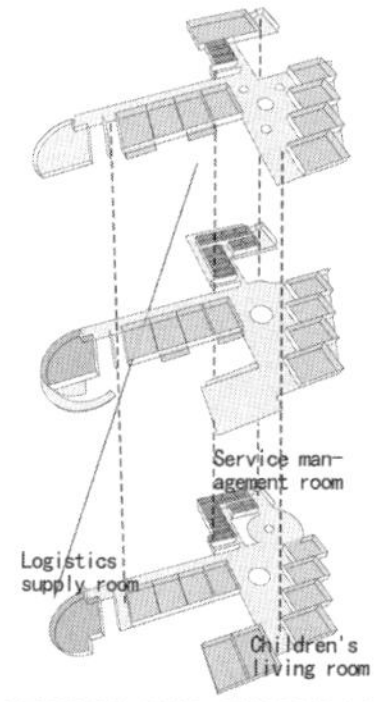
Service management room
Logistics supply room
Children's living room

STREAMLINE ANALYSIS

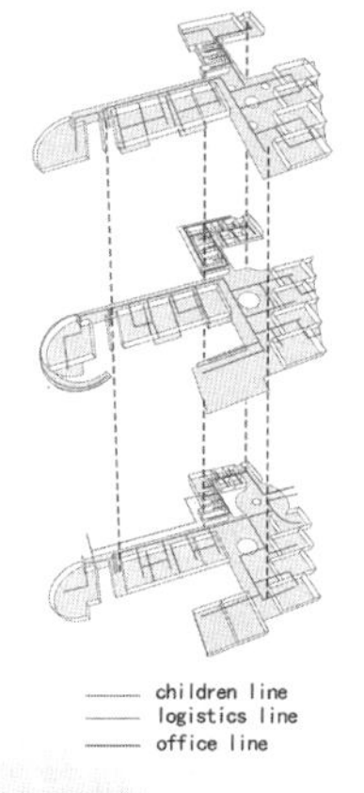
children line
logistics line
office line

14.400
12.000
8.000
4.000

NORTH FACADE site plan 1：250

向阳 · 梦园 Sunny · Dream Garden 04

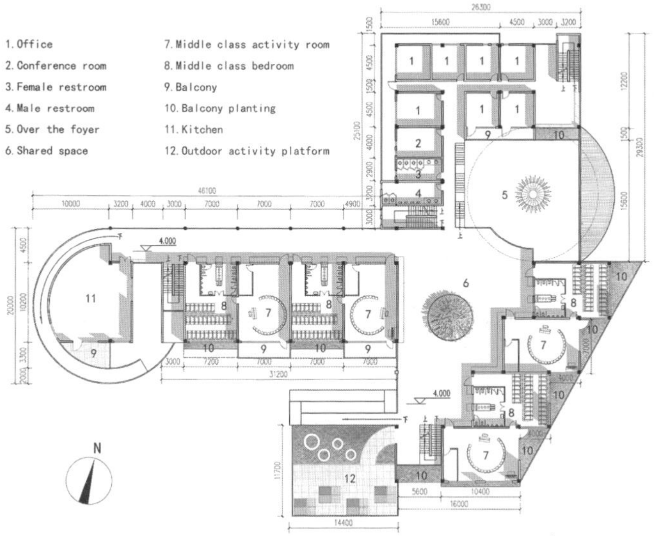
1. Office
2. Conference room
3. Female restroom
4. Male restroom
5. Over the foyer
6. Shared space
7. Middle class activity room
8. Middle class bedroom
9. Balcony
10. Balcony planting
11. Kitchen
12. Outdoor activity platform
N

SECOND FLOOR PLAN site plan 1:250

CORRIDOR SPACE

SHARED SPACE

FOYER SPACE

READING SPACE

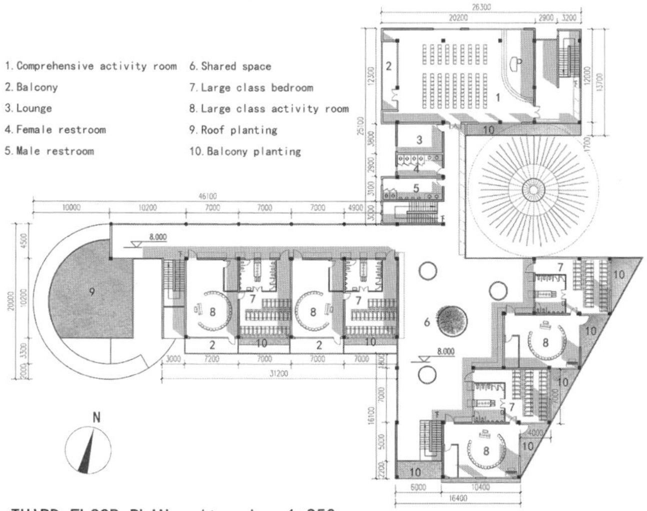
1. Comprehensive activity room
2. Balcony
3. Lounge
4. Female restroom
5. Male restroom
6. Shared space
7. Large class bedroom
8. Large class activity room
9. Roof planting
10. Balcony planting
N

THIRD FLOOR PLAN site plan 1:250

DRIP IRRIGATION SYSTEM

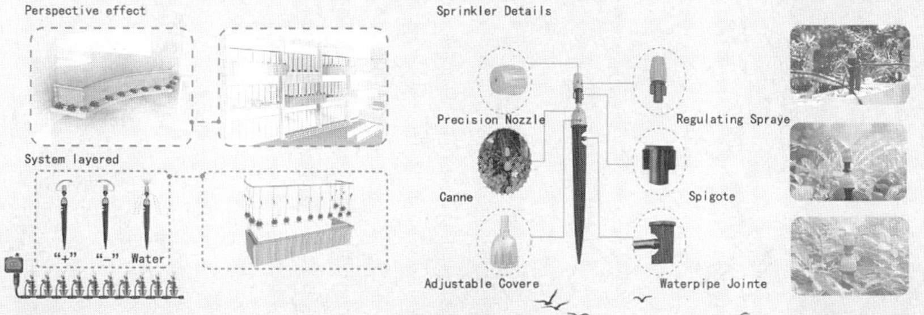
Perspective effect
System layered
"+"
"-"
Water
Sprinkler Details
Precision Nozzle
Regulating Spraye
Canne
Spigote
Adjustable Covere
Waterpipe Jointe

GROUND SOURCE HEAT PUMP

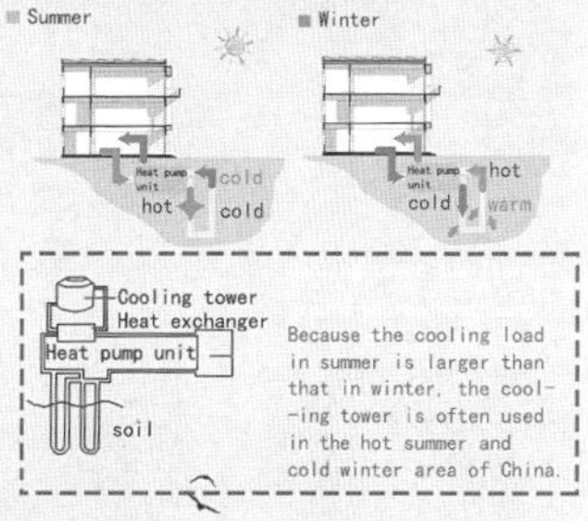
Summer
Winter
cold
hot
cold
hot
cold
warm
Cooling tower
Heat exchanger
Heat pump unit
soil
Because the cooling load in summer is larger than that in winter, the cool-ing tower is often used in the hot summer and cold winter area of China.

14.400
12.000
8.000
4.000

SOUTH FACADE site plan 1:250

向阳 · 梦园 Sunny · Dream Garden 05

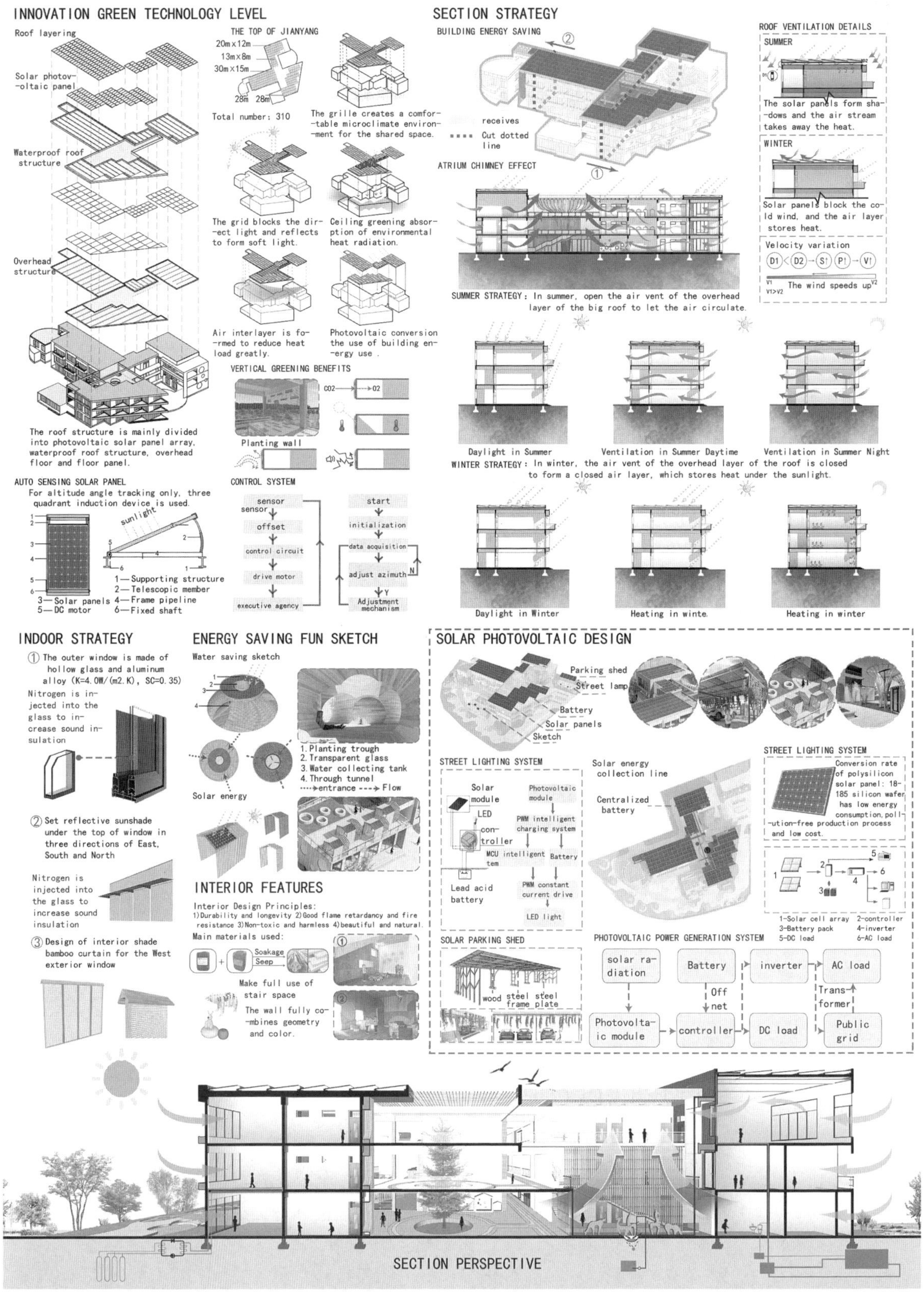
INNOVATION GREEN TECHNOLOGY LEVEL
Roof layering
Solar photov-
-oltaic panel
Waterproof roof
structure
Overhead
structure
The roof structure is mainly divided into photovoltaic solar panel array, waterproof roof structure, overhead floor and floor panel.
THE TOP OF JIANYANG
20m×12m
13m×8m
30m×15m
28m 28m
Total number: 310
The grille creates a comfor-
-table microclimate environ-
-ment for the shared space.
The grid blocks the dir-
-ect light and reflects
to form soft light.
Ceiling greening absor-
ption of environmental
heat radiation.
Air interlayer is fo-
-rmed to reduce heat
load greatly.
Photovoltaic conversion
the use of building en-
-ergy use .
VERTICAL GREENING BENEFITS
CO2 O2
Planting wall
AUTO SENSING SOLAR PANEL
For altitude angle tracking only, three quadrant induction device is used.
sunlight
1—Supporting structure
2—Telescopic member
3—Solar panels
4—Frame pipeline
5—DC motor
6—Fixed shaft
CONTROL SYSTEM
sensor
offset
control circuit
drive motor
executive agency
start
initialization
data acquisition
adjust azimuth
N
Y
Adjustment mechanism
SECTION STRATEGY
BUILDING ENERGY SAVING
receives
Cut dotted line
ATRIUM CHIMNEY EFFECT
SUMMER STRATEGY: In summer, open the air vent of the overhead layer of the big roof to let the air circulate.
ROOF VENTILATION DETAILS
SUMMER
The solar panels form shadows and the air stream takes away the heat.
WINTER
Solar panels block the cold wind, and the air layer stores heat.
Velocity variation
D1 D2 S P V
V1>V2 The wind speeds up
Daylight in Summer
Ventilation in Summer Daytime
Ventilation in Summer Night
WINTER STRATEGY: In winter, the air vent of the overhead layer of the roof is closed to form a closed air layer, which stores heat under the sunlight.
Daylight in Winter
Heating in winte
Heating in winter
INDOOR STRATEGY
① The outer window is made of hollow glass and aluminum alloy (K=4.0W/(m2.K), SC=0.35)
Nitrogen is injected into the glass to increase sound insulation
② Set reflective sunshade under the top of window in three directions of East, South and North
Nitrogen is injected into the glass to increase sound insulation
③ Design of interior shade bamboo curtain for the West exterior window
ENERGY SAVING FUN SKETCH
Water saving sketch
1.Planting trough
2.Transparent glass
3.Water collecting tank
4.Through tunnel
entrance Flow
Solar energy
INTERIOR FEATURES
Interior Design Principles:
1)Durability and longevity 2)Good flame retardancy and fire resistance 3)Non-toxic and harmless 4)beautiful and natural.
Main materials used:
Soakage
Seep
Make full use of stair space
The wall fully combines geometry and color.
SOLAR PHOTOVOLTAIC DESIGN
Parking shed
Street lamp
Battery
Solar panels
Sketch
STREET LIGHTING SYSTEM
Solar module
LED
controller
Lead acid battery
Photovoltaic module
PWM intelligent charging system
MCU intelligent system
Battery
PWM constant current drive
LED light
Solar energy collection line
Centralized battery
STREET LIGHTING SYSTEM
Conversion rate of polysilicon solar panel: 18-185 silicon wafer has low energy consumption, pollution-free production process and low cost.
1-Solar cell array 2-controller
3-Battery pack 4-inverter
5-DC load 6-AC load
SOLAR PARKING SHED
wood steel frame steel plate
PHOTOVOLTAIC POWER GENERATION SYSTEM
solar radiation
Battery
inverter
AC load
Off net
Transformer
Photovoltaic module
controller
DC load
Public grid
SECTION PERSPECTIVE

向阳 · 梦园 Sunny · Dream Garden

06

MATERIAL ANALYSIS

The building materials are made of bamboo board which is abundant in the area, and the building enclosure is well insulated

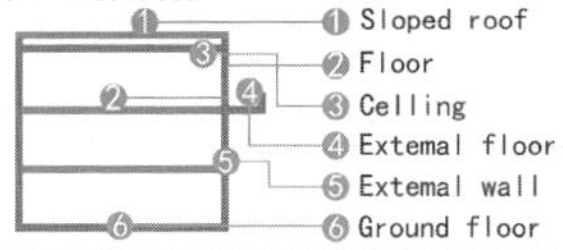

1 Sloped roof
2 Floor
3 Celling
4 Extemal floor
5 Extemal wall
6 Ground floor

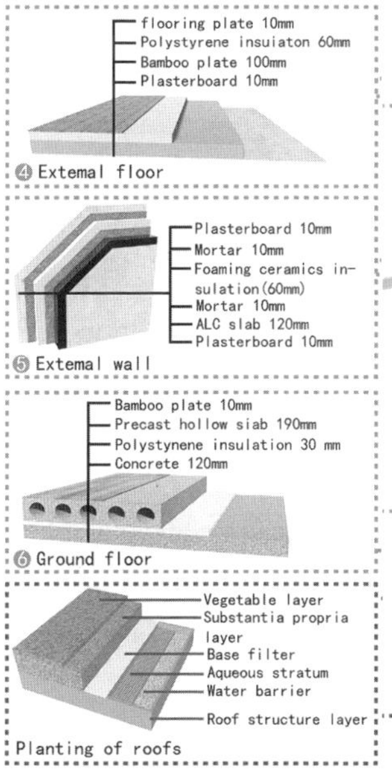

flooring plate 10mm
Polystyrene insuiaton 60mm
Bamboo plate 100mm
Plasterboard 10mm
4 Extemal floor
Plasterboard 10mm
Mortar 10mm
Foaming ceramics insulation(60mm)
Mortar 10mm
ALC slab 120mm
Plasterboard 10mm
5 Extemal wall
Bamboo plate 10mm
Precast hollow slab 190mm
Polystynene insulation 30 mm
Concrete 120mm
6 Ground floor
Vegetable layer
Substantia propria layer
Base filter
Aqueous stratum
Water barrier
Roof structure layer
Planting of roofs

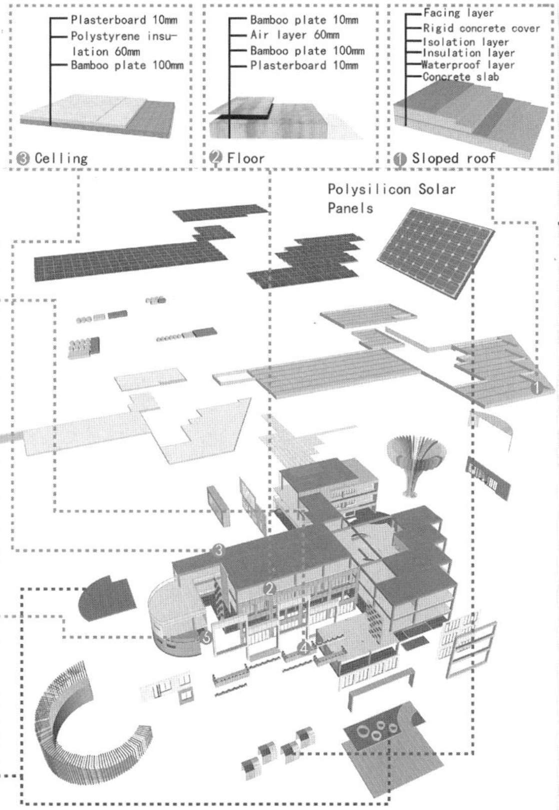

Plasterboard 10mm
Polystyrene insulation 60mm
Bamboo plate 100mm
3 Celling
Bamboo plate 10mm
Air layer 60mm
Bamboo plate 100mm
Plasterboard 10mm
2 Floor
Facing layer
Rigid concrete cover
Isolation layer
Insulation layer
Waterproof layer
Concrete slab
1 Sloped roof
Polysilicon Solar Panels

STORM-WATER SYSTEM

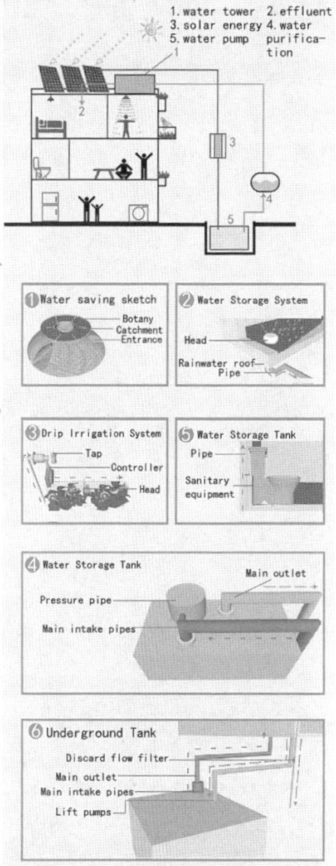

1.water tower 2.effluent
3.solar energy 4.water purification
5.water pump
1 Water saving sketch
Botany
Catchment
Entrance
2 Water Storage System
Head
Rainwater roof
Pipe
3 Drip Irrigation System
Tap
Controller
Head
5 Water Storage Tank
Pipe
Sanitary equipment
4 Water Storage Tank
Main outlet
Pressure pipe
Main intake pipes
6 Underground Tank
Discard flow filter
Main outlet
Main intake pipes
Lift pumps

AERIAL VIEW OF ENERGY CONSERVATION

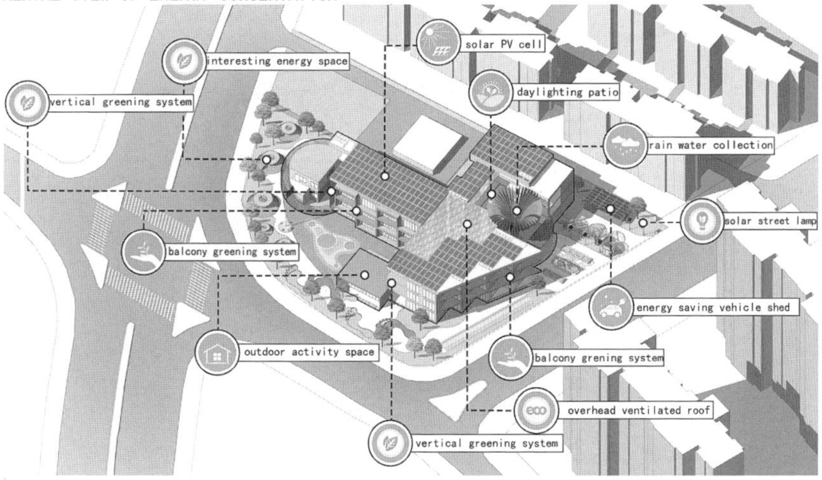

interesting energy space
vertical greening system
solar PV cell
daylighting patio
rain water collection
solar street lamp
balcony greening system
energy saving vehicle shed
outdoor activity space
balcony grening system
overhead ventilated roof
vertical greening system

BAMBOO STRUCTURE

Planting Wall Details

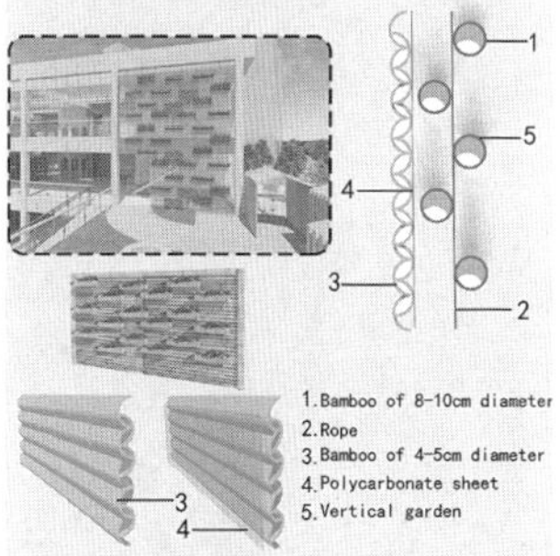

1.Bamboo of 8-10cm diameter
2.Rope
3.Bamboo of 4-5cm diameter
4.Polycarbonate sheet
5.Vertical garden

BAMBOO COMPOSITE STRUCTURE

Bamboo requirements:
1) Use dried bamboo; 2) insect proof treatment is required to prevent; 3) do not use bamboo with longitudinal cracks

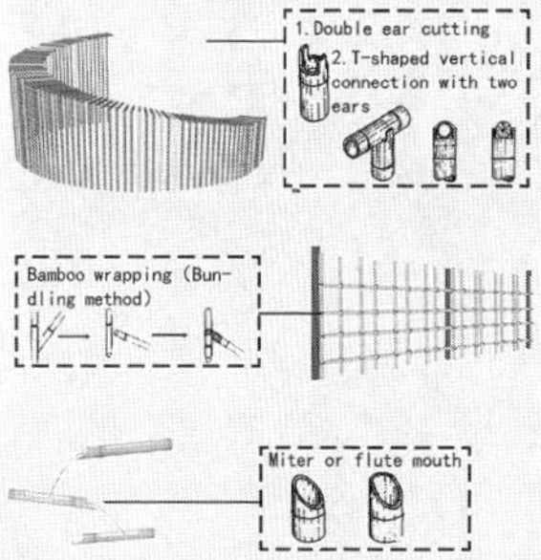

1.Double ear cutting
2.T-shaped vertical connection with two ears
Bamboo wrapping (Bundling method)
Miter or flute mouth

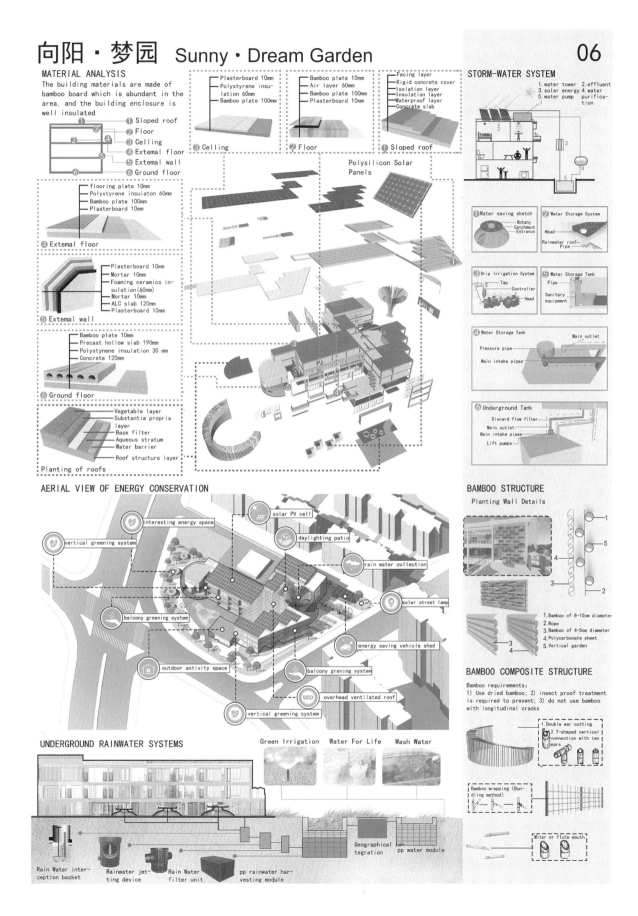

UNDERGROUND RAINWATER SYSTEMS

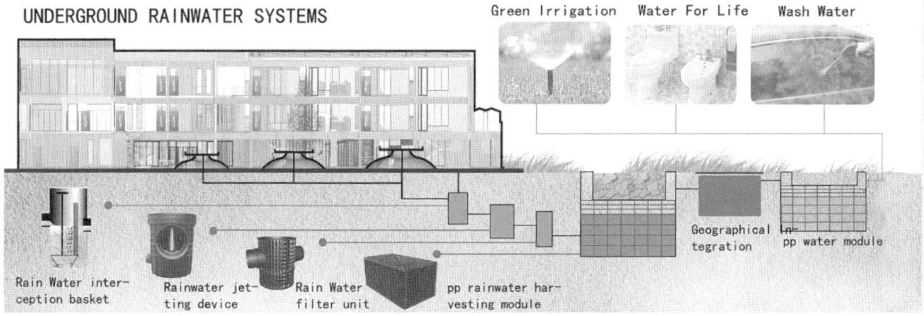

Green Irrigation
Water For Life
Wash Water
Geographical integration
pp water module
Rain Water interception basket
Rainwater jetting device
Rain Water filter unit
pp rainwater harvesting module

综合奖 · 优秀奖
General Prize Awarded · Honorable Mention Prize

注 册 号：7406

项目名称：“小世界”环游记（南平）
Dream Quest Travel (Nanping)

作　　者：周扬空、李景秀、苗振轩、赵晓彤、金勇运、许仁杰、李正洁、甄　成

参赛单位：天津大学

指导教师：郭娟利

“小世界”环游记
Dream Quest Travel 01
2020台达杯国际太阳能建筑设计竞赛
International Solar Building Design Competition 2020

Local Feature

Southern Foot of Wuyi Mountain　Kaoting Academy　Ocean of Forest and Town of Bamboo　Tea and Fruit Base　Longyao Kiln Site　Jianzhan Porcelain

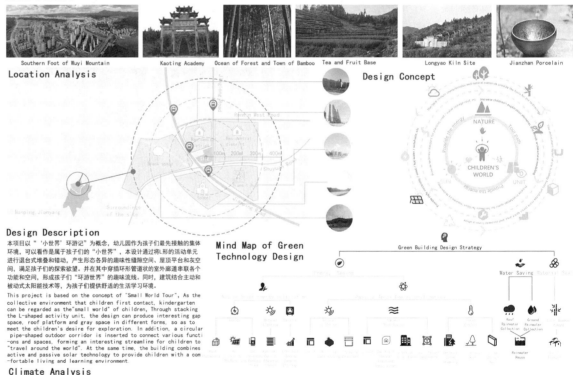

Location Analysis

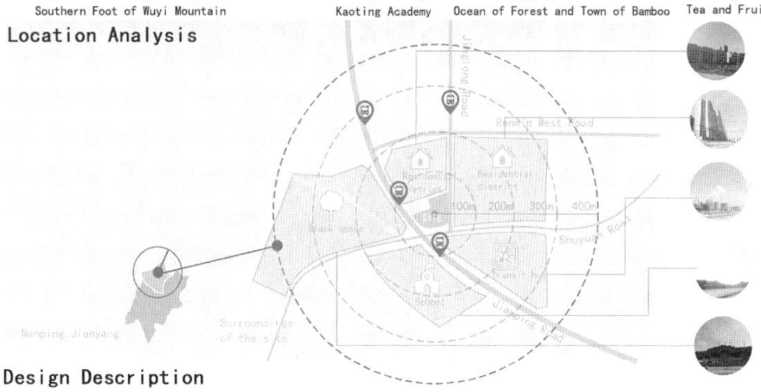

Design Concept

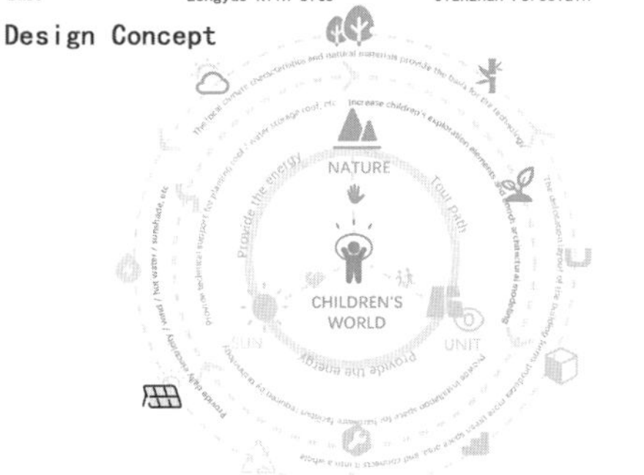
NATURE
CHILDREN'S WORLD
UNIT

Design Description

本项目以“‘小世界’环游记”为概念，幼儿园作为孩子们最先接触的集体环境，可以看作是属于孩子们的“小世界”，本设计通过将L形的活动单元进行退台式堆叠和错动，产生形态各异的趣味性缝隙空间、屋顶平台和灰空间，满足孩子们的探索欲望。并在其中穿插环形管道状的室外廊道串联各个功能和空间，形成孩子们“环游世界”的趣味流线。同时，建筑结合主动和被动式太阳能技术等，为孩子们提供舒适的生活学习环境。

This project is based on the concept of "Small World Tour", As the collective environment that children first contact, kindergarten can be regarded as the"small world" of children, Through stacking the L-shaped activity unit, the design can produce interesting gap space, roof platform and gray space in different forms, so as to meet the children's desire for exploration. In addition, a circular pipe-shaped outdoor corridor is inserted to connect various functi-ons and spaces, forming an interesting streamline for children to "travel around the world". At the same time, the building combines active and passive solar technology to provide children with a com-fortable living and learning environment.

Mind Map of Green Technology Design

Green Building Design Strategy
Water Saving
Roof Rainwater Collection
Ground Rainwater Collection
Rainwater Reuse

Climate Analysis

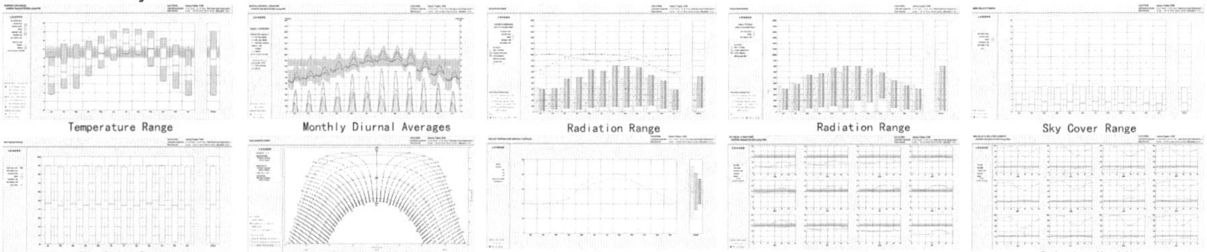
Temperature Range　Monthly Diurnal Averages　Radiation Range　Radiation Range　Sky Cover Range
Sky Cover Range　Sun Shading Chart　Ground Temperature　Dew Point　Relative Humidity

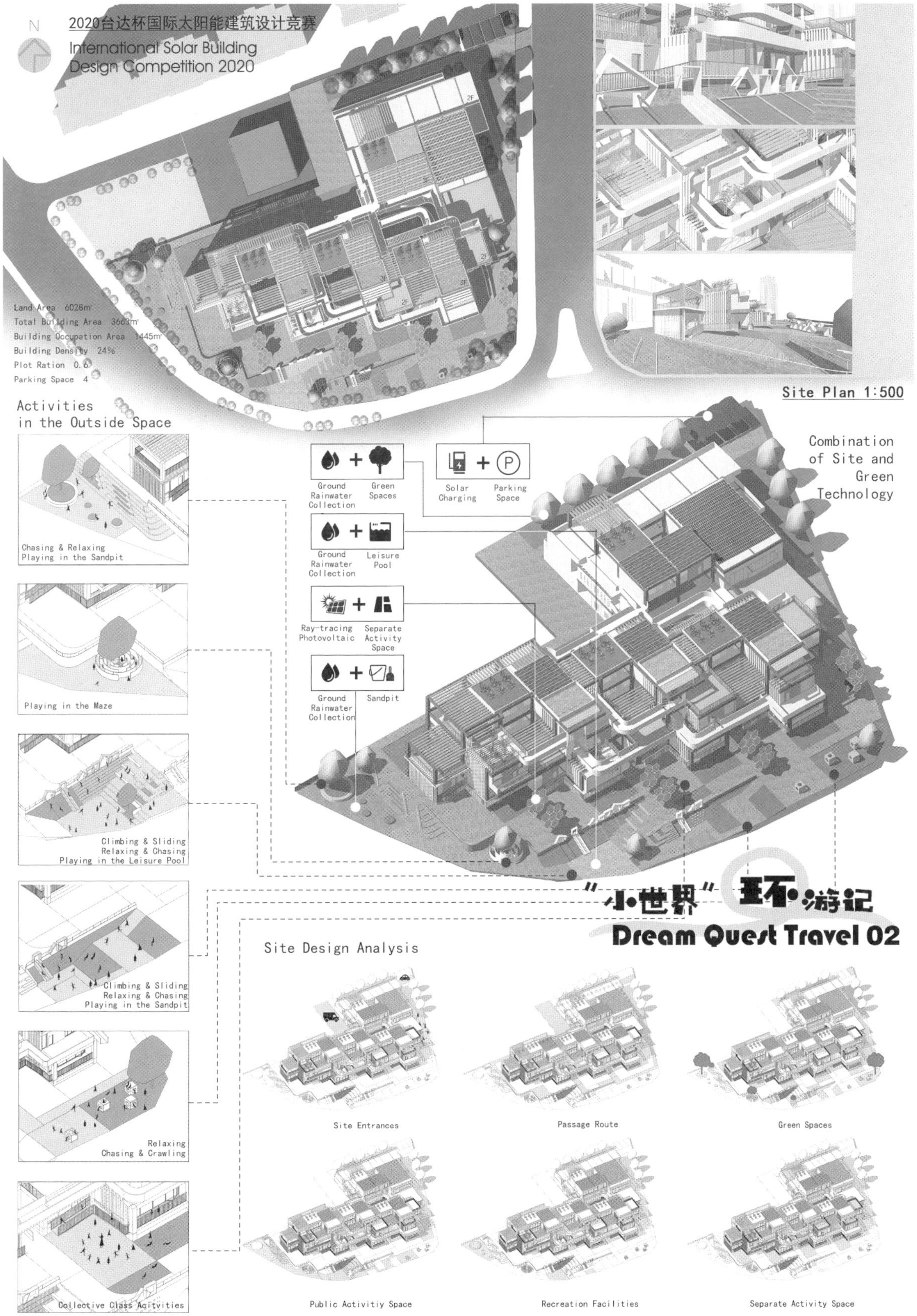
2020台达杯国际太阳能建筑设计竞赛
International Solar Building Design Competition 2020
N
Land Area 6028㎡
Total Building Area 3603㎡
Building Occupation Area 1445㎡
Building Density 24%
Plot Ration 0.6
Parking Space 4
Site Plan 1:500
Activities in the Outside Space
Chasing & Relaxing
Playing in the Sandpit
Playing in the Maze
Climbing & Sliding
Relaxing & Chasing
Playing in the Leisure Pool
Climbing & Sliding
Relaxing & Chasing
Playing in the Sandpit
Relaxing
Chasing & Crawling
Collective Class Acitvities
Ground Rainwater Collection
Green Spaces
Solar Charging
Parking Space
Ground Rainwater Collection
Leisure Pool
Ray-tracing Photovoltaic
Separate Activity Space
Ground Rainwater Collection
Sandpit
Combination of Site and Green Technology
"小世界"环游记
Dream Quest Travel 02
Site Design Analysis
Site Entrances
Passage Route
Green Spaces
Public Activitiy Space
Recreation Facilities
Separate Activity Space

“小世界”环游记
Dream Quest Travel 03

2020台达杯国际太阳能建筑设计竞赛
International Solar Building Design Competition 2020

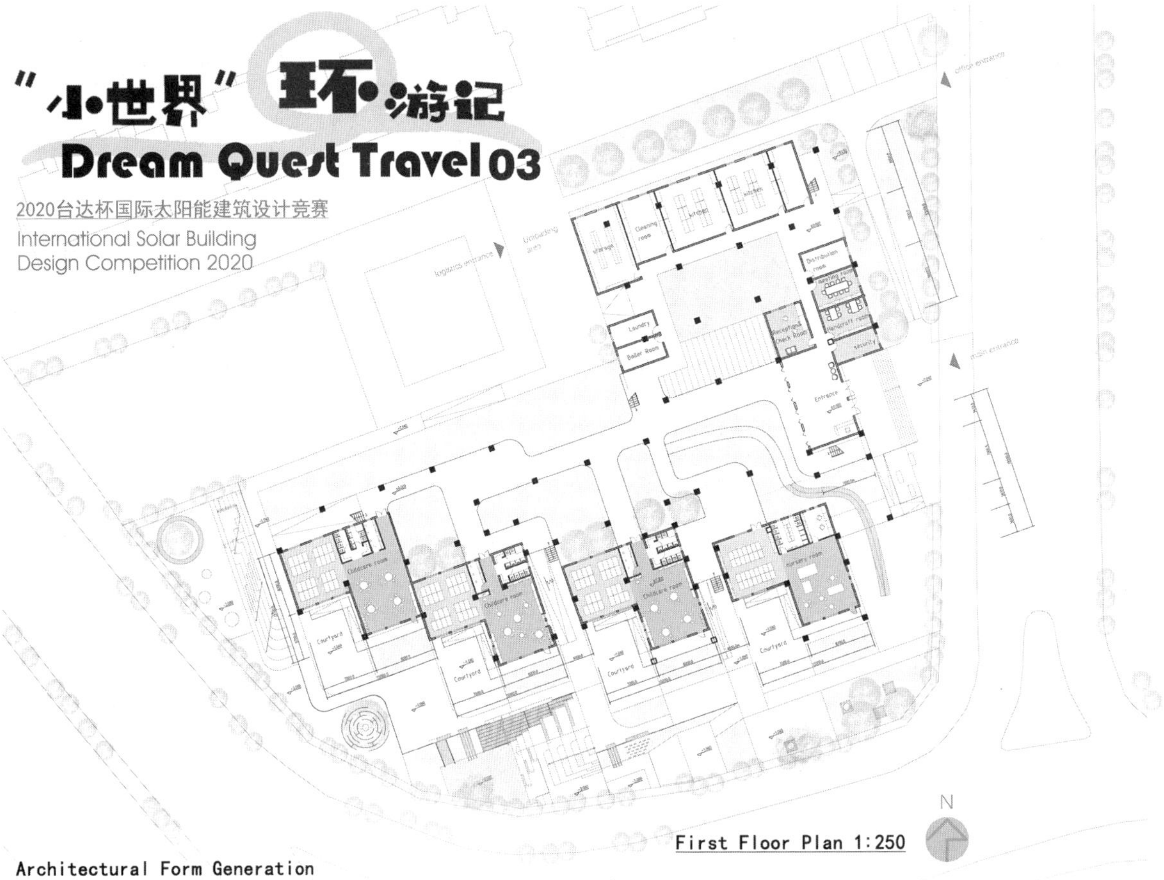
N

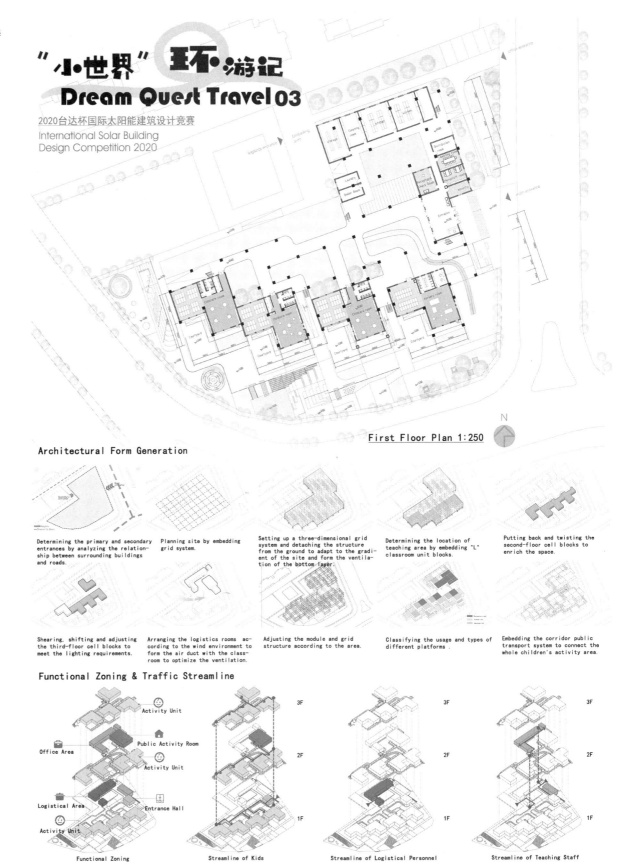

First Floor Plan 1:250

Architectural Form Generation

Determining the primary and secondary entrances by analyzing the relationship between surrounding buildings and roads.

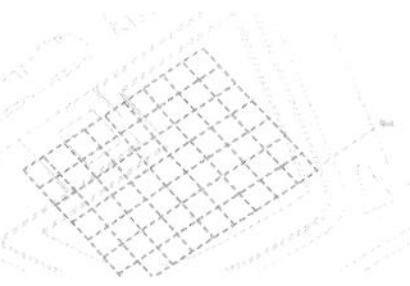

Planning site by embedding grid system.

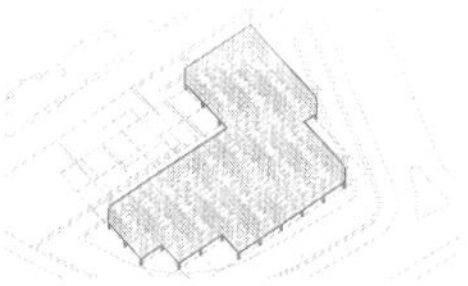

Setting up a three-dimensional grid system and detaching the structure from the ground to adapt to the gradient of the site and form the ventilation of the bottom layer.

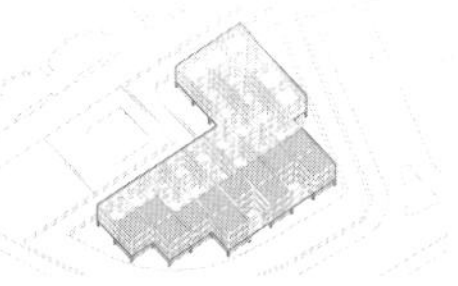

Determining the location of teaching area by embedding "L" classroom unit blocks.

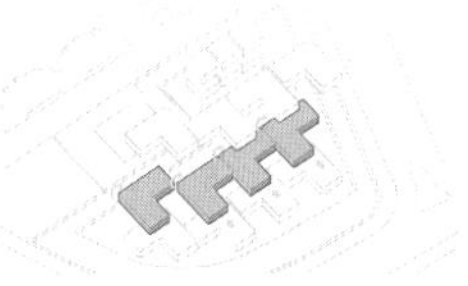

Putting back and twisting the second-floor cell blocks to enrich the space.

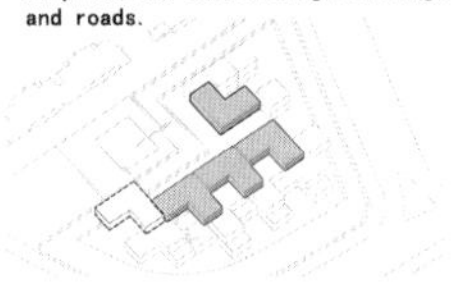

Shearing, shifting and adjusting the third-floor cell blocks to meet the lighting requirements.

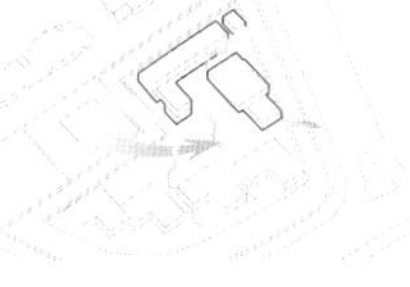

Arranging the logistics rooms according to the wind environment to form the air duct with the classroom to optimize the ventilation.

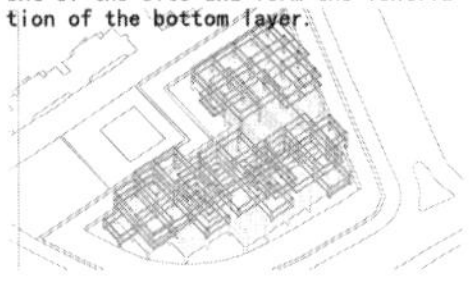

Adjusting the module and grid structure according to the area.

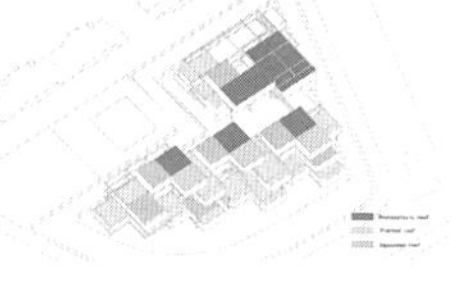

Classifying the usage and types of different platforms .

Embedding the corridor public transport system to connect the whole children's activity area.

Functional Zoning & Traffic Streamline

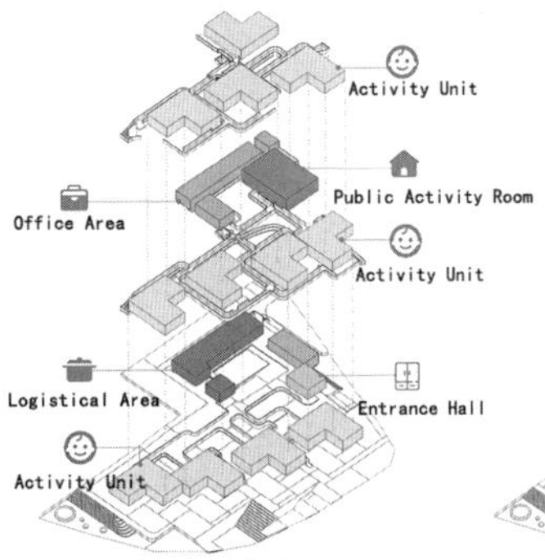
Activity Unit
Public Activity Room
Office Area
Activity Unit
Logistical Area
Entrance Hall
Activity Unit

Functional Zoning

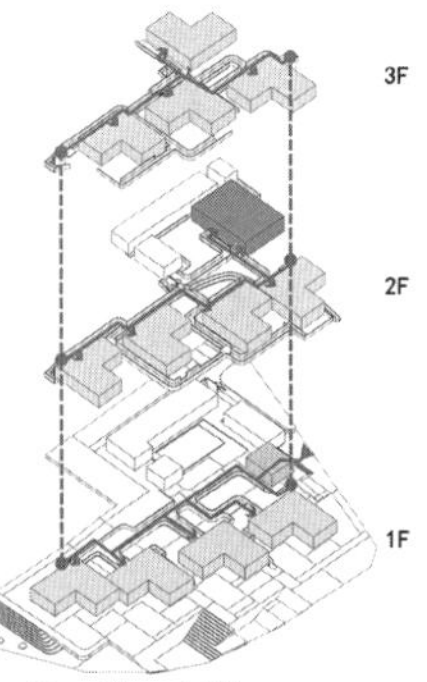
3F
2F
1F

Streamline of Kids

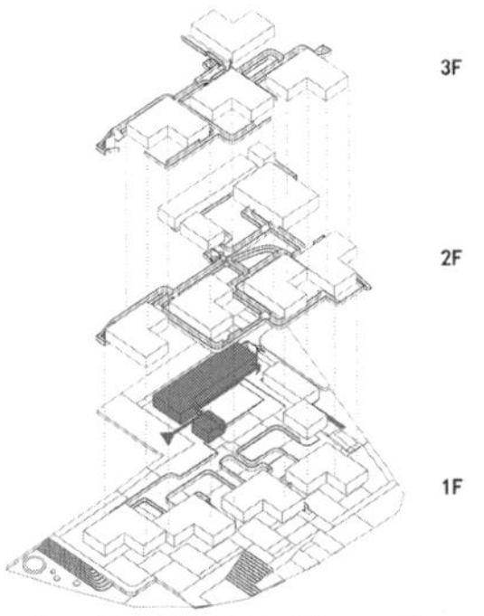
3F
2F
1F

Streamline of Logistical Personnel

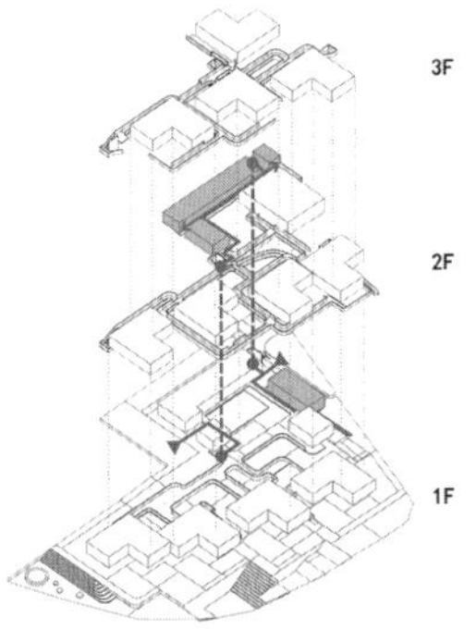
3F
2F
1F

Streamline of Teaching Staff

2020台达杯国际太阳能建筑设计竞赛
International Solar Building
Design Competition 2020

"小世界"环游记
Dream Quest Travel 04

Building Block Design Analysis

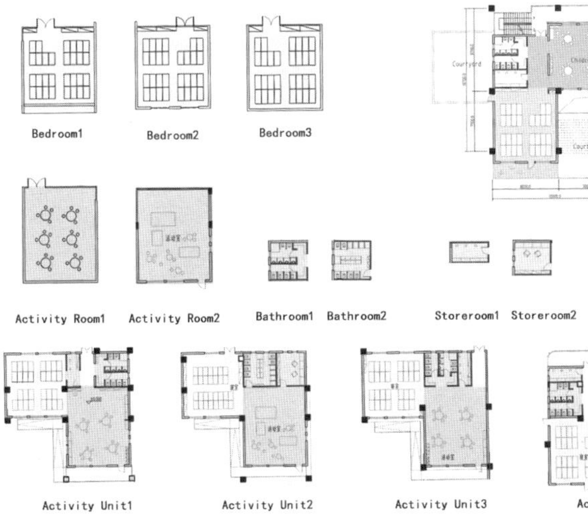
Bedroom1
Bedroom2
Bedroom3
Activity Room1
Activity Room2
Bathroom1
Bathroom2
Storeroom1
Storeroom2
Activity Unit1
Activity Unit2
Activity Unit3
Activity Unit4
Activity Unit5

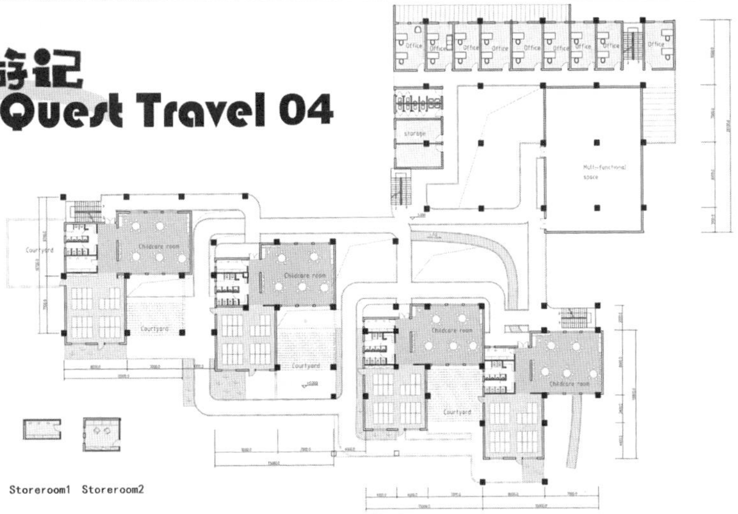

Second Floor Plan 1 : 250

Corridor Public Transport System Analysis

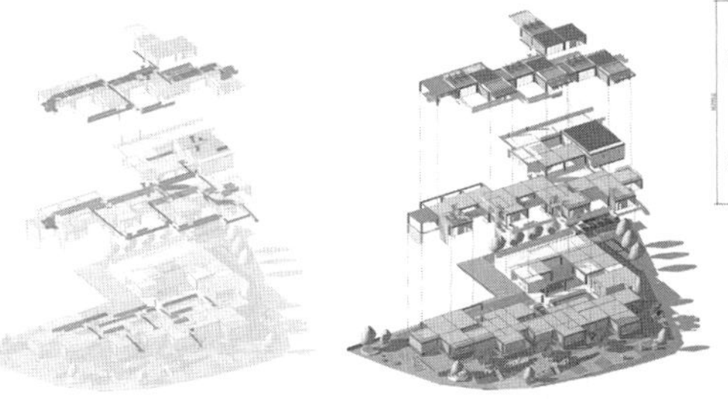

Third Floor Plan 1 : 250

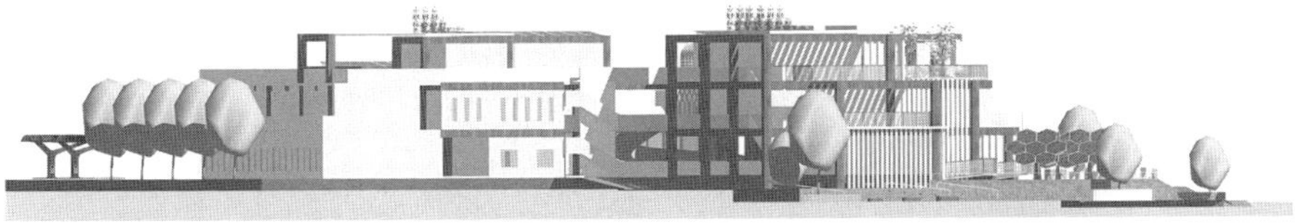

West Elevation 1 : 250

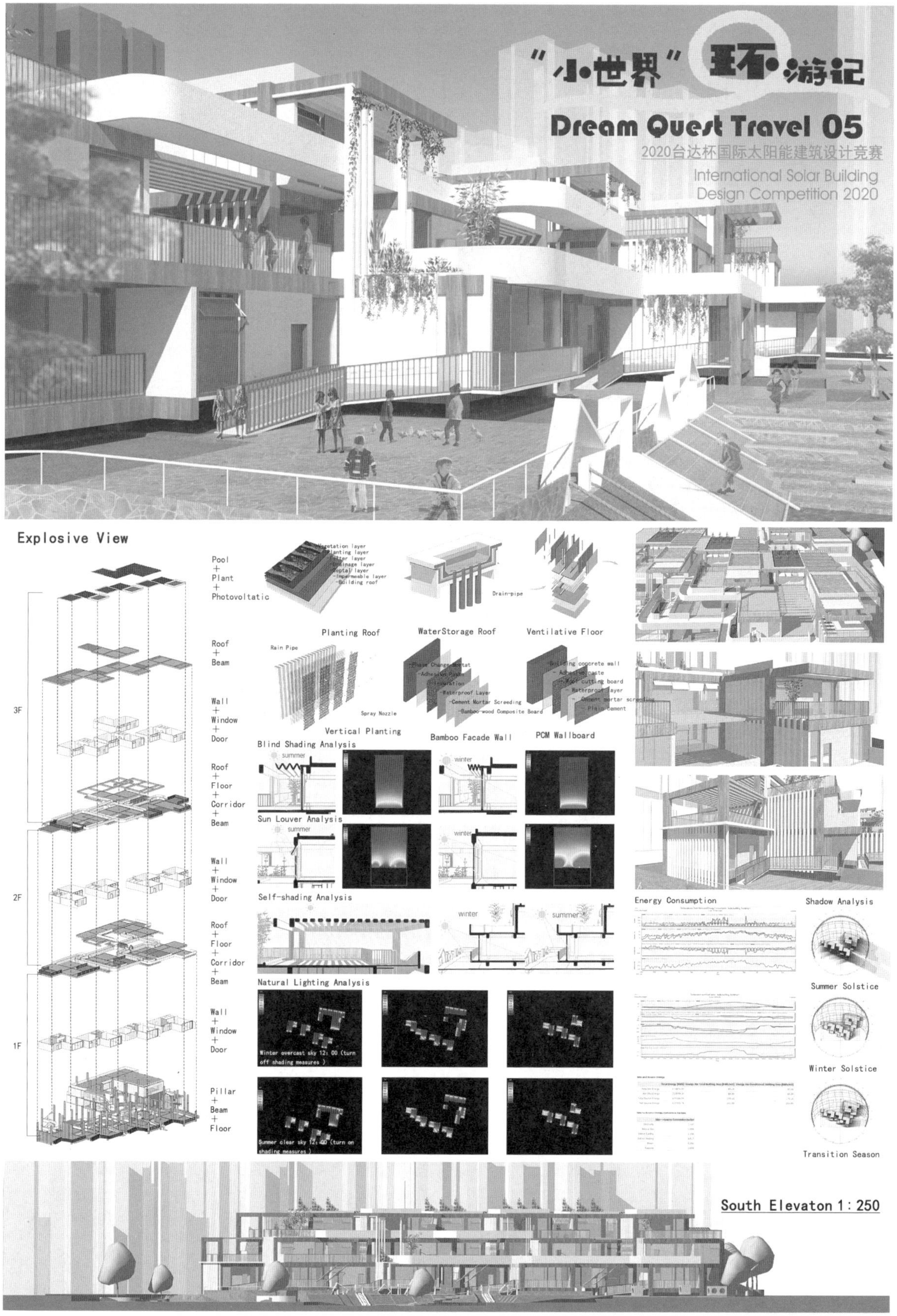

“小世界”环游记
Dream Quest Travel 05
2020台达杯国际太阳能建筑设计竞赛
International Solar Building
Design Competition 2020
Explosive View
3F
2F
1F
Pool + Plant + Photovoltatic
Roof + Beam
Wall + Window + Door
Roof + Floor + Corridor + Beam
Wall + Window + Door
Roof + Floor + Corridor + Beam
Wall + Window + Door
Pillar + Beam + Floor
Planting Roof
WaterStorage Roof
Ventilative Floor
Drain-pipe
Rain Pipe
Spray Nozzle
Vertical Planting
Bamboo Facade Wall
PCM Wallboard
Blind Shading Analysis
Sun Louver Analysis
Self-shading Analysis
Natural Lighting Analysis
summer
winter
Winter overcast sky 12:00 (turn off shading measures)
Summer clear sky 12:00 (turn on shading measures)
Energy Consumption
Shadow Analysis
Summer Solstice
Winter Solstice
Transition Season
South Elevaton 1:250

2020台达杯国际太阳能建筑设计竞赛
International Solar Building Design Competition 2020
"小世界"环·游记
Dream Quest Travel 06

Integrated Intelligent Building System

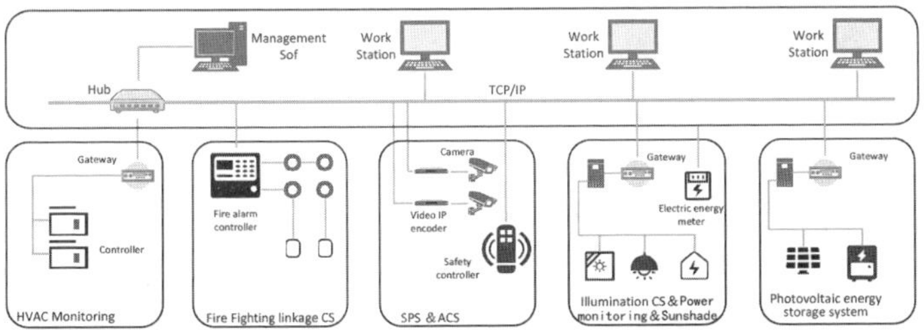
Management Sof
Work Station
Work Station
Work Station
Hub
TCP/IP
Gateway
Controller
HVAC Monitoring
Fire alarm controller
Fire Fighting linkage CS
Camera
Video IP encoder
Safety controller
SPS & ACS
Gateway
Electric energy meter
Illumination CS & Power monitoring & Sunshade
Gateway
Photovoltaic energy storage system

Control Level of Air Conditioning System

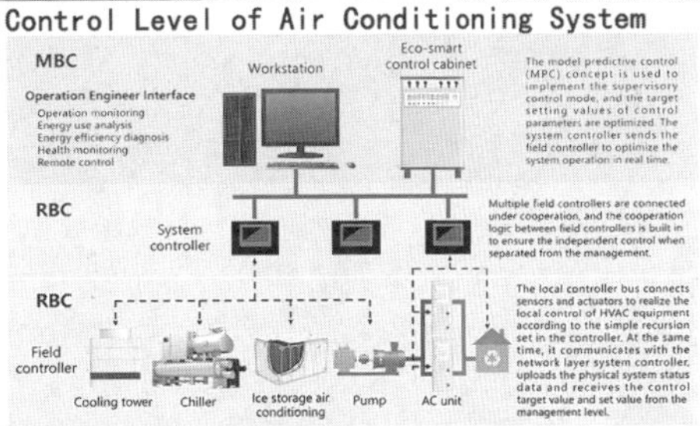
MBC
Operation Engineer Interface
Workstation
Eco-smart control cabinet
RBC
System controller
RBC
Field controller
Cooling tower
Chiller
Ice storage air conditioning
Pump
AC unit

Optimization Logic of Air Conditioning System Control

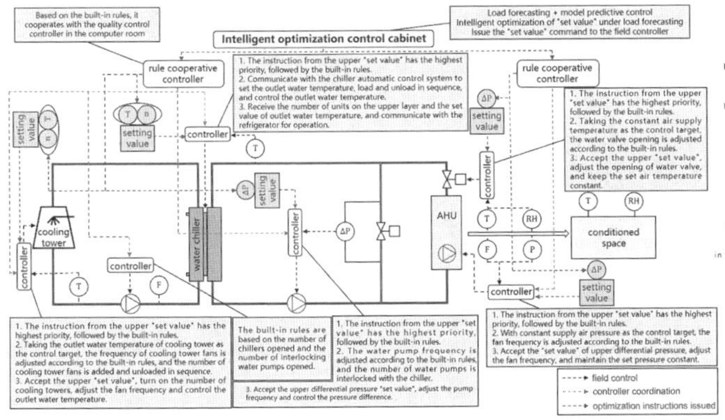
Intelligent optimization control cabinet
rule cooperative controller
rule cooperative controller
controller
setting value
cooling tower
water chiller
AHU
conditioned space
field control
controller coordination
optimization instructions issued

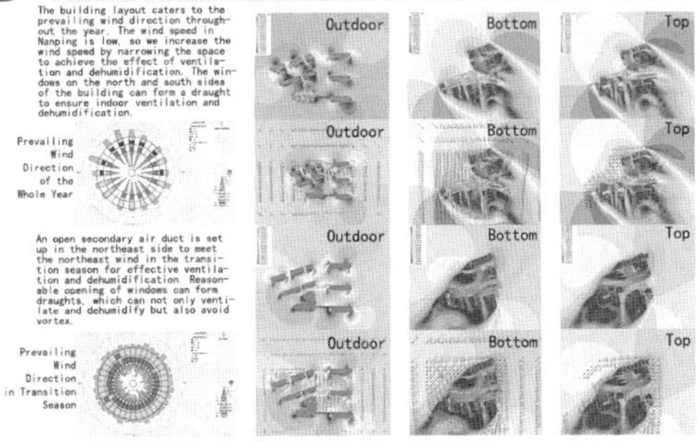
Outdoor
Bottom
Top
Prevailing Wind Direction of the Whole Year
Prevailing Wind Direction in Transition Season

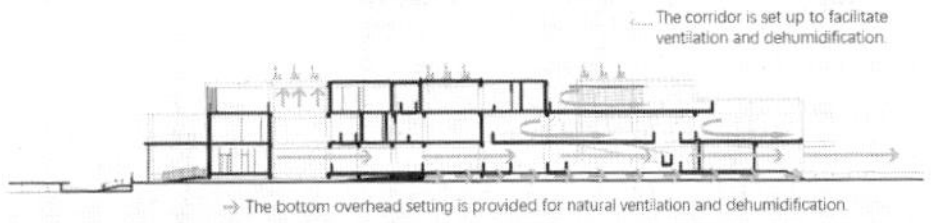
The corridor is set up to facilitate ventilation and dehumidification
The bottom overhead setting is provided for natural ventilation and dehumidification

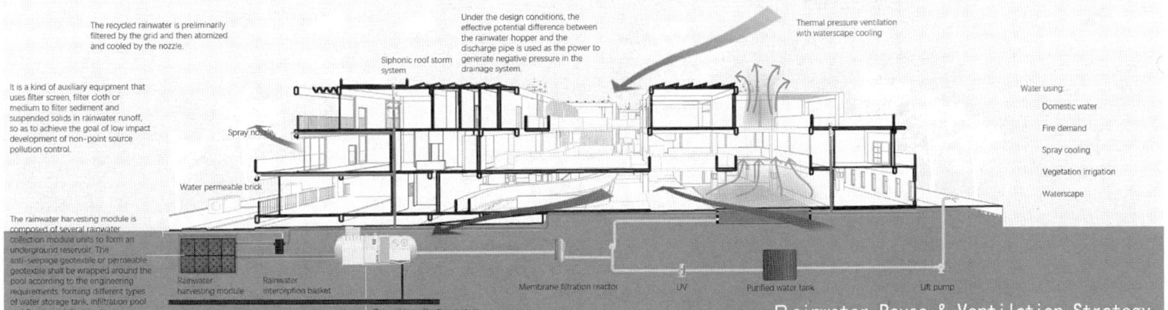
Siphonic roof storm system
Thermal pressure ventilation with waterscape cooling
Spray nozzle
Water permeable brick
Water using:
Domestic water
Fire demand
Spray cooling
Vegetation irrigation
Waterscape
Rainwater harvesting module
Rainwater interception basket
Membrane filtration reactor
UV
Purified water tank
Lift pump
Municipal sewerage pipe network
Rainwater purification system

Rainwater Reuse & Ventilation Strategy

综合奖 · 优秀奖
General Prize Awarded · Honorable Mention Prize

注 册 号：7408

项目名称：趣园（南平）

Interesting Garden(Nanping)

作　　者：崔林森、薛荣骅、张丛媛

参赛单位：西安科技大学

指导教师：孙倩倩

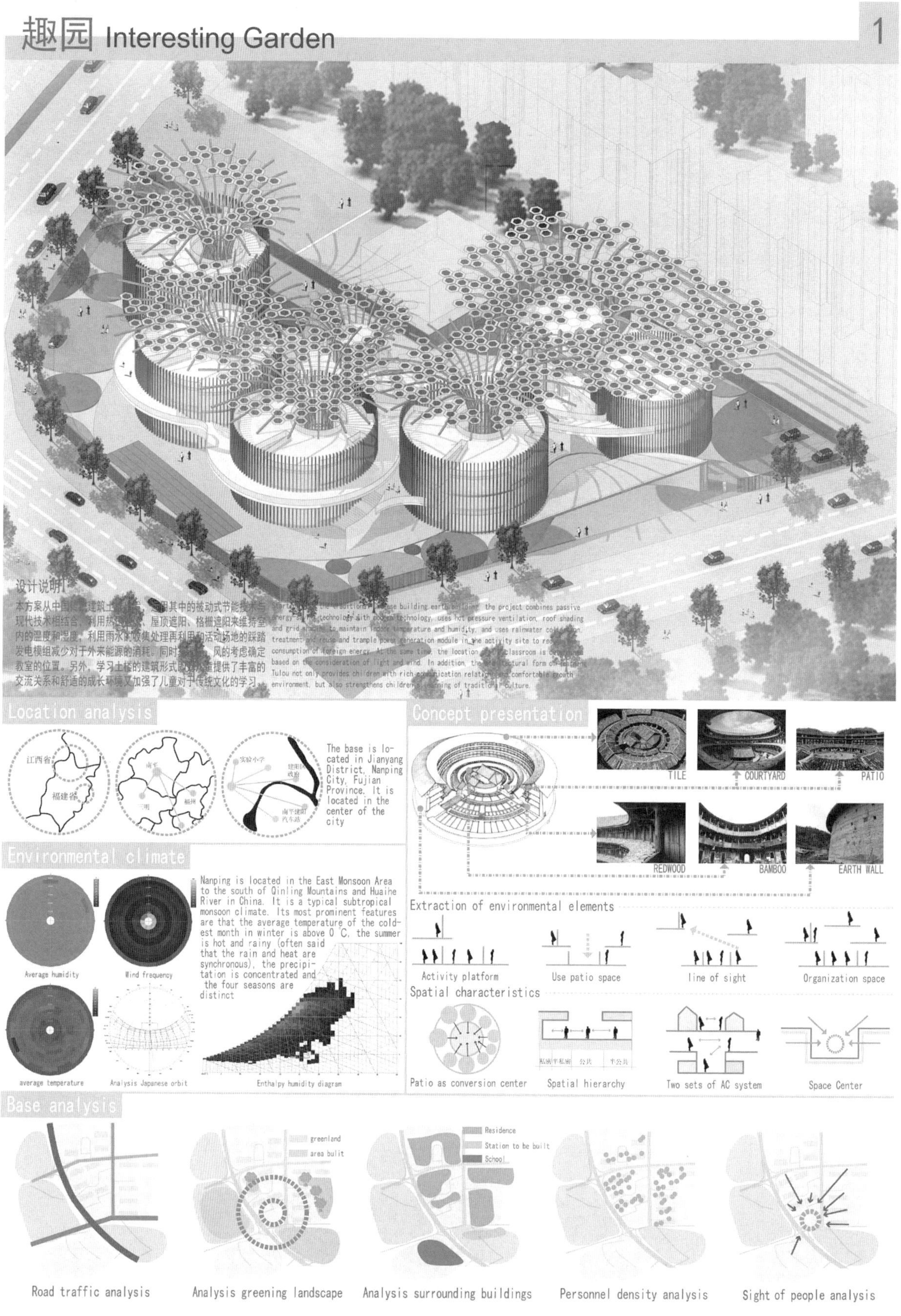
趣园 Interesting Garden
1
设计说明
本方案从中国 建筑土 其中的被动式节能技 与
现代技术相结合，利用热 、屋顶遮阳、格栅遮阳来维持室
内的温度和湿度，利用雨水的收集处理再利用和活动场地的踩踏
发电模组减少对于外来能源的消耗。同时 风的考虑确定
教室的位置。另外，学习土楼的建筑形式 童提供了丰富的
交流关系和舒适的成长环境又加强了儿童对于传统文化的学习。
the project combines passive energy technology with technology, uses hot pressure ventilation, roof shading and grid to maintain indoor temperature and humidity, and uses rainwater treatment and reuse and trample power generation module in the activity site to consumption of foreign energy. the location classroom is based on the consideration of light and wind. In addition, the form of Tulou not only provides children with rich communication relat comfortable growth environment, but also strengthens children of traditional culture.
Location analysis
江西省
福建省
The base is located in Jianyang District, Nanping City, Fujian Province. It is located in the center of the city
Concept presentation
TILE
COURTYARD
PATIO
REDWOOD
BAMBOO
EARTH WALL
Extraction of environmental elements
Activity platform
Use patio space
line of sight
Organization space
Spatial characteristics
Patio as conversion center
Spatial hierarchy
Two sets of AC system
Space Center
Environmental climate
Nanping is located in the East Monsoon Area to the south of Qinling Mountains and Huaihe River in China. It is a typical subtropical monsoon climate. Its most prominent features are that the average temperature of the coldest month in winter is above 0°C, the summer is hot and rainy (often said that the rain and heat are synchronous), the precipitation is concentrated and the four seasons are distinct
Average humidity
Wind frequency
average temperature
Analysis Japanese orbit
Enthalpy humidity diagram
Base analysis
greenland
area bulit
Residence
Station to be built
School
Road traffic analysis
Analysis greening landscape
Analysis surrounding buildings
Personnel density analysis
Sight of people analysis

趣园 Interesting Garden

2

N
Logistics entrance
Main entrance
Economic and technical indicators
Total
area
6 8m²
24 55m²
construction
4514.32m²
Plot ratio
0.74
Building density
40.1%
Green space
rate
32.7%
space
dividual

General layout 1:200

Classroom monomer

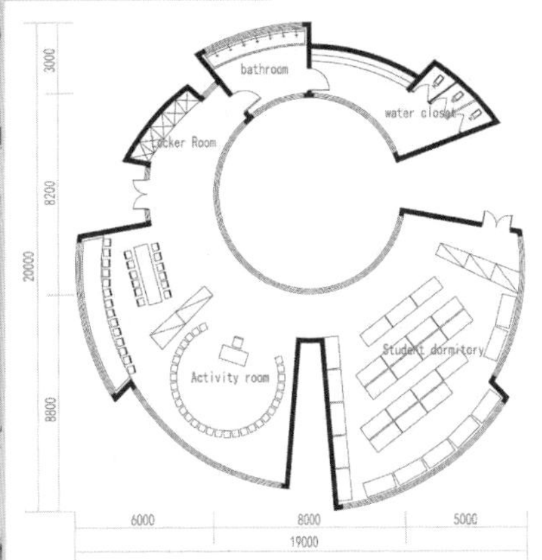
bathroom
water closet
Locker Room
Activity room
Student dormitory
3000
8200
20000
8800
6000
8000
5000
19000

Classroom monomer plane 1:100

Block generation analysis

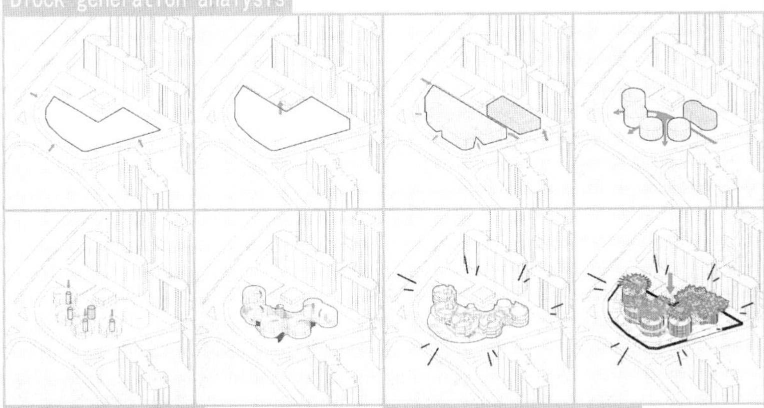

General plane analysis

Entrance square
Park parking
Park greening
Park public activities
Sand pit in the park
Park Pool
Classroom area
Administrative Area
Back up area
Public activity area
Student streamline
Teacher streamline
Streamline of office staff
Logistics personnel streamline

Sunshine analysis

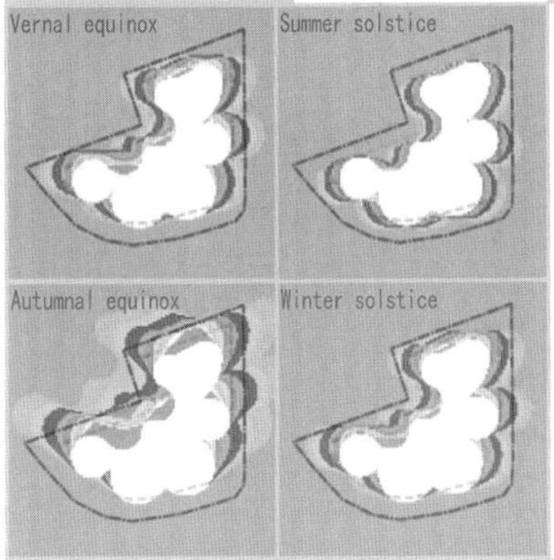
Vernal equinox
Summer solstice
Autumnal equinox
Winter solstice

Green toilet structure

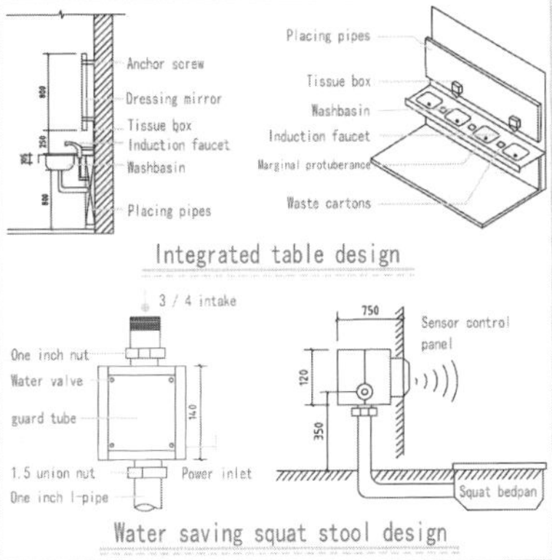
Anchor screw
Dressing mirror
Tissue box
Induction faucet
Washbasin
Placing pipes
Placing pipes
Tissue box
Washbasin
Induction faucet
Marginal protuberance
Waste cartons
Integrated table design
3 / 4 intake
One inch nut
Water valve
guard tube
1.5 union nut
One inch I-pipe
Power inlet
750
Sensor control panel
120
350
140
Squat bedpan
Water saving squat stool design

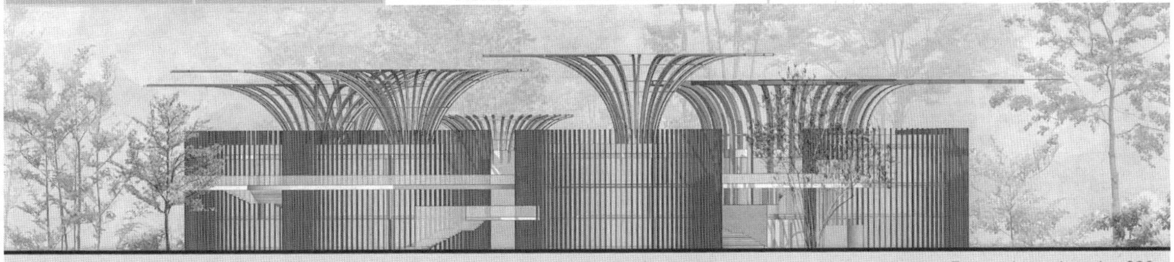

East elevation 1:200

趣园 Interesting Garden

3

1 Student dormitory
2 Activity room
3 Student toilet
4 bathroom
5 Locker Room
6 Cortile
7 Traffic core
8 Staff toilet
9 Storeroom
10 Kitchen processing room
11 Kitchen pantry
12 Kitchen washroom
13 Kitchen cutting room
14 Back bedroom changing room
15 Duty Room
16 Lobby
17 Morning examination room
Logistics entrance
Office entrance
Main entrance
N

First floor plan 1：250

Classroom profile analysis

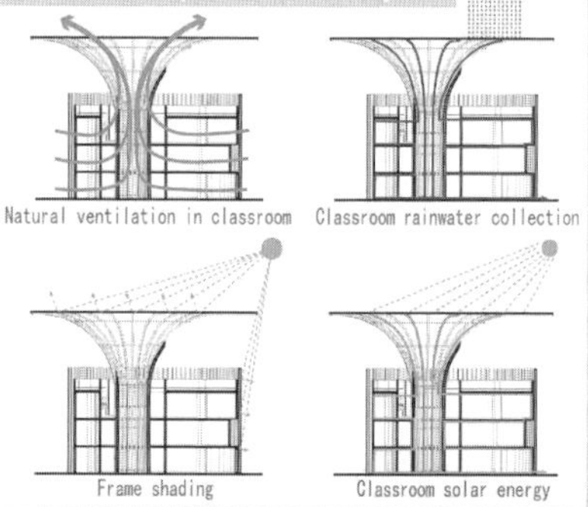
Natural ventilation in classroom
Classroom rainwater collection
Frame shading
Classroom solar energy

Solar energy utilization

The outer enclosure of the building is made of solar panels, which can not only guide the wind inside the building through the rotation angle, but also play the role of converting solar energy into electric energy when shielding the sun.

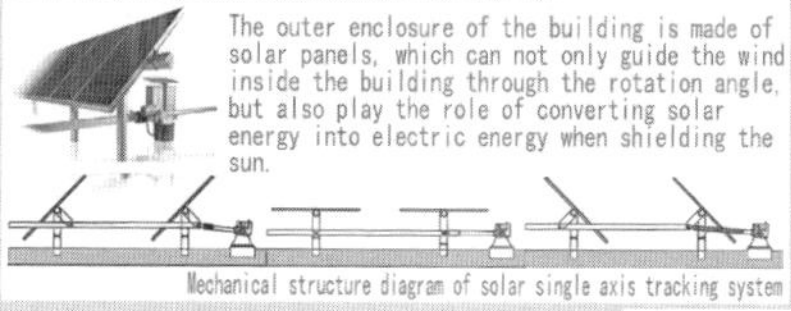

Mechanical structure diagram of solar single axis tracking system

Kinetic energy conversion system

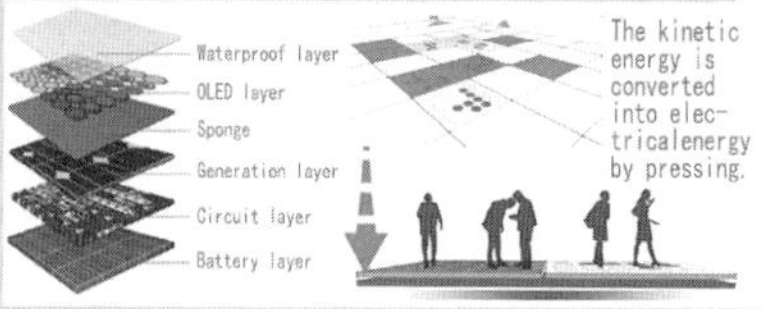
Waterproof layer
OLED layer
Sponge
Generation layer
Circuit layer
Battery layer

The kinetic energy is converted into electricalenergy by pressing.

Rainwater collection system

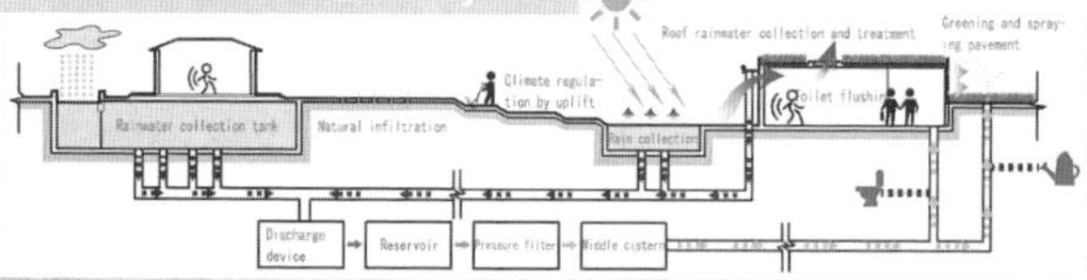
Roof rainwater collection and treatment
Greening and spraying pavement
Climate regulation by uplift
Rainwater collection tank
Natural infiltration
Rain collection
Toilet flushing
Discharge device
Reservoir
Pressure filter
Middle system

Rainwater collection is mainly divided into roof and ground collection. Rainwater is collected through diversion channels to the collection pipe network. Filters and infiltration facilities are laid on the site. Then the rainwater is collected into the reservoir for treatment and use.

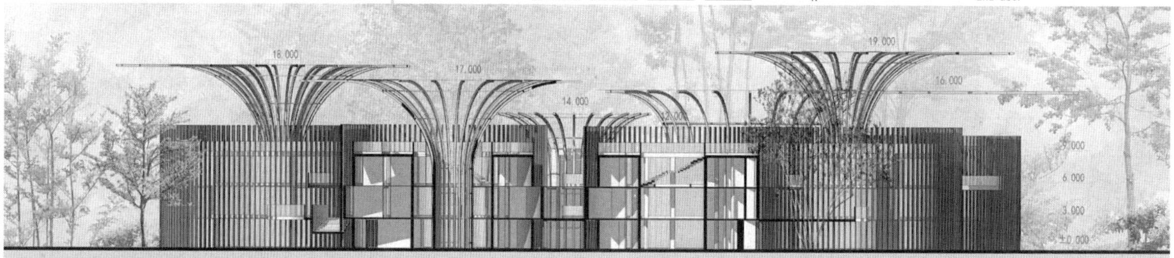
18.000
17.000
14.000
19.000
16.000
9.000
6.000
3.000
±0.000

1-1 section 1：200

趣园 Interesting Garden

4

Exploded view

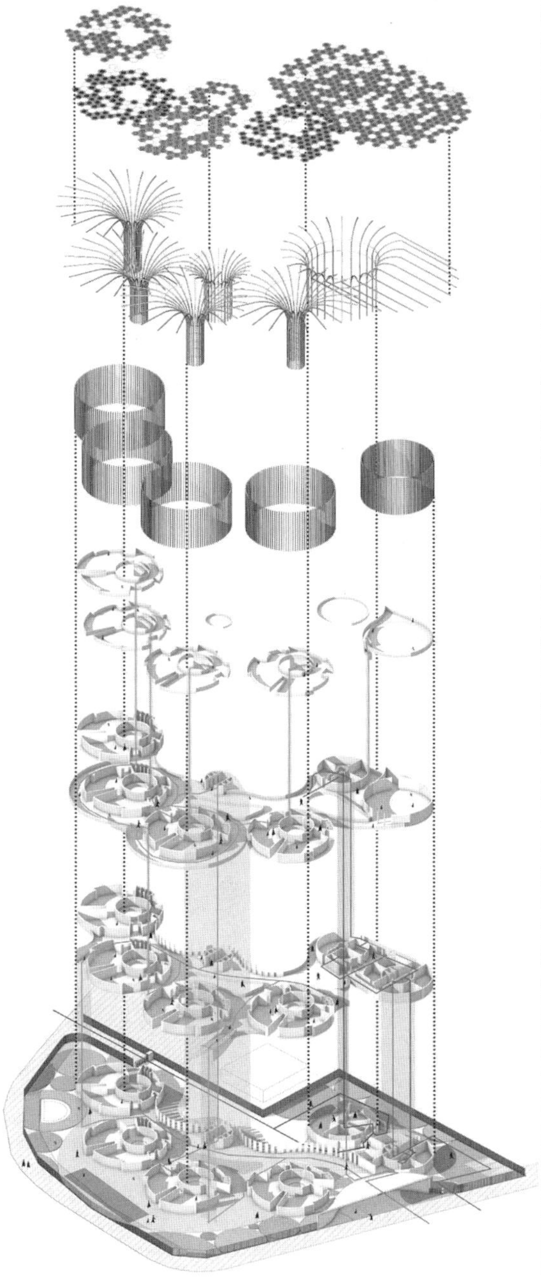

Second floor plan 1：250

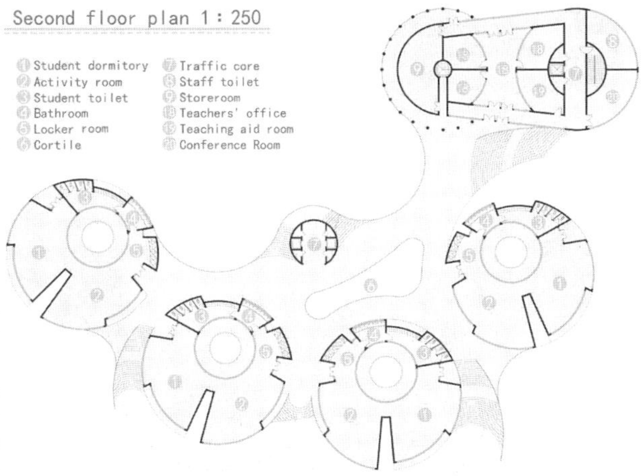
① Student dormitory
② Activity room
③ Student toilet
④ Bathroom
⑤ Locker room
⑥ Cortile
⑦ Traffic core
⑧ Staff toilet
⑨ Storeroom
⑩ Teachers' office
⑪ Teaching aid room
⑳ Conference Room

Third floor plan 1：250

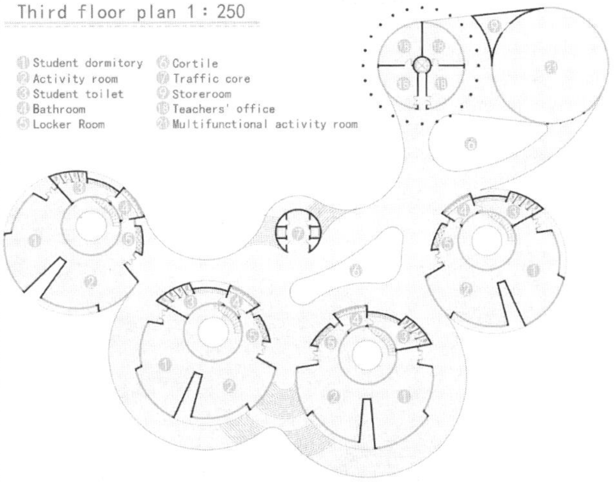
① Student dormitory
② Activity room
③ Student toilet
④ Bathroom
⑤ Locker Room
⑥ Cortile
⑦ Traffic core
⑨ Storeroom
⑩ Teachers' office
㉑ Multifunctional activity room

Features of regional rchitecyural

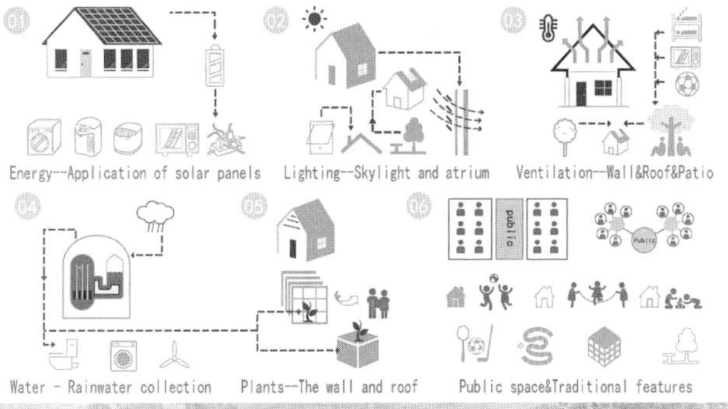
01
Energy--Application of solar panels
02
Lighting--Skylight and atrium
03
Ventilation--Wall&Roof&Patio
04
Water - Rainwater collection
05
Plants--The wall and roof
06
public
Public space&Traditional features

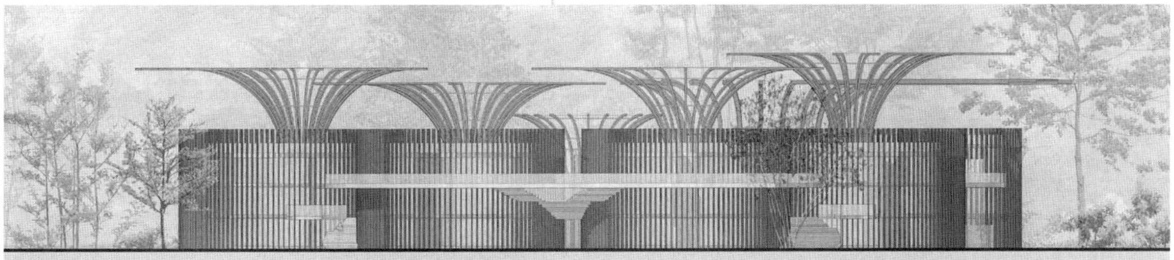

South elevation 1：200

综合奖 · 优秀奖
General Prize Awarded · Honorable Mention Prize

注 册 号：7422

项目名称：院 · 积木之家（南平）

Courtyard · House of Building Blocks（Nanping）

作　　者：梁钟琪、朱游学、秦瑞炜

参赛单位：南京工业大学

指导教师：叶起瑾、郗皎如

院 · 积木之家

FUJIAN NANPING KINDERGARTEN DESIGN　1

LOCATION ANALYSIS

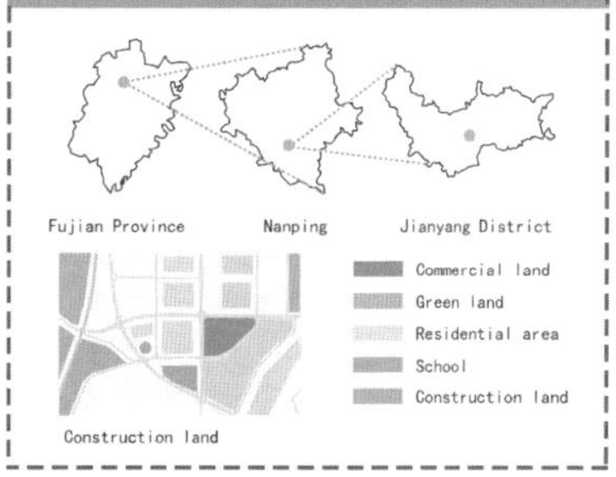
Fujian Province
Nanping
Jianyang District
Commercial land
Green land
Residential area
School
Construction land
Construction land

THCHNICS OF BUILDING DESIGN

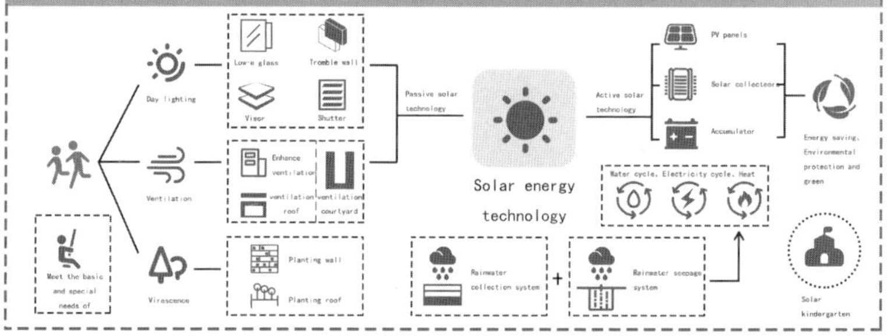
Day lighting
Ventilation
Virescence
Low-e glass
Tromble wall
Visor
Shutter
Planting wall
Planting roof
Passive solar technology
Solar energy technology
Active solar technology
PV panels
Solar collector
Accumulator
Rainwater collection system
Solar kindergarten

CLIMATE ANALYSIS

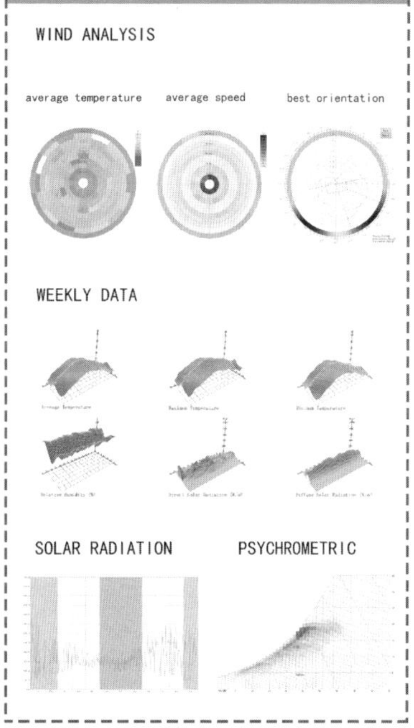
WIND ANALYSIS
average temperature
average speed
best orientation
WEEKLY DATA
SOLAR RADIATION
PSYCHROMETRIC

EXTRCTION OF TRADITIONAL ARCHITECTURAL ELEMENTS

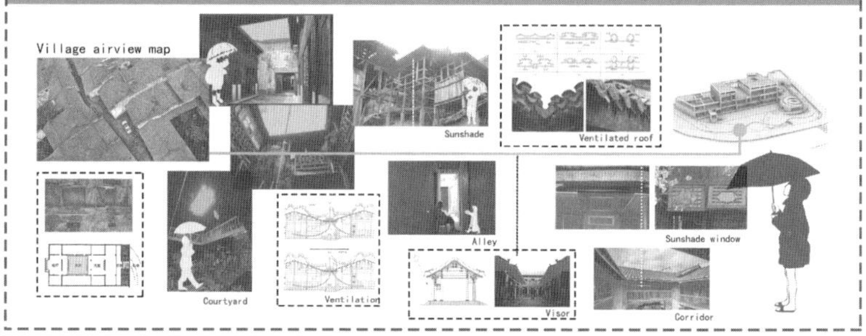
Village airview map
Sunshade
Ventilated roof
Alley
Sunshade window
Courtyard
Ventilation
Visor
Corridor

DESIGN NOTES 设计说明

天井、院落作为福建传统建筑中不可或缺的元素，在建筑的通风、采光方面起到了重要的作用。该设计提取“院落”的要素，并结合孩子们最喜欢的玩具——积木，将幼儿活动单元以积木的形式进行堆叠，同时利用一些鲜明的色块丰富建筑色彩。室外活动场地、屋顶花园、圆形综合活动室、植物廊道等等满足了儿童的成长需要，也在设计上提供了更多的可能。

技术方面，采用通风隔热屋面、植物墙、拔风庭院、low-e玻璃和遮阳板解决夏季通风隔热问题。采用雨水收集和渗漏等水循环系统、太阳能板电热循环系统实现节能，并根据环境特点改进了特朗勃墙。

As an indispensable element in traditional Fujian architecture, patios and courtyards have played an important role in the ventilation and lighting of buildings. The design extracts the elements of the "yard" and combines the children's favorite toys-building blocks, stacking the children's activity units in the form of building blocks, and enriching the architectural colors with some bright color blocks.

In terms of technology, ventilation and heat insulation roofs, plant walls, wind-pulling courtyards, low-e glass and sun visors are used to solve summer ventilation and heat insulation problems. Water circulation systems such as rainwater collection and leakage, and solar panel electrothermal circulation systems have achieved energy savings, and improved the Tromble Wall according to environmental characteristics.

2 FUJIAN NANPING KINDERGARTEN DESIGN

院·积木之家

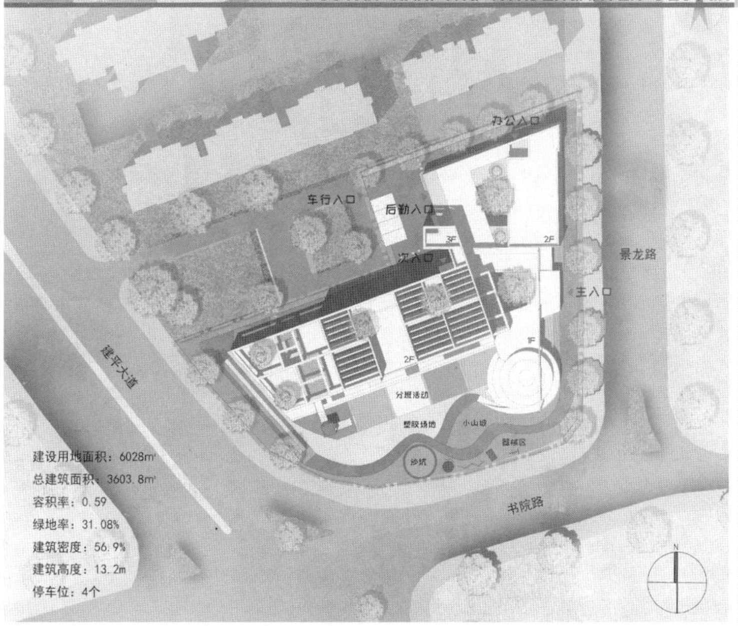
办公入口
车行入口
后勤入口
次入口
主入口
景龙路
建平大道
书院路
3F
2F
1F
分散活动
小山坡
种植区
沙坑
建设用地面积：6028m²
总建筑面积：3603.8m²
容积率：0.59
绿地率：31.08%
建筑密度：56.9%
建筑高度：13.2m
停车位：4个
N

VENTILATED & INSULATED ROOF

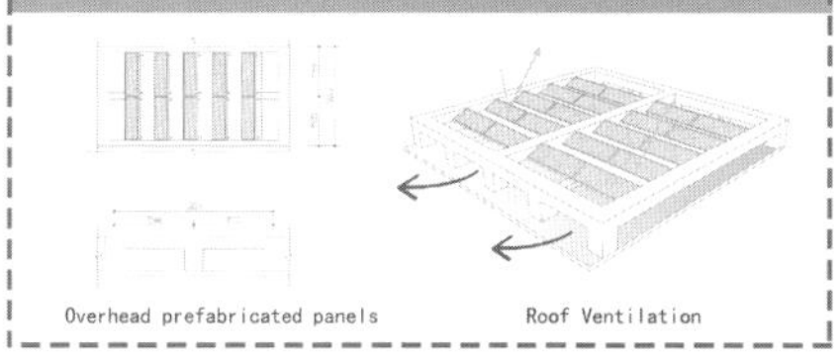
Overhead prefabricated panels
Roof Ventilation

PLANTING ROOF & WALL

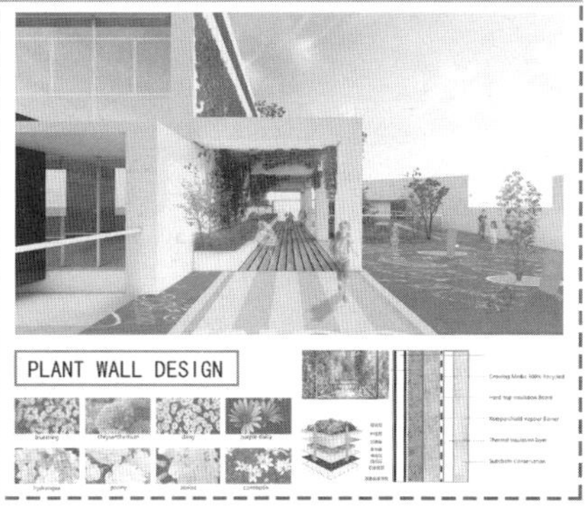
PLANT WALL DESIGN

SITE ANALYSIS

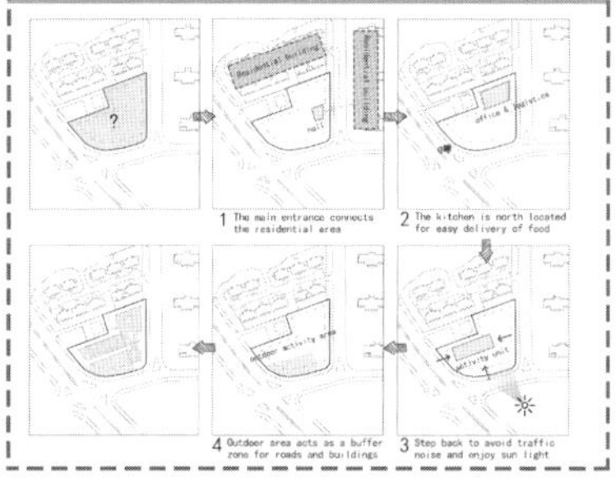
1 The main entrance connects the residential area
2 The kitchen is north located for easy delivery of food
3 Step back to avoid traffic noise and enjoy sun light
4 Outdoor area acts as a buffer zone for roads and buildings

FORM-CREATION

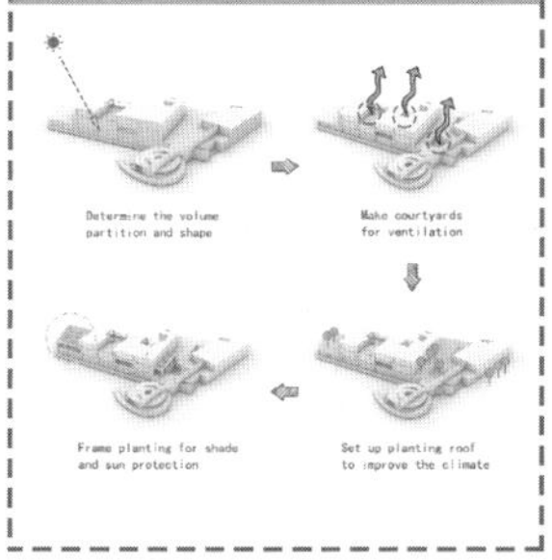
Determine the volume partition and shape
Make courtyards for ventilation
Set up planting roof to improve the climate
Frame planting for shade and sun protection

TROMBLE WALL

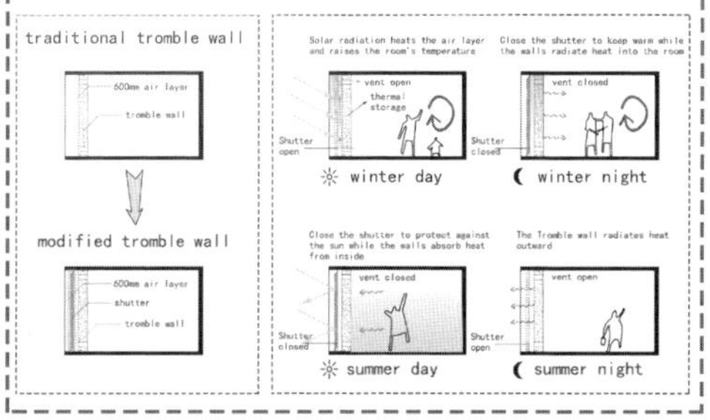
traditional tromble wall
600mm air layer
tromble wall
modified tromble wall
600mm air layer
shutter
tromble wall
vent open
vent closed
Shutter open
Shutter closed
winter day
winter night
summer day
summer night

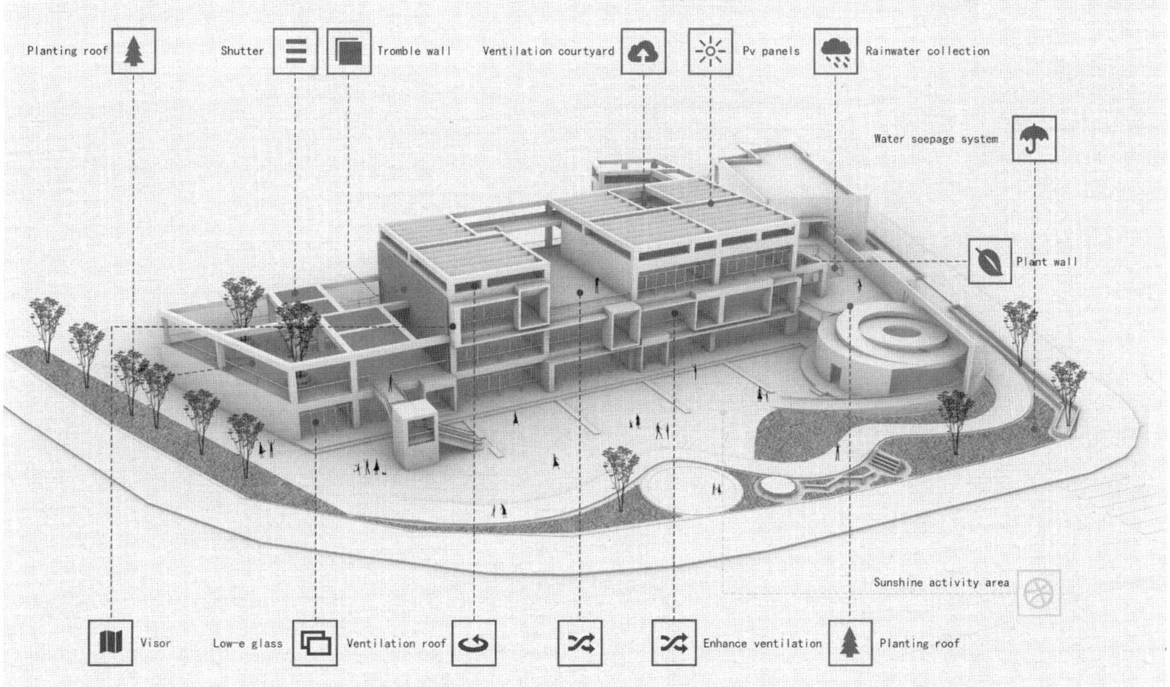
Planting roof
Shutter
Tromble wall
Ventilation courtyard
Pv panels
Rainwater collection
Water seepage system
Plant wall
Sunshine activity area
Visor
Low-e glass
Ventilation roof
Enhance ventilation
Planting roof

院·积木之家
FUJIAN NANPING KINDERGARTEN DESIGN
3

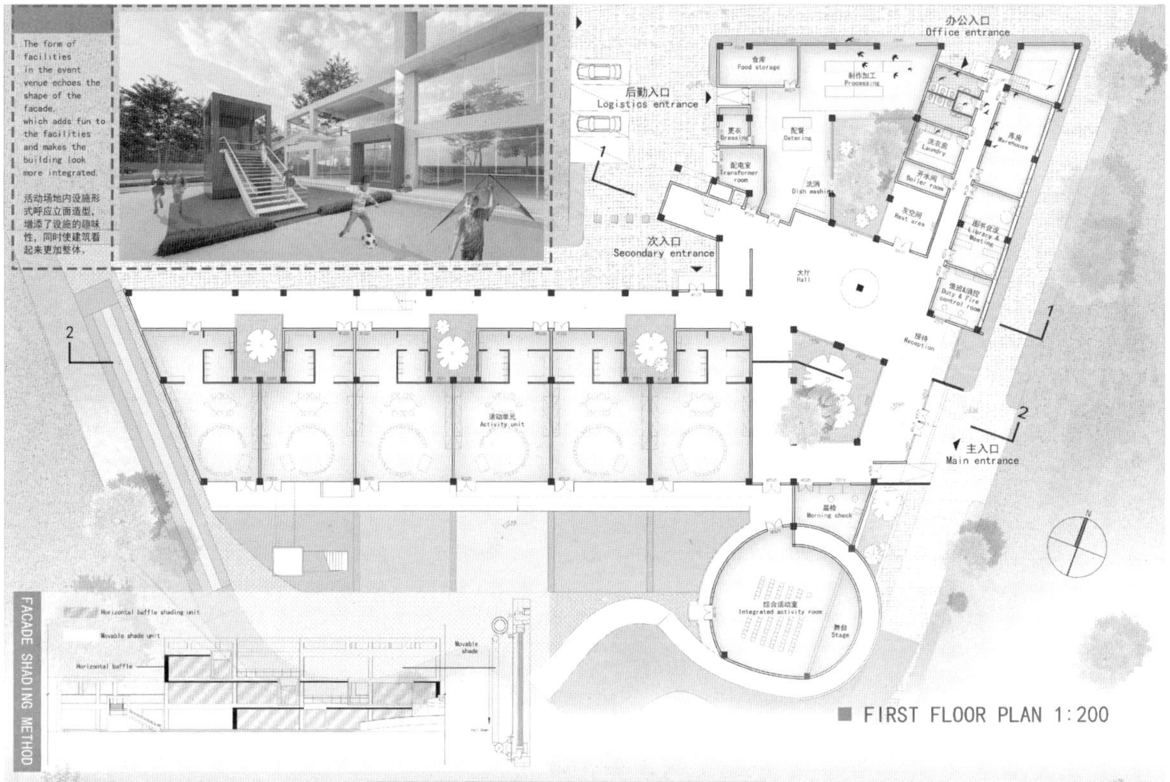
The form of facilities in the event venue echoes the shape of the facade, which adds fun to the facilities and makes the building look more integrated.
活动场地内设施形式呼应立面造型，增添了设施的趣味性，同时使建筑看起来更加整体。
办公入口
Office entrance
食库
Food storage
制作加工
Processing
后勤入口
Logistics entrance
更衣
Dressing
配餐
Catering
配电室
Transformer room
洗消
Dish washing
洗衣房
Laundry
库房
Warehouse
开水间
Boiler room
休息间
Rest area
图书会议
Library & Meeting
值班&消控
Duty & Fire control room
次入口
Secondary entrance
大厅
Hall
接待
Reception
活动单元
Activity unit
主入口
Main entrance
晨检
Morning check
综合活动室
Integrated activity room
舞台
Stage
FACADE SHADING METHOD
Horizontal baffle shading unit
Movable shade unit
Horizontal baffle
Movable shade
FIRST FLOOR PLAN 1:200

SOUTH ELEVATION 1:200

院·积木之家

FUJIAN NANPING KINDERGARTEN DESIGN

4

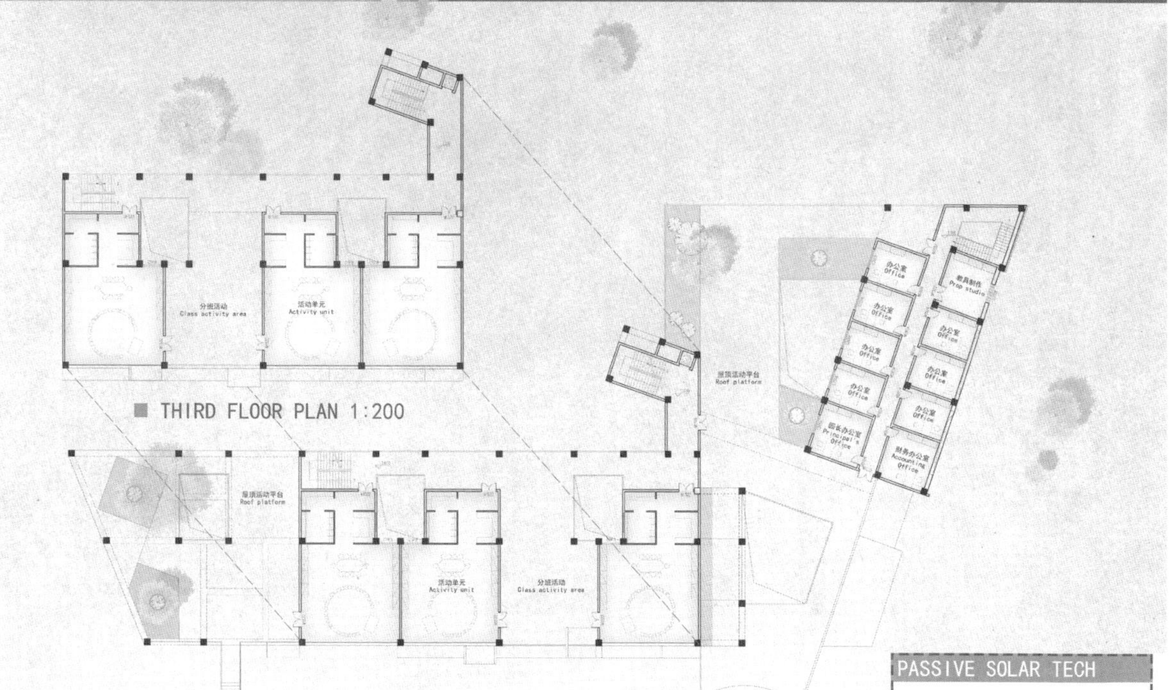
分班活动
Class activity area
活动单元
Activity unit
屋顶活动平台
Roof platform
办公室
Office
教具制作
Prop studio
园长办公室
Principal's Office
财务办公室
Accounting Office

THIRD FLOOR PLAN 1:200

SECOND FLOOR PLAN 1:200

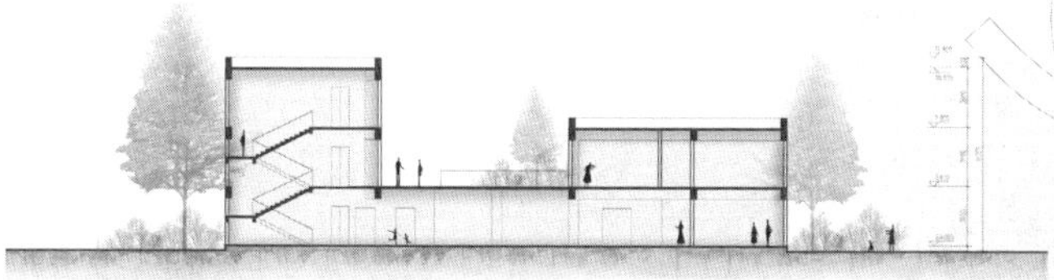

1-1 SECTION 1:200

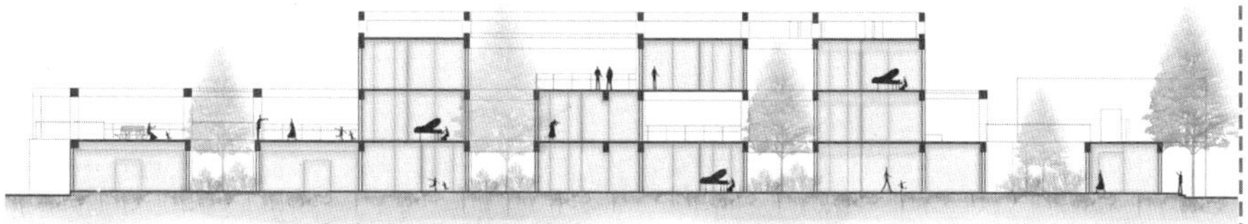

2-2 SECTION 1:200

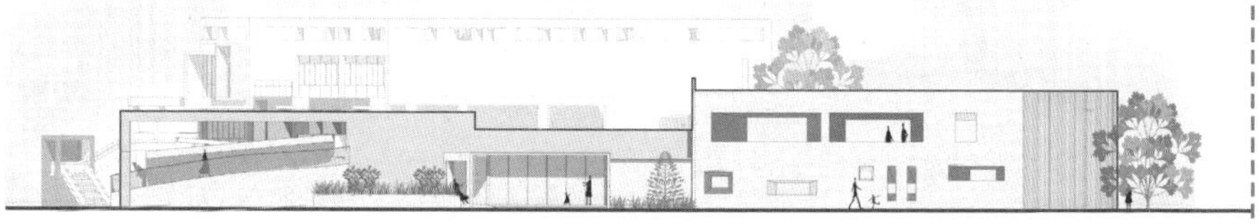

EAST ELEVATION 1:200

SUNLIGHT ANALYSIS

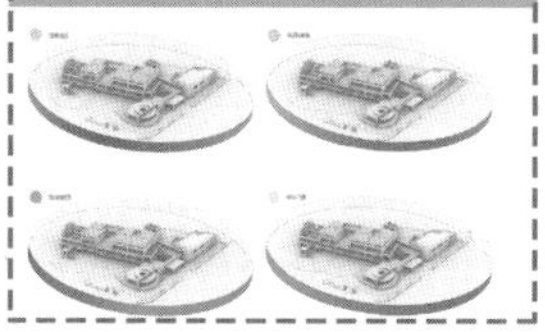

PERFORMANCE INSPECTION

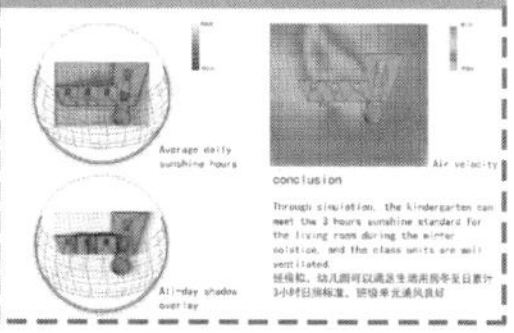
Average daily sunshine hours
All-day shadow overlay
Air velocity

conclusion

Through simulation, the kindergarten can meet the 3 hours sunshine standard for the living room during the winter solstice, and the class units are well ventilated.

PASSIVE SOLAR TECH

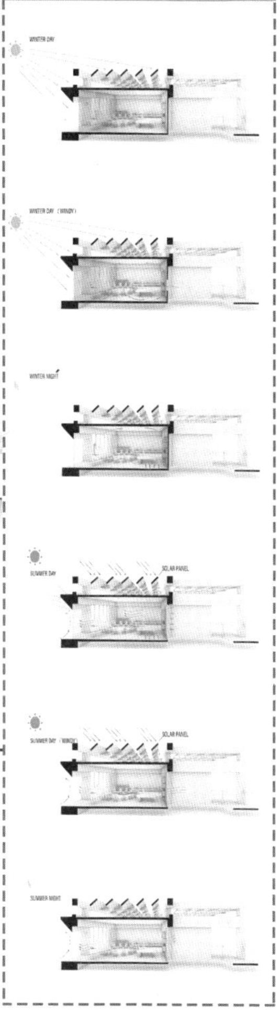

5 FUJIAN NANPING KINDERGARTEN DESIGN 院 · 积木之家

GREEN TECHNICS OF ACTIVITY UNIT

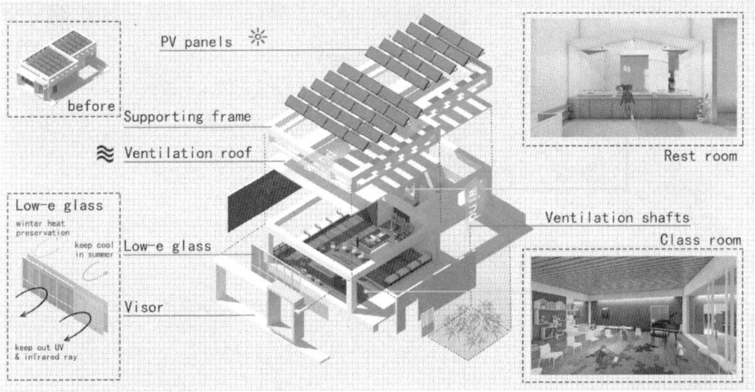

before
PV panels
Supporting frame
Ventilation roof
Low-e glass
winter heat preservation
keep cool in summer
keep out UV & infrared ray
Low-e glass
Visor
Ventilation shafts
Rest room
Class room

PLANE ARRANGEMENT

UNIT PLAN 1:200

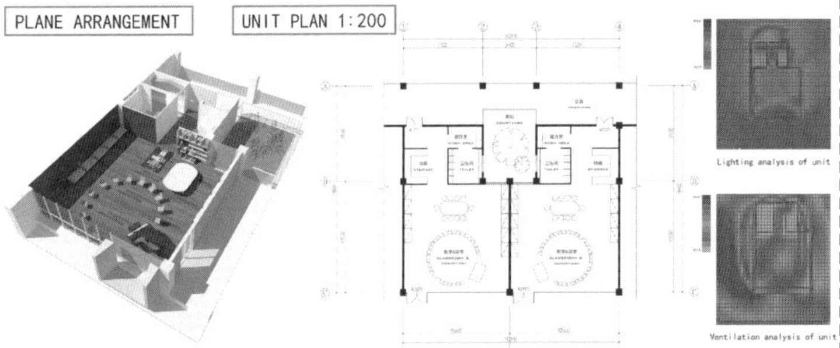

Lighting analysis of unit
Ventilation analysis of unit

NATURAL WATER SEEPAGE SYSTEM

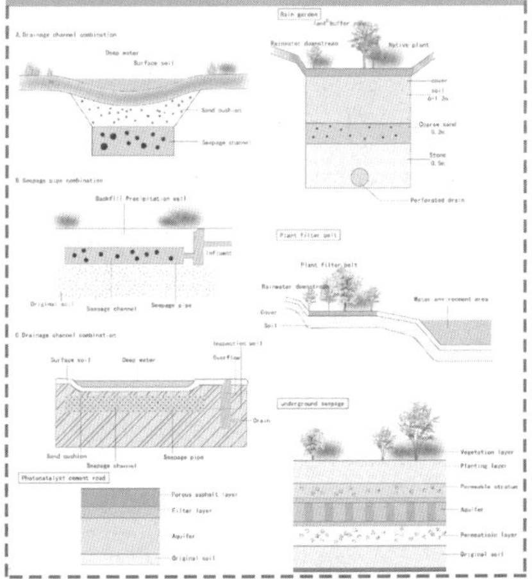

Rain garden

OVERVIEW PLAN OF GREEN TECHNICS

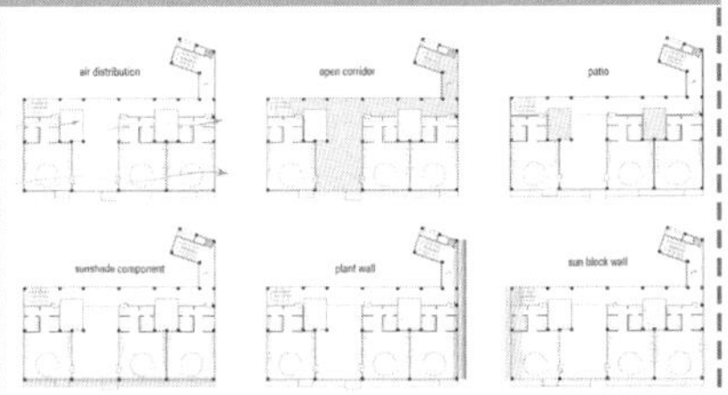

air distribution
open corridor
patio
sunshade component
plant wall
sun block wall

SOLAR PANEL TECHNICS

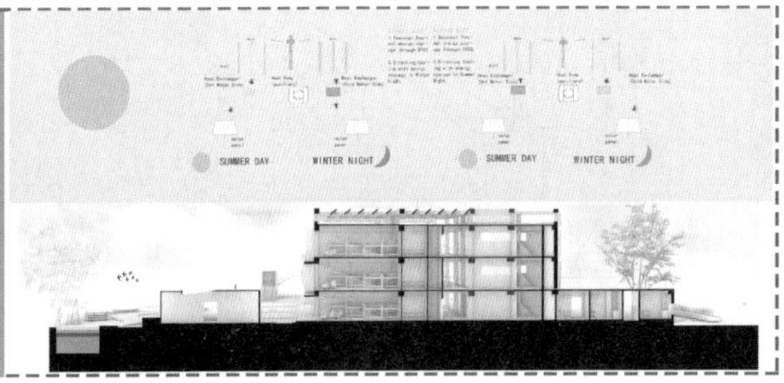

SUMMER DAY
WINTER NIGHT
SUMMER DAY
WINTER NIGHT

LIGHTING & RAIN COLLECTION SYSTEM

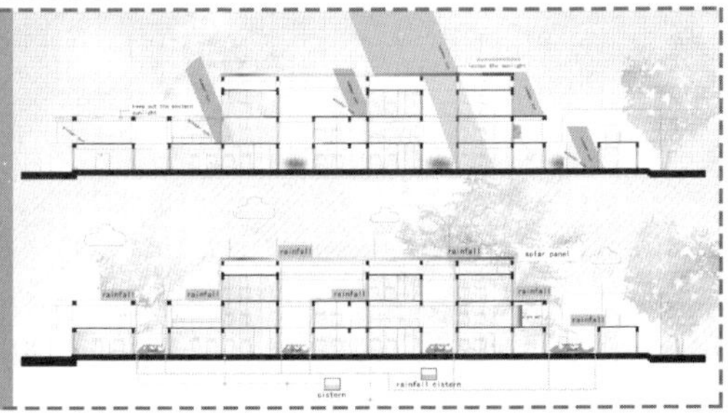

rainfall

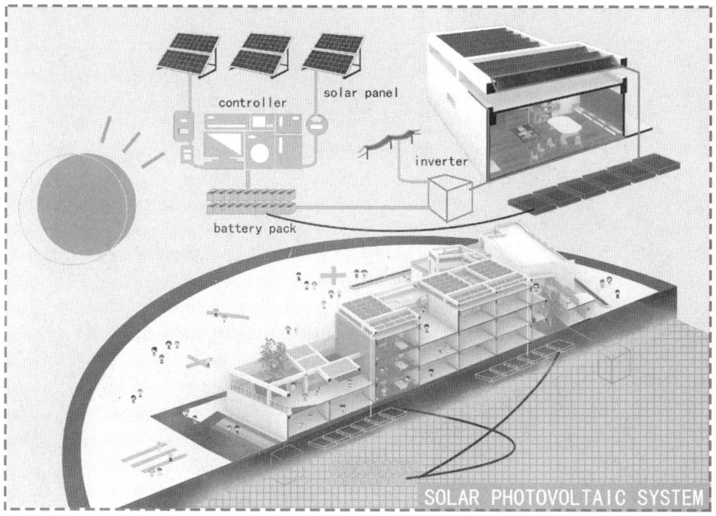
controller
solar panel
inverter
battery pack
SOLAR PHOTOVOLTAIC SYSTEM

PLANTING ANALYSIS

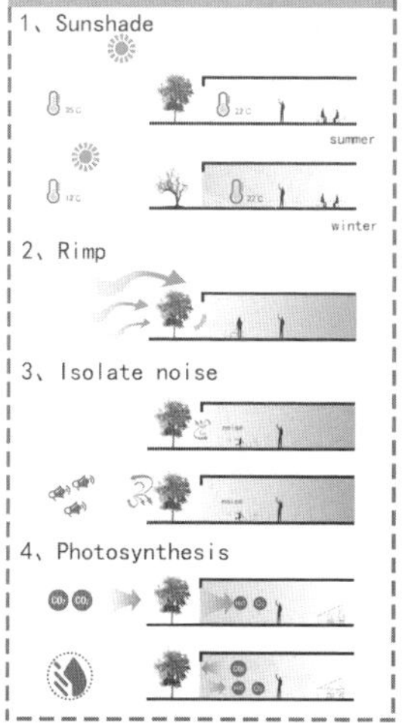
1、Sunshade
summer
winter
2、Rimp
3、Isolate noise
4、Photosynthesis

ENERGY SAVING CALCULATION

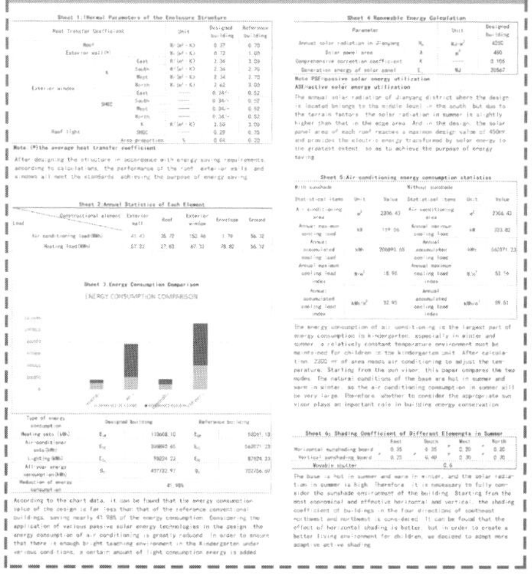
Sheet 1.Thermal Parameters of the Enclosure Structure
Sheet 2.Annual Statistics of Each Element
Sheet 3.Energy Consumption Comparison
Sheet 4.Renewable Energy Calculation
Sheet 5.Air conditioning energy consumption statistics
Sheet 6.Shading Coefficient of Different Elements in Summer

STRUCTURE

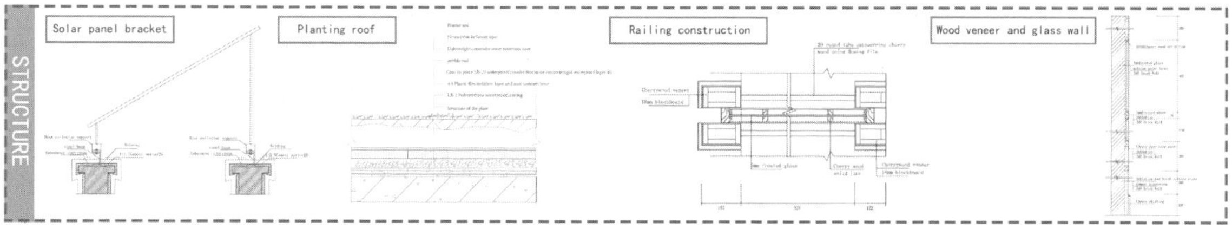
Solar panel bracket
Planting roof
Railing construction
Wood veneer and glass wall

SCHEMATIC DIAGRAM OF PATIO VENTILATION

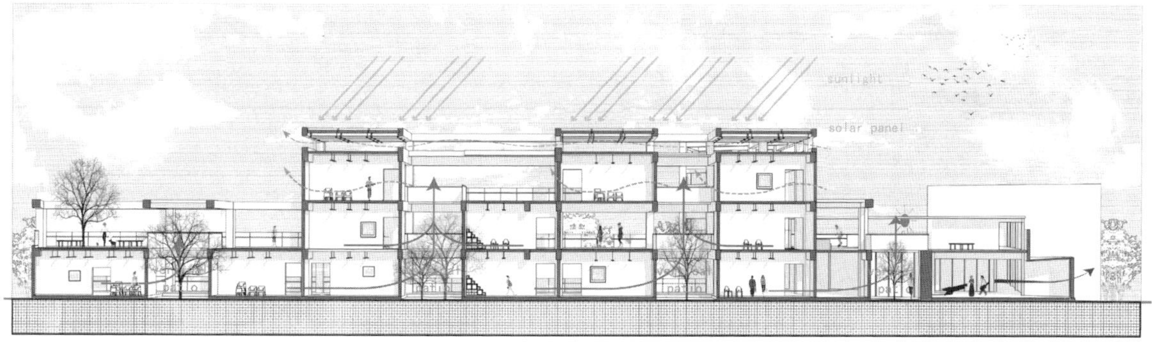
sunlight
solar panel

综合奖·优秀奖
General Prize Awarded · Honorable Mention Prize

注 册 号：7489

项目名称：巷院·阡陌（南平）
Lane · Compound（Nanping）

作　　者：王笑涵、沈一飞、段昭丞、
陈　晴

参赛单位：重庆大学

指导教师：张海滨、周铁军

巷院·阡陌 Lane · Compound 1

Site Location

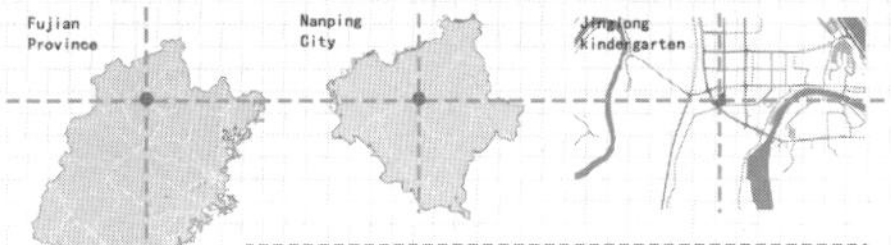
Fujian
Province
Nanping
City
Kindergarten

Nanping City has a typical subtropical humid monsoon climate.
The annual average temperature is 18-20 ℃.
and the monthly average temperature is 28-29 ℃.
the coldest month is 9-10 ℃.
Rainfall 1650 mm. frost free period 250-300 days.

Field Background

/ Traditional Residence / Lounge Bridge / Folk Culture / Detail Decoration

Climate Analysis

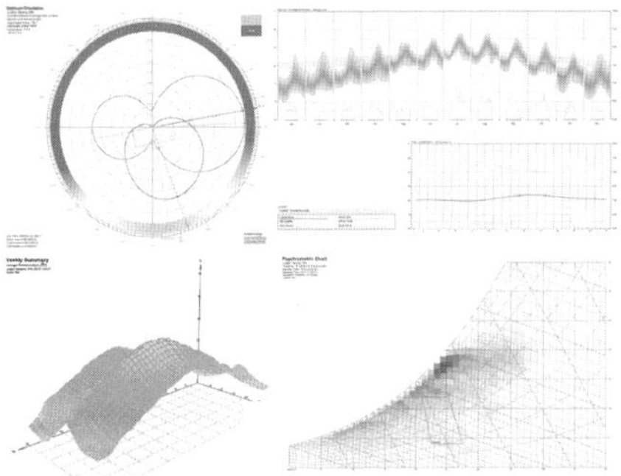

Nanping belongs to the subtropical monsoon humid climate, and the local mountainous area is the subtropical mountain climate, forming a landform dominated by hills and mountains. Belongs to the subtropical monsoon humid climate, the local mountainous area is the subtropical mountain climate, known as "Fujian granary", "Southern forest", "China's bamboo township". Known as "Fujian granary", "Southern Forest sea", "China's bamboo township"

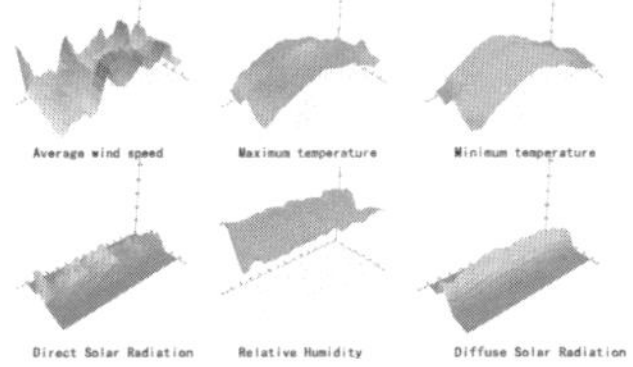
Average wind speed
Maximum temperature
Minimum temperature
Direct Solar Radiation
Relative Humidity
Diffuse Solar Radiation

设计说明

设计从被动节能技术出发，结合场地特征、当地文化风俗及气候特点，突出冷巷和院落的当地建筑特色，以院为点，连巷成线，阡陌交通，院落与冷巷以地方性建筑语汇从孩提时代培养儿童的乡土感情。

在院落布局中，小院落形成局部拔风井，冷巷与幼儿单元相连发挥热压通风效果，冷巷既在功能上满足了交通需求，又充分利用风形成原理，安排相对低温空间，制造微风循环，以达到降温目的，从源头节约能源。

Design instruction

Starting from the passive energy-saving technology, combining with the site characteristics, local cultural customs and climate characteristics, the design highlights the local architectural features of cold alleys and courtyards. With the courtyard as the point, the alleys form a line, and the courtyard and cold lane use local architectural vocabulary to cultivate children's local feelings from childhood.

In the courtyard layout, the small courtyard forms a local ventilation shaft, and the cold lane is connected with the children's unit to play the effect of hot pressure ventilation. The cold Lane not only meets the traffic demand in function, but also makes full use of the wind formation principle to arrange relatively low temperature space and create breeze circulation, so as to achieve the cooling purpose and save energy from the source.

巷院 · 阡陌 Lane · Compound 2

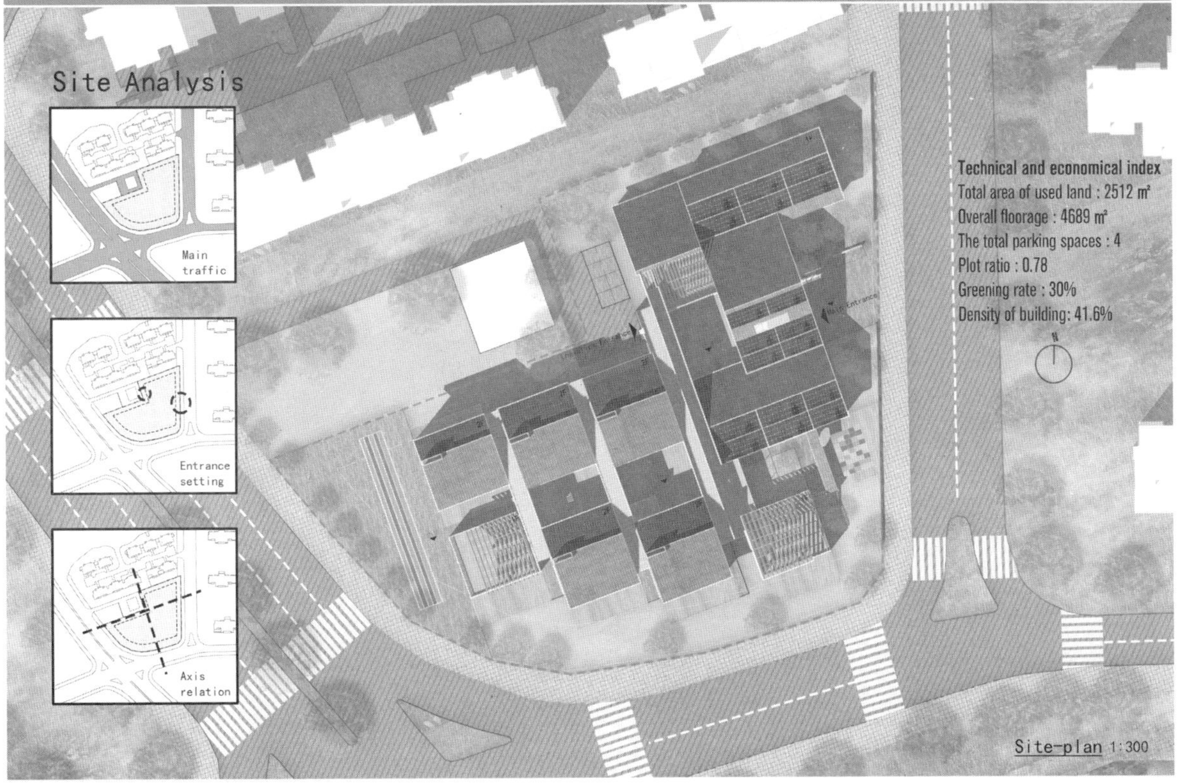
Site Analysis
Main traffic
Entrance setting
Axis relation
Technical and economical index
Total area of used land : 2512 m²
Overall floorage : 4689 m²
The total parking spaces : 4
Plot ratio : 0.78
Greening rate : 30%
Density of building: 41.6%
Site-plan 1:300

Function Partition

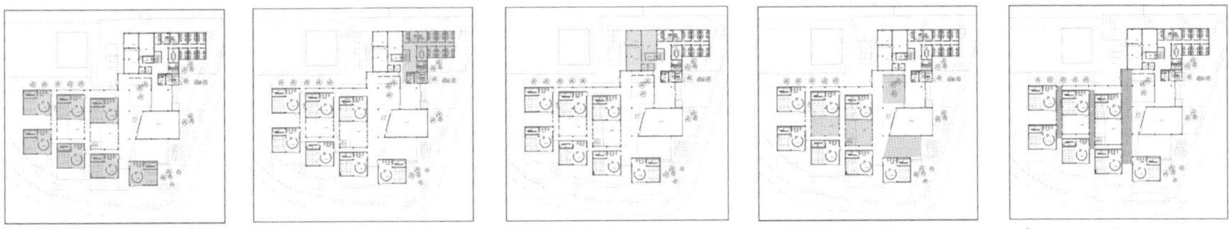

/ Class unit　/ office zone　/ Back up area　/ courtyard　/ tunnel space

Morphogenesis

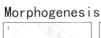

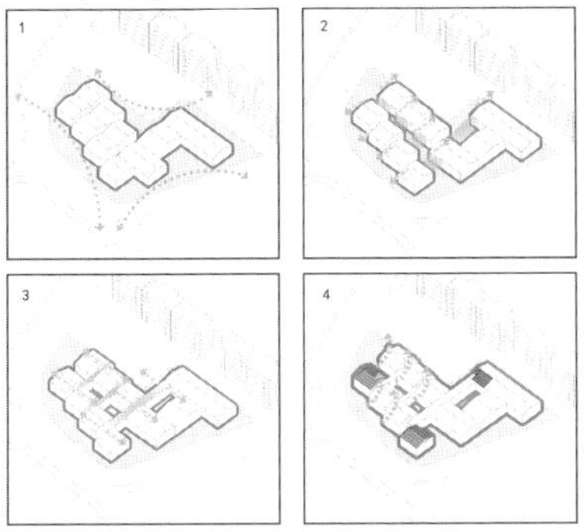

1. Step back from the site relationship.
2. The monomer gives way to the yard.
3. Block combination generates cold lanes to adapt to local climate.
4. Monomer deepens the influence of technological core on the morphological relationship.

Axonometric Decomposition

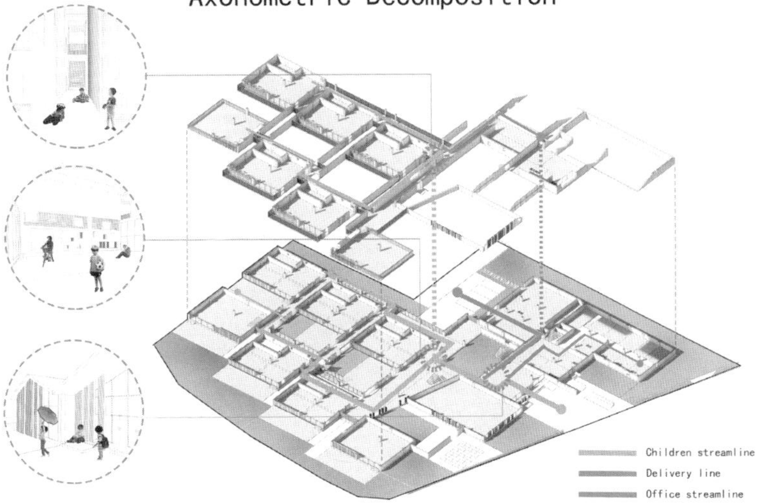
Children streamline
Delivery line
Office streamline

巷院 · 阡陌 Lane • Compound 3

Food Warehouse
kitchen operator
prepare room
process room
locker room
wash room
principal office
financial office
office
conference room
duty room
Logistics Entrance
courtyard
lobby
Main Entrance
Lorem Ipsum
Music room
Cloakroom
Classroom
courtyard
N
6100
6100
2800
6100
6100
2800
6100
6100
5000
17000
5200
4000
4000
4000
81400

First Floor Plan 1:200

The Shadow Range

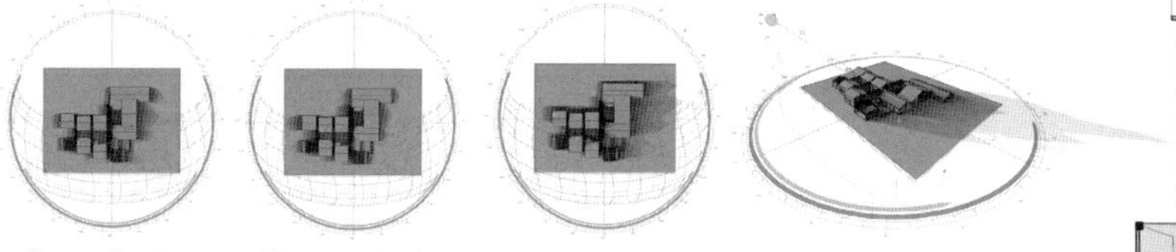

The vernal equinox　The summer slostice　The autumnal equinox　The winter slostice

The Air-flow Analysis

5.6m/s
The South-east wind

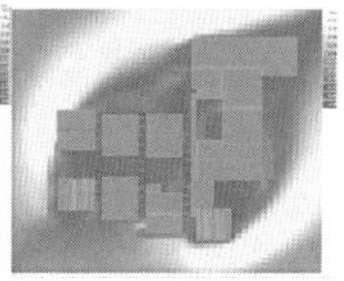

2.8m/s
The North-east wind

7.8m/s
The South-west wind

7.8m/s
The South-west wind

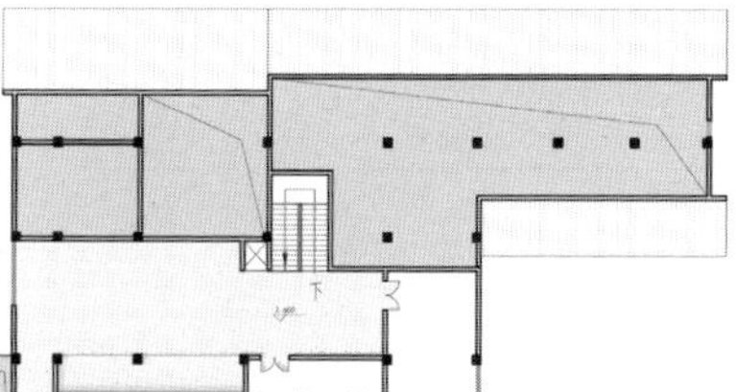
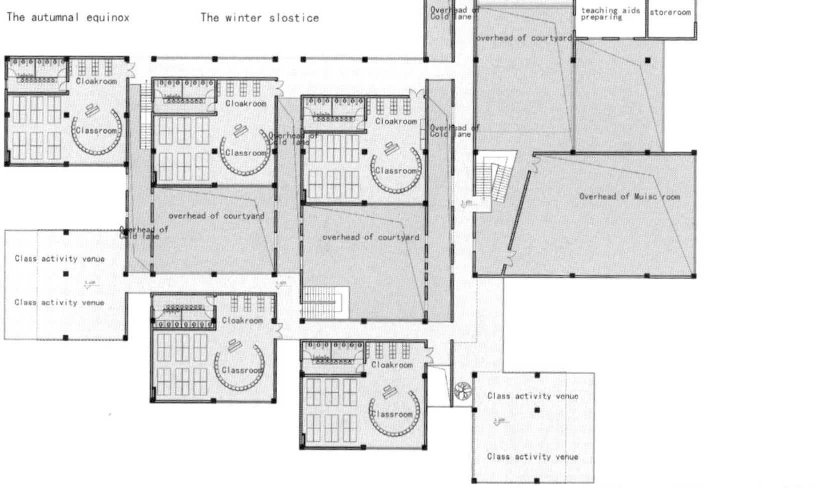
teaching aids preparing
storeroom
Overhead of Cold lane
overhead of courtyard
Cloakroom
Classroom
Overhead of Muisc room
Class activity venue

Second Floor Plan 1:200

巷院 · 阡陌 Lane · Compound 4

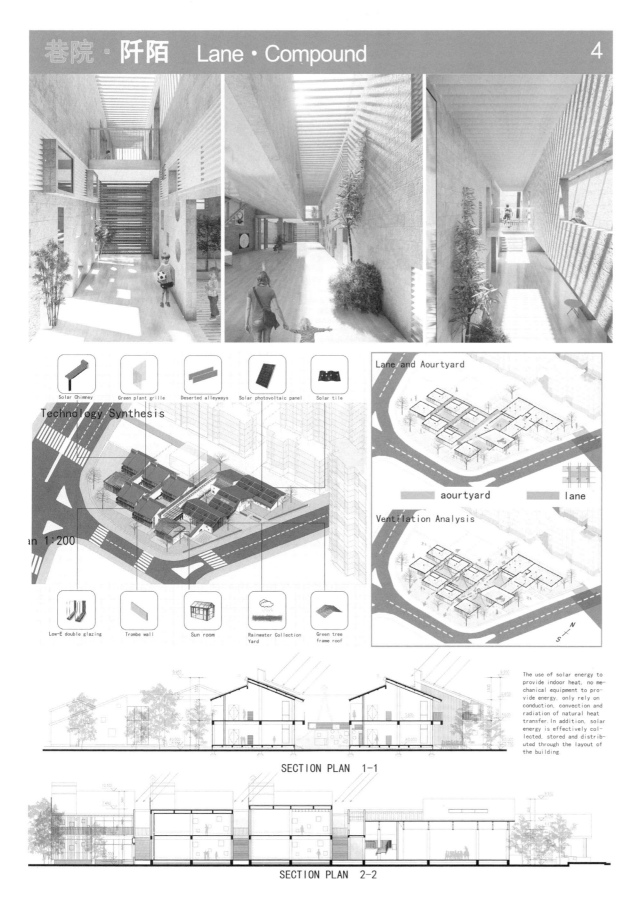

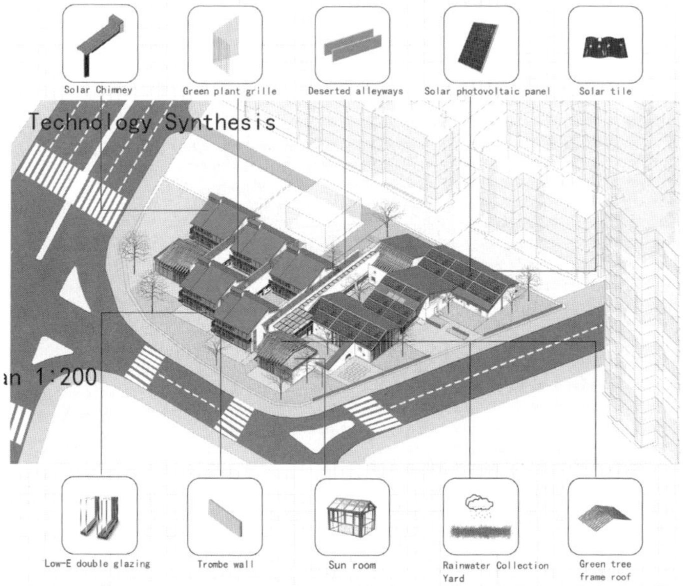
Solar Chimney
Green plant grille
Deserted alleyways
Solar photovoltaic panel
Solar tile
Technology Synthesis
an 1:200
Low-E double glazing
Trombe wall
Sun room
Rainwater Collection Yard
Green tree frame roof

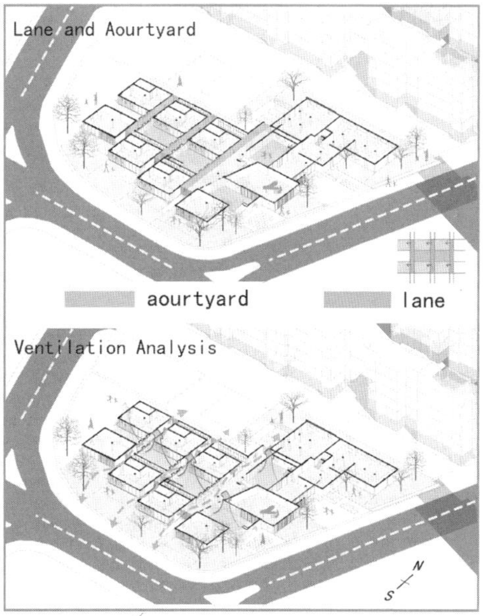
Lane and Aourtyard
aourtyard
lane
Ventilation Analysis
N
S

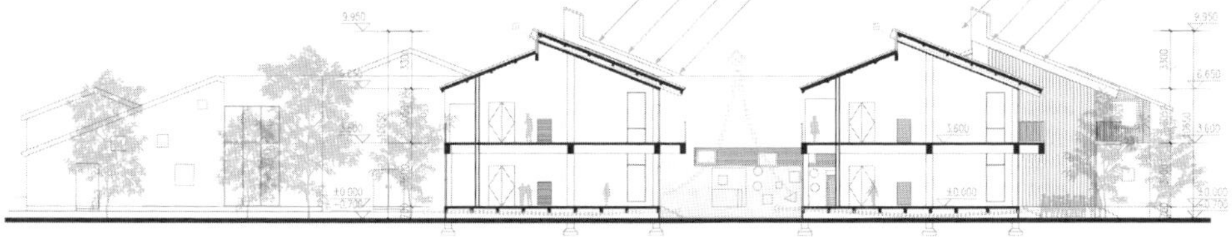

The use of solar energy to provide indoor heat, no mechanical equipment to provide energy, only rely on conduction, convection and radiation of natural heat transfer. In addition, solar energy is effectively collected, stored and distributed through the layout of the building.

SECTION PLAN 1-1

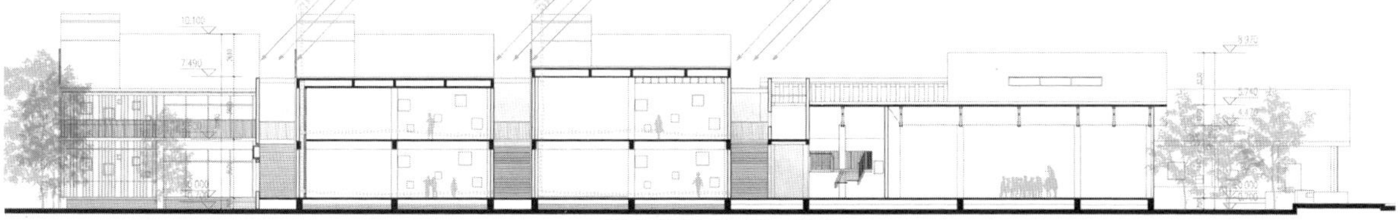

SECTION PLAN 2-2

巷院·阡陌 Lane · Compound 5

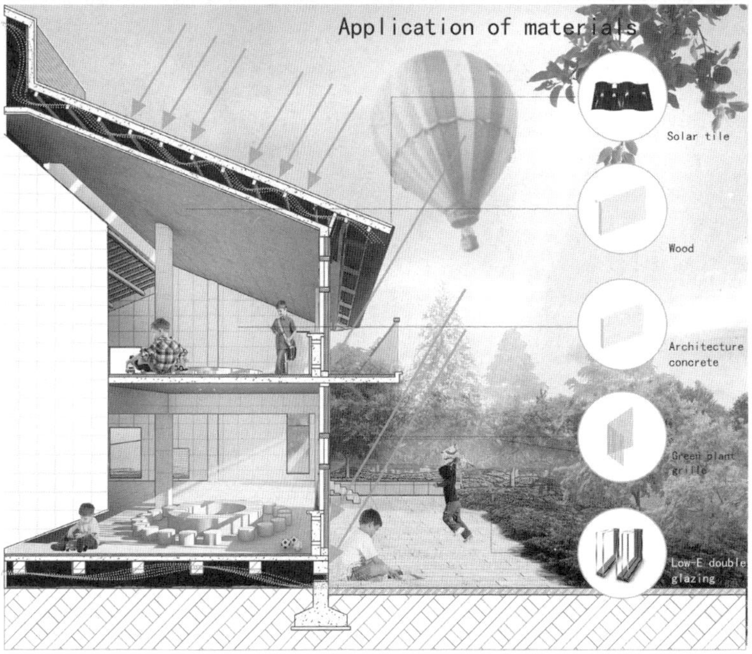

Application of materials
Solar tile
Wood
Architecture concrete
Green plant grille
Low-E double glazing

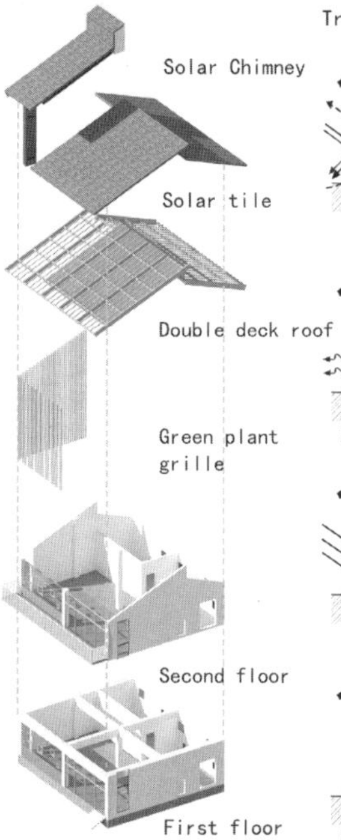

Solar Chimney
Solar tile
Double deck roof
Green plant grille
Second floor
First floor

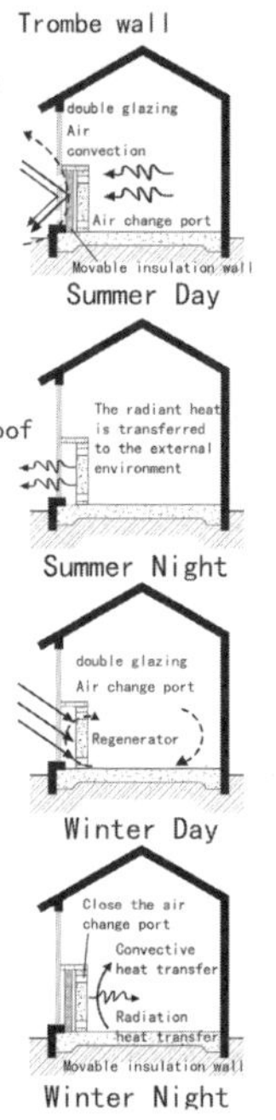

Trombe wall
double glazing
Air convection
Air change port
Movable insulation wall
Summer Day
The radiant heat is transferred to the external environment
Summer Night
double glazing
Air change port
Regenerator
Winter Day
Close the air change port
Convective heat transfer
Radiation heat transfer
Movable insulation wall
Winter Night

Technical Details Analysis (single model)

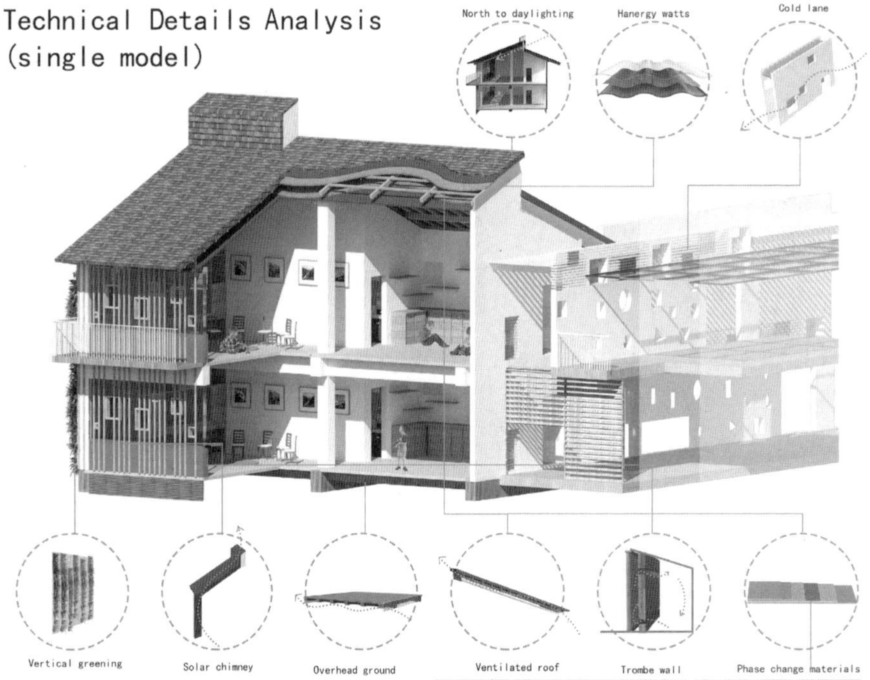

North to daylighting
Hanergy watts
Cold lane
Vertical greening
Solar chimney
Overhead ground
Ventilated roof
Trombe wall
Phase change materials

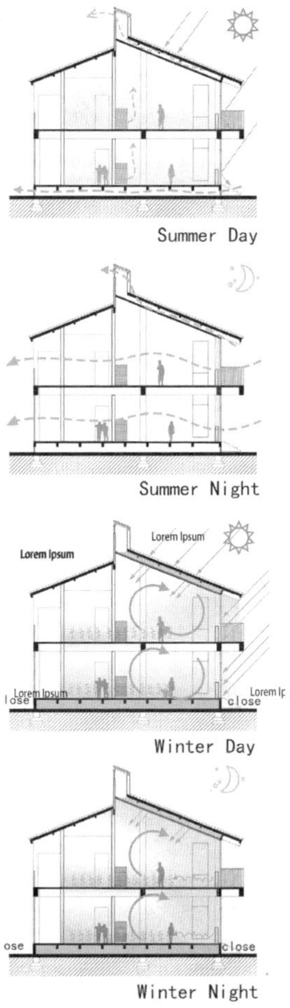

Summer Day
Summer Night
Lorem Ipsum
Winter Day
close
Winter Night
close

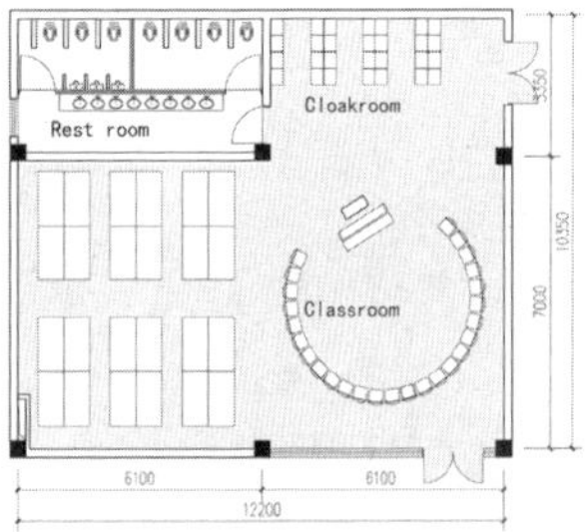

Rest room
Cloakroom
Classroom
6100
6100
12200

巷院 · 阡陌 Lane • Compound 6

Solar Radiation analysis

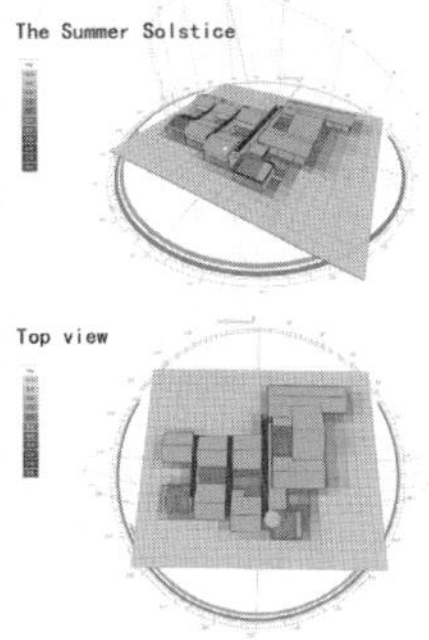
The Summer Solstice
Top view

Indoor Thermoanalysis

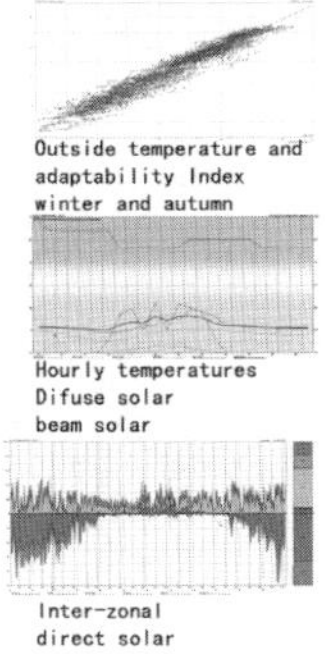
Outside temperature and adaptability Index winter and autumn
Hourly temperatures
Difuse solar
beam solar
Inter-zonal direct solar

Type of energy comsumption	Designed building		Reference building	
Heating sets(kwh)	E	55304.07	E	26530.20
Air-conditioner sets(Kwh)	E	7766.45	E	211003.50
Lighting (Kwh)	E	55864.39	E	54245.61
All-year energy consumption(kWh)	B	188835.92	B	291788.68
Redution of energy consumption	34.21%			

Power generation of Unit's solar panels

	Generated energy [kWh]	Number [Piece]	Area [㎡]	Average output [kWh/piece]	Average output [kWh/㎡]	Average efficiency
Year	32328.72	108	99.6	299.34	324.58	16.41%
Summer(7、8)	6696.6	108	99.6	62	67.23	19.61%
Winter (11、12、1)	4714.2	108	99.6	43.65	47.30	13.20%

Renewable Energy Calcutation

Parameter	Unit	Designed building
Annual solar radiation in Fujian H_A	MJ/㎡	4700
Solar panel area A	㎡	800
Comprehensive correction coefficient K	—	0.105
Generation energy of solar panel E_p	MJ	11638

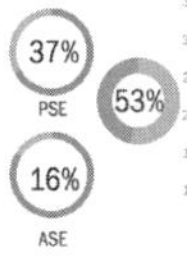
37%
PSE
53%
16%
ASE

Note: PSE=passive solar energy utilization
ASE=Active solar energy utilization

Power supply and demand relationship

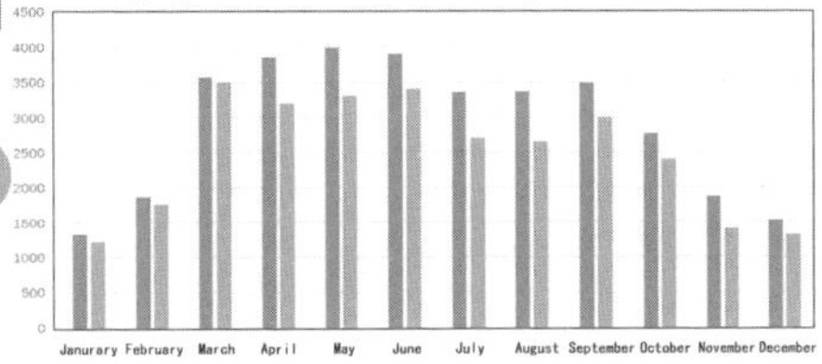
Power supply
Power demand
4500
4000
3500
3000
2500
2000
1500
1000
500
0
Janurary
February
March
April
May
June
July
August
September
October
November
December

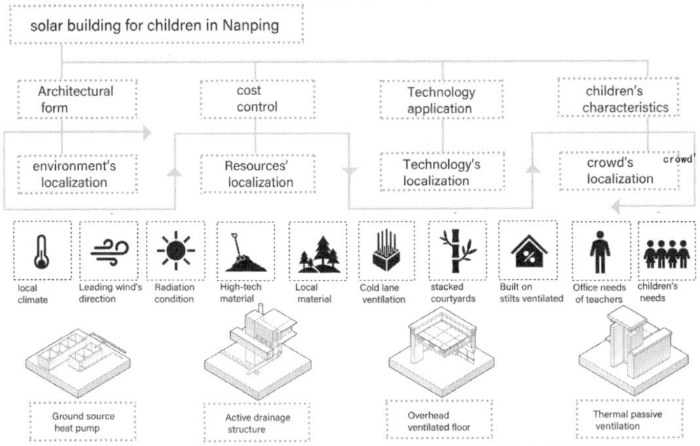
solar building for children in Nanping
Architectural form
cost control
Technology application
children's characteristics
environment's localization
Resources' localization
Technology's localization
crowd's localization
crowd's
local climate
Leading wind's direction
Radiation condition
High-tech material
Local material
Cold lane ventilation
stacked courtyards
Built on stilts ventilated
Office needs of teachers
children's needs
Ground source heat pump
Active drainage structure
Overhead ventilated floor
Thermal passive ventilation

Capillary network radiant heating system

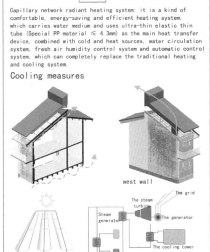

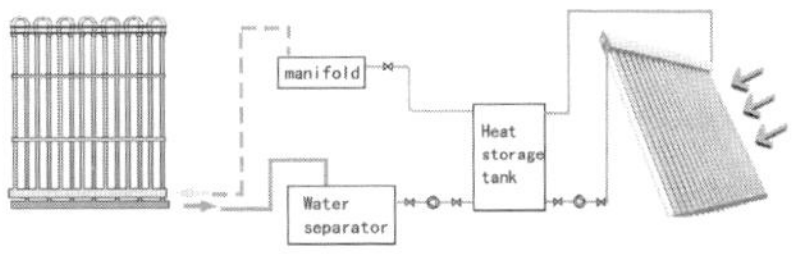
manifold
Heat storage tank
Water separator

Capillary network radiant heating system: it is a kind of comfortable, energy-saving and efficient heating system, which carries water medium and uses ultra-thin elastic thin tube (Special PP material ≤ 4.3mm) as the main heat transfer device, combined with cold and heat sources, water circulation system, fresh air humidity control system and automatic control system, which can completely replace the traditional heating and cooling system.

Cooling measures

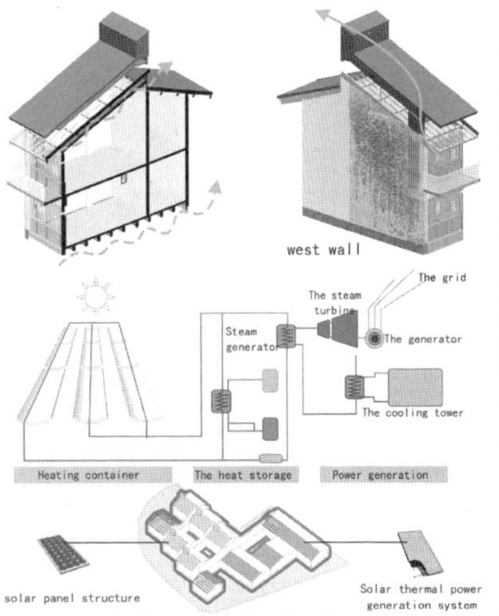
west wall
The grid
The steam turbine
Steam generator
The generator
The cooling tower
Heating container
The heat storage
Power generation
solar panel structure
Solar thermal power generation system

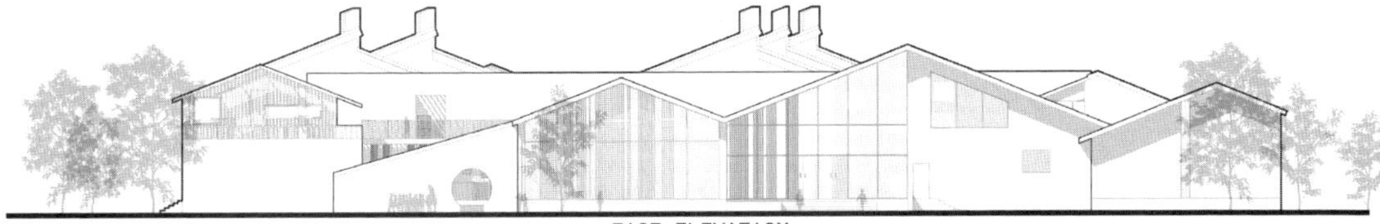

EAST ELEVATION

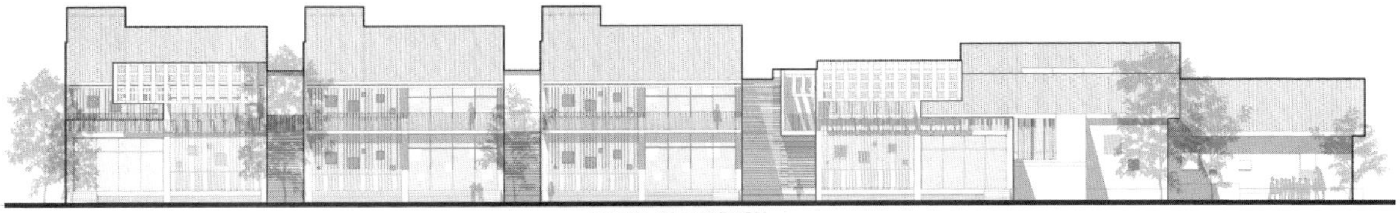

SOUTH ELEVATION

综合奖 · 优秀奖

General Prize Awarded · Honorable Mention Prize

注 册 号： 7608

项目名称： 稚巷 · 拾光（南平）

Children Alley & Chasing Childlight（Nanping）

作　　者： 牛静仁、刘宸溪、张清亮、高　力、付浩然、谭若晨

参赛单位： 石家庄铁道大学、中铁建安工程设计有限公司

指导教师： 樊海彬、高力强

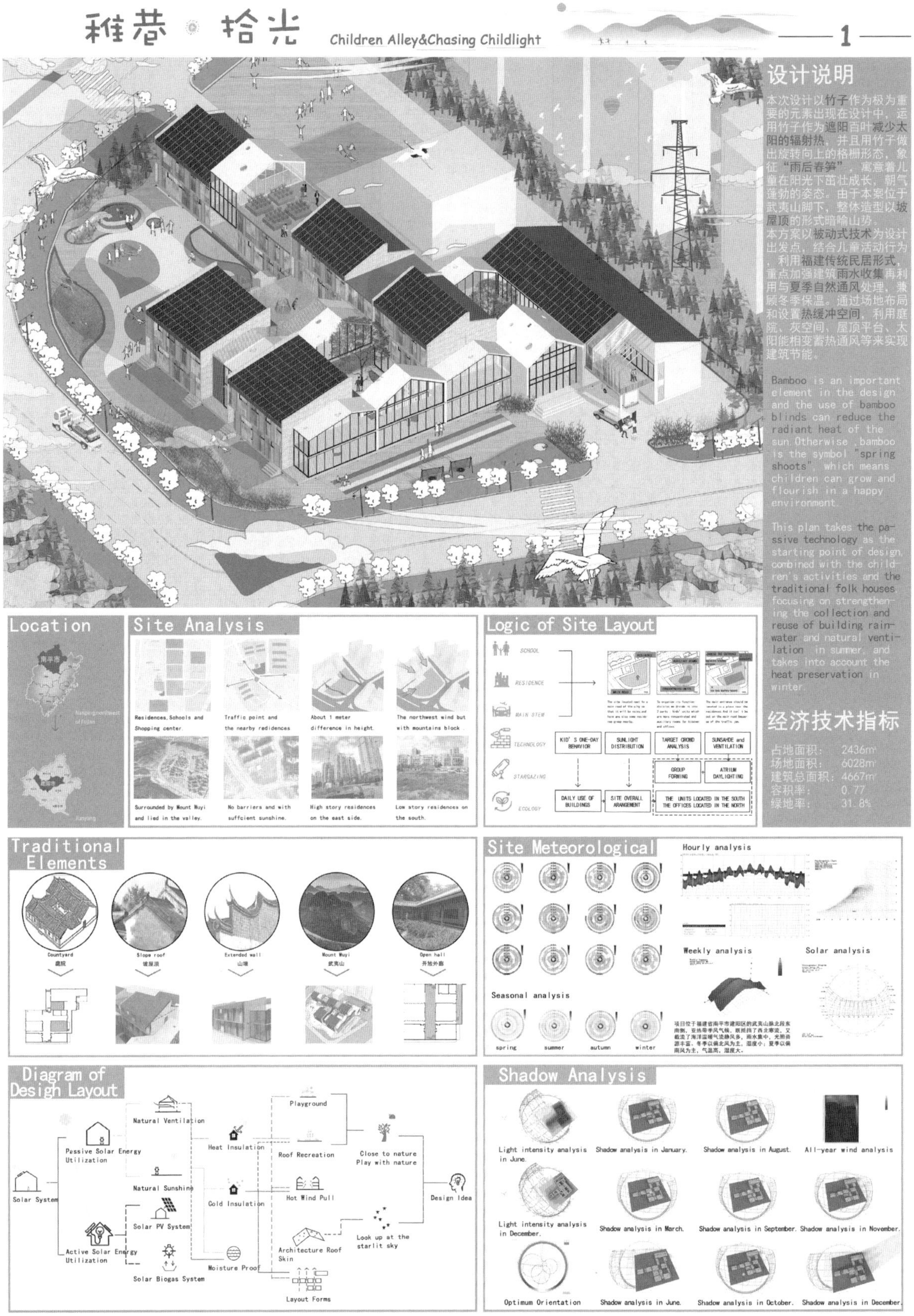

稚巷 · 拾光
Children Alley&Chasing Childlight
1
设计说明
本次设计以竹子作为极为重要的元素出现在设计中，运用竹子作为遮阳百叶减少太阳的辐射热，并且用竹子做出旋转向上的格栅形态，象征"雨后春笋"，寓意着儿童在阳光下茁壮成长，朝气蓬勃的姿态。由于本案位于武夷山脚下，整体造型以坡屋顶的形式暗喻山势。
本方案以被动式技术为设计出发点，结合儿童活动行为，利用福建传统民居形式，重点加强建筑雨水收集再利用与夏季自然通风处理，兼顾冬季保温。通过场地布局和设置热缓冲空间，利用庭院、灰空间、屋顶平台、太阳能相变蓄热通风等来实现建筑节能。
Bamboo is an important element in the design and the use of bamboo blinds can reduce the radiant heat of the sun.Otherwise ,bamboo is the symbol "spring shoots", which means children can grow and flourish in a happy environment.
This plan takes the passive technology as the starting point of design, combined with the children's activities and the traditional folk houses, focusing on strengthening the collection and reuse of building rainwater and natural ventilation in summer, and takes into account the heat preservation in winter.
经济技术指标
占地面积： 2436m²
场地面积： 6028m²
建筑总面积： 4667m²
容积率： 0.77
绿地率： 31.8%
Location
Site Analysis
Logic of Site Layout
Traditional Elements
Site Meteorological
Hourly analysis
Weekly analysis
Solar analysis
Seasonal analysis
Diagram of Design Layout
Shadow Analysis

稚巷 · 拾光 Children Alley&Chasing Childlight

—— 2 ——

Green Behaviors

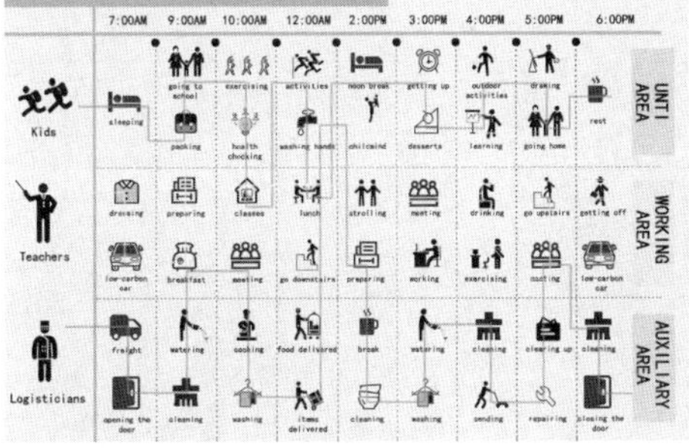
7:00AM
9:00AM
10:00AM
12:00AM
2:00PM
3:00PM
4:00PM
5:00PM
6:00PM
Kids
Teachers
Logisticians
UNIT AREA
WORKING AREA
AUXILIARY AREA

Logical Generation

STEP1:
The land area is 6028m' and the boundary line of the building is irregular.

STEP2:
According to the conditions, we divide the functional partition into 2 parts.

STEP3:
Refine the block and divide 6 units in detail. Traffic cores are formed in the middle.

STEP4:
One of the units is overheaded so that wind can be easily ventilated. And a room setted in the middle.

STEP5:
Accord the corridor to make 'cold lane'. Then a slide connect the traffic flow.

STEP6:
Finally, the courtyards are formed in the middle of the corridor, and trim the details. Here is the new spatial architecture.

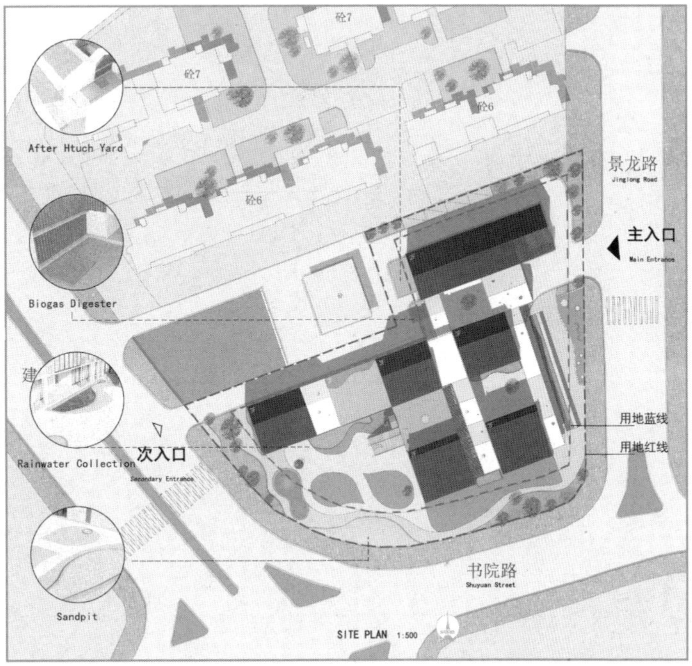
砼7
砼6
After Htuch Yard
Biogas Digester
Rainwater Collection
Sandpit
建
景龙路
Jinglong Road
主入口
Main Entrance
次入口
Secondary Entrance
用地蓝线
用地红线
书院路
Shuyuan Street
SITE PLAN 1:500

Environmental Analysis

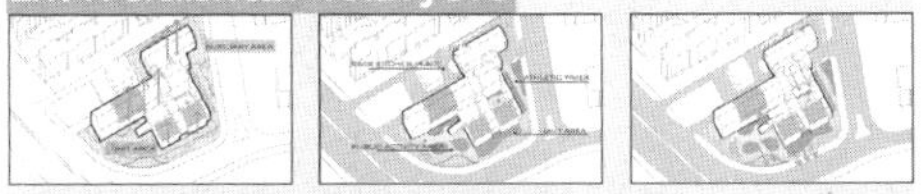

NO.1
Concentrate on the use of land resourses.

NO.2
Reasonable layout of green landscape for kids.

NO.3
Plan stormwater runoff properly.

NO.4
To assure that there is enough sunshine in the site.

NO.5
Provide public services, such as activity yard and sandpit.

NO.6
Reasonable rainwater collection system and biogas system.

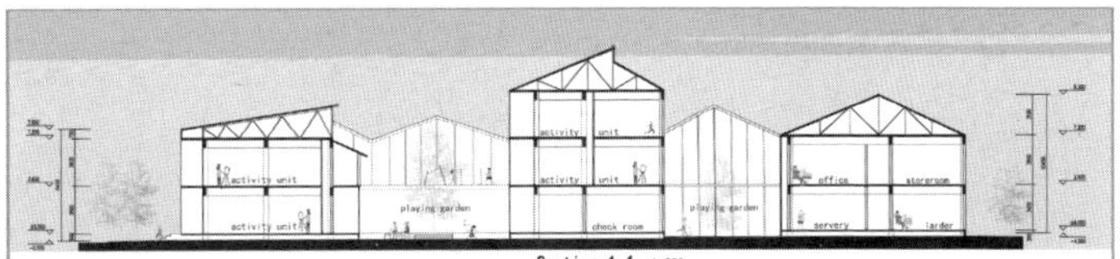
activity unit
playing garden
check room
office
storeroom
servery
larder

Section 1-1 1:250

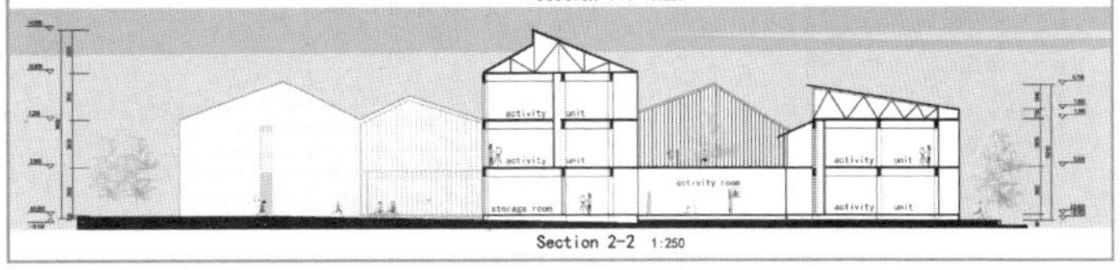
activity
unit
activity room
storage room

Section 2-2 1:250

稚巷 · 拾光 Children Alley&Chasing Childlight

3

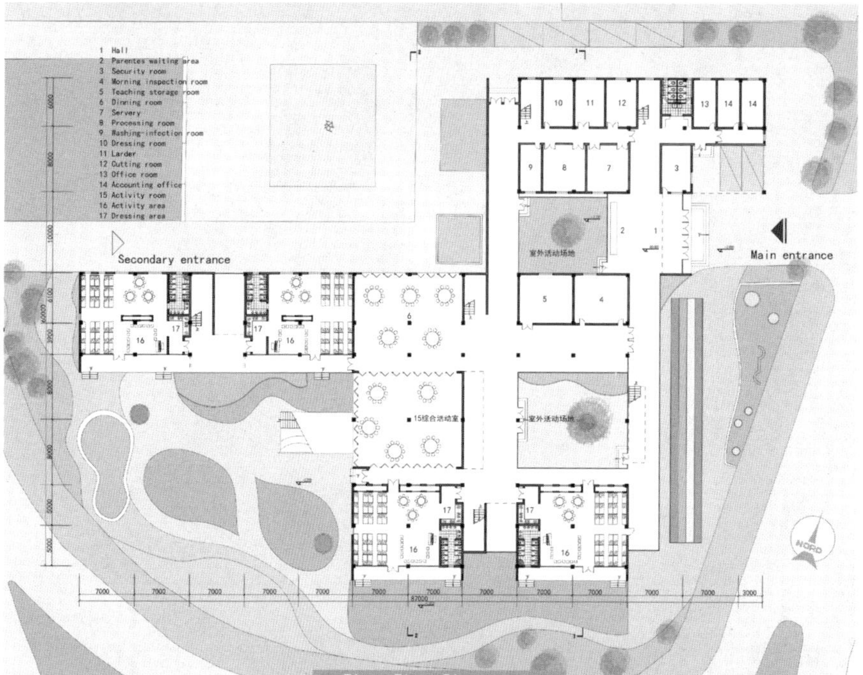
1 Hall
2 Parentes waiting area
3 Security room
4 Morning inspection room
5 Teaching storage room
6 Dinning room
7 Servery
8 Processing room
9 Washing-infection room
10 Dressing room
11 Larder
12 Cutting room
13 Office room
14 Accounting office
15 Activity room
16 Activity area
17 Dressing area
Secondary entrance
Main entrance
室外活动场地
15综合活动室
NORD

First Floor Plan 1:250

Weekly Time Line

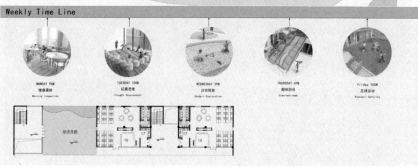

MONDAY 9AM
健康晨检
Morning Inspection
TUESDAY 10AM
拓展思维
Thought Development
WEDNESDAY 3PM
沙坑探索
Sandpit Exploration
THURSDAY 4PM
趣味游戏
Hide-and-seek
Friday 10AM
足球活动
Football Activity

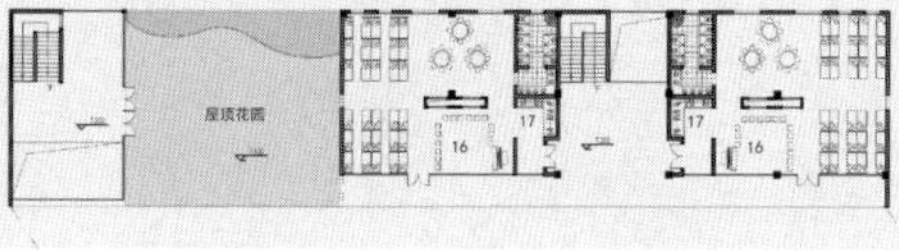
屋顶花园

Third Floor Plan 1:250

13 Office
14 Accounting office
16 Activity area
17 Dressing area
18 Meeting room
19 Switch box room
20 Storeroom
21 Washroom
22 Dean' s office
屋顶花园

Second Floor Plan 1:250

Strategy of Plan

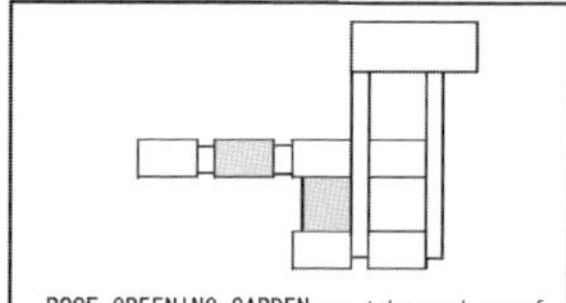

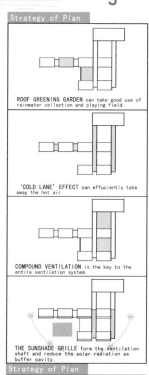

ROOF GREENING GARDEN can take good use of rainwater collection and playing field.

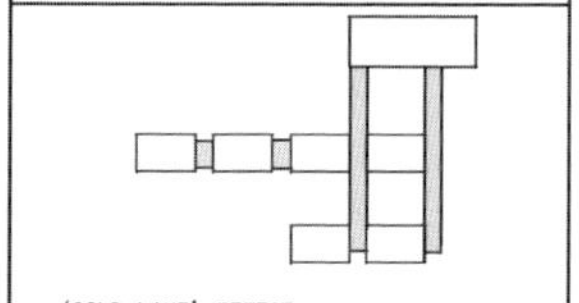

'COLD LANE' EFFECT can effuciently take away the hot air.

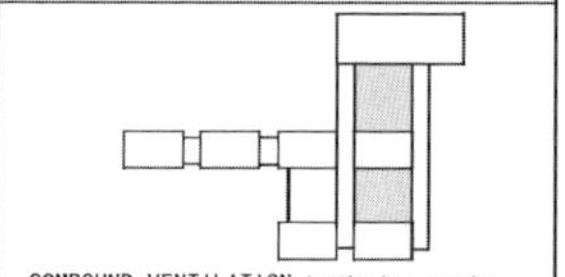

COMPOUND VENTILATION is the key to the entire ventilation system.

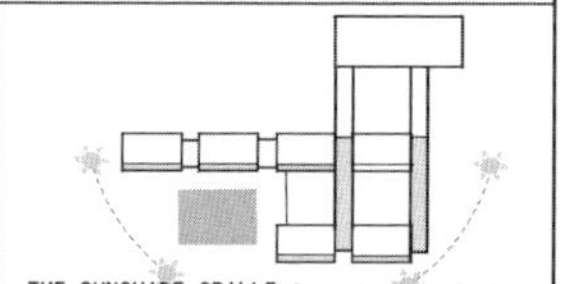

THE SUNSHADE GRILLE form the ventilation shaft and reduce the solar radiation as buffer cavity.

Strategy of Plan

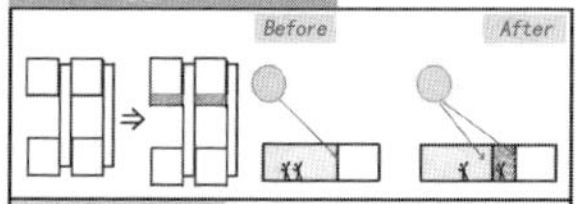
Before
After

+SOLAR CORRIDOR form several adumbral area, which provide kids indoor and outdoor sun bath respectively.

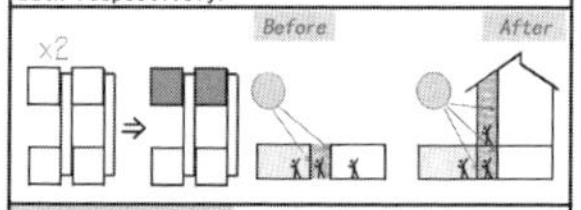
x2
Before
After

+BUILDING VOLUME is the best way to utillize solar, we increase the volume of playing area and day-light area to collect more solar energy.

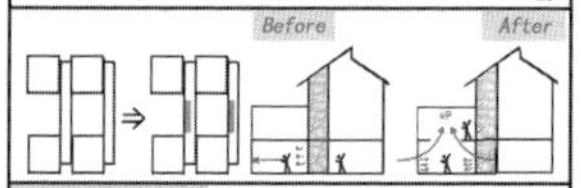
Before
After

+WIND EVULSION may provide kids a more fresh and cool place to play in summer, adding a space trifling change gout.

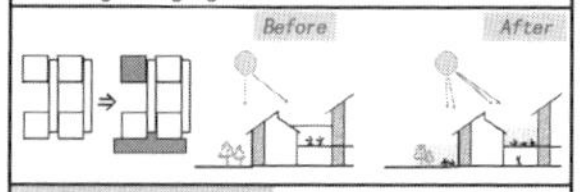
Before
After

+GARDEN BEFORE YARD may provide a shareing between people and nature. The sequence is enhanced from open to private simultaneously.

Ventilation and Heat

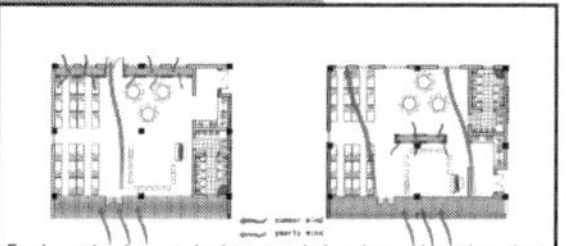

Each unit has a balcony with visor. It also has good ventilation. In summer, the chimney window will open to increase ventilation.

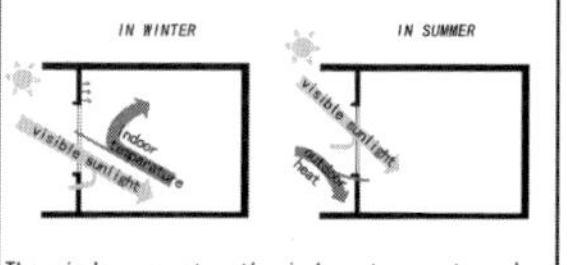
IN WINTER
IN SUMMER

The window can stop the indoor temperature lose in winter and can prevent outdoor heat from entering in summer.

稚巷 · 拾光 Children Alley&Chasing Childlight 4

Analasis of Activity Room

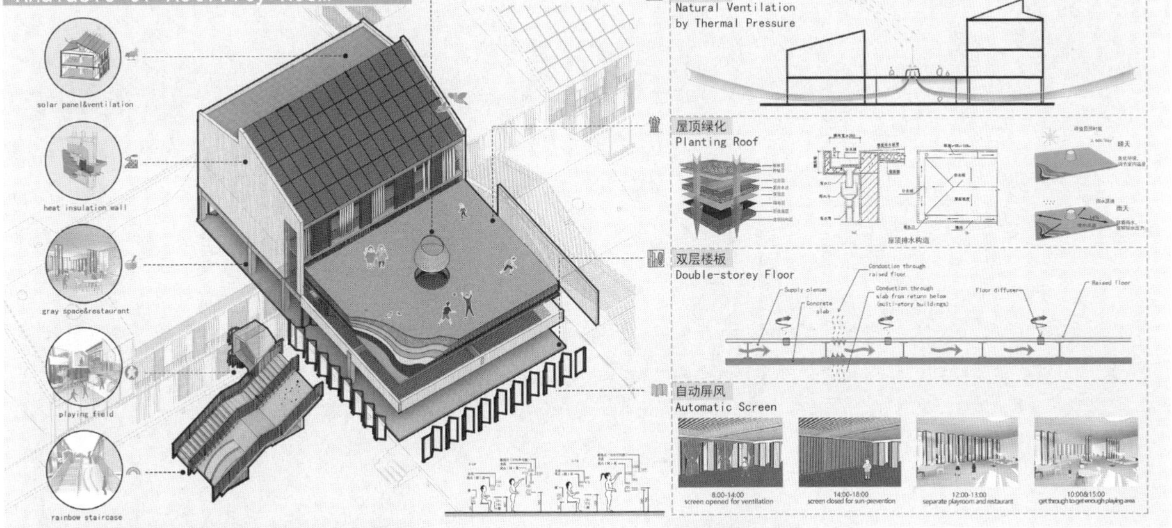
solar panel&ventilation
heat insulation wall
gray space&restaurant
playing field
rainbow staircase
热压通风
Natural Ventilation by Thermal Pressure
屋顶绿化
Planting Roof
双层楼板
Double-storey Floor
自动屏风
Automatic Screen
8:00-14:00 screen opened for ventilation
14:00-18:00 screen closed for sun-prevention
12:00-13:00 separate playroom and restaurant
10:00&15:00 get though to get enough playing area

Unit Plan

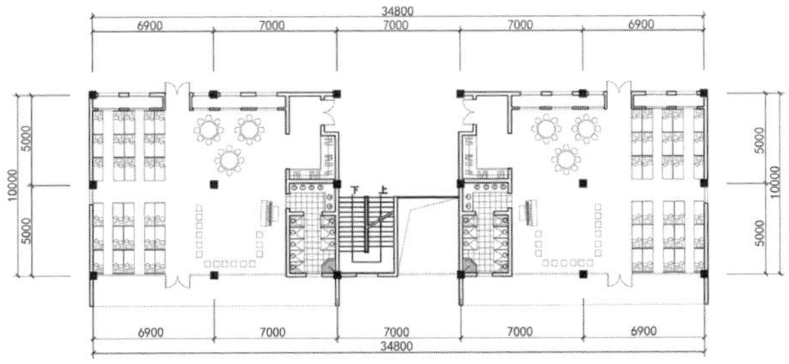

单元 大样1 1：150

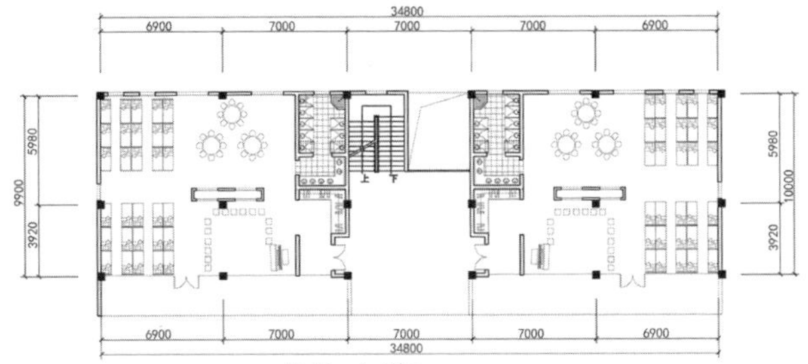

单元 大样2 1：150

Analasis of Single Architecture

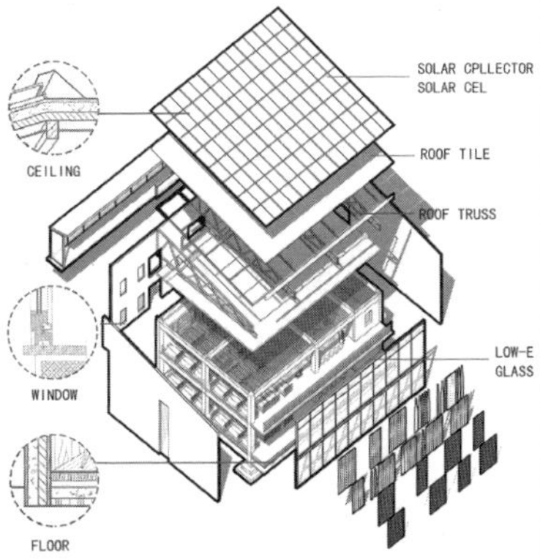
SOLAR CPLLECTOR
SOLAR CEL
ROOF TILE
ROOF TRUSS
LOW-E GLASS
CEILING
WINDOW
FLOOR

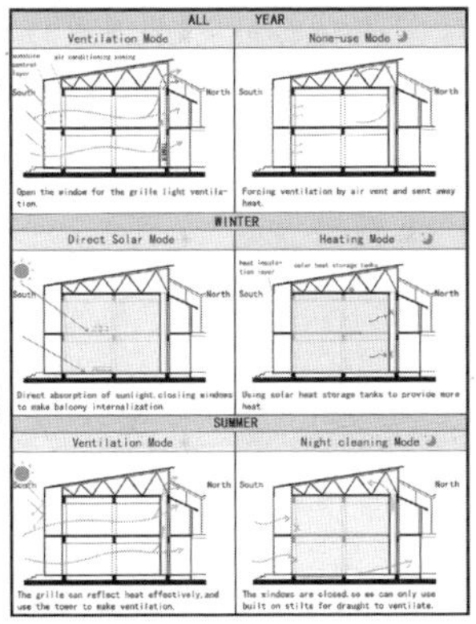
ALL YEAR
Ventilation Mode
None-use Mode
WINTER
Direct Solar Mode
Heating Mode
SUMMER
Ventilation Mode
Night cleaning Mode

Technics

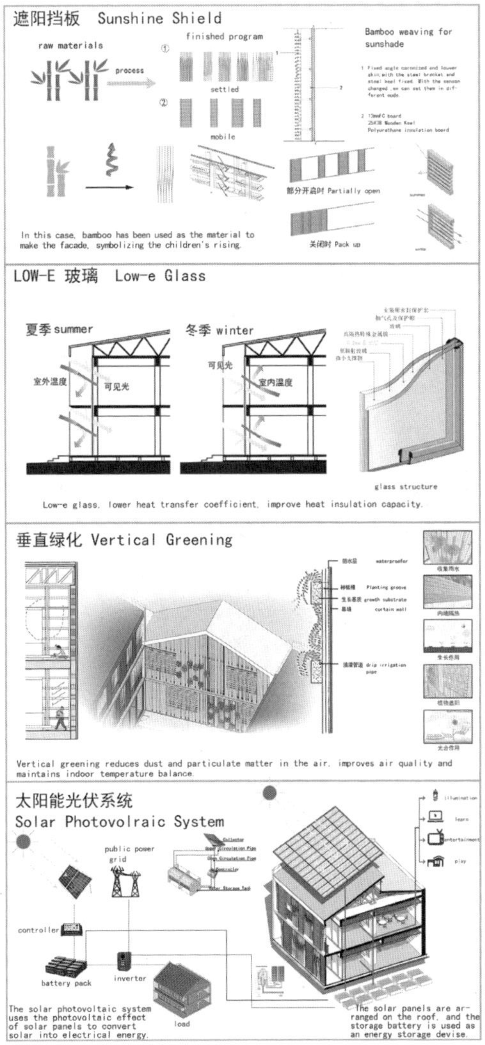
遮阳挡板 Sunshine Shield
raw materials
process
finished program
settled
mobile
Bamboo weaving for sunshade
部分开启时 Partially open
关闭时 Pack up
In this case, bamboo has been used as the material to make the facade, symbolizing the children's rising.
LOW-E 玻璃 Low-e Glass
夏季 summer
冬季 winter
室外温度
可见光
室内温度
glass structure
Low-e glass, lower heat transfer coefficient, improve heat insulation capacity.
垂直绿化 Vertical Greening
Vertical greening reduces dust and particulate matter in the air, improves air quality and maintains indoor temperature balance.
太阳能光伏系统
Solar Photovolraic System
public power grid
controller
battery pack
inverter
load
The solar photovoltaic system uses the photovoltaic effect of solar panels to convert solar into electrical energy.
The solar panels are arranged on the roof, and the storage battery is used as an energy storage devise.

稚巷·拾光 Children Alley&Chasing Childlight

5

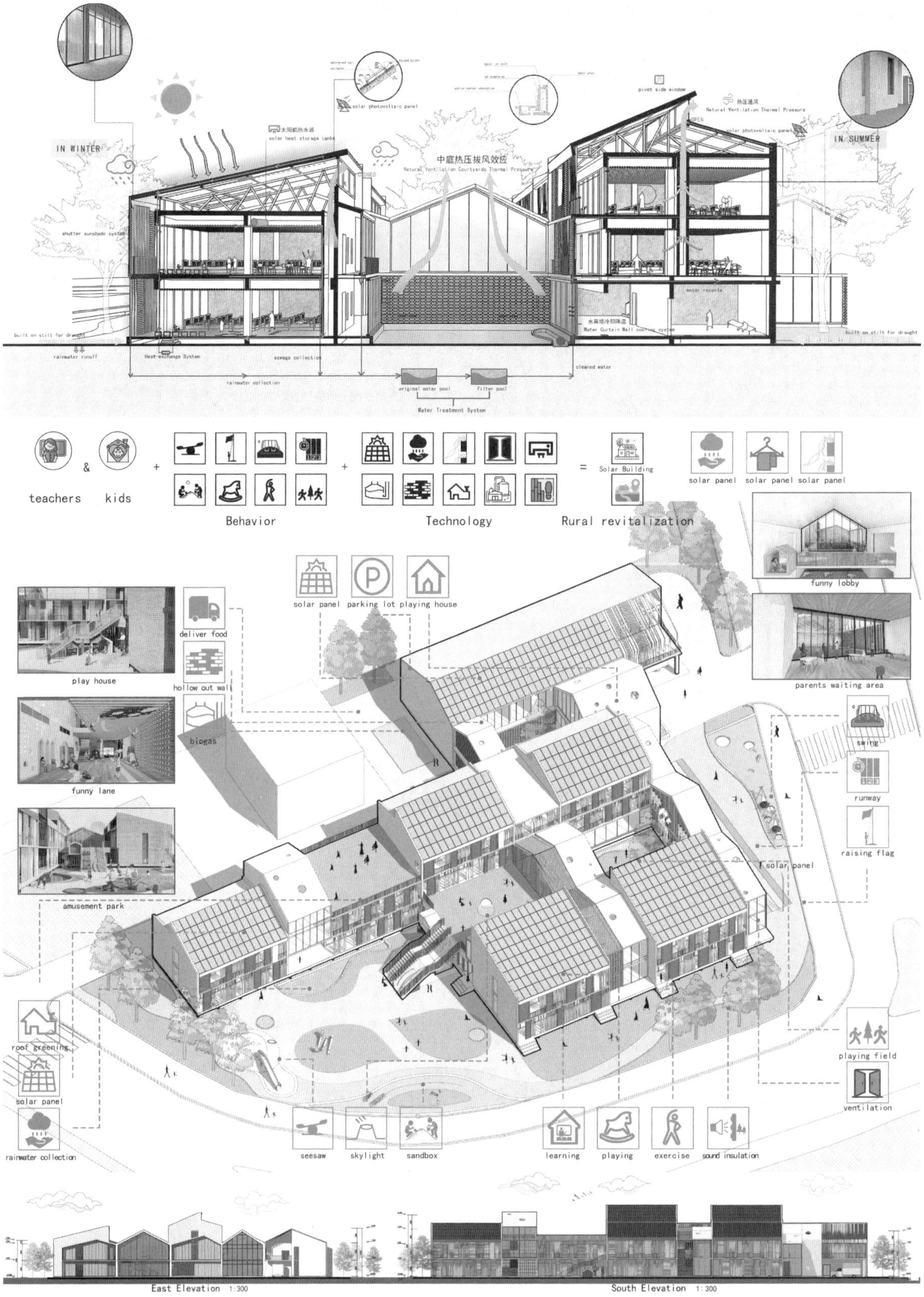
IN WINTER
IN SUMMER
中庭热压拔风效应
Natural Ventilation Courtyards Thermal Pressure
solar photovoltaic panel
pivot side window
热压通风
太阳能热水器
solar heat storage tanks
shutter sunshade system
水幕墙冷却降温
water recycle
built on stilt for drought
rainwater runoff
Heat-exchange System
sewage collection
rainwater collection
original water pool
filter pool
cleaned water
Water Treatment System
teachers
&
kids
+
Behavior
+
Technology
=
Solar Building
Rural revitalization
solar panel
funny lobby
parents waiting area
parking lot
playing house
deliver food
hollow out wall
biogas
play house
funny lane
amusement park
swing
runway
raising flag
roof greening
playing field
ventilation
rainwater collection
seesaw
skylight
sandbox
learning
playing
exercise
sound insulation
East Elevation 1:300
South Elevation 1:300

稚巷 · 拾光 Children Alley&Chasing Childlight

—— 6 ——

Sheet1:

Thermal Parameter of the Enclosure Structure

Heat transfer coeffident			Unit	Designed building	Reference building
Roof			W/(m²/K)	0.75	0.82
Exterior wall			W/(m²/K)	1.32	1.45
Exterior window	K	East	W/(m²/K)	2.40	3
		South	W/(m²/K)	2.40	2.5
		West	W/(m²/K)	2.40	2.5
		North	W/(m²/K)	2.40	3.8
	SHGC	East	---	0.32/---	0.32/---
		South	---	0.32/---	0.32/---
		West	---	0.32/---	0.32/---
		North	---	0.30	0.40

Note: (*) the averge heat transfer coefficient

Sheet2:Heat transfer coeffident

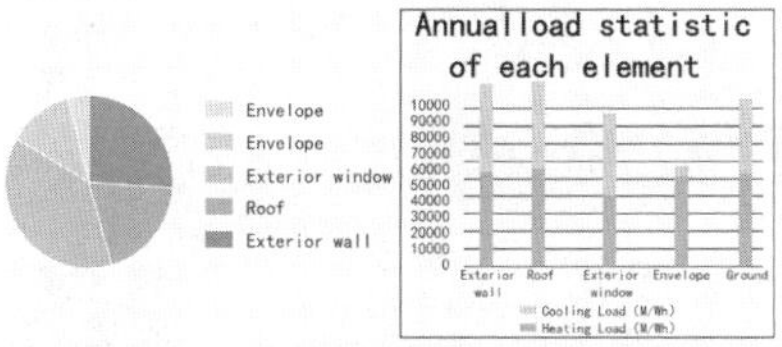
Envelope
Envelope
Exterior window
Roof
Exterior wall
Annualload statistic of each element
100000
90000
80000
70000
60000
50000
40000
30000
20000
10000
0
Exterior wall
Roof
Exterior window
Envelope
Ground
Cooling Load (M/Wh)
Heating Load (M/Wh)

Constructional element / Load	Exterior wall	Roof	Exterior window	Envelope	Ground
Cooling Load (M/Wh)	54.12	54.21	59.88	2.43	63.21
Heating Load (M/Wh)	67.43	74.30	46.90	60.33	63.21

Sheet3:Energy consumption comparison

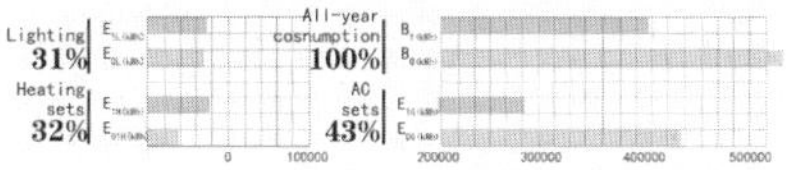
Lighting
31%
Heating sets
32%
All-year cosnumption
100%
AC sets
43%
0
100000
200000
300000
400000
500000

Type of Energy Consumption	Designed Building		Reference Building	
Heating sets (kWh)	E_{1H}	92532.40	E_{0H}	44325.62
Cooling sets (kWh)	E_{1Q}	130026.51	E_{0Q}	367746.46
Lighting (kWh)	E_{1L}	91003.72	E_{0L}	897549.74
All-year energy consumption (kWh)	B_1	315342.65	B_0	503633.07
Redction of energy consumption (kWh)	37.4%			

Sheet4:Renewable energy calculation

Parameter		Unit	Designed building
Annual solar radiation in JianYang	H_A	MJ/m²	4800
Solar panel area	A	m²	420
Comprehensive correction coefficient	K	—	0.201
Generation energy of solar panel	E_A	MJ/m²	24100

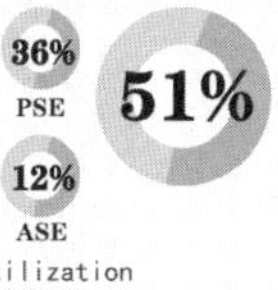
36%
PSE
12%
ASE
51%

Note: PSE=Passive solar energy utilization
ASE=Active solar energy utilization

Biogas utilization System

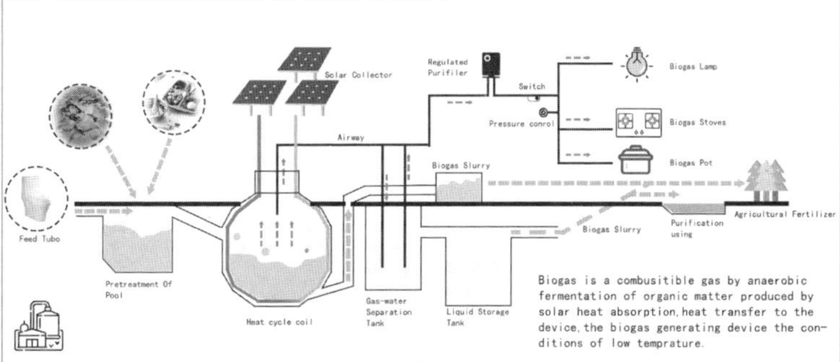
Solar Collector
Regulated Purifier
Switch
Pressure conrol
Airway
Biogas Lamp
Biogas Stoves
Biogas Pot
Biogas Slurry
Agricultural Fertilizer
Purification using
Feed Tubo
Pretreatment Of Pool
Heat cycle coil
Gas-water Separation Tank
Liquid Storage Tank
Biogas is a combusitible gas by anaerobic fermentation of organic matter produced by solar heat absorption,heat transfer to the device,the biogas generating device the con-ditions of low temprature.

The Rainwater Collection System

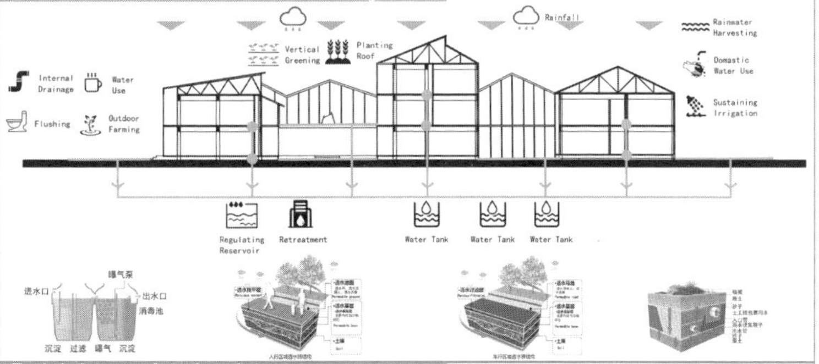
Rainfall
Rainwater Harvesting
Vertical Greening
Planting Roof
Domastic Water Use
Sustaining Irrigation
Internal Drainage
Water Use
Flushing
Outdoor Farming
Regulating Reservoir
Retreatment
Water Tank
Water Tank
Water Tank
进水口
曝气泵
出水口
消毒池
沉淀 过滤 曝气 沉淀

Cold Lane Analysis

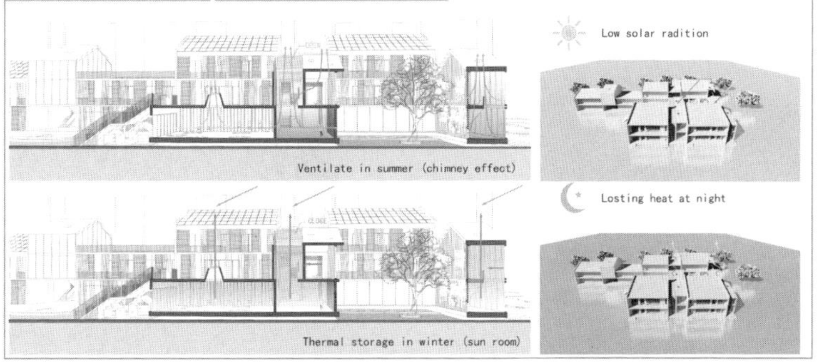
Low solar radition
Ventilate in summer (chimney effect)
Losting heat at night
Thermal storage in winter (sun room)

综合奖·优秀奖
General Prize Awarded · Honorable Mention Prize

注 册 号：7612
项目名称：楼台·织梦（南平）
Platforms of Dream（Nanping）

作　　者：李奇芫、于汉泽、李　洁、经翔宇、高皓然、陈柯欣
参赛单位：天津大学、天津大学建筑设计规划研究总院有限公司
指导教师：杨　崴、刘　刚、贡小雷

楼台·织梦 Platforms of Dream 01

Site Analysis

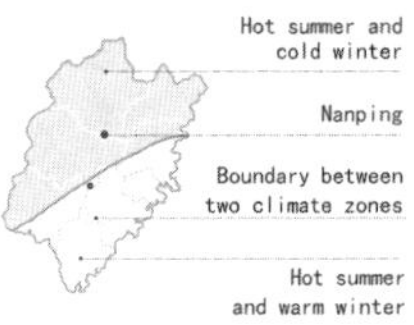
Hot summer and cold winter
Nanping
Boundary between two climate zones
Hot summer and warm winter

基于南平地区夏季湿热漫长、冬季寒冷短暂的气候特征，提取地域建筑的适宜性技术，如底层架空、天井、露台、冷巷、捕风窗，形成体块错落的建筑方案。风和光在错落的建筑体量之间穿梭，加强室内自然通风和采光。同时，错落的建筑体量形成自遮阳，保证室外活动平台的全天候的舒适性。通过双层表皮、相变蓄冷等，最大限度地利用被动式太阳能技术实现不同工况下的室内热舒适需求。此外，利用光伏屋面平衡全年约1/3的能耗需求。

Based on Nanping's climatic characteristics of a long-term hot and humid summer with a short-term cold winter, the appropriate technologies for local buildings, such as ground floor overhead, patios, terraces, cold lanes, and wind windows, are extracted to form a scattered architectural plan. Wind and light are roaming between the scattered building volumes to enhance indoor natural ventilation and lighting. At the same time, the scattered building volumes form self-shading to ensure the all-weather comfort of the outdoor activity platforms. At the structural side, double skins, sunshade louvers, phase change cold storage material, wind catchers, air supply floor slabs, etc., passive solar technology are used to the greatest extent to achieve indoor thermal comfort requirements under different working conditions. Besides, photo-voltaic roofs are used to balance about 1/3 of the overall annual energy consumption demand.

Regional Culture

Stilted buildings

High wall narrow lanes

Settlement platform

Patio space

Space under the eaves

Flexible envelope

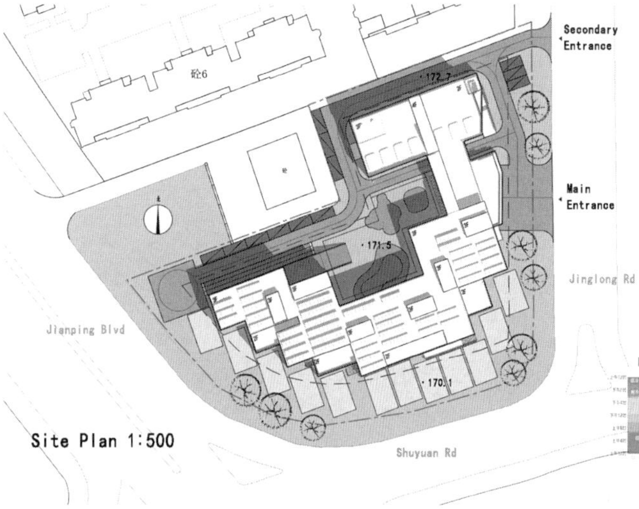
Secondary Entrance
Main Entrance
172.7
171.5
170.1
Jinglong Rd
Jianping Blvd
Shuyuan Rd
Site Plan 1:500

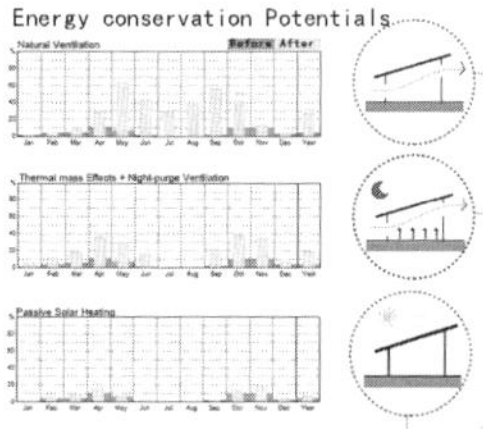
Energy conservation Potentials
Before
After

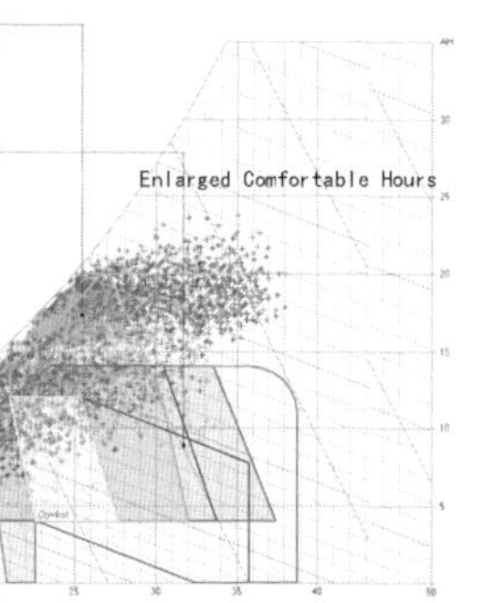
Passive strategy analysis
Enlarged Comfortable Hours

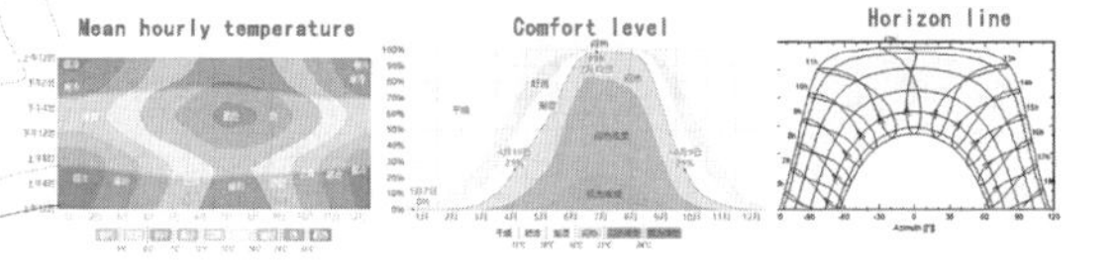
Mean hourly temperature
Comfort level
Horizon line

楼台·织梦

Platforms of Dream 02

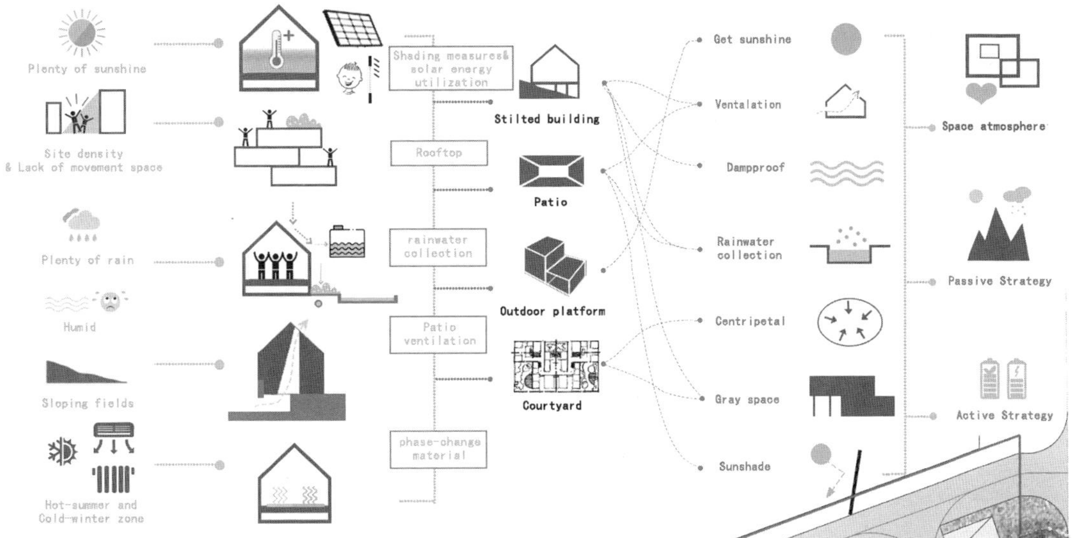

Problems in Site
Strategy
Looking for Prototype
Ecological& Space Significance
Plenty of sunshine
Site density & Lack of movement space
Plenty of rain
Humid
Sloping fields
Hot-summer and Cold-winter zone
Shading measures& solar energy utilization
Rooftop
rainwater collection
Patio ventilation
phase-change material
Stilted building
Patio
Outdoor platform
Courtyard
Get sunshine
Ventalation
Dampproof
Rainwater collection
Centripetal
Gray space
Sunshade
Space atmosphere
Passive Strategy
Active Strategy

Problem Analysis and Response

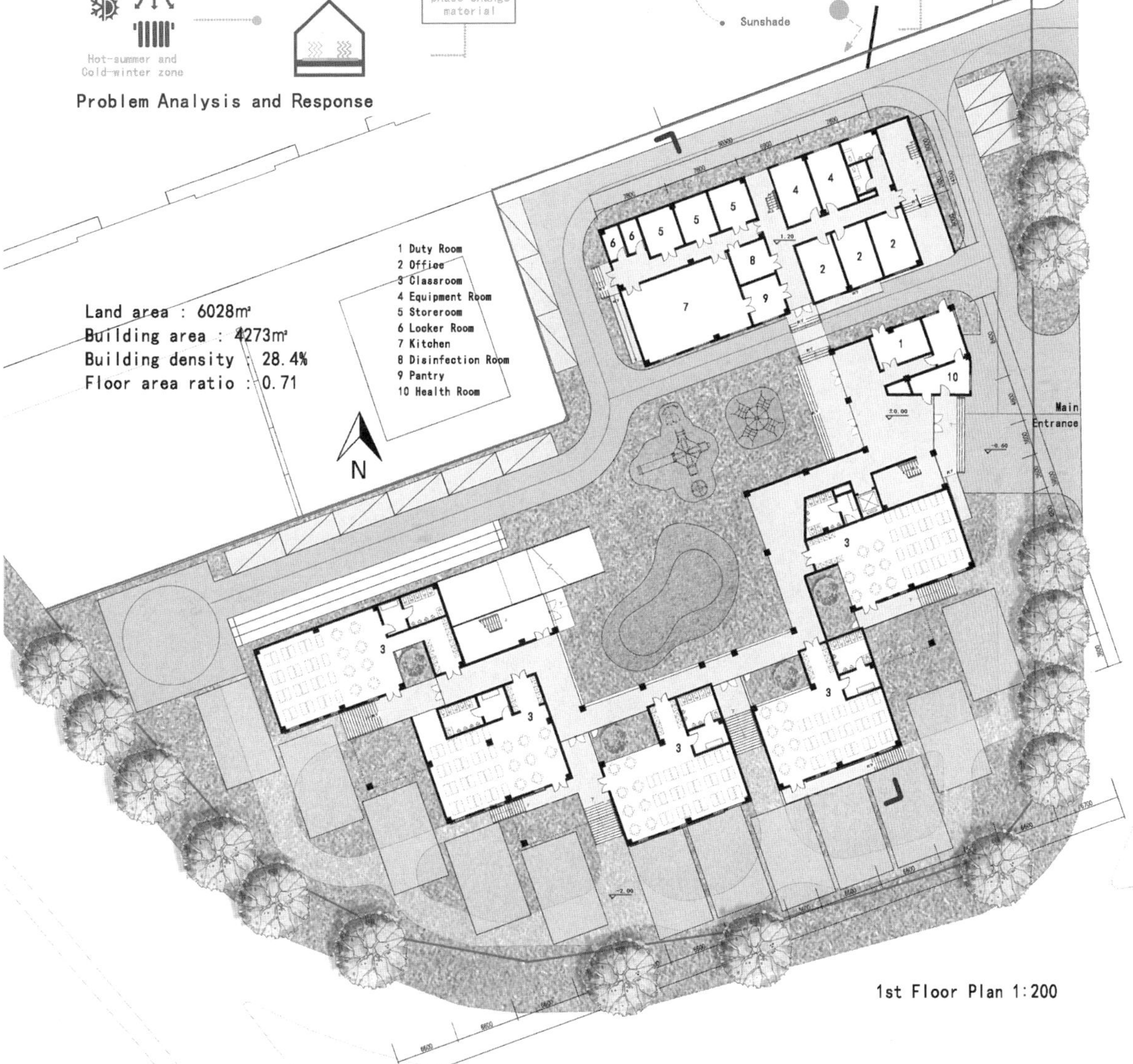

Land area : 6028m²
Building area : 4273m²
Building density : 28.4%
Floor area ratio : 0.71
1 Duty Room
2 Office
3 Classroom
4 Equipment Room
5 Storeroom
6 Locker Room
7 Kitchen
8 Disinfection Room
9 Pantry
10 Health Room
N
Main Entrance

1st Floor Plan 1:200

楼台 · 织梦 Platforms of Dream 03

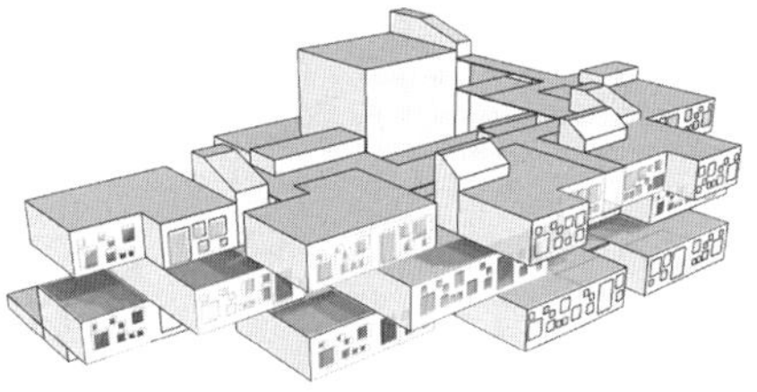

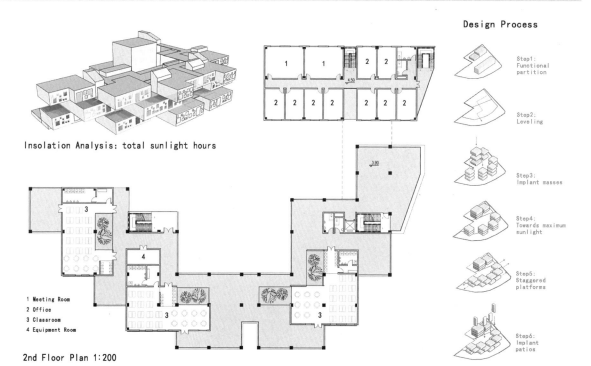

Insolation Analysis: total sunlight hours

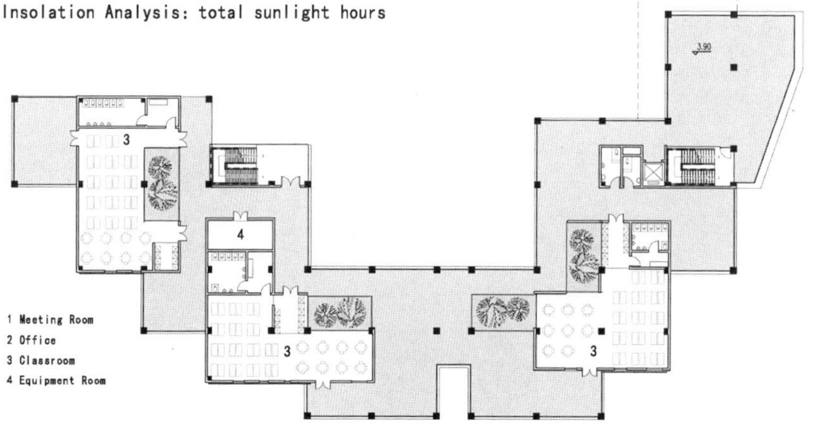
1 Meeting Room
2 Office
3 Classroom
4 Equipment Room
4.50
3.90

2nd Floor Plan 1:200

Design Process

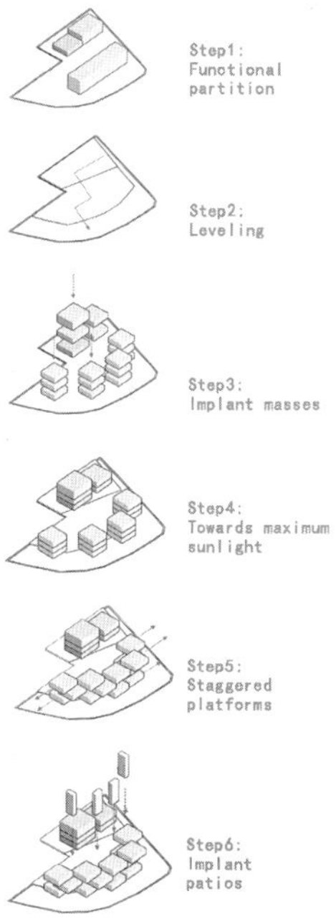
Step1: Functional partition
Step2: Leveling
Step3: Implant masses
Step4: Towards maximum sunlight
Step5: Staggered platforms
Step6: Implant patios

Details of Activity Unit

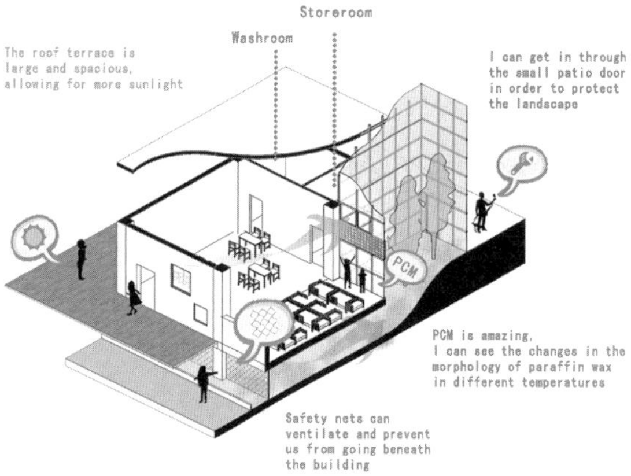
Storeroom
Washroom
The roof terrace is large and spacious, allowing for more sunlight
I can get in through the small patio door in order to protect the landscape
PCM
PCM is amazing, I can see the changes in the morphology of paraffin wax in different temperatures
Safety nets can ventilate and prevent us from going beneath the building

Function Block Composition Method

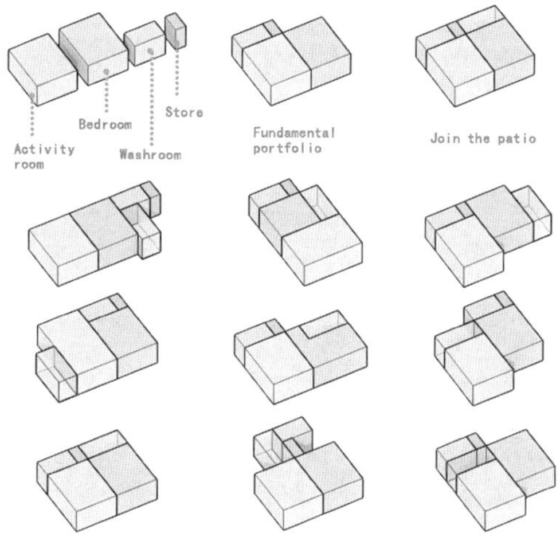
Activity room
Bedroom
Washroom
Store
Fundamental portfolio
Join the patio

South Elevation 1:200

楼台 · 织梦 Platforms of Dream 04

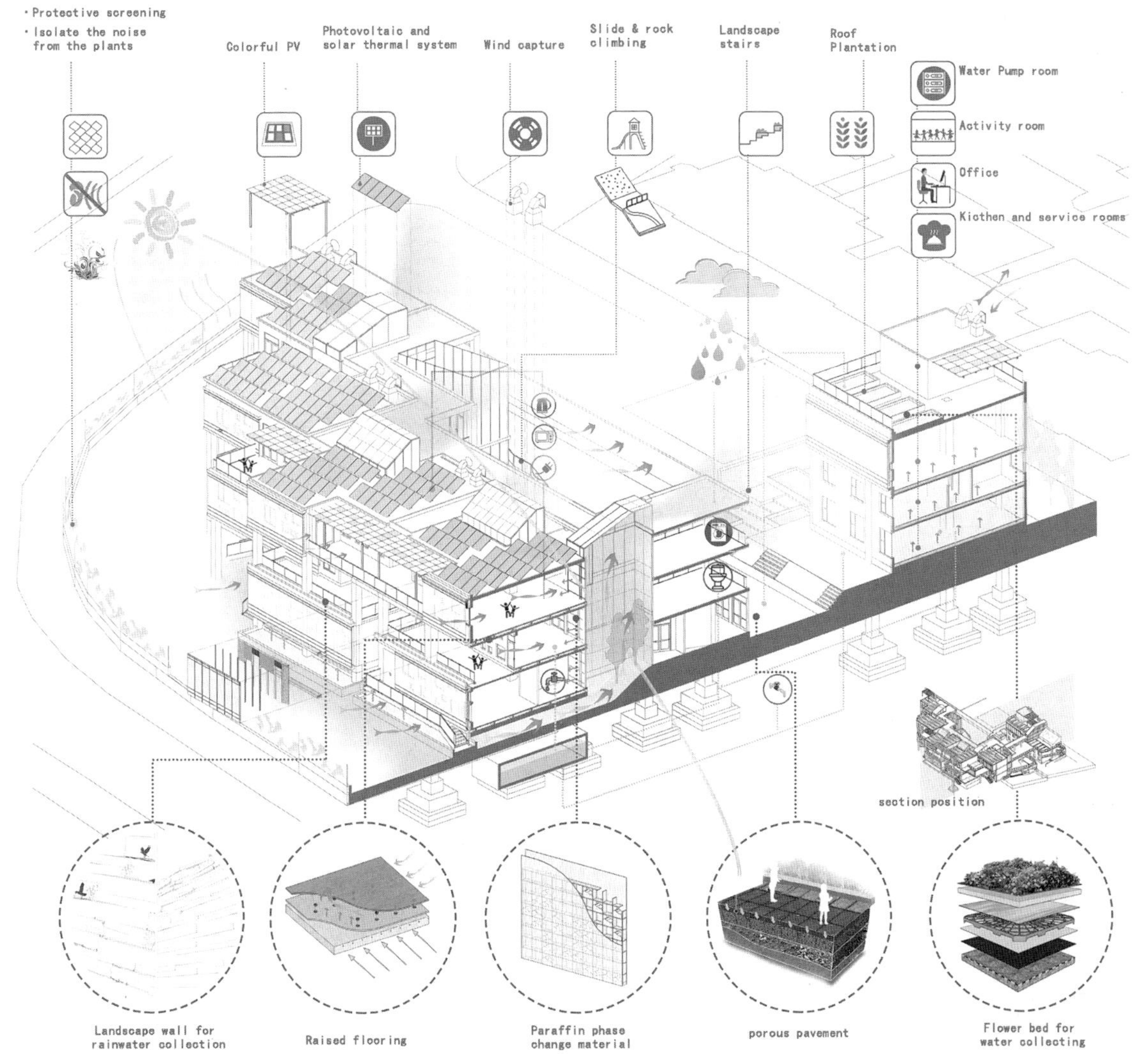

· Protective screening
· Isolate the noise from the plants
Colorful PV
Photovoltaic and solar thermal system
Wind capture
Slide & rock climbing
Landscape stairs
Roof Plantation
Water Pump room
Activity room
Office
Kitchen and service rooms
section position
Landscape wall for rainwater collection
Raised flooring
Paraffin phase change material
porous pavement
Flower bed for water collecting

Wind environment analysis diagram

According to Assessment standard for green building GB/T 50378-2019.

In Winter

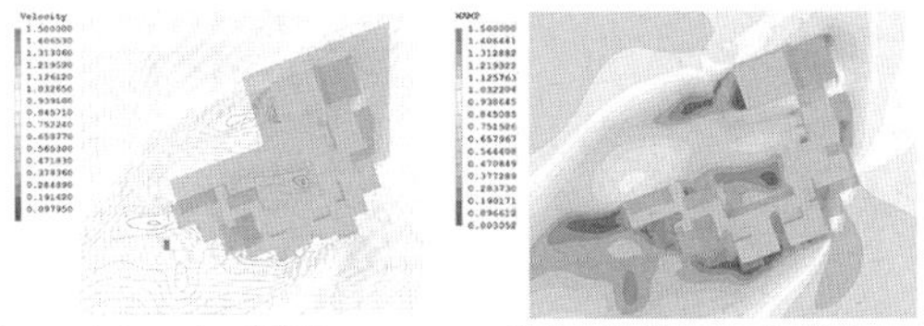

Velocity
WAMP

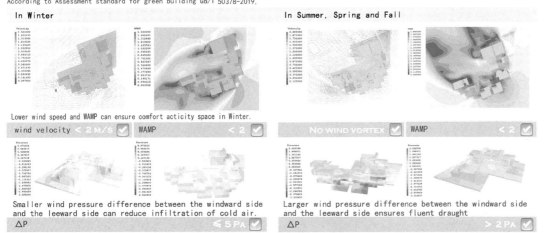

Lower wind speed and WAMP can ensure comfort acticity space in Winter.

wind velocity < 2 M/S ✓ | WAMP < 2 ✓

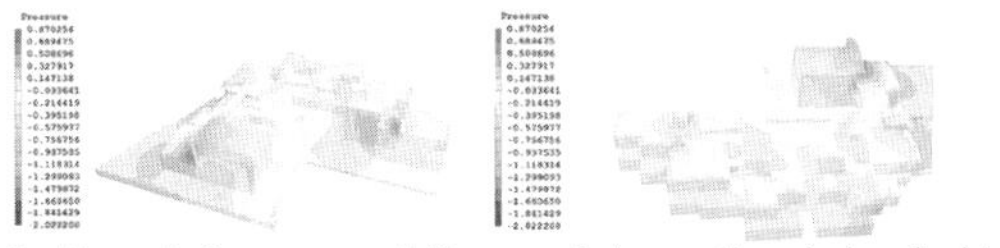

Pressure
Pressure

Smaller wind pressure difference between the windward side and the leeward side can reduce infiltration of cold air.

ΔP ≤ 5 PA ✓

In Summer, Spring and Fall

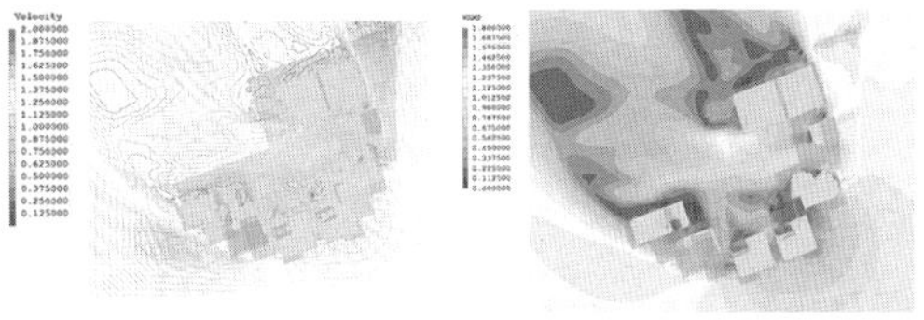

Velocity
WAMP

NO WIND VORTEX ✓ | WAMP < 2 ✓

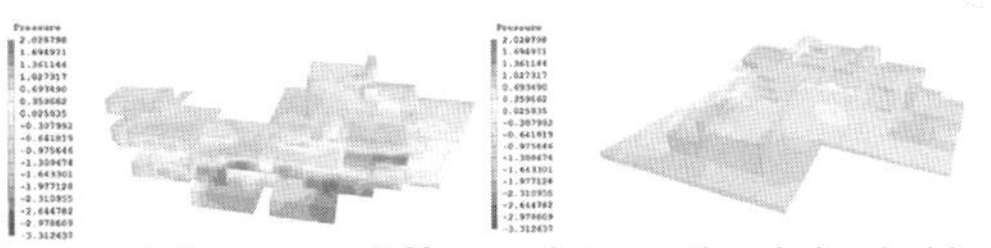

Pressure
Pressure

Larger wind pressure difference between the windward side and the leeward side ensures fluent draught

ΔP > 2 PA ✓

楼台·织梦 Platforms of Dream 05

Fresh air supply through double floors

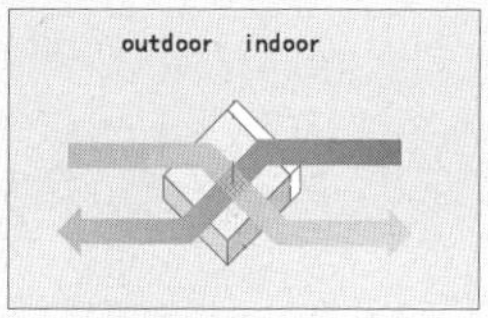
outdoor
indoor

Heat exchanger

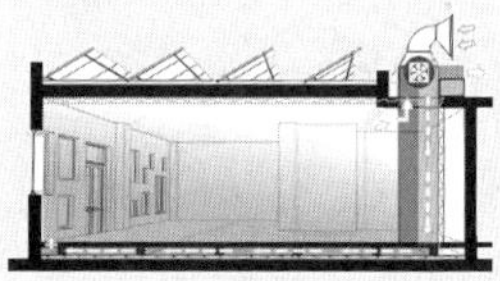

· Captured Wind exchange heat with stale air ouput, and then supplied into rooms through double floors.

· In Summer, fresh air is cooled.

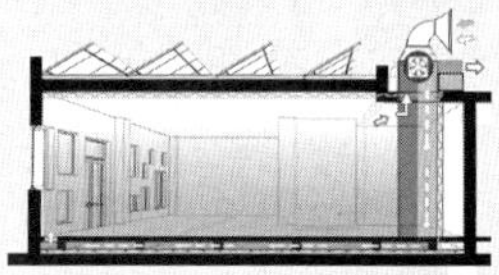

· Captured Wind exchange heat with stale air ouput, and then supplied into rooms through double floors.

· In Winter, fresh air is heated.

Chimney effect optimization

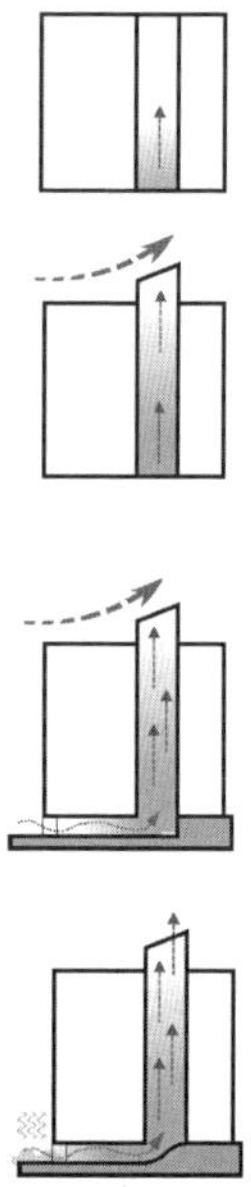

Patio work mode in Spring and Fall

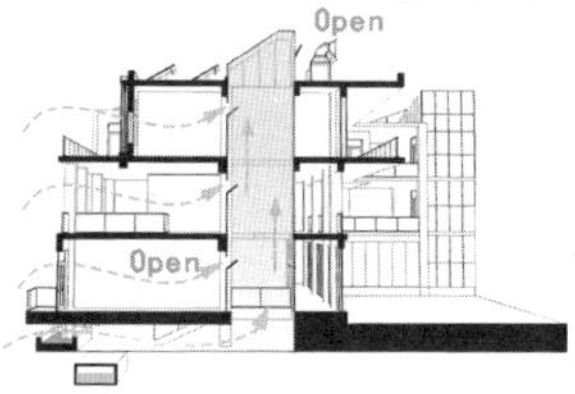
Open
Open

Patio work mode in Summer

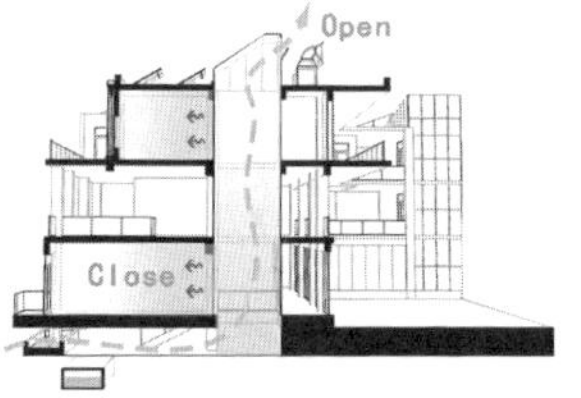
Open
Close

Patio work mode in Winter

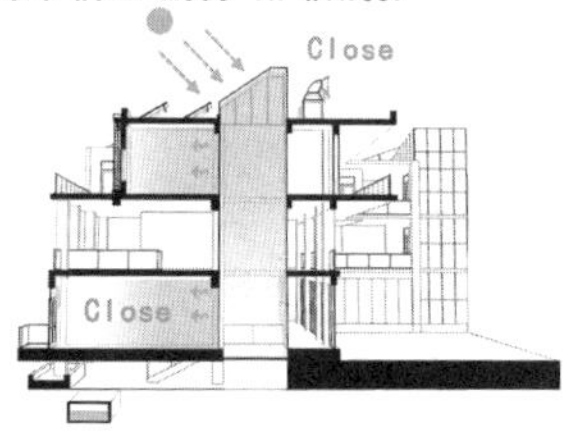
Close
Close

According to *Assessment standard for green building GB/T 50378-2019.*

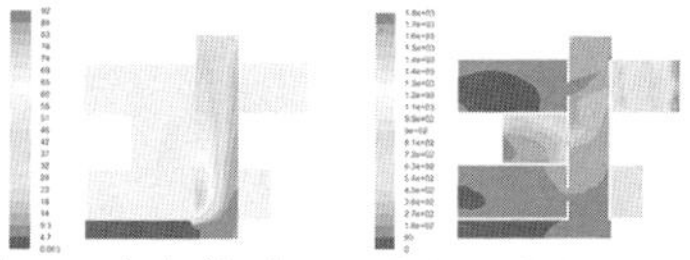

In summer, air should exchange more than 2 tiames.

Air Age > 1800s ✓

In Spring and Fall, Open the high windows to promote draught.

GREAT DRAUGHT ✓

According to Design code for heating ventilation and air conditioning of civil buildings GB 50736-2012.

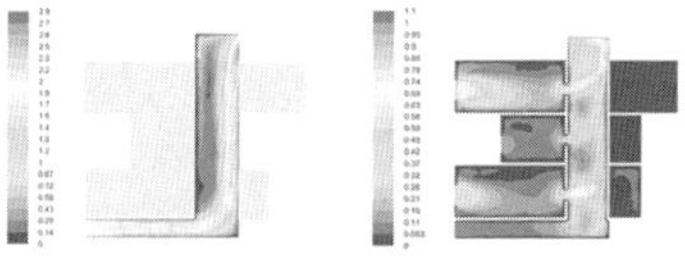

In summer, the higher the wind speed in the patio, the more heat is removed.

Wind Velocity > 2 M/S ✓

In Spring and Fall, All wind speeds in the room are below 0.25.

Wind Velocity < 0.25 M/S ✓

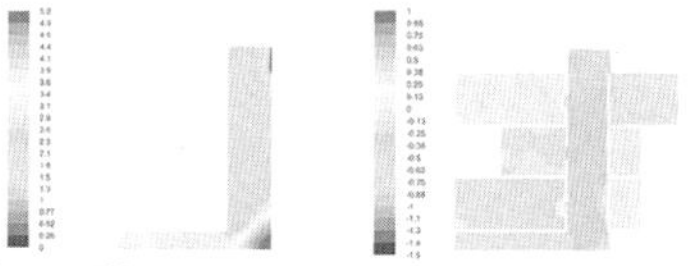

Pressure difference in vertion can promote hot pressure ventilation

Vertical Pressure Stratification < 0.25 M/S ✓

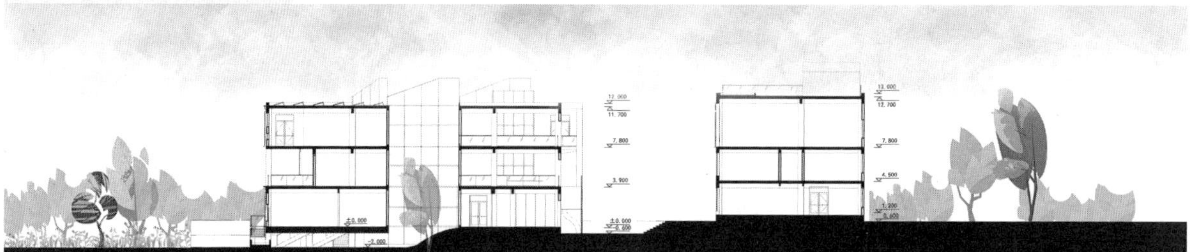

Section 1:200

楼台 · 织梦 Platforms of Dream 06

Main materials' Life Cycle Analysis

Element	materials	front of the life cycle	back of the life cycle	lifetime
Roof	Concrete	1/2 dust and 1/2 Construction waste and recycling of industrial waste and aggregate	Shredding for reuse as aggregates	20 years
Roof	bamboo	native plants	Dismantled and reused as wall coverings or degraded into the soil	20-30 years
External wall	foamed concrete	Construction waste and recycled industrial waste	Shredding for reuse as aggregates	60 years
interior wall	reed board	Plants around Dianchi	Degradation to soil	30 years
Window	glass	New production materials	Recycled for glaze	40 years
Floor	Concrete	1/2 dust and 1/2 Construction waste and recycling of industrial waste	Shredding for reuse as aggregates	60 years

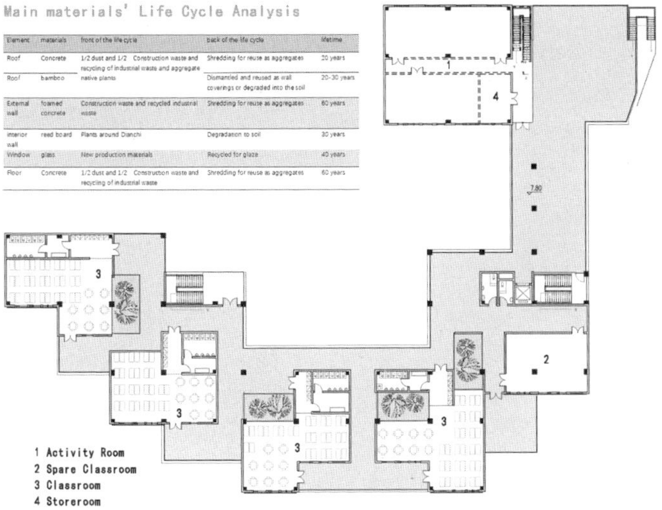
1 Activity Room
2 Spare Classroom
3 Classroom
4 Storeroom

3rd Floor Plan 1:250

According to the Standard for daylighting Design of Buildings GB 50033-2013,

Scattered building volumes can enhance natural ventilation with enough shading.
Simulations are performed to ensure comfortable daylighting for the 1st and 2nd floor.

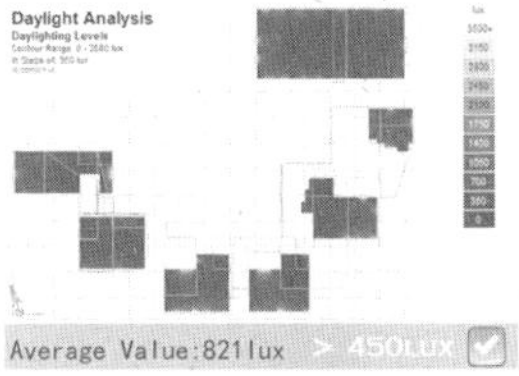
Daylight Analysis
Average Value:821lux
> 450LUX

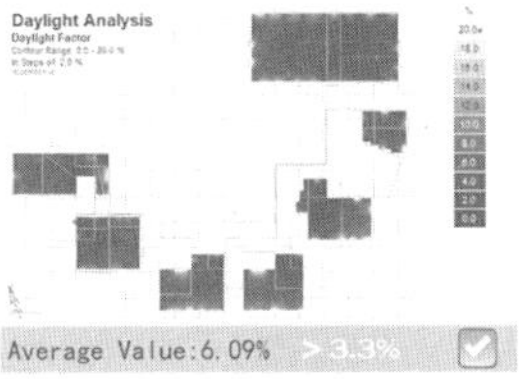
Daylight Analysis
Average Value:6.09%
> 3.3%

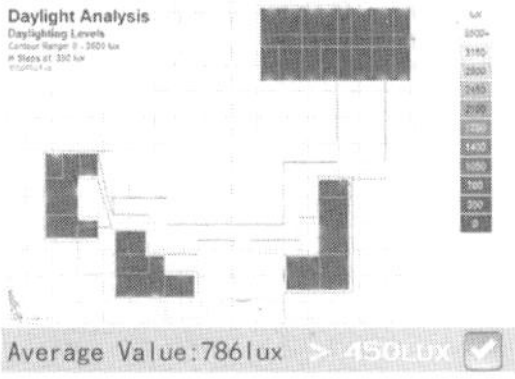
Daylight Analysis
Average Value:786lux
> 450LUX

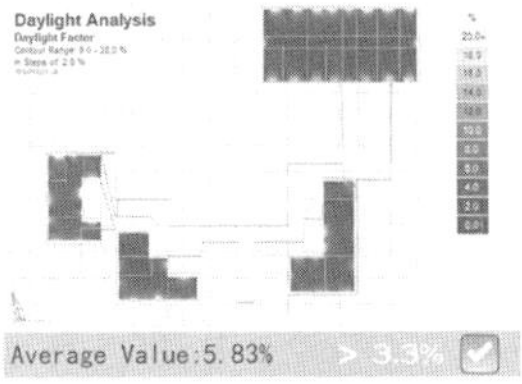
Daylight Analysis
Average Value:5.83%
> 3.3%

Nanping, located in hot summer and cold winter zone, is required to meet the winter heating demand through passive strategies due to the lack of district heating in winter.
Double Skin and Colorful Louvers are installed to fit various requirements among the seasons.

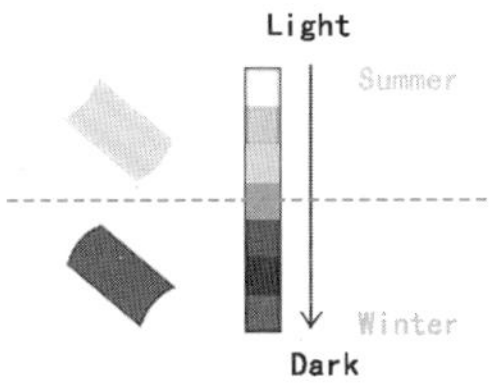
Light
Summer
Winter
Dark

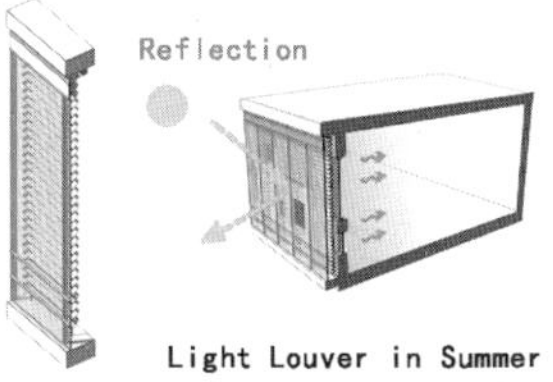
Reflection
Light Louver in Summer

One side of each piece is coated with light color while the other side with dark. In Summer, light colored louver reflects solar. In Winter, Dark colored Louver absorbs heat then transfers into the inside of rooms.

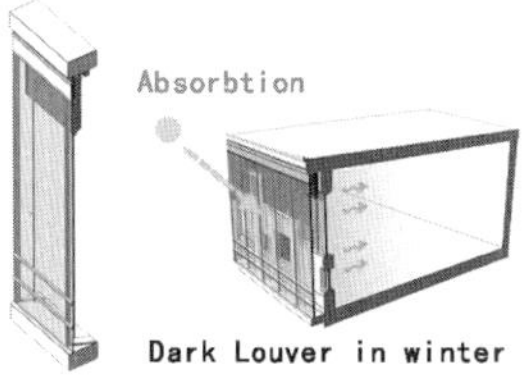
Absorbtion
Dark Louver in winter

Energy Consumption Comparison

	Baseline	Designed Building	Reduction
Cooling Load (kW·h)	92.5	86.0	6.98%
Heating Load (kW·h)	25.0	22.3	10.87%
Lighting (kW·h)	0.8	0.8	3.25%
Equipment (kW·h)	0.3	0.3	1.98%

Roof PV system

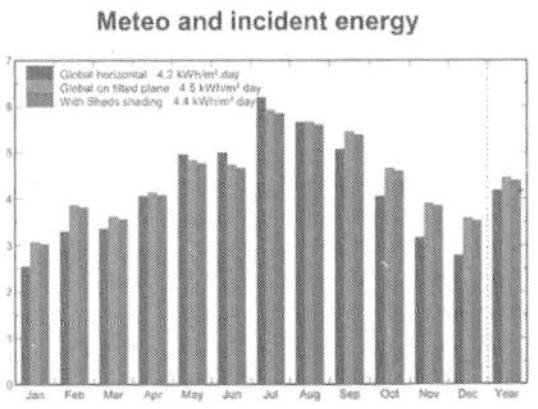
Meteo and incident energy
Jan Feb Mar Apr May Jun Jul Aug Sep Oct Nov Dec Year

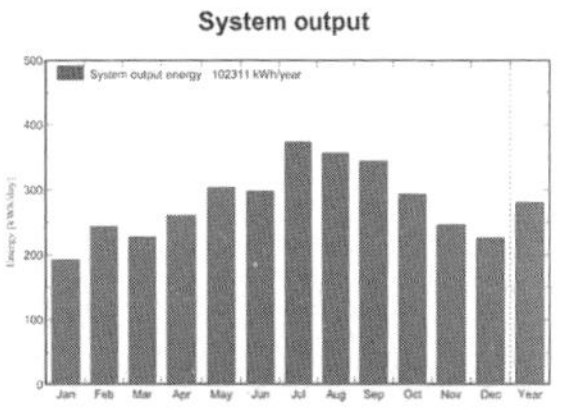
System output
Jan Feb Mar Apr May Jun Jul Aug Sep Oct Nov Dec Year

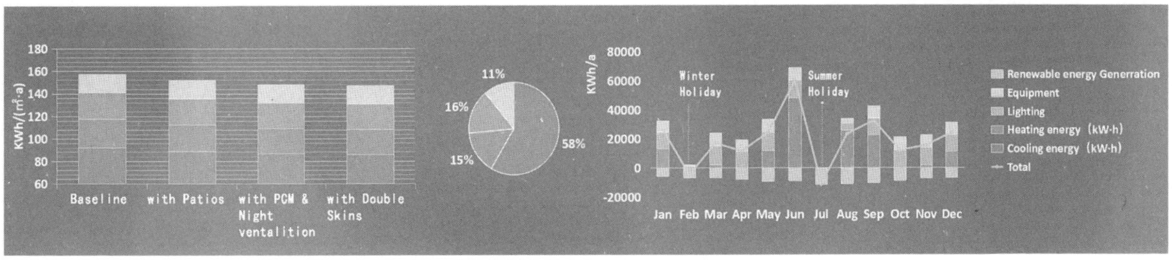
180
160
140
120
100
80
60
KWh/(m²·a)
Baseline
with Patios
with PCM & Night ventalition
with Double Skins
11%
16%
15%
58%
KWh/a
80000
60000
40000
20000
0
-20000
Winter Holiday
Summer Holiday
Jan Feb Mar Apr May Jun Jul Aug Sep Oct Nov Dec
Renewable energy Generation
Equipment
Lighting
Heating energy (kW·h)
Cooling energy (kW·h)
Total

综合奖·优秀奖
General Prize Awarded · Honorable Mention Prize

注 册 号：7630

项目名称：巷·间（南平）

Cold Lane·Interspace (Nanping)

作　　者：徐衍新、郑仲意、李迪萌、相　楠、刘权仪、张鹏娜

参赛单位：山东建筑大学、北京师范大学珠海分校

指导教师：薛一冰、房　涛、何文晶

巷·间 | COLD LANE · INTERSPACE |

LOCATION ANALYSIS

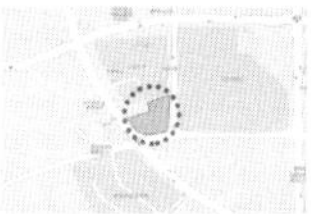

Nan Ping——Site

CLIMATE ANALYSIS

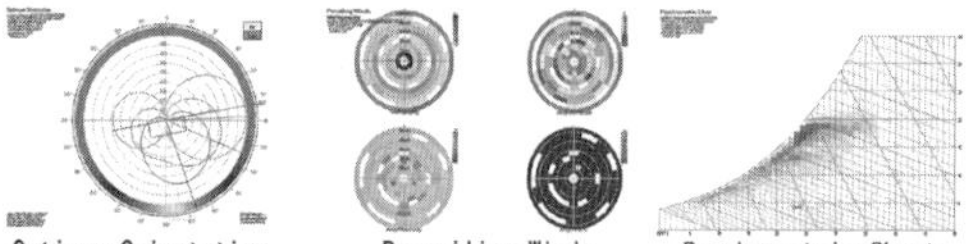

Optimum Orientation　Prevailing Winds　Psychrometric Chart

REGIONAL FEATURE

Traditional Roof　Narrow courtyard　Open Hall　Eaves Gallery

SITE ANALYSIS

基地

The texture of buildings around the site

基地
场地入口

The traffic around the site

基地

Landscape green space around the site

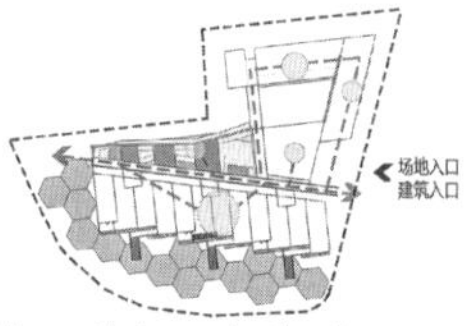
场地入口
建筑入口

The spatial organization in the site

DESCRIPTION OF DESIGN

This project takes "cold lane", a traditional ventilation and cooling method for residential buildings in Nanping city, Fujian Province, as the entry point, and combines the hot summer and cold winter areas. It also takes into full consideration the design key points of the building types of the kindergarten and the perceived needs of the school-age children to the space site.

At the same time, in the active technology, the solar photovoltaic system and solar photovoltaic system are set up integrating design of the building roof. Rainwater harvesting systems have also been added to use rainwater for landscape and other non-domestic water.

本方案以福建省南平市的传统民居通风降温手段“冷巷”为切入点，结合了夏热冬冷地区的气候特点，并充分考虑了幼儿园建筑类型的设计要点和适龄儿童对空间场地的感知需求。

同时在主动式技术上，设置太阳能光伏系统和太阳能光热系统，并于建筑屋顶一体化设计。此外，还加入了雨水收集系统，将雨水用于景观环境和其他非生活用水。

COLD LANE STRATEGY

Villages in Fujian Province

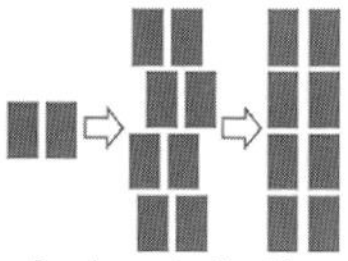

Development of cold alleys in villages

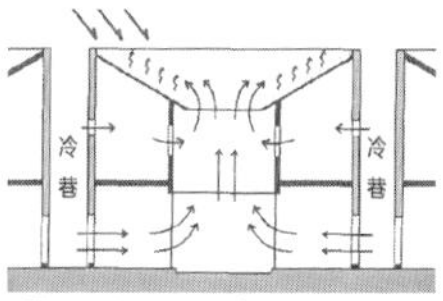
冷巷
冷巷

The combination of cold lane and courtyard is the traditional folk house in Fujian Province. Scheme design from Passive energy saving technology starts with the use of traditional cold lanes. With the courtyard to formthe ventilation cooling principle to form the building buffer layer.

The scene of Village cold lane

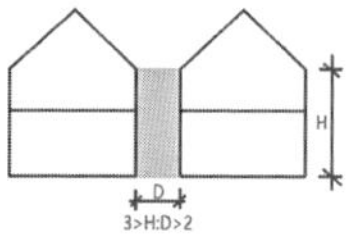
H
D
3>H:D>2

Comfortable value

APPLICATION OF COLD LANE

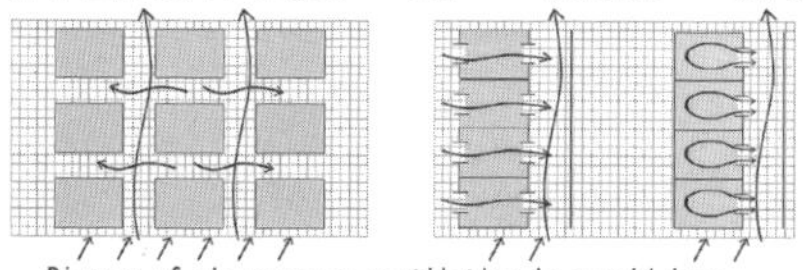

Diagram of air pressure ventilation in a cold lane

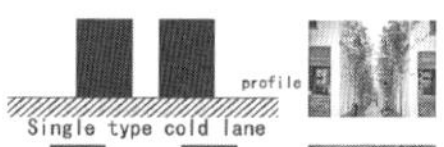
profile

Single type cold lane

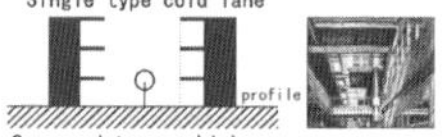
profile

Compound type cold lane

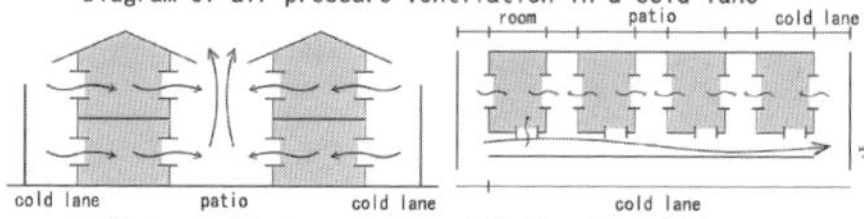
room
patio
cold lane
cold lane
patio
cold lane
cold lane

Diagram of hot pressure ventilation in cold lane

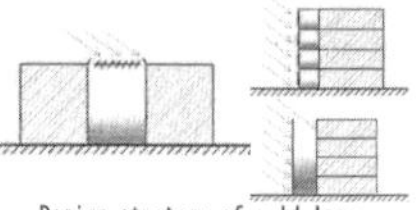

Design strategy of cold lane artificial shading

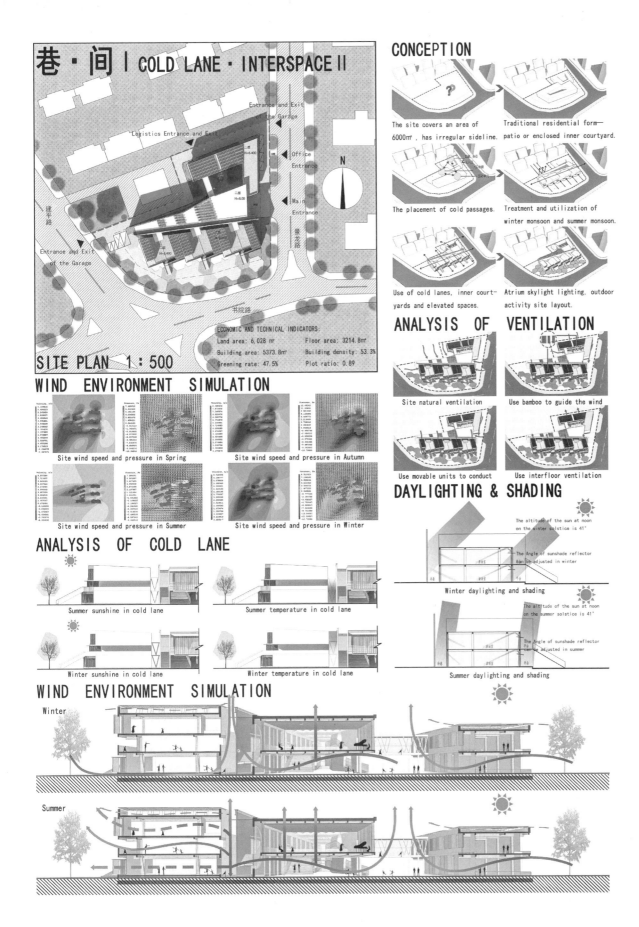

巷 · 间 | COLD LANE · INTERSPACE II

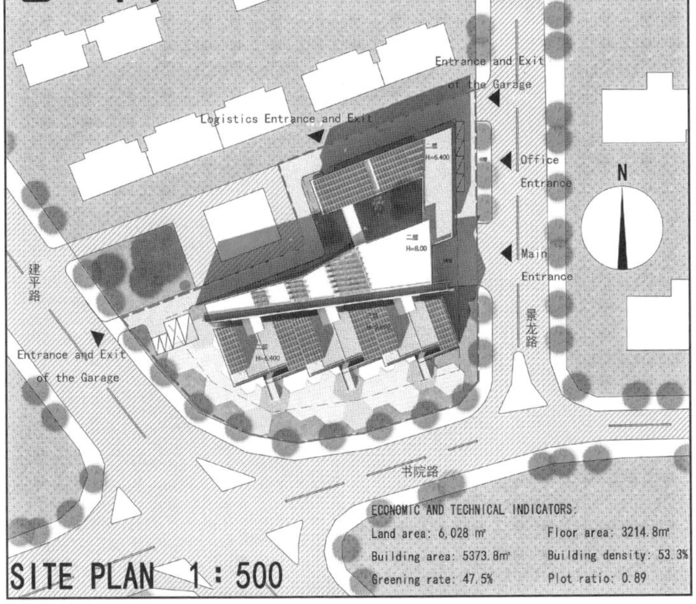

Entrance and Exit of the Garage
Logistics Entrance and Exit
Office Entrance
Main Entrance
N
建平路
景龙路
书院路
Entrance and Exit of the Garage

SITE PLAN 1:500

ECONOMIC AND TECHNICAL INDICATORS:

Land area: 6,028 m²　Floor area: 3214.8m²

Building area: 5373.8m²　Building density: 53.3%

Greening rate: 47.5%　Plot ratio: 0.89

WIND ENVIRONMENT SIMULATION

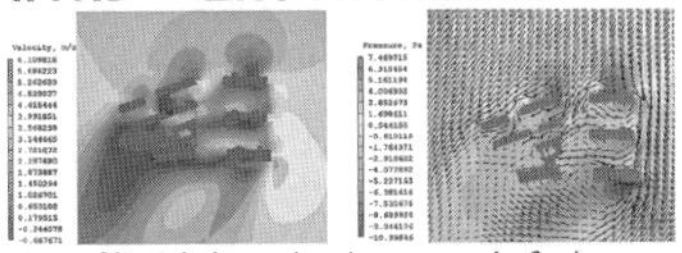

Site wind speed and pressure in Spring

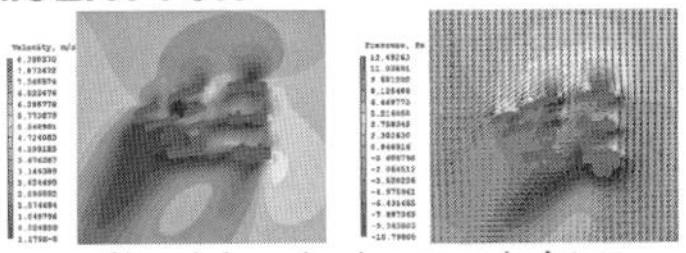

Site wind speed and pressure in Autumn

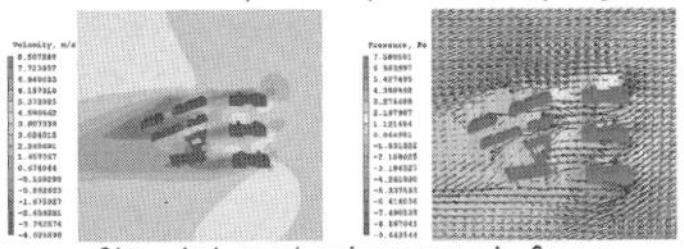

Site wind speed and pressure in Summer

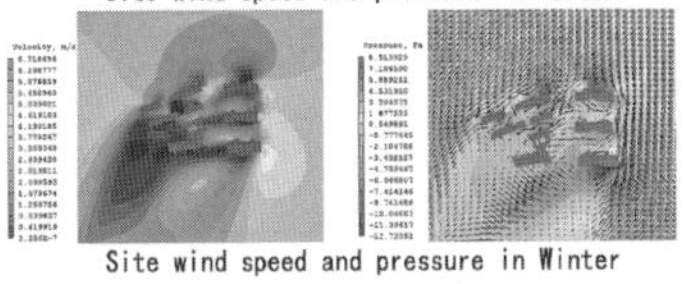

Site wind speed and pressure in Winter

ANALYSIS OF COLD LANE

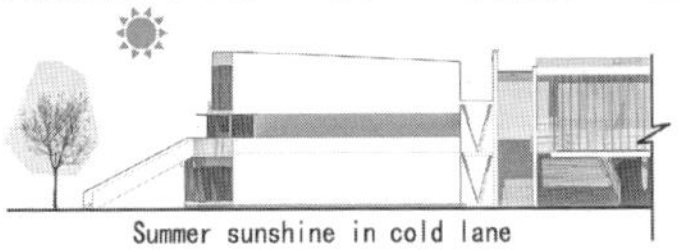

Summer sunshine in cold lane

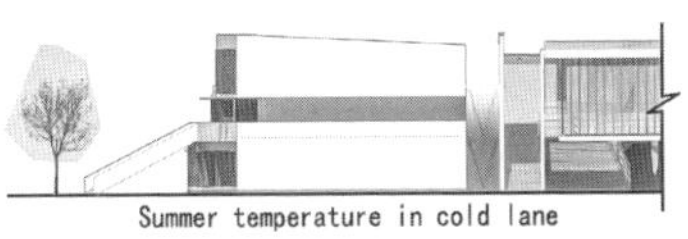

Summer temperature in cold lane

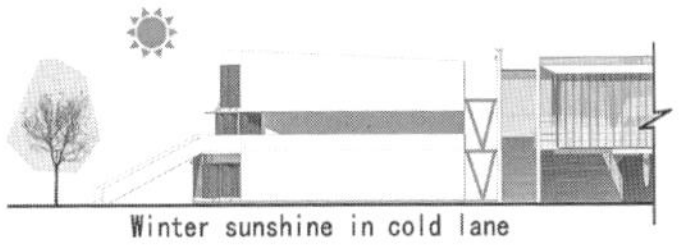

Winter sunshine in cold lane

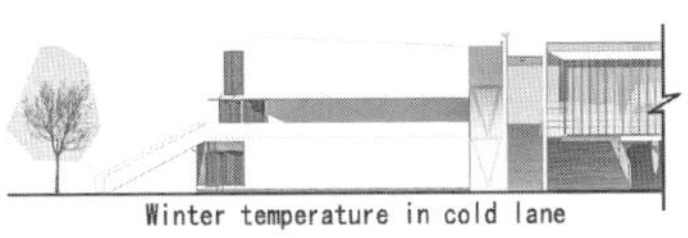

Winter temperature in cold lane

CONCEPTION

The site covers an area of 6000m², has irregular sideline.

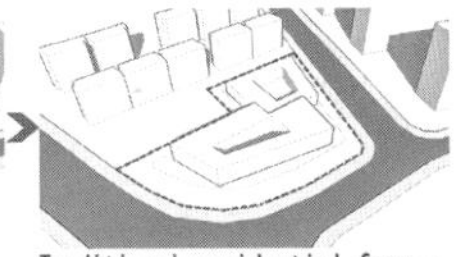

Traditional residential form—patio or enclosed inner courtyard.

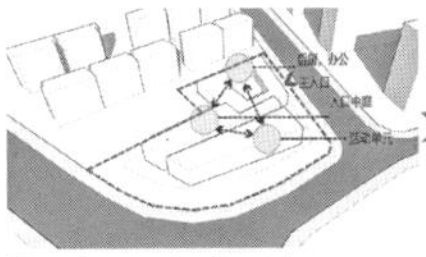

The placement of cold passages.

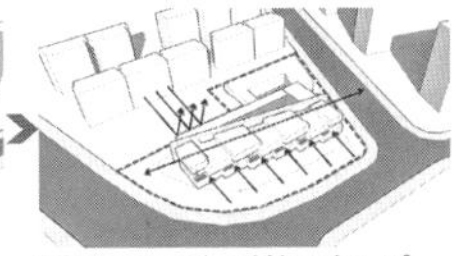

Treatment and utilization of winter monsoon and summer monsoon.

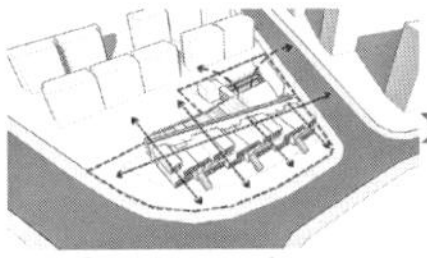

Use of cold lanes, inner courtyards and elevated spaces.

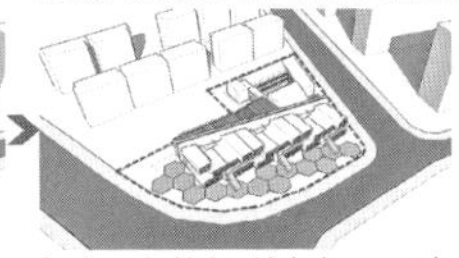

Atrium skylight lighting, outdoor activity site layout.

ANALYSIS OF VENTILATION

Site natural ventilation

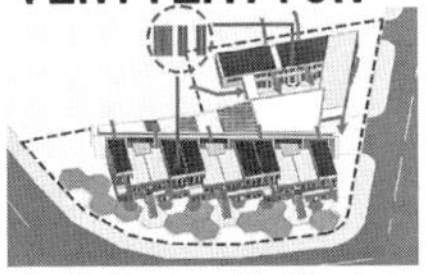

Use bamboo to guide the wind

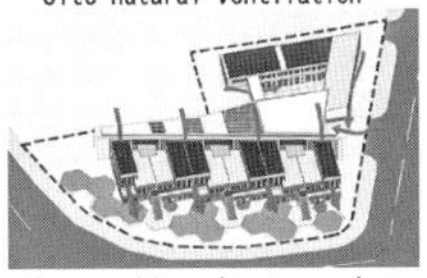

Use movable units to conduct

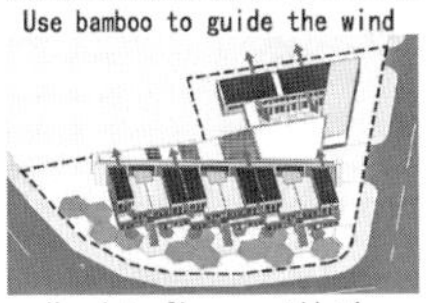

Use interfloor ventilation

DAYLIGHTING & SHADING

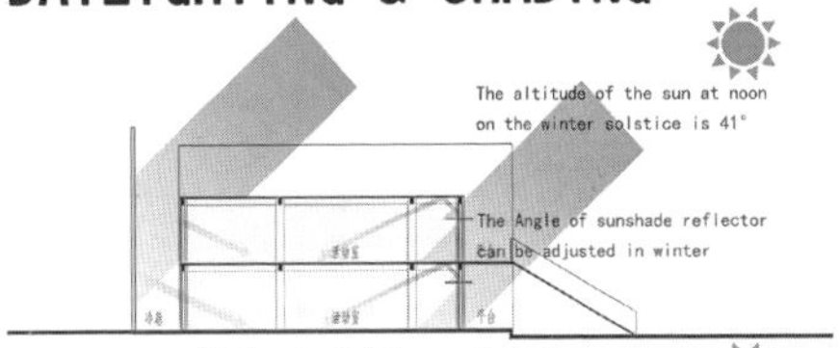

The altitude of the sun at noon on the winter solstice is 41°
The Angle of sunshade reflector can be adjusted in winter

Winter daylighting and shading

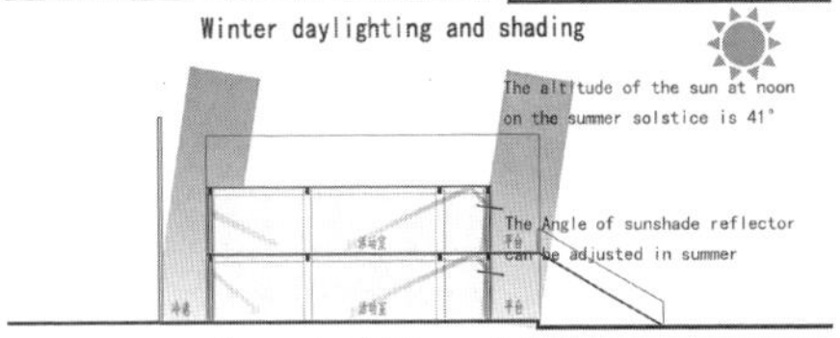

The altitude of the sun at noon on the summer solstice is 41°
The Angle of sunshade reflector can be adjusted in summer

Summer daylighting and shading

WIND ENVIRONMENT SIMULATION

Winter

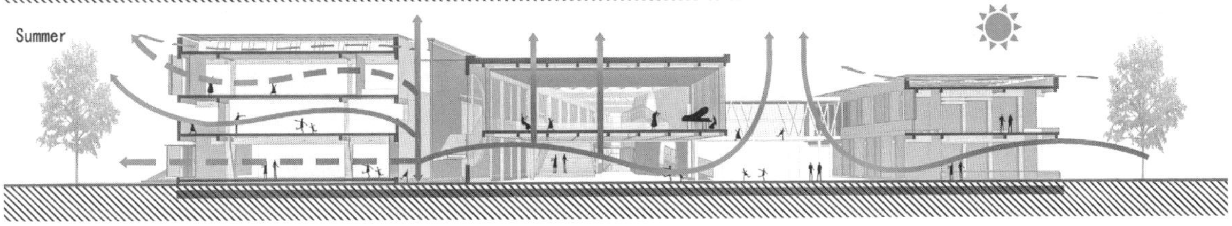

Summer

巷·间 | COLD LANE · INTERSPACE |

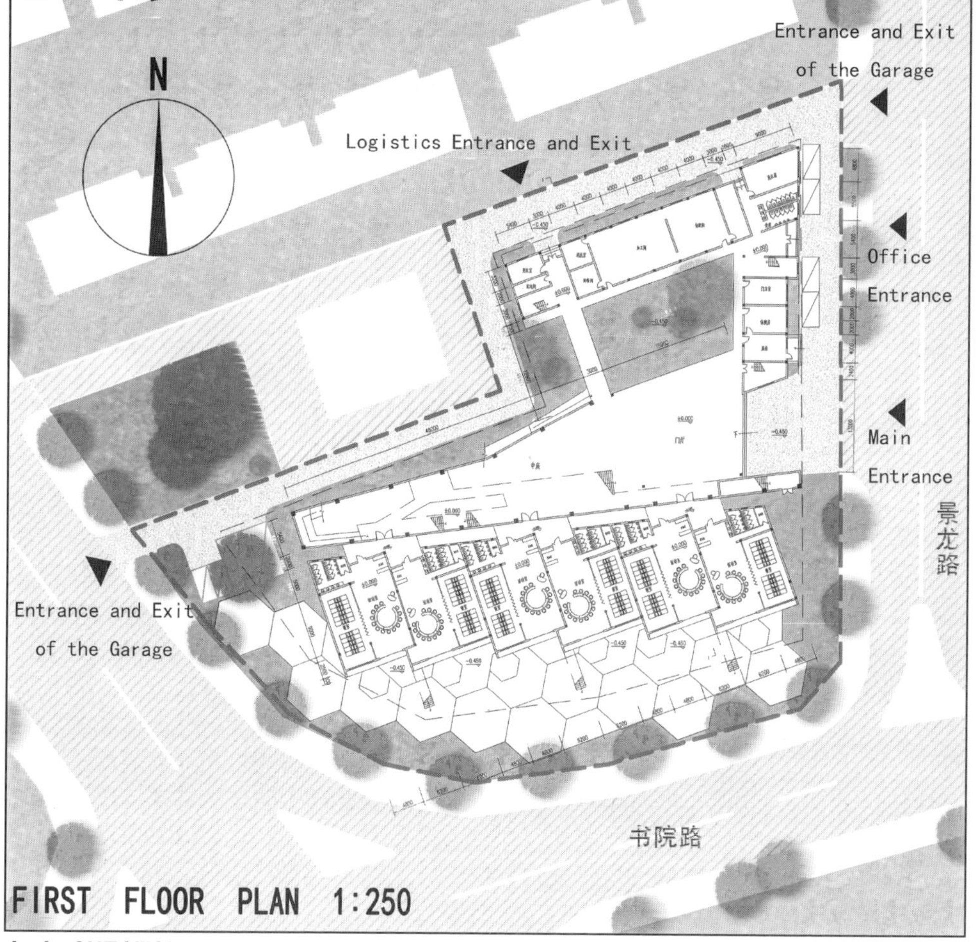
N
Entrance and Exit
of the Garage
Logistics Entrance and Exit
Office
Entrance
Main
Entrance
景龙路
Entrance and Exit
of the Garage
书院路
FIRST FLOOR PLAN 1:250

Morning check
Indoor teaching
Outdoor activities
Lunch time
Lunch break
Outdoor activities
Indoor teaching
Home time
At home

A-A CUTAWAY VIEW 1:250

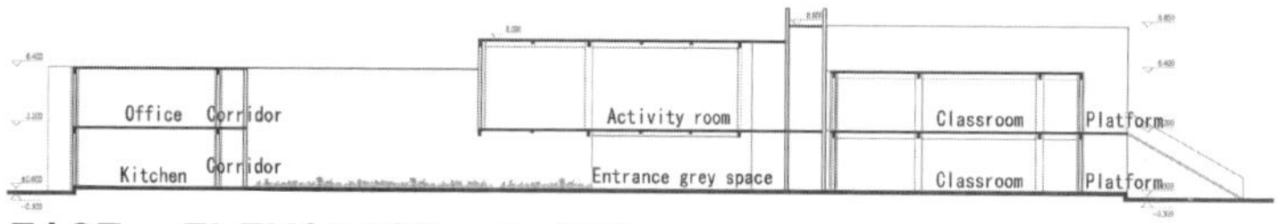
Office
Corridor
Kitchen
Corridor
Activity room
Entrance grey space
Classroom
Classroom
Platform
Platform

SPATIAL FUNCTION ORGANIZES

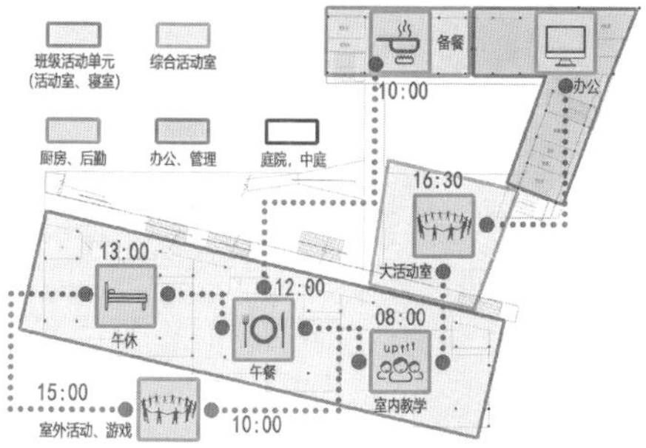
班级活动单元
(活动室、寝室)
综合活动室
厨房、后勤
办公、管理
庭院、中庭
备餐
10:00
办公
16:30
大活动室
13:00
午休
12:00
午餐
08:00
室内教学
15:00
室外活动、游戏
10:00

EAST ELEVATION 1:250

巷·间 | COLD LANE · INTERSPACE Ⅳ

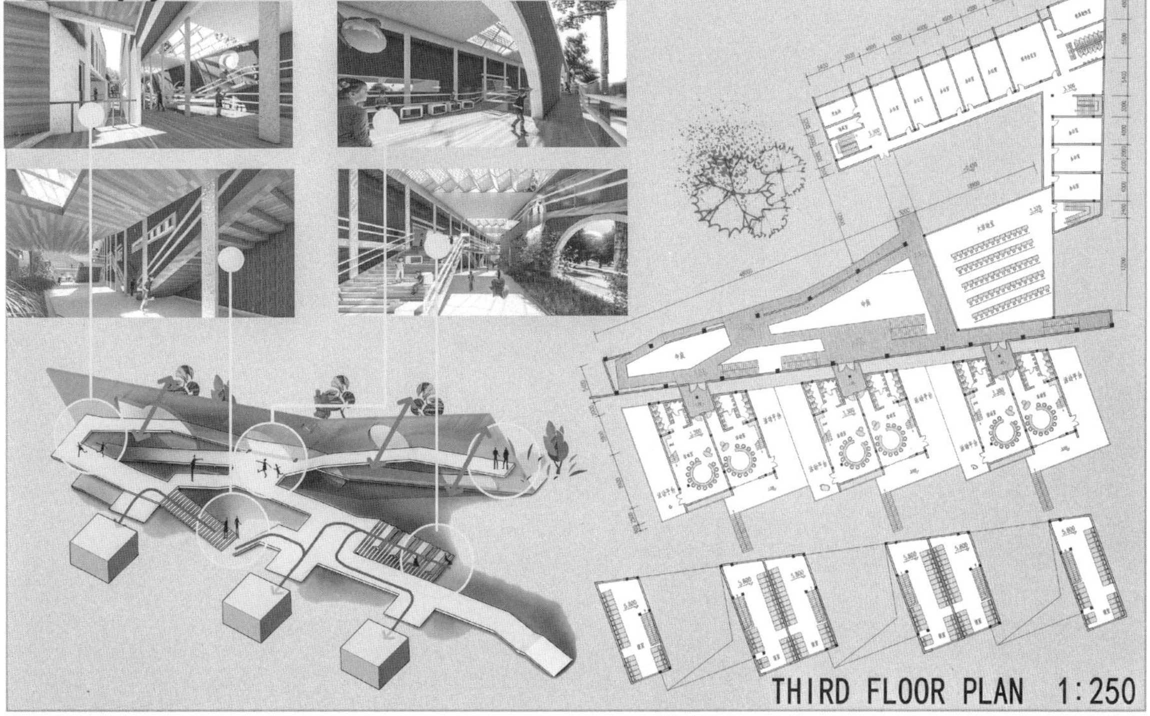
SECOND FLOOR PLAN 1:250
THIRD FLOOR PLAN 1:250

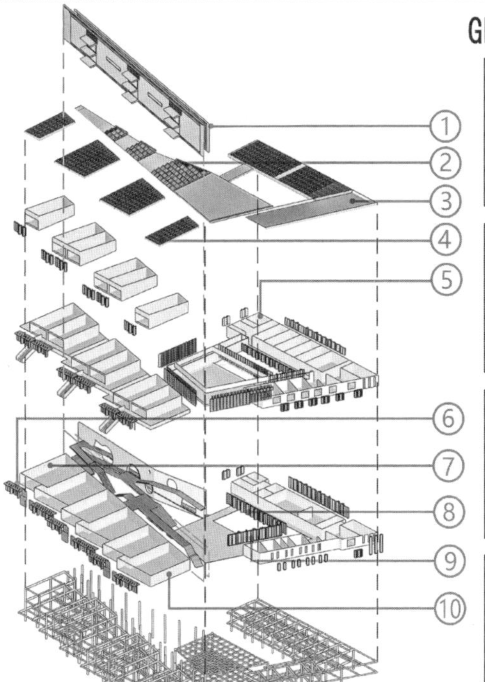
①
②
③
④
⑤
⑥
⑦
⑧
⑨
⑩

GREEN BUILDING TECHNOLOGY ORGANIZATION

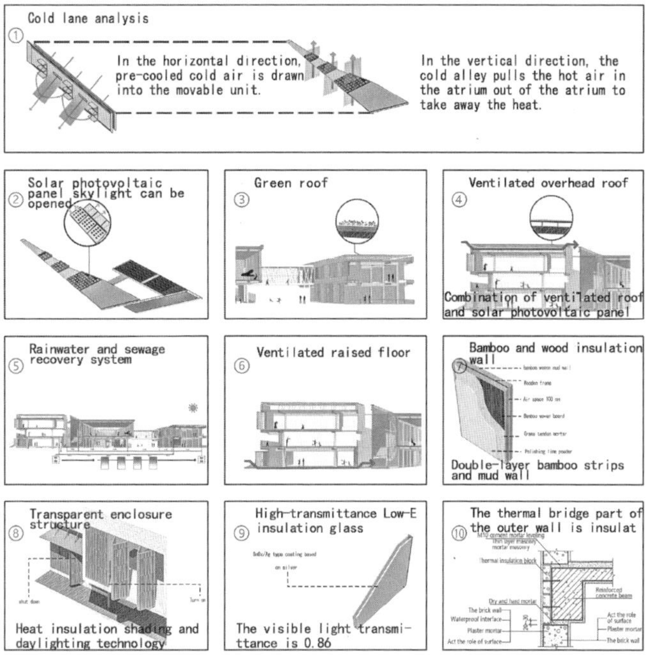
Cold lane analysis
①
In the horizontal direction, pre-cooled cold air is drawn into the movable unit.
In the vertical direction, the cold alley pulls the hot air in the atrium out of the atrium to take away the heat.
② Solar photovoltaic panel skylight can be opened
③ Green roof
④ Ventilated overhead roof
Combination of ventilated roof and solar photovoltaic panel
⑤ Rainwater and sewage recovery system
⑥ Ventilated raised floor
⑦ Bamboo and wood insulation wall
Double-layer bamboo strips and mud wall
⑧ Transparent enclosure structure
Heat insulation shading and daylighting technology
⑨ High-transmittance Low-E insulation glass
The visible light transmittance is 0.86
⑩ The thermal bridge part of the outer wall is insulat
Thermal insulation block
Reinforced concrete beam
The brick wall
Plaster mortar
Act the role of surface

NORTH ELEVATION 1:250

综合奖 · 优秀奖
General Prize Awarded · Honorable Mention Prize

注 册 号：6804

项目名称：瑶域 · 疆来（新疆）

Yao Yu · Xin Jiang Future (Xinjiang)

作　　者：韩欣琰、李文珠、杨　阳、谭　雯

参赛单位：大连大学

指导教师：黄世岩

瑶域 · 疆来 1 Yao Yu · Xin Jiang Future

Climate

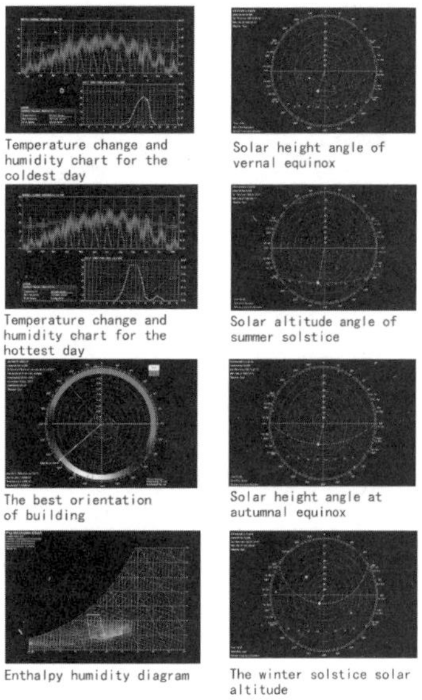
Temperature change and humidity chart for the coldest day
Solar height angle of vernal equinox
Temperature change and humidity chart for the hottest day
Solar altitude angle of summer solstice
The best orientation of building
Solar height angle at autumnal equinox
Enthalpy humidity diagram
The winter solstice solar altitude

Analysis of Historical Context

Sustainable and stable development towards a better future

The development of reclamation promoted the economic recovery and cultural development of Xinjiang

Establishment of production and Construction Corps of Xinjiang Military Region of Chinese people's Liberation Army

Location Analysis Diagram

42 ° 58 ′ N and 86 ° 33 ′ E

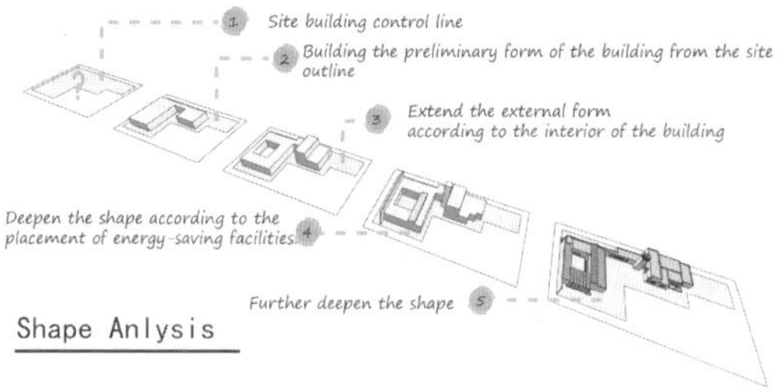
1 Site building control line
2 Building the preliminary form of the building from the site outline
3 Extend the external form according to the interior of the building
Deepen the shape according to the placement of energy-saving facilities 4
Further deepen the shape 5

Shape Anlysis

Regional Feature

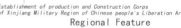

Tall towers
Small humanly scaled
Retaining wall and the slope roof
The stone wall
A thick earthen wall
The local situation

设计说明（Design Notes）

建筑风格源于当地民居，建筑形体围绕建设用地生成。在被动式技术上我们使用了烟囱效应，嵌入式阳光房，热压通风，双层屋面降温和可调节百叶窗及空调效应等技术，并应用光伏电池板，太阳能集热装置，导光管，生态污水处理装置，实现了被动式与主动式相结合的采暖、隔热和通风，力图给儿童及公职人员提供一个舒适健康的环境，在改善当地居民生活的同时，为国际节能事业贡献一份微薄之力。

The architectural style originates from the local dwellings, and the architectural form is generated around the construction land. We use the stack effect on passive technology, embedded sun room, thermal pressure ventilation, and adjustable shutter double-layer roof cooling and air conditioning effect such as technology, and application of photovoltaic panels and solar heating device, light pipe, ecological sewage treatment plant, the combination of passive and active heating, insulation and ventilation, to give children and office staff to provide a comfortable and healthy environment, in improving the life of local residents to make our contribution to the international energy saving at the same time.

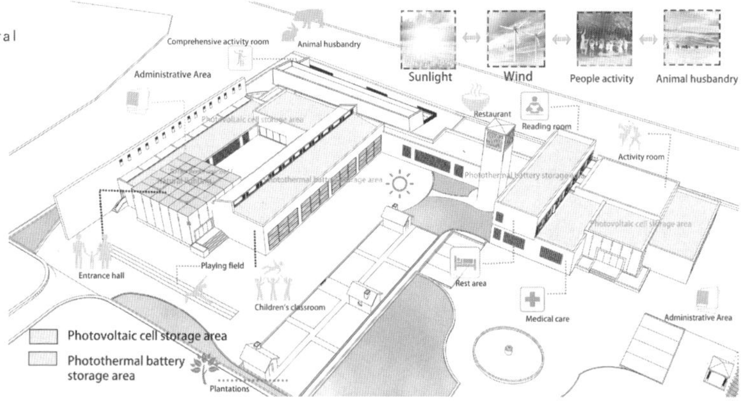
Sunlight
Wind
People activity
Animal husbandry
Comprehensive activity room
Animal husbandry
Administrative Area
Restaurant
Reading room
Activity room
Entrance hall
Playing field
Children's classroom
Rest area
Medical care
Administrative Area
Plantations
Photovoltaic cell storage area
Photothermal battery storage area

Model Photos

Overview of Solar Technology

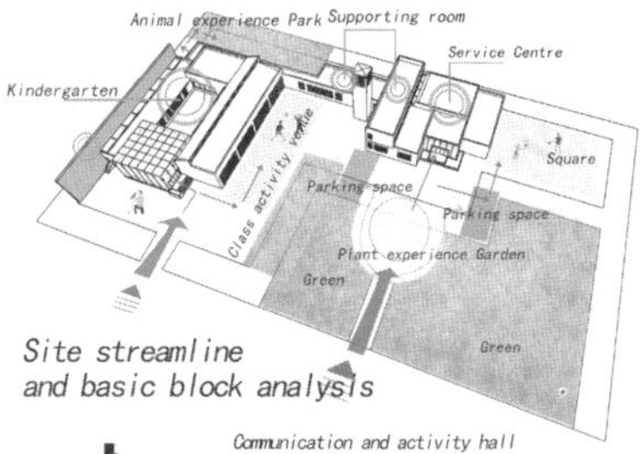
Animal experience Park
Supporting room
Service Centre
Kindergarten
Square
Parking space
Class activity venue
Plant experience Garden
Green
Site streamline
and basic block analysis

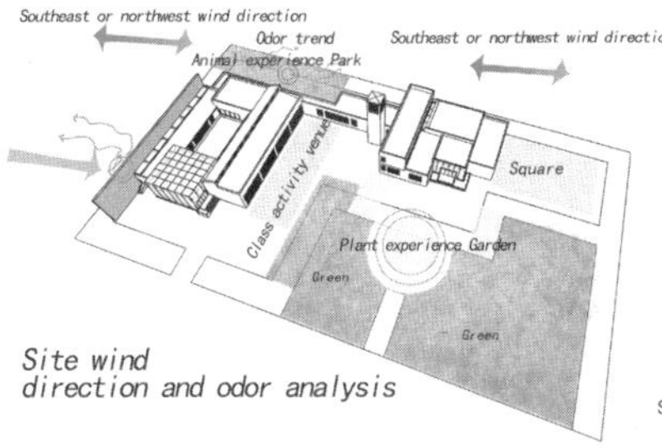
Southeast or northwest wind direction
Odor trend
Animal experience Park
Square
Class activity venue
Plant experience Garden
Green
Site wind
direction and odor analysis

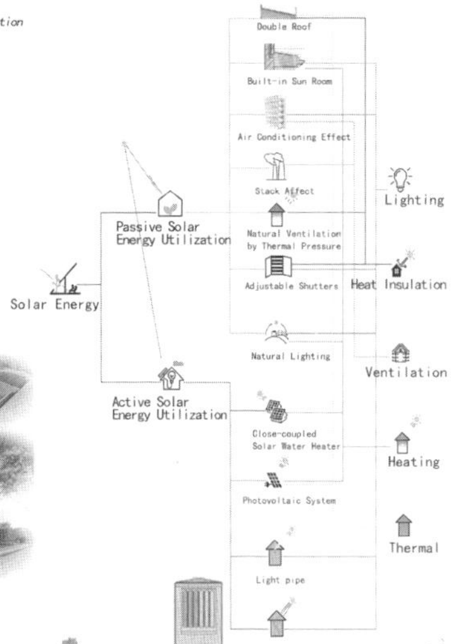
Double Roof
Built-in Sun Room
Air Conditioning Effect
Stack Affect
Natural Ventilation by Thermal Pressure
Adjustable Shutters
Natural Lighting
Close-coupled Solar Water Heater
Photovoltaic System
Light pipe
Passive Solar Energy Utilization
Active Solar Energy Utilization
Solar Energy
Lighting
Heat Insulation
Ventilation
Heating
Thermal

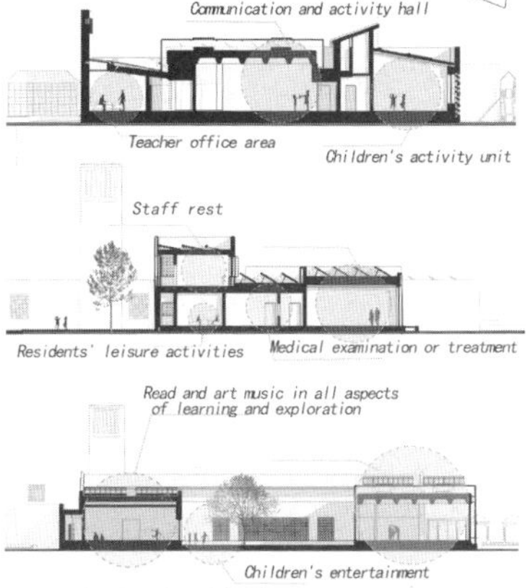
Communication and activity hall
Teacher office area
Children's activity unit
Staff rest
Residents' leisure activities
Medical examination or treatment
Read and art music in all aspects
of learning and exploration
Children's entertainment

Analysis Of Character scenes

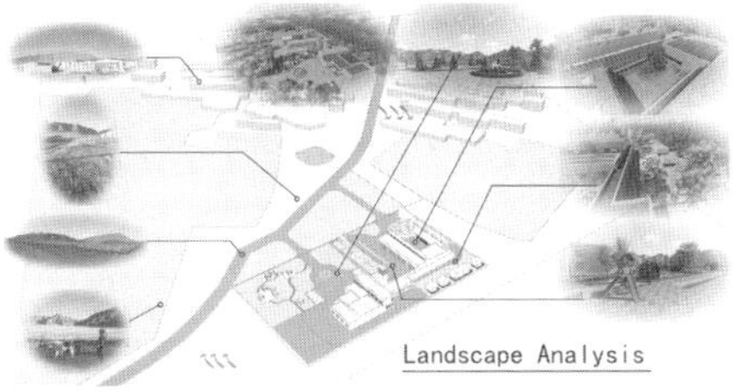
Landscape Analysis

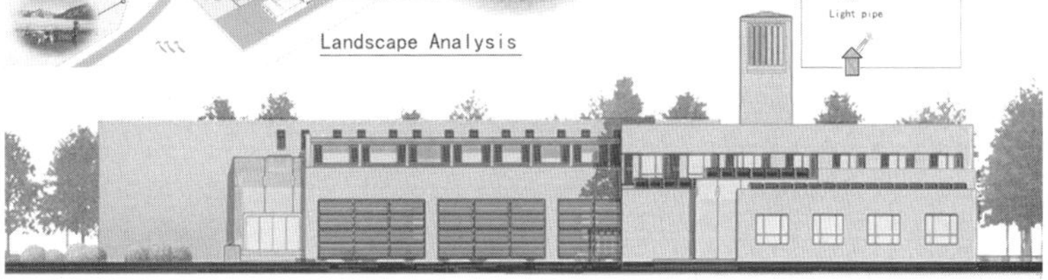

Southeast Evelation 1：200

瑶域·疆来 2

Yao Yu · Xin Jiang Future

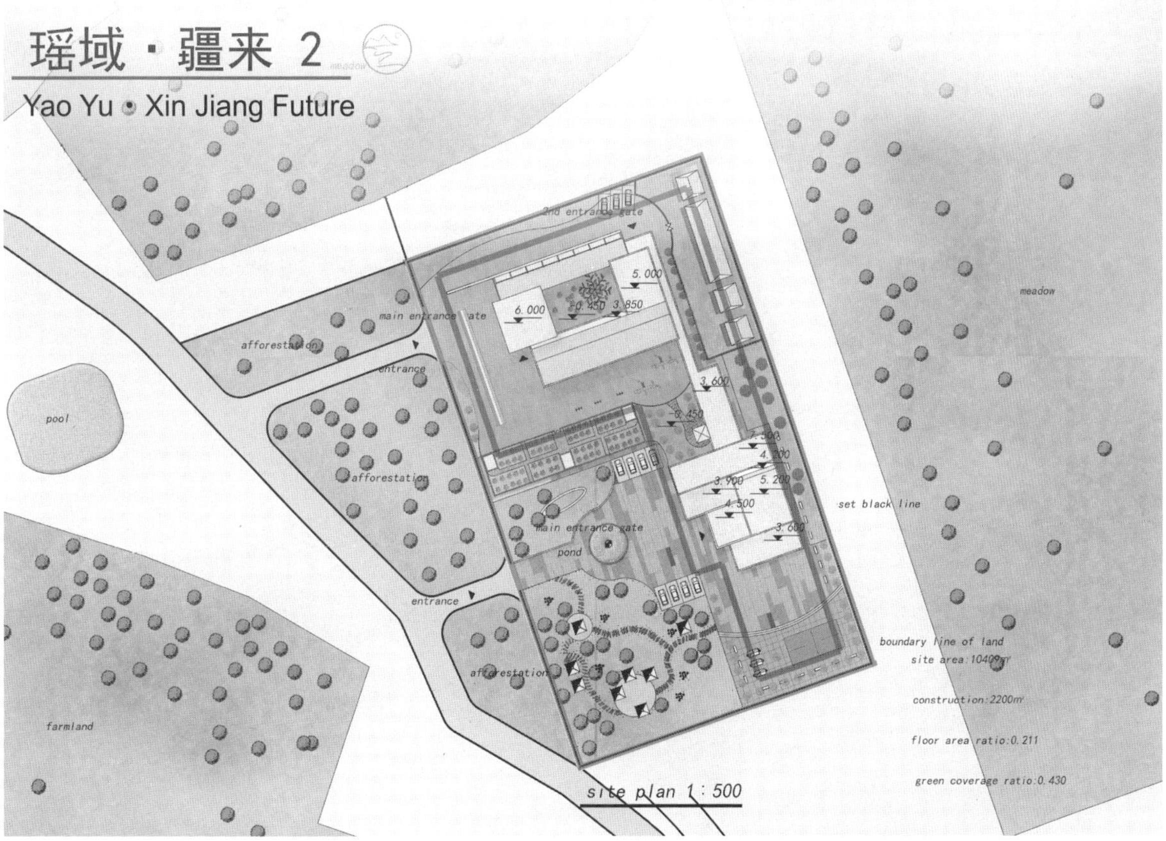
meadow
pool
farmland
afforestation
entrance
main entrance gate
2nd entrance gate
pond
6.000
5.000
3.850
0.450
3.600
-0.450
3.900
4.500
5.200
3.600
set black line
boundary line of land
site area:10409㎡
construction:2200㎡
floor area ratio:0.211
green coverage ratio:0.430

site plan 1：500

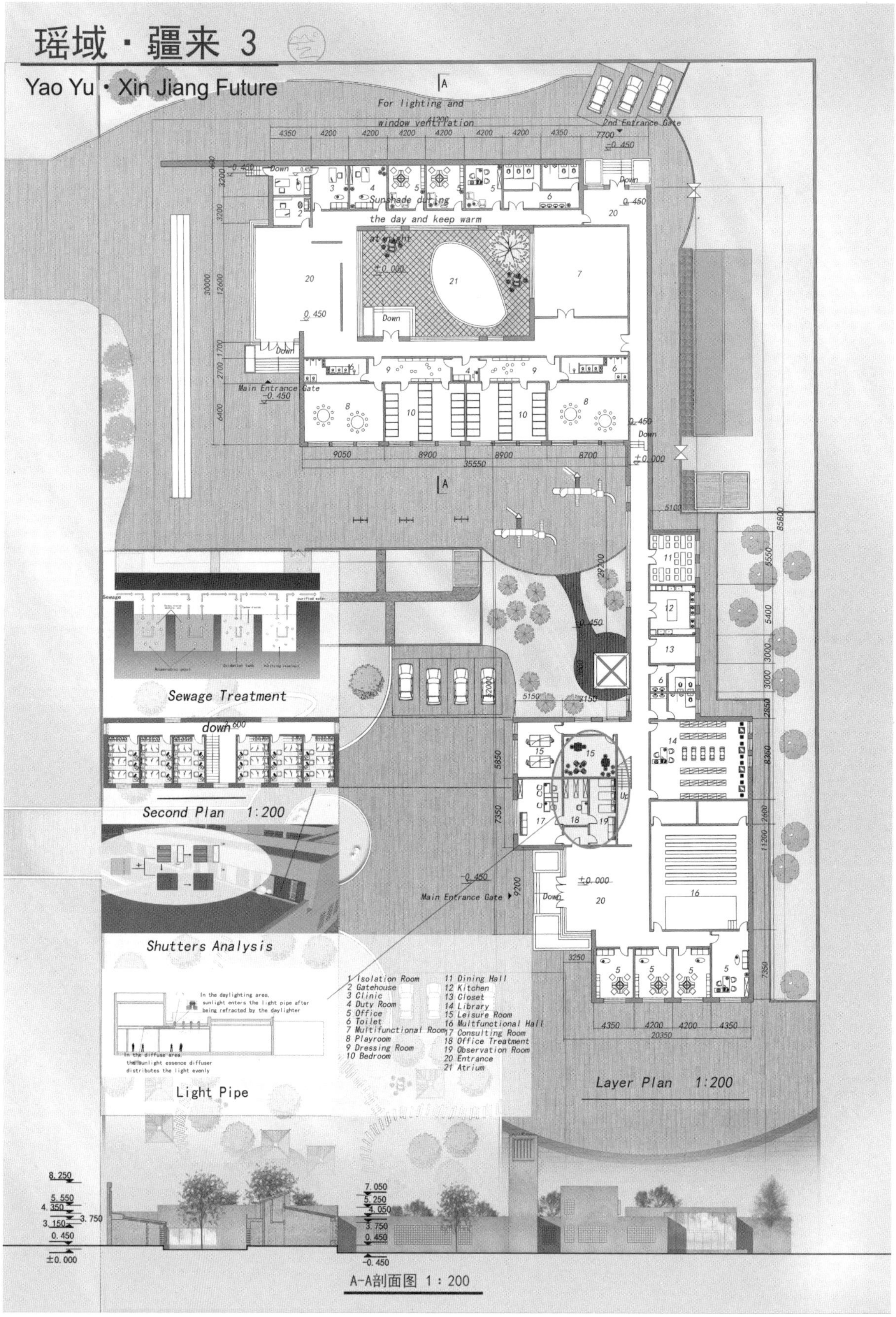
瑶域·疆来 3
Yao Yu · Xin Jiang Future
For lighting and window ventilation
Sunshade during the day and keep warm at night
2nd Entrance Gate
Main Entrance Gate
Sewage Treatment
Second Plan 1:200
Shutters Analysis
Light Pipe
In the daylighting area, sunlight enters the light pipe after being refracted by the daylighter
In the diffuse area, the sunlight essence diffuser distributes the light evenly
1 Isolation Room
2 Gatehouse
3 Clinic
4 Duty Room
5 Office
6 Toilet
7 Multifunctional Room
8 Playroom
9 Dressing Room
10 Bedroom
11 Dining Hall
12 Kitchen
13 Closet
14 Library
15 Leisure Room
16 Multifunctional Hall
17 Consulting Room
18 Office Treatment
19 Observation Room
20 Entrance
21 Atrium
Layer Plan 1:200
A-A剖面图 1：200

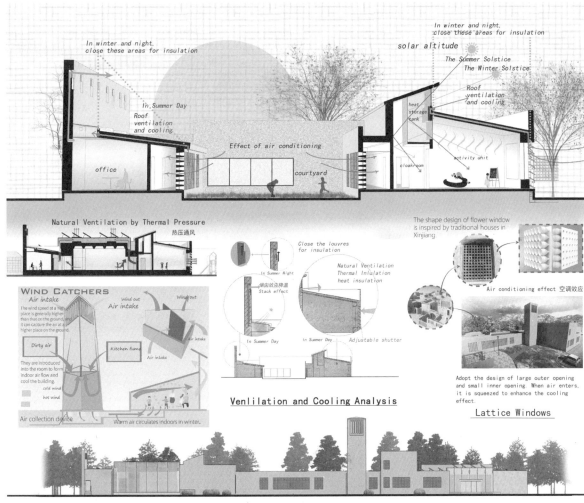

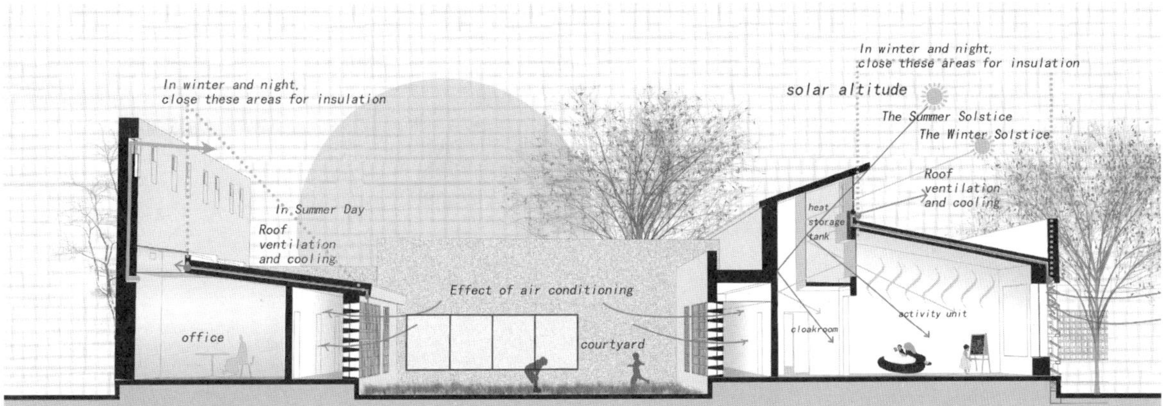
In winter and night,
close these areas for insulation
In Summer Day
Roof
ventilation
and cooling
office
Effect of air conditioning
courtyard
In winter and night,
close these areas for insulation
solar altitude
The Summer Solstice
The Winter Solstice
Roof
ventilation
and cooling
heat
storage
tank
cloakroom
activity unit

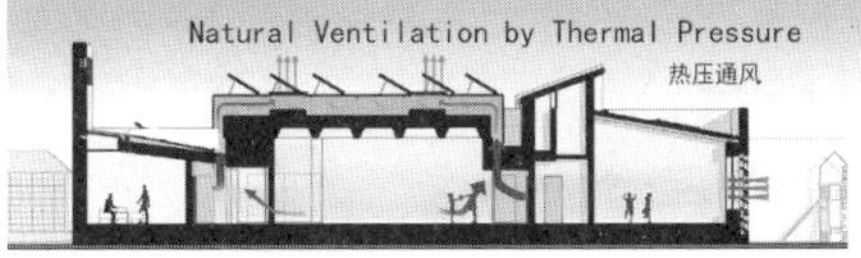
Natural Ventilation by Thermal Pressure
热压通风

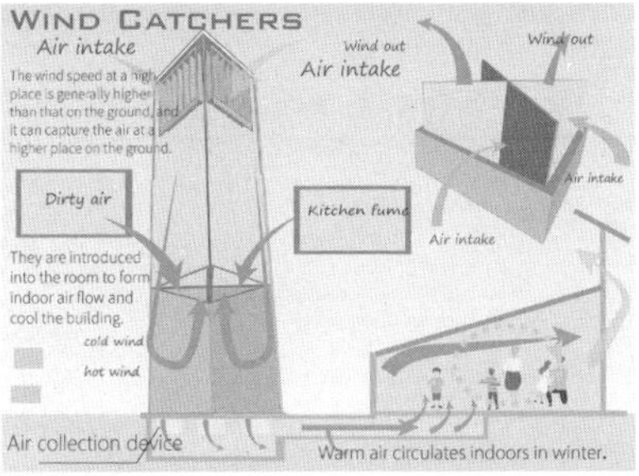
WIND CATCHERS
Air intake
Wind out
Air intake
Wind out
Dirty air
Kitchen fume
Air intake
Air intake
They are introduced into the room to form indoor air flow and cool the building.
cold wind
hot wind
Air collection device
Warm air circulates indoors in winter.

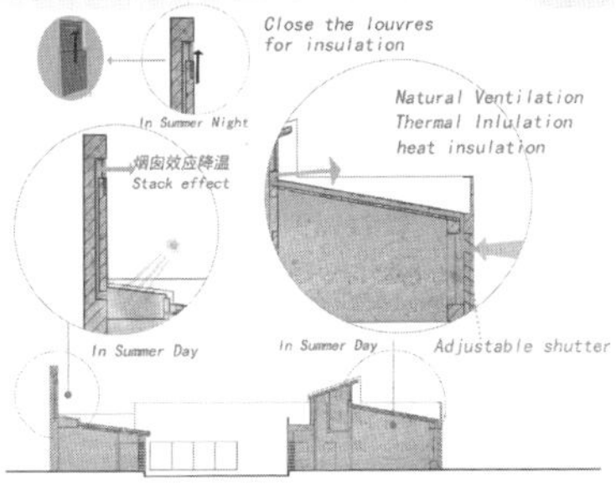
Close the louvres
for insulation
In Summer Night
烟囱效应降温
Stack effect
Natural Ventilation
Thermal Inlulation
heat insulation
In Summer Day
In Summer Day
Adjustable shutter

Venlilation and Cooling Analysis

The shape design of flower window is inspired by traditional houses in Xinjiang.

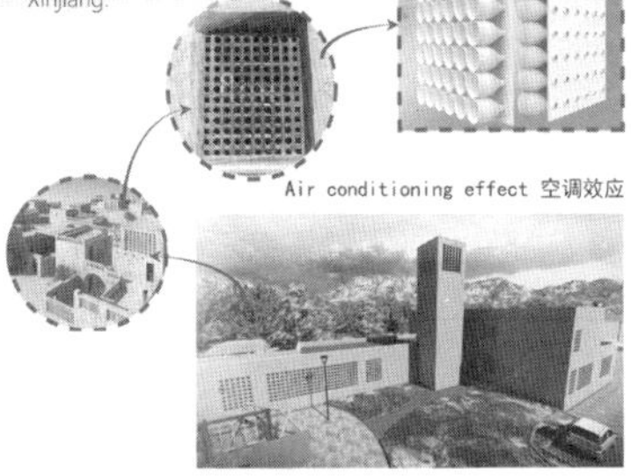
Air conditioning effect 空调效应

Adopt the design of large outer opening and small inner opening. When air enters, it is squeezed to enhance the cooling effect.

Lattice Windows

Southwest Evelation 1：200

瑶域·疆来 4
Yao Yu · Xin Jiang Future

综合奖·优秀奖
General Prize Awarded · Honorable Mention Prize

注 册 号：6817

项目名称：晨兴日起巴伦台（新疆）
Bring the Sunrise to BLUNTAI
(Xinjiang)

作　　者：郭嘉钰、张玉琪、王松瑞

参赛单位：天津大学、哈尔滨工业大学、上海同济规划院设计研究院

指导教师：刘家韦华

BRING THE SUNRISE TO BLUNTAI

晨兴日起巴伦台——幼儿园·牧场服务中心联合体设计 01

CHAPTER1: BACKGROND

LOCATION ANALYSIS

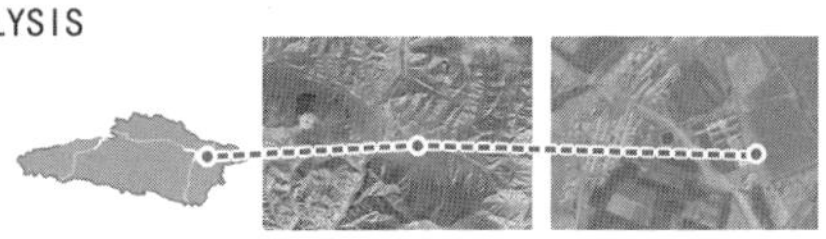

CLIMATE ANALYSIS

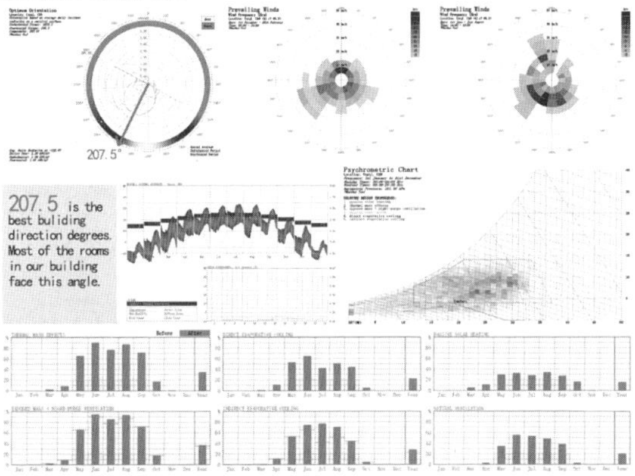
207.5 is the best builiding direction degrees. Most of the rooms in our building face this angle.
Psychrometric Chart

DESIGN DESCRIPTION 设计说明

“晨兴日起巴伦台”为幼儿园与牧场服务中心的综合体，提供幼托、办公、活动、医疗等服务。本方案中，幼儿园同牧场服务中心在功能与流线上完全分隔，保障幼儿园活动场地的安全性。建筑综合考虑了低能耗、环境友好、人体舒适等技术要求，尽可能地利用了可再生能源，并且使用多项主被动技术提高室内环境舒适性。

BRING THE SUNRISE TO BLUNTAI combined Kindergarten and Ranch Service Center, provide child care, office, activities, medical services.The kindergarten and the ranch service center are completely separated in function and streamline, and ensured the safety of kindergarten playground.This building use low energy consumption technology, environment-friendly technology and human comfort technology. Using renewable energy as much as possible, and a number of active and passive technologies has been used to improve indoor's environmental comfort.

LOCAL CHARACTERISTIC

CHALLENGE 1 Accommodate 40 Kids + 100 Adult

40 + 9 + 10 + 90 + ···

CHALLENGE 2 How to Let Everyone Not Interfered by Each Other

CRITERION: ENERGY RECYCLING

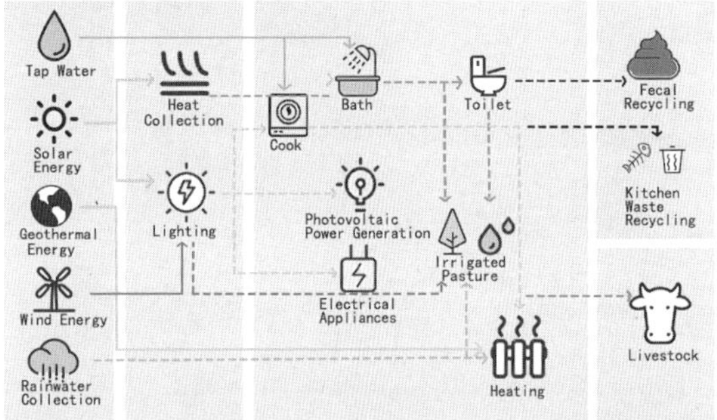
Tap Water
Solar Energy
Geothermal Energy
Wind Energy
Rainwater Collection
Heat Collection
Cook
Bath
Toilet
Fecal Recycling
Kitchen Waste Recycling
Lighting
Photovoltaic Power Generation
Irrigated Pasture
Electrical Appliances
Heating
Livestock

TRY TO SOLVE ↑ PROBLEM : FORMATION ANALYSIS

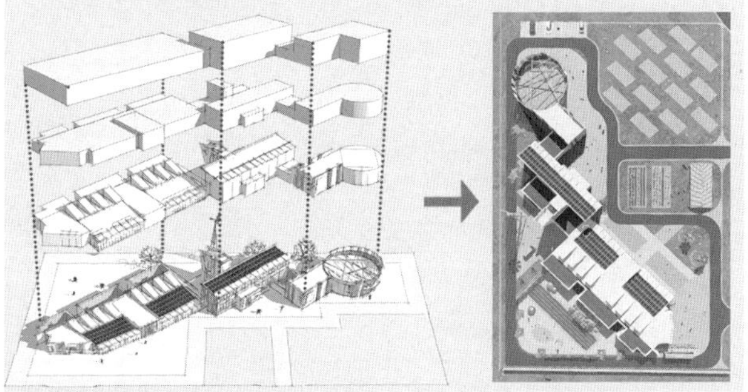

SHOW OUR ARCHITECTURE

BRING THE SUNRISE TO BLUNTAI

晨兴日起巴伦台 ——幼儿园·牧场服务中心联合体设计

02

CHAPTER2:DESIGN AND ANALYSIS

GREEN TECHNOLOGY OVERVIEW

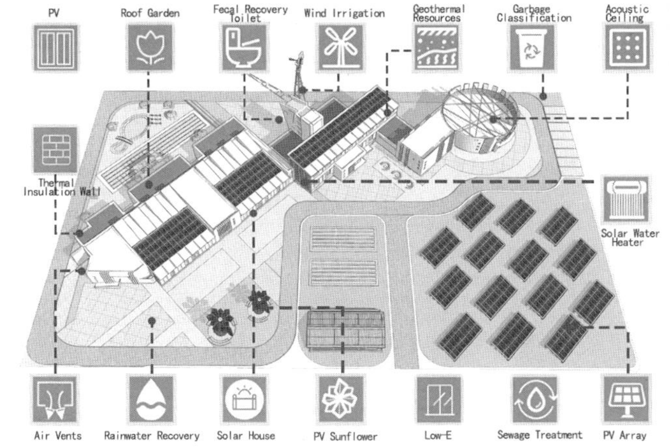
PV
Roof Garden
Fecal Recovery Toilet
Wind Irrigation
Geothermal Resources
Garbage Classification
Acoustic Ceiling
Thermal Insulation Wall
Solar Water Heater
Air Vents
Rainwater Recovery
Solar House
PV Sunflower
Low-E
Sewage Treatment
PV Array

DECOMPOSITION OF ENERGY-SAVING TECHNOLOGY

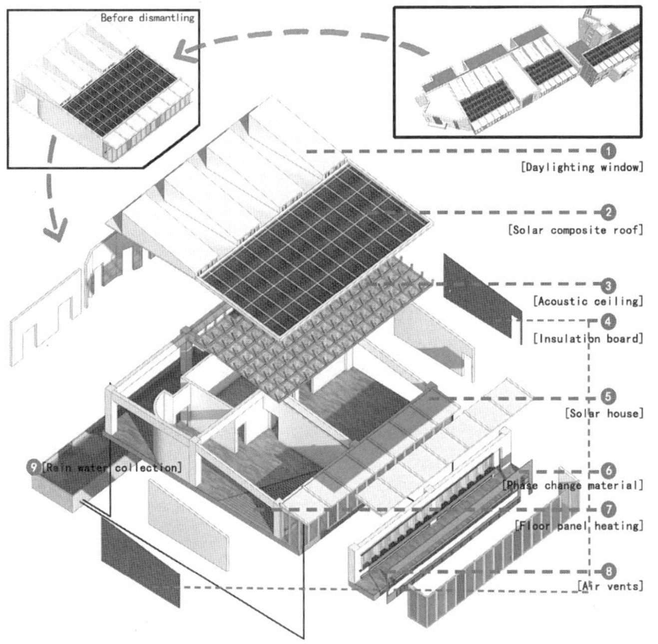
Before dismantling
1 [Daylighting window]
2 [Solar composite roof]
3 [Acoustic ceiling]
4 [Insulation board]
5 [Solar house]
6 [Phase change material]
7 [Floor panel heating]
8 [Air vents]
9 [Rain water collection]

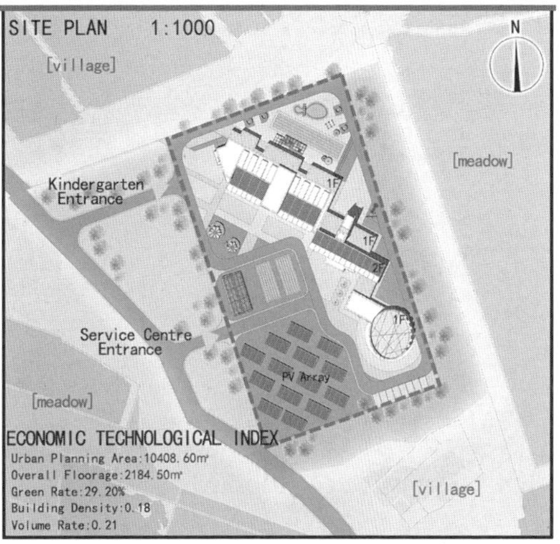
SITE PLAN 1:1000
[village]
[meadow]
Kindergarten Entrance
Service Centre Entrance
PV Array
[meadow]
[village]
ECONOMIC TECHNOLOGICAL INDEX
Urban Planning Area:10408.60㎡
Overall Floorage:2184.50㎡
Green Rate:29.20%
Building Density:0.18
Volume Rate:0.21

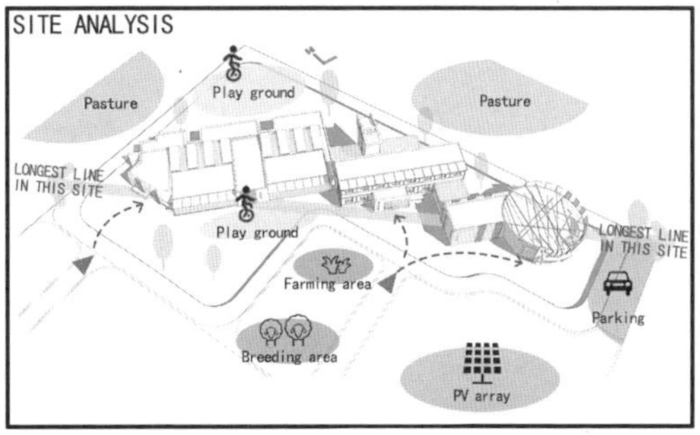
SITE ANALYSIS
Pasture
Play ground
Pasture
LONGEST LINE IN THIS SITE
Play ground
Farming area
LONGEST LINE IN THIS SITE
Parking
Breeding area
PV array

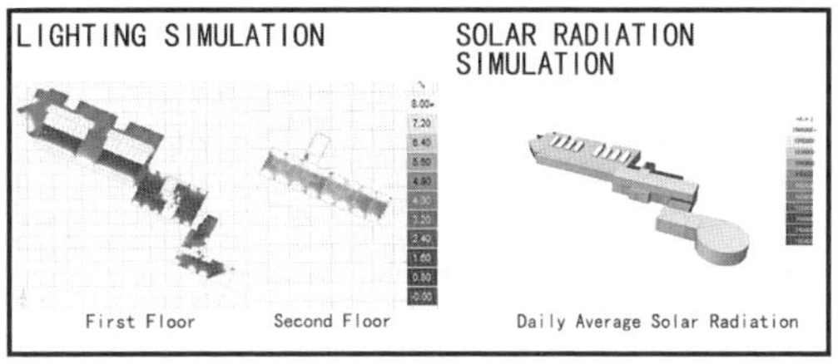
LIGHTING SIMULATION
SOLAR RADIATION SIMULATION
First Floor
Second Floor
Daily Average Solar Radiation

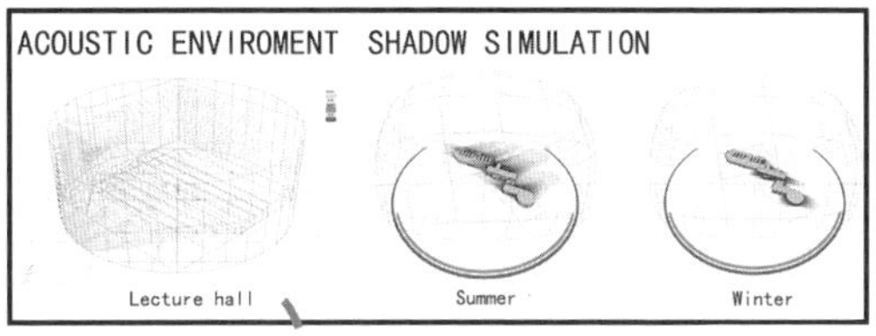
ACOUSTIC ENVIROMENT
SHADOW SIMULATION
Lecture hall
Summer
Winter

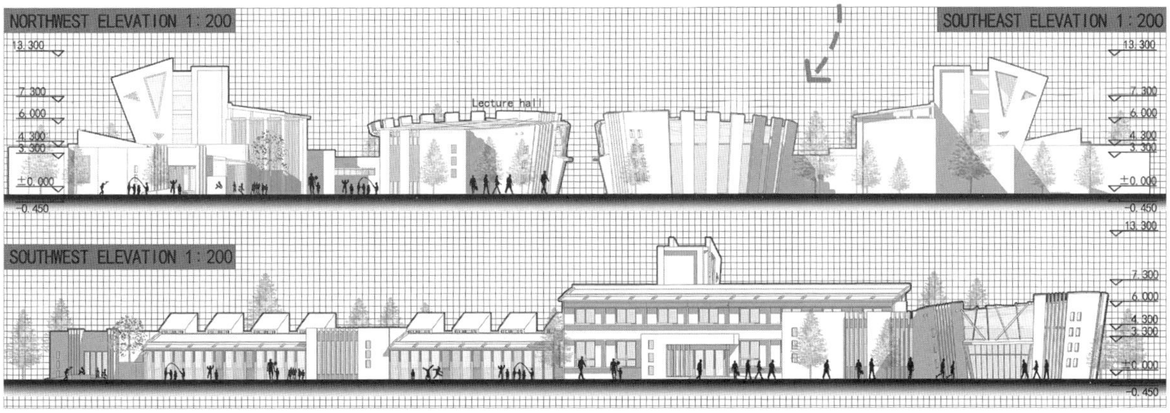
NORTHWEST ELEVATION 1:200
SOUTHEAST ELEVATION 1:200
13.300
7.300
6.000
4.300
3.300
±0.000
-0.450
Lecture hall
SOUTHWEST ELEVATION 1:200

BRING THE SUNRISE TO BLUNTAI

晨兴日起巴伦台 ——幼儿园·牧场服务中心联合体设计

03

CHAPTER3: USERS' ACTIVITY AND PLAN

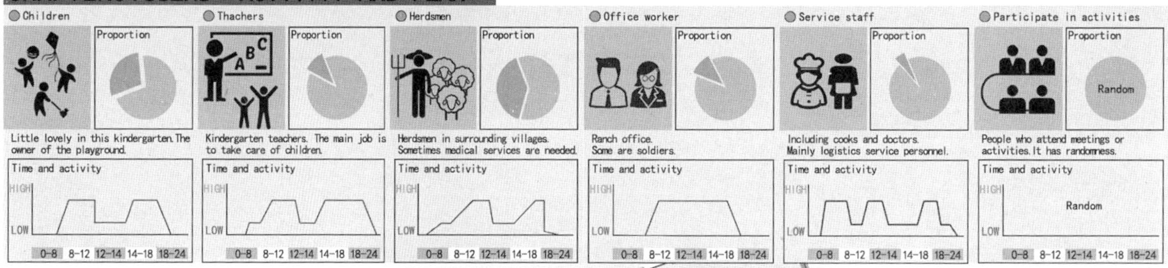
Children
Proportion
Little lovely in this kindergarten. The owner of the playground.
Time and activity
HIGH
LOW
0-8 8-12 12-14 14-18 18-24
Thachers
Proportion
Kindergarten teachers. The main job is to take care of children.
Time and activity
HIGH
LOW
0-8 8-12 12-14 14-18 18-24
Herdsmen
Proportion
Herdsmen in surrounding villages. Sometimes medical services are needed.
Time and activity
HIGH
LOW
0-8 8-12 12-14 14-18 18-24
Office worker
Proportion
Ranch office. Some are soldiers.
Time and activity
HIGH
LOW
0-8 8-12 12-14 14-18 18-24
Service staff
Proportion
Including cooks and doctors. Mainly logistics service personnel.
Time and activity
HIGH
LOW
0-8 8-12 12-14 14-18 18-24
Participate in activities
Proportion
Random
People who attend meetings or activities. It has randomness.
Time and activity
HIGH
Random
LOW
0-8 8-12 12-14 14-18 18-24

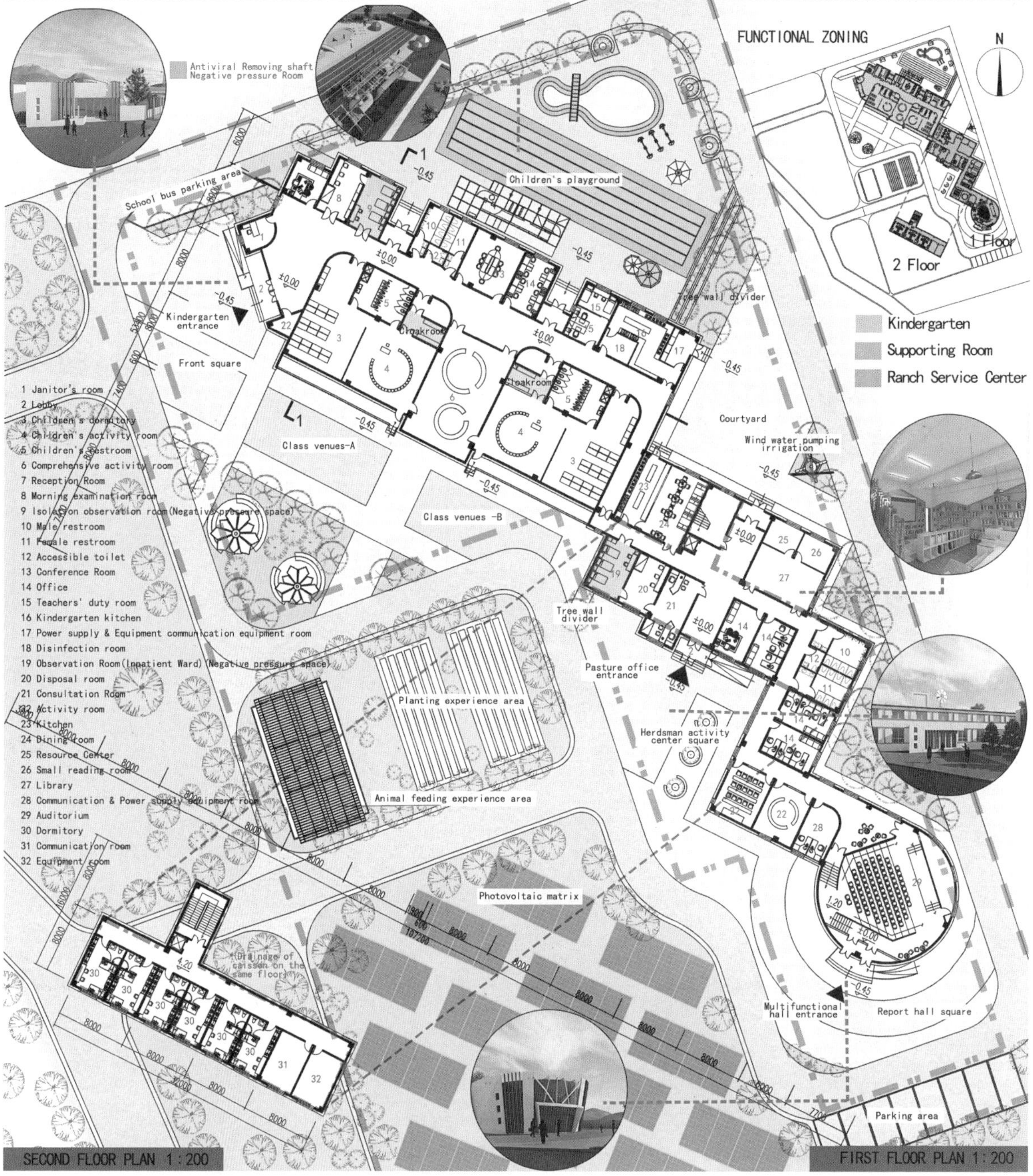
Antiviral Removing shaft
Negative pressure Room
FUNCTIONAL ZONING
N
1 Floor
2 Floor
Kindergarten
Supporting Room
Ranch Service Center
Children's playground
School bus parking area
Kindergarten entrance
Front square
Cloakroom
Tree wall divider
Courtyard
Wind water pumping irrigation
Class venues-A
Class venues -B
Tree wall divider
Pasture office entrance
Herdsman activity center square
Planting experience area
Animal feeding experience area
Photovoltaic matrix
Drainage of caisson on the same floor
Multifunctional hall entrance
Report hall square
Parking area
1 Janitor's room
2 Lobby
3 Children's dormitory
4 Children's activity room
5 Children's restroom
6 Comprehensive activity room
7 Reception Room
8 Morning examination room
9 Isolation observation room(Negative pressure space)
10 Male restroom
11 Female restroom
12 Accessible toilet
13 Conference Room
14 Office
15 Teachers' duty room
16 Kindergarten kitchen
17 Power supply & Equipment communication equipment room
18 Disinfection room
19 Observation Room(Inpatient Ward)(Negative pressure space)
20 Disposal room
21 Consultation Room
22 Activity room
23 Kitchen
24 Dining room
25 Resource Center
26 Small reading room
27 Library
28 Communication & Power supply equipment room
29 Auditorium
30 Dormitory
31 Communication room
32 Equipment room
SECOND FLOOR PLAN 1:200
FIRST FLOOR PLAN 1:200

BRING THE SUNRISE TO BLUNTAI

晨兴日起巴伦台——幼儿园·牧场服务中心联合体设计

04

CHAPTER4: GREEN BUILDING TECHNOLOGY

PV PANEL ENERGY EFFICIENCY SIMULATION

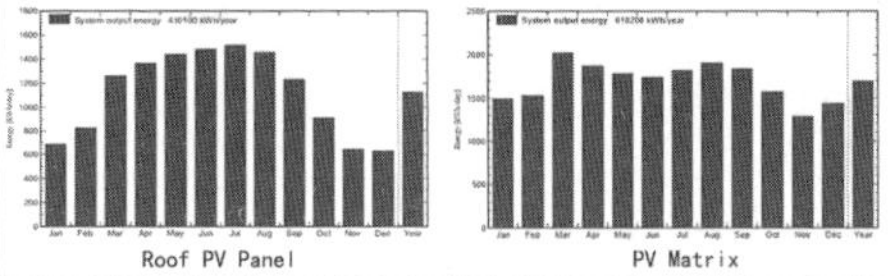
Roof PV Panel
PV Matrix

DRY SEALED TOILET

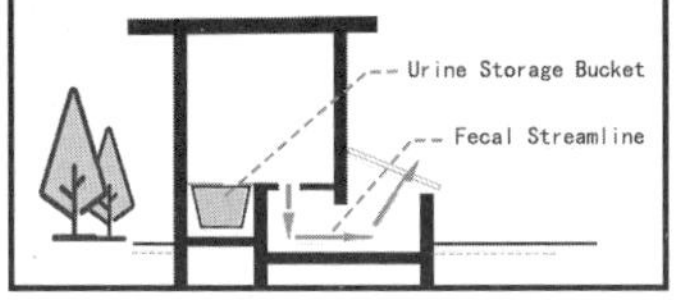
Urine Storage Bucket
Fecal Streamline

SOLAR HOUSE

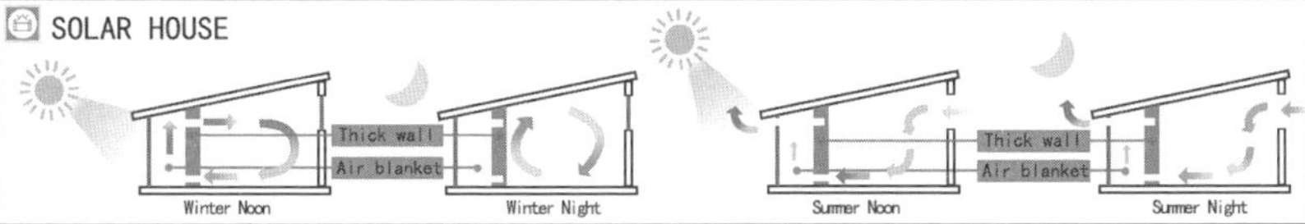
Thick wall
Air blanket
Winter Noon
Winter Night
Summer Noon
Summer Night

WIND ENVIRONMENT SIMULATION

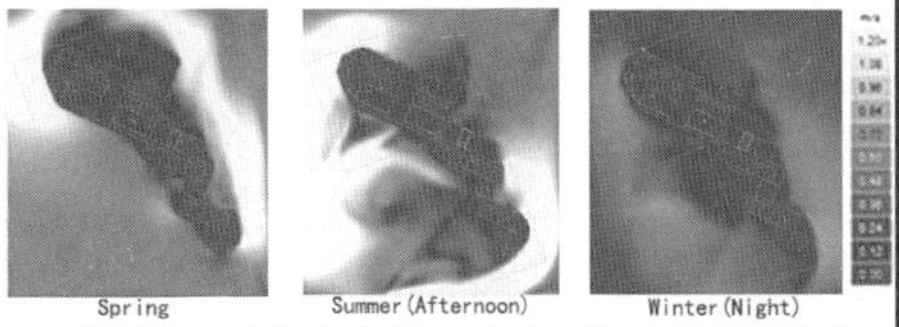
Spring
Summer (Afternoon)
Winter (Night)

Spring have the highest wind speed of the hole year. Trees wall can protect the playground against wind's attack. Building forms effectively blocks the cold wind in winter. In summer's afternoon, appropriate wind speed helping ventilate and cooling.

AIR VENTS

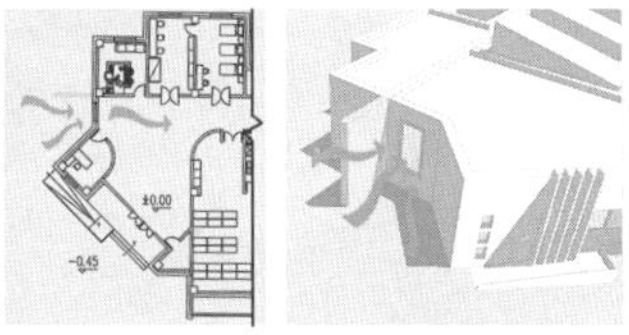

The wall is used to catch the wind around. It can effectively improve ventilation efficiency.

PHOTOVOLTAIC DIRECT DRIVE AIR CINDITIONER

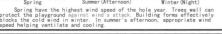

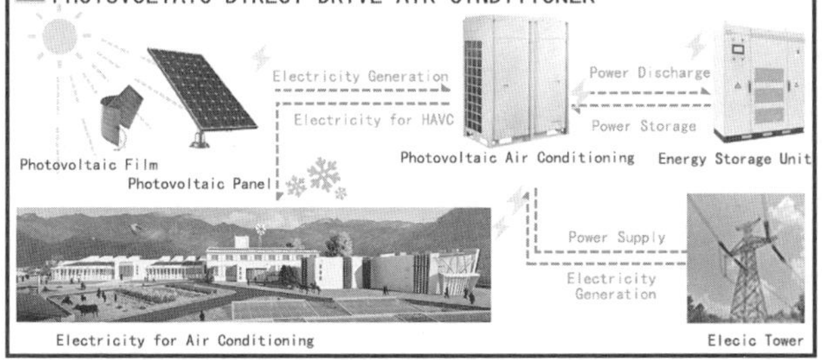
Electricity Generation
Electricity for HAVC
Power Discharge
Power Storage
Photovoltaic Film
Photovoltaic Panel
Photovoltaic Air Conditioning
Energy Storage Unit
Power Supply
Electricity Generation
Electricity for Air Conditioning
Elecic Tower

PV SUNFLOWER

ALL SEASONS

Spring(raining)
Summer
Fall
Winter(snowy day)

SECTIONAL VIEW 1-1

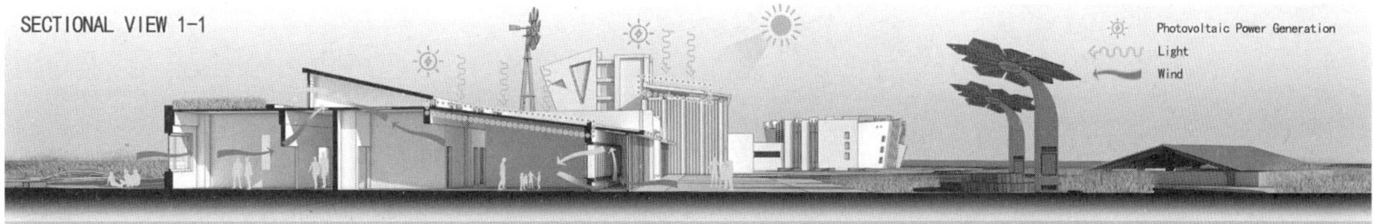
Photovoltaic Power Generation
Light
Wind

SHOW OUR ARCHITECTURE

综合奖 · 优秀奖

General Prize Awarded · Honorable Mention Prize

注 册 号：6986

项目名称：风 · 院（新疆）

Wind · Yard（Xinjiang）

作　　者：周之恒、潘嘉豪、贺子琦、沈　聪、叶慧瑶、沈立一、曹鑫媛

参赛单位：浙江理工大学

指导教师：文　强

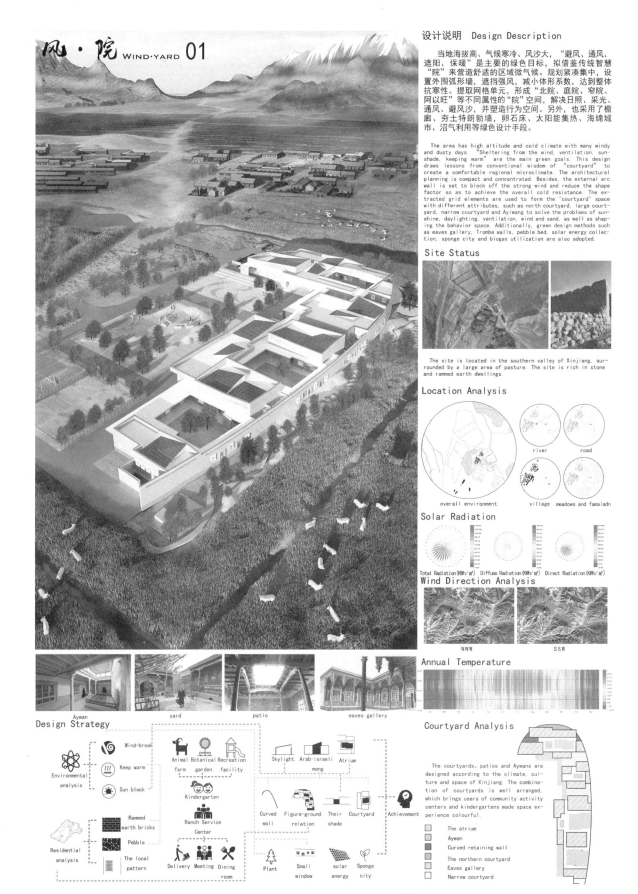

风 · 院 WIND·YARD 01

设计说明 Design Description

当地海拔高、气候寒冷、风沙大，“避风、通风、遮阳、保暖”是主要的绿色目标，拟借鉴传统智慧“院”来营造舒适的区域微气候。规划紧凑集中，设置外围弧形墙，遮挡强风，减小体形系数，达到整体抗寒性。提取网格单元，形成“北院、庭院、窄院、阿以旺”等不同属性的“院”空间，解决日照、采光、通风、避风沙，并塑造行为空间。另外，也采用了檐廊、夯土特朗勃墙，卵石床、太阳能集热、海绵城市、沼气利用等绿色设计手段。

The area has high altitude and cold climate with many windy and dusty days. “Sheltering from the wind, ventilation, sunshade, keeping warm” are the main green goals. This design draws lessons from conventional wisdom of “courtyard” to create a comfortable regional microclimate. The architectural planning is compact and concentrated. Besides, the external arc wall is set to block off the strong wind and reduce the shape factor so as to achieve the overall cold resistance. The extracted grid elements are used to form the "courtyard" space with different attributes, such as north courtyard, large courtyard, narrow courtyard and Ayiwang to solve the problems of sunshine, daylighting, ventilation, wind and sand, as well as shaping the behavior space. Additionally, green design methods such as eaves gallery, Trombe walls, pebble bed, solar energy collection, sponge city and biogas utilization are also adopted.

Site Status

The site is located in the southern valley of Xinjiang, surrounded by a large area of pasture. The site is rich in stone and rammed earth dwellings.

Location Analysis

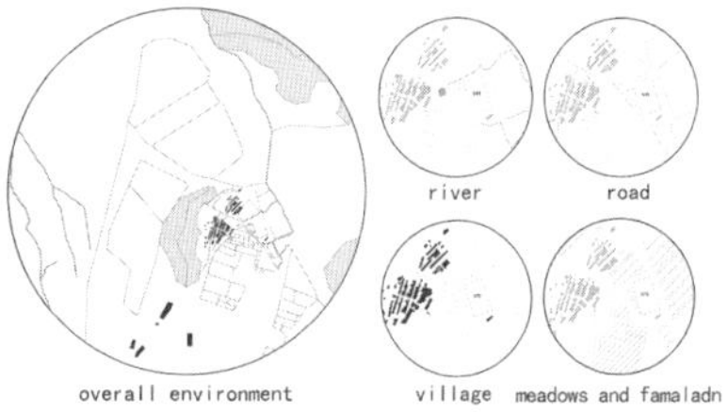
river
road
overall environment
village
meadows and famaladn

Solar Radiation

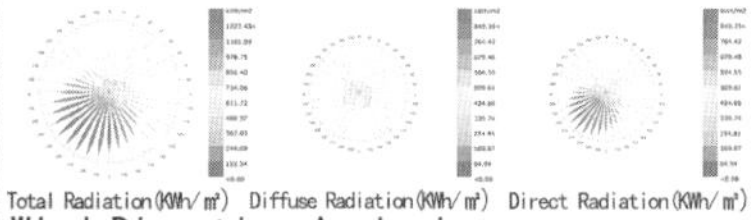
Total Radiation(KWh/㎡)
Diffuse Radiation(KWh/㎡)
Direct Radiation(KWh/㎡)

Wind Direction Analysis

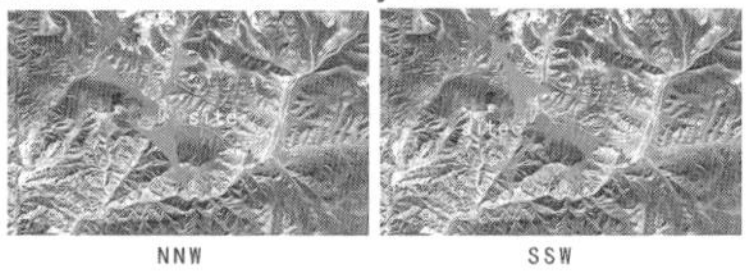
NNW
SSW

Aywan
yard
patio
eaves gallery

Annual Temperature

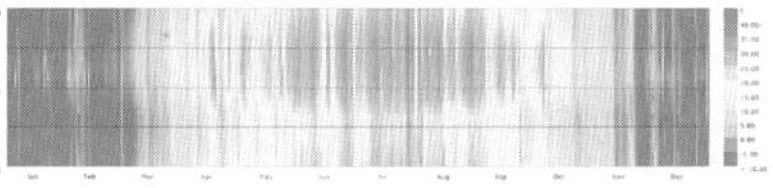

Design Strategy

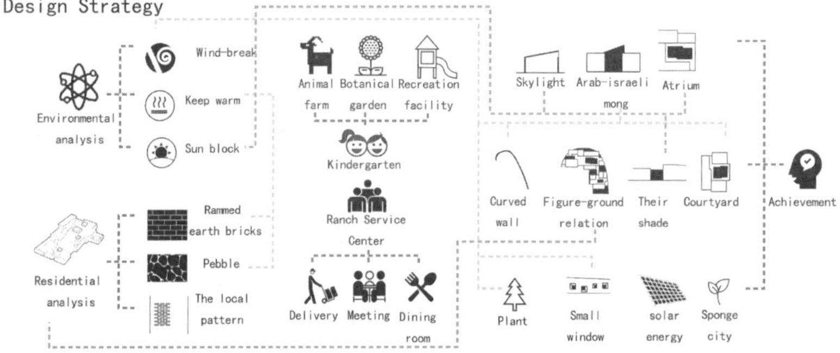
Environmental analysis
Wind-break
Keep warm
Sun block
Animal farm
Botanical garden
Recreation facility
Kindergarten
Ranch Service Center
Delivery
Meeting
Dining room
Skylight
Arab-israeli mong
Atrium
Curved wall
Figure-ground relation
Their shade
Courtyard
Achievement
Residential analysis
Rammed earth bricks
Pebble
The local pattern
Plant
Small window
solar energy
Sponge city

Courtyard Analysis

The courtyards、patios and Aywans are designed according to the climate, culture and space of Xinjiang. The combination of courtyards is well arranged, which brings users of community activity centers and kindergartens made space experience colourful.

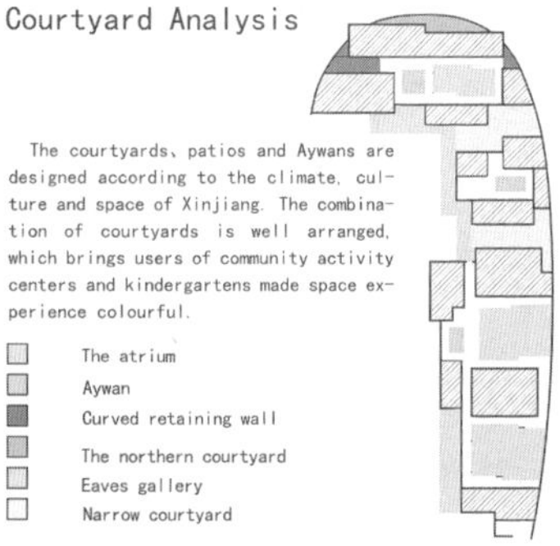
The atrium
Aywan
Curved retaining wall
The northern courtyard
Eaves gallery
Narrow courtyard

风·院 WIND·YARD 02

幼儿园
Kindergarten Entrance Perspective

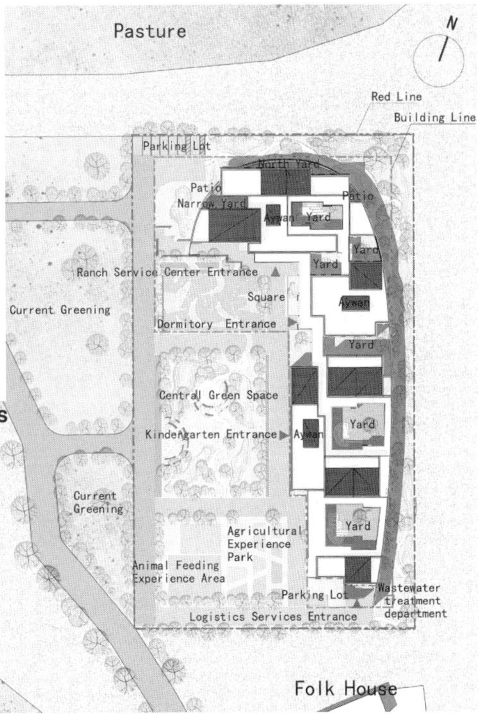
Pasture
N
Red Line
Building Line
Parking Lot
North Yard
Patio
Narrow Yard
Patio
Aywan
Yard
Yard
Yard
Ranch Service Center Entrance
Square
Aywan
Current Greening
Dormitory Entrance
Yard
Central Green Space
Kindergarten Entrance
Aywan
Yard
Current Greening
Agricultural Experience Park
Yard
Animal Feeding Experience Area
Parking Lot
Wastewater treatment department
Logistics Services Entrance
Folk House

Ranch Service Center Entrance Perspective

Economic Indicators

Land Area: 10408.6m²

Building Area: 2084.5m²

Service Center: 687.7m²

Dormitory: 371.9m²

Kindergarten: 994.9m²

Building Density: 20%

Building Floors: 1

Greening Rate: 57.8%

Plot Ratio: 0.20

Parking Spaces: 8

SITE PLAN 1:500

Generate analysis

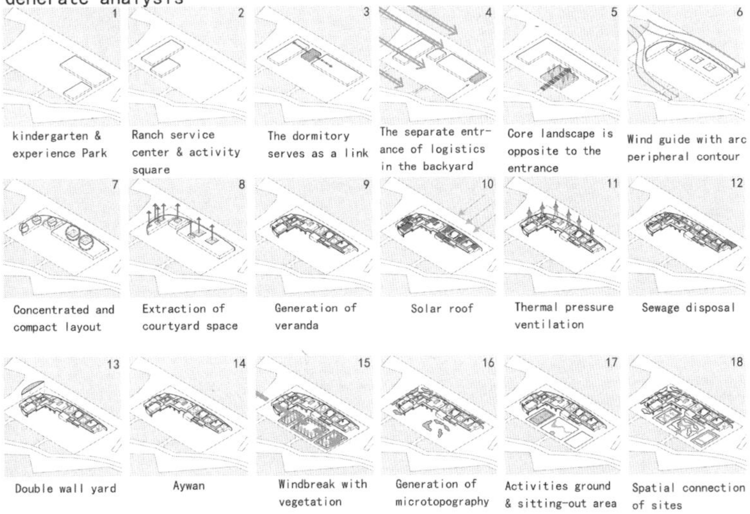
1 kindergarten & experience Park
2 Ranch service center & activity square
3 The dormitory serves as a link
4 The separate entrance of logistics in the backyard
5 Core landscape is opposite to the entrance
6 Wind guide with arc peripheral contour
7 Concentrated and compact layout
8 Extraction of courtyard space
9 Generation of veranda
10 Solar roof
11 Thermal pressure ventilation
12 Sewage disposal
13 Double wall yard
14 Aywan
15 Windbreak with vegetation
16 Generation of microtopography
17 Activities ground & sitting-out area
18 Spatial connection of sites

Aywan Internal Perspective

员工宿舍
Main Entrance Perspective

风·院 WIND·YARD 03

FIRST FLOOR PLAN 1：200

1 Entrance Hall
2 Function Hall
3 Library
4 Activity Room
5 Consulting Room
6 Observation Ward
7 Treatment Room
8 Office
9 Toilet
10 Dormitory
11 Kitchen
12 Food Bank
13 Restaurant
14 Storage Room
15 Guard Room
16 Comprehensive Activity Room
17 Children's Activity Room
18 Children's Toilet
19 Bathroom
20 Cloakroom
21 Yard
22 Health Center
23 Isolation Room
24 Kindergarten Office
25 Interesting Gallery
26 Duty Room
27 Disinfection Room
28 Kindergarten Kitchen
29 Locker Room
30 Backyard
31 Animal Feeding Experience Area
32 Agricultural Experience Park
33 Sewage treatment Division
34 Microtopography
35 Square
36 Basketball Court
37 Equipment Room
38 Hot Water Room
39 Sun Lawn

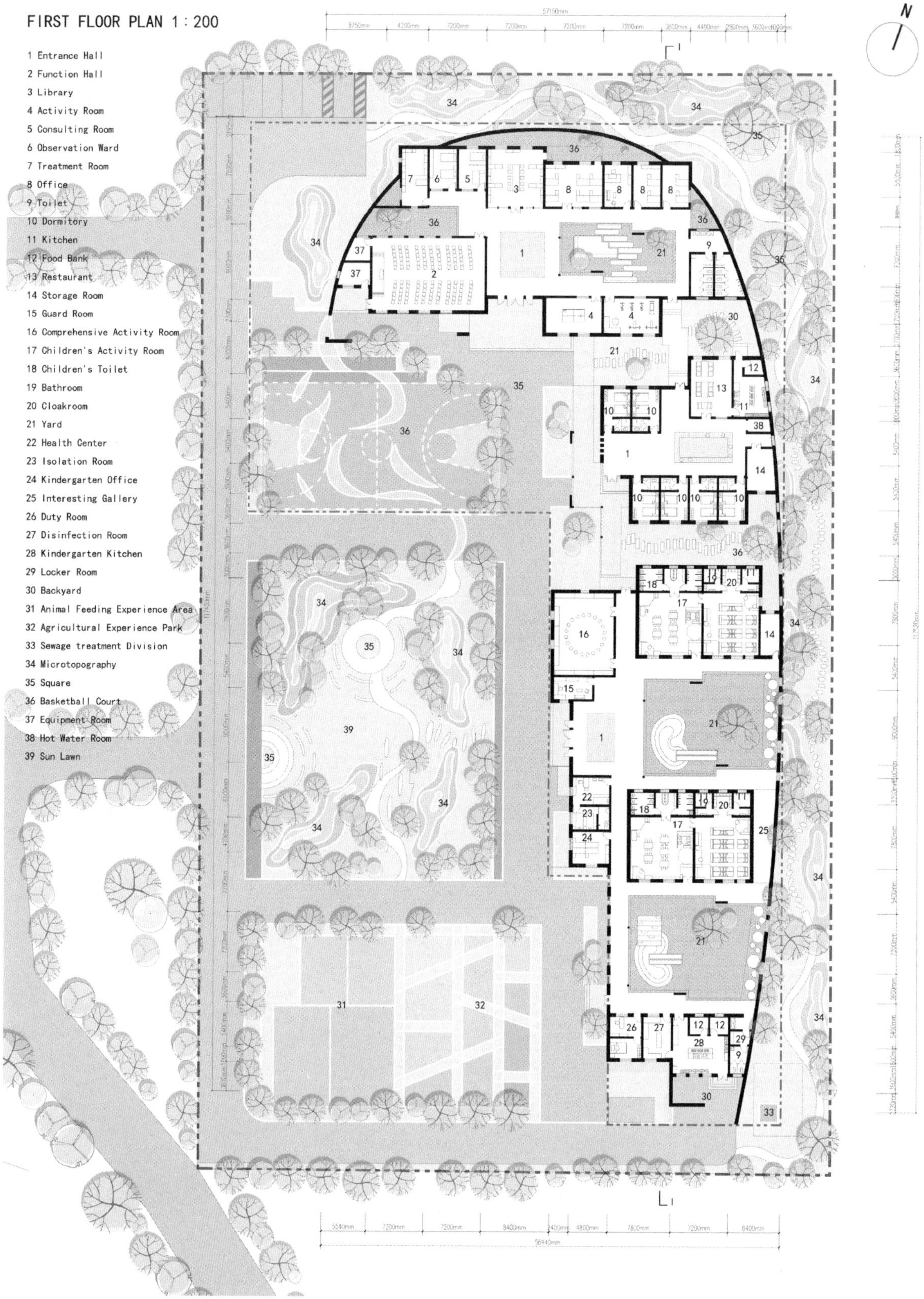

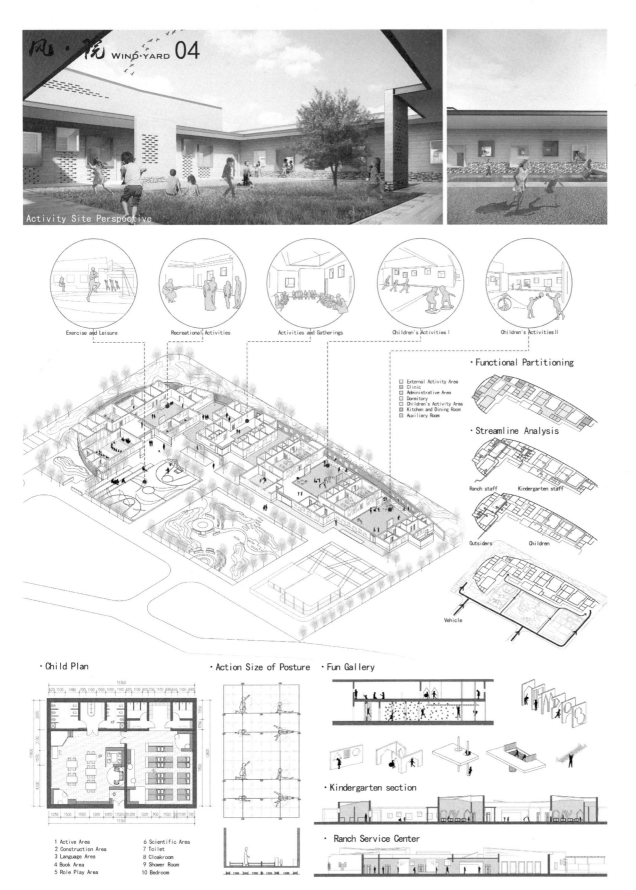

风·院 WIND·YARD 04
Activity Site Perspective

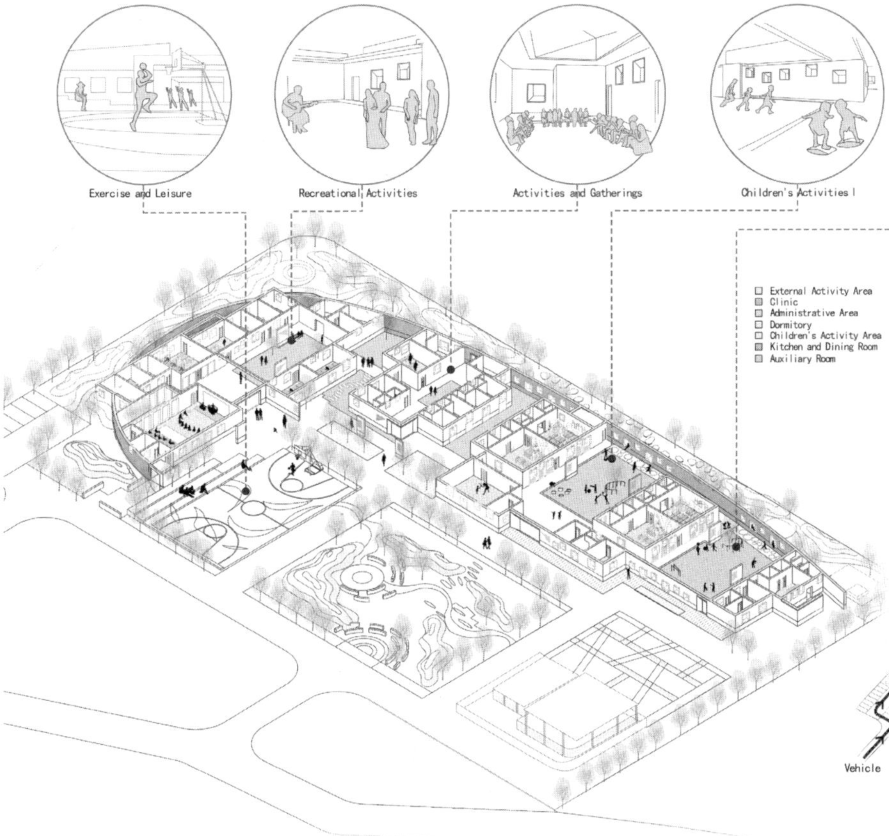
Exercise and Leisure
Recreational Activities
Activities and Gatherings
Children's Activities I
Children's Activities II
External Activity Area
Clinic
Administrative Area
Dormitory
Children's Activity Area
Kitchen and Dining Room
Auxiliary Room

· Functional Partitioning

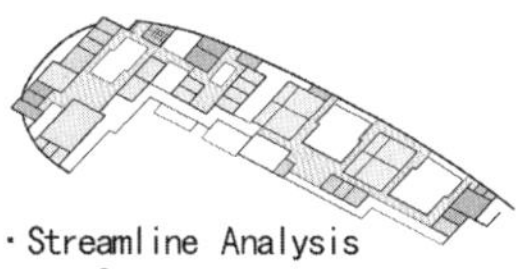

· Streamline Analysis

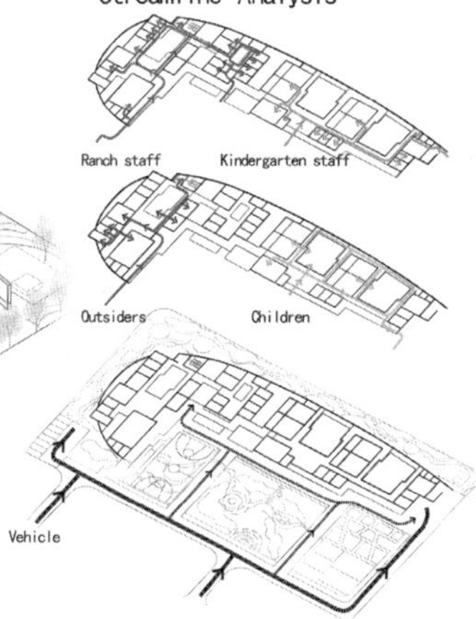
Ranch staff
Kindergarten staff
Outsiders
Children
Vehicle

· Child Plan

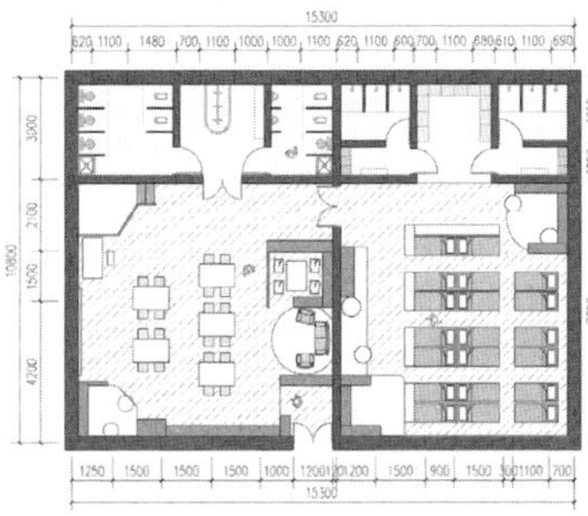

1 Active Area
2 Construction Area
3 Language Area
4 Book Area
5 Role Play Area
6 Scientific Area
7 Toilet
8 Cloakroom
9 Shower Room
10 Bedroom

· Action Size of Posture

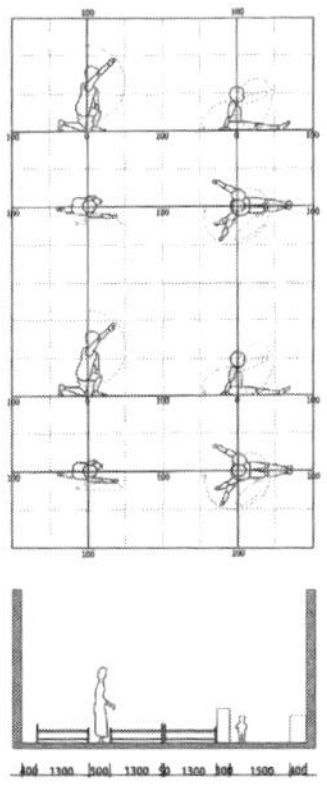

· Fun Gallery

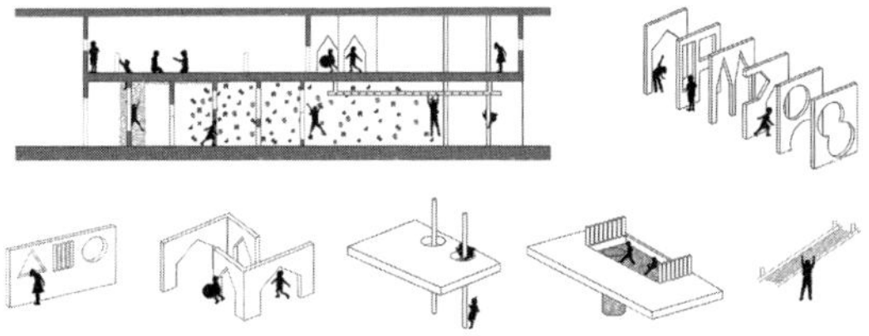

· Kindergarten section

· Ranch Service Center

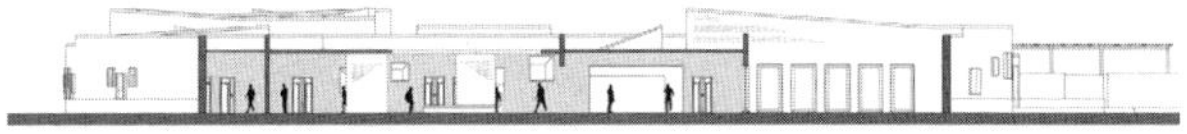

风·院 WIND·YARD 05

Courtyard draught
Solar panels
Pebble bed
Courtyard draught
Courtyard draught
High skylight lighting

· Passive Insulation Strategies

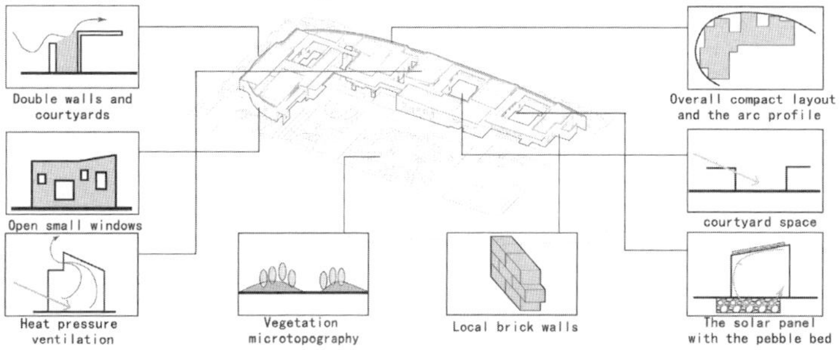
Double walls and courtyards
Open small windows
Heat pressure ventilation
Vegetation microtopography
Local brick walls
Overall compact layout and the arc profile
courtyard space
The solar panel with the pebble bed

· Passive Windbreak Strategies

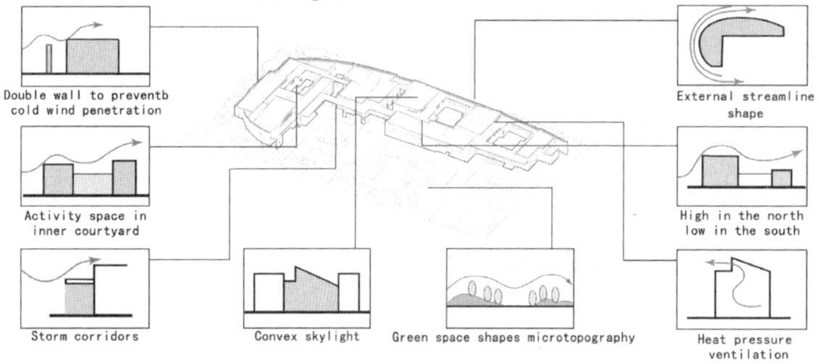
Double wall to preventb cold wind penetration
Activity space in inner courtyard
Storm corridors
Convex skylight
Green space shapes microtopography
External streamline shape
High in the north low in the south
Heat pressure ventilation

· Passive Shading and Sun Protection Strategies

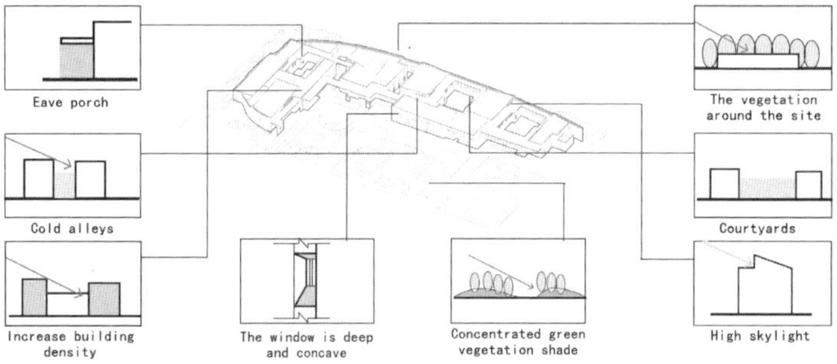
Eave porch
Cold alleys
Increase building density
The window is deep and concave
Concentrated green vegetation shade
The vegetation around the site
Courtyards
High skylight

· Ranch Service Center

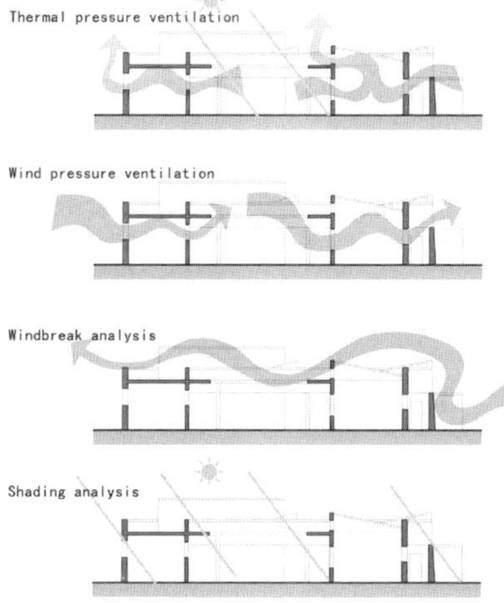
Thermal pressure ventilation
Wind pressure ventilation
Windbreak analysis
Shading analysis

· Kindergarten

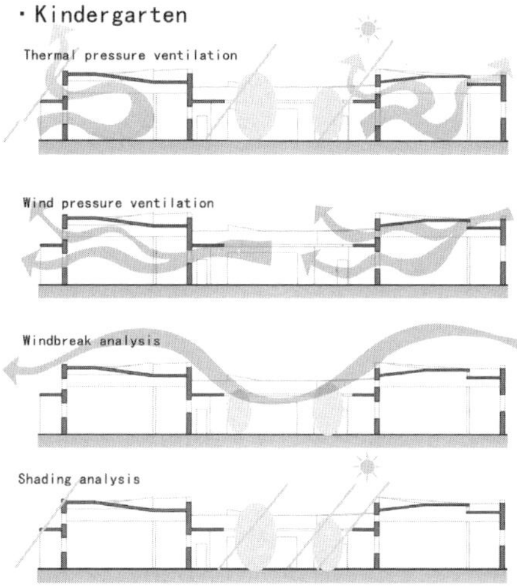
Thermal pressure ventilation
Wind pressure ventilation
Windbreak analysis
Shading analysis

Section Ⅰ-Ⅰ 1：250

West Elevation Ⅰ-Ⅰ 1：250

风·院 WIND·YARD 06

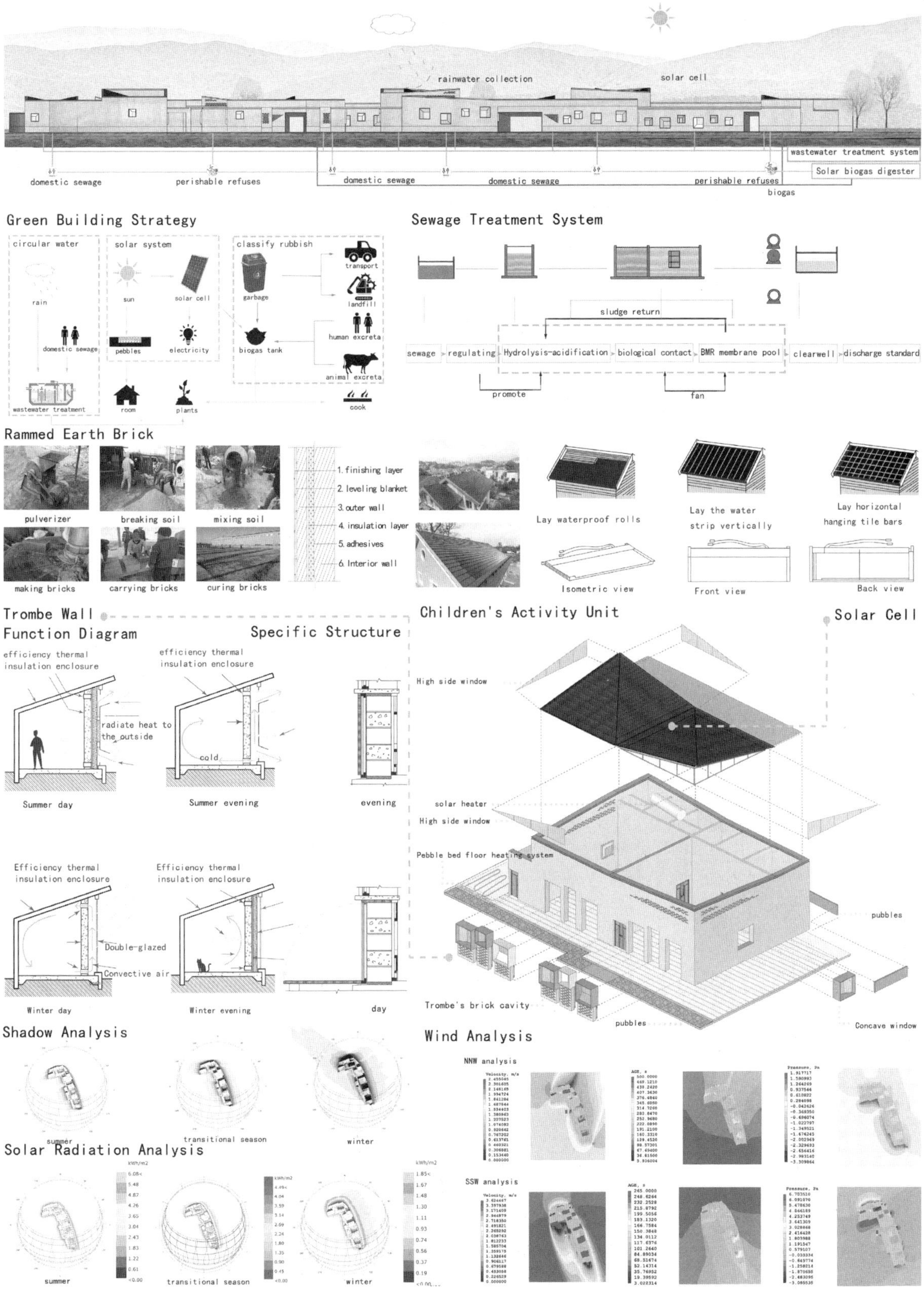
rainwater collection
solar cell
wastewater treatment system
Solar biogas digester
domestic sewage
perishable refuses
domestic sewage
domestic sewage
perishable refuses
biogas
Green Building Strategy
circular water
rain
domestic sewage
wastewater treatment
solar system
sun
solar cell
pebbles
electricity
room
plants
classify rubbish
garbage
transport
landfill
human excreta
biogas tank
animal excreta
cook
Sewage Treatment System
sludge return
sewage
regulating
Hydrolysis-acidification
biological contact
BMR membrane pool
clearwell
discharge standard
promote
fan
Rammed Earth Brick
pulverizer
breaking soil
mixing soil
making bricks
carrying bricks
curing bricks
1. finishing layer
2. leveling blanket
3. outer wall
4. insulation layer
5. adhesives
6. Interior wall
Lay waterproof rolls
Lay the water strip vertically
Lay horizontal hanging tile bars
Isometric view
Front view
Back view
Trombe Wall
Function Diagram
Specific Structure
efficiency thermal insulation enclosure
radiate heat to the outside
Summer day
efficiency thermal insulation enclosure
cold
Summer evening
evening
Efficiency thermal insulation enclosure
Double-glazed
Convective air
Winter day
Efficiency thermal insulation enclosure
Winter evening
day
Children's Activity Unit
Solar Cell
High side window
solar heater
High side window
Pebble bed floor heating system
pubbles
Trombe's brick cavity
pubbles
Concave window
Shadow Analysis
summer
transitional season
winter
Solar Radiation Analysis
summer
transitional season
winter
Wind Analysis
NNW analysis
SSW analysis

综合奖 · 优秀奖
General Prize Awarded · Honorable Mention Prize

注 册 号：7068
项目名称：向阳而生（新疆）
Grow in the Sun（Xinjiang）
作　　者：古丽・玉素甫、阿依旦、冀雯思、刘　静、刘梦君、刘　帆、廖小苗
参赛单位：新疆大学
指导教师：王万江、滕树勤、樊　辉

向阳而生 1
GROW IN THE SUN

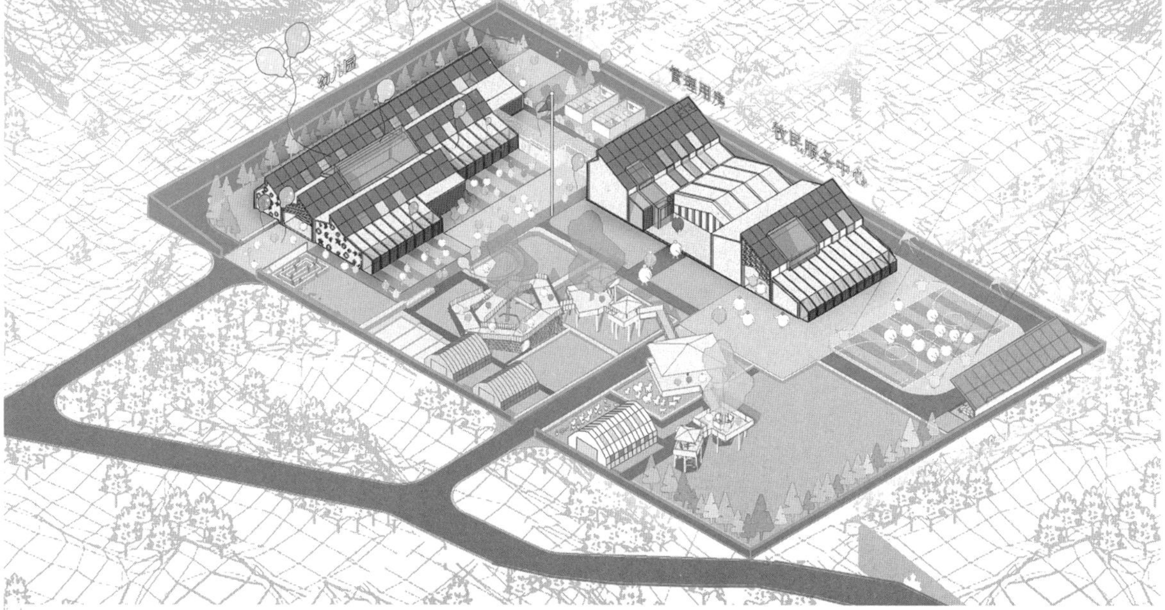

Climate

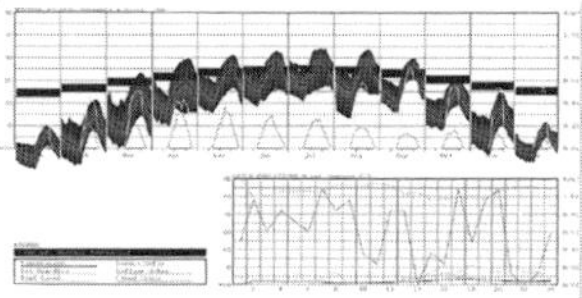

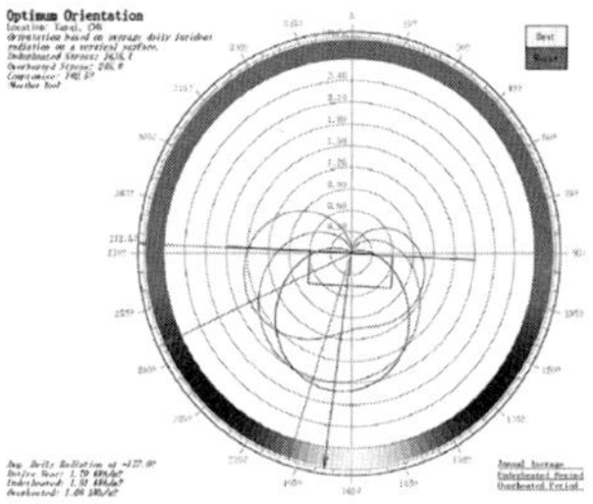

Best Orientation

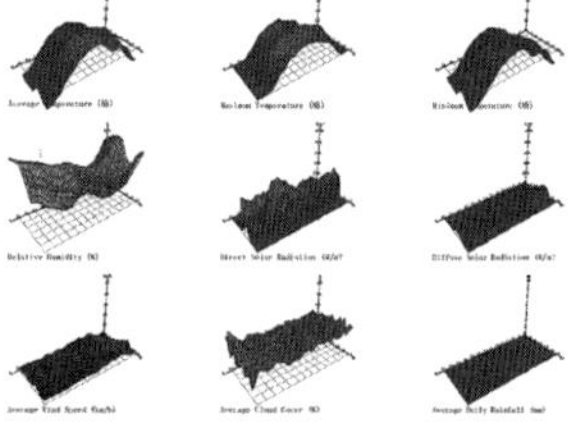

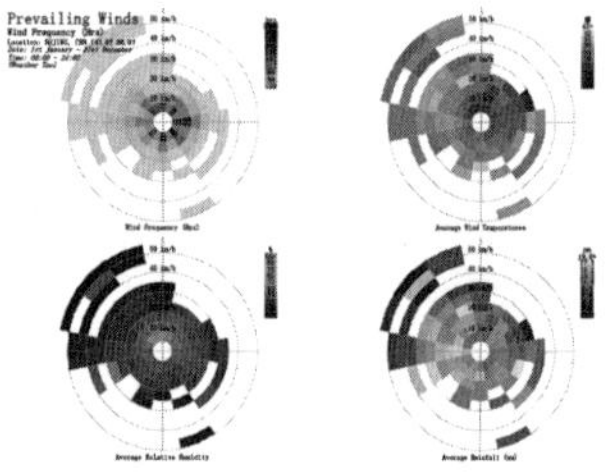

Prevailing Winds

Design Desription

本方案以“向阳而生”为主题，一方面，希望孩子们在阳光下快乐自由的成长；另一方面，在建筑空间上，每个体块设置了阳光大厅赋予建筑明亮的光环境，夏季利于通风，冬季形成暖房。此外，通过场地布局，入口灰空间，太阳房，特朗勃墙体，双层窗，百叶遮阳等实现建筑节能。建筑立面则提取周围山峰的天际线形成多折坡屋顶，实现太阳能光伏一体化的同时利于雪水与雨水的收集。此外，在主动式技术上还利用沼气，地源热泵与主动式太阳能实现一体化供给能量并循环使用。

The theme of project is “Grow In The Sun”, On the one hand, it is hoped that children will grow up freely and happily in the sun. In terms of space, each building block is equipped with a sunny hall to give the building a bright light environment, which is good for ventilation in summer and forms a warm room in winter;In addition, building energy saving can be achieved through the layout of the site, entrance gray space, solar room, Trumb wall, LOW-E glass, and Louver shading system;The façade of the building extracts the skyline of the surrounding mountains to form a multi-fold slope roof, which realizes the integration of solar energy and photovoltaic and is conducive to the collection of snow and rain water. In addition, biogas, ground source heat pumps and active solar energy are also used in active technology to provide integrated energy supply and recycle.

Lacation

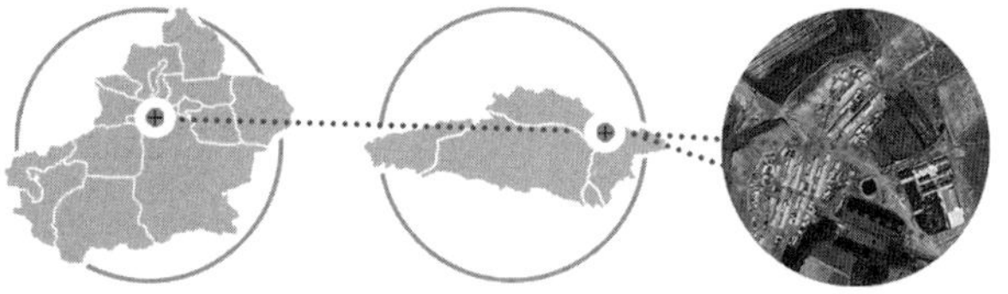

Site Analysis

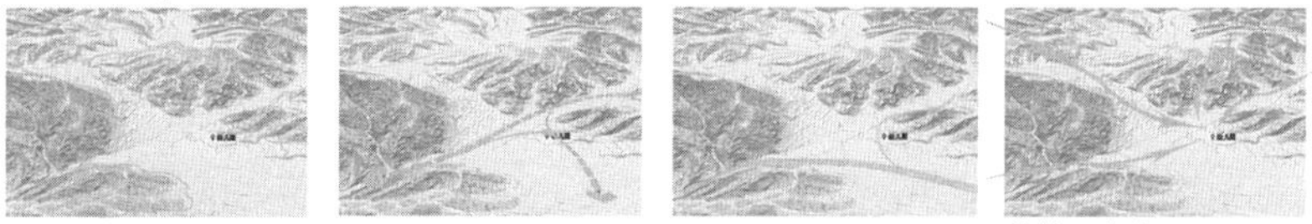

Terrain　Rivers　Roads　Valley wind, mainly northwest wind.

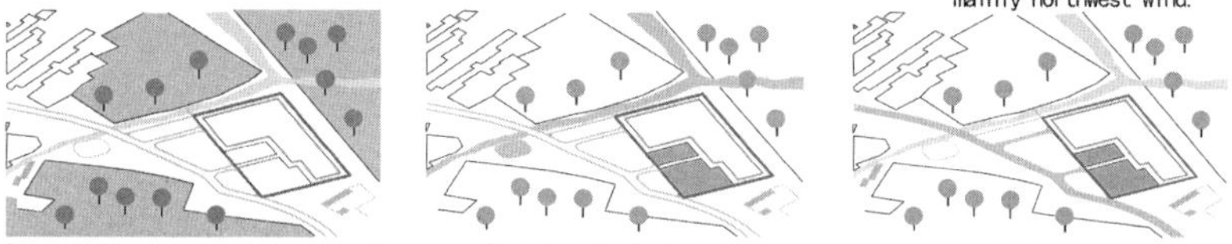

Surrounded by pastures around the site　Streams around the site　The road to the site

Regional Feature

向阳而生 2
GROW IN THE SUN

1. Outdoor Playground For Children
2. Children's Outdoor Activity Unit
3. Sandpit
4. Pool
5. Outdoor Planting Site
6. Greenhouse
7. Sky Corridor
8. Outdoor Animal Breeding Experience Area
9. Animal Pen
10. Outdoor Venue For Herders
11. Parking Lot
12. Green Area
13. Yard
14. Garbage Disposal Pool

Economic Analysis

Site Area: 10409m²

Kindergarten Building Area: 983m²

Total building area: 1979m²

Greening Rate: 31.8%

Floor Area Ratio: 0.39

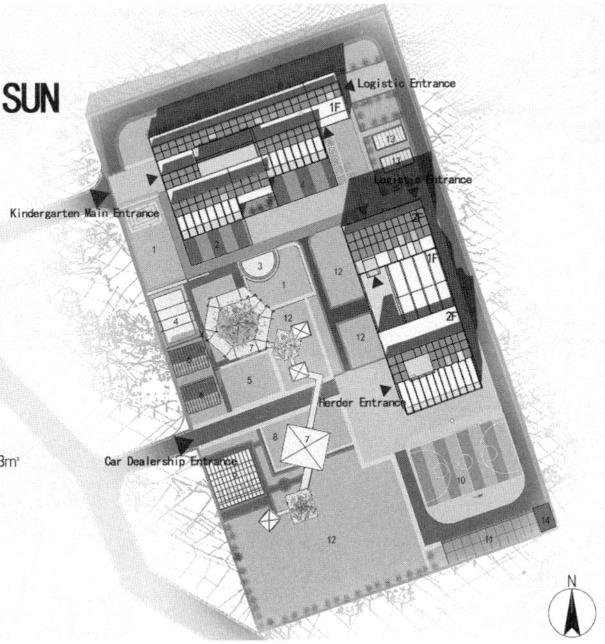
Logistic Entrance
Logistic Entrance
Kindergarten Main Entrance
Herder Entrance
Car Dealership Entrance
1F
2F
N

Site-Plan 1 : 500

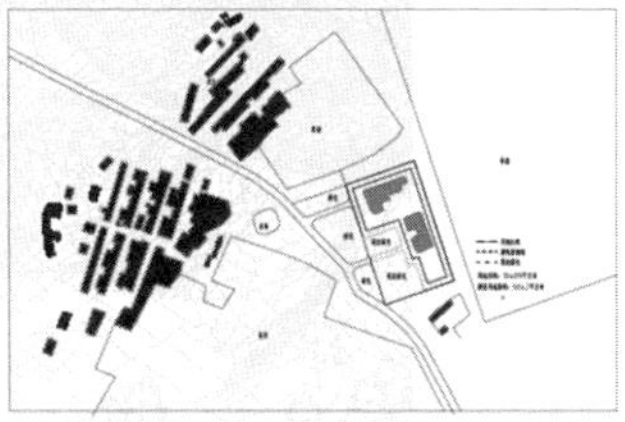

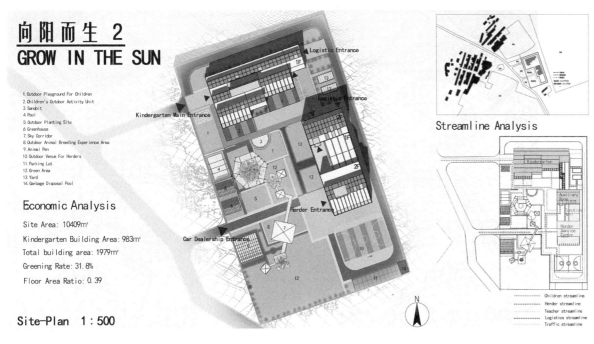

Streamline Analysis

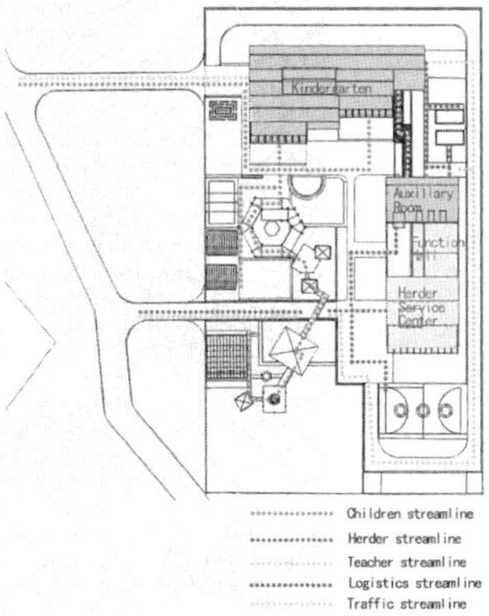
Kindergarten
Auxiliary Room
Function Hall
Herder Service Center
Children streamline
Herder streamline
Teacher streamline
Logistics streamline
Traffic streamline

Concept Generation

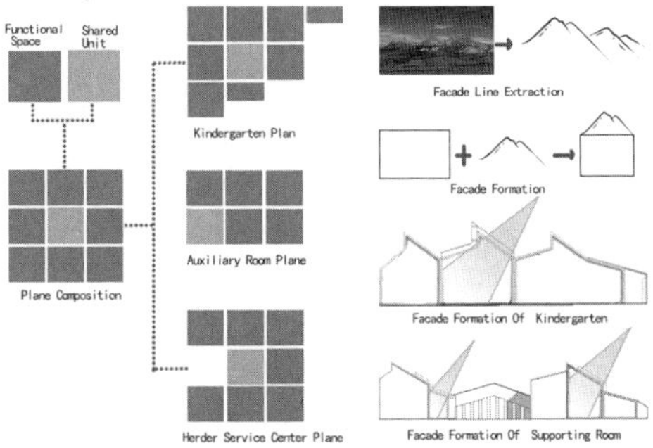
Functional Space
Shared Unit
Plane Composition
Kindergarten Plan
Auxiliary Room Plane
Herder Service Center Plane
Facade Line Extraction
Facade Formation
Facade Formation Of Kindergarten
Facade Formation Of Supporting Room

Logical Generation

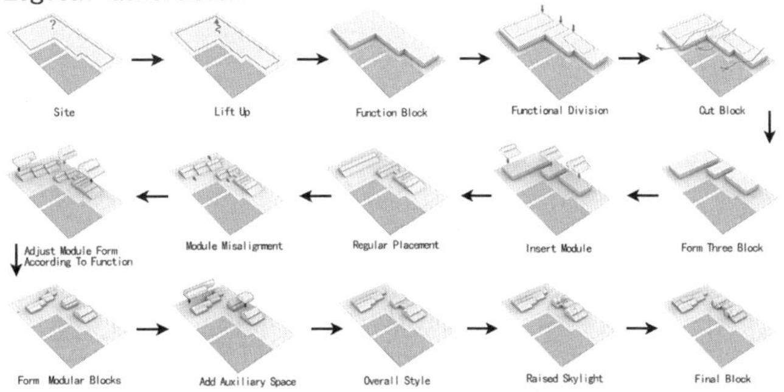
Site
Lift Up
Function Block
Functional Division
Cut Block
Form Three Block
Insert Module
Regular Placement
Module Misalignment
Adjust Module Form According To Function
Form Modular Blocks
Add Auxiliary Space
Overall Style
Raised Skylight
Final Block

Wind Simulation

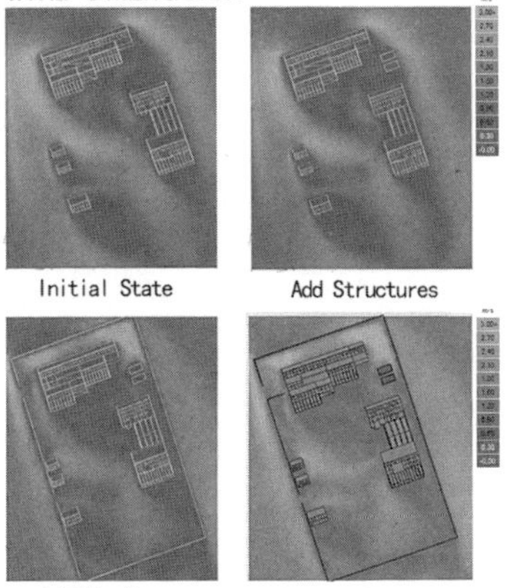

Initial State

Add Structures

Add The Fence

Optimal Solution

Shadow Simulation

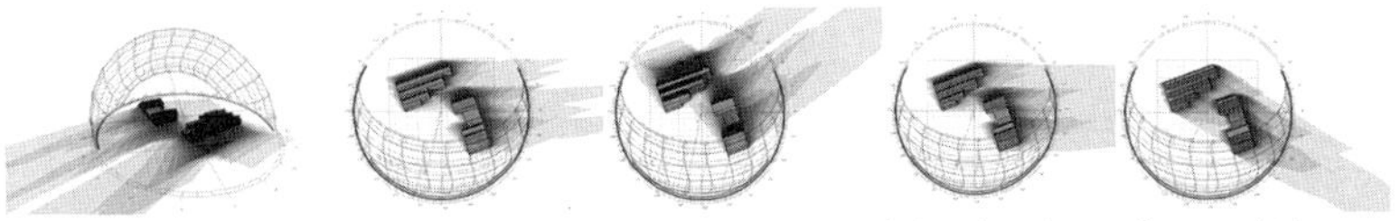

3D of Wind Solatice

spring breeze

Winter solstice

Autumnal equinox

Summer solstice

Lighting Simulation

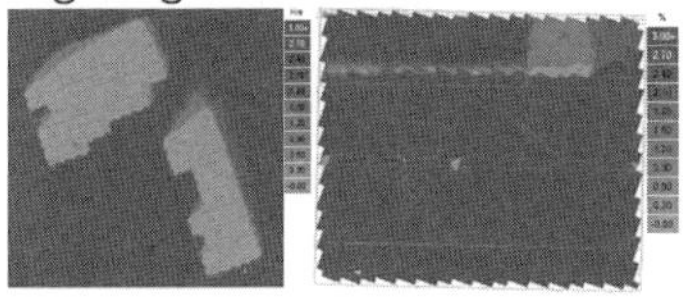

Sunshine Duration

Daylighting Coefficient

Technology Strategy

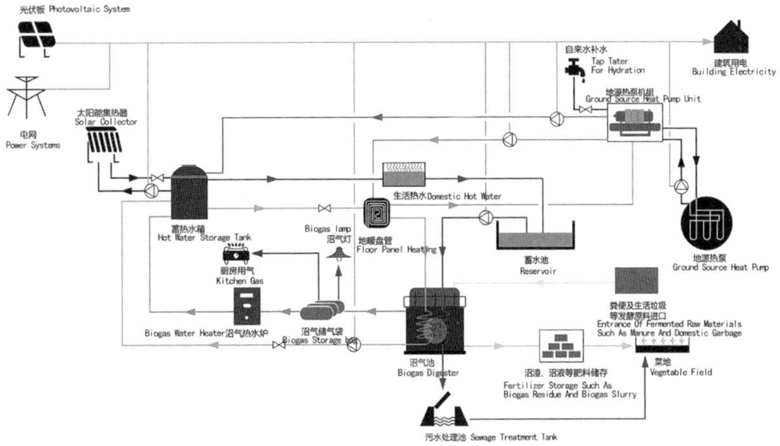
光伏板 Photovoltaic System
电网 Power Systems
太阳能集热器 Solar Collector
自来水补水 Tap Tater For Hydration
建筑用电 Building Electricity
地源热泵机组 Ground Source Heat Pump Unit
蓄热水箱 Hot Water Storage Tank
生活热水 Domestic Hot Water
地暖盘管 Floor Panel Heating
Biogas lamp 沼气灯
厨房用气 Kitchen Gas
蓄水池 Reservoir
地源热泵 Ground Source Heat Pump
Biogas Water Heater沼气热水炉
沼气储气袋 Biogas Storage bag
沼气池 Biogas Digester
粪便及生活垃圾等发酵原料进口 Entrance Of Fermented Raw Materials Such As Manure And Domestic Garbage
沼渣、沼液等肥料储存 Fertilizer Storage Such As Biogas Residue And Biogas Slurry
菜地 Vegetable Field
污水处理池 Sewage Treatment Tank

向阳而生 3
GROW IN THE SUN

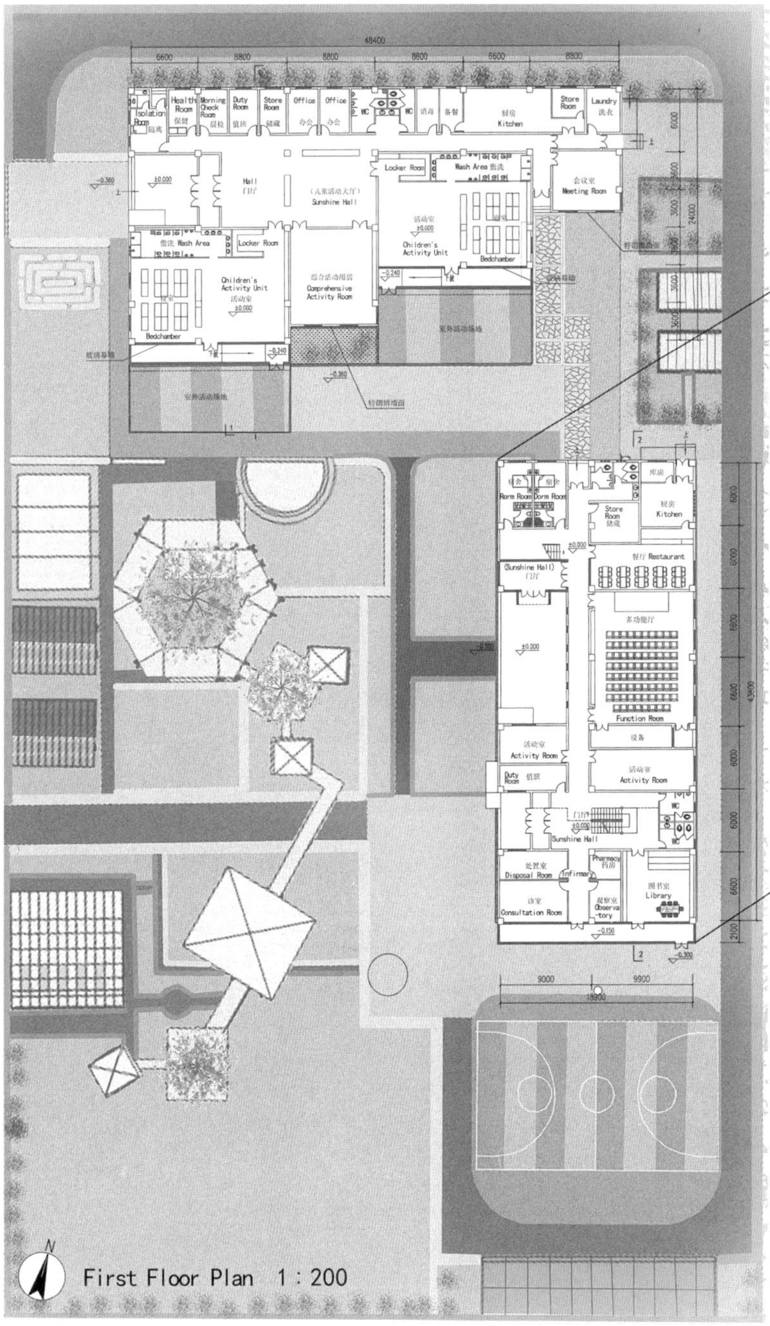
48400
Isolation Room
Health Room
Morning Check Room
Duty Room
Store Room
Office
Office
WC
WC
Kitchen
Store Room
Laundry
Hall
Sunshine Hall
Locker Room
Wash Area
Meeting Room
Wash Area
Locker Room
Children's Activity Unit
Comprehensive Activity Room
Children's Activity Unit
Bedchamber
Bedchamber
Dorm Room
Kitchen
Restaurant
Sunshine Hall
Function Room
Activity Room
Activity Room
Duty Room
Sunshine Hall
Pharmacy
Disposal Room
Infirmary
Library
Consultation Room
Observatory
18900

First Floor Plan 1：200

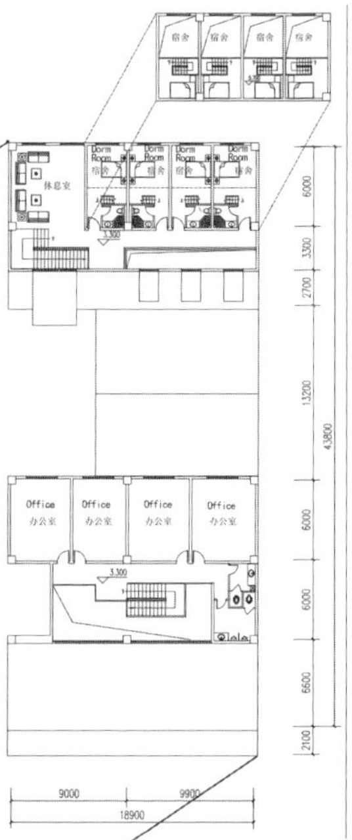
Dorm Room
Office
Office
Office
Office
9000
9900
18900
43800

Second Floor Plan 1：200

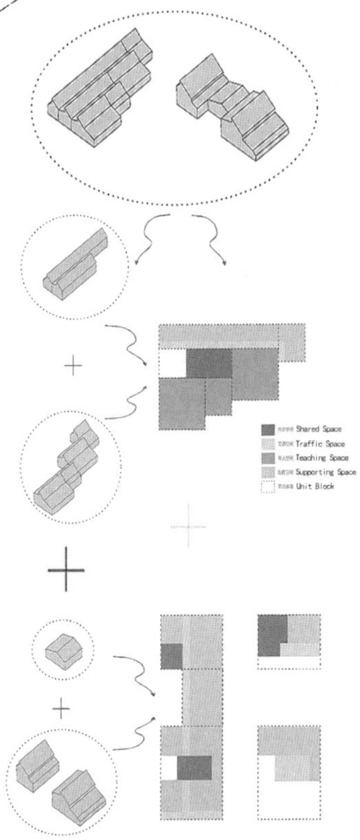
Shared Space
Traffic Space
Teaching Space
Supporting Space
Unit Block

Building Block Analysis

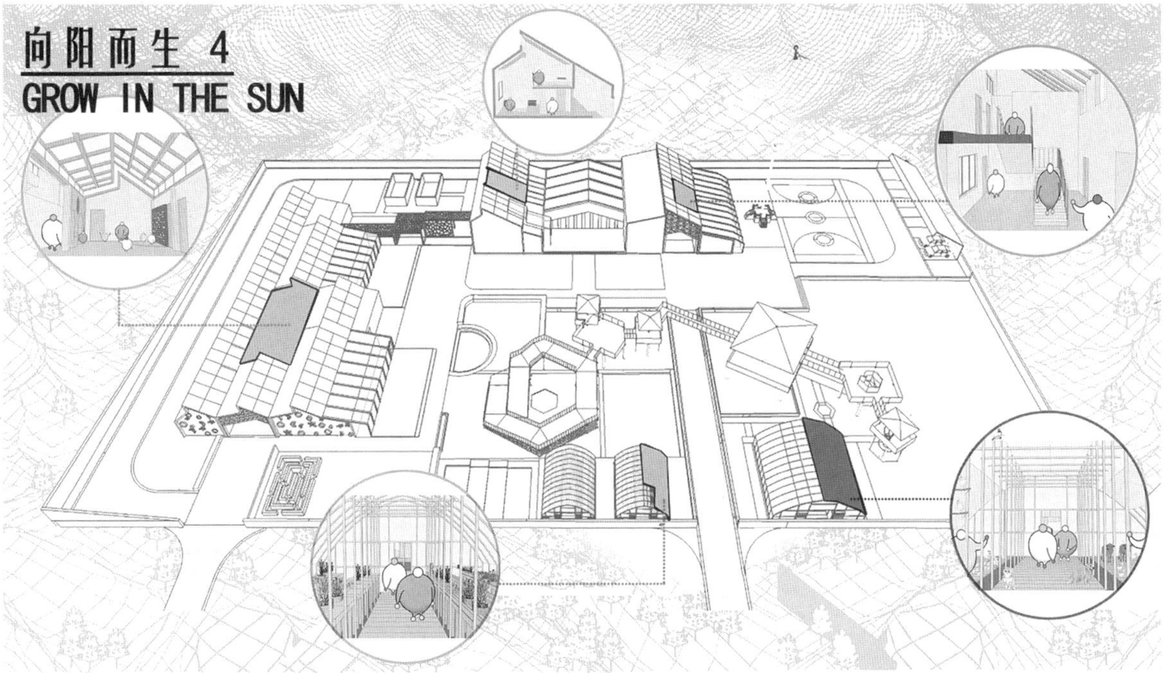
向阳而生 4
GROW IN THE SUN

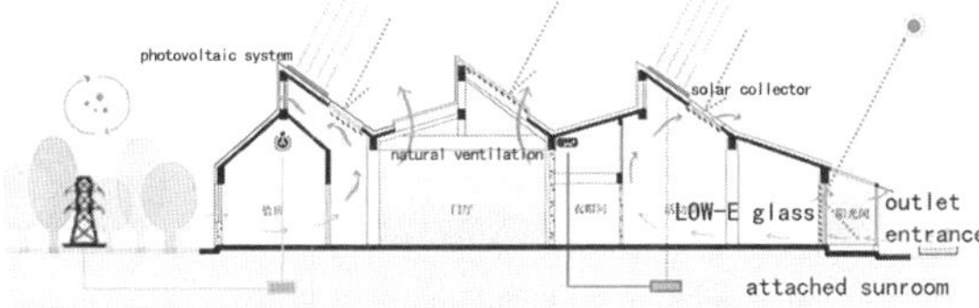
photovoltaic system
natural ventilation
solar collector
LOW-E glass
outlet
entrance
attached sunroom

kindergarten
Summer Ventilation and Daylight

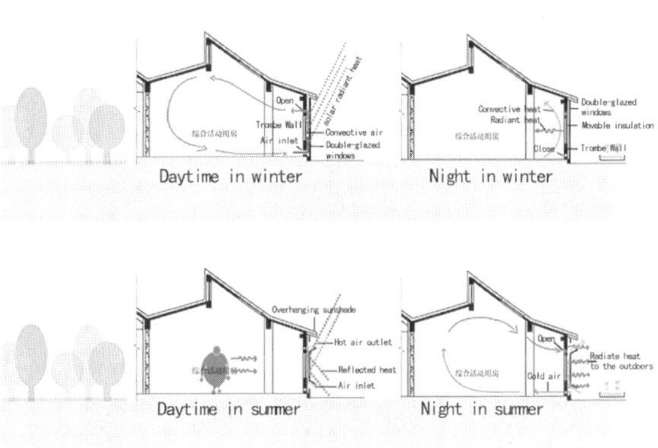
Daytime in winter
Night in winter
Daytime in summer
Night in summer

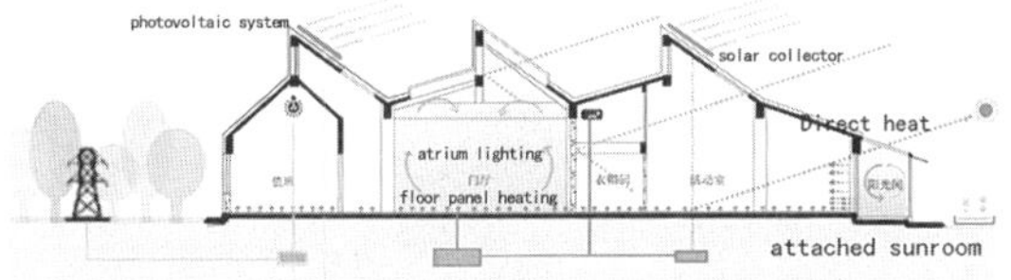
photovoltaic system
solar collector
atrium lighting
floor panel heating
Direct heat
attached sunroom

kindergarten
Winter Ventilation and Daylight

Comprehensive Activity Room

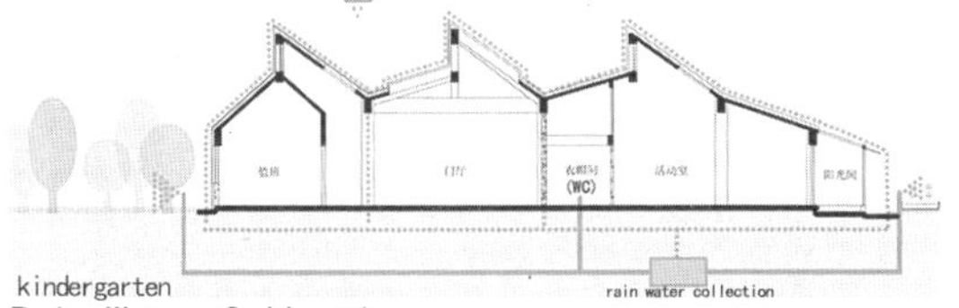
rain water collection

kindergarten
Rain Water Collection

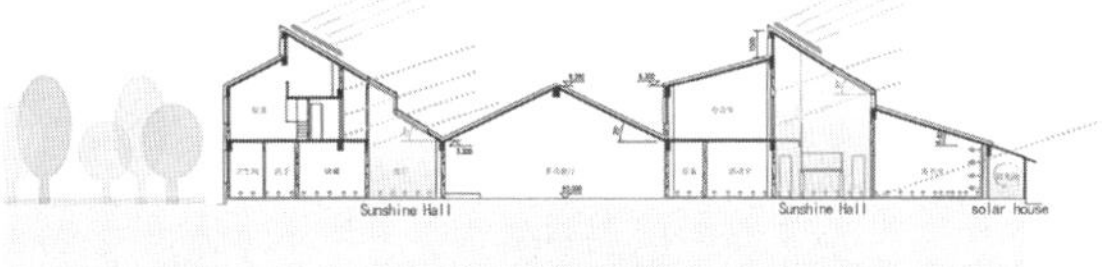
Sunshine Hall
solar house

Herdsman Service Center
Winter Ventilation and Daylight

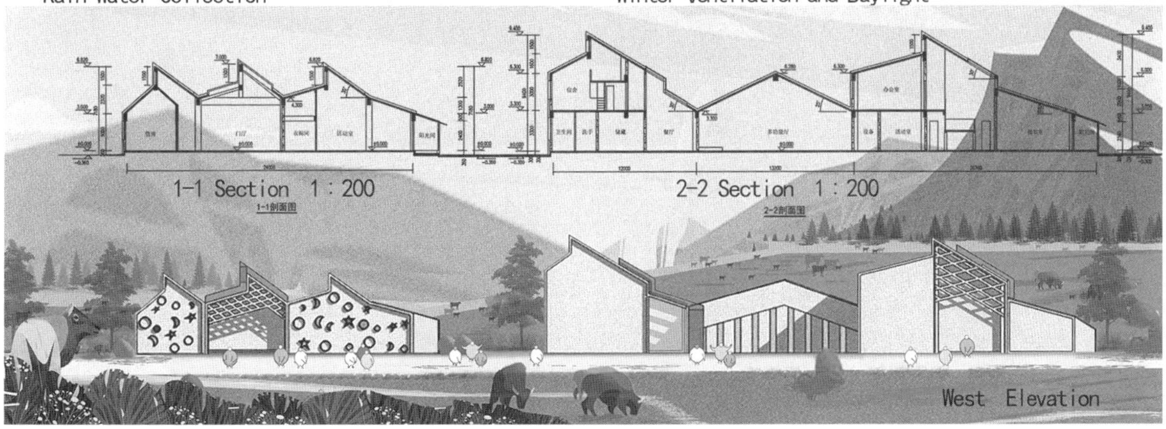
1-1 Section 1 : 200
1-1剖面图
2-2 Section 1 : 200
2-2剖面图
West Elevation

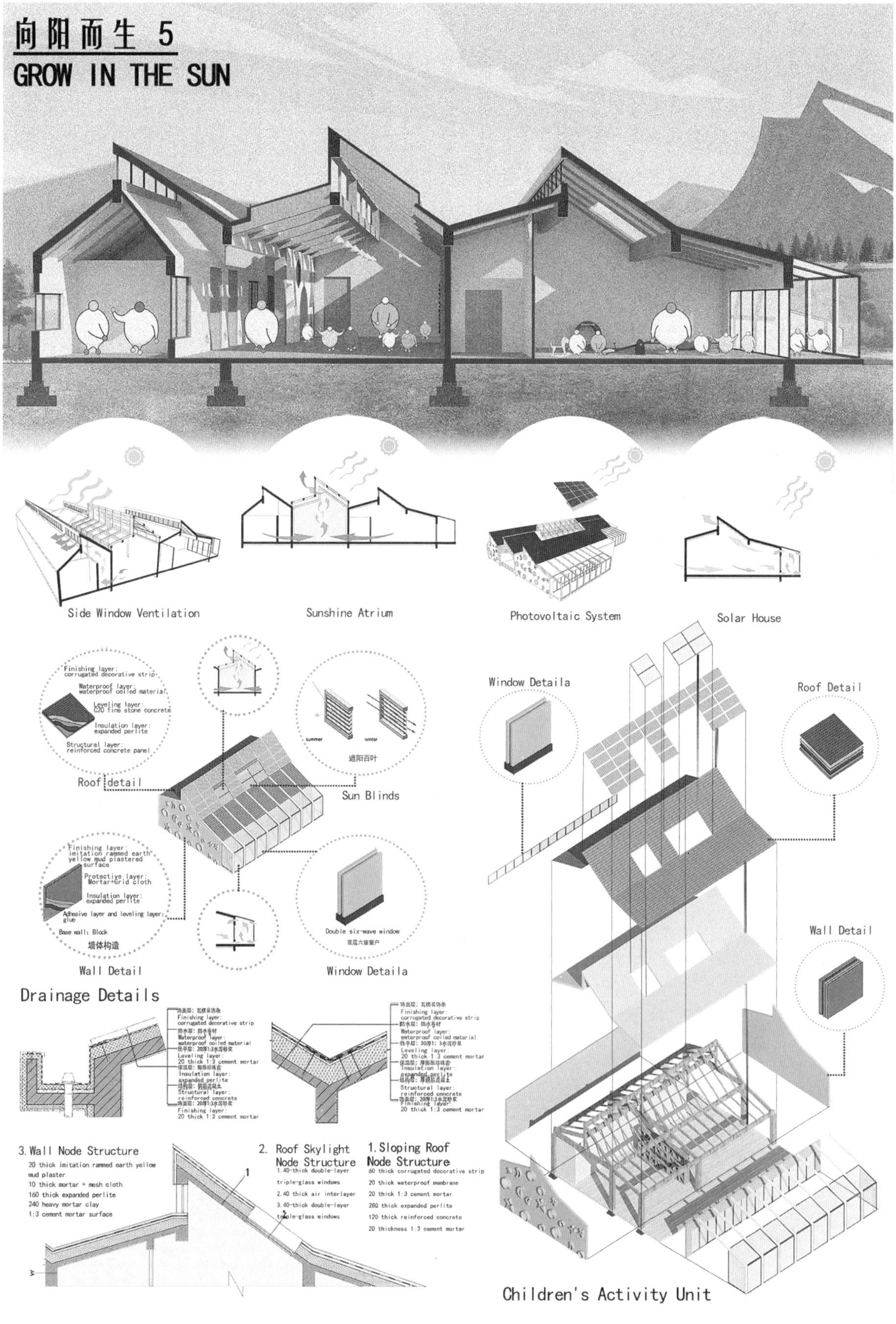

向阳而生 5
GROW IN THE SUN
Side Window Ventilation
Sunshine Atrium
Photovoltaic System
Solar House
Finishing layer: corrugated decorative strip
Waterproof layer: waterproof coiled material
Leveling layer: C20 fine stone concrete
Insulation layer: expanded perlite
Structural layer: reinforced concrete panel
Roof detail
summer
winter
遮阳百叶
Sun Blinds
Finishing layer: imitation rammed earth yellow mud plastered surface
Protective layer: Mortar+Grid cloth
Insulation layer: expanded perlite
Adhesive layer and leveling layer: glue
Base wall: Block
墙体构造
Wall Detail
Double six-wave window
双层六玻窗户
Window Detaila
Drainage Details
Finishing layer: corrugated decorative strip
Waterproof layer: waterproof coiled material
Leveling layer: 20 thick 1:3 cement mortar
Insulation layer: expanded perlite
Structural layer: reinforced concrete
Finishing layer: 20 thick 1:3 cement mortar
3. Wall Node Structure
20 thick imitation rammed earth yellow mud plaster
10 thick mortar + mesh cloth
160 thick expanded perlite
240 heavy mortar clay
1:3 cement mortar surface
2. Roof Skylight Node Structure
1. 40-thick double-layer triple-glass windows
2. 40 thick air interlayer
3. 40-thick double-layer triple-glass windows
1. Sloping Roof Node Structure
60 thick corrugated decorative strip
20 thick waterproof membrane
20 thick 1:3 cement mortar
280 thick expanded perlite
120 thick reinforced concrete
20 thickness 1:3 cement mortar
Window Detaila
Roof Detail
Wall Detail
Children's Activity Unit

向阳而生 6

GROW IN THE SUN

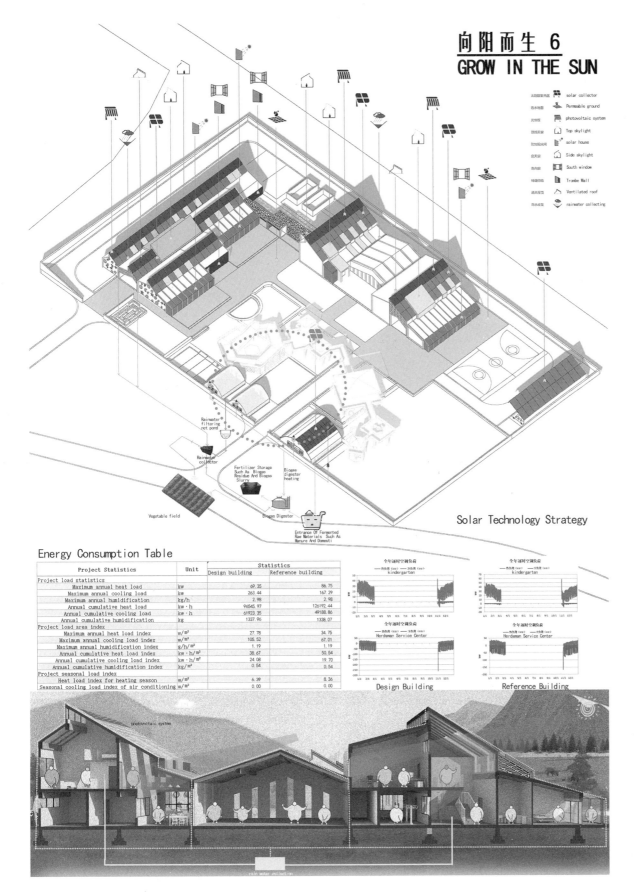

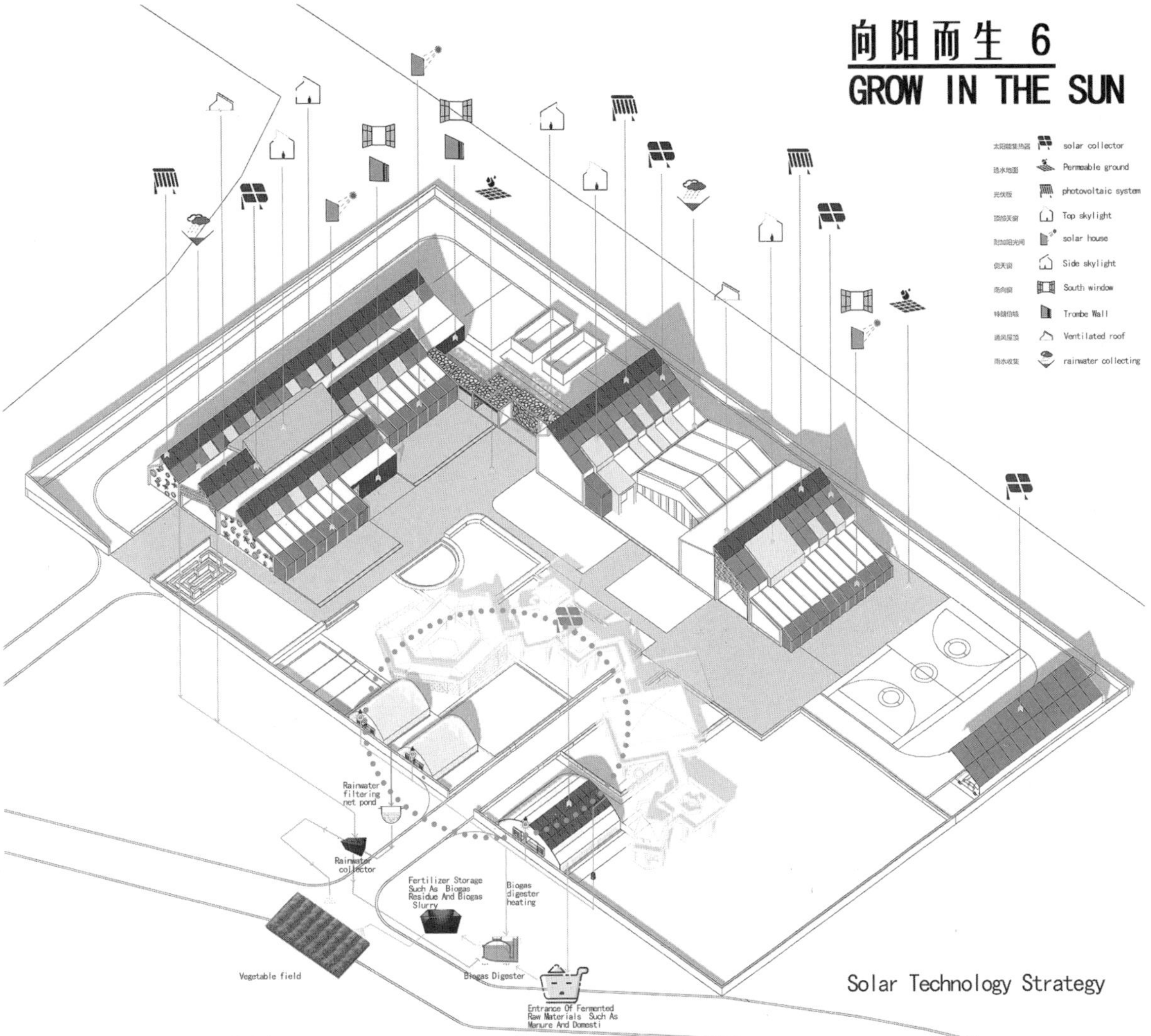
solar collector
Permeable ground
photovoltaic system
Top skylight
solar house
Side skylight
South window
Trombe Wall
Ventilated roof
rainwater collecting
Rainwater filtering net pond
Rainwater collector
Fertilizer Storage Such As Biogas Residue And Biogas Slurry
Biogas digester heating
Vegetable field
Biogas Digester
Entrance Of Fermented Raw Materials Such As Manure And Domesti

Solar Technology Strategy

Energy Consumption Table

Project Statistics	Unit	Statistics	
		Design building	Reference building
Project load statistics			
Maximum annual heat load	kw	69.35	86.75
Maximum annual cooling load	kw	263.44	167.29
Maximum annual humidification	kg/h	2.98	2.98
Annual cumulative heat load	kw·h	96545.97	126192.44
Annual cumulative cooling load	kw·h	61923.35	49188.86
Annual cumulative humidification	kg	1337.96	1338.07
Project load area index			
Maximum annual heat load index	w/m²	27.78	34.75
Maximum annual cooling load index	w/m²	105.52	67.01
Maximum annual humidification index	g/h/m²	1.19	1.19
Annual cumulative heat load index	kw·h/m²	38.67	50.54
Annual cumulative cooling load index	kw·h/m²	24.08	19.70
Annual cumulative humidification index	kg/m²	0.54	0.54
Project seasonal load index			
Heat load index for heating season	w/m²	6.39	8.36
Seasonal cooling load index of air conditioning	w/m²	0.00	0.00

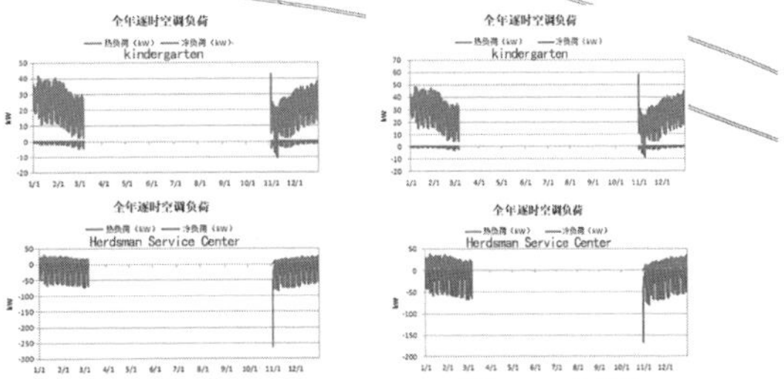
kindergarten
kindergarten
Herdsman Service Center
Herdsman Service Center

Design Building　　Reference Building

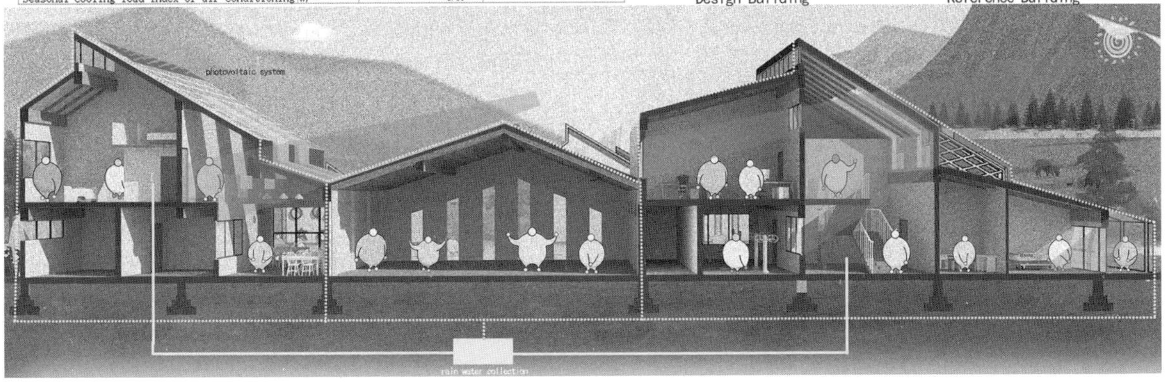
photovoltaic system
rain water collection

综合奖 · 优秀奖
General Prize Awarded · Honorable Mention Prize

注 册 号：7080

项目名称：童旭 · 土生（新疆）
“Raising” Children | Vernacular Homeland（Xinjiang）

作　　者：艾晓雄、方振威、李成成、齐　放、吴　昊、张　波

参赛单位：内蒙古工业大学

指导教师：贾晓浒、常泽辉

童旭 · 土生
“Raising” Children | Vernacular Homeland
新疆巴音郭楞州和静县建设兵团牧场幼儿园及服务中心

Site Location Analysis

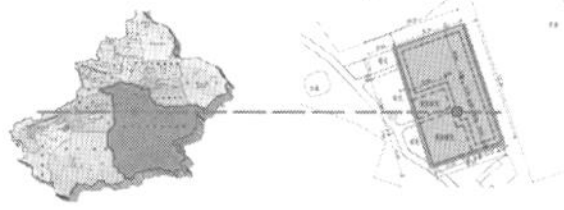

The geographical position

Dilapidated Rammed Earth Wall & Dilapidated Road

Slooping Roof　Surrounding Villages　Residents' Lives

Climate Analysis

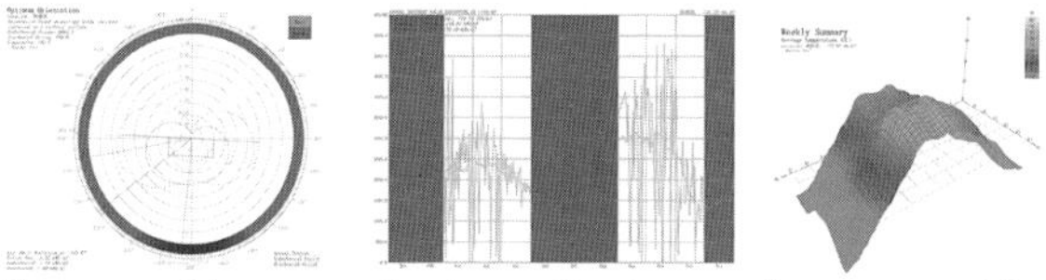

Best sunshine　Sun's radiation　Average weekly

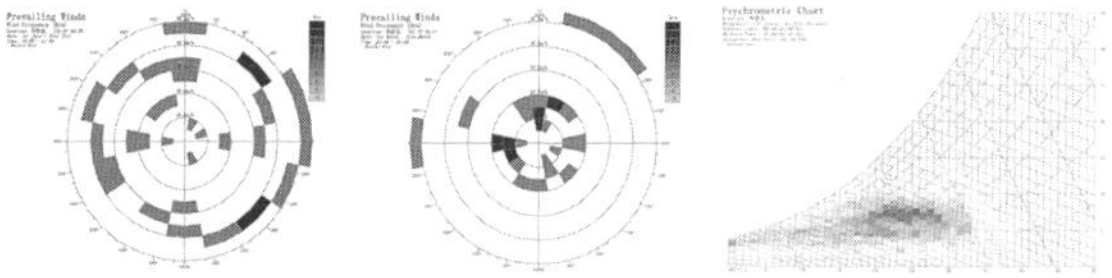

Summer　Winter　Psychrometric chart

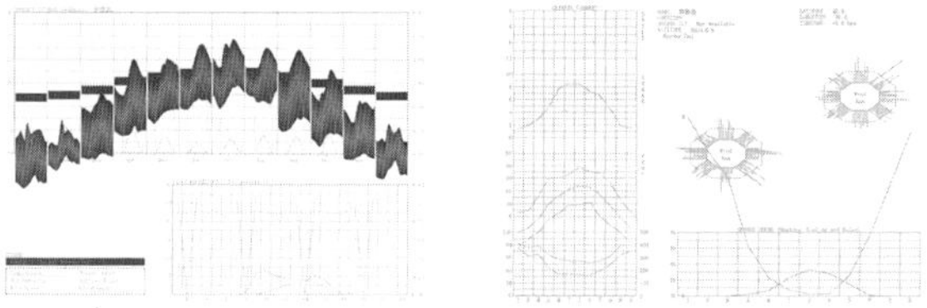

Monthly average　Days of the sun

Technology System

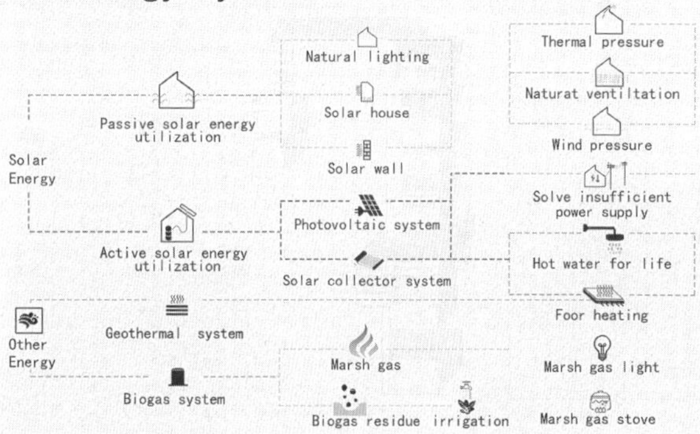
Solar Energy
Passive solar energy utilization
Active solar energy utilization
Natural lighting
Solar house
Solar wall
Photovoltaic system
Solar collector system
Thermal pressure
Naturat ventilation
Wind pressure
Solve insufficient power supply
Hot water for life
Foor heating
Other Energy
Geothermal system
Biogas system
Marsh gas
Biogas residue irrigation
Marsh gas light
Marsh gas stove

Design Specification

本方案通过结合各功能部分的设计来布置多个室内外趣味空间来满足儿童好奇心与活跃性强的特点。基于光、风环境模拟等措施，从主被动技术与建筑设计的融洽联系出发，通过热压通风、阳光房结合百叶窗与采用当地生土构造做法的围护结构以兼顾冬季保温与夏季遮阳通风。采用光伏发电系统与新型太阳能热泵系统，解决当地供电不足与采暖供冷的需求问题。同时通过污水处理等措施合理利用不可再生能源，增强方案的可实施性与经济效益。

This scheme combines the design of each functional part to arrange a number of indoor and outdoor interesting space to meet the children's curiosity and active characteristics. Based on the light and wind environment simulation and other measures, starting from the harmonious relationship between active and passive technology and architectural design, the thermal pressure ventilation, the sunlight room combined with louvers and the enclosure structure adopting the local soil structure, can give consideration to the heat preservation in winter and shading ventilation in summer. Photovoltaic power generation system and new solar heat pump system are adopted to solve the problem of insufficient local power supply and heating and cooling demand. At the same time, the non-renewable energy can be rationally utilized through sewage treatment and other measures to enhance the feasibility and economic benefits of the scheme.

童旭·土生 “Raising” Children | Vernacular Homeland 02

新疆巴音郭楞州和静县建设兵团牧场幼儿园及服务中心

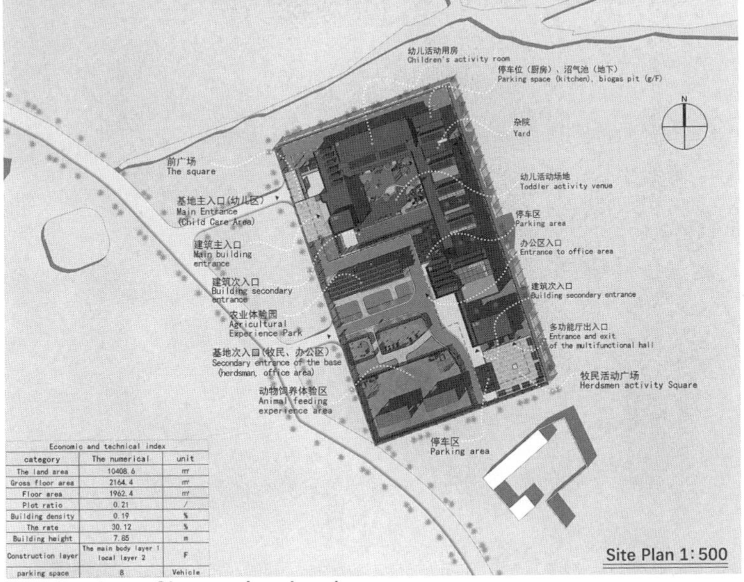

幼儿活动用房
Children's activity room
停车位（厨房）、沼气池（地下）
Parking space (kitchen), biogas pit (g/F)
杂院
Yard
前广场
The square
基地主入口(幼儿区)
Main Entrance
(Child Care Area)
幼儿活动场地
Toddler activity venue
停车区
Parking area
建筑主入口
Main building entrance
办公区入口
Entrance to office area
建筑次入口
Building secondary entrance
建筑次入口
Building secondary entrance
农业体验园
Agricultural Experience Park
多功能厅出入口
Entrance and exit of the multifunctional hall
基地次入口(牧民、办公区)
Secondary entrance of the base (herdsman, office area)
牧民活动广场
Herdsmen activity Square
动物饲养体验区
Animal feeding experience area
停车区
Parking area
N

Site Plan 1:500

Economic and technical index

category	The numerical	unit
The land area	10408.6	㎡
Gross floor area	2164.4	㎡
Floor area	1962.4	㎡
Plot ratio	0.21	/
Building density	0.19	%
The rate	30.12	%
Building height	7.85	m
Construction layer	The main body layer 1 local layer 2	F
parking space	8	Vehicle

Site Environment Analysis

Speed of cloud

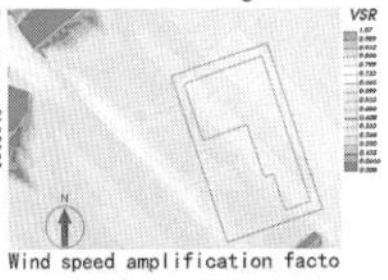

Wind speed amplification facto

The pressure of cloud

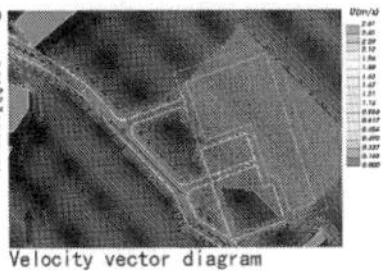

Velocity vector diagram

The Form Generated

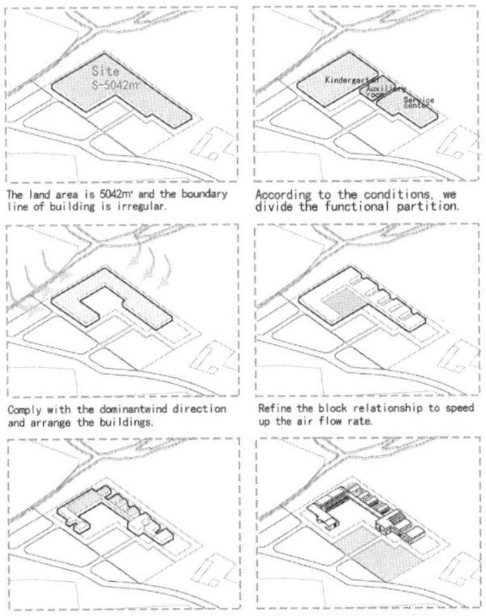

Site
S=5042㎡
Kindergarten
Service Center

The land area is 5042㎡ and the boundary line of building is irregular.

According to the conditions, we divide the functional partition.

Comply with the dominantwind direction and arrange the buildings.

Refine the block relationship to speed up the air flow rate.

Optimal tilt angle andSolar photovoltaic panel.

Architecture is integrated with agriculture and animal experience garden.

Explosion Chart Analysis

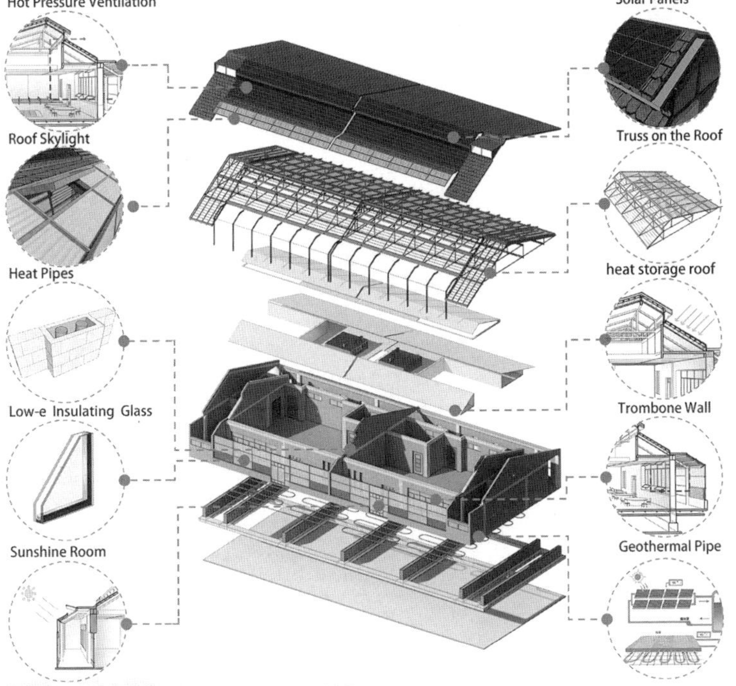

Hot Pressure Ventilation
Roof Skylight
Heat Pipes
Low-e Insulating Glass
Sunshine Room
Solar Panels
Truss on the Roof
heat storage roof
Trombone Wall
Geothermal Pipe

Detail Structure

Floor Radiant Heating System

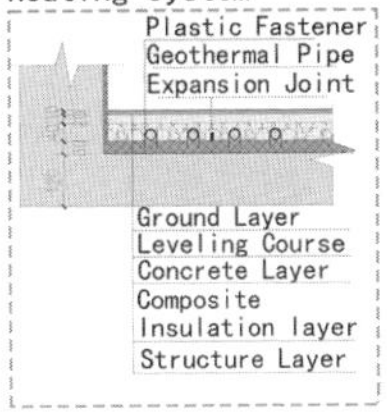

Plastic Fastener
Geothermal Pipe
Expansion Joint
Ground Layer
Leveling Course
Concrete Layer
Composite
Insulation layer
Structure Layer

Solar Collector Structure Detail

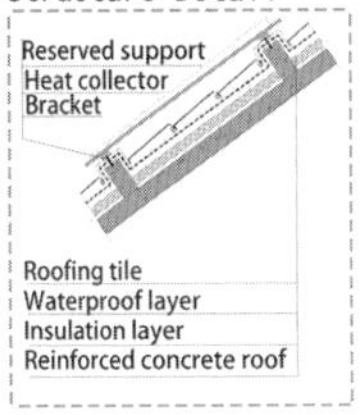

Reserved support
Heat collector
Bracket
Roofing tile
Waterproof layer
Insulation layer
Reinforced concrete roof

Shutter Structure

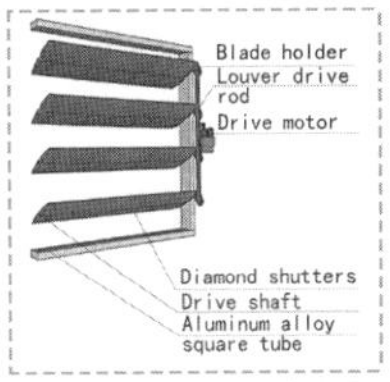

Blade holder
Louver drive rod
Drive motor
Diamond shutters
Drive shaft
Aluminum alloy square tube

Phase Change Thermal Storage Wall Structure

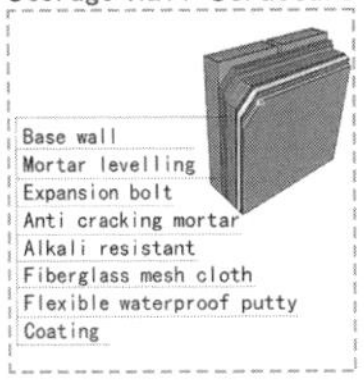

Base wall
Mortar levelling
Expansion bolt
Anti cracking mortar
Alkali resistant
Fiberglass mesh cloth
Flexible waterproof putty
Coating

Profile Perspective Effect Drawing

童旭 · 土生 "Raising" Children | Vernacular Homeland 03

新疆巴音郭楞州和静县建设兵团牧场幼儿园及服务中心

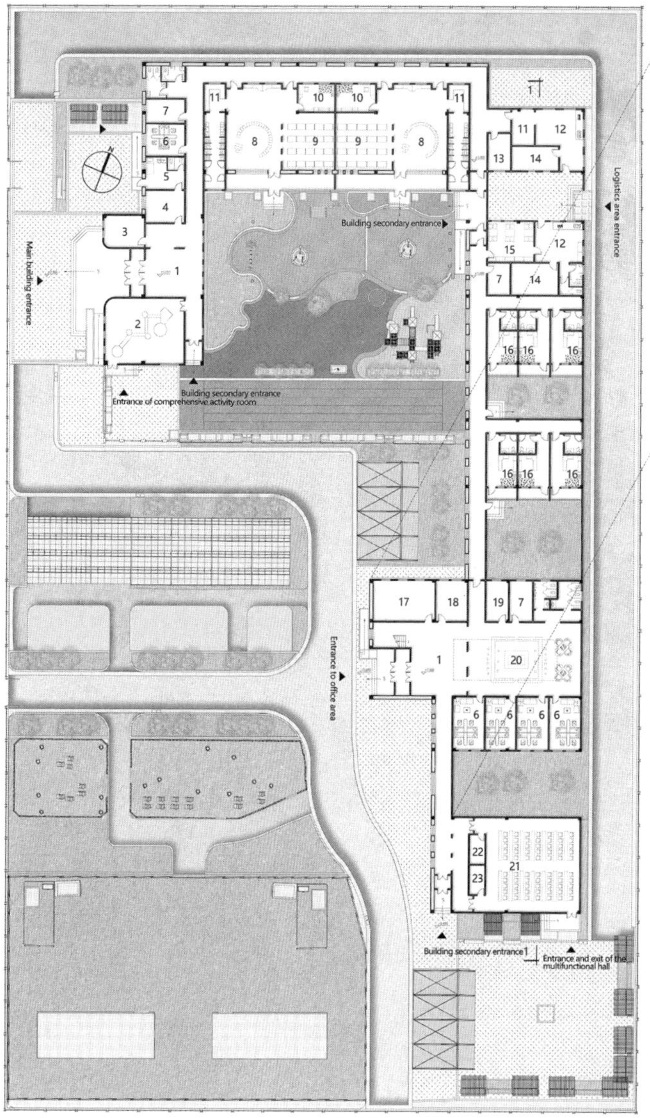

Building secondary entrance
Logistics area entrance
Main building entrance
Entrance of comprehensive activity room
Entrance to office area
Building secondary entrance
Entrance and exit of the multifunctional hall

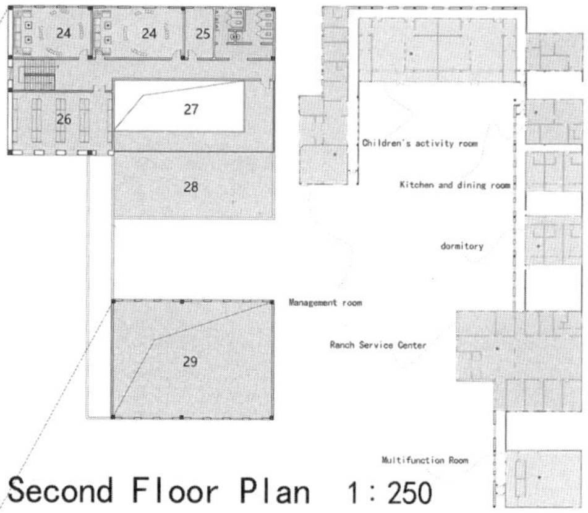

Children's activity room
Kitchen and dining room
dormitory
Management room
Ranch Service Center
Multifunction Room

Second Floor Plan 1：250

Kindergarten Analysis

Kindergarten bed-room section

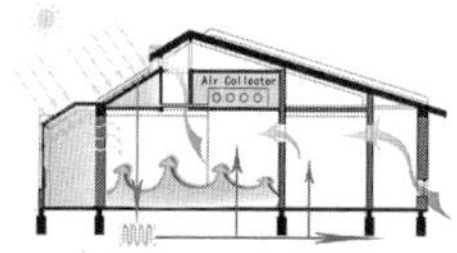

Air Collector

In winter, solar energy is converted into heat energy

Section of Kinder-garten Activity Room

Summer ventilation, solar energy is converted into electricity

Biochemical Analysis

Solar Energy Combine Bioga Boiler For Heating up Biogas Plant

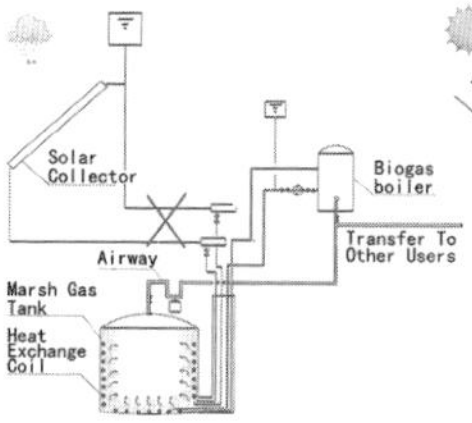

Solar Collector
Biogas boiler
Transfer To Other Users
Airway
Marsh Gas Tank
Heat Exchange Coil

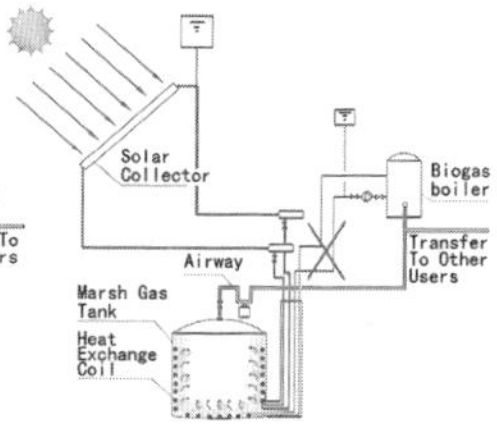

Solar Collector
Biogas boiler
Transfer To Other Users
Airway
Marsh Gas Tank
Heat Exchange Coil

1 Entrance hall
2 Comprehensive activity room
3 Guard
4 Healthcare room
5 Isolation room
6 Office
7 Equipment room
8 Activity room
9 Dorm
10 Teachers' duty room
11 Cloakroom
11+ Pantry
12 Kitchen
13 Disinfection room
14 Storeroom
15 Restaurant
16 Dormitory
17 Disposal room
18 Consultation room
19 Observation room
20 Indoor leisure area
21 Multifunction room
22 Preparation room
23 Equipment room
24 Activity room
25 Storeroom
26 The library
27 Above the indoor leisure area
28 Terrace
29 Above the multipurpose hall

First Floor Plan 1：250

In order to maintain the temperature of the biogas digester above ten degrees, the solar water system is used to maintain the temperature during sunny days in winter. In rainy or snowy weather, the biogas boiler is turned on to ensure the temperaturefluctuation is less than three degrees.

South Elevation

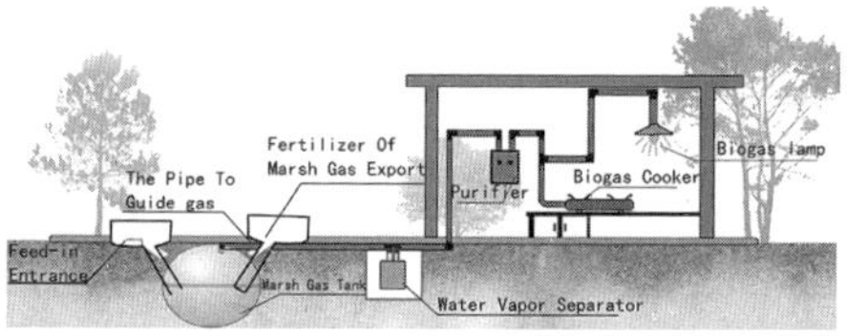

Fertilizer Of Marsh Gas Export
The Pipe To Guide gas
Purifier
Biogas Cooker
Biogas lamp
Feed-in Entrance
Marsh Gas Tank
Water Vapor Separator

West Elevation

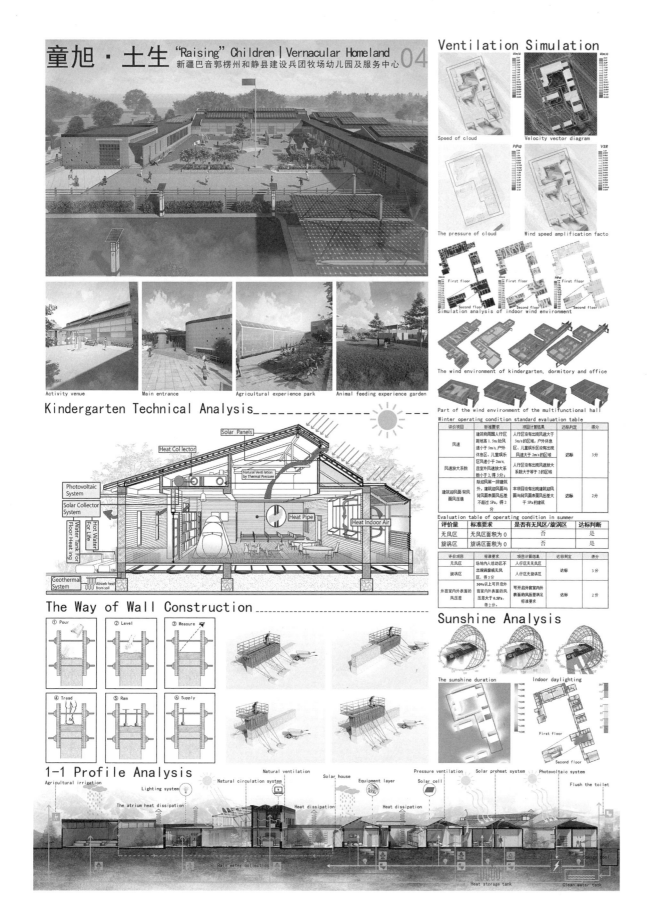

童旭·土生 “Raising” Children | Vernacular Homeland 04

新疆巴音郭楞州和静县建设兵团牧场幼儿园及服务中心

Activity venue

Main entrance

Agricultural experience park

Animal feeding experience garden

Kindergarten Technical Analysis

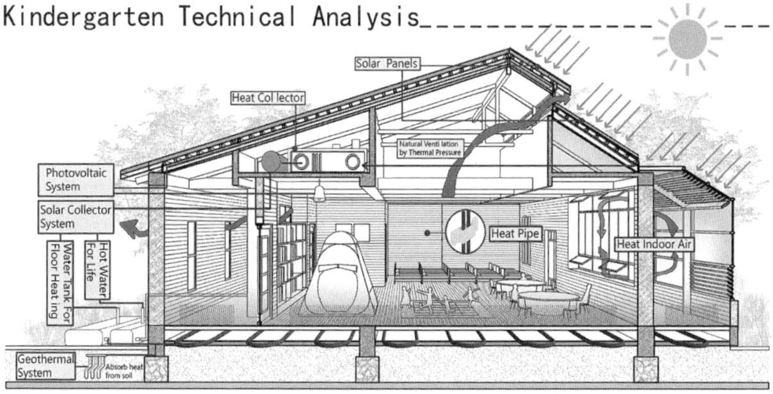

Solar Panels
Heat Collector
Natural Ventilation by Thermal Pressure
Photovoltaic System
Solar Collector System
Hot Water For Life
Water Tank For Floor Heating
Heat Pipe
Heat Indoor Air
Geothermal System
Absorb heat from soil

The Way of Wall Construction

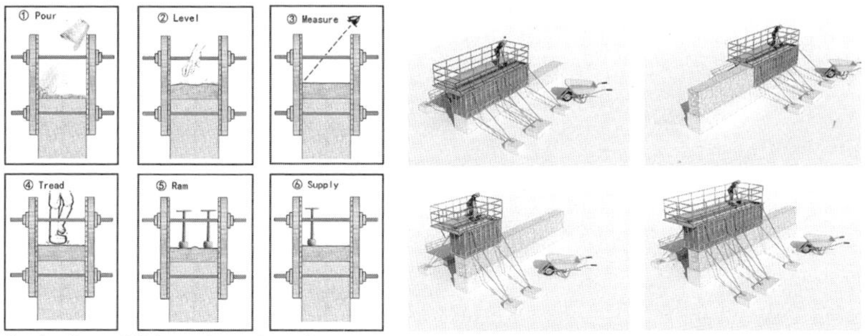

① Pour
② Level
③ Measure
④ Tread
⑤ Ram
⑥ Supply

1-1 Profile Analysis

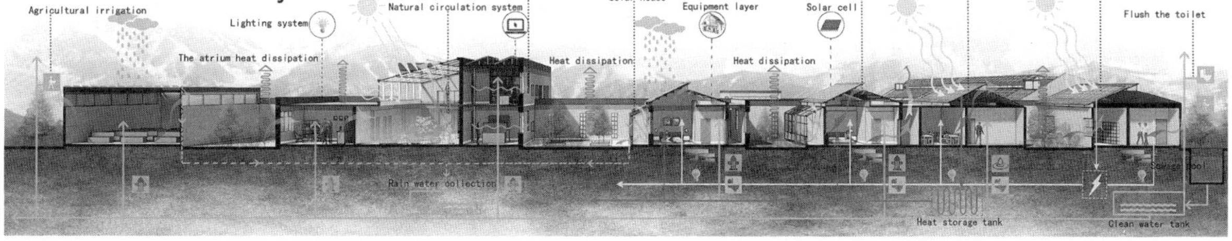

Agricultural irrigation
Lighting system
The atrium heat dissipation
Natural circulation system
Natural ventilation
Solar house
Equipment layer
Heat dissipation
Heat dissipation
Pressure ventilation
Solar cell
Solar preheat system
Photovoltaic system
Flush the toilet
Rain water collection
Heat storage tank
Clean water tank

Ventilation Simulation

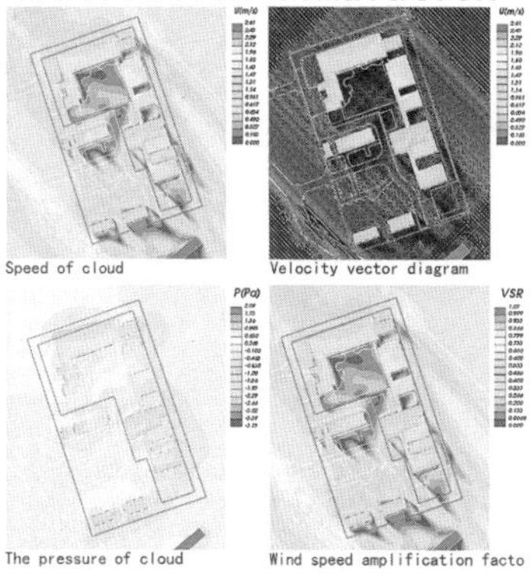
Speed of cloud
Velocity vector diagram
The pressure of cloud
Wind speed amplification facto

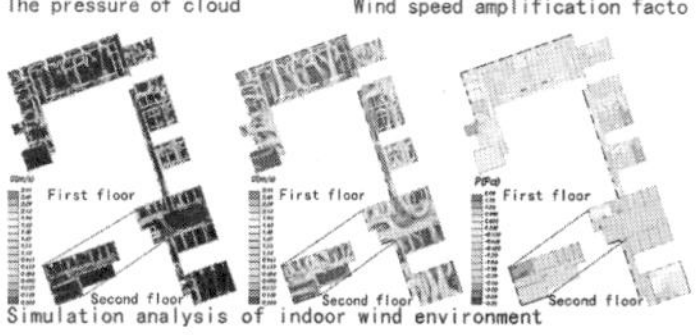

First floor
Second floor
First floor
Second floor
First floor
Second floor

Simulation analysis of indoor wind environment

The wind environment of kindergarten, dormitory and office

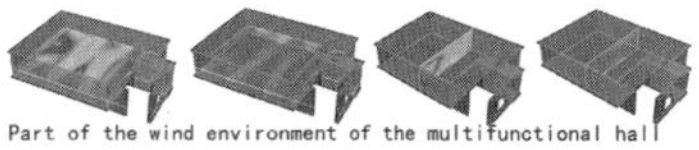
Part of the wind environment of the multifunctional hall

Winter operating condition standard evaluation table

评价项目	标准要求	项目计算结果	达标判定	得分
风速	建筑物周围人行区距地高1.5m处风速小于5m/s,户外休息区、儿童娱乐区风速小于2m/s,且室外风速放大系数小于2,得3分;	人行区没有出现风速大于5m/s的区域,户外休息区、儿童娱乐区没有出现风速大于2m/s的区域	达标	3分
风速放大系数		人行区没有出现风速放大系数大于等于2的区域		
建筑迎风面/背风面风压值	除迎风第一排建筑外,建筑迎风面与背风面表面风压差不超过5Pa,得2分	本项目没有出现建筑迎风面与背风面表面风压差大于5Pa的建筑	达标	2分

Evaluation table of operating condition in summer

评价量	标准要求	是否有无风区/旋涡区	达标判断
无风区	无风区面积为0	否	是
旋涡区	旋涡区面积为0	否	是

评价项目	标准要求	项目计算结果	达标判定	得分
无风区	场地内人活动区不出现涡旋或无风区,得3分	人行区无无风区	达标	3分
旋涡区		人行区无旋涡区		
外窗室内外表面的风压差	50%以上可开启外窗室内外表面的风压差大于0.5Pa,得2分。	可开启外窗室内外表面的风压差满足标准要求	达标	2分

Sunshine Analysis

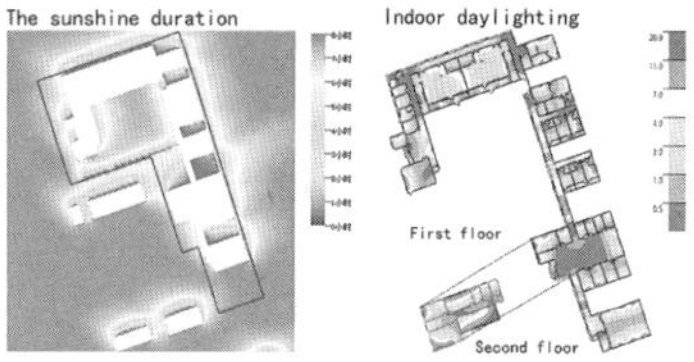

First floor
Second floor

The sunshine duration
Indoor daylighting

综合奖·优秀奖
General Prize Awarded · Honorable Mention Prize

注 册 号：7129

项目名称：朝晖·光场（新疆）

Haven（Xinjiang）

作　　者：黄　瑞、沈　焜、熊旎颖、高　青、卢惠阳、徐艺倩、杰　布、李永晖、刘　奎

参赛单位：湖南省建筑设计院有限公司、上迈新能源（北京）公司

指导教师：蔡　屹、Tim Mason

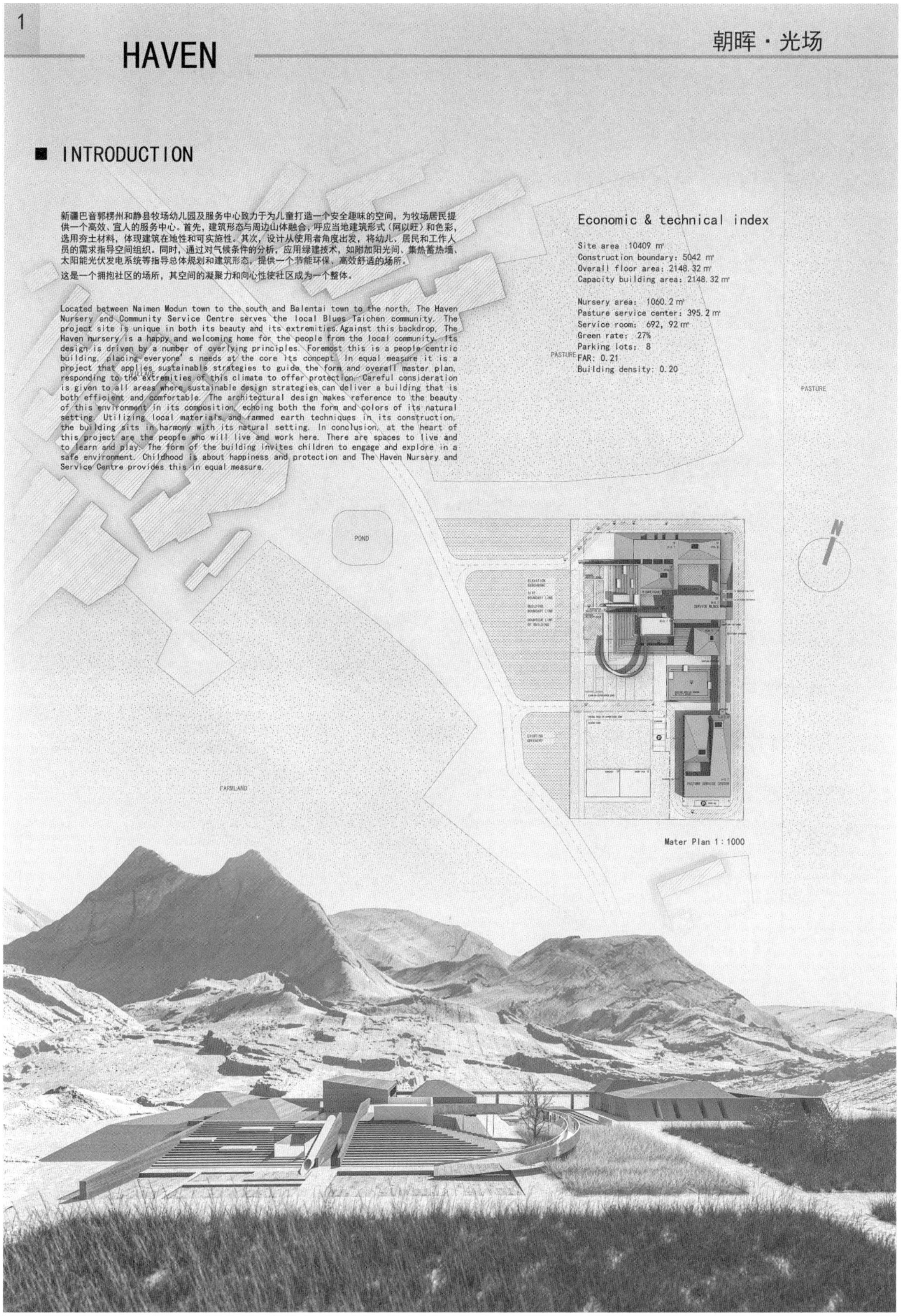
1
HAVEN
朝晖 · 光场
INTRODUCTION
新疆巴音郭楞州和静县牧场幼儿园及服务中心致力于为儿童打造一个安全趣味的空间，为牧场居民提供一个高效、宜人的服务中心。首先，建筑形态与周边山体融合，呼应当地建筑形式（阿以旺）和色彩，选用夯土材料，体现建筑在地性和可实施性。其次，设计从使用者角度出发，将幼儿、居民和工作人员的需求指导空间组织。同时，通过对气候条件的分析，应用绿建技术，如附加阳光间、集热蓄热墙、太阳能光伏发电系统等指导总体规划和建筑形态，提供一个节能环保、高效舒适的场所。
这是一个拥抱社区的场所，其空间的凝聚力和向心性使社区成为一个整体。
Located between Naimen Modun town to the south and Balentai town to the north. The Haven Nursery and Community Service Centre serves the local Blues Taichen community. The project site is unique in both its beauty and its extremities.Against this backdrop, The Haven nursery is a happy and welcoming home for the people from the local community. Its design is driven by a number of overlying principles. Foremost this is a people centric building, placing everyone' s needs at the core its concept. In equal measure it is a project that applies sustainable strategies to guide the form and overall master plan, responding to the extremities of this climate to offer protection. Careful consideration is given to all areas where sustainable design strategies can deliver a building that is both efficient and comfortable. The architectural design makes reference to the beauty of this environment in its composition, echoing both the form and colors of its natural setting. Utilizing local materials and rammed earth techniques in its construction, the building sits in harmony with its natural setting. In conclusion, at the heart of this project are the people who will live and work here. There are spaces to live and to learn and play. The form of the building invites children to engage and explore in a safe environment. Childhood is about happiness and protection and The Haven Nursery and Service Centre provides this in equal measure.
Economic & technical index
Site area :10409 m²
Construction boundary: 5042 m²
Overall floor area: 2148.32 m²
Capacity building area: 2148.32 m²
Nursery area: 1060.2 m²
Pasture service center: 395.2 m²
Service room: 692, 92 m²
Green rate: 27%
Parking lots: 8
FAR: 0.21
Building density: 0.20
PASTURE
PASTURE
POND
FARMLAND
Mater Plan 1:1000

2

HAVEN

朝晖 · 光场

■ DESIGN PRINCIPLE

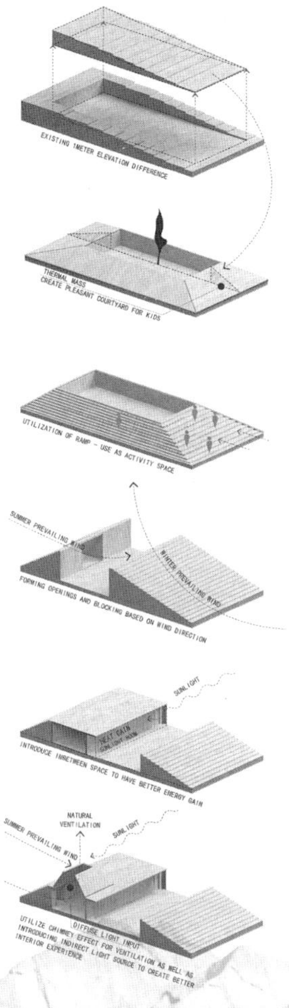
EXISTING 1METER ELEVATION DIFFERENCE
THERMAL MASS
CREATE PLEASANT COURTYARD FOR KIDS
UTILIZATION OF RAMP - USE AS ACTIVITY SPACE
SUMMER PREVAILING WIND
WINTER PREVAILING WIND
FORMING OPENINGS AND BLOCKING BASED ON WIND DIRECTION
SUNLIGHT
HEAT GAIN
INTRODUCE INBETWEEN SPACE TO HAVE BETTER ENERGY GAIN
NATURAL VENTILATION
SUNLIGHT
SUMMER PREVAILING WIND
DIFFUSE LIGHT INPUT
UTILIZE CHIMNEY EFFECT FOR VENTILATION AS WELL AS INTRODUCING INDIRECT LIGHT SOURCE TO CREATE BETTER INTERIOR EXPERIENCE

■ SITE RESEARCH

□ LOCAL BUILDINGS

□ LOCAL CLIMATE ANALYSIS

1. Annual Temperature

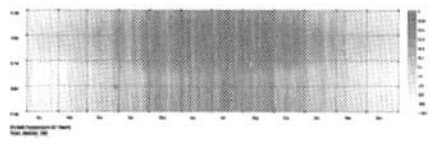

2. Annual Wind Speed

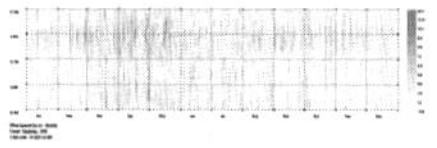

3. Annual Raletive Humidity

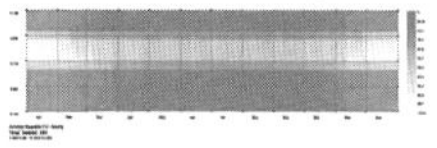

4. Global Horizontal Radiation

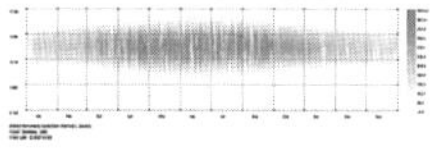

5. Outdoor Comfort Analysis

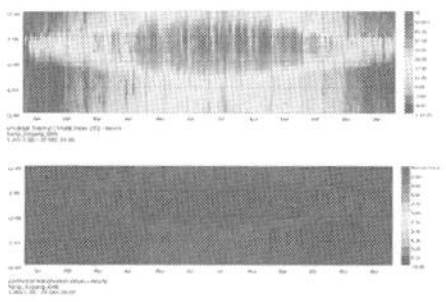

6. Wind Rose

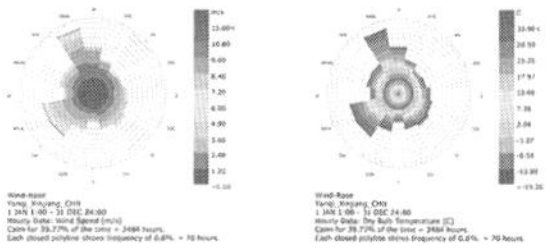

7. Solar Coronal Map

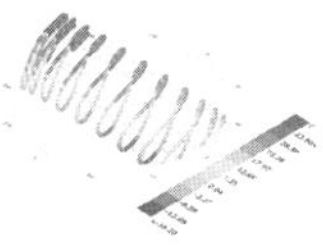

8. Comfortable Hours Affected by Different Green Building Strategies

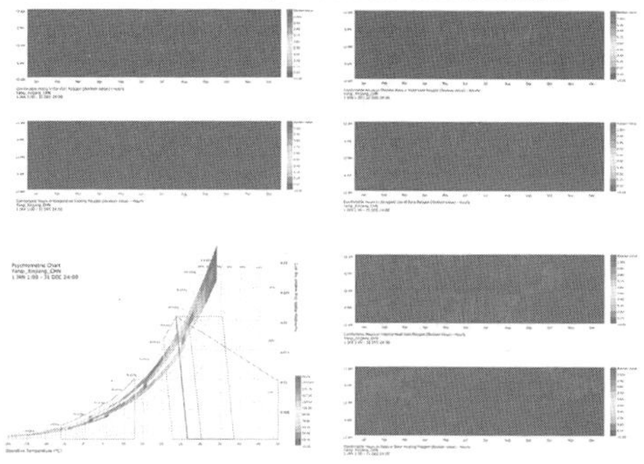

Climate Analysis Summary:

1. Majority of the year is cold. However, summertime is comfortable.
2. Summer's prevailing wind is from the northwest, while in Winter the wind is mainly from the southwest.
3. The radiation is strong and mainly from the southwest.
4. The wind environment has a positive influence in summers however acting the opposite in winters in terms of humidity and temperature.

Green Design Stratigies:

1. Based on the climate analysis, our design aim is to lower the body temperature during the day in Summer as well as keep the indoor warm during the Winter.
2. There are mainly four stratigies in Summer including the Evaporative Cooling System, Thermal Mass +Night Vent System, Occupant Use of Fans and Internal Heat Gain. Meanwhile the passive Solar Heating System has a big role in Improving the comfort in Winter.
3. According to the Psychrometric Chart, using the stratigies of the evaporative Cooling System, Thermal Mass +Night Vent System, Occupant Use of Fans can lower the temperature during the day in Summer. Also using the stratigy of the Internal Heat Gain System can improve the comfort during the night in Summer.

3

HAVEN

朝晖 · 光场

WEST ELEVATION 1 : 250

■ RAMMED EARTH

□ RAMMED EARTH CONSTRUCTION

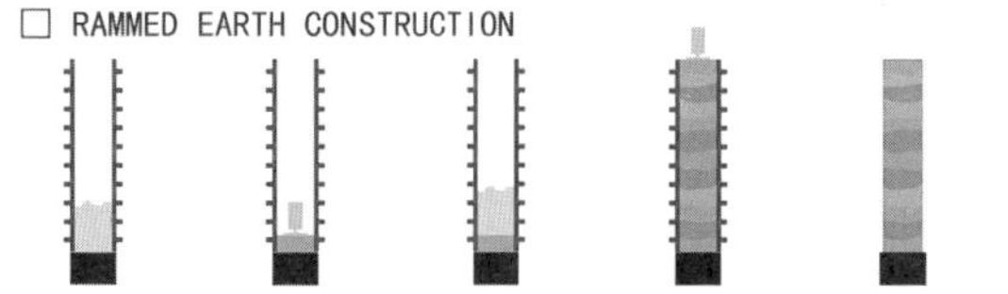
Framework & Filling
Rammed
Filling
Compression
Framework removal

□ BUILDING PROCESS

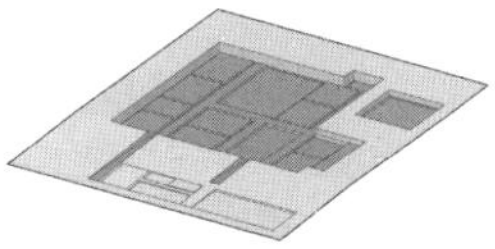

1. Excavation of foundation pit

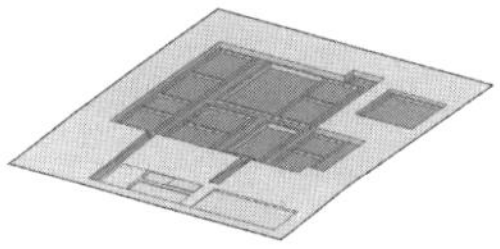

2. Concrete cast ring beam

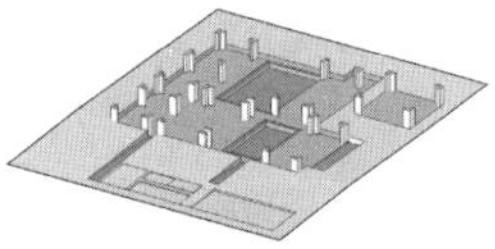

3. Concrete cast floor slab and shear wall

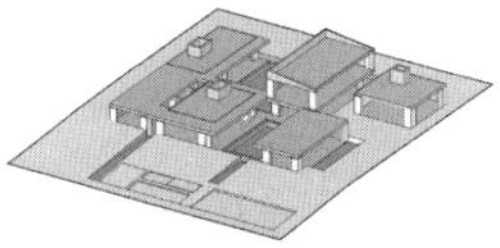

4. Concrete cast the second floor and roof

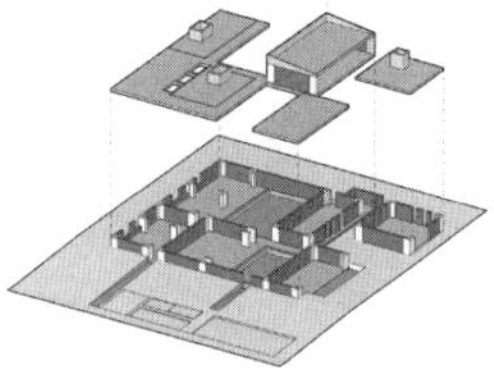

5. Rammed earth wall

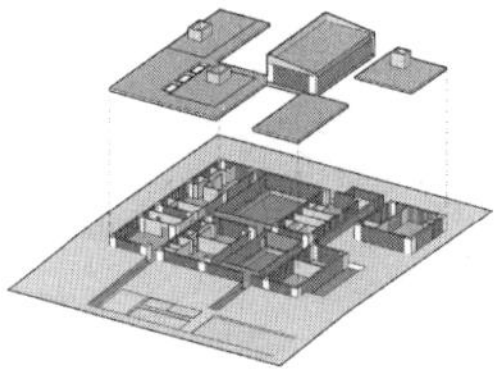

6. Build internal partitions, install doors and windows

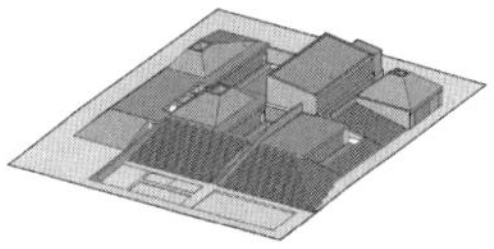

7. Turns the soil on roofs and surrounding

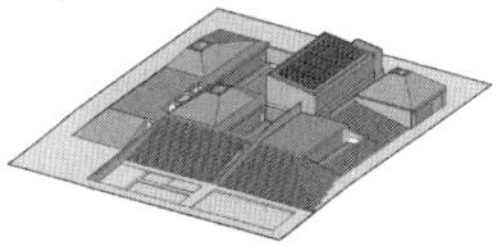

8. Install the solar panels

□ CONSTRUCTION DETAILS

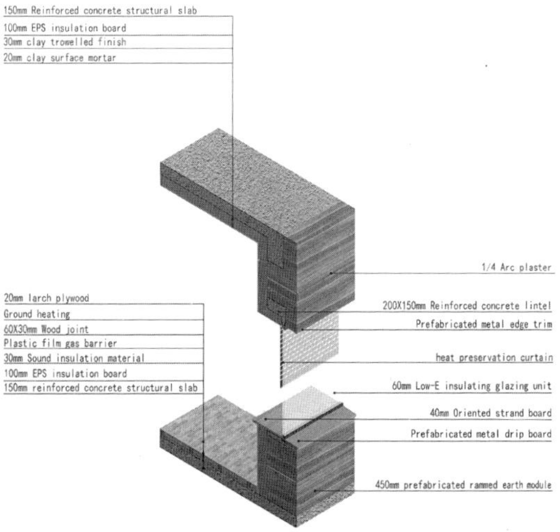
150mm Reinforced concrete structural slab
100mm EPS insulation board
30mm clay trowelled finish
20mm clay surface mortar
1/4 Arc plaster
20mm larch plywood
Ground heating
60X30mm Wood joint
Plastic film gas barrier
30mm Sound insulation material
100mm EPS insulation board
150mm reinforced concrete structural slab
200X150mm Reinforced concrete lintel
Prefabricated metal edge trim
heat preservation curtain
60mm Low-E insulating glazing unit
40mm Oriented strand board
Prefabricated metal drip board
450mm prefabricated rammed earth module

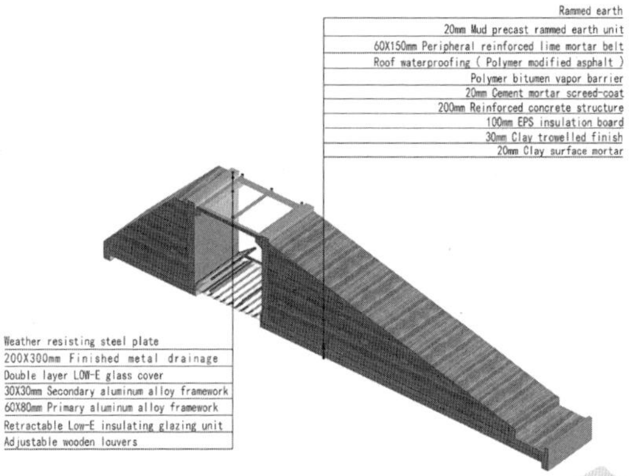
Rammed earth
20mm Mud precast rammed earth unit
60X150mm Peripheral reinforced lime mortar belt
Roof waterproofing (Polymer modified asphalt)
Polymer bitumen vapor barrier
20mm Cement mortar screed-coat
200mm Reinforced concrete structure
100mm EPS insulation board
30mm Clay trowelled finish
20mm Clay surface mortar
Weather resisting steel plate
200X300mm Finished metal drainage
Double layer LOW-E glass cover
30X30mm Secondary aluminum alloy framework
60X80mm Primary aluminum alloy framework
Retractable Low-E insulating glazing unit
Adjustable wooden louvers

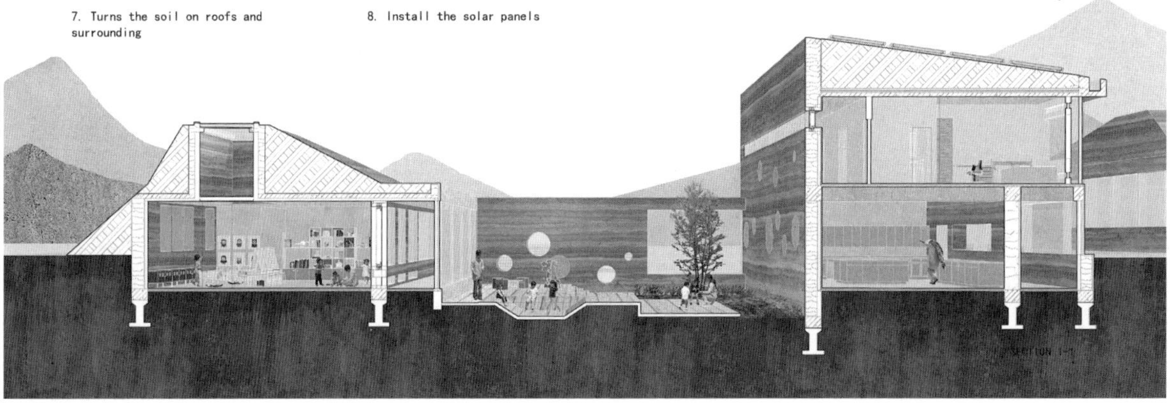

SECTION 1-1

4

HAVEN

朝晖 · 光场

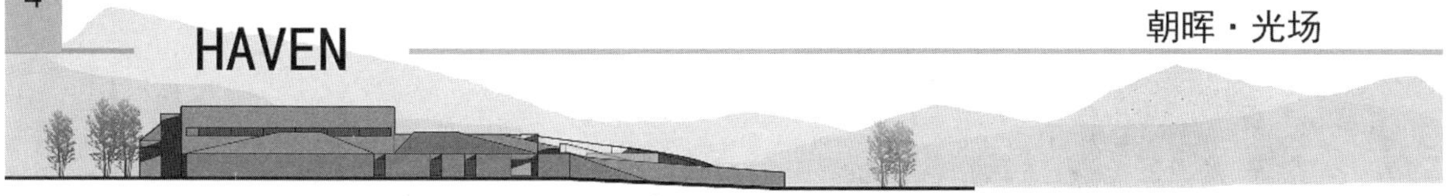

NORTH ELEVATION 1 : 250

□ ENERGY CONSUMPTION AND EFFICIENCY ANALYSIS

Overall energy consumption and energy efficiency — Without applying new energy technologies, the annual energy consumption rate of the kindergarten and service center is 14MWh and the heating energy consumption is 55% (7.7mWh i.e 77000kwh). Utilizing the solar thermal system and the water source heat pump generates 70mWh (7000kwh) and provides winter heating consumption of 20kwh/hr. The heating energy consumption adds up to 3.6MWh (36000kwh). Water source heat pump can achieve energy saving up to 4.1MWh (41000kwh). Reducing the overall energy consumption rate of the building to 11.1 MWH (110000kwh) and energy saving rate to 79%.

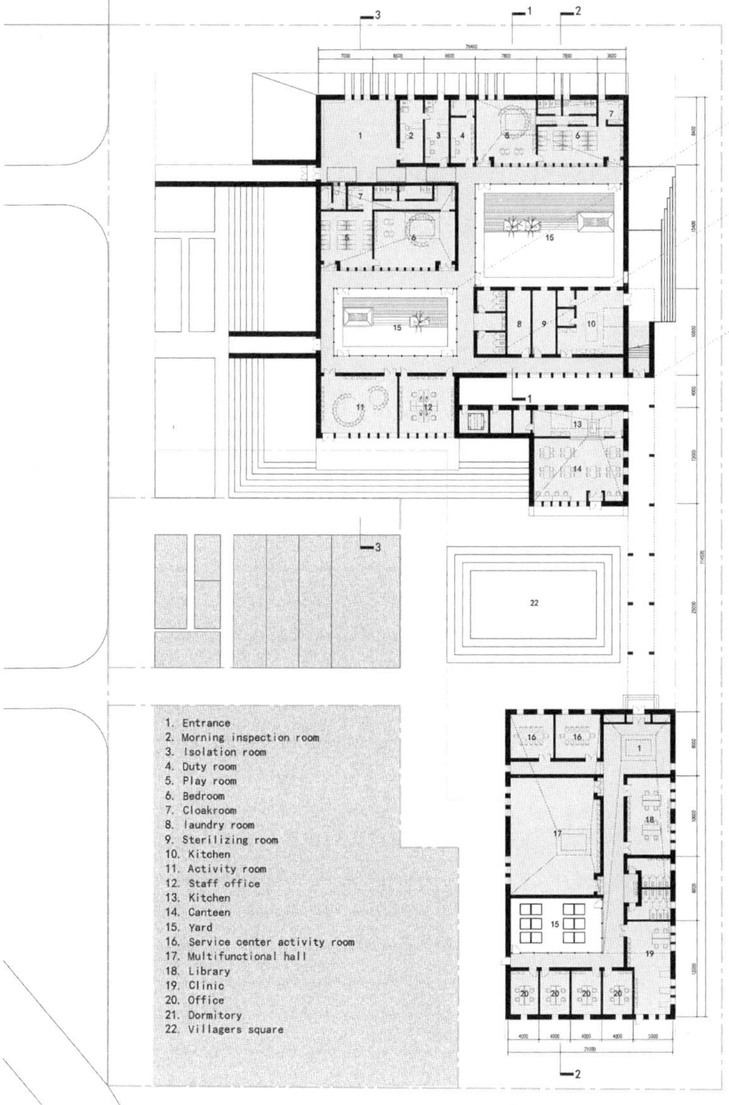
1. Entrance
2. Morning inspection room
3. Isolation room
4. Duty room
5. Play room
6. Bedroom
7. Cloakroom
8. laundry room
9. Sterilizing room
10. Kitchen
11. Activity room
12. Staff office
13. Kitchen
14. Canteen
15. Yard
16. Service center activity room
17. Multifunctional hall
18. Library
19. Clinic
20. Office
21. Dormitory
22. Villagers square

1F PLAN 1 : 250

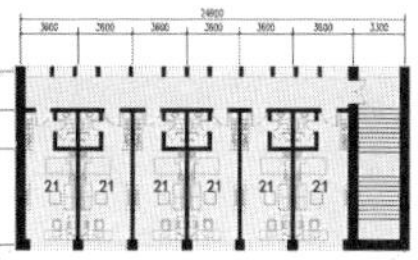

2F PLAN 1 : 250

□ ESTIMATED BUILDING ENERGY CONSUMPTION

- Electricity Consumption (kWh)

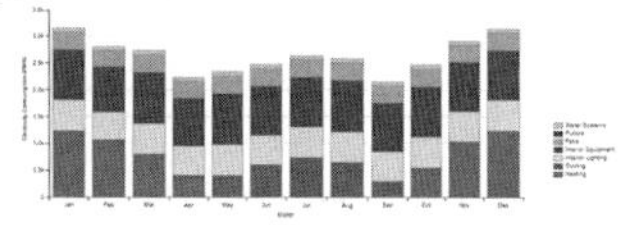

- Electricity Peak Demand (kW)

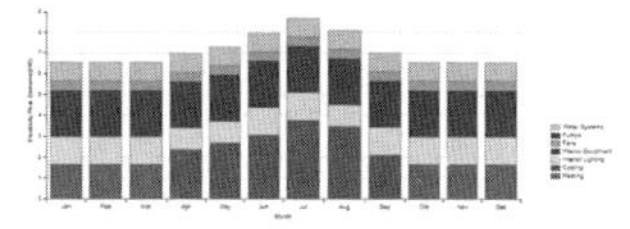

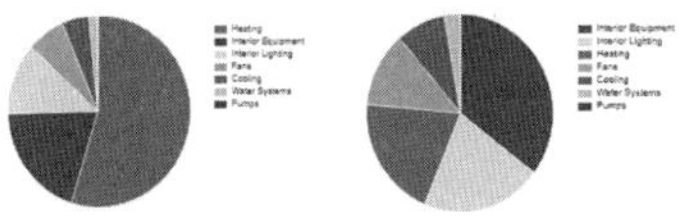

- HVAC Load Profiles

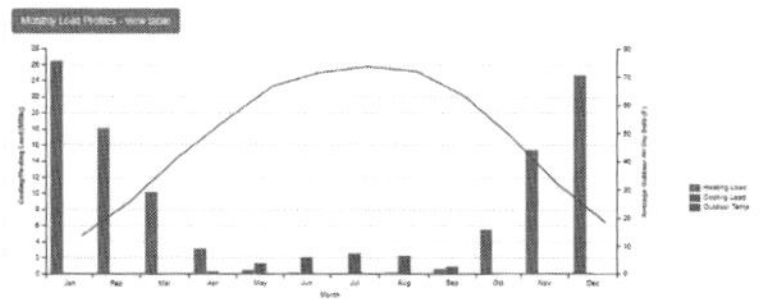

- Building Water Use

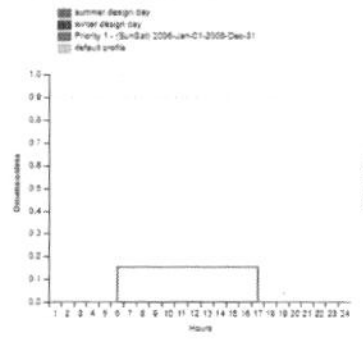

- Core Zone Occupancy

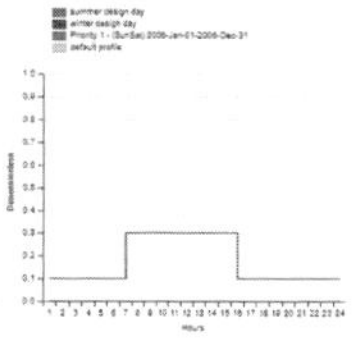

- Building Light

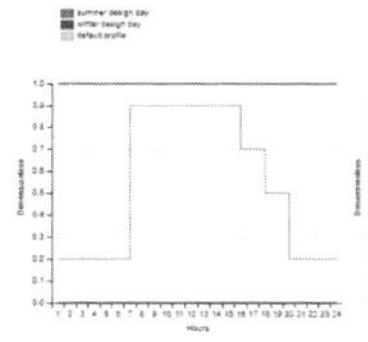

- Building Plug Load

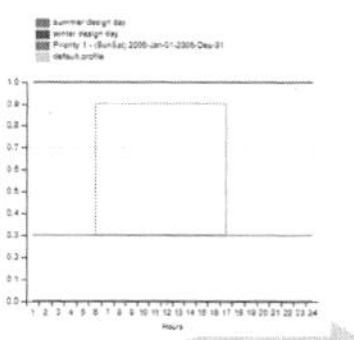

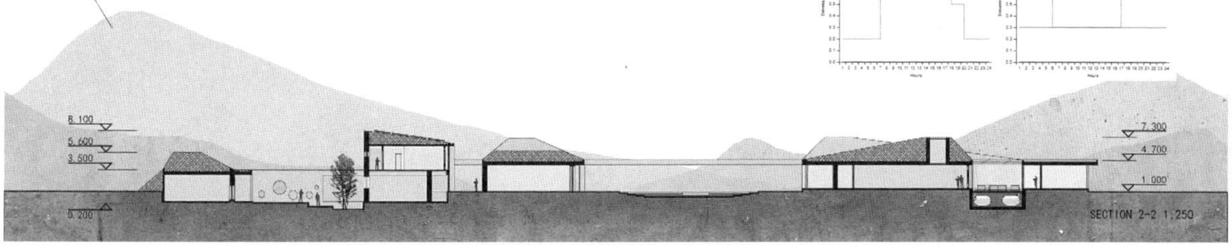
8.100
5.600
3.500
0.200
7.300
4.700
1.000

SECTION 2-2 1:250

5

HAVEN

朝晖 · 光场

■ SUSTAINABLE STRATEGIES

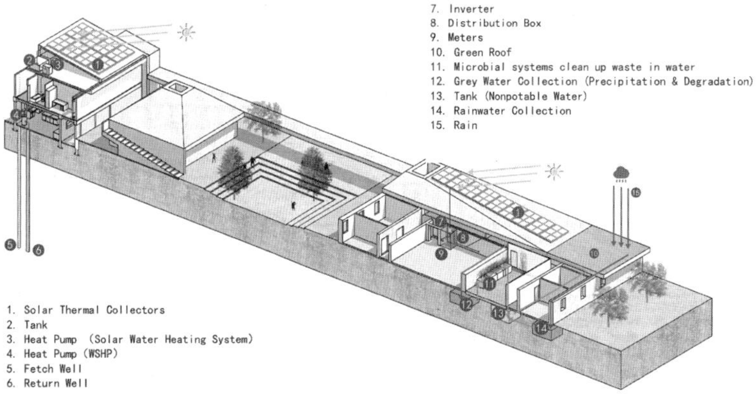
7. Inverter
8. Distribution Box
9. Meters
10. Green Roof
11. Microbial systems clean up waste in water
12. Grey Water Collection (Precipitation & Degradation)
13. Tank (Nonpotable Water)
14. Rainwater Collection
15. Rain
1. Solar Thermal Collectors
2. Tank
3. Heat Pump (Solar Water Heating System)
4. Heat Pump (WSHP)
5. Fetch Well
6. Return Well

Green Strategies

- Passive Solar Energy
  - Site Design
    - ■ Site Ventilation Design
  - Architecture Design
    - ■ Direct-gain windows
    - ■ Additional sunroom
    - ■ Heat collection and storage wall
    - ■ Patio ventilation skylight
    - ■ Architectural shading
    - ■ Roof Greening
- Active Solar Energy
  - ■ Solar photothermal system
  - ■ WSHP + Solar thermal heating
  - ■ CSPV
  - ■ Efficient fresh air system
- Other Technologies
  - ■ Rainwater collection
  - ■ Living Machine (sewage disposal)

□ ENERGY ANALYSIS

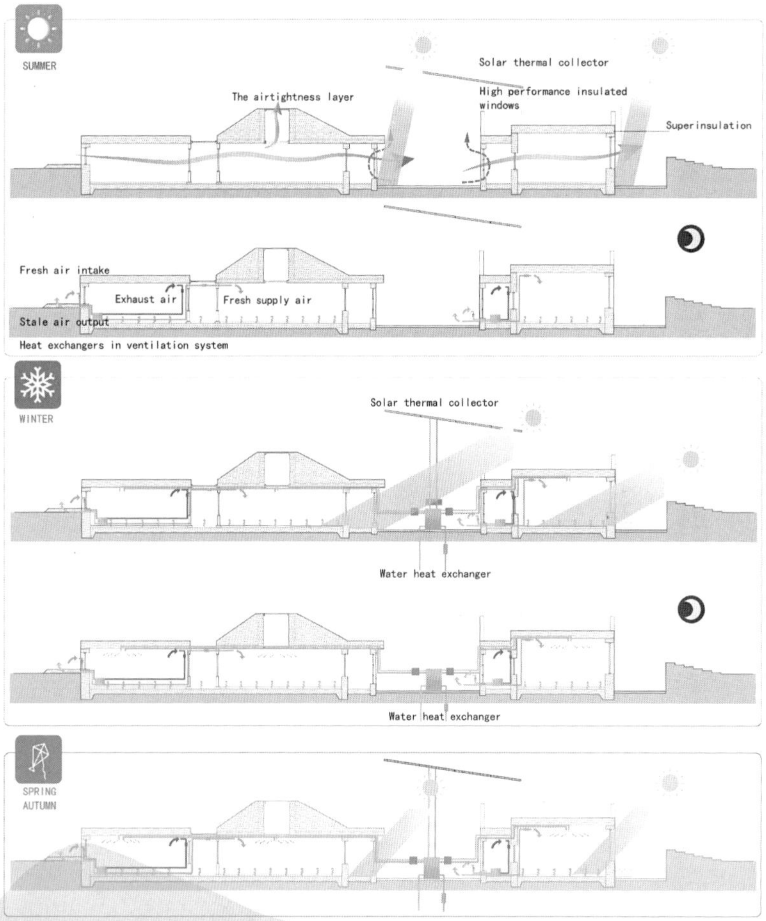
SUMMER
Solar thermal collector
The airtightness layer
High performance insulated windows
Superinsulation
Fresh air intake
Exhaust air
Fresh supply air
Stale air output
Heat exchangers in ventilation system
WINTER
Solar thermal collector
Water heat exchanger
Water heat exchanger
SPRING AUTUMN

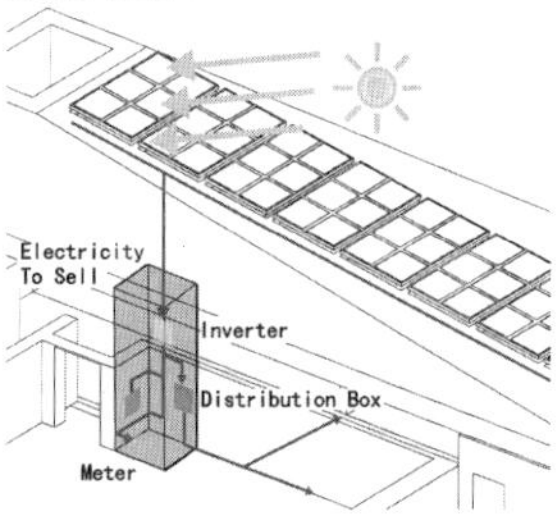
Solar PV system
Electricity To Sell
Inverter
Distribution Box
Meter

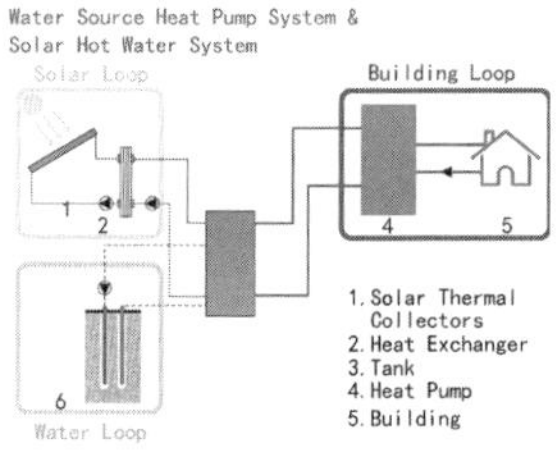
Water Source Heat Pump System & Solar Hot Water System
Solar Loop
Building Loop
Water Loop
1. Solar Thermal Collectors
2. Heat Exchanger
3. Tank
4. Heat Pump
5. Building

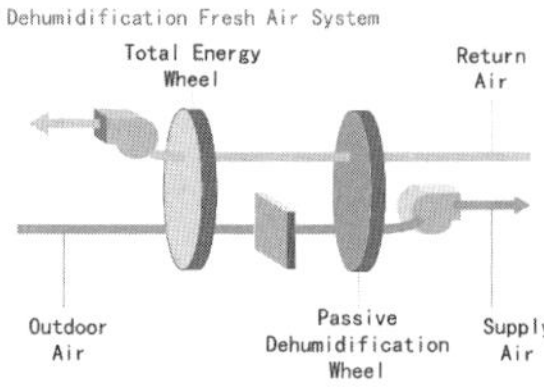
Dehumidification Fresh Air System
Total Energy Wheel
Return Air
Outdoor Air
Passive Dehumidification Wheel
Supply Air

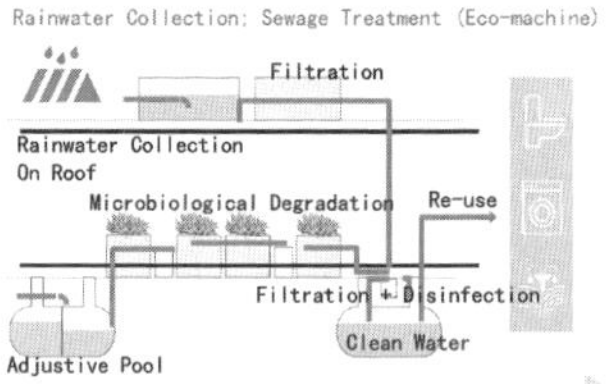
Rainwater Collection: Sewage Treatment (Eco-machine)
Filtration
Rainwater Collection On Roof
Microbiological Degradation
Re-use
Filtration + Disinfection
Clean Water
Adjustive Pool

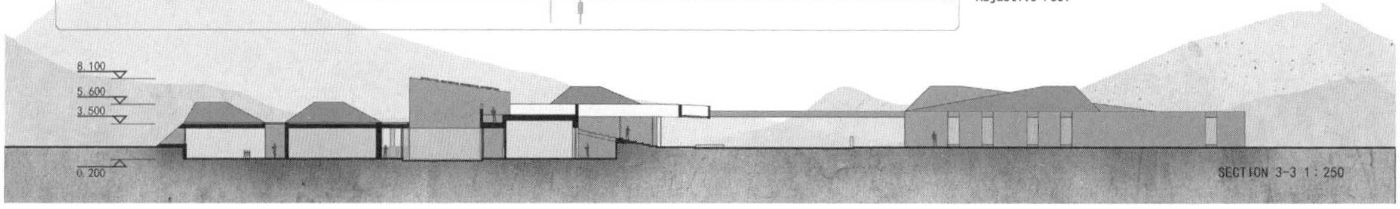
8.100
5.600
3.500
0.200
SECTION 3-3 1:250

6

朝晖·光场

HAVEN

☐ PRE-ASSESSMENT of NATIONAL GREEN BUILDING CERTIFICATION

	PRE-ASSESSMENT INDIVIDUAL VALUE						CREATIVITY AND INNOVATION SCORE
	INDIVIDUAL PRE-ASSESSMENT SCORE	SAFETY AND DURABILITY	HEALTH AND COMFORT	LIFE CON-VENIENCE	SAVE RESOURCES	ENVIRONMENT LIVABILITY	
TOTAL SCORE OF PRE-ASSESSMENT	400	100	100	70	200	100	100
ATTAINED SCORE	400	75	87	42	160	78	15
MINIMUM SCORE	—	30	30	21	60	30	—
INDIVIDUAL SCORE	40	7.5	8.7	4.2	16	7.8	1.5
TOTAL SCORE							

☐ SUSTAINABLE SYSTEMS DIAGRAM

1 RAMMED EARTH STRUCTURE
Use local materials and work as a thermal mass, which could stabilize the temperature

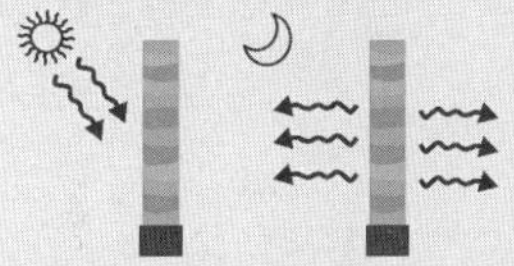

2 LIVING SKYLIGHT
During the summer daytime, keep the window open, stack effect boosted by wind pressure to enhance the ventilation

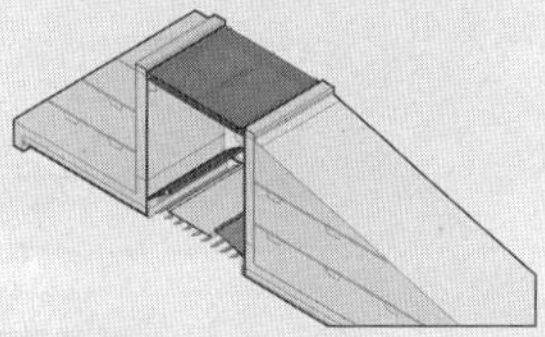

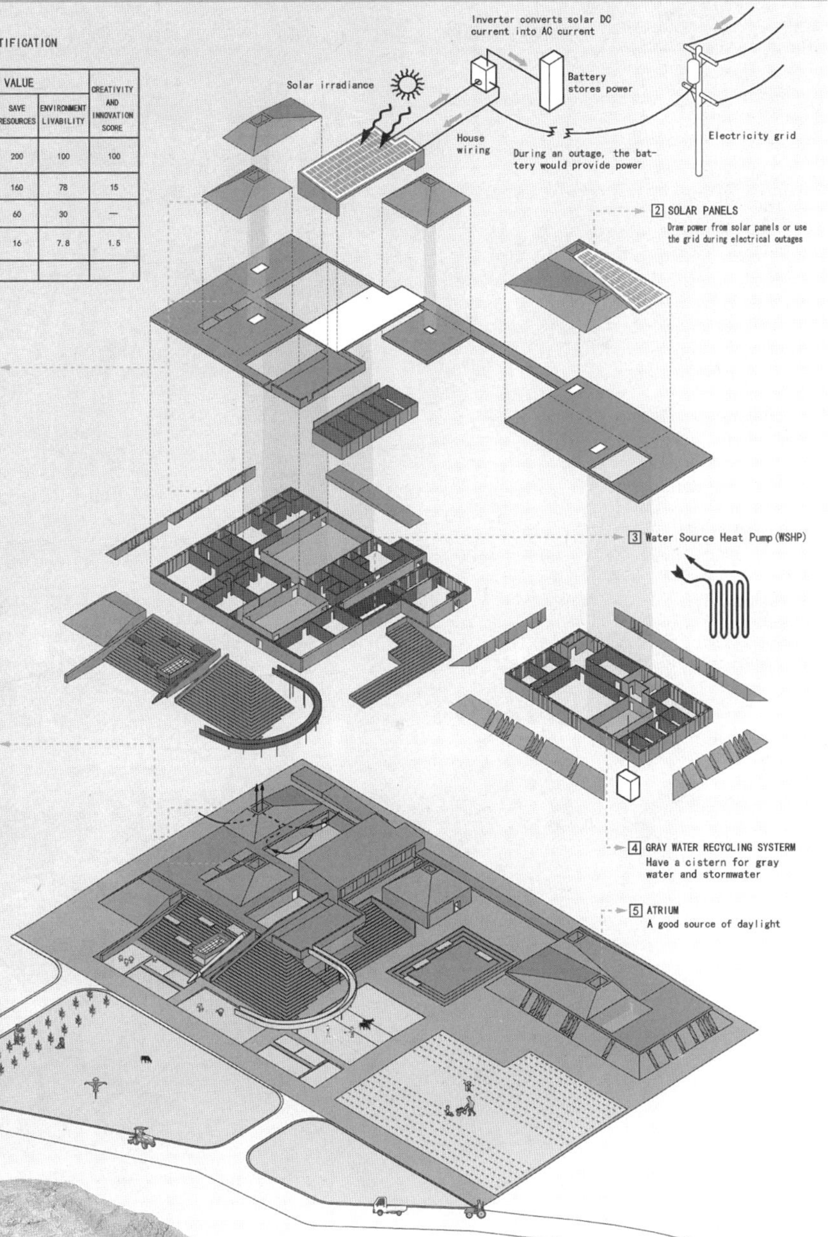

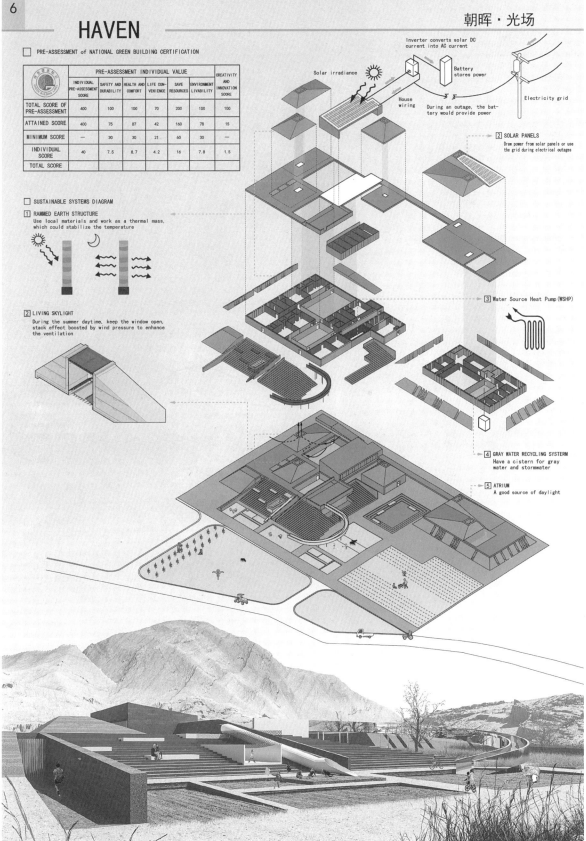

Inverter converts solar DC current into AC current
Solar irradiance
Battery stores power
House wiring
During an outage, the battery would provide power
Electricity grid
2 SOLAR PANELS
Draw power from solar panels or use the grid during electrical outages
3 Water Source Heat Pump (WSHP)
4 GRAY WATER RECYCLING SYSTERM
Have a cistern for gray water and stormwater
5 ATRIUM
A good source of daylight

COLLECTING SUNSHINE 1

综合奖 · 优秀奖
General Prize Awarded · Honorable Mention Prize

注册号：7145
项目名称：光之幼儿园（新疆）
Collecting Sunshine（Xinjiang）
作　　者：陈文静、董春朝、白一伟
参赛单位：西北工业大学
指导教师：李　静、毕景龙

LOCATION ANALYSIS

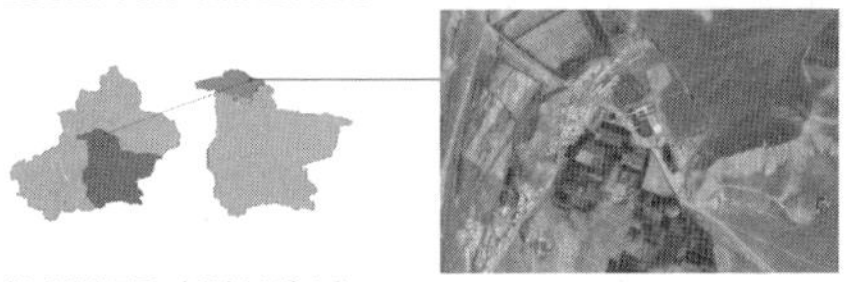

CLIMATE ANALYSIS

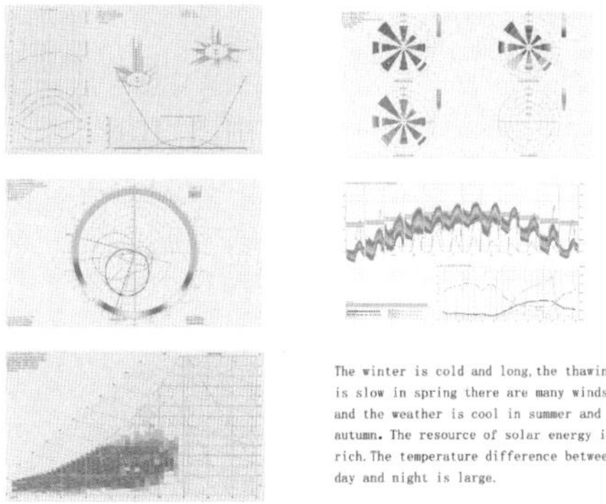

The winter is cold and long, the thawing is slow in spring there are many winds, and the weather is cool in summer and in autumn. The resource of solar energy is rich. The temperature difference between day and night is large.

SITE ENVIRONMENT ANALYSIS

Village

The west of the project site is the village.

Road

The roads inside the village are rib like.

Water Resource

The north is separated from the pasture by a stream, and the East is a pool.

Green Belt

There are two parts of grassland in the north and the south, and less greening in the village.

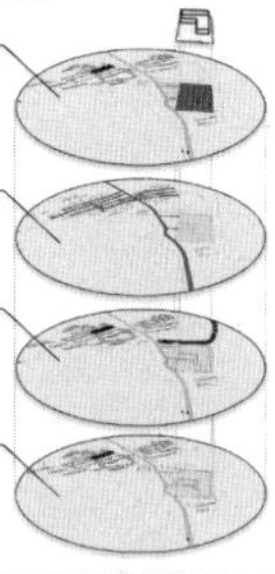

CURRENT SITUATION ANALYSIS

Village houses　The road outside the project site

Current project land　Current land Eentrance

The old houses are dilapidated, which need to be rebuilt. The infrastructure in the base is poor, no drainage measures, and unstable communication, but with running water system, poor power supply stability. The village is in the northwest side of the land, with a distance of about 100 meters. The land is surrounded by pastures.

DESIGN CONCEPT

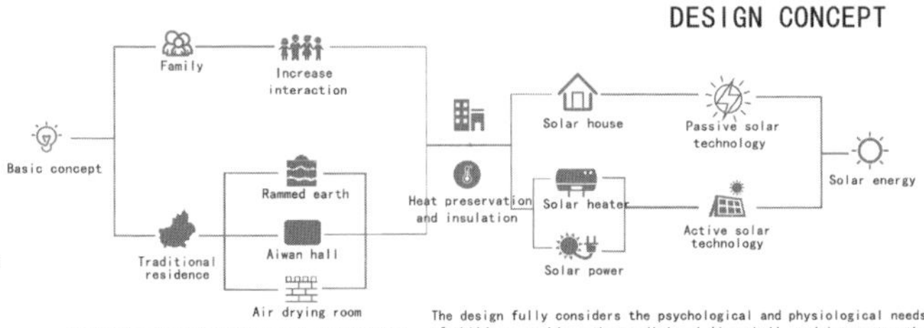
Family
Increase interaction
Basic concept
Rammed earth
Aiwan hall
Traditional residence
Air drying room
Heat preservation and insulation
Solar house
Solar heater
Solar power
Passive solar technology
Active solar technology
Solar energy

本设计充分考虑幼儿的心理和生理需求，将新疆传统民中的阿以往大厅形式与整体建筑结合，在形成过程中将主动式与被动式太阳能技术、A2/O污水处理系统、风力发电技术等纳入建筑之中，形成完整的节能系统。建筑建造采用适应当地气候特征的夯土结构来减少建造成本和后期维护成本，更好达到保温隔热的效果，为孩子、村民提供温馨的生活、娱乐场所。

The design fully considers the psychological and physiological needs of children, combines the traditional Aiwan hall, and integrates the active and passive solar energy technology, A2 / O sewage treatment system, wind power generation technology, etc. into the building during the formation process, forming a complete energy-saving system. The rammed earth structure adapted to the local climate features is adopted in the construction to reduce the construction cost and later maintenance cost, better achieve the effect of heat preservation and insulation, and form a warm living and entertainment place for children and villagers.

TIME OF DIFFERENT PEOPLE IN DIFFERENT PLACES

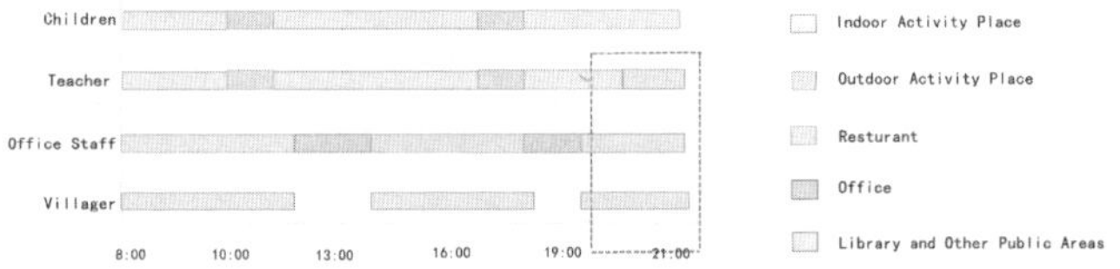
Children
Teacher
Office Staff
Villager
8:00
10:00
13:00
16:00
19:00
21:00
Indoor Activity Place
Outdoor Activity Place
Resturant
Office
Library and Other Public Areas

COLLECTING SUNSHINE 2

SITE PLANTING/LOGICAL GENERATION

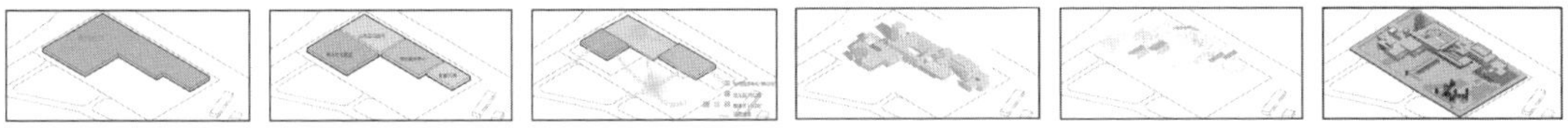

1. The land boundary is irregular.
2. According to the outline of building land, fit the outline of building and divide the area reasonably
3. Two Aiwan living units form the overall Aiwan pattern and create a flow line of activities
4. Refine the connection of partition blocks and deepen the modeling design.
5. Deepen the details
6. Appropriate landscape layout adjusts microclimate, deepens square and strengthens blocks connection.

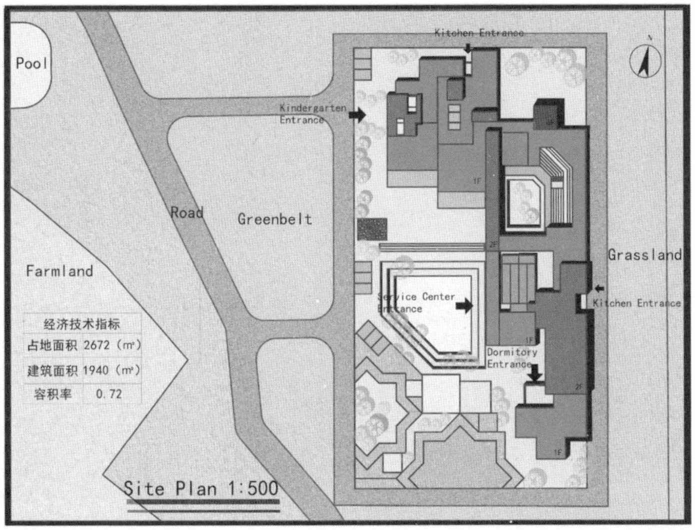

Pool
Kitchen Entrance
Kindergarten Entrance
Road
Greenbelt
Farmland
Grassland
Service Center Entrance
Kitchen Entrance
Dormitory Entrance
Site Plan 1:500

经济技术指标	
占地面积	2672（m²）
建筑面积	1940（m²）
容积率	0.72

TRAFFIC ANALYSIS

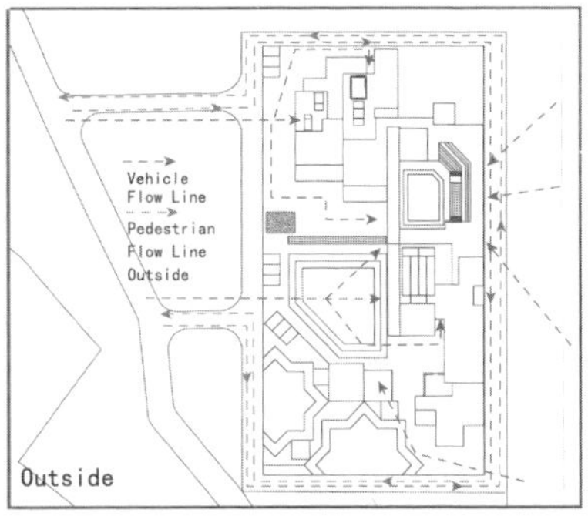

Vehicle Flow Line
Pedestrian Flow Line
Outside
Outside

SUNSHINE ANALYSIS

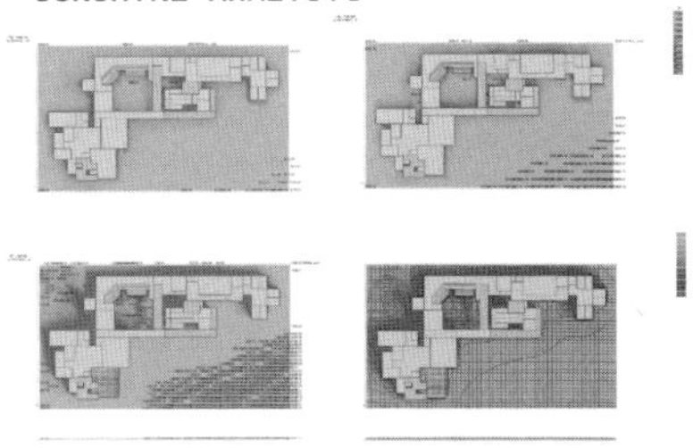

SHADOW ANALYSIS

WIND POWER SYSTEM

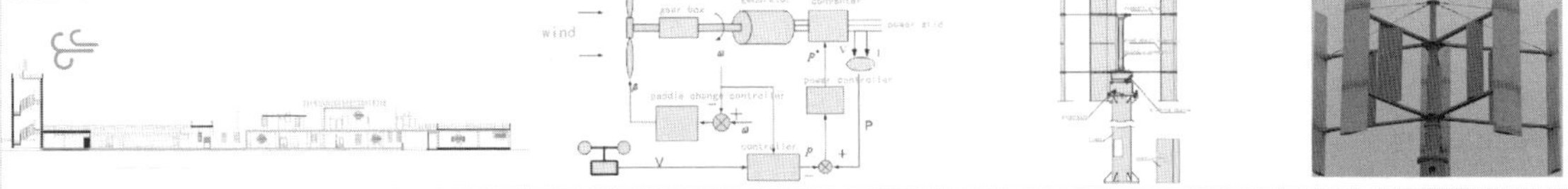

wind

SOLAR WATER HEATER SYSTEM

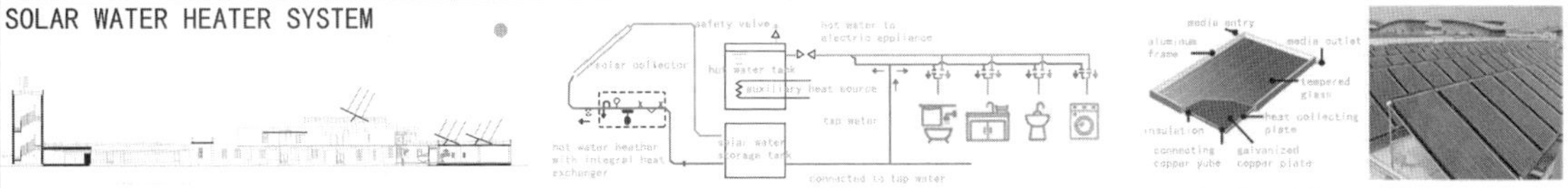

SOLAR PHOTOVOLTAIC SYSTEM

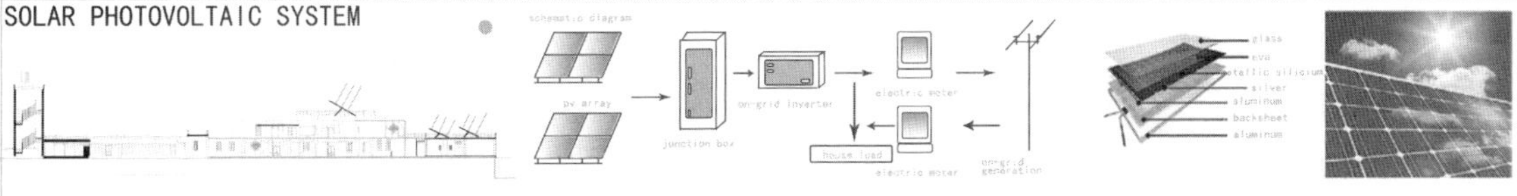

1-1 SECTION

COLLECTING SUNSHINE 3

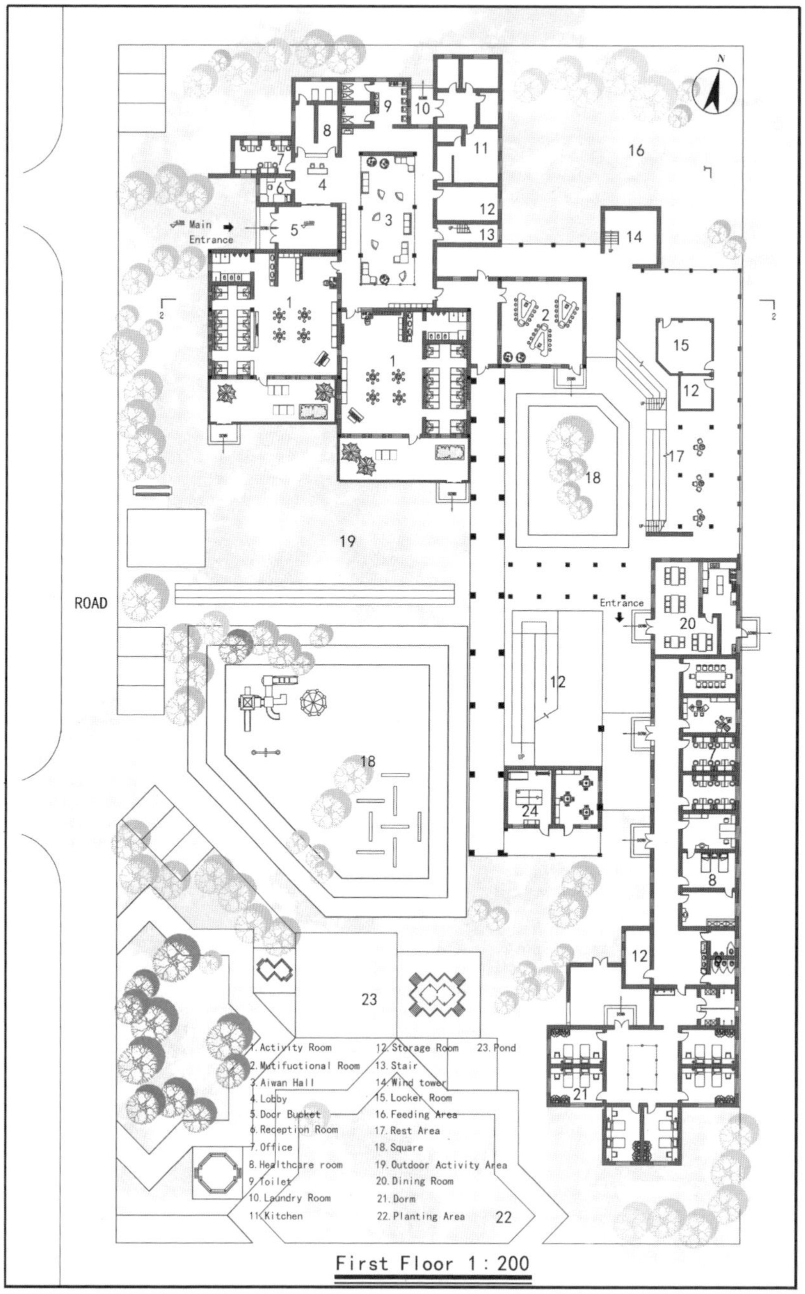
ROAD
Main Entrance
Entrance
1. Activity Room
2. Mutifuctional Room
3. Aiwan Hall
4. Lobby
5. Door Bucket
6. Reception Room
7. Office
8. Healthcare room
9. Toilet
10. Laundry Room
11. Kitchen
12. Storage Room
13. Stair
14. Wind tower
15. Locker Room
16. Feeding Area
17. Rest Area
18. Square
19. Outdoor Activity Area
20. Dining Room
21. Dorm
22. Planting Area
23. Pond

First Floor 1 : 200

PEOPLE' S FLOW ANALYSIS

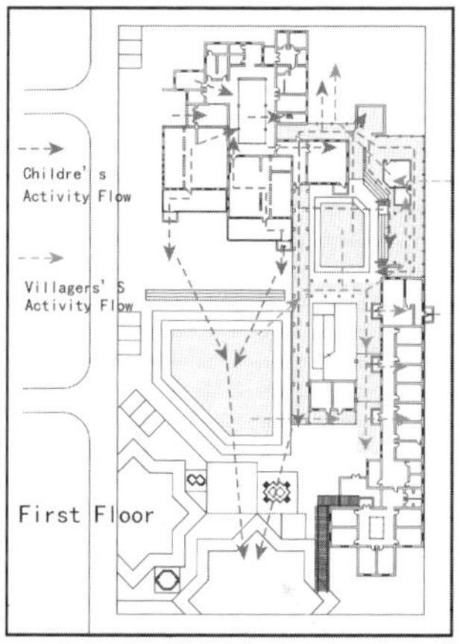
Childre' s Activity Flow
Villagers' S Activity Flow

First Floor

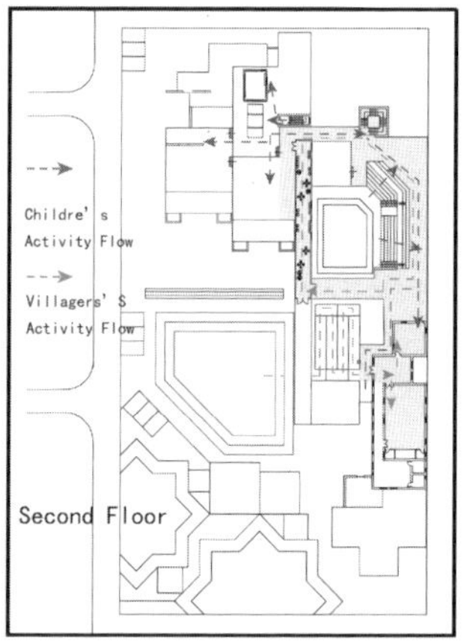
Childre' s Activity Flow
Villagers' S Activity Flow

Second Floor

THREE FROMS OF ACTIVITY ROOM

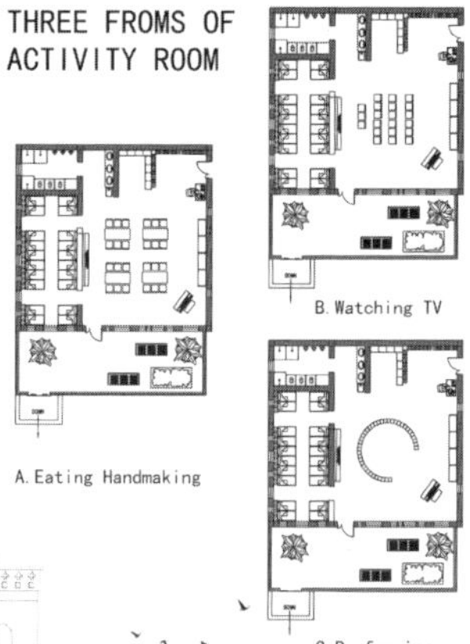

A. Eating Handmaking

B. Watching TV

C. Performing

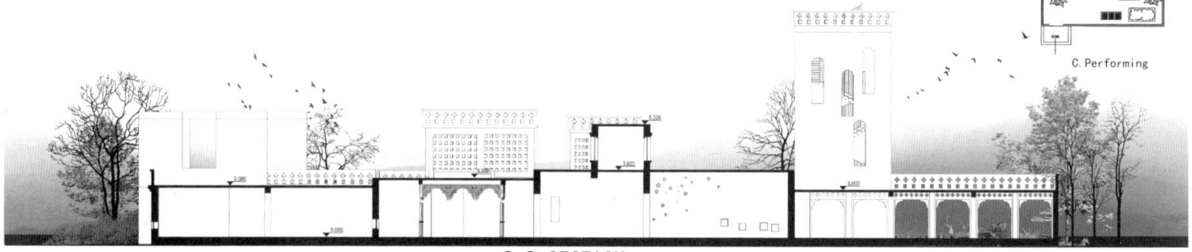

2-2 SECTION

COLLECTING SUNSHINE 4

SEWAGE TREATMENT SYSTEM

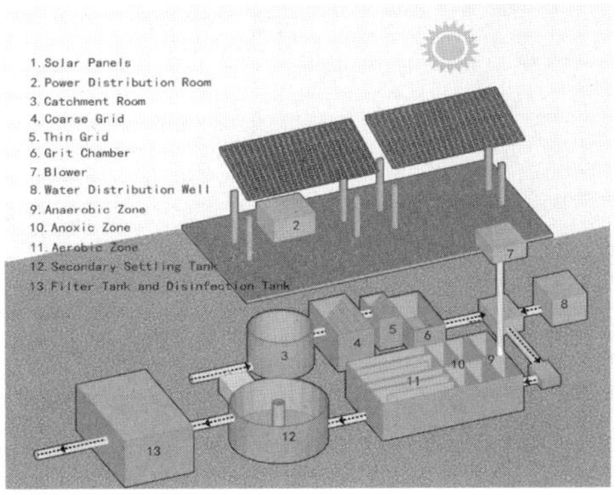
1. Solar Panels
2. Power Distribution Room
3. Catchment Room
4. Coarse Grid
5. Thin Grid
6. Grit Chamber
7. Blower
8. Water Distribution Well
9. Anaerobic Zone
10. Anoxic Zone
11. Aerobic Zone
12. Secondary Settling Tank
13. Filter Tank and Disinfection Tank

PASSIVE SOLAR HOUSE

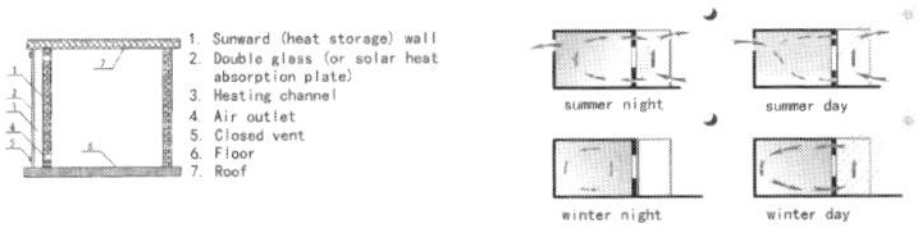
1. Sunward (heat storage) wall
2. Double glass (or solar heat absorption plate)
3. Heating channel
4. Air outlet
5. Closed vent
6. Floor
7. Roof
summer night
summer day
winter night
winter day

renderings of the solar room

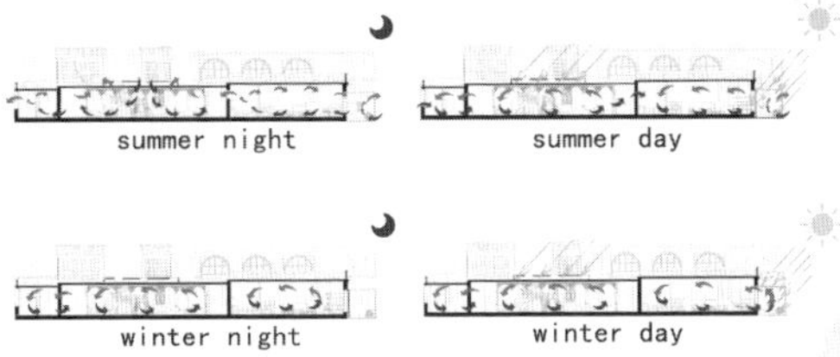
summer night
summer day
winter night
winter day

HOT PRESSURE VENTILATION IN SUMMER

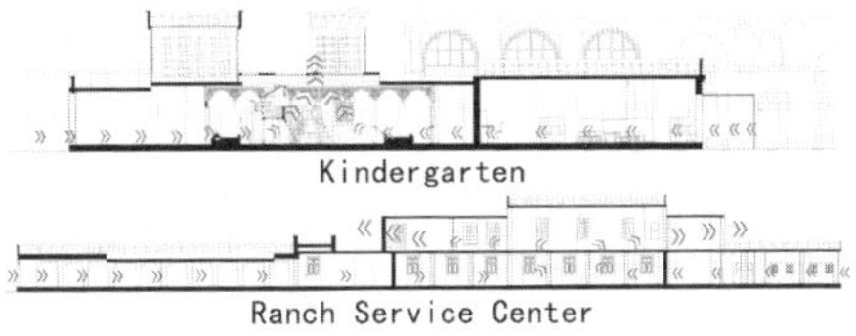
Kindergarten
Ranch Service Center

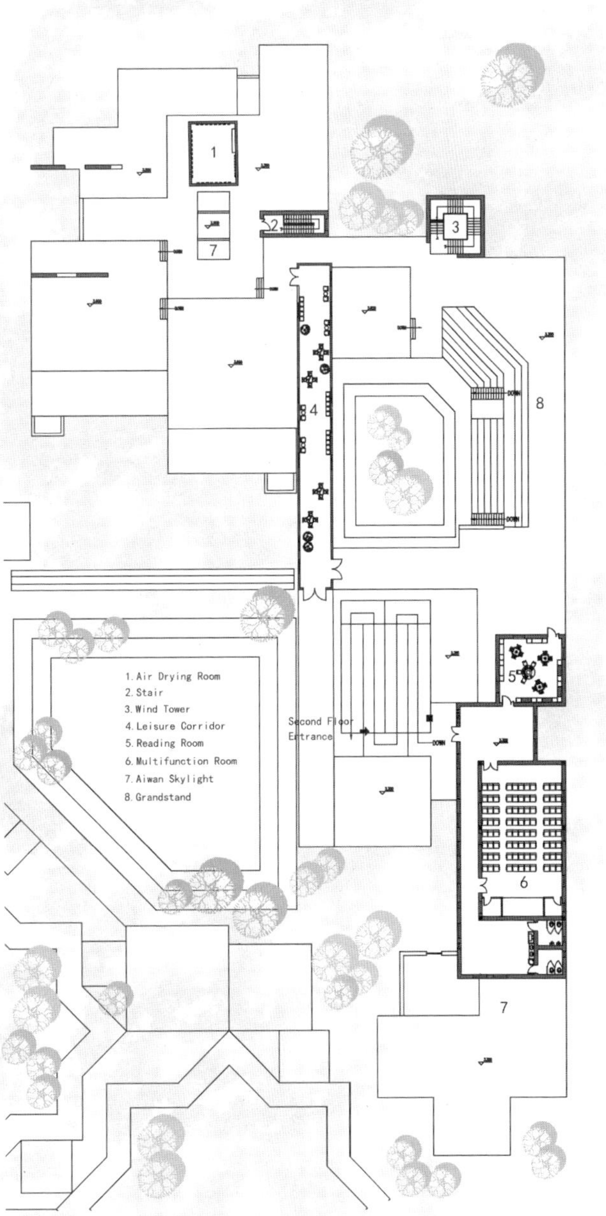
1. Air Drying Room
2. Stair
3. Wind Tower
4. Leisure Corridor
5. Reading Room
6. Multifunction Room
7. Aiwan Skylight
8. Grandstand
Second Floor Entrance

Second Floor 1:200

WEST ELEBATION

COLLECTING SUNSHINE 5

Building Decomposition Diagram

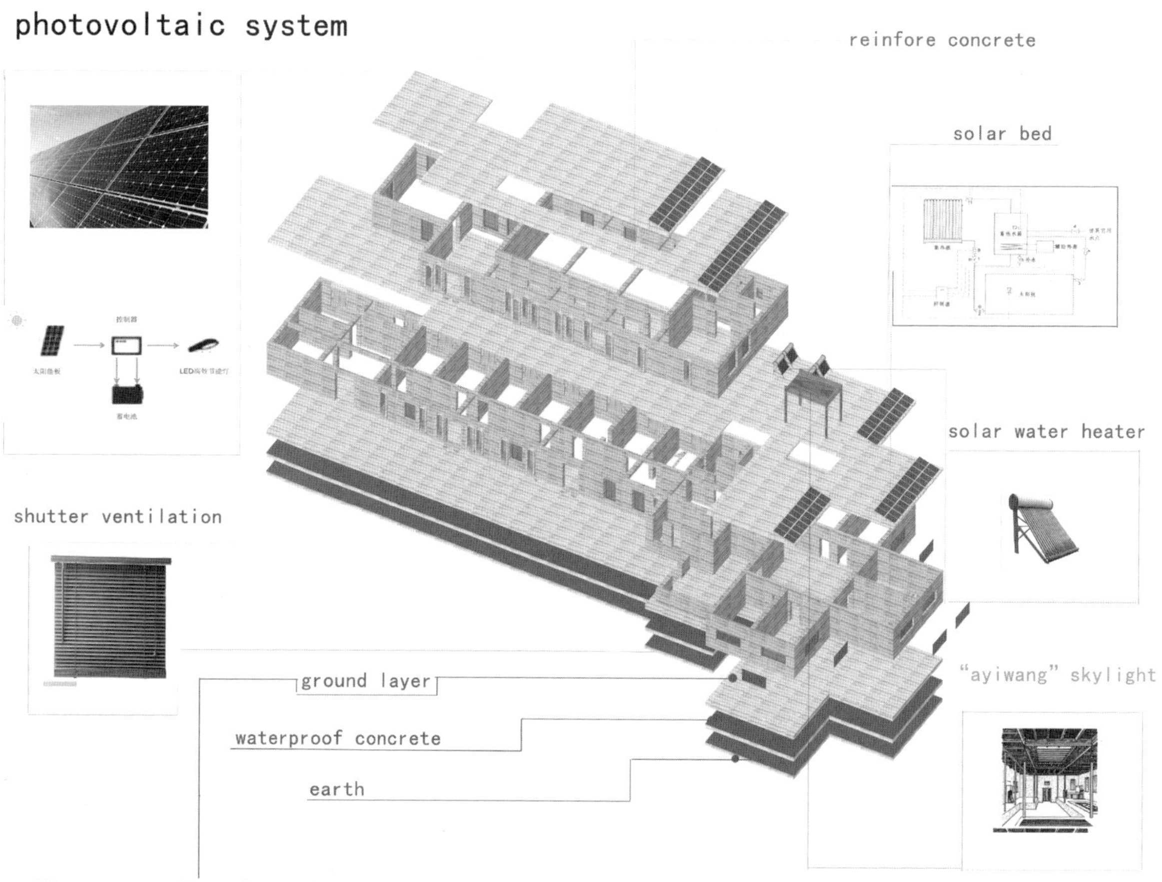
photovoltaic system
reinfore concrete
solar bed
solar water heater
shutter ventilation
ground layer
waterproof concrete
earth
"ayiwang" skylight

floor radiation heating system

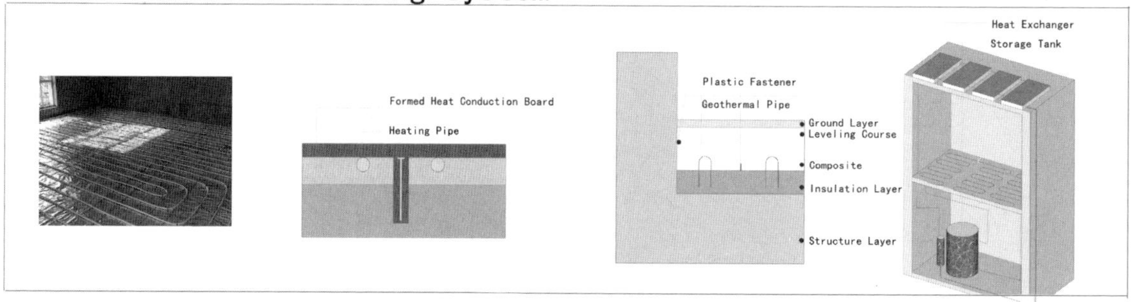
Formed Heat Conduction Board
Heating Pipe
Plastic Fastener
Geothermal Pipe
Ground Layer
Leveling Course
Composite
Insulation Layer
Structure Layer
Heat Exchanger
Storage Tank

WALL CONSTRUCYION

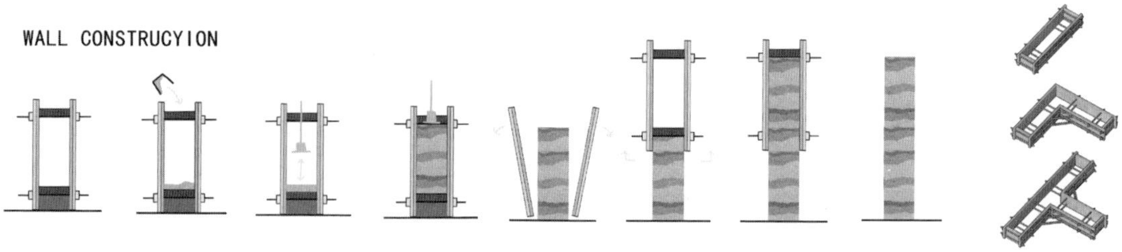

COLLECTING SUNSHINE 6

SCENCE CONSTRUCTION

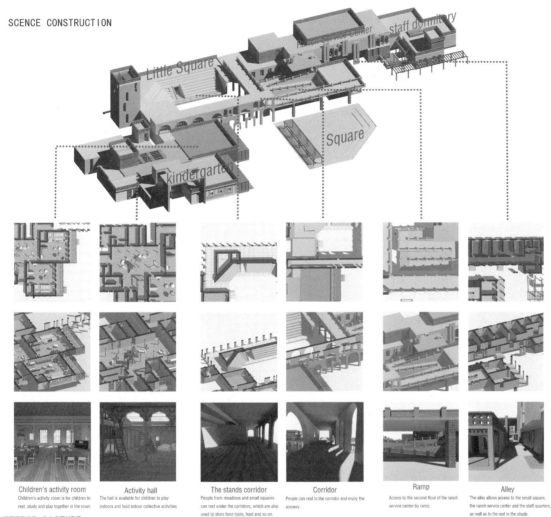

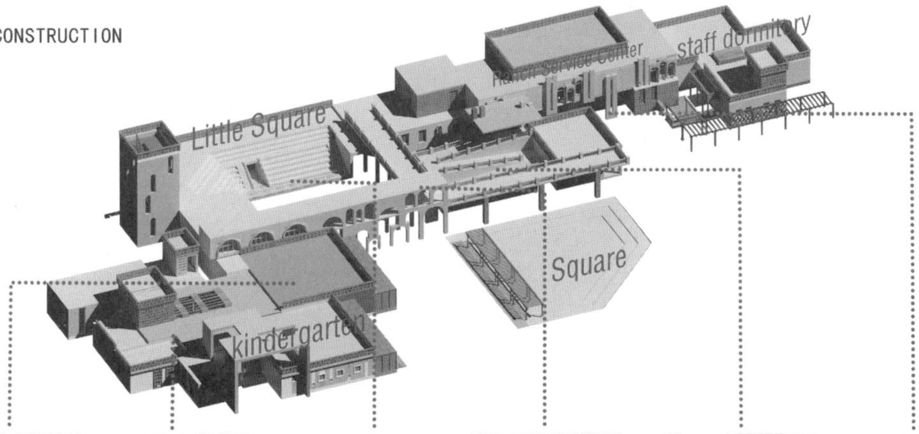
Little Square
staff dormitory
Square
kindergarten

Children's activity room

Children's activity room is for children to rest, study and play together in the room

Activity hall

The hall is available for children to play indoors and hold indoor collective activities

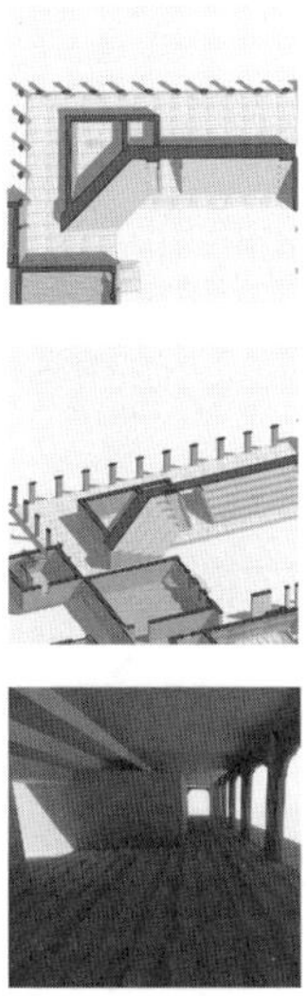

The stands corridor

People from meadows and small squares can rest under the corridors, which are also used to store farm tools, feed and so on.

Corridor

People can rest in the corridor and enjoy the scenery.

Ramp

Access to the second floor of the ranch service center by ramp.

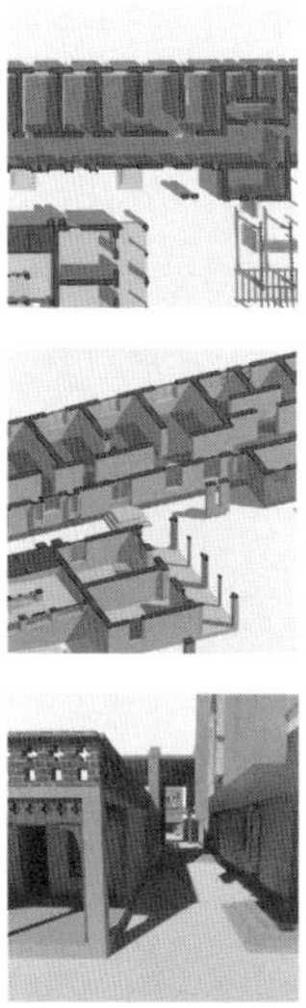

Alley

The alley allows access to the small square, the ranch service center and the staff quarters, as well as to the rest in the shade.

EFFECT PICTURE

Square

Alley

Acitivity room

Aiwan hall

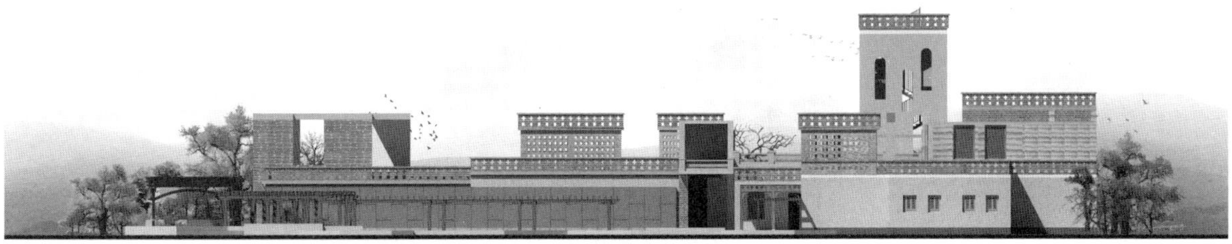

EAST ELEVATION

综合奖 · 优秀奖
General Prize Awarded · Honorable Mention Prize

注 册 号：7182

项目名称：阳光托马斯（新疆）

Sunny Thomas（Xinjiang）

作　　者：张　琳、黄和荣、张金锦

参赛单位：重庆大学

指导教师：周铁军、张海滨

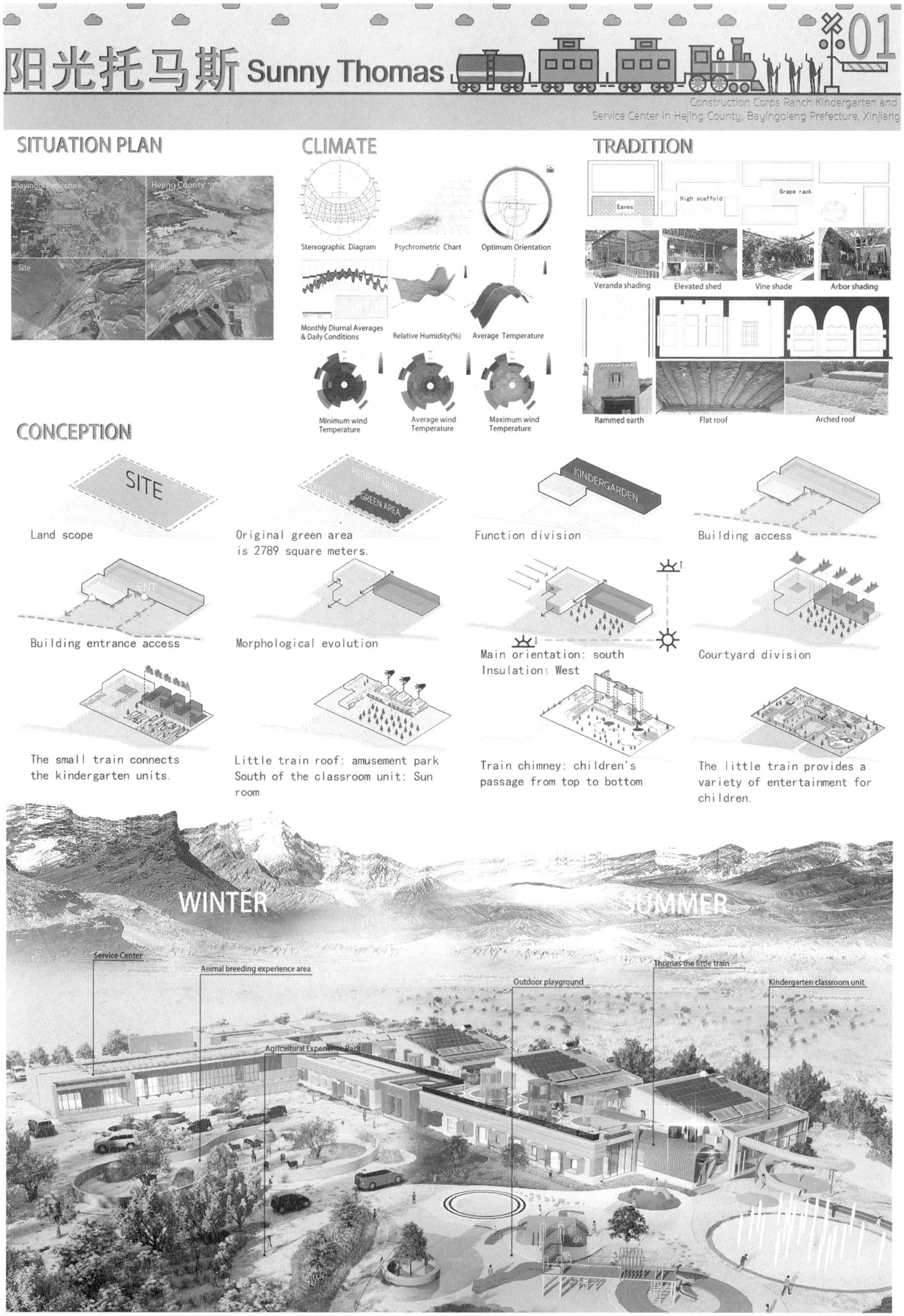
阳光托马斯 Sunny Thomas
01
Construction Corps Ranch Kindergarten and
Service Center in Hejing County, Bayingoleng Prefecture, Xinjiang
SITUATION PLAN
Bayingol Prefecture
Hejing County
Site
Building site
CLIMATE
Stereographic Diagram
Psychrometric Chart
Optimum Orientation
Monthly Diurnal Averages & Daily Conditions
Relative Humidity(%)
Average Temperature
Minimum wind Temperature
Average wind Temperature
Maximum wind Temperature
TRADITION
Eaves
High scaffold
Grape rack
Veranda shading
Elevated shed
Vine shade
Arbor shading
Rammed earth
Flat roof
Arched roof
CONCEPTION
SITE
Land scope
GREEN AREA
Original green area
is 2789 square meters.
KINDERGARDEN
Function division
Building access
Building entrance access
Morphological evolution
Main orientation: south
Insulation: West
Courtyard division
The small train connects
the kindergarten units.
Little train roof: amusement park
South of the classroom unit: Sun
room
Train chimney: children's
passage from top to bottom
The little train provides a
variety of entertainment for
children.
WINTER
SUMMER
Service Center
Animal breeding experience area
Agricultural Experience Park
Outdoor playground
Thomas the little train
Kindergarten classroom unit

阳光托马斯 Sunny Thomas

02

Construction Corps Ranch Kindergarten and Service Center in Hejing County, Bayingoleng Prefecture, Xinjiang

设计说明

项目的主题来自于托马斯小火车，用小火车将幼儿园单元连接起来，同时利用小火车的烟囱和屋顶创造出一个上下贯通的游戏空间，给孩子们更多样的空间选择。在服务中心和幼儿园设计部分，运用了主动式和被动式太阳能技术，包括太阳能供暖系统、光伏系统以及太阳能烟囱地道风耦合系统，帮助解决整个建筑的用电，采暖和通风问题。值得注意的是设计巧妙地结合小火车的烟囱进行技术和功能的再创造，这些五颜六色的烟囱既是联系屋顶和室内游戏场所的桥梁又是用于通风和发电的太阳能烟囱，将太阳能技术和建筑设计结合成了一体。

Design Description

The theme of the project comes from the Thomas Train. The little train connects the kindergarten units. At the same time, The chimney and roof of the small train create a play space through up and down, giving children more choices of space.

In the design of the service center and kindergarten, active and passive solar technoloies are used, including solar heating systems, photovoltaic systems, and solar chimney tunnel wind coupling systems to help solve the electricity, heating and ventilation problems of the entire building.

It is worth noting that the design cleverly combines the chimneys of the small trains to recreate the technology and functions. These colorful chimneys are not only a bridge connecting the roof and indoor playgrounds, but also used for Solar chimneys for ventilation and power generation, which combines solar technology and architectural design into one.

Sunshine and shadow analysis

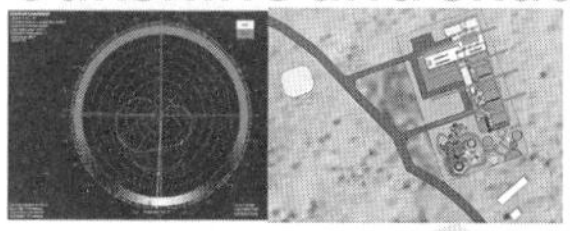

analysed by ecotect,2.5° is the best angle.

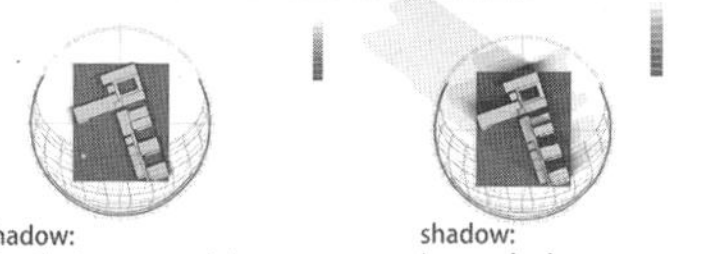

shadow: The shadows are mainly concentrated in the east and north in summer.

shadow: Larger shade range in winter.

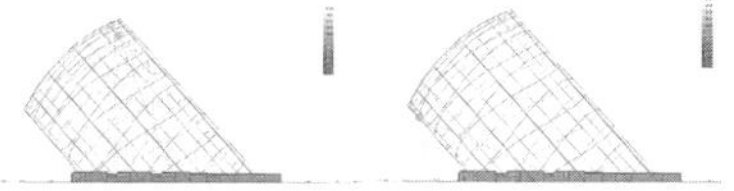

orientation: 15° inclined to the up side for solar roof.

TECHICAL - ECONOMIC INDEX IN CONSTRUCTION SITE	
CONSTRUCTION SITE AREA	10409 SQM.
FLOOR AREA	5042.6 SQM.
BUILDING AREA	2981.3 SQM.
BUILDING HEIGHT	7300 mm.
FAR	2.587
SITE COVERAGE	800 mm.
TOTAL HOSEHOLD	6
TOTAL POPPULATION	N/A
GREENING RATE	52.6 %
GROUND PARKING NUMBER	8
ENERGY RECEIVING AREA RATE	64.3%

Site plan

Perspective about kindergarten in a snowy day. On the side of the train

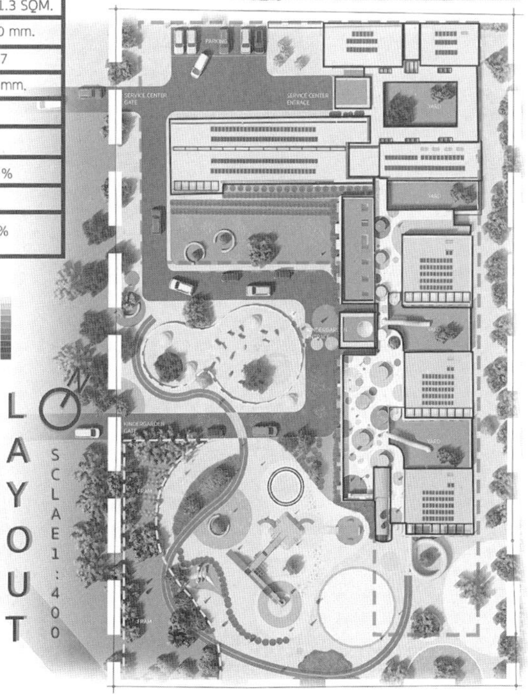
LAYOUT
SCLAE1:400

Temperature & thermal radiation analysis

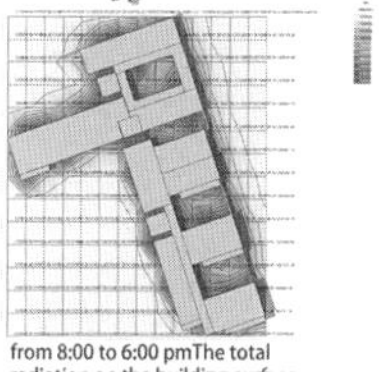

from 8:00 to 6:00 pmThe total radiation on the building surface

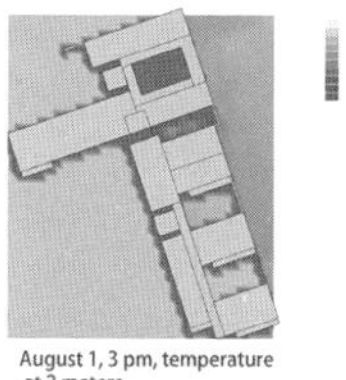

August 1, 3 pm, temperature at 2 meters

Wind speed at 3pm on August 1

The building's style for the most comfortable conditions arises from the integration of vernacular architecture with existing architectural technology. To provide a model for creating a sustainable community.

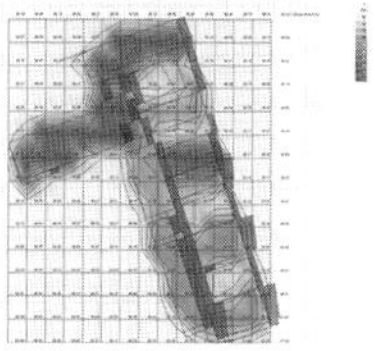

Daylighting factor map on August 1
The average daylighting factor is 80.66%

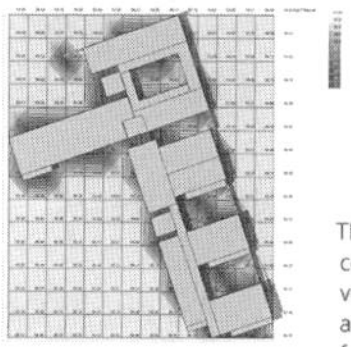

Space comfort analysis

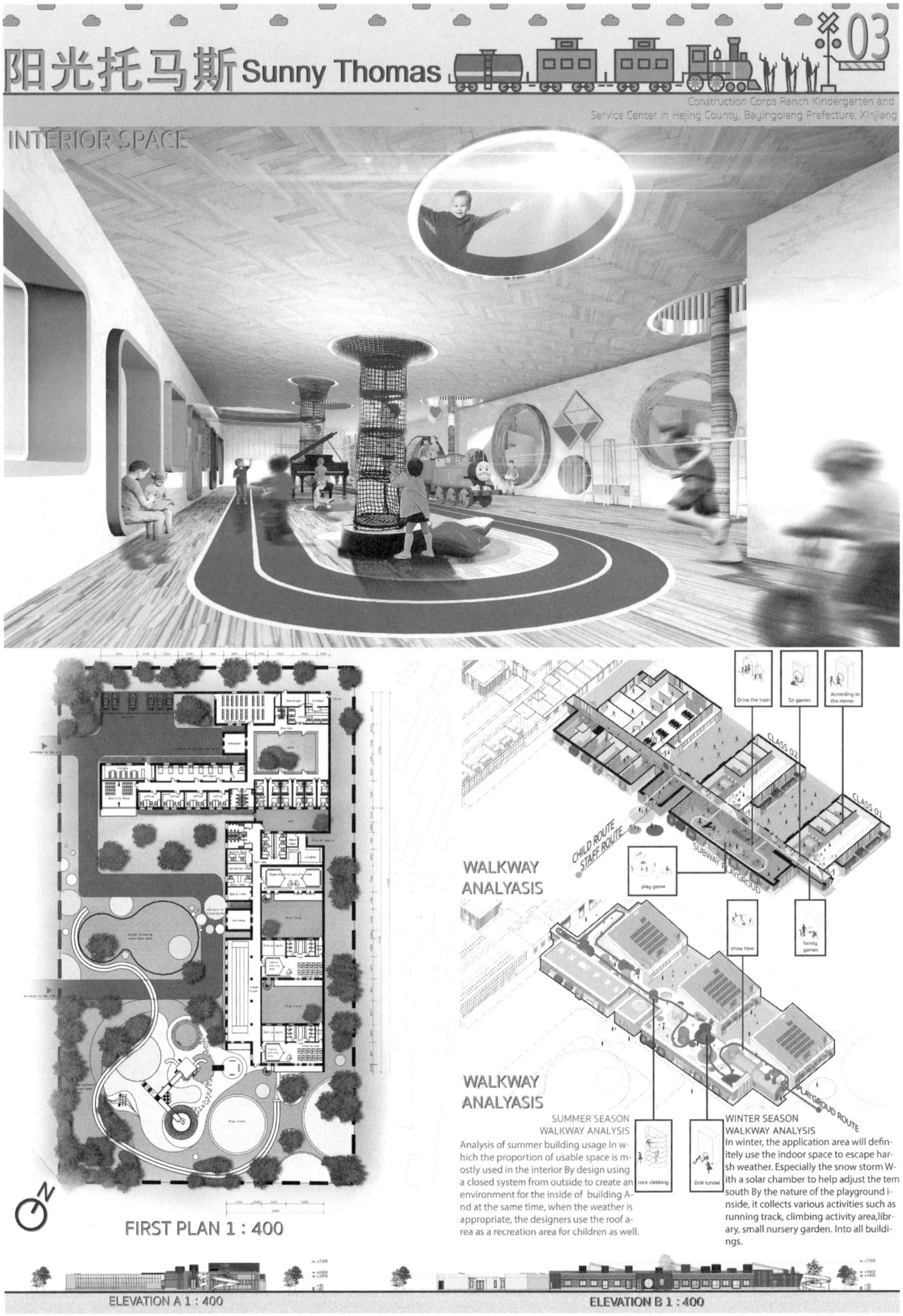
03
阳光托马斯 Sunny Thomas
Construction Corps Ranch Kindergarten and
Service Center in Hejing County, Bayingoleng Prefecture, Xinjiang
INTERIOR SPACE
FIRST PLAN 1 : 400
WALKWAY
ANALYASIS
CLASS 02
CLASS 01
CHILD ROUTE
STAFF ROUTE
SUBWAY PLAYGROUD
Drive the train
Sit games
According to the mirror
play game
show time
family games
WALKWAY
ANALYASIS
PLAYGROUD ROUTE
SUMMER SEASON
WALKWAY ANALYSIS
Analysis of summer building usage In w-
hich the proportion of usable space is m-
ostly used in the interior By design using
a closed system from outside to create an
environment for the inside of building A-
nd at the same time, when the weather is
appropriate, the designers use the roof a-
rea as a recreation area for children as well.
rock climbing
Drill tunnel
WINTER SEASON
WALKWAY ANALYSIS
In winter, the application area will defin-
itely use the indoor space to escape har-
sh weather. Especially the snow storm W-
ith a solar chamber to help adjust the tem
south By the nature of the playground i-
nside, it collects various activities such as
running track, climbing activity area,libr-
ary, small nursery garden. Into all buildings.
ELEVATION A 1 : 400
ELEVATION B 1 : 400

阳光托马斯 Sunny Thomas

04

Construction Corps Ranch Kindergarten and Service Center in Hejing County, Bayingoleng Prefecture, Xinjiang

Technology integration map

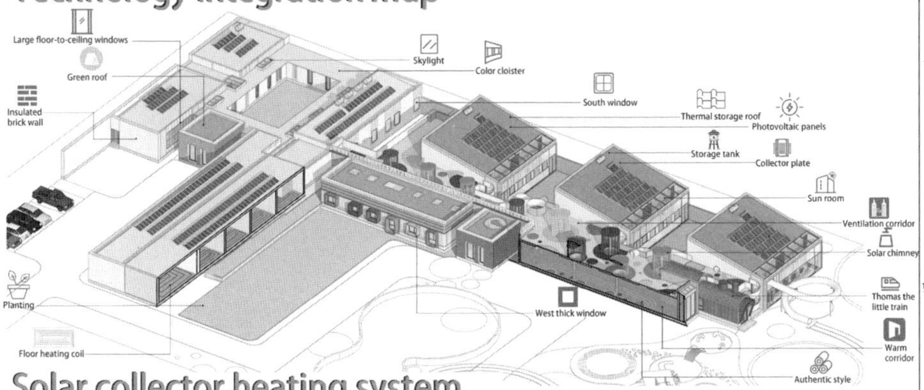
Large floor-to-ceiling windows
Green roof
Insulated brick wall
Skylight
Color cloister
South window
Thermal storage roof
Photovoltaic panels
Storage tank
Collector plate
Sun room
Ventilation corridor
Solar chimney
Thomas the little train
Warm corridor
Authentic style
West thick window
Planting
Floor heating coil

Solar photovoltaic

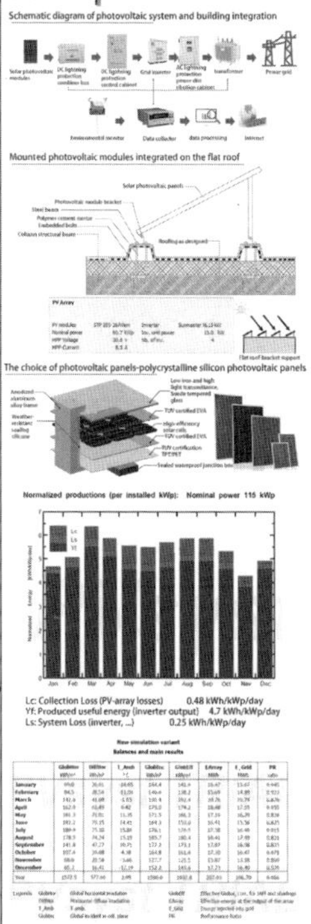
Schematic diagram of photovoltaic system and building integration
Mounted photovoltaic modules integrated on the flat roof
The choice of photovoltaic panels-polycrystalline silicon photovoltaic panels
Normalized productions (per installed kWp): Nominal power 115 kWp
Lc: Collection Loss (PV-array losses) 0.48 kWh/kWp/day
Yf: Produced useful energy (inverter output) 4.7 kWh/kWp/day
Ls: System Loss (inverter, ...) 0.25 kWh/kWp/day

Solar collector heating system

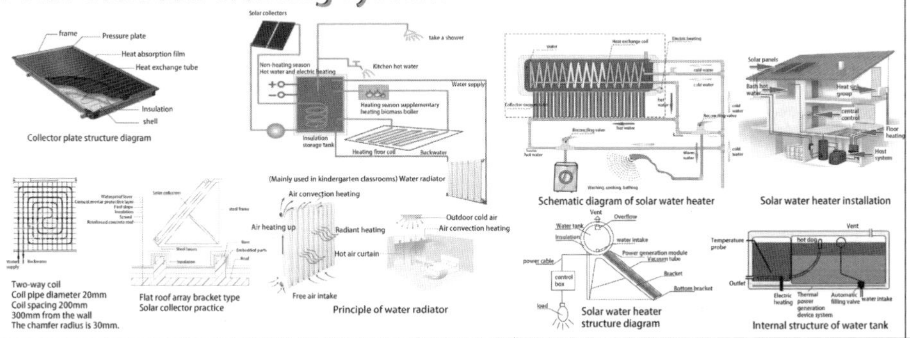
frame
Pressure plate
Heat absorption film
Heat exchange tube
Insulation
shell
Collector plate structure diagram
Solar collectors
take a shower
Kitchen hot water
Water supply
Heating season supplementary heating biomass boiler
Heating floor coil
Backwater
Schematic diagram of solar water heater
Solar water heater installation
Two-way coil
Coil pipe diameter 20mm
Coil spacing 200mm
300mm from the wall
The chamfer radius is 30mm.
Flat roof array bracket type
Solar collector practice
(Mainly used in kindergarten classrooms) Water radiator
Air convection heating
Air heating up
Radiant heating
Hot air curtain
Free air intake
Outdoor cold air
Air convection heating
Principle of water radiator
Vent
Overflow
Water tank
Insulation
Power generation module
Vacuum tube
Bracket
Bottom bracket
power cable
control box
load
Solar water heater structure diagram
Temperature probe
hot dog
Outlet
Electric heating
Thermal power generation device system
Automatic filling valve
water intake
Internal structure of water tank

Solar chimney-tunnel wind composite system

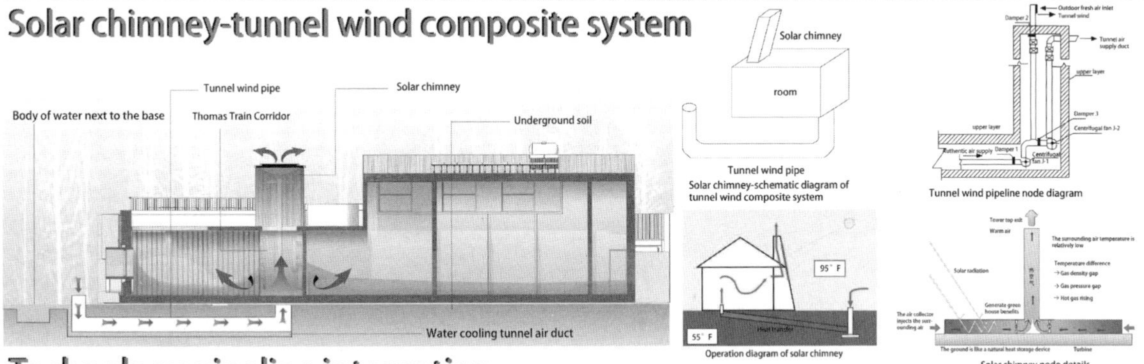
Tunnel wind pipe
Solar chimney
Body of water next to the base
Thomas Train Corridor
Underground soil
Water cooling tunnel air duct
Solar chimney
room
Tunnel wind pipe
Solar chimney-schematic diagram of tunnel wind composite system
95° F
55° F
Operation diagram of solar chimney
Outdoor fresh air inlet
Tunnel wind
Tunnel air supply duct
upper layer
Damper 3
Centrifugal fan 3-2
Damper 1
Centrifugal fan 3-1
Tunnel wind pipeline node diagram
Solar chimney node details

Technology pipeline integration

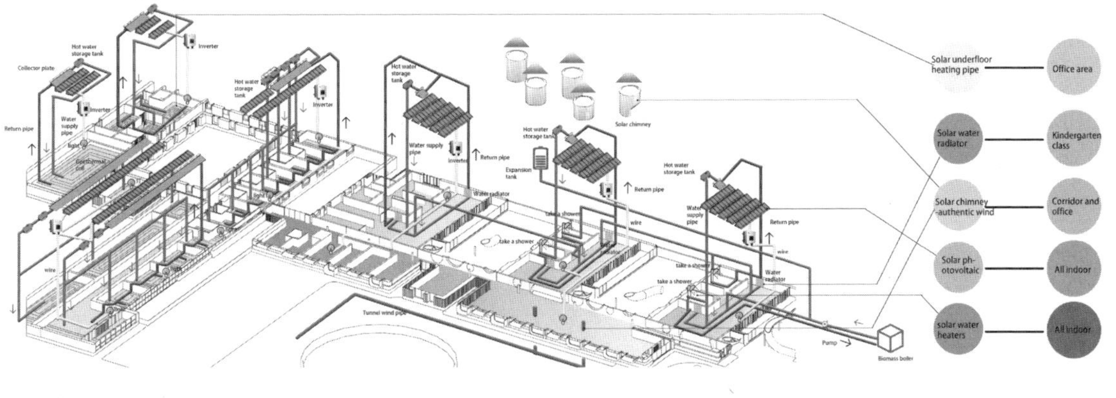
Collector plate
Hot water storage tank
Inverter
Return pipe
Water supply pipe
Solar chimney
Expansion tank
Water radiator
take a shower
wire
Tunnel wind pipe
Pump
Biomass boiler
Solar underfloor heating pipe
Office area
Solar water radiator
Kindergarten class
Solar chimney-authentic wind
Corridor and office
Solar photovoltaic
All indoor
solar water heaters
All indoor

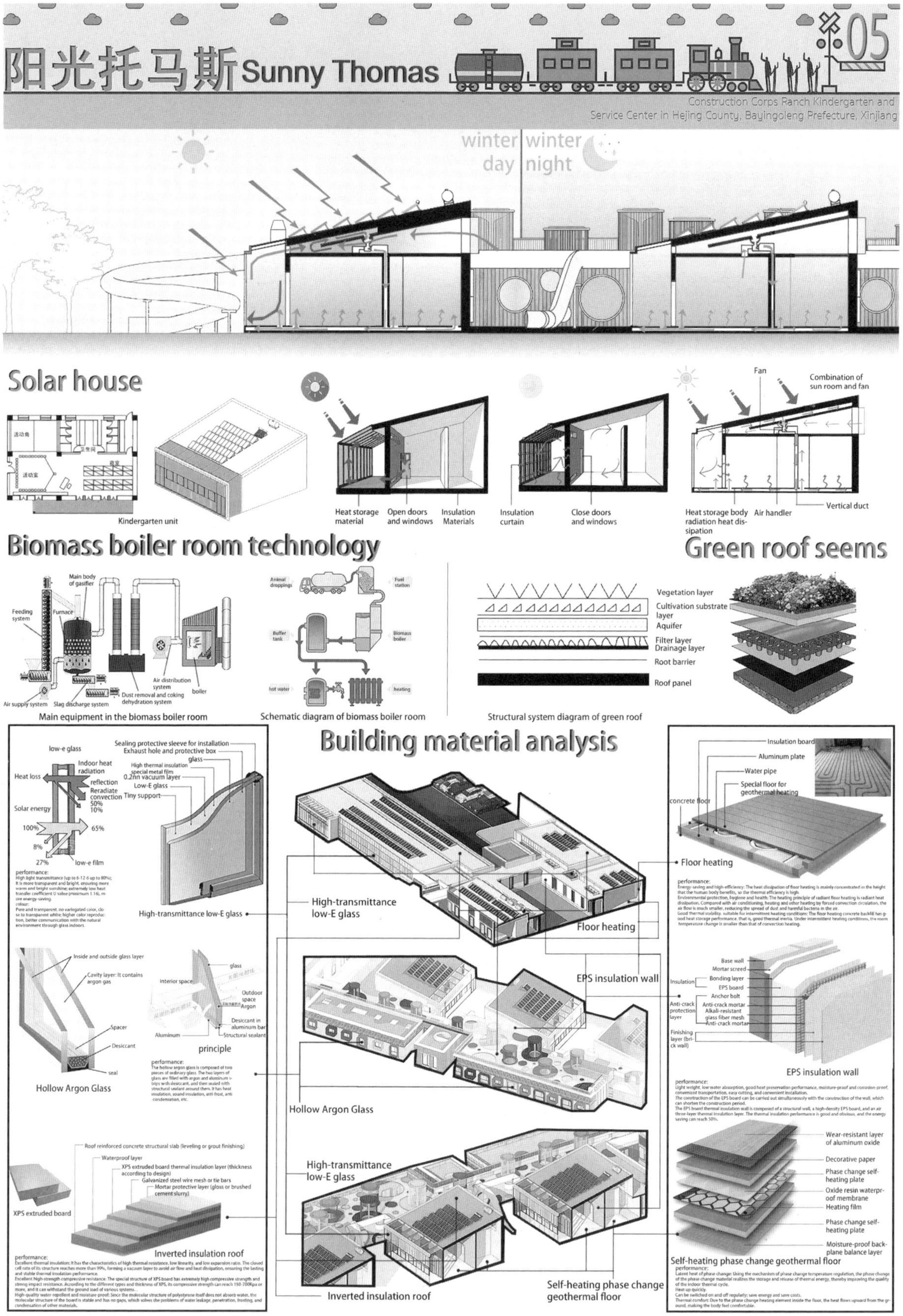
05
阳光托马斯 Sunny Thomas
Construction Corps Ranch Kindergarten and Service Center in Hejing County, Bayingoleng Prefecture, Xinjiang
winter day
winter night
Solar house
Kindergarten unit
Heat storage material
Open doors and windows
Insulation Materials
Insulation curtain
Close doors and windows
Fan
Combination of sun room and fan
Heat storage body radiation heat dissipation
Air handler
Vertical duct
Biomass boiler room technology
Main equipment in the biomass boiler room
Schematic diagram of biomass boiler room
Green roof seems
Vegetation layer
Cultivation substrate layer
Aquifer
Filter layer
Drainage layer
Root barrier
Roof panel
Structural system diagram of green roof
Building material analysis
High-transmittance low-E glass
Hollow Argon Glass
Inverted insulation roof
Floor heating
EPS insulation wall
Self-heating phase change geothermal floor
XPS extruded board
principle

06

阳光托马斯 Sunny Thomas

Construction Corps Ranch Kindergarten and Service Center in Hejing County, Bayingoleng Prefecture, Xinjiang

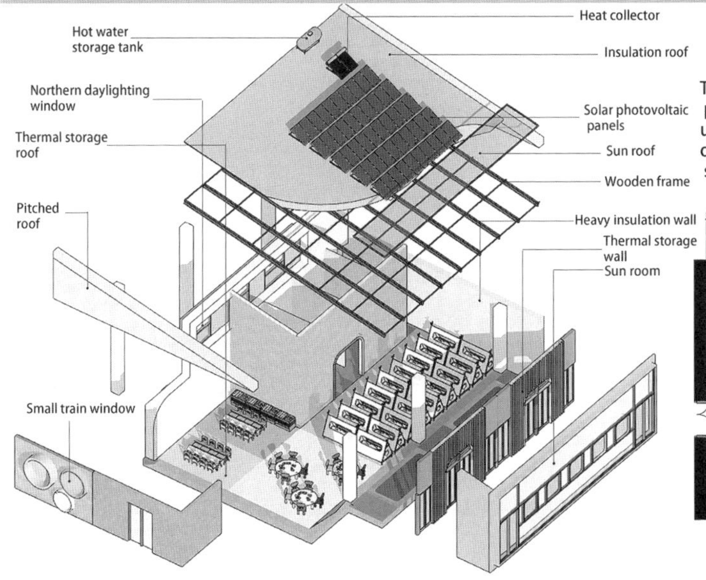
Hot water storage tank
Heat collector
Insulation roof
Northern daylighting window
Solar photovoltaic panels
Thermal storage roof
Sun roof
Wooden frame
Pitched roof
Heavy insulation wall
Thermal storage wall
Sun room
Small train window

Kindergarten unit

The kindergarten unit uses both active solar technology and passive solar technology.Passive technology uses some insulation and heat storage enclosure structures.Active technology uses solar water radiators and mechanical ventilation systems to heat buildings.

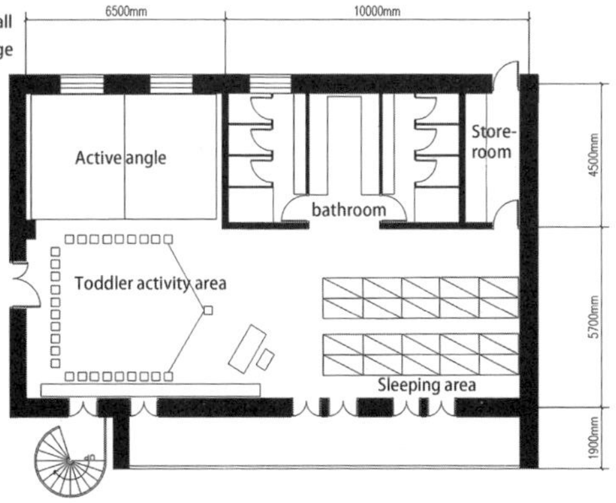
6500mm
10000mm
Active angle
Store-room
bathroom
Toddler activity area
Sleeping area
4500mm
5700mm
1900mm

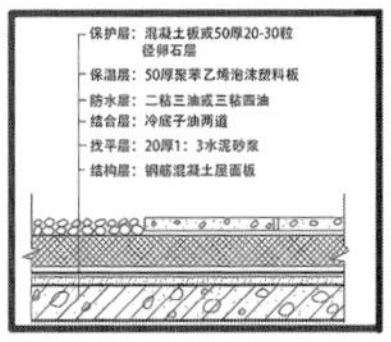

Insulated roof

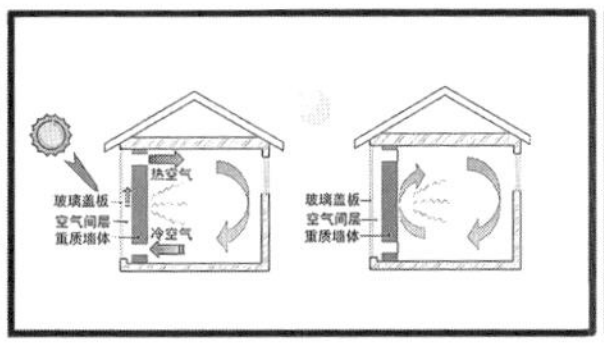

Thermal storage wall

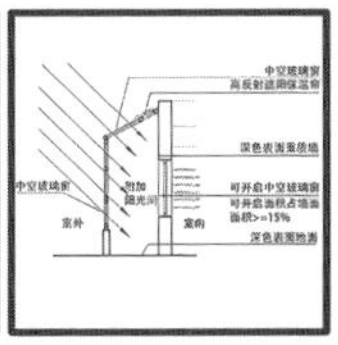

Sun room

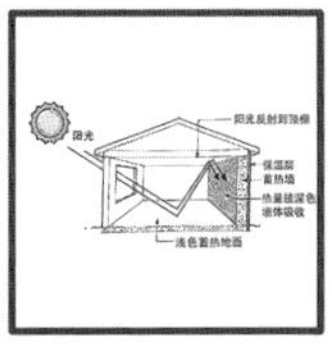

Schematic diagram of heat storage body

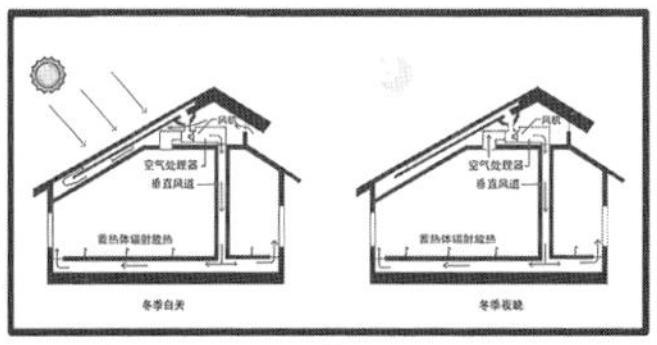

Active ventilation system

Physical analysis

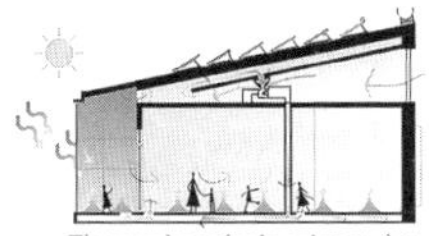

Thermal analysis-winter day

In winter, the sun room and the mechanical ventilation system work together to heat the room.

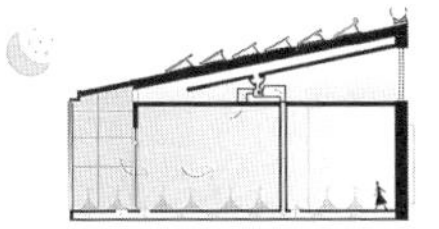

Thermal analysis-winter night

At night, close the heat preservation curtain, the heat storage body in the room will release heat to heat the room.

Thermal analysis-summer day

The hollow roof and sun room jointly ventilate and cool the room.

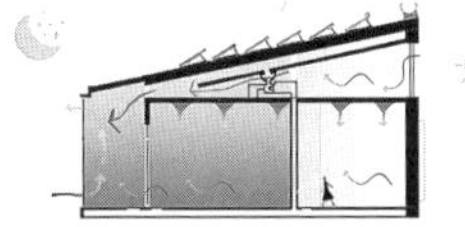

Thermal analysis-summer night

At night, the windows of the sun room are opened, and together with the hollow roof, the heat in the house is discharged.

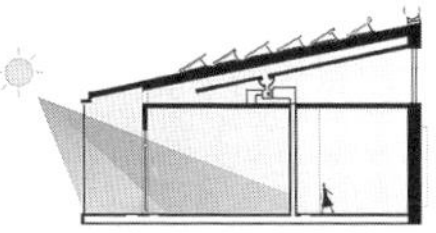

Light analysis-winter day

In winter, the sun's altitude angle is low and it can shine into a large part of the room.

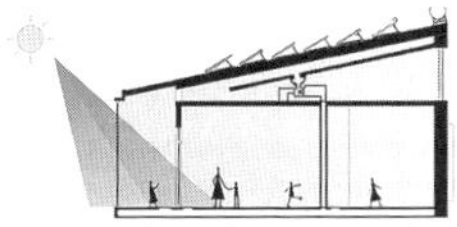

Light analysis-summer day

In summer, the sun has a high altitude angle, which mainly shines into the south-facing sun room, which can ensure good light in the room.

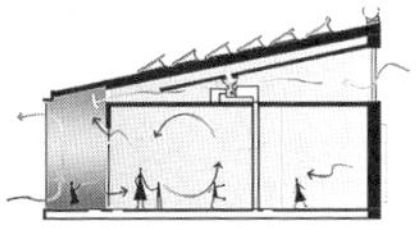

Ventilation analysis-summer

Open doors and windows in summer, and the sun room and hollow roof together help the room ventilate.

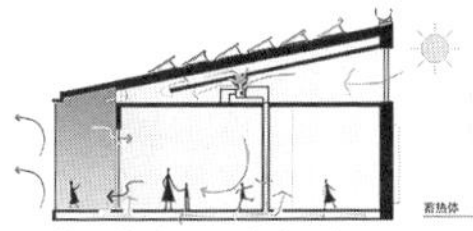

Ventilation analysis-winter

In winter, the sun room and mechanical ventilation are combined to provide hot air to the room.

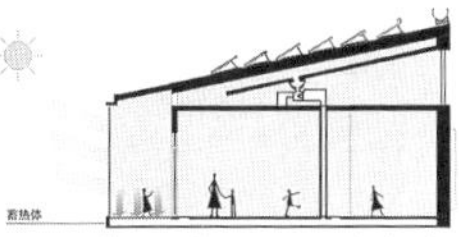

Thermal storage analysis-day

During the day, the heat storage ground in the sun room absorbs heat radiation and stores heat.

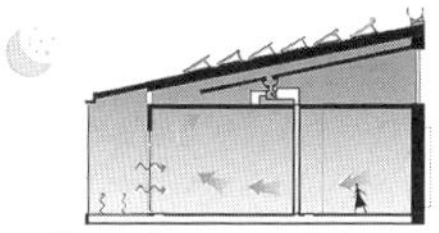

Thermal storage analysis-night

At night, the heat storage body emits heat to heat the room.

Energy consumption analysis

power consumption per month

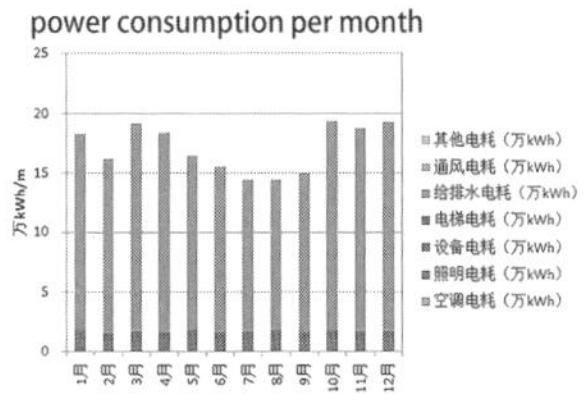
其他电耗（万kWh）
通风电耗（万kWh）
给排水电耗（万kWh）
电梯电耗（万kWh）
设备电耗（万kWh）
照明电耗（万kWh）
空调电耗（万kWh）

Hourly humidification throughout the year

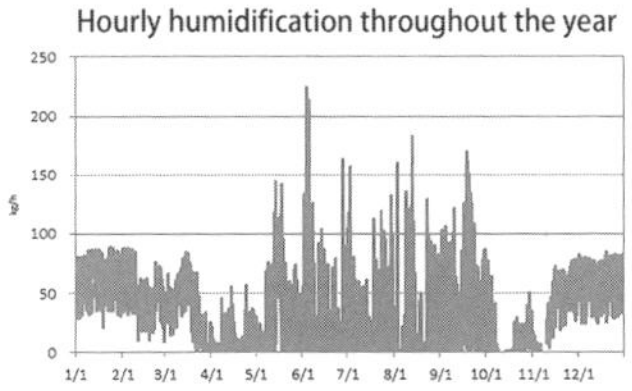

Hourly air conditioning load

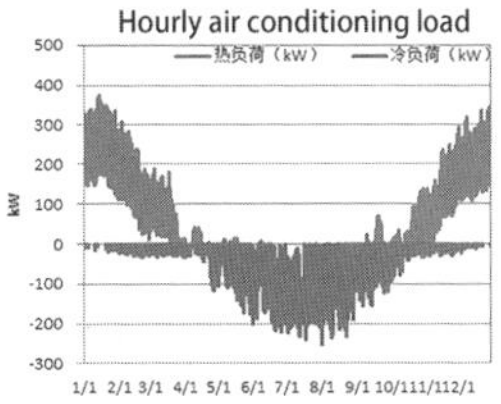
热负荷（kW）
冷负荷（kW）
kW

Energy load table

Project Statistics	unit	Statistics
Total building air-conditioning area	m²	2518.89
Project load statistics		
Maximum annual heat load	kw	374.20
Maximum annual cooling load	kw	248.15
Maximum annual humidification	kg/h	224.15
Annual cumulative heat load	kw.h	314795.97
Annual cumulative cooling load	kw.h	245497.84
Maximum annual heat load index	w/m.	148.56
Maximum annual heat load index	w/m²	98.51
Annual cumulative heat load index	kw.h/m²	124.97
Annual cumulative cooling load index	kw.h/m²	97.46
Project seasonal load index		
Heat load index for heating season	w/m²	39.21
Seasonal cooling load index of air conditioning	w/m²	30.91

综合奖 · 优秀奖
General Prize Awarded · Honorable Mention Prize

注 册 号：7212

项目名称：杲杲日光（新疆）

High in the Sun（Xinjiang）

作　　者：闫鑫懿、杨承超、高　力、张清亮

参赛单位：石家庄铁道大学、中铁建安工程设计有限公司

指导教师：樊海彬、高力强

杲杲日光——HIGH IN THE SUN 1

□ Desciiption of design

“杲杲冬日光，明暖真可爱。” 本次设计源于白居易的这一句诗，理念在于在寒冷的时候为孩子们提供温暖的阳光，同时结合新疆当地的建筑特色“阿以旺”，主要利用被动式太阳能，配合太阳能光伏技术，地窖燃烧秸秆等技术最大限度地为建筑提供热量，同时尽可能地减少了北侧的开窗，为建筑抵挡了寒冷的北风。本次设计基于就地取材经济优先的策略，主要利用了高蓄热夯土墙。同时，多处种树以及草帘的运用起到了夏天防晒的作用。

“There were bright winter lights, bright and sweet." From bai juyi's poem, the design idea is to provide for the children in the cold warm sunshine, the combination of xinjiang local architectural features “Ayiwang”, the main use of passive solar, solar photovoltaic technology, burning straw cellar technology such as maximum provides heat for the building, at the same time, as far as possible to reduce the north side of the window, for buildings against the cold north wind. This design is based on the strategy of economic priority of local materials, mainly using rammed earth wall with high storage heat. At the same time many plant trees and the application of straw curtain solved the summer bask in the action.

□ 经济技术指标

用地面积：10409㎡　容积率：0.2

总建筑面积：2052㎡

建筑占地面积：1582㎡

建筑密度：15.19%

Economic and technical norms

Land area:10409㎡　Plot ratio:0.2

Total building area:2052㎡

Building occupation area:1582㎡

Building density:15.19%

□ Inspiration source

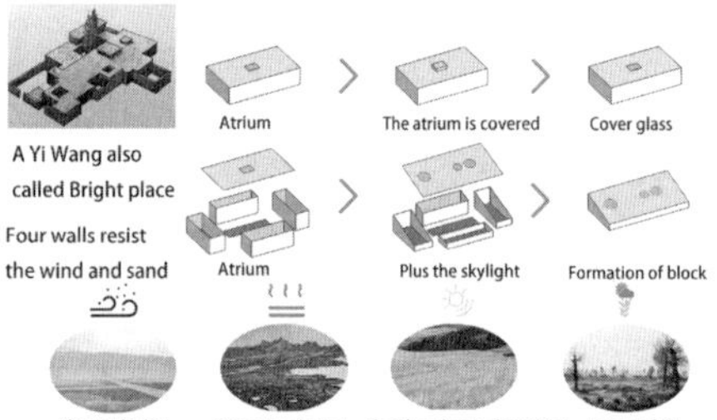
A Yi Wang also called Bright place
Four walls resist the wind and sand
Atrium
The atrium is covered
Cover glass
Atrium
Plus the skylight
Formation of block
The sand is big
Evaporated water
There' s plenty of sunshine
Lass rainfall

□ Sunlight analysis

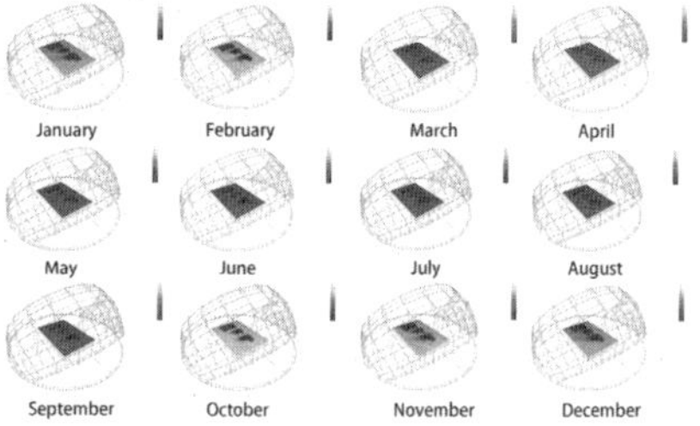
January
February
March
April
May
June
July
August
September
October
November
December

Temperature change at twelve months

□ Surroundings

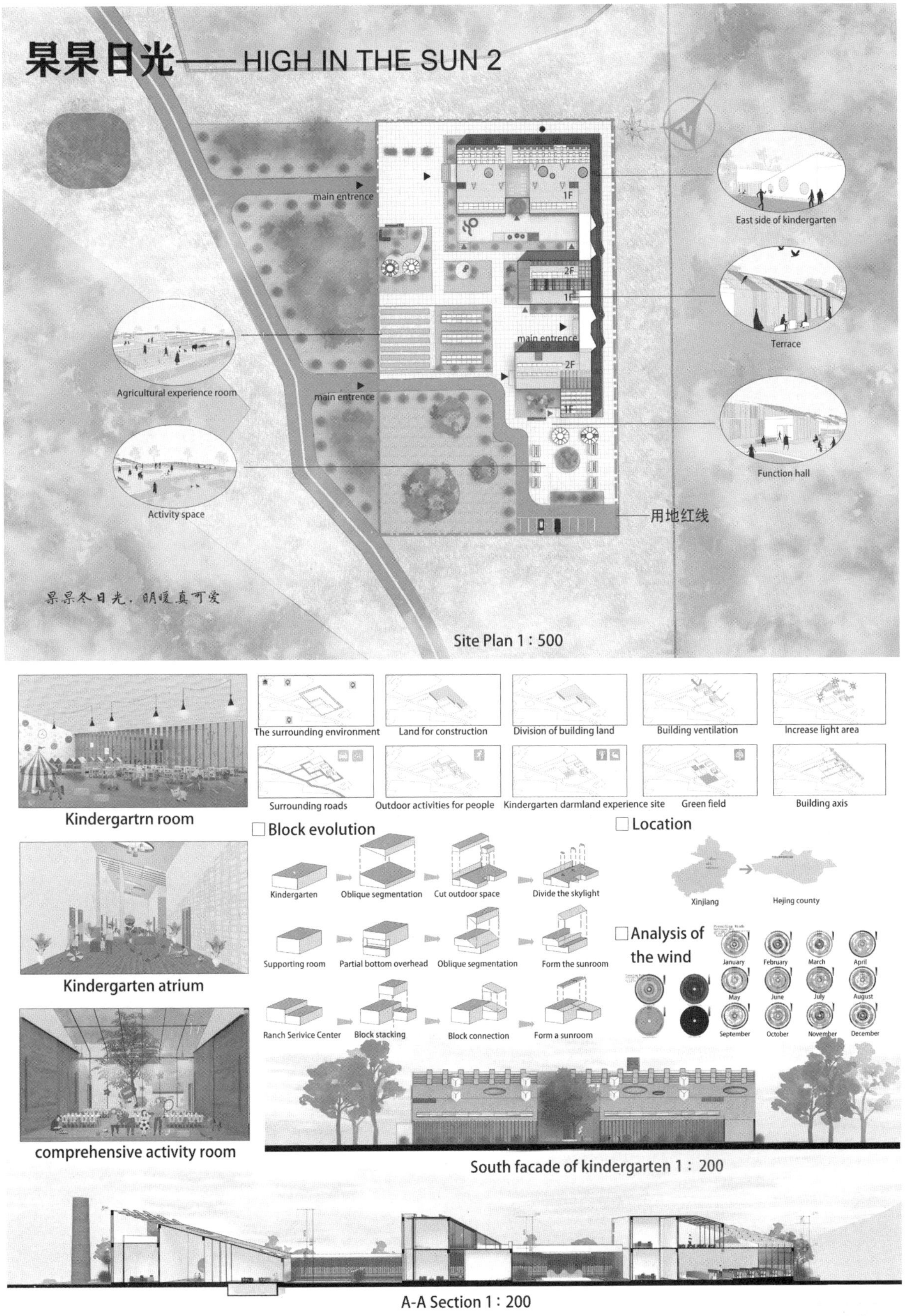
杲杲日光——HIGH IN THE SUN 2
main entrence
main entrence
main entrence
1F
2F
1F
2F
1F
East side of kindergarten
Terrace
Function hall
Agricultural experience room
Activity space
用地红线
杲杲冬日光，明暖真可爱
Site Plan 1：500
Kindergartrn room
Kindergarten atrium
comprehensive activity room
The surrounding environment
Land for construction
Division of building land
Building ventilation
Increase light area
Surrounding roads
Outdoor activities for people
Kindergarten darmland experience site
Green field
Building axis
Block evolution
Kindergarten
Oblique segmentation
Cut outdoor space
Divide the skylight
Supporting room
Partial bottom overhead
Oblique segmentation
Form the sunroom
Ranch Serivice Center
Block stacking
Block connection
Form a sunroom
Location
Xinjiang
Hejing county
Analysis of the wind
January
February
March
April
May
June
July
August
September
October
November
December
South facade of kindergarten 1：200
A-A Section 1：200

杲杲日光——HIGH IN THE SUN 3

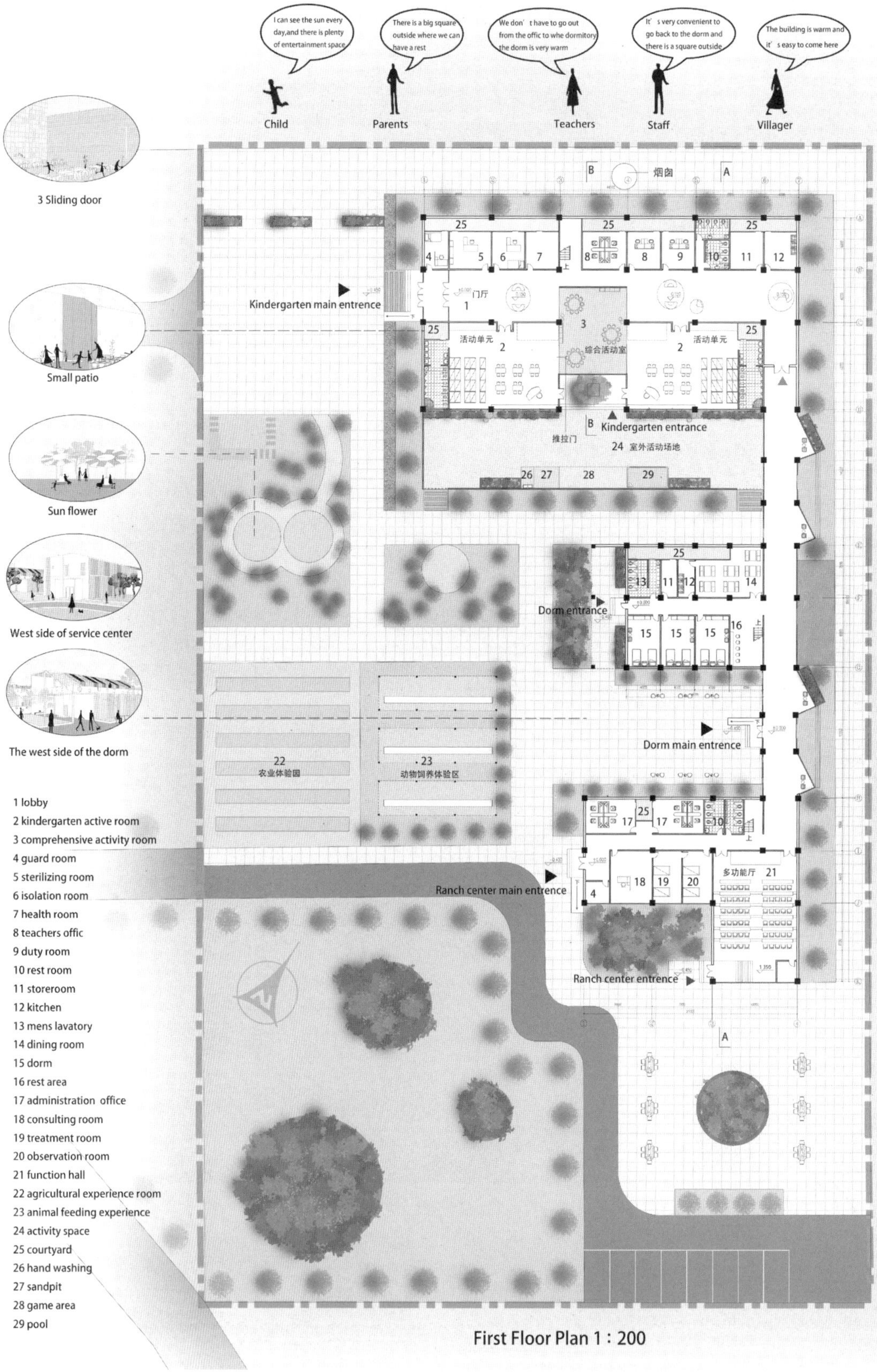
I can see the sun every day,and there is plenty of entertainment space
There is a big square outside where we can have a rest
We don' t have to go out from the offic to whe dormitory the dorm is very warm
It' s very convenient to go back to the dorm and there is a square outside
The building is warm and it' s easy to come here
Child
Parents
Teachers
Staff
Villager
3 Sliding door
Small patio
Sun flower
West side of service center
The west side of the dorm
烟囱
Kindergarten main entrence
门厅
活动单元
综合活动室
Kindergarten entrance
推拉门
24 室外活动场地
Dorm entrance
Dorm main entrence
22 农业体验园
23 动物饲养体验区
Ranch center main entrence
多功能厅 21
Ranch center entrence
1 lobby
2 kindergarten active room
3 comprehensive activity room
4 guard room
5 sterilizing room
6 isolation room
7 health room
8 teachers offic
9 duty room
10 rest room
11 storeroom
12 kitchen
13 mens lavatory
14 dining room
15 dorm
16 rest area
17 administration office
18 consulting room
19 treatment room
20 observation room
21 function hall
22 agricultural experience room
23 animal feeding experience
24 activity space
25 courtyard
26 hand washing
27 sandpit
28 game area
29 pool

First Floor Plan 1：200

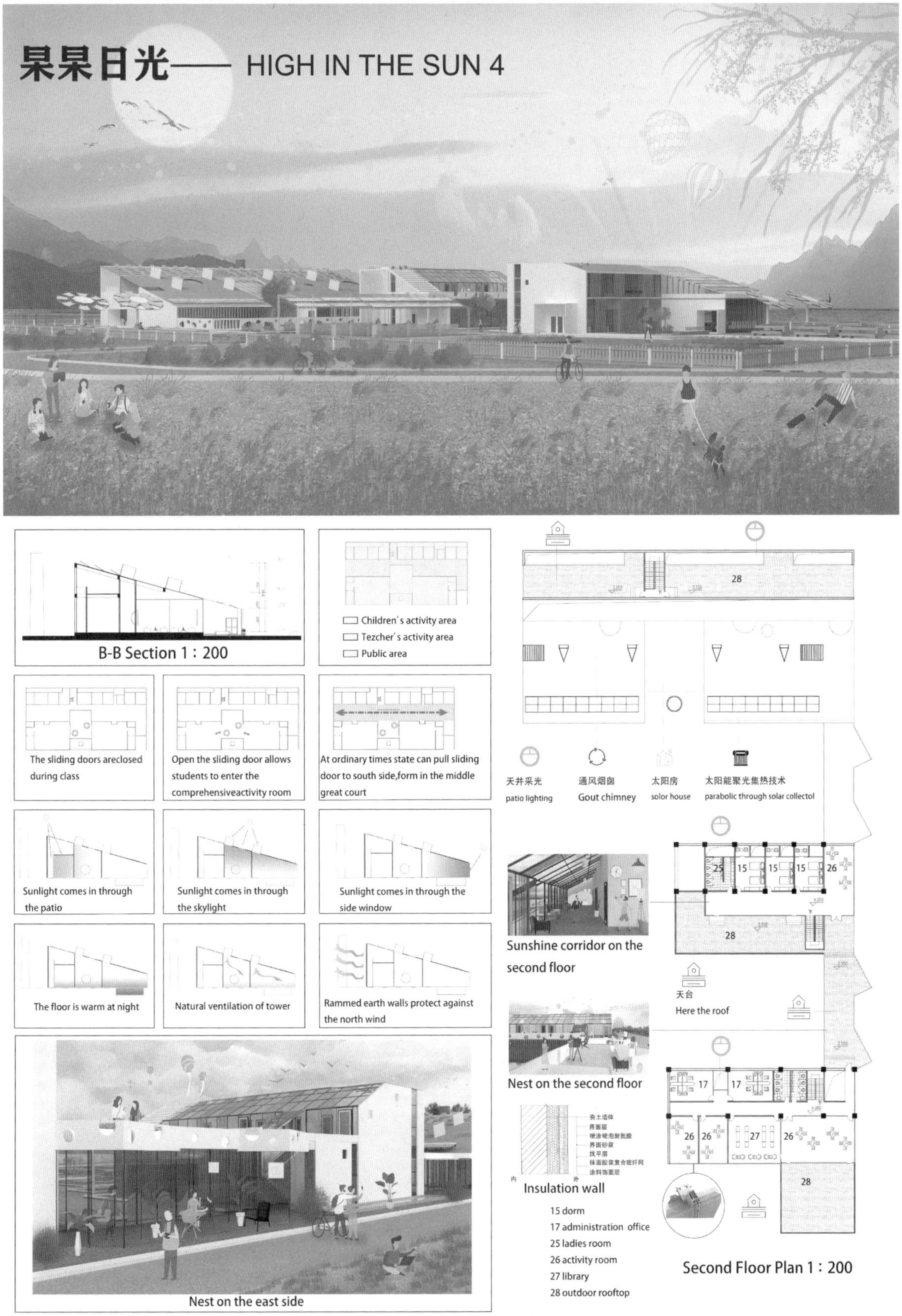
杲杲日光—— HIGH IN THE SUN 4
B-B Section 1：200
Children' s activity area
Tezcher' s activity area
Public area
The sliding doors areclosed during class
Open the sliding door allows students to enter the comprehensiveactivity room
At ordinary times state can pull sliding door to south side,form in the middle great court
Sunlight comes in through the patio
Sunlight comes in through the skylight
Sunlight comes in through the side window
The floor is warm at night
Natural ventilation of tower
Rammed earth walls protect against the north wind
Nest on the east side
天井采光
patio lighting
通风烟囱
Gout chimney
太阳房
solor house
太阳能聚光集热技术
parabolic through solar collectol
Sunshine corridor on the second floor
天台
Here the roof
Nest on the second floor
Insulation wall
15 dorm
17 administration office
25 ladies room
26 activity room
27 library
28 outdoor rooftop
Second Floor Plan 1：200

杲杲日光——HIGH IN THE SUN 5

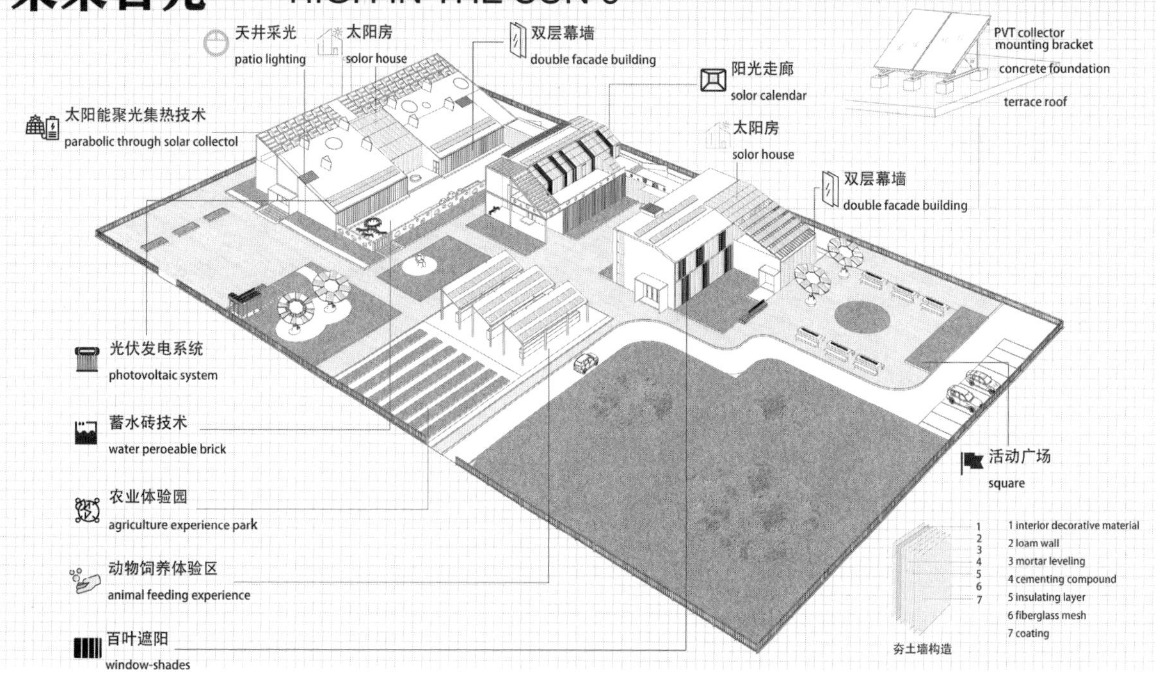

天井采光
patio lighting
太阳房
solor house
双层幕墙
double facade building
阳光走廊
solor calendar
太阳房
solor house
双层幕墙
double facade building
PVT collector mounting bracket
concrete foundation
terrace roof
太阳能聚光集热技术
parabolic through solar collectol
光伏发电系统
photovoltaic system
蓄水砖技术
water peroeable brick
农业体验园
agriculture experience park
动物饲养体验区
animal feeding experience
百叶遮阳
window-shades
活动广场
square
1 interior decorative material
2 loam wall
3 mortar leveling
4 cementing compound
5 insulating layer
6 fiberglass mesh
7 coating
夯土墙构造

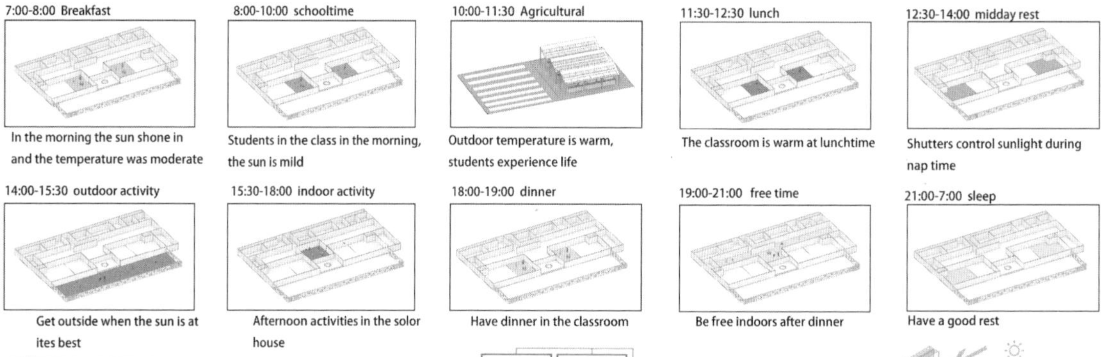

7:00-8:00 Breakfast
In the morning the sun shone in and the temperature was moderate
8:00-10:00 schooltime
Students in the class in the morning, the sun is mild
10:00-11:30 Agricultural
Outdoor temperature is warm, students experience life
11:30-12:30 lunch
The classroom is warm at lunchtime
12:30-14:00 midday rest
Shutters control sunlight during nap time
14:00-15:30 outdoor activity
Get outside when the sun is at ites best
15:30-18:00 indoor activity
Afternoon activities in the solor house
18:00-19:00 dinner
Have dinner in the classroom
19:00-21:00 free time
Be free indoors after dinner
21:00-7:00 sleep
Have a good rest

Outside the kindergarten

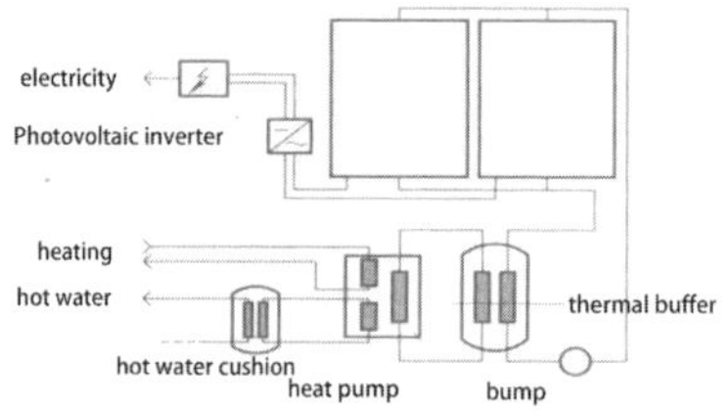

electricity
Photovoltaic inverter
heating
hot water
thermal buffer
hot water cushion
heat pump
bump

Energy conversion diagram

The interior of the corridor

• Natural light heating

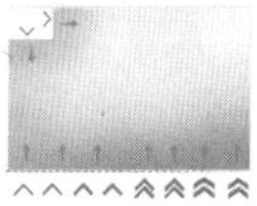

Control the lighting of the kindergarten through louvers

• The dormitory shutter

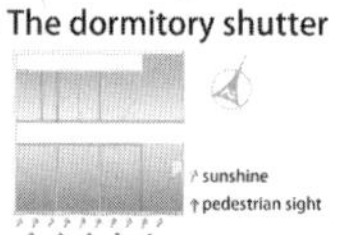

sunshine
pedestrian sight

Block the pedestrians' view without affecting the sunlight

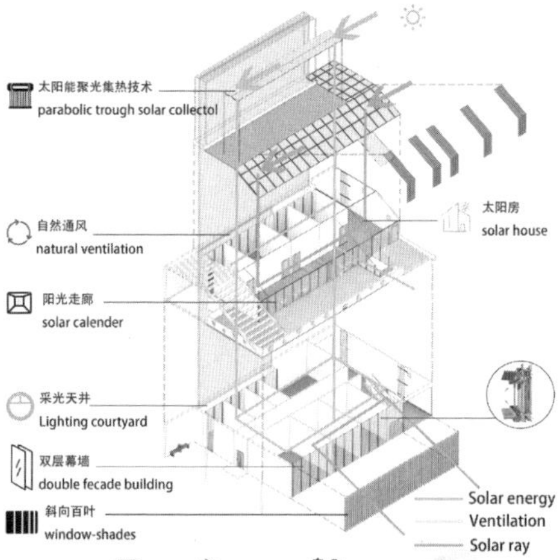

太阳能聚光集热技术
parabolic trough solar collectol
太阳房
solar house
自然通风
natural ventilation
阳光走廊
solar calender
采光天井
Lighting courtyard
双层幕墙
double fecade building
斜向百叶
window-shades
Solar energy
Ventilation
Solar ray

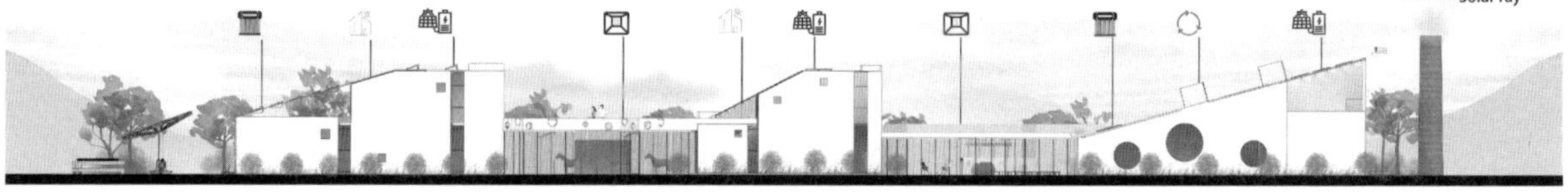

East elevation 1：200

杲杲日光——HIGH IN THE SUN 6

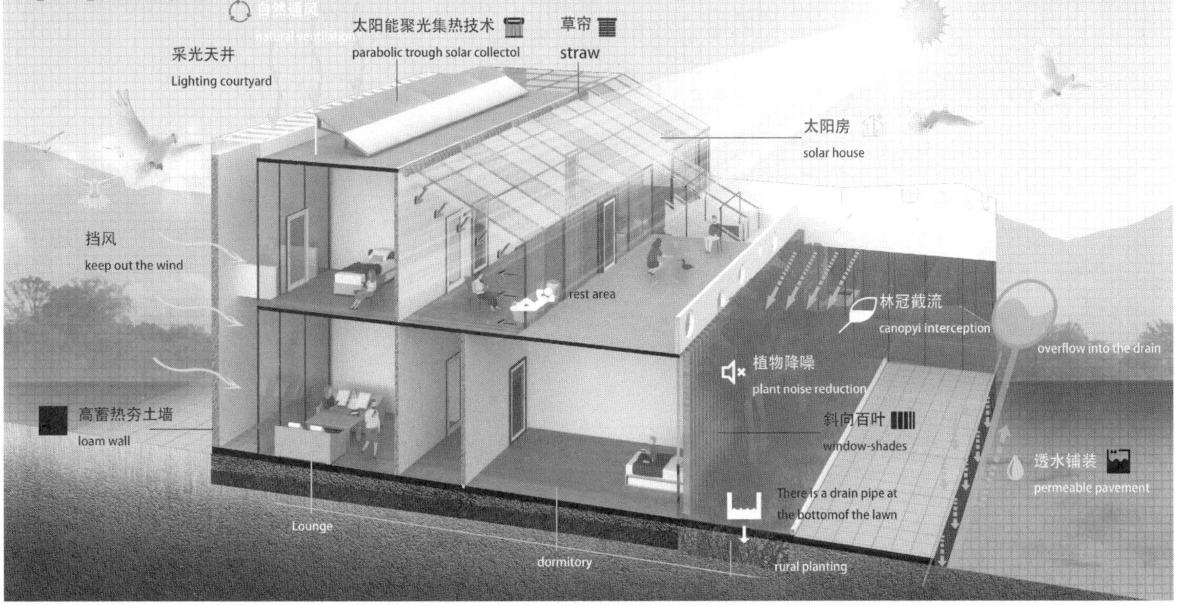
自然通风
natural ventilation
太阳能聚光集热技术
parabolic trough solar collectol
草帘
straw
采光天井
Lighting courtyard
太阳房
solar house
挡风
keep out the wind
rest area
林冠截流
canopyi interception
overflow into the drain
植物降噪
plant noise reduction
高蓄热夯土墙
loam wall
斜向百叶
window-shades
透水铺装
permeable pavement
There is a drain pipe at
the bottomof the lawn
Lounge
dormitory
rural planting

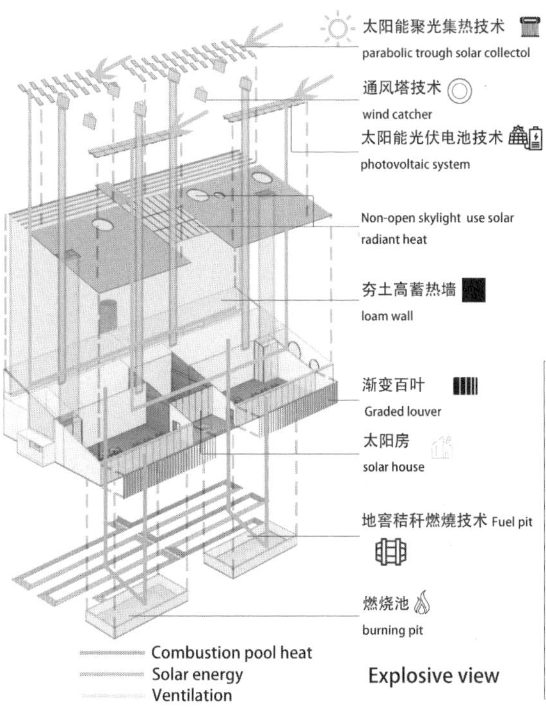
太阳能聚光集热技术
parabolic trough solar collectol
通风塔技术
wind catcher
太阳能光伏电池技术
photovoltaic system
Non-open skylight use solar
radiant heat
夯土高蓄热墙
loam wall
渐变百叶
Graded louver
太阳房
solar house
地窖秸秆燃烧技术 Fuel pit
燃烧池
burning pit
Combustion pool heat
Solar energy
Ventilation

Explosive view

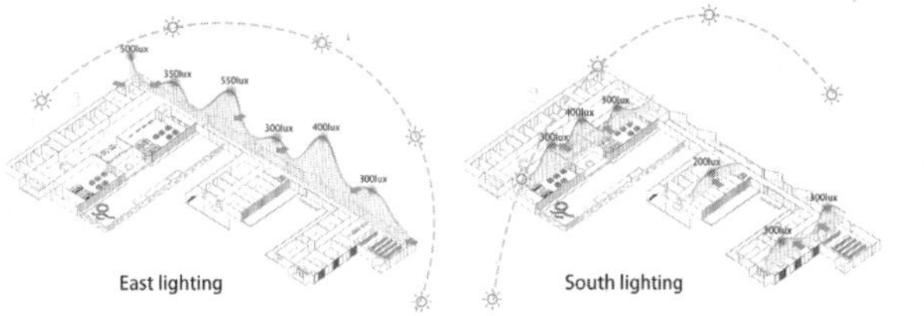
East lighting
South lighting

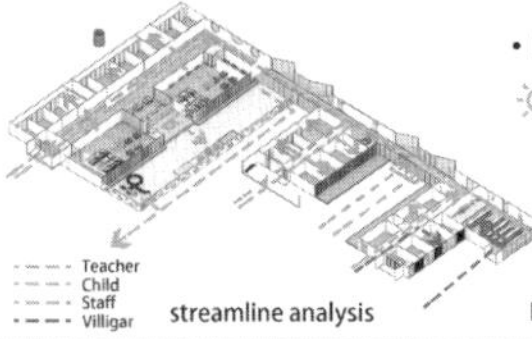
Teacher
Child
Staff
Villigar
streamline analysis

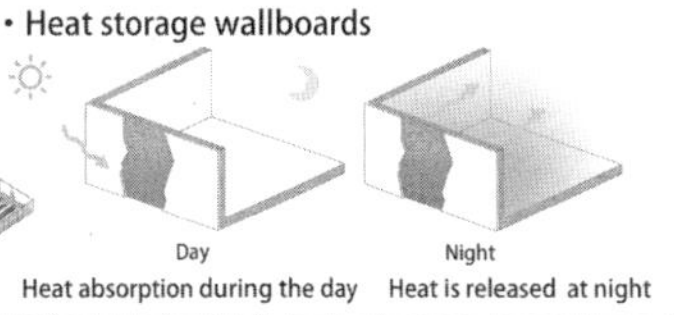
• Heat storage wallboards
Day
Night
Heat absorption during the day
Heat is released at night

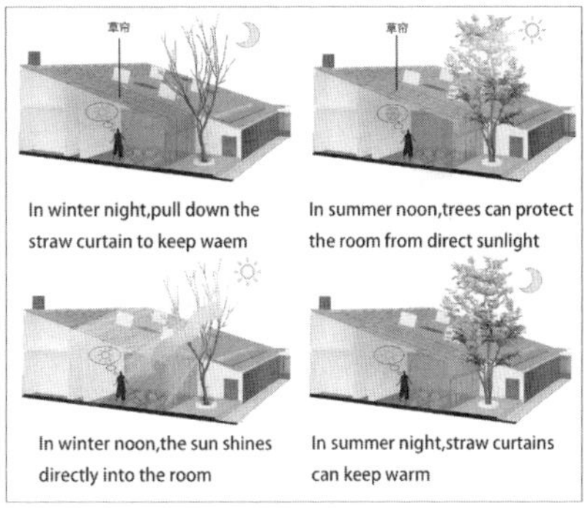
草帘
草帘
In winter night,pull down the
straw curtain to keep waem
In summer noon,trees can protect
the room from direct sunlight
In winter noon,the sun shines
directly into the room
In summer night,straw curtains
can keep warm

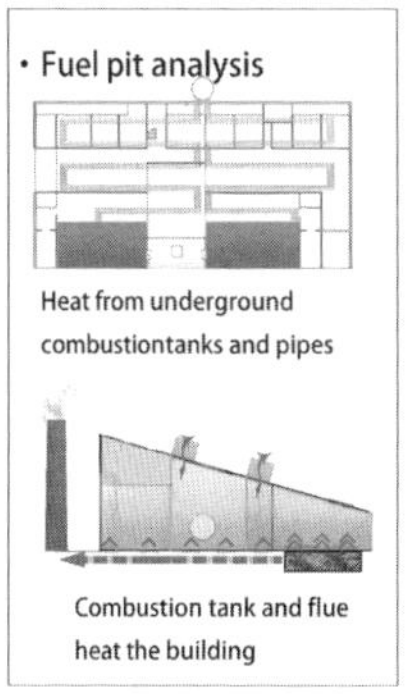
• Fuel pit analysis
Heat from underground
combustiontanks and pipes
Combustion tank and flue
heat the building

West elevation 1：200

综合奖 · 优秀奖
General Prize Awarded · Honorable Mention Prize

注 册 号：7277
项目名称：煦阳 · 和风（新疆）
Warm Sunshine · Gentle Wind（Xinjiang）
作 者：蔡 双、胡诗雨、王叶涵
参赛单位：重庆大学
指导教师：周铁军、张海滨

煦阳 · 和风 1
WARM SUNSHINE · GENTLE WIND
2020台达杯国际太阳能建筑设计竞赛
International Solar Building Design Competition 2020
Aerial View

Climate

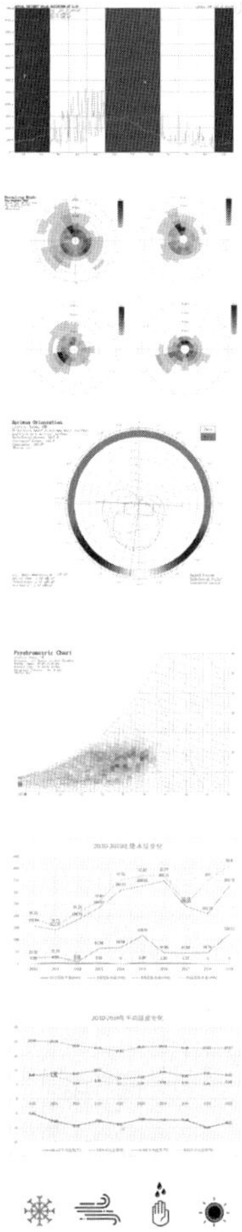
Cold + cool + dry + solar

Regional Feature

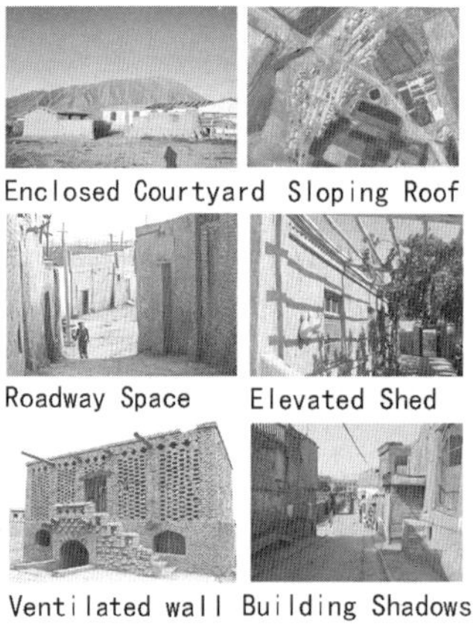
Enclosed Courtyard
Sloping Roof
Roadway Space
Elevated Shed
Ventilated wall
Building Shadows

Geographical Location

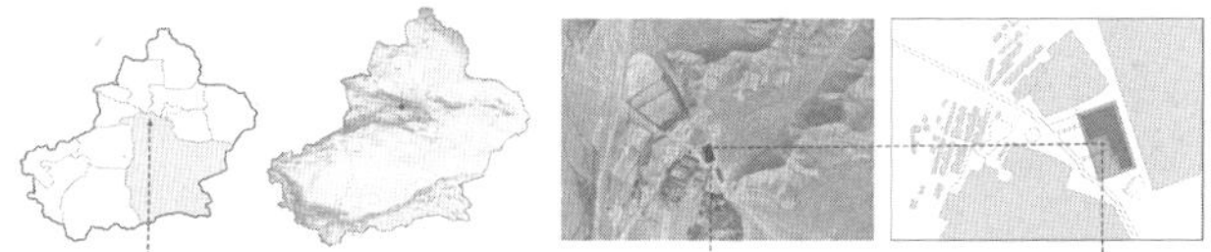

The site is located between the mountains, between the north and the south. Belongs to severe cold area.

The site is located on the plain, surrounded by grasslands, and there are villages along the mountains to the northwest.

Site Analysis

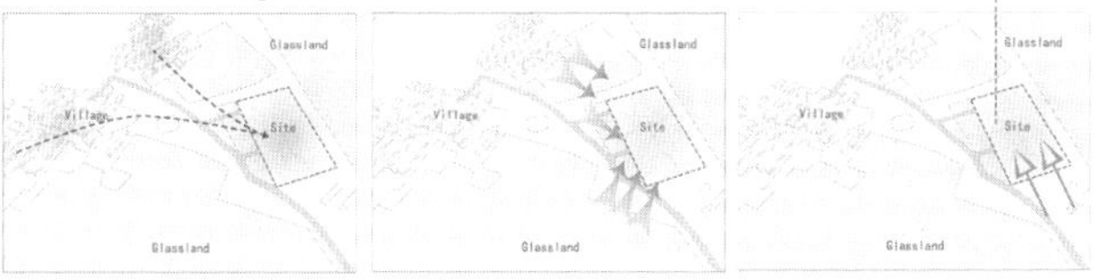

The flow of people on the site mainly comes from the villages in the northwest of the site.

The northwest wind prevails in winter and the southwest wind prevails in summer.

Cool in summer and cold in winter, with plenty of sunshine to the south throughout the year

Design Description

本方案以"煦阳和风"为主题，主要采用被动技术，利用太阳能、风能实现节能和舒适环境，再整合可再生能源技术实现产能建筑。

结合儿童作息时间，当地民居形式，在南向附加大进深活动阳光房，冬季关闭保温、夏季开敞通风，儿童活动不受气候影响。此外，包含相变蓄热墙体和材料、西向遮阳通风间墙、利用原蓄水池加湿空气的地道风系统等。主动技术上，除PVT系统外，利用太阳能沼气池，解决部分垃圾处理问题。

The theme of this project is "the warm sun and gentle wind", which means useing solar energy and wind energy to achieve energy saving and comfort by passive technology, and then integrating appropriate renewable energy technology to realize Energy plus house.

Combined with children's time table and local residential form, adding a large deep sunroom in the south of the room. The sunroom is closed in winter and open in summer, providing the most important outdoor activity space for children's growth and leisure area. In addition, there are the phase change heat storage wall and material, the sunshade ventilation wall facing west and the tunnel air system using original pool to humidify air. In terms of active technology, in addition to PVT system, there are solar biogas digesters to solve some garbage disposal problems and bring sustainable energy to self sufficiency of Architecture

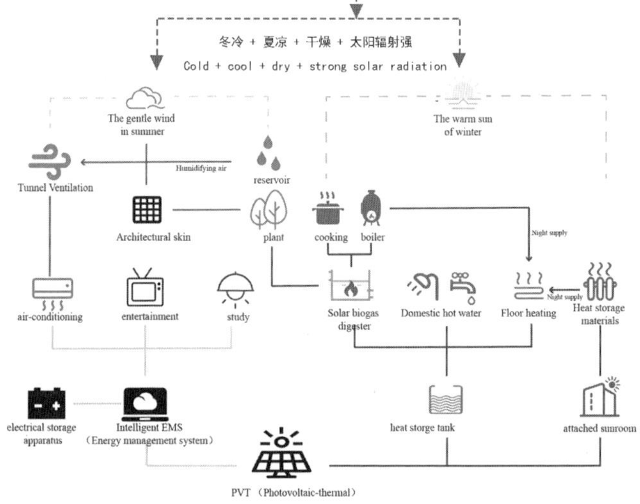
冬冷 + 夏凉 + 干燥 + 太阳辐射强
Cold + cool + dry + strong solar radiation
The gentle wind in summer
The warm sun of winter
Tunnel Ventilation
Humidifying air
reservoir
Architectural skin
plant
cooking
boiler
Night supply
air-conditioning
entertainment
study
Solar biogas digester
Domestic hot water
Floor heating
Heat storage materials
electrical storage apparatus
Intelligent EMS (Energy management system)
heat storge tank
attached sunroom
PVT (Photovoltaic-thermal)

煦阳 · 和风 2

WARM SUNSHINE · GENTLE WIND

2020台达杯国际太阳能建筑设计竞赛
International Solar Building Design Competition 2020

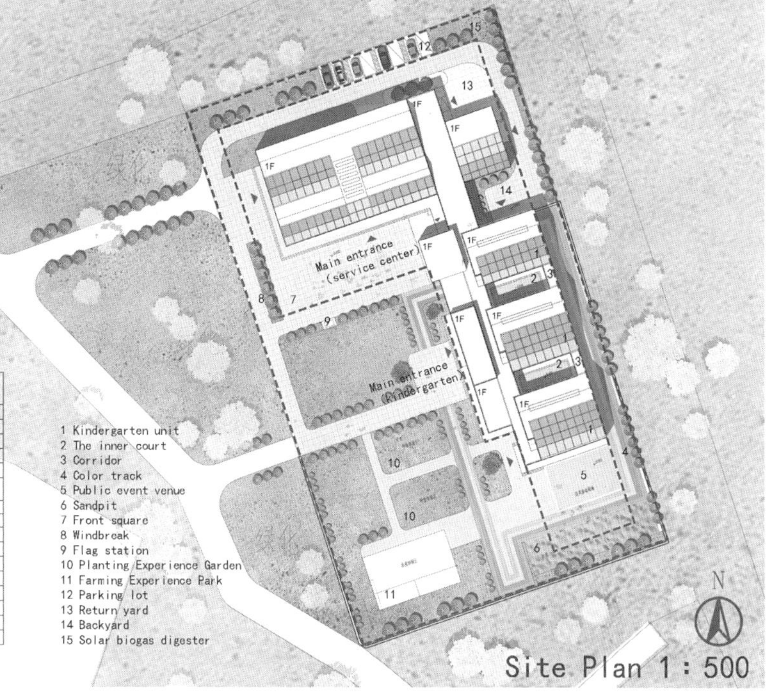

水池
Main entrance (service center)
Main entrance (Kindergarten)
1 Kindergarten unit
2 The inner court
3 Corridor
4 Color track
5 Public event venue
6 Sandpit
7 Front square
8 Windbreak
9 Flag station
10 Planting Experience Garden
11 Farming Experience Park
12 Parking lot
13 Return yard
14 Backyard
15 Solar biogas digester
N

Site Plan 1 : 500

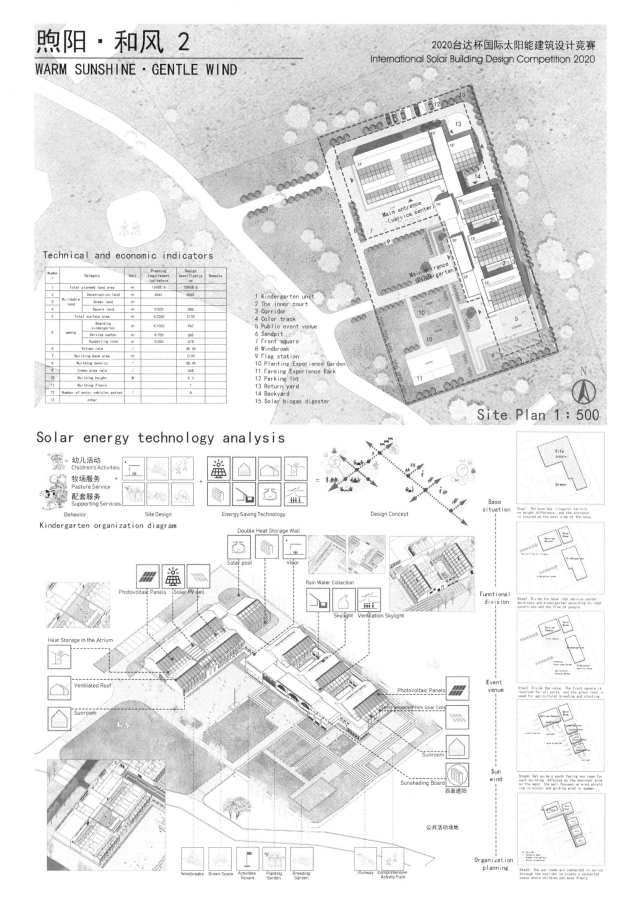

Technical and economic indicators

Number	Category		Unit	Planning requirement indicators	Design specifications	Remarks
1	Total planned land area		㎡	10408.6	10408.6	
2	Buildable land	Construction land	㎡	5541	5541	
3		Green land	㎡			
4		Square land	㎡	≥500	580	
5	Total surface area		㎡	≤2200	2130	
5	among	Boarding kindergarten	㎡	≤1000	967	
		Service center	㎡	≤700	685	
		Supporting room	㎡	≤500	478	
6	Volume rate		/		38.4%	
7	Building base area		㎡		2130	
8	Building density		/		38.4%	
9	Green area rate		/		36%	
10	Building height		M		6.3	
11	Building floors				1	
12	Number of motor vehicles parked		/		8	
13	other					

Solar energy technology analysis

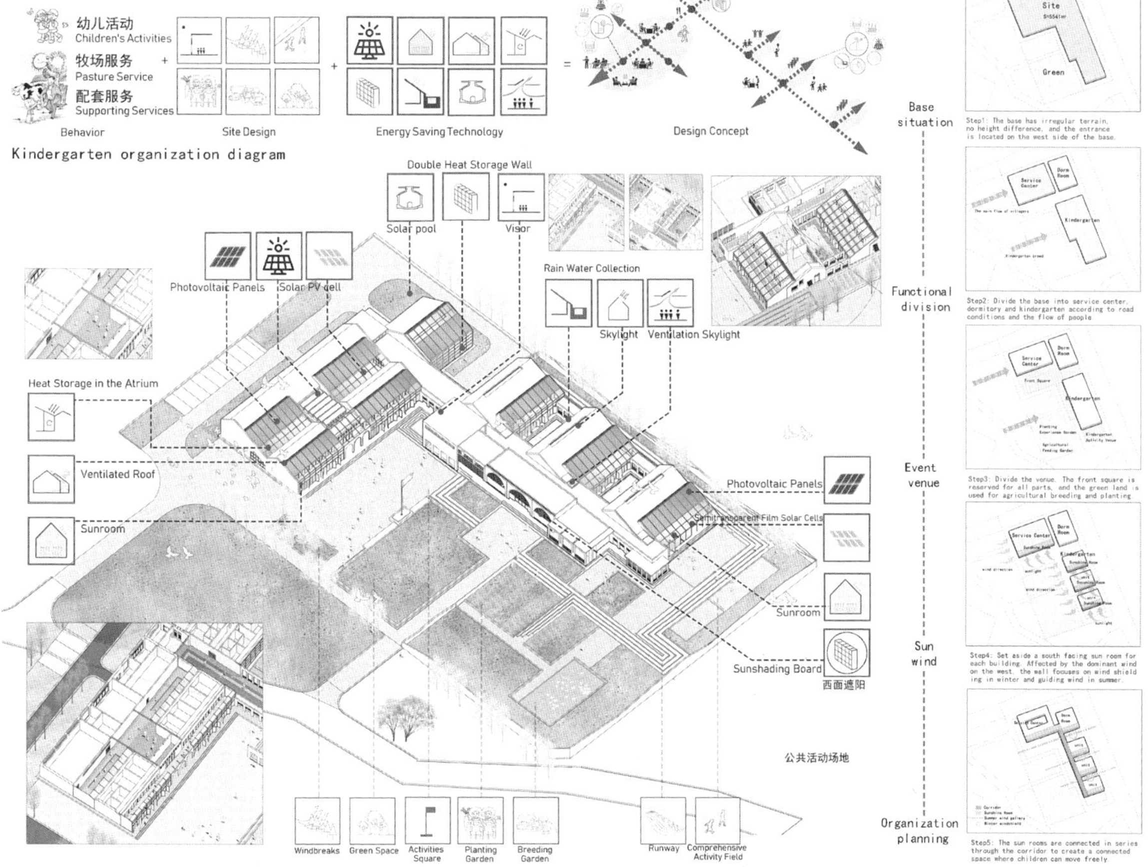

幼儿活动
Children's Activities
牧场服务
Pasture Service
配套服务
Supporting Services
Behavior
Site Design
Energy Saving Technology
Design Concept
Kindergarten organization diagram
Double Heat Storage Wall
Solar pool
Visor
Photovoltaic Panels
Solar PV cell
Rain Water Collection
Skylight
Ventilation Skylight
Heat Storage in the Atrium
Ventilated Roof
Sunroom
Photovoltaic Panels
Semitransparent Film Solar Cells
Sunroom
Sunshading Board
西面遮阳
公共活动场地
Windbreaks
Green Space
Activities Square
Planting Garden
Breeding Garden
Runway
Comprehensive Activity Field
Base situation
Functional division
Event venue
Sun wind
Organization planning
Green
Step1: The base has irregular terrain, no height difference, and the entrance is located on the west side of the base.
Service Center
Dorm Room
Kindergarten
Step2: Divide the base into service center, dormitory and kindergarten according to road conditions and the flow of people
Step3: Divide the venue. The front square is reserved for all parts, and the green land is used for agricultural breeding and planting
Step4: Set aside a south facing sun room for each building. Affected by the dominant wind on the west, the wall focuses on wind shielding in winter and guiding wind in summer.
Step5: The sun rooms are connected in series through the corridor to create a connected space where children can move freely.

煦阳 · 和风 3

WARM SUNSHINE · GENTLE WIND

2020台达杯国际太阳能建筑设计竞赛
International Solar Building Design Competition 2020

1 Activity room
2 Bathroom
3 Infant bedroom
4 Class activity area
5 Comprehensive activity room
6 Kindergarten kitchen
7 Health room
8 Sterilization chamber
9 Laundry room
10 Public toilet
11 Guard room
12 Equipment room
13 Centralized bathroom
14 Teacher Office
15 Teacher duty room
16 Unit corridor
17 Southbound corridor
18 Entrance foyer
19 Passing hall
20 Staff dormitory
21 Staff canteen
22 Staff kitchen
23 Public room
24 Function Room
25 Library
26 Consultation room
27 Heating atrium

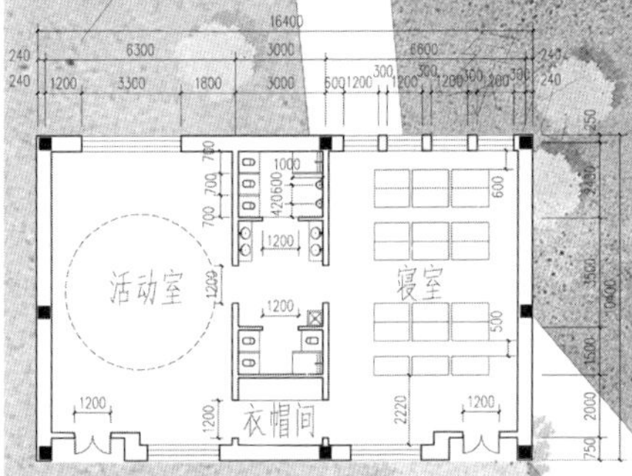
活动室
寝室
衣帽间

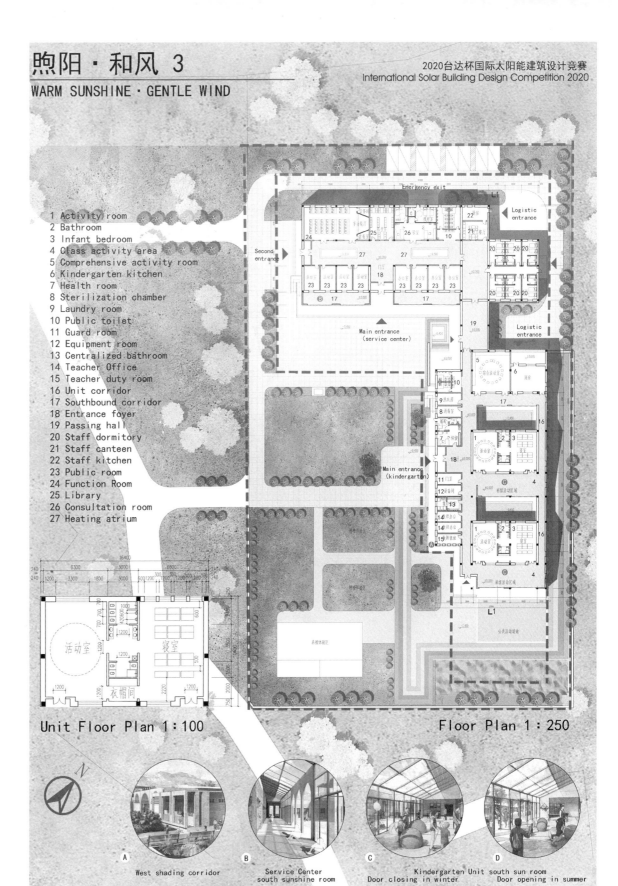

Unit Floor Plan 1：100

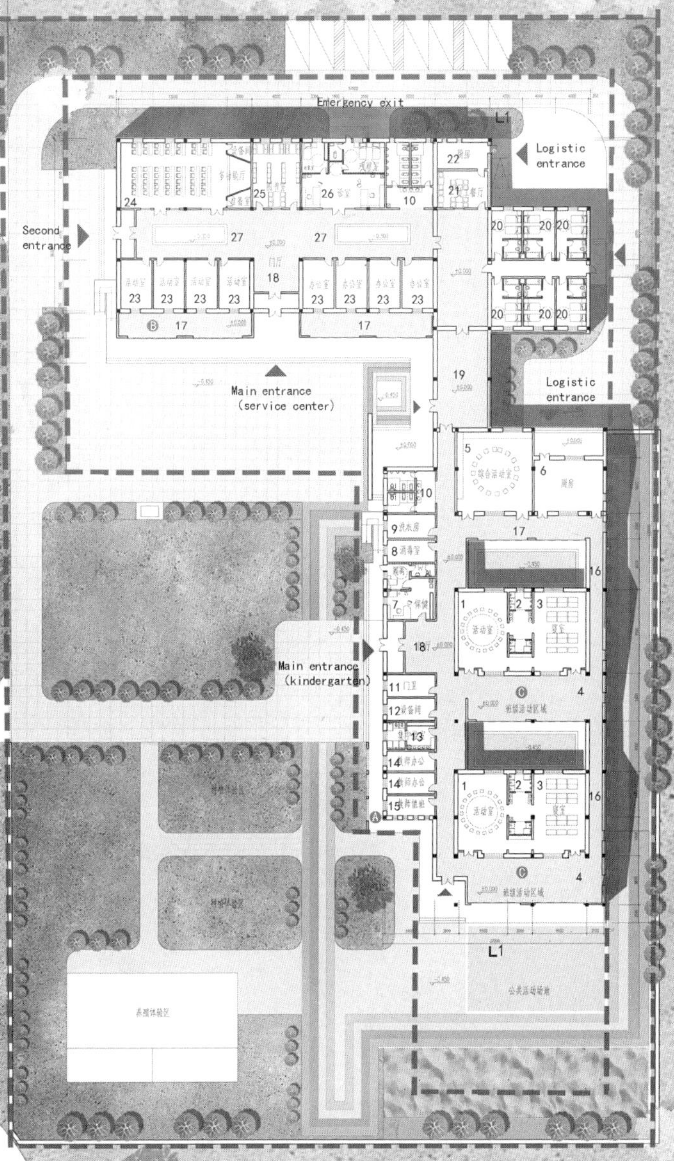
Emergency exit
Logistic entrance
Second entrance
Main entrance (service center)
Logistic entrance
Main entrance (kindergarten)
L1

Floor Plan 1：250

A West shading corridor

B Service Center south sunshine room

C Kindergarten Unit south sun room Door closing in winter

D Door opening in summer

煦阳·和风 4

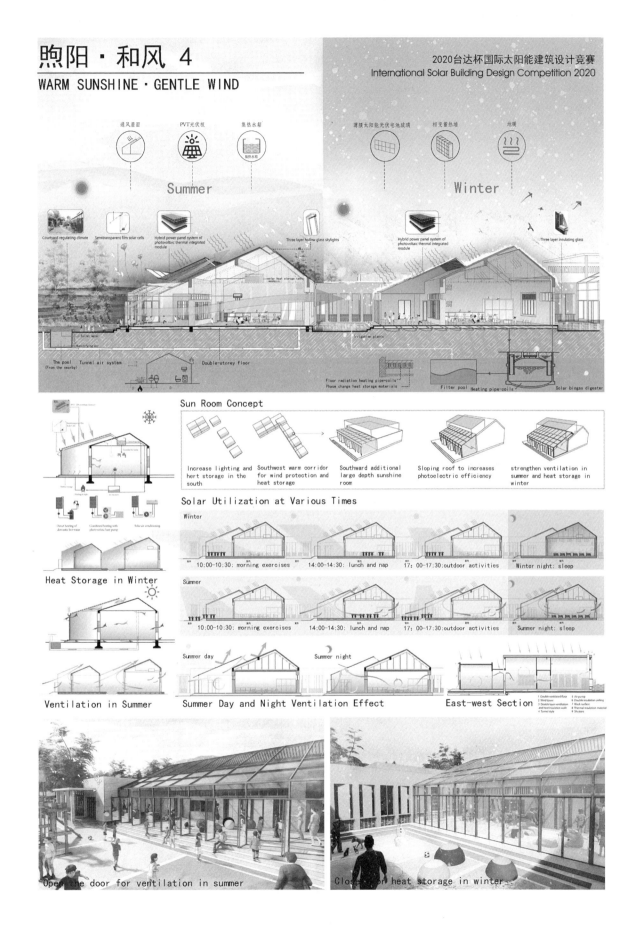

WARM SUNSHINE · GENTLE WIND

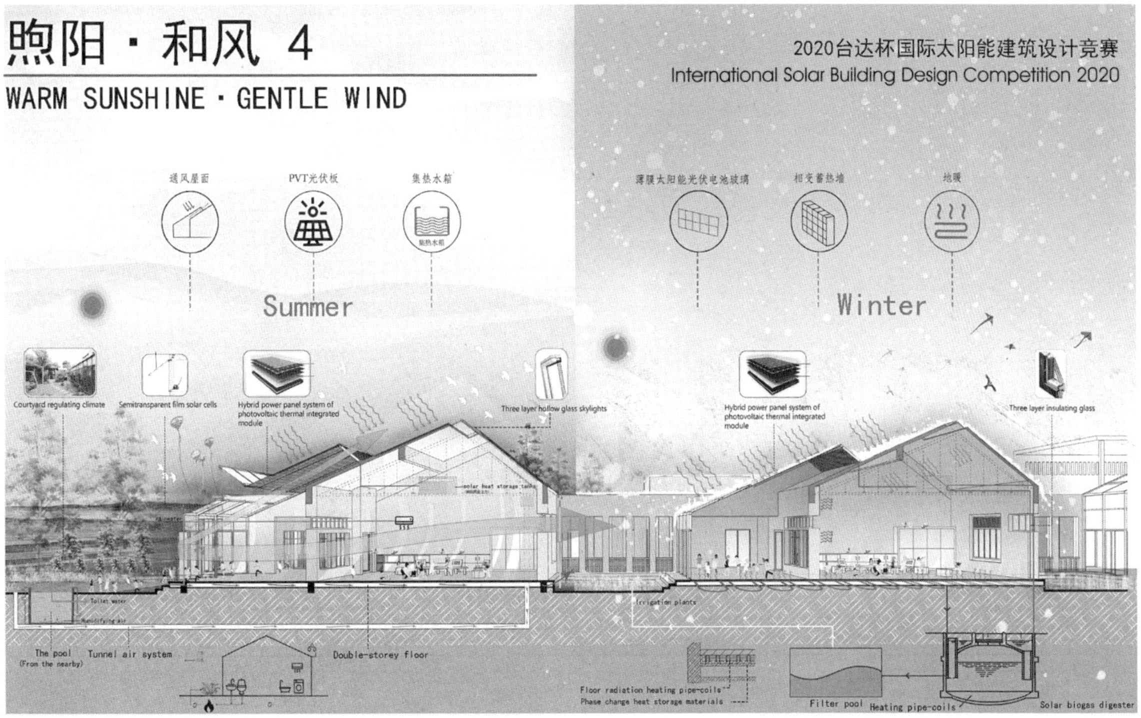
2020台达杯国际太阳能建筑设计竞赛
International Solar Building Design Competition 2020
通风屋面
PVT光伏板
集热水箱
薄膜太阳能光伏电池玻璃
相变蓄热墙
地暖
Summer
Winter
Courtyard regulating climate
Semitransparent film solar cells
Hybrid power panel system of photovoltaic thermal integrated module
Three layer hollow glass skylights
Three layer insulating glass
The pool (From the nearby)
Tunnel air system
Double-storey floor
Floor radiation heating pipe-coils
Phase change heat storage materials
Filter pool
Heating pipe-coils
Solar biogas digester

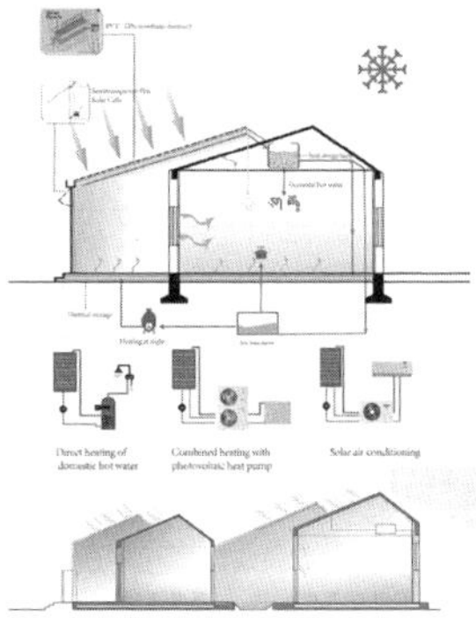

Heat Storage in Winter

Ventilation in Summer

Sun Room Concept

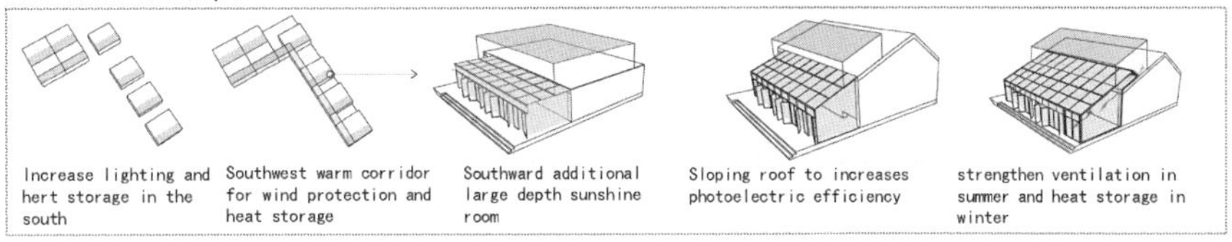
Increase lighting and hert storage in the south
Southwest warm corridor for wind protection and heat storage
Southward additional large depth sunshine room
Sloping roof to increases photoelectric efficiency
strengthen ventilation in summer and heat storage in winter

Solar Utilization at Various Times

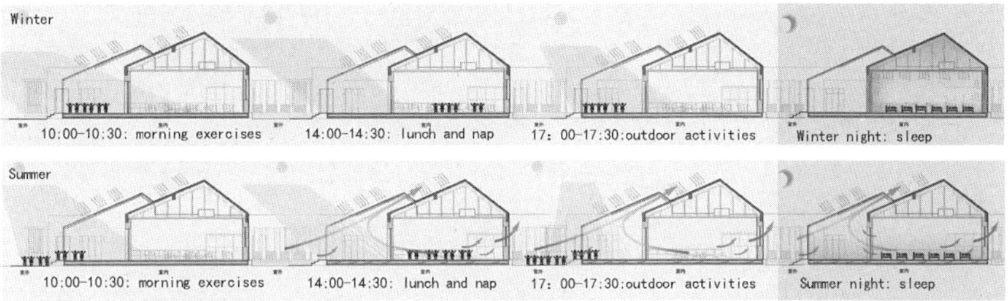
Winter
10:00-10:30: morning exercises
14:00-14:30: lunch and nap
17: 00-17:30:outdoor activities
Winter night: sleep
Summer
10:00-10:30: morning exercises
14:00-14:30: lunch and nap
17: 00-17:30:outdoor activities
Summer night: sleep

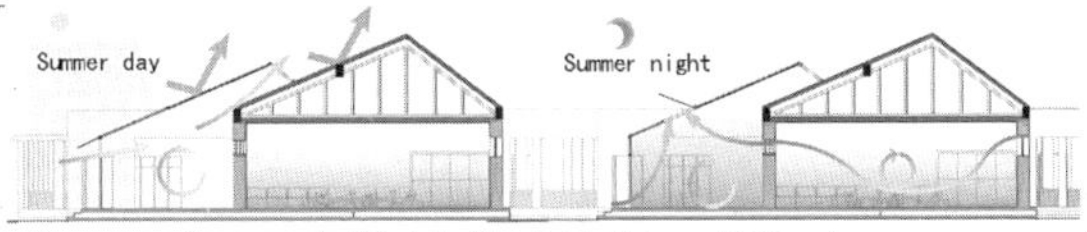
Summer day
Summer night

Summer Day and Night Ventilation Effect

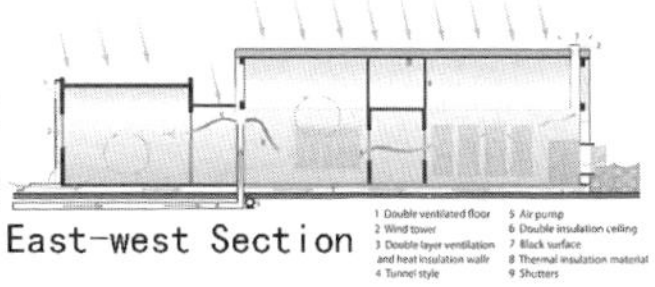

East-west Section

Open the door for ventilation in summer

Close for heat storage in winter

煦阳 · 和风 5

WARM SUNSHINE · GENTLE WIND

2020台达杯国际太阳能建筑设计竞赛
International Solar Building Design Competition 2020

Kindergarten Units and Public Venues

Unit Structure

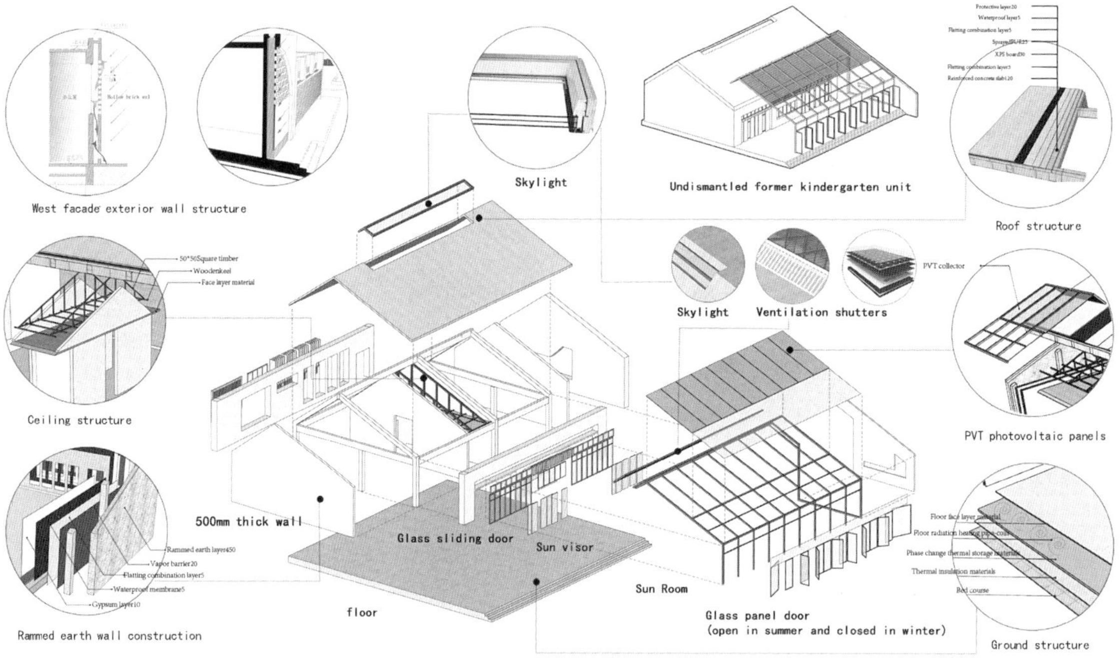
West facade exterior wall structure
Skylight
Undismantled former kindergarten unit
Roof structure
50*50Square timber
Woodenkeel
Face layer material
Skylight
Ventilation shutters
PVT collector
Ceiling structure
PVT photovoltaic panels
500mm thick wall
Glass sliding door
Sun visor
Sun Room
Rammed earth layer450
Vapor barrier20
Waterproof membrane5
Gypsum layer10
Thermal insulation materials
floor
Glass panel door
(open in summer and closed in winter)
Rammed earth wall construction
Ground structure

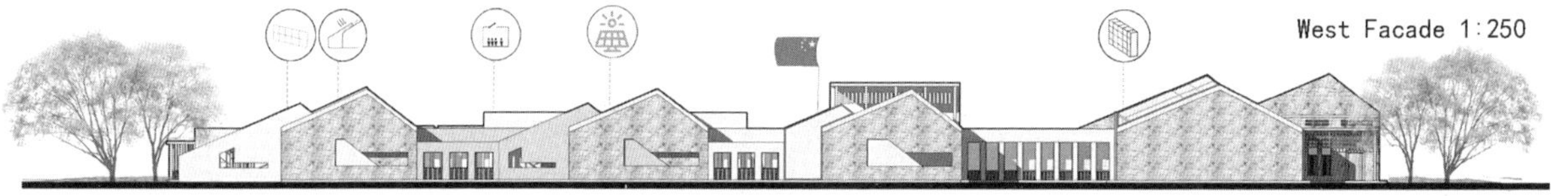
West Facade 1:250

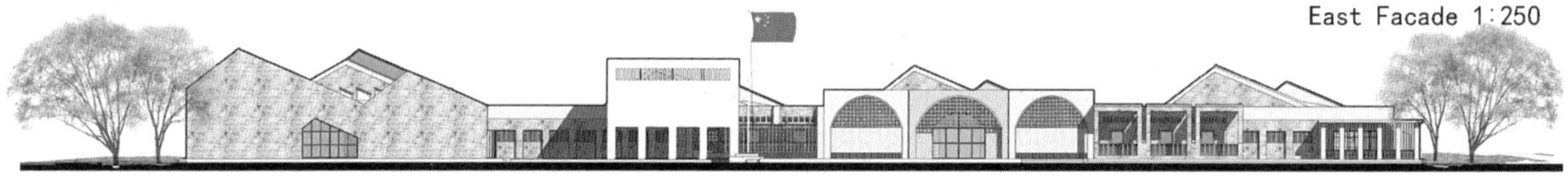
East Facade 1:250

煦阳·和风 6

WARM SUNSHINE · GENTLE WIND

2020台达杯国际太阳能建筑设计竞赛
International Solar Building Design Competition 2020

Service Center Entrance Plaza

Unit Analysis

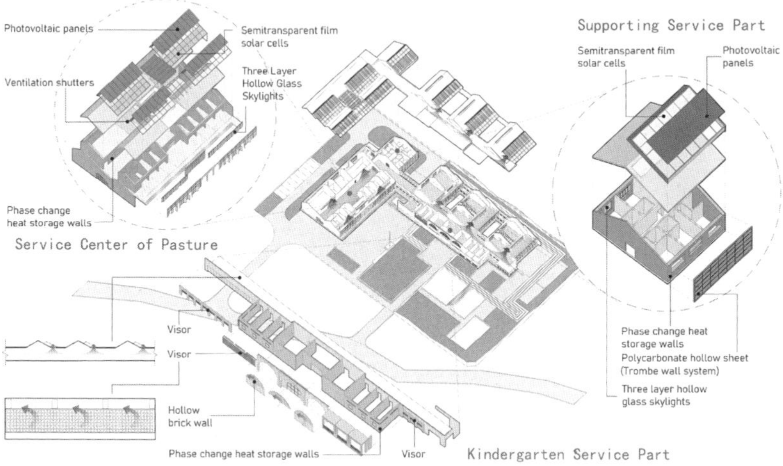
Photovoltaic panels
Semitransparent film solar cells
Ventilation shutters
Three Layer Hollow Glass Skylights
Phase change heat storage walls
Service Center of Pasture
Supporting Service Part
Semitransparent film solar cells
Photovoltaic panels
Phase change heat storage walls
Polycarbonate hollow sheet (Trombe wall system)
Three layer hollow glass skylights
Visor
Visor
Hollow brick wall
Phase change heat storage walls
Visor
Kindergarten Service Part

Wind-heat Environment Simulation

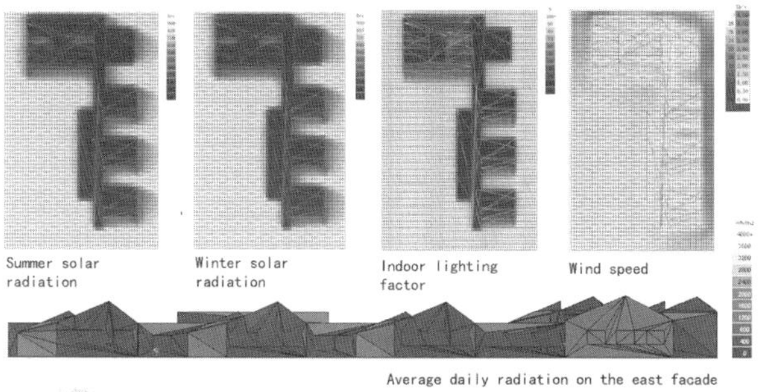
Summer solar radiation
Winter solar radiation
Indoor lighting factor
Wind speed
Average daily radiation on the east facade

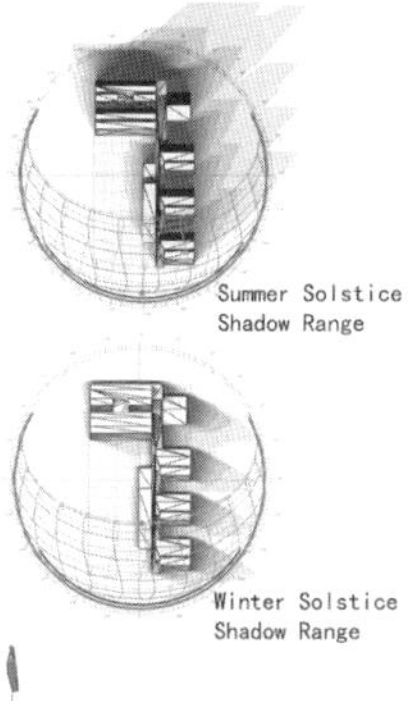
Summer Solstice Shadow Range
Winter Solstice Shadow Range

Sheet1: Thermal Parameters of the Enclosure Structure

Heat transfer coefficient	Factor		Unit	Designed building	Type
Roof	R		m²·k/W	3.56	水泥珍珠岩保温屋面（未考虑隔热吊顶）
Exterior wall	K		W/(m²·K)	0.550	500厚草泥土坯墙+聚苯板内保温
Exterior window	K	East	W/(m²·K)	[illegible]	三玻两腔单Low-E中空玻璃（双银）+PVC塑料窗框
		South	W/(m²·K)	1.581	三玻两腔单Low-E中空玻璃（双银）+PVC塑料窗框
		West	W/(m²·K)	[illegible]	三玻两腔单Low-E中空玻璃（双银）+PVC塑料窗框
		North	W/(m²·K)	1.581	三玻两腔单Low-E中空玻璃（双银）+PVC塑料窗框
	SHGC	East	W/(m²·K)	[illegible]	
		South	W/(m²·K)	0.395	
		West	W/(m²·K)	[illegible]	
		North	W/(m²·K)	0.399	
Roof light	K		W/(m²·K)	[illegible]	
	SHGC		W/(m²·K)	0.209	

Sheet2: Room Temperature Contrast of Activity Room
(compared with no solar house)

Close Sun Room Windows In Winter				
hours		BEFORE	AFTER	Outdoor temperature
≤16℃	Activity Room1	3604	1825	5775
	Activity Room2	3331	1292	
	Activity Room3	3670	1287	
16℃~22℃	Activity Room1	1379	1750	1432
	Activity Room2	1543	1912	
	Activity Room3	1340	1911	
22℃~29℃	Activity Room1	2744	2136	1200
	Activity Room2	2754	2074	
	Activity Room3	2646	2077	

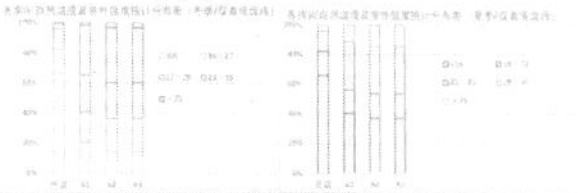

Open Sun Room Windows In Summer				
22℃~29℃	Activity Room1	2744	2332	1200
	Activity Room2	2754	2313	
	Activity Room3	2648	2304	
29℃~35℃	Activity Room1	1033	996	331
	Activity Room2	1222	1139	
	Activity Room3	1092	1056	
>35℃	Activity Room1	0	0	22
	Activity Room2	0	0	
	Activity Room3	0	0	

Sheet3: Energy Consumption Comparison

type of energy consumption		building without passive measures	designed building	reduction of energy consumption
Energy consumption of cold and heat sources	Air-conditioner sets (Kwh/a)	20.23	16.03	4.2
	Tunnel style Cooling (Kwh/a)	——	——	1.603(10%Air-conditioner sets energy consumption)
	Heating sets (natural gas) (万Nm³/a)	4.85	3.5	1.35
	Total energy consumption (equivalent electricity)(万kWh/a)	54.94	41.06	28%
other energy consumption	Lighting (万Kwh/a)	9.79		
	Domestic hot water (万kwh/a)	0.722		

Sheet4: Renewable Energy Calculation

Parameter		Unit	designed building
Slope daily radiation in Urumqi		MJ/m²	16594
Solar panel area		m²	500
PVT system	electric energy (140-160W/m²)	万kwh/a	21.6
	heat energy (200-500 W/m²)	万kwh/a	32.4

PSE:24%

ASE:82%

Energy plus house

Note: PSE=Passive solar energy utilization
ASE=Active solar energy utilization

Activity Room
Activity Room
Public Room
6.600
5.600
3.900
3.300
±0.000
-0.450

1-1 Section elevation 1 : 250

综合奖·优秀奖
General Prize Awarded · Honorable Mention Prize

注 册 号：7321

项目名称：腔间光语（新疆）
The Melody of Sunlight in the Chamber（Xinjiang）

作　　者：刘静怡、马宝裕、刘晓俐、王邑心、俞欣妤

参赛单位：华南理工大学、浙江大学、南加州建筑学院

指导教师：史劲松、赵立华

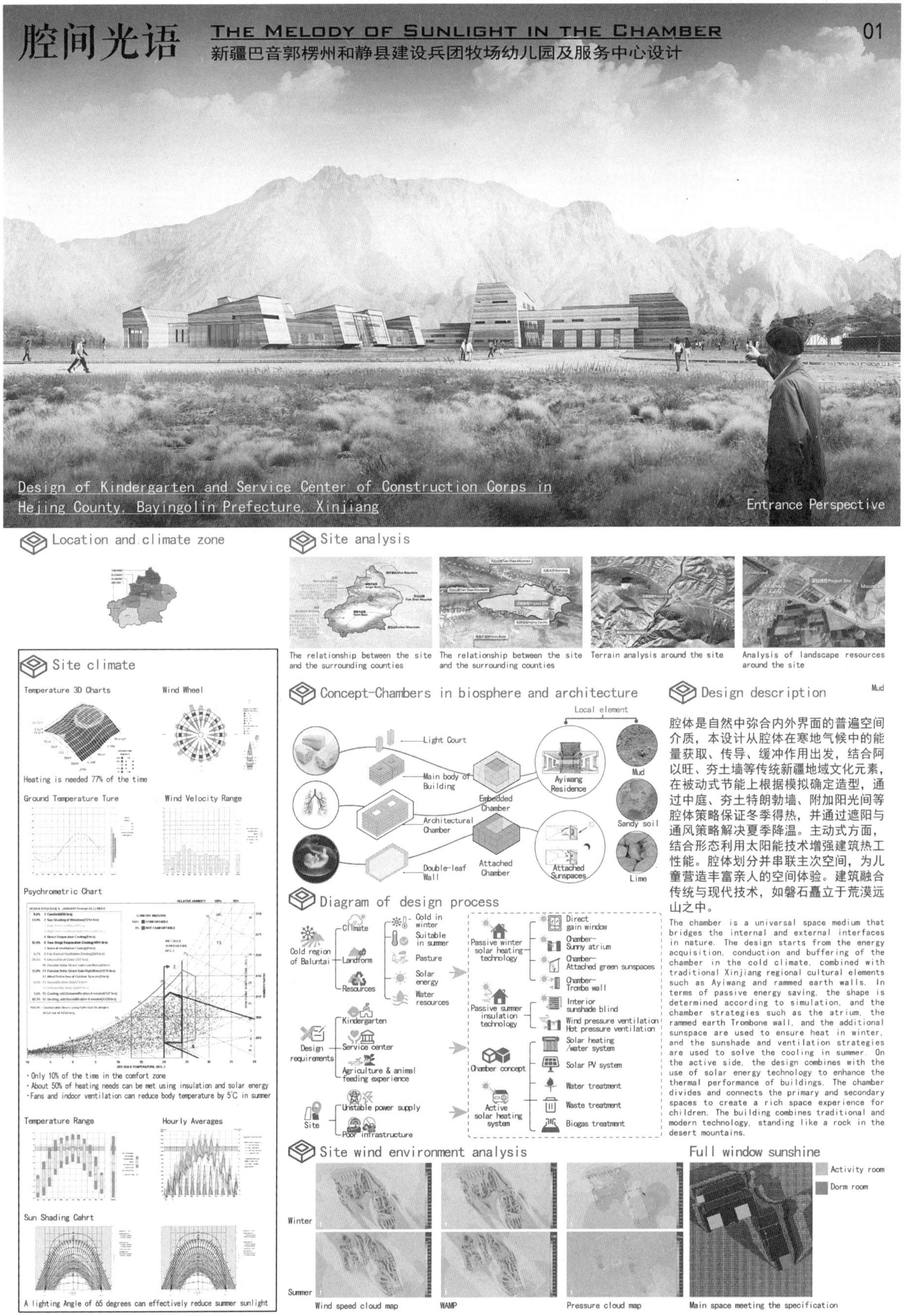
腔间光语
THE MELODY OF SUNLIGHT IN THE CHAMBER
新疆巴音郭楞州和静县建设兵团牧场幼儿园及服务中心设计
01
Design of Kindergarten and Service Center of Construction Corps in Hejing County, Bayingolin Prefecture, Xinjiang
Entrance Perspective
Location and climate zone
Site analysis
The relationship between the site and the surrounding counties
The relationship between the site and the surrounding counties
Terrain analysis around the site
Analysis of landscape resources around the site
Site climate
Temperature 3D Charts
Wind Wheel
Heating is needed 77% of the time
Ground Temperature Ture
Wind Velocity Range
Psychrometric Chart
·Only 10% of the time in the comfort zone
·About 50% of heating needs can be met using insulation and solar energy
·Fans and indoor ventilation can reduce body temperature by 5°C in summer
Temperature Range
Hourly Averages
Sun Shading Cahrt
A lighting Angle of 65 degrees can effectively reduce summer sunlight
Concept-Chambers in biosphere and architecture
Local element
Light Court
Main body of Building
Embedded Chamber
Aylwang Residence
Mud
Architectural Chamber
Sandy soil
Double-leaf Wall
Attached Chamber
Attached Sunspaces
Lime
Design description
腔体是自然中弥合内外界面的普遍空间介质，本设计从腔体在寒地气候中的能量获取、传导、缓冲作用出发，结合阿以旺、夯土墙等传统新疆地域文化元素，在被动式节能上根据模拟确定造型，通过中庭、夯土特朗勃墙、附加阳光间等腔体策略保证冬季得热，并通过遮阳与通风策略解决夏季降温。主动式方面，结合形态利用太阳能技术增强建筑热工性能。腔体划分并串联主次空间，为儿童营造丰富亲人的空间体验。建筑融合传统与现代技术，如磐石矗立于荒漠远山之中。
The chamber is a universal space medium that bridges the internal and external interfaces in nature. The design starts from the energy acquisition, conduction and buffering of the chamber in the cold climate, combined with traditional Xinjiang regional cultural elements such as Ayiwang and rammed earth walls. In terms of passive energy saving, the shape is determined according to simulation, and the chamber strategies such as the atrium, the rammed earth Trombone wall, and the additional sunspace are used to ensure heat in winter, and the sunshade and ventilation strategies are used to solve the cooling in summer. On the active side, the design combines with the use of solar energy technology to enhance the thermal performance of buildings. The chamber divides and connects the primary and secondary spaces to create a rich space experience for children. The building combines traditional and modern technology, standing like a rock in the desert mountains.
Diagram of design process
Cold region of Baluntai
Climate
Landform
Resources
Cold in winter
Suitable in summer
Pasture
Solar energy
Water resources
Design requirements
Kindergarten
Service center
Agriculture & animal feeding experience
Site
Unstable power supply
Poor infrastructure
Passive winter solar heating technology
Passive summer insulation technology
Chamber concept
Active solar heating system
Direct gain window
Chamber-Sunny atrium
Chamber-Attached green sunspaces
Chamber-Trombe wall
Interior sunshade blind
Wind pressure ventilation Hot pressure ventilation
Solar heating /water system
Solar PV system
Water treatment
Waste treatment
Biogas treatment
Site wind environment analysis
Winter
Summer
Wind speed cloud map
WAMP
Pressure cloud map
Full window sunshine
Activity room
Dorm room
Main space meeting the specification

腔间光语 THE MELODY OF SUNLIGHT IN THE CHAMBER

02

新疆巴音郭楞州和静县建设兵团牧场幼儿园及服务中心设计

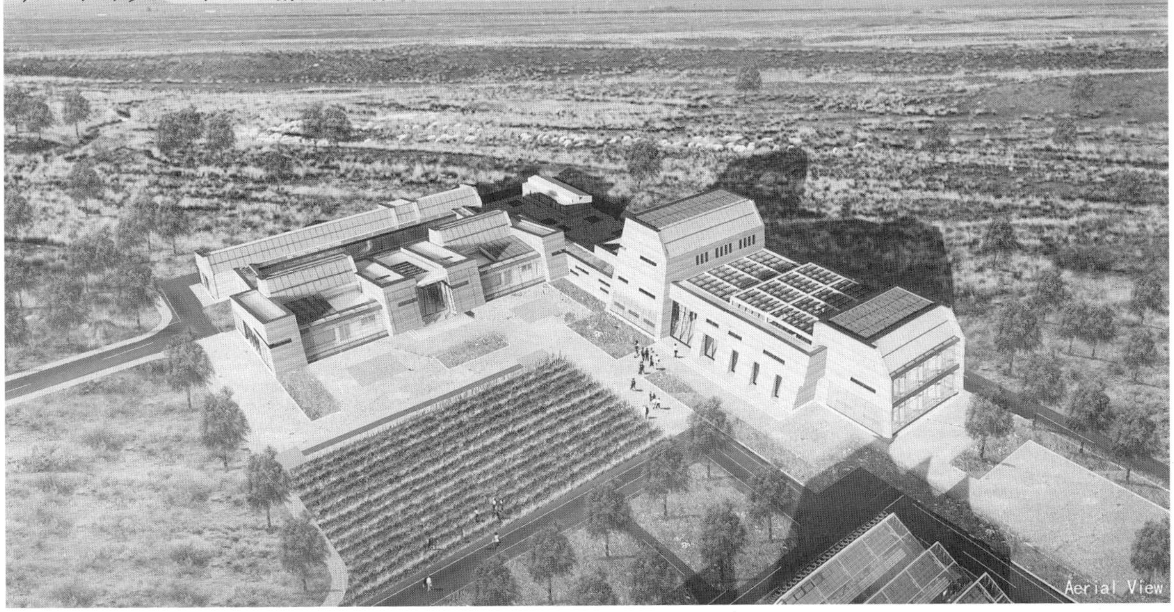
Aerial View

Form Generation Analysis

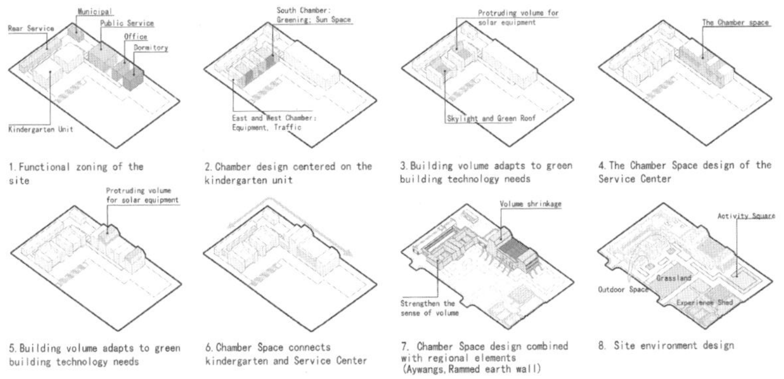
Rear Service
Municipal
Public Service
Office
Dormitory
Kindergarten Unit
1. Functional zoning of the site
South Chamber: Greening; Sun Space
East and West Chamber: Equipment, Traffic
2. Chamber design centered on the kindergarten unit
Protruding volume for solar equipment
Skylight and Green Roof
3. Building volume adapts to green building technology needs
The Chamber space
4. The Chamber Space design of the Service Center
Protruding volume for solar equipment
5. Building volume adapts to green building technology needs
6. Chamber Space connects kindergarten and Service Center
Volume shrinkage
Strengthen the sense of volume
7. Chamber Space design combined with regional elements (Aywangs, Rammed earth wall)
Activity Square
Grassland
Outdoor Space
Experience Shed
8. Site environment design

Kindergarten Foyer

Kindergarten Sunshine Room

Service Center Atrium

Service Center Corridor

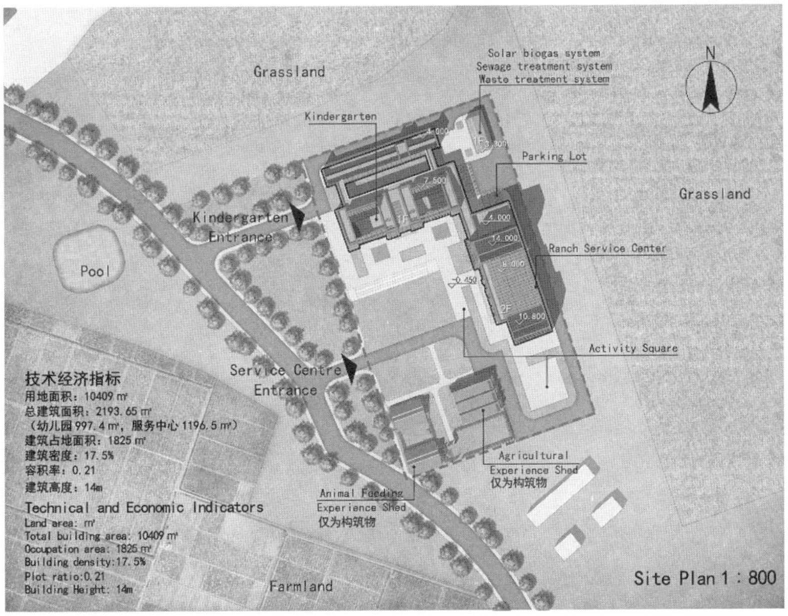
Grassland
Solar biogas system
Sewage treatment system
Waste treatment system
N
Kindergarten
Parking Lot
Grassland
Kindergarten Entrance
Ranch Service Center
Pool
Activity Square
Service Centre Entrance
Agricultural Experience Shed
仅为构筑物
Animal Feeding Experience Shed
仅为构筑物
Farmland
技术经济指标
用地面积：10409 m²
总建筑面积：2193.65 m²
（幼儿园997.4 m²，服务中心1196.5 m²）
建筑占地面积：1825 m²
建筑密度：17.5%
容积率：0.21
建筑高度：14m
Technical and Economic Indicators
Land area: m²
Total building area: 10409 m²
Occupation area: 1825 m²
Building density:17.5%
Plot ratio:0.21
Building Height: 14m

Site Plan 1：800

腔间光语 THE MELODY OF SUNLIGHT IN THE CHAMBER

新疆巴音郭楞州和静县建设兵团牧场幼儿园及服务中心设计

03

Kindergarten Entrance Perspective

Strategy overview

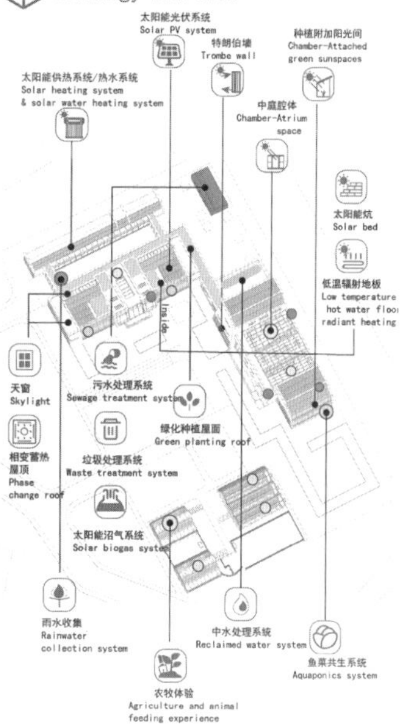
太阳能光伏系统 Solar PV system
特朗伯墙 Trombe wall
种植附加阳光间 Chamber-Attached green sunspaces
太阳能供热系统/热水系统 Solar heating system & solar water heating system
中庭腔体 Chamber-Atrium space
太阳能炕 Solar bed
低温辐射地板 Low temperature hot water floor radiant heating
天窗 Skylight
污水处理系统 Sewage treatment system
绿化种植屋面 Green planting roof
相变蓄热屋顶 Phase change roof
垃圾处理系统 Waste treatment system
太阳能沼气系统 Solar biogas system
雨水收集 Rainwater collection system
中水处理系统 Reclaimed water system
鱼菜共生系统 Aquaponics system
农牧体验 Agriculture and animal feeding experience

First Floor Plan 1：250

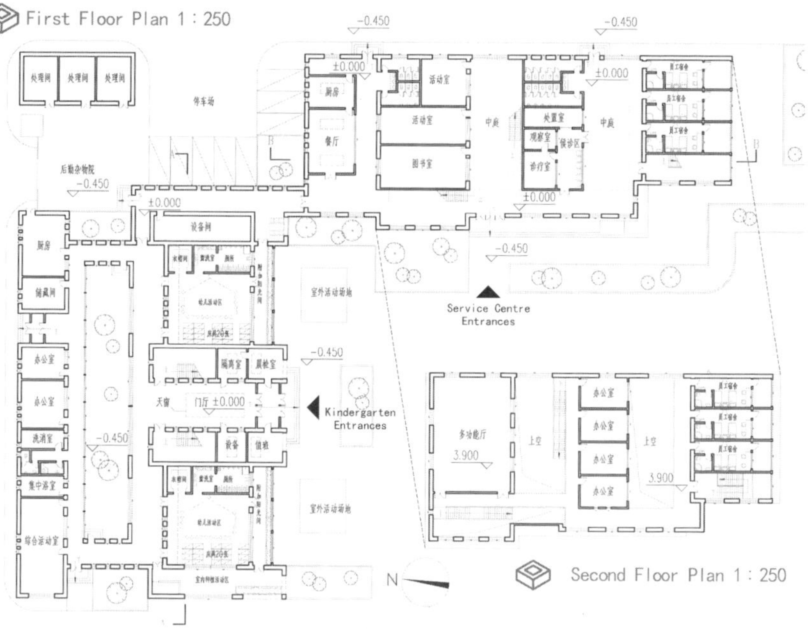
Service Centre Entrances
Kindergarten Entrances
N

Second Floor Plan 1：250

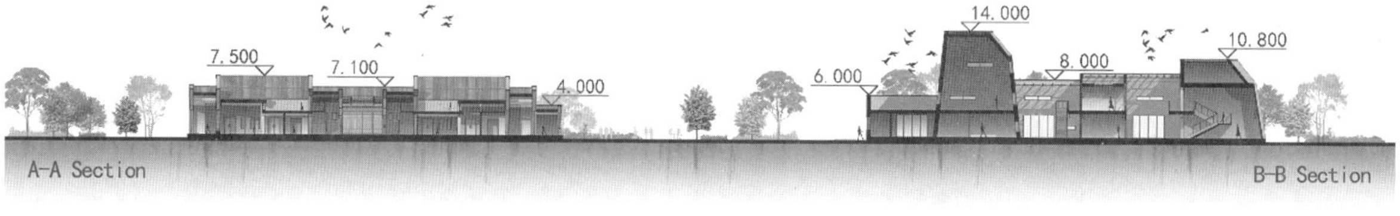
7.500
7.100
4.000
14.000
6.000
8.000
10.800

A-A Section

B-B Section

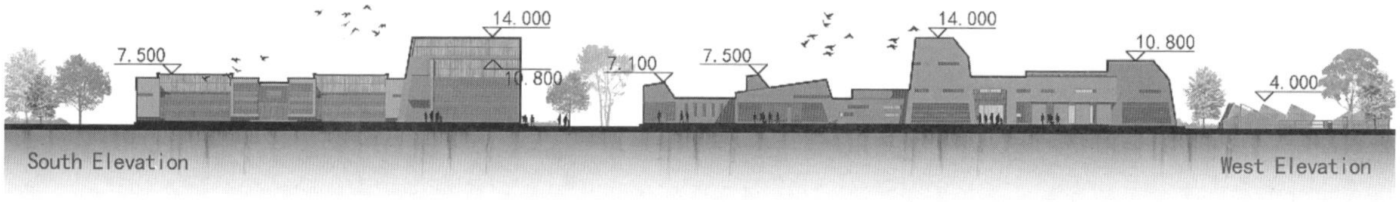
7.500
14.000
10.800
7.100
7.500
14.000
10.800
4.000

South Elevation

West Elevation

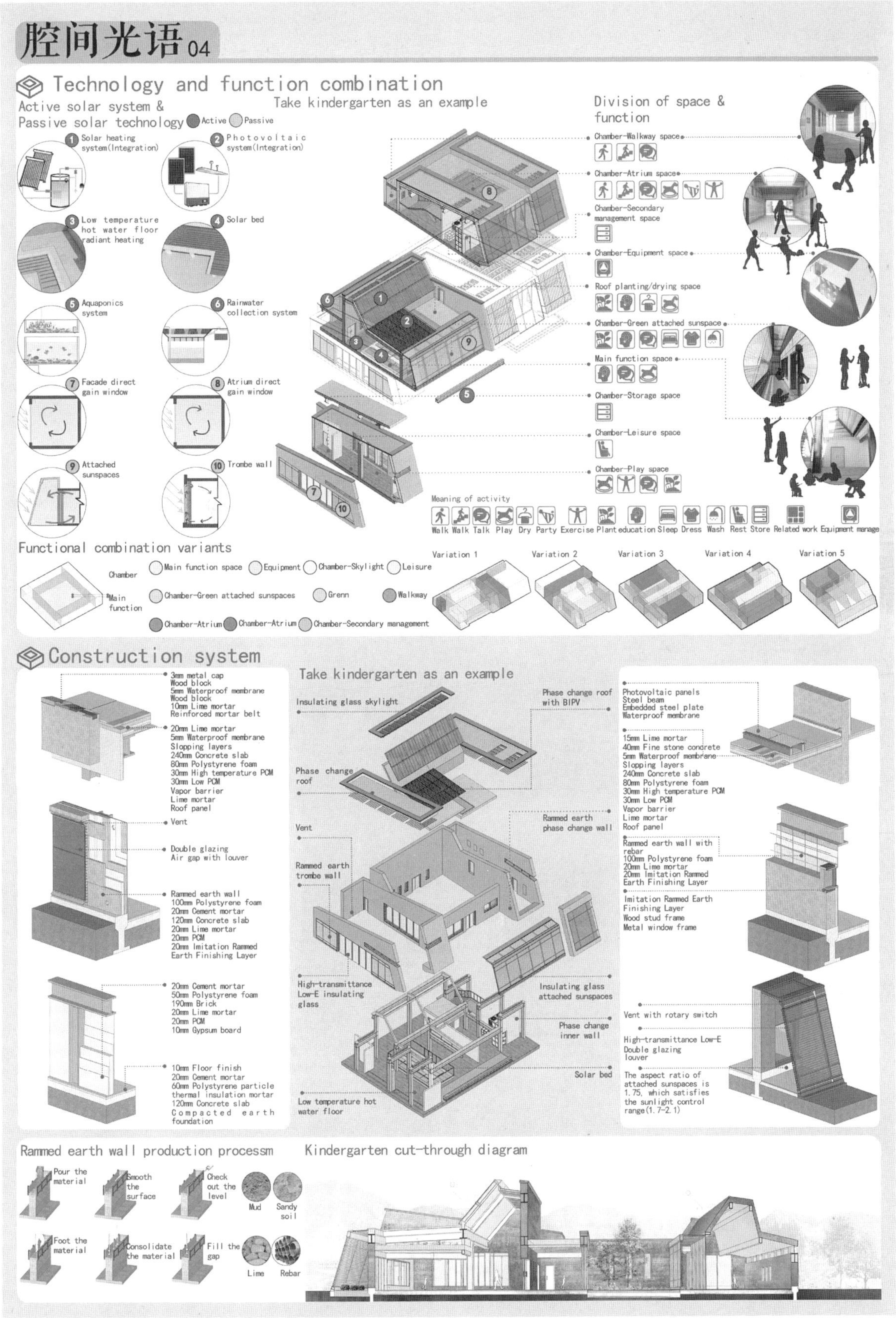

腔间光语 04
Technology and function combination
Active solar system &
Passive solar technology
Active
Passive
Take kindergarten as an example
1 Solar heating system(Integration)
2 Photovoltaic system(Integration)
3 Low temperature hot water floor radiant heating
4 Solar bed
5 Aquaponics system
6 Rainwater collection system
7 Facade direct gain window
8 Atrium direct gain window
9 Attached sunspaces
10 Trombe wall
Division of space & function
Chamber-Walkway space
Chamber-Atrium space
Chamber-Secondary management space
Chamber-Equipment space
Roof planting/drying space
Chamber-Green attached sunspace
Main function space
Chamber-Storage space
Chamber-Leisure space
Chamber-Play space
Meaning of activity
Walk Walk Talk Play Dry Party Exercise Plant education Sleep Dress Wash Rest Store Related work Equipment manage
Functional combination variants
Chamber
Main function
Main function space
Equipment
Chamber-Skylight
Leisure
Chamber-Green attached sunspaces
Grenn
Walkway
Chamber-Atrium
Chamber-Atrium
Chamber-Secondary management
Variation 1
Variation 2
Variation 3
Variation 4
Variation 5
Construction system
Take kindergarten as an example
3mm metal cap
Wood block
5mm Waterproof membrane
Wood block
10mm Lime mortar
Reinforced mortar belt
20mm Lime mortar
5mm Waterproof membrane
Slopping layers
240mm Concrete slab
80mm Polystyrene foam
30mm High temperature PCM
30mm Low PCM
Vapor barrier
Lime mortar
Roof panel
Vent
Double glazing
Air gap with louver
Rammed earth wall
100mm Polystyrene foam
20mm Cement mortar
120mm Concrete slab
20mm Lime mortar
20mm PCM
20mm Imitation Rammed Earth Finishing Layer
20mm Cement mortar
50mm Polystyrene foam
190mm Brick
20mm Lime mortar
20mm PCM
10mm Gypsum board
10mm Floor finish
20mm Cement mortar
60mm Polystyrene particle thermal insulation mortar
120mm Concrete slab
Compacted earth foundation
Insulating glass skylight
Phase change roof with BIPV
Phase change roof
Vent
Rammed earth trombe wall
Rammed earth phase change wall
High-transmittance Low-E insulating glass
Insulating glass attached sunspaces
Phase change inner wall
Solar bed
Low temperature hot water floor
Photovoltaic panels
Steel beam
Embedded steel plate
Waterproof membrane
15mm Lime mortar
40mm Fine stone concrete
5mm Waterproof membrane
Slopping layers
240mm Concrete slab
80mm Polystyrene foam
30mm High temperature PCM
30mm Low PCM
Vapor barrier
Lime mortar
Roof panel
Rammed earth wall with rebar
100mm Polystyrene foam
20mm Lime mortar
20mm Imitation Rammed Earth Finishing Layer
Imitation Rammed Earth Finishing Layer
Wood stud frame
Metal window frame
Vent with rotary switch
High-transmittance Low-E Double glazing louver
The aspect ratio of attached sunspaces is 1.75, which satisfies the sunlight control range(1.7-2.1)
Rammed earth wall production processm
Pour the material
Smooth the surface
Check out the level
Mud
Sandy soil
Foot the material
Consolidate the material
Fill the gap
Lime
Rebar
Kindergarten cut-through diagram

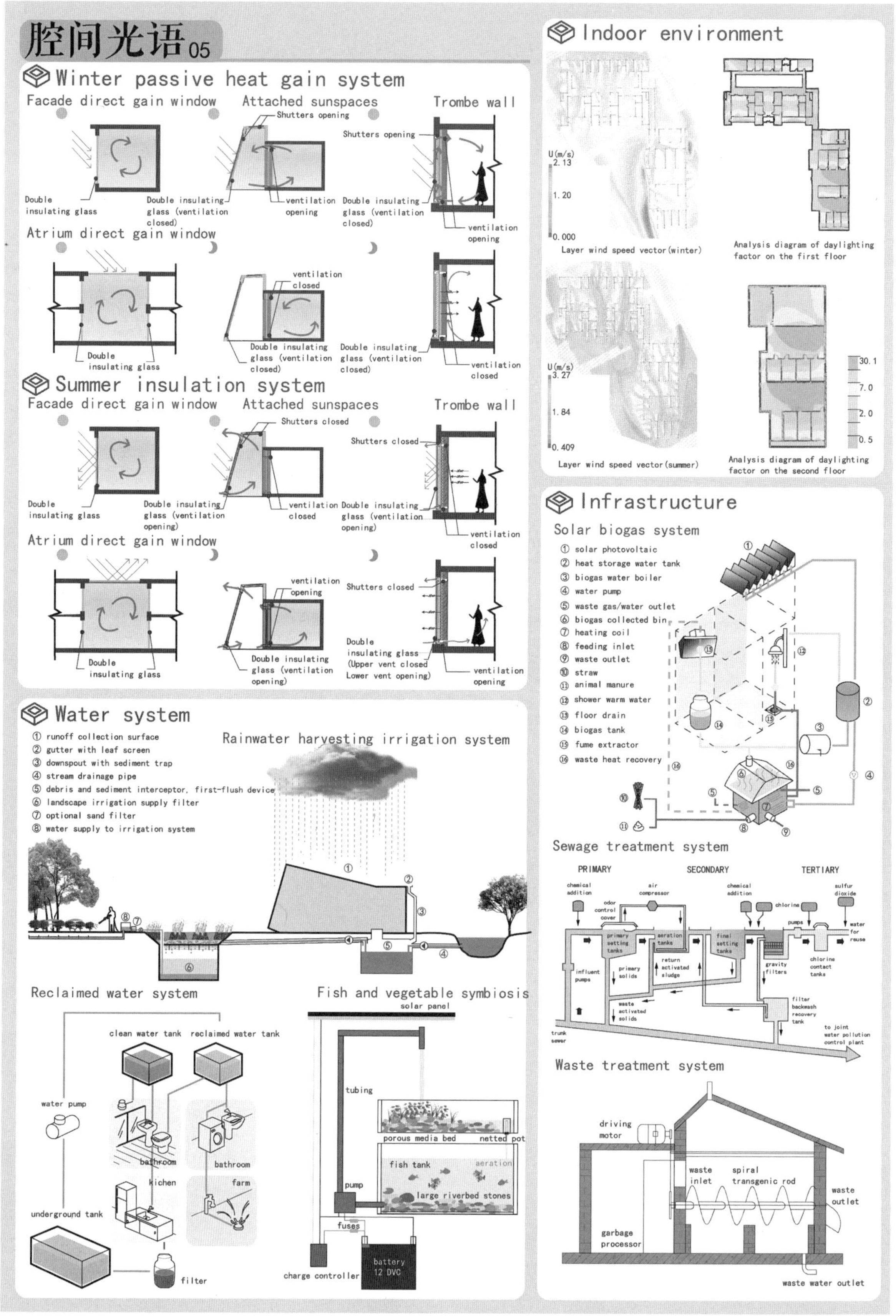

腔间光语 05
Winter passive heat gain system
Facade direct gain window
Attached sunspaces
Trombe wall
Shutters opening
Double insulating glass
Double insulating glass (ventilation closed)
ventilation opening
Atrium direct gain window
ventilation closed
Summer insulation system
Shutters closed
Double insulating glass (ventilation opening)
ventilation closed
Double insulating glass (Upper vent closed Lower vent opening)
Water system
① runoff collection surface
② gutter with leaf screen
③ downspout with sediment trap
④ stream drainage pipe
⑤ debris and sediment interceptor, first-flush device
⑥ landscape irrigation supply filter
⑦ optional sand filter
⑧ water supply to irrigation system
Rainwater harvesting irrigation system
Reclaimed water system
clean water tank
reclaimed water tank
water pump
bathroom
kichen
farm
underground tank
filter
Fish and vegetable symbiosis
solar panel
tubing
porous media bed
netted pot
fish tank
aeration
pump
large riverbed stones
fuses
charge controller
battery 12 DVC
Indoor environment
U(m/s)
2.13
1.20
0.000
Layer wind speed vector(winter)
Analysis diagram of daylighting factor on the first floor
3.27
1.84
0.409
Layer wind speed vector(summer)
30.1
7.0
2.0
0.5
Analysis diagram of daylighting factor on the second floor
Infrastructure
Solar biogas system
① solar photovoltaic
② heat storage water tank
③ biogas water boiler
④ water pump
⑤ waste gas/water outlet
⑥ biogas collected bin
⑦ heating coil
⑧ feeding inlet
⑨ waste outlet
⑩ straw
⑪ animal manure
⑫ shower warm water
⑬ floor drain
⑭ biogas tank
⑮ fume extractor
⑯ waste heat recovery
Sewage treatment system
PRIMARY
SECONDARY
TERTIARY
Waste treatment system
driving motor
waste inlet
spiral transgenic rod
waste outlet
garbage processor
waste water outlet

腔间光语 06

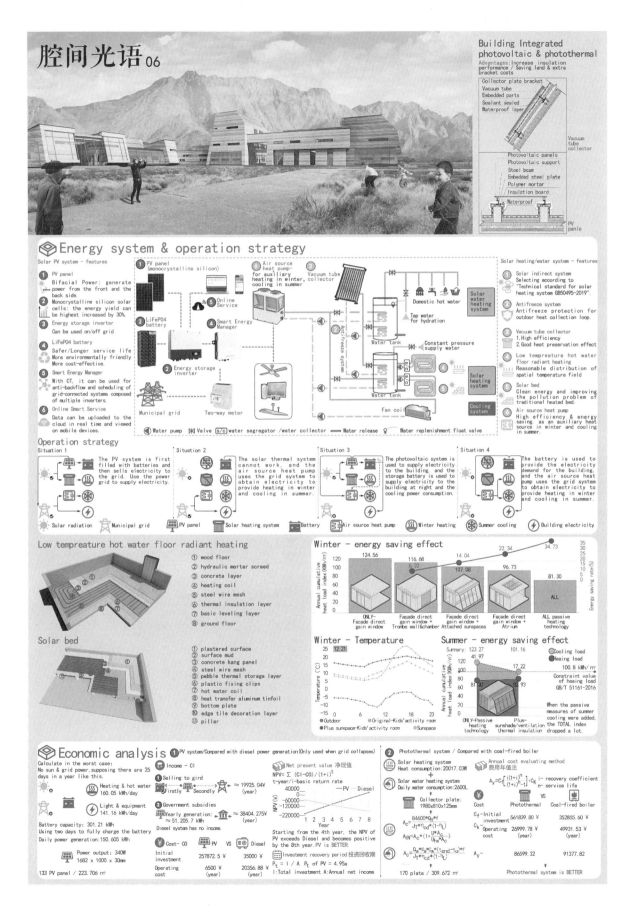

Building Integrated photovoltaic & photothermal

Advantages: Increase insulation performance / Saving land & extra bracket costs

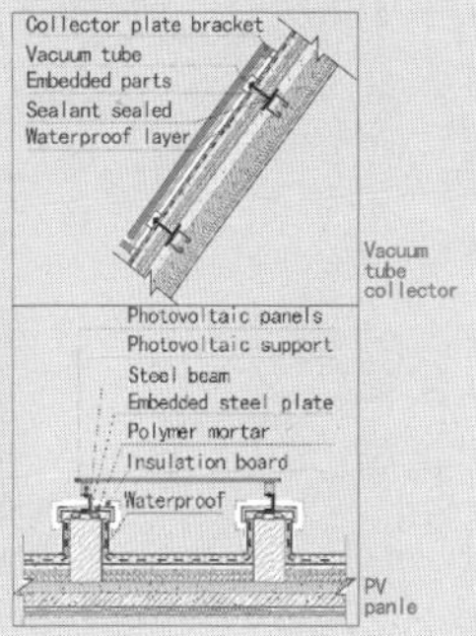

Collector plate bracket
Vacuum tube
Embedded parts
Sealant sealed
Waterproof layer
Vacuum tube collector
Photovoltaic panels
Photovoltaic support
Steel beam
Embedded steel plate
Polymer mortar
Insulation board
Waterproof
PV panle

Energy system & operation strategy

Solar PV system - features

1. PV panel
   Bifacial Power: generate power from the front and the back side.
2. Monocrystalline silicon solar cells: the energy yield can be highest increased by 30%.
3. Energy storage inverter
   Can be used on/off grid
4. LiFePO4 battery
   Safer/Longer service life
   More environmentally friendly
   More cost-effective.
5. Smart Energy Manager
   With CT, it can be used for anti-backflow and scheduling of grid-connected systems composed of multiple inverters.
6. Online Smart Service
   Data can be uploaded to the cloud in real time and viewed on mobile devices.

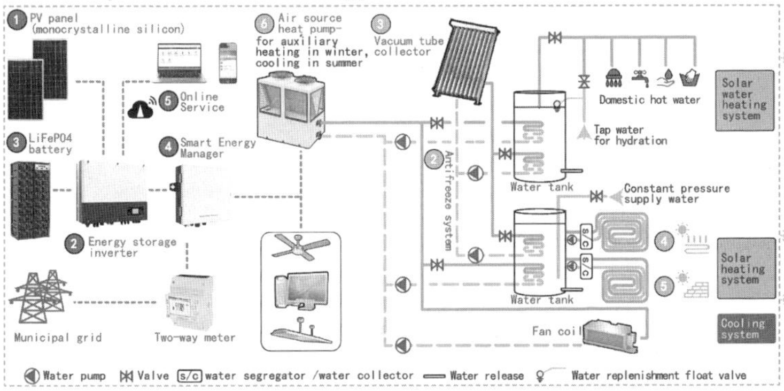

① PV panel (monocrystalline silicon)
⑥ Air source heat pump- for auxiliary heating in winter, cooling in summer
③ Vacuum tube collector
⑤ Online Service
③ LiFePO4 battery
④ Smart Energy Manager
② Energy storage inverter
Municipal grid
Two-way meter
② Antifreeze system
Domestic hot water
Tap water for hydration
Water tank
Constant pressure supply water
Solar water heating system
Solar heating system
Cooling system
Fan coil
Water pump
Valve
S/C water segregator /water collector
Water release
Water replenishment float valve

Solar heating/water system - features

1. Solar indirect system
   Selecting according to "Technical standard for solar heating system GB50495-2019".
2. Antifreeze system
   Antifreeze protection for outdoor heat collection loop.
3. Vacuum tube collector
   1. High efficiency
   2. Good heat preservation effect
4. Low tempreature hot water floor radiant heating
   Reasonable distribution of spatial temperature field
5. Solar bed
   Clean energy and improving the pollution problem of traditional heated bed.
6. Air source heat pump
   High efficiency & energy saving, as an auxiliary heat source in winter and cooling in summer.

Operation strategy

Situation 1

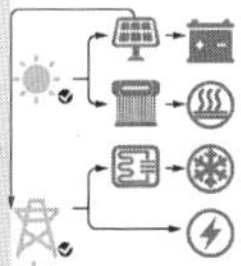

The PV system is first filled with batteries and then sells electricity to the grid. Use the power grid to supply electricity.

Situation 2

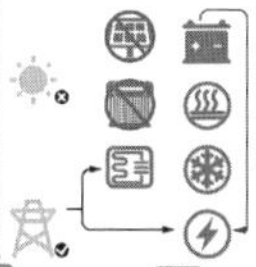

The solar thermal system cannot work, and the air source heat pump uses the grid system to obtain electricity to provide heating in winter and cooling in summer.

Situation 3

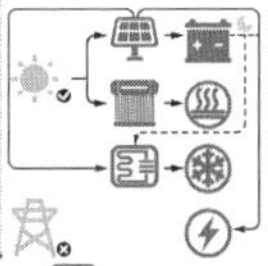

The photovoltaic system is used to supply electricity to the building, and the storage battery is used to supply electricity to the building at night and the cooling power consumption.

Situation 4

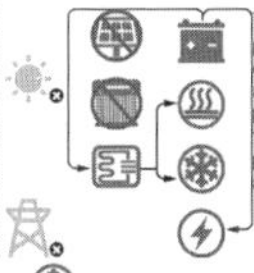

The battery is used to provide the electricity demand for the building, and the air source heat pump uses the grid system to obtain electricity to provide heating in winter and cooling in summer.

Solar radiation | Municipal grid | PV panel | Solar heating system | Battery | Air source heat pump | Winter heating | Summer cooling | Building electricity

Low tempreature hot water floor radiant heating

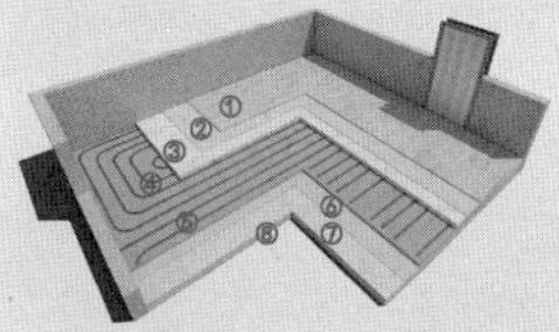

① wood floor
② hydraulic mortar screed
③ concrete layer
④ heating coil
⑤ steel wire mesh
⑥ thermal insulation layer
⑦ basic leveling layer
⑧ ground floor

Solar bed

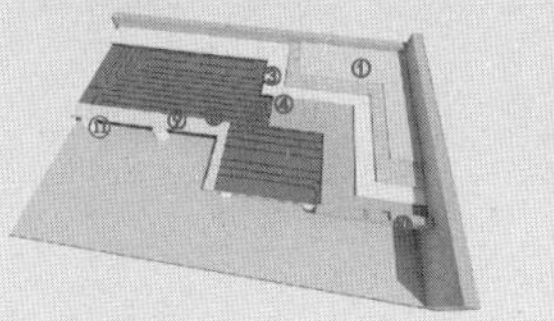

① plastered surface
② surface mud
③ concrete kang panel
④ steel wire mesh
⑤ pebble thermal storage layer
⑥ plastic fixing clips
⑦ hot water coil
⑧ heat transfer aluminum tinfoil
⑨ bottom plate
⑩ edge tile decoration layer
⑪ pillar

Winter - energy saving effect

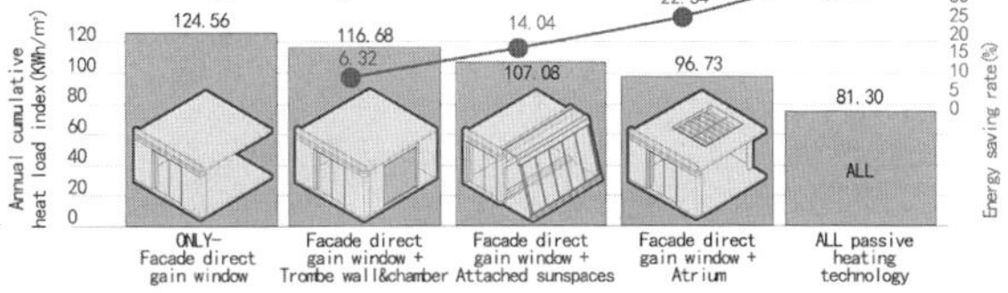

Annual cumulative heat load index (KWh/m²)
Energy saving rate(%)
124.56
116.68
107.08
96.73
81.30
6.32
14.04
22.34
34.73
ALL
ONLY-Facade direct gain window
Facade direct gain window + Trombe wall&chamber
Facade direct gain window + Attached sunspaces
Facade direct gain window + Atrium
ALL passive heating technology

Winter - Temperature

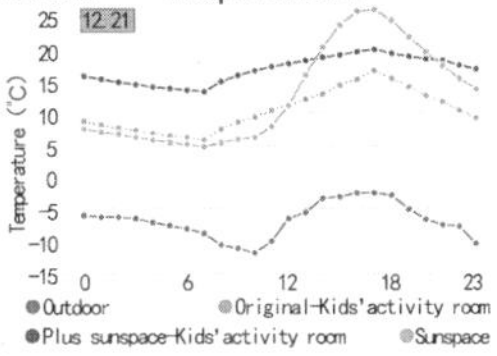

Temperature (°C)
12.21
Outdoor
Original-Kids'activity room
Plus sunspace-Kids'activity room
Sunspace

Summer - energy saving effect

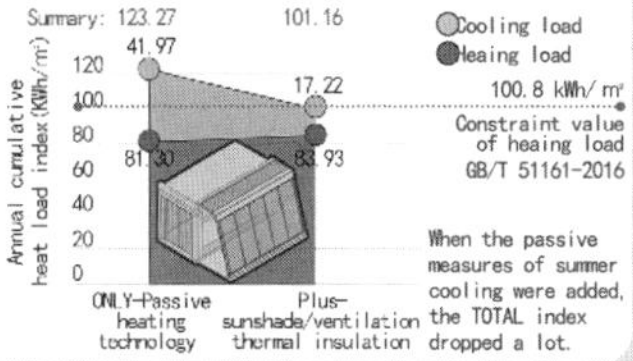

Summary: 123.27 101.16
Cooling load
Heaing load
41.97
17.22
100.8 kWh/m²
Constraint value of heaing load GB/T 51161-2016
81.30
83.93
Annual cumulative heat load index (KWh/m²)
ONLY-Passive heating technology
Plus-sunshade/ventilation thermal insulation

When the passive measures of summer cooling were added, the TOTAL index dropped a lot.

Economic analysis

① PV system/Compared with diesel power generation(Only used when grid collapses)

Calculate in the worst case:
No sun & grid power, supposing there are 25 days in a year like this.

Heating & hot water 160.05 kWh/day
Light & equipment 141.16 kWh/day

Battery capacity: 301.21 kWh
Using two days to fully charge the battery
Daily power generation: 150.605 kWh

Power output: 340W
1682 x 1000 x 30mm

133 PV panel / 223.706 m²

Income - CI

① Selling to gird
Firstly → Secondly → ≈ 19925.04¥ (year)

② Government subsidies
Yearly generation: ≈ 51,205.7 kWh → ≈ 38404.275¥ (year)

Diesel system has no income.

Cost - CO

	PV	VS	Diesel
Initial investment	257872.5 ¥		35000 ¥
Operating cost	6500 ¥ (year)		20356.88 ¥ (year)

Net present value 净现值

$NPV=\sum (CI-CO)/(1+i)^t$

t-year / i-basic return rate

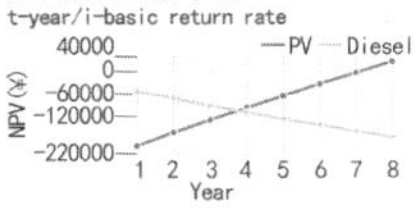

PV
Diesel
NPV (¥)
40000
0
-60000
-120000
-220000
1 2 3 4 5 6 7 8
Year

Starting from the 4th year, the NPV of PV exceeds Diesel and becomes positive by the 8th year. PV is BETTER.

Investment recovery period 投资回收期

$P_t = I / A$ P_t of PV = 4.95a

I: Total investment A: Annual net income

② Photothermal system / Compared with coal-fired boiler

Solar heating system
Heat consumption: 20017.03W
+
Solar water heating system
Daily water consumption: 2600L

Collector plate:
1980x810x125mm

$A_c=\frac{84600*Q_H*f}{J_T*\eta_{cd}*(1-\eta_L)}$

$A_{IN}=A_c*(1+\frac{U*A_c}{U_{hx}*A_c})$

$A_c=\frac{Q_w*\rho_w*C_w*(t_{end}-t_o)*f}{J_T*\eta_{cd}*(1-\eta_L)}$

170 plate / 309.672 m²

Annual cost evaluating method 费用年值法

$A_y=C_f\left[\frac{i(1+i)^n}{(1+i)^n-1}\right]+C_k$ i- recovery coefficient, n- service life

Cost	Photothermal	VS	Coal-fired boiler
C_f-Initial investment	561839.80 ¥		352855.60 ¥
C_k-Operating cost	26999.78 ¥ (year)		49931.53 ¥ (year)
A_y-	86599.32		91377.82

Photothermal system is BETTER

专项奖 · 技术奖
Special Prize Awarded · Technology Prize

注 册 号：6984

项目名称：记忆中的糖块（南平）

Candy in Memory（Nanping）

作　　者：丁海忠、潘兴清、施鹏仙、李松博、刘　杰

参赛单位：昆明学院

指导教师：陈虹羽

区位分析 Locationg Analysis

Nanping
Jianyang
Jinglong

Base location

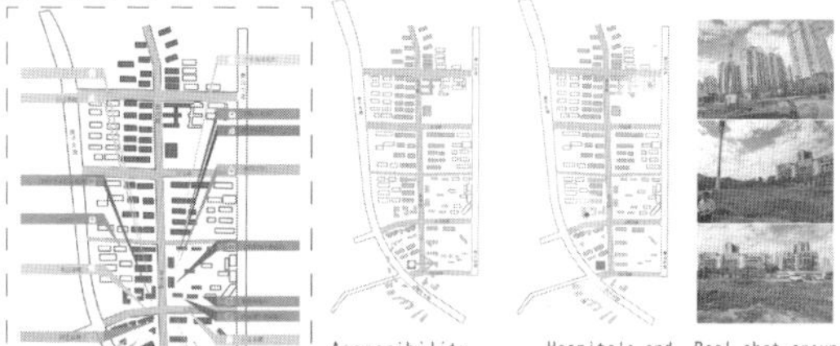

Environmental situation　　Accessibility　　Hospitals and schools　　Real shot around the base

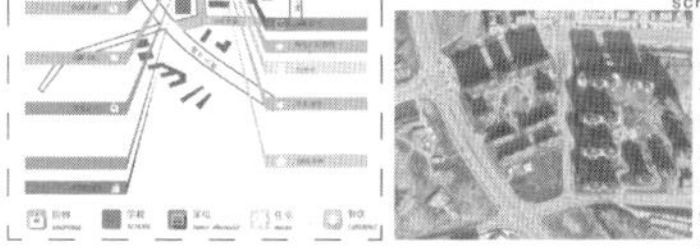

Location map of base area

The land for this project is adjacent to Jing long Road in the east, Shuyuan South Road in the south and West. Jianping Avenue, a city-level trunk road, is close to the landscape residential area in the north, with a land area of 6028m. Site The northwest side is a waterpump houseand open space, with entrances and exits leading to the site.

气候分析 Climate Analysis

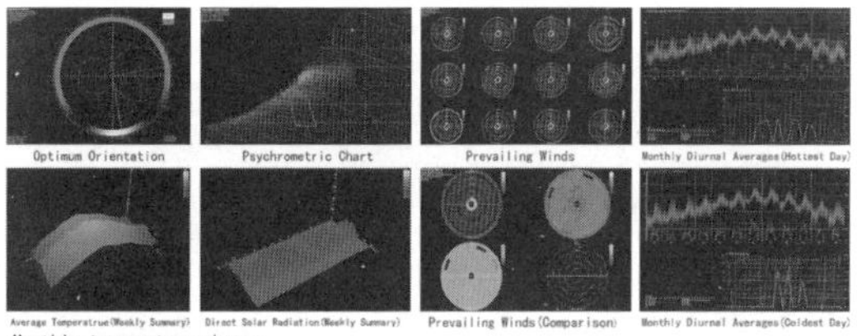
Optimum Orientation
Psychrometric Chart
Prevailing Winds
Monthly Diurnal Averages(Hottest Day)
Average Temperature(Weekly Summary)
Direct Solar Radiation(Weekly Summary)
Prevailing Winds(Comparison)
Monthly Diurnal Averages(Coldest Day)

Monthly temperature change

It is hot in summer and cold in winter at the junction with the southeast region, with no hot summer and no severe cold in winter. There are abundant sunshine, humid climate and abundant rain.

文化底蕴 Cultural Heritage

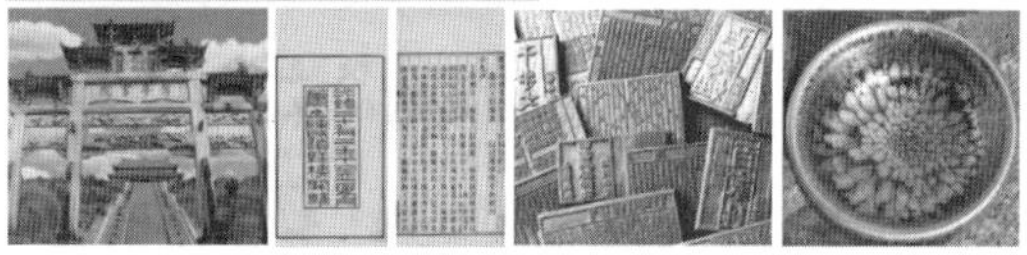

Vietnam and Zhu Ziwen　　The oldest county　　Book edition　　Royal tea lamp

幼儿全天活动及太阳能 Activities and Solar Energy

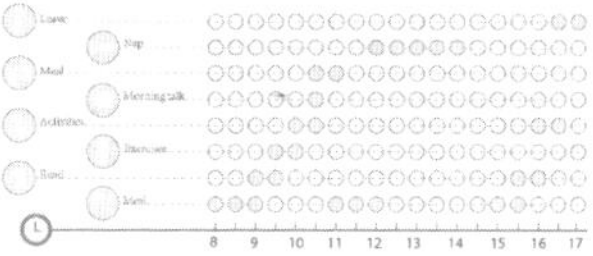

A happy day for children

序号	BIPV形式	光伏组件	建筑要求	类型
1	Solar collector	Ordinary photovoltaic module	Architectural effect, Daylighting, structural strength	combine
2	Solar photovoltaic panel	Photovoltaic roof tile	Architectural effect, Daylighting	Integration
3	Photovoltaic tree	Ordinary photovoltaic module	Daylighting	Integration
4	Photovoltaic parking shed	Ordinary photovoltaic module	Daylighting	Integration
5	Photovoltaic stool	Ordinary photovoltaic module	Daylighting	Integration

Installation form of building photovoltaic

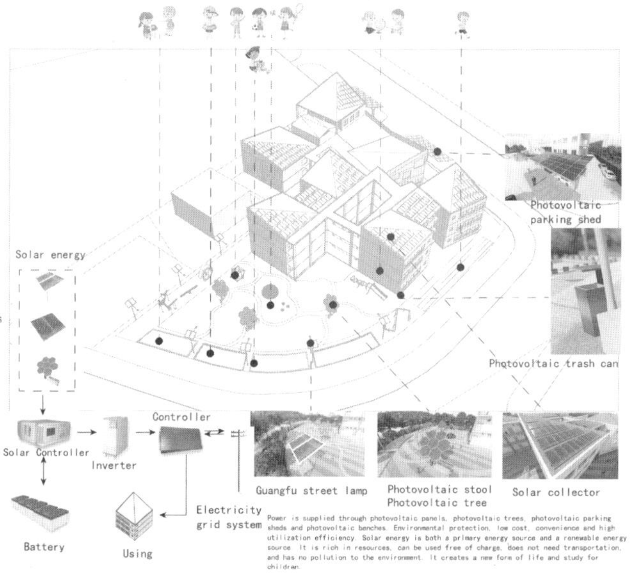
Photovoltaic parking shed
Solar energy
Photovoltaic trash can
Controller
Solar Controller
Inverter
Battery
Using
Electricity grid system
Guangfu street lamp
Photovoltaic stool
Photovoltaic tree
Solar collector

Power is supplied through photovoltaic panels, photovoltaic trees, photovoltaic parking sheds and photovoltaic benches. Environmental protection, low cost, convenience and high utilization efficiency. Solar energy is both a primary energy source and a renewable energy source. It is rich in resources, can be used free of charge, does not need transportation, and has no pollution to the environment. It creates a new form of life and study for children.

设计说明 Design Description

幼儿园基于“本然教育"思想，立足本土，我们将建盏文化内涵渗透到幼儿园的办园理念和培养目标中。借以"建盏”烧制过程所需的13道工序中的4道工序——揉泥、拉坯、上釉、窑烧，以寓儿童成长过程，形成了“揉捏成形,坯中记忆，加甜为釉，出窑展翅”的园本教育理念。并于幼儿园创“陶艺坊”给儿童们用橡皮泥制作建盏，旨在尊重儿童，顺应天性，营造丰富多彩的童年文化，希望儿童们长大后还能记得这个“记忆中的糖块”。

Kindergarten is based on the idea of "natural education" and based on the local, and we will infiltrate the cultural connotation of Jianzhan into the kindergarten Ideas and training objectives. Four of the 13 processes needed in the firing process of "building a lamp"-kneading mud, drawing billet, Glazing, kiln burning, in order to accommodate children's growth process, formed a "kneading molding, memory in the blank, adding sweetness to glaze, and spreading wings out of the kiln." Garden-based education concept. And create a "pottery workshop" in kindergarten to make lanterns with plasticine for children, aiming at respecting children. Comply with nature, create a colorful childhood culture, and hope that children can remember this "candy bar in memory" when they grow up.

1 记忆中的糖块 Candy in Memory

福建南平建阳景龙幼儿园设计

The Design of Jianyang Jinglong Kindergarten in Nanping Fujian

2 记忆中的糖块 Candy in Memory

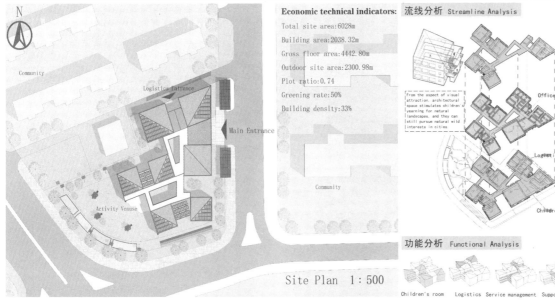

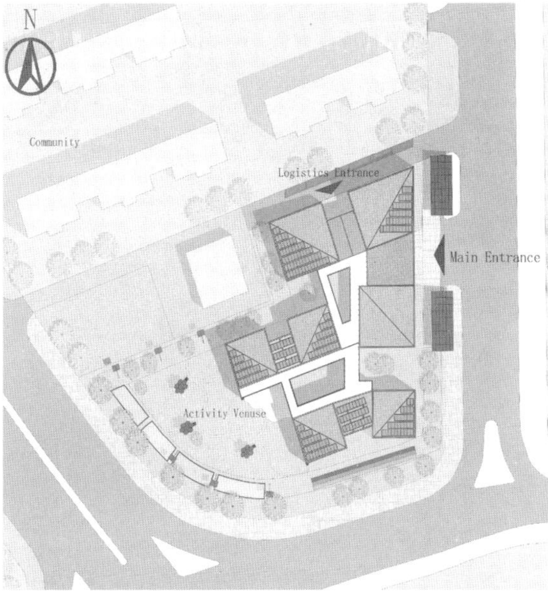
N
Community
Logistics Entrance
Main Entrance
Activity Venue
Community

Site Plan 1：500

Economic technical indicators:

Total site area:6028m

Building area:2038.32m

Gross floor area:4442.80m

Outdoor site area:2300.98m

Plot ratio:0.74

Greening rate:50%

Building density:33%

流线分析 Streamline Analysis

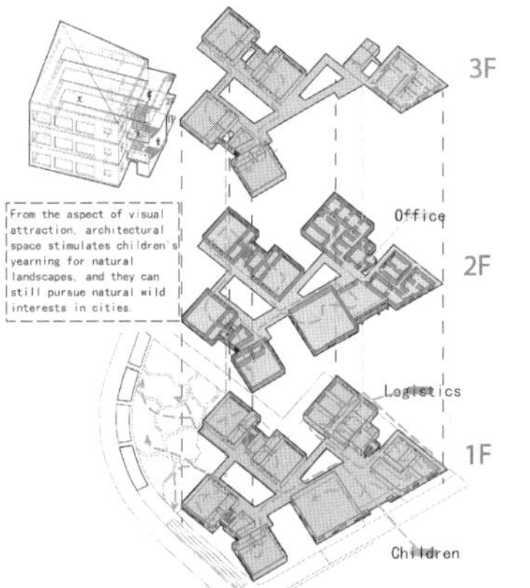
3F
From the aspect of visual attraction, architectural space stimulates children's yearning for natural landscapes, and they can still pursue natural wild interests in cities.
Office
2F
Logistics
1F
Children

功能分析 Functional Analysis

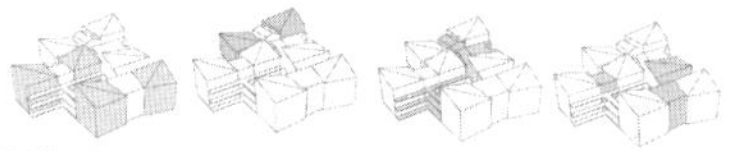

Children's room　Logistics　Service management　Supporting room

软件模拟分析 Software Simulation Analysis

Shadow Analysis

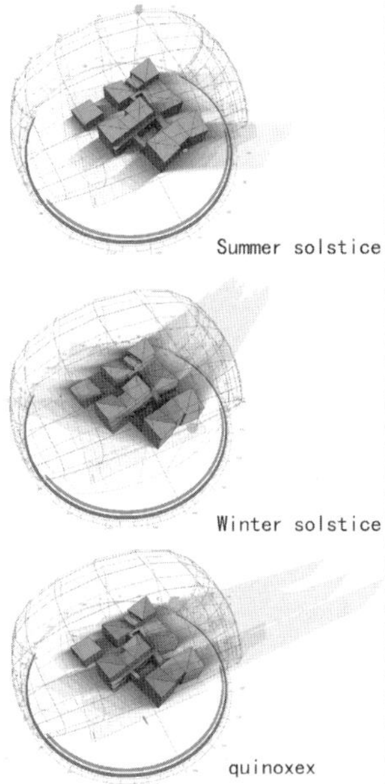
Summer solstice
Winter solstice
quinoxex

Daylight Analysis

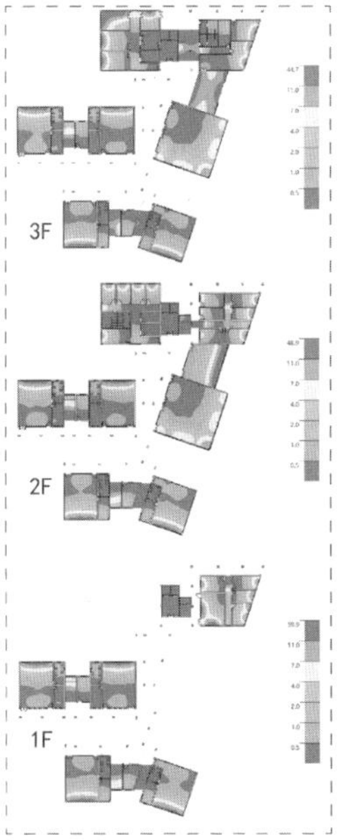
3F
2F
1F

概念分析 Concept Analysis

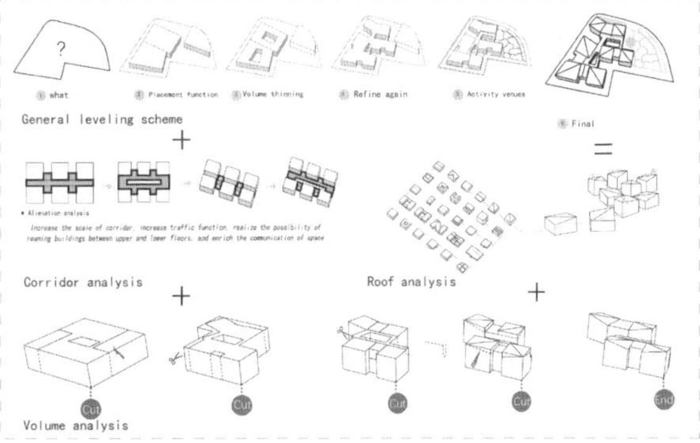
?
General leveling scheme
Corridor analysis
Roof analysis
Volume analysis
Cut
End

Activity room renderings

Corridor renderings

建筑声环境 Architectural Acoustic Environment

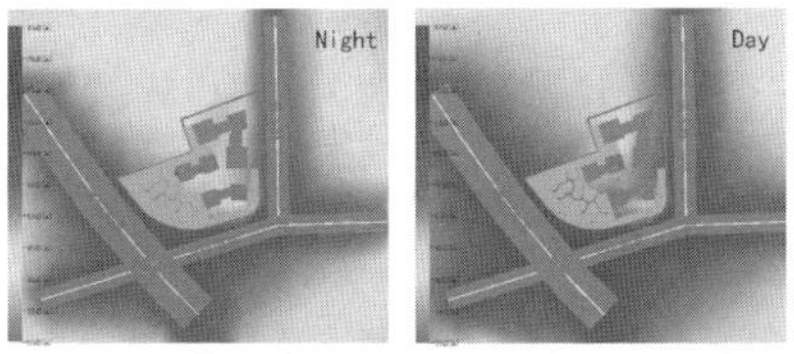
Night
Day

1.5m sound pressure level

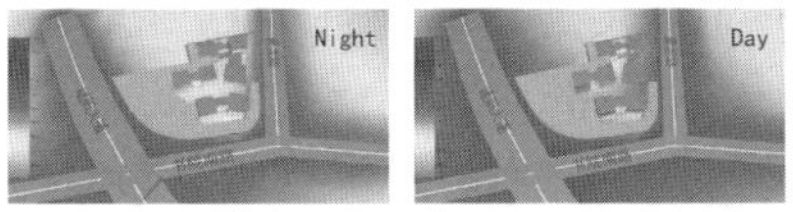
Night
Day

Noise distribution

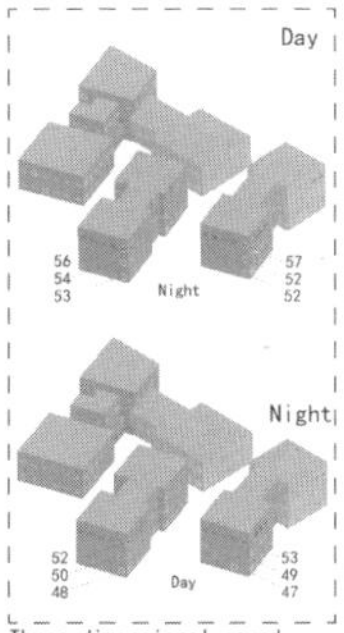
Day
Night

Three dimensional sound pressure

建筑节能 Building Energy Conservation

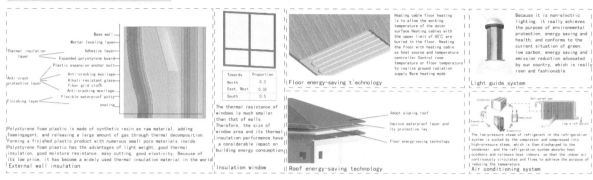

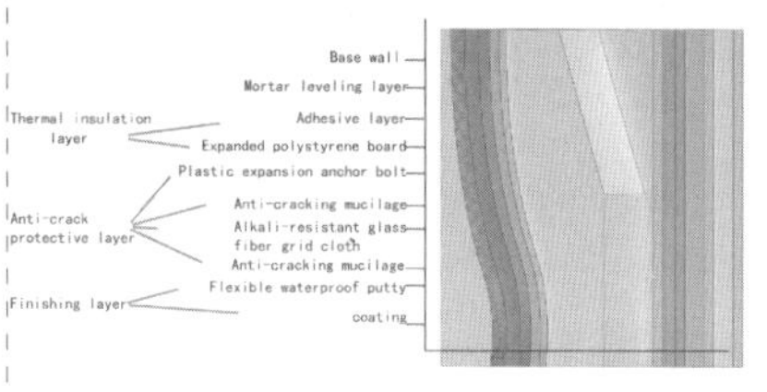
Base wall
Mortar leveling layer
Thermal insulation layer
Adhesive layer
Expanded polystyrene board
Plastic expansion anchor bolt
Anti-crack protective layer
Anti-cracking mucilage
Alkali-resistant glass fiber grid cloth
Anti-cracking mucilage
Finishing layer
Flexible waterproof putty
coating

Polystyrene foam plastic is made of synthetic resin as raw material, adding foamingagent, and releasing a large amount of gas through thermal decomposition. Forming a finished plastic product with numerous small pore materials inside. Polystyrene foam plastic has the advantages of light weight, good thermal insulation, good moisture resistance, easy cutting, good elasticity. Because of its low price, it has become a widely used thermal insulation material in the world

External wall insulation

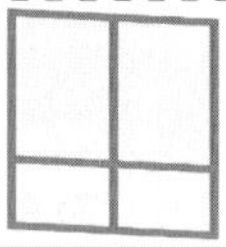

Towards	Proportion
North	0.3
East, West	0.35
South	0.5

The thermal resistance of windows is much smaller than that of walls. Therefore, the size of window area and its thermal insulation performance have a considerable impact on building energy consumption.

Insulation window

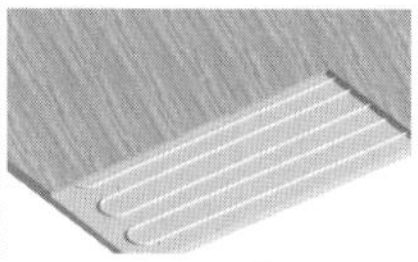

Heating cable floor heating is to allow the working temperature of the outer surface Heating cables with the upper limit of 65℃ are buried in the floor. Heating the floor with heating cable as heat source and temperature controller Control room temperature or floor temperature to realize ground radiation supply Warm heating mode.

Floor energy-saving t`echnology

Because it is non-electric lighting, it really achieves the purpose of environmental protection, energy saving and health, and conforms to the current situation of green, low carbon, energy saving and emission reduction advocated by our country, which is really reen and fashionable.

Light guide system

Adopt sloping roof
Improve waterproof layer and its protective lay
Floor energy-saving technology

Roof energy-saving technology

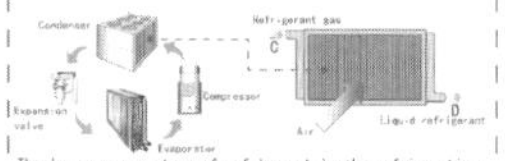
Condenser
Refrigerant gas
C
Compressor
Expansion valve
Evaporator
Air
D
Liquid refrigerant

The low-pressure steam of refrigerant in the refrigeration system is sucked by the compressor and compressed into high-pressure steam, which is then discharged to the condenser, and the refrigeration system absorbs heat outdoors and releases heat indoors, so that the indoor air continuously circulates and flows to achieve the purpose of reducing the temperature

Air conditioning system

幼儿用房分解 Decomposition of Children's House

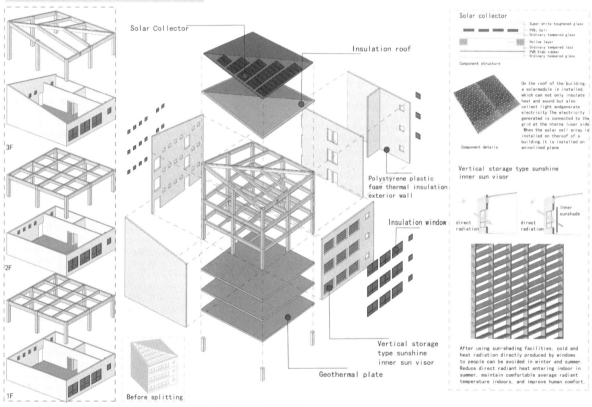

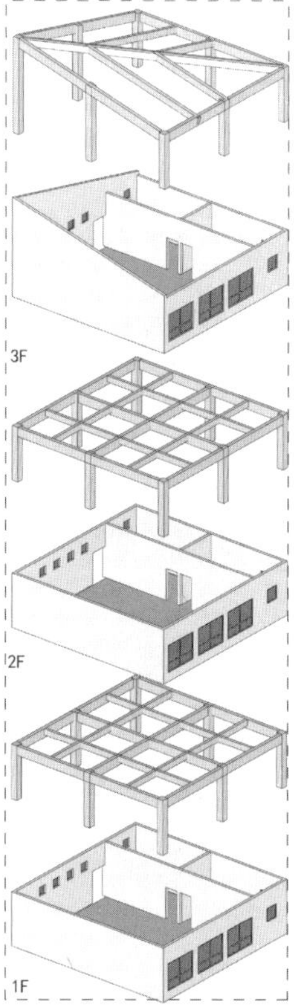
3F
2F
1F

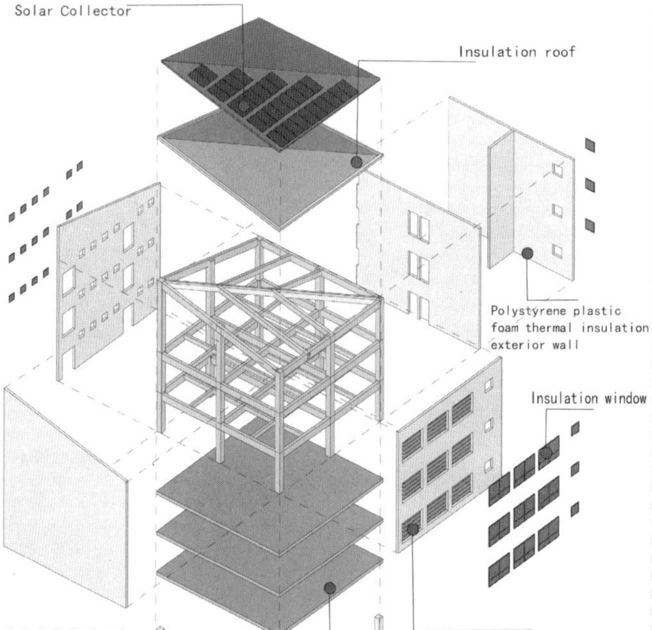
Solar Collector
Insulation roof
Polystyrene plastic foam thermal insulation exterior wall
Insulation window
Vertical storage type sunshine inner sun visor
Geothermal plate

Before splitting

Solar collector

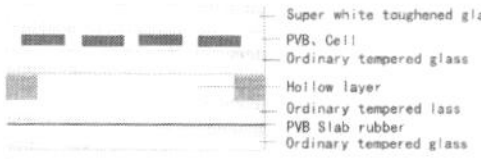
Super white toughened glass
PVB、Cell
Ordinary tempered glass
Hollow layer
Ordinary tempered lass
PVB Slab rubber
Ordinary tempered glass

Component structure

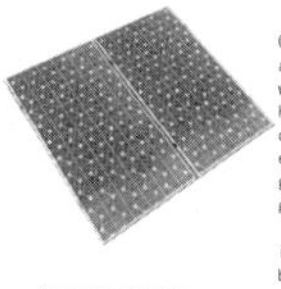

Component details

On the roof of the building, a solarmodule is installed, which can not only insulate heat and sound, but also collect light andgenerate electricity. The electricity generated is connected to the grid at the nterna luser side. When the solar cell array is installed on theroof of a building, it is installed on aninclined plane.

Vertical storage type sunshine inner sun visor

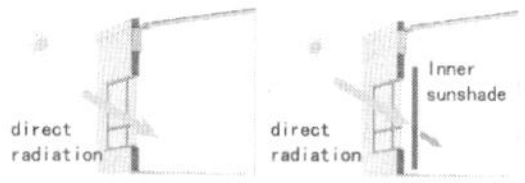
direct radiation
direct radiation
Inner sunshade

After using sun-shading facilities, cold and heat radiation directly produced by windows to people can be avoided in winter and summer. Reduce direct radiant heat entering indoor in summer, maintain comfortable average radiant temperature indoors, and improve human comfort.

3 记忆中的

4 记忆中的糖块 Candy In Memory

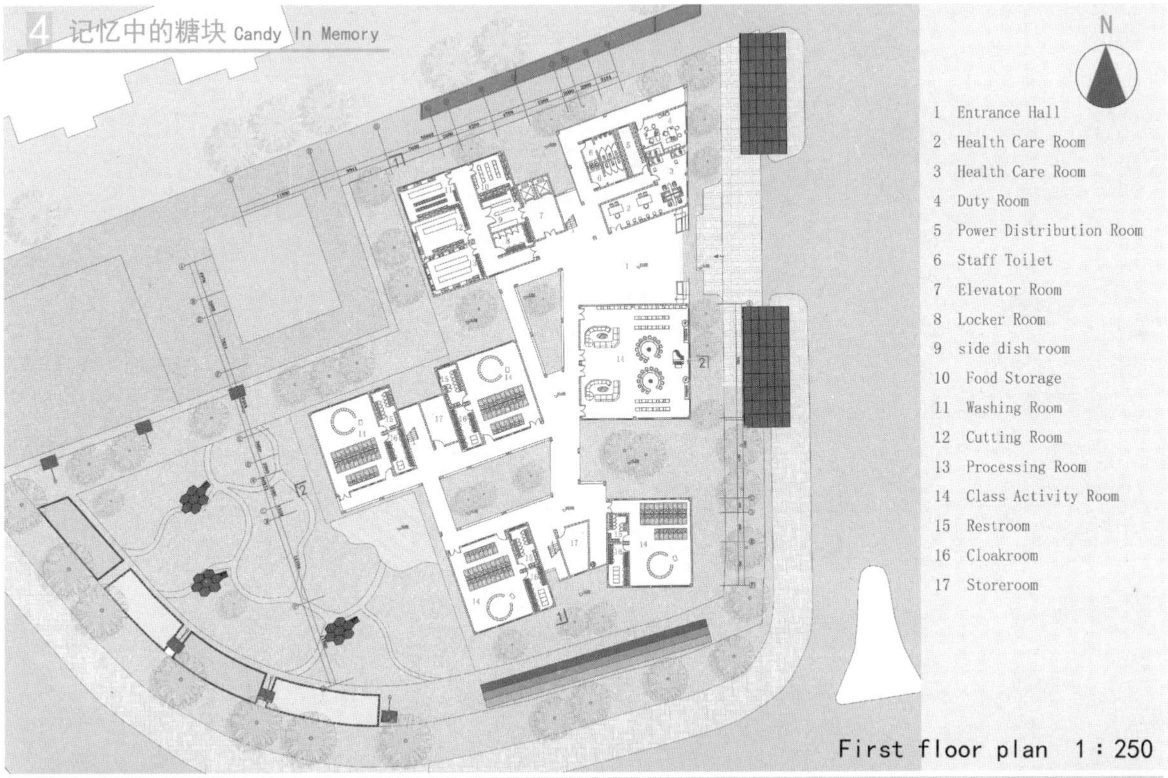
N
1 Entrance Hall
2 Health Care Room
3 Health Care Room
4 Duty Room
5 Power Distribution Room
6 Staff Toilet
7 Elevator Room
8 Locker Room
9 side dish room
10 Food Storage
11 Washing Room
12 Cutting Room
13 Processing Room
14 Class Activity Room
15 Restroom
16 Cloakroom
17 Storeroom

First floor plan 1∶250

软件模拟分析 Software Simulation Analysis

场地日照分析 Sunshine Analysis

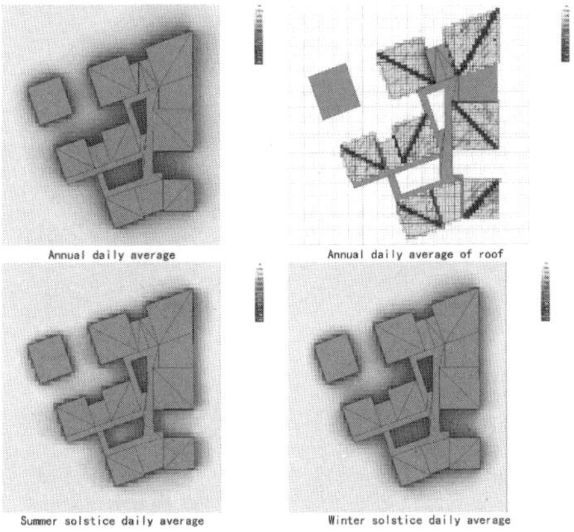
Annual daily average
Annual daily average of roof
Summer solstice daily average
Winter solstice daily average

建筑通风 Building Ventilation

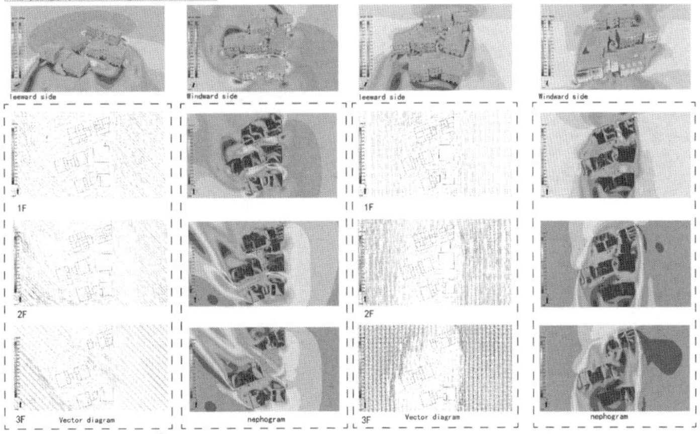
leeward side
Windward side
leeward side
Windward side
1F
2F
3F
Vector diagram
nephogram
1F
2F
3F
Vector diagram
nephogram
Summer
Winter

雨水收集与利用 Rainwater Collection and Utilization

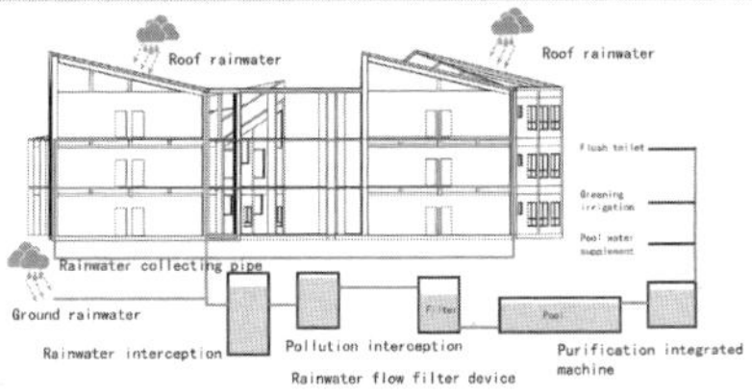
Roof rainwater
Roof rainwater
Flush toilet
Greening irrigation
Pool water supplement
Rainwater collecting pipe
Ground rainwater
Rainwater interception
Pollution interception
Rainwater flow filter device
Purification integrated machine

Rainwater collection can save energy, reduce emission, protect the environment, reduce the amount of rainwater, and make wateravailable in drought and emergency (such as fire). In addition, it can use miscellaneous water in life, save tap water and reduce the cost of water treatment.

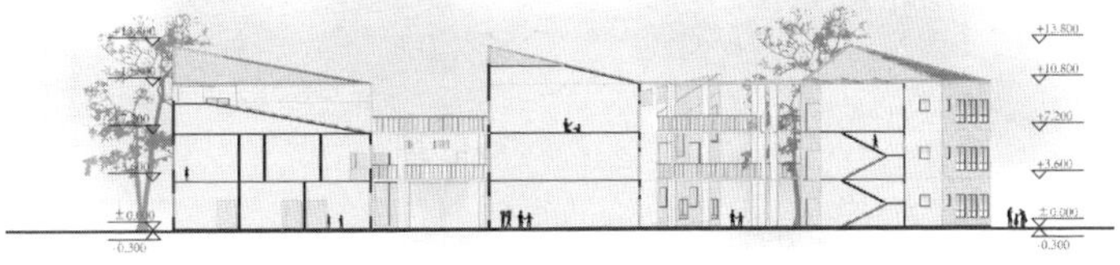

Section 1-1 1∶250

风向和热量 Wind Direction and Energy

Ventilation
Lighting
Heat cycle

Through natural ventilation, the dirty air in the building is directly or purified and discharged to the outside, and then the fresh air is added in, so as to keep the indoor air environment in line with the hygienic standard. Buildings provide a comfortable environment for children through indoor heating equipment and the heat supplied by solar energy.

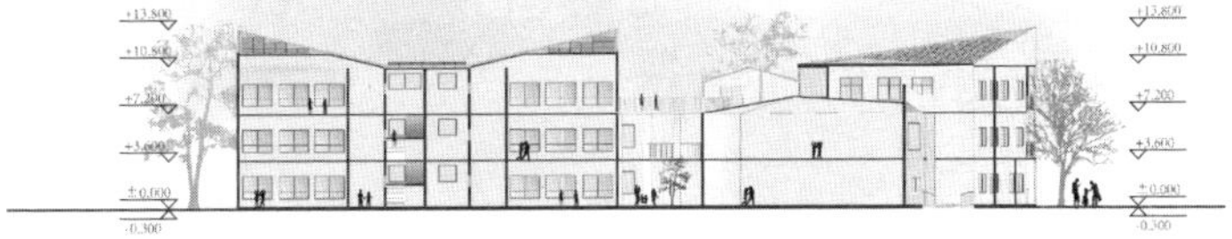

Section 2-2 1∶250

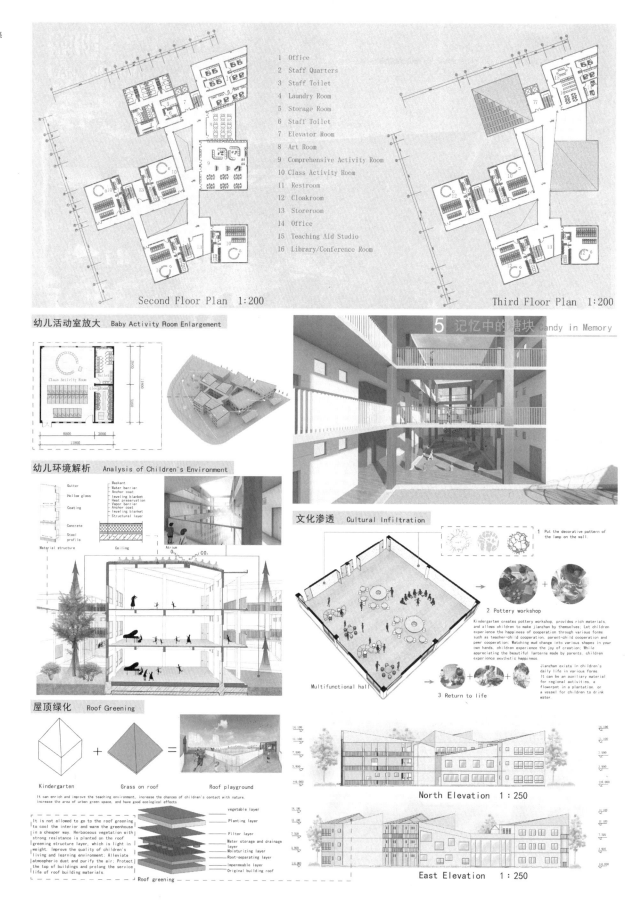

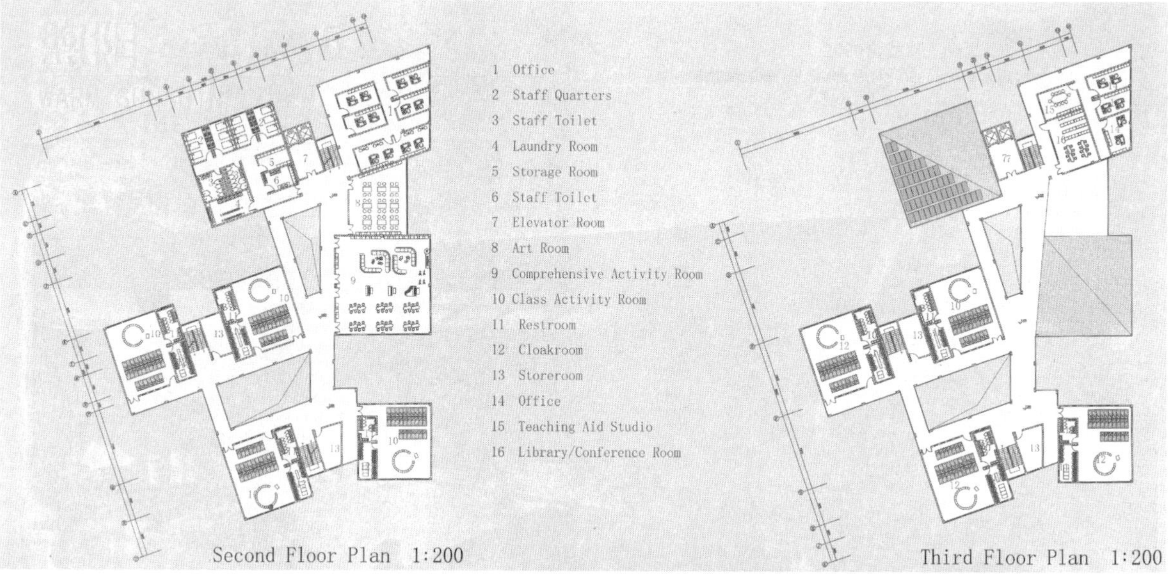
1 Office
2 Staff Quarters
3 Staff Toilet
4 Laundry Room
5 Storage Room
6 Staff Toilet
7 Elevator Room
8 Art Room
9 Comprehensive Activity Room
10 Class Activity Room
11 Restroom
12 Cloakroom
13 Storeroom
14 Office
15 Teaching Aid Studio
16 Library/Conference Room
Second Floor Plan 1:200
Third Floor Plan 1:200

幼儿活动室放大 Baby Activity Room Enlargement

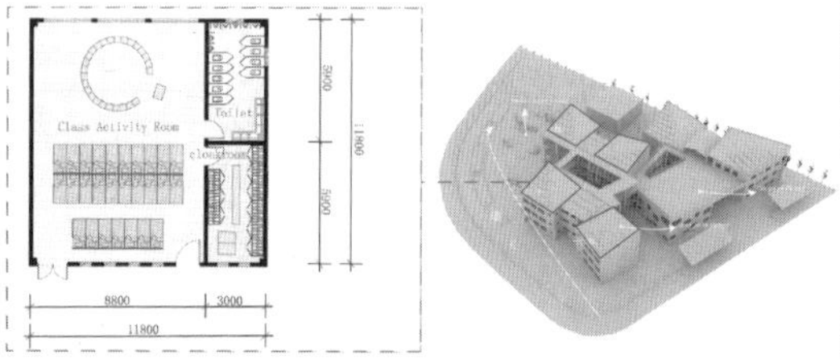
Class Activity Room
toilet
8800
3000
11800

5 记忆中的糖块 Candy in Memory

幼儿环境解析 Analysis of Children's Environment

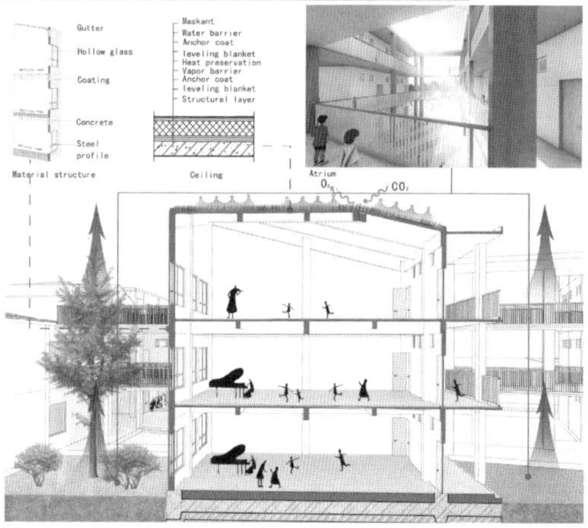
Gutter
Hollow glass
Coating
Concrete
Steel profile
Material structure
Ceiling
Atrium
O_2
CO_2

文化渗透 Cultural Infiltration

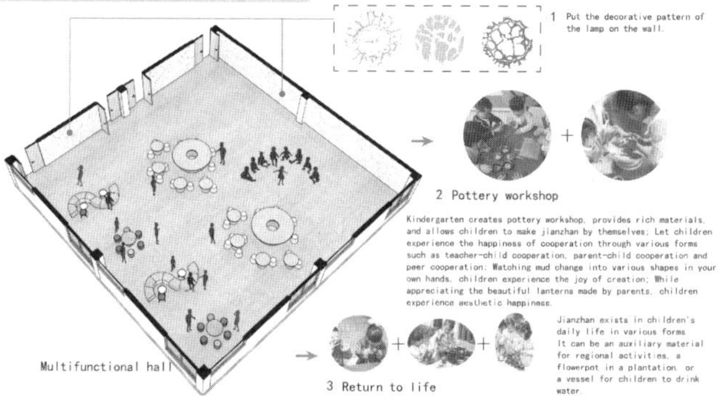
1 Put the decorative pattern of the lamp on the wall.
2 Pottery workshop
Kindergarten creates pottery workshop, provides rich materials, and allows children to make jianzhan by themselves; Let children experience the happiness of cooperation through various forms such as teacher-child cooperation, parent-child cooperation and peer cooperation; Watching mud change into various shapes in your own hands, children experience the joy of creation; While appreciating the beautiful lanterns made by parents, children experience aesthetic happiness.
Multifunctional hall
3 Return to life
Jianzhan exists in children's daily life in various forms. It can be an auxiliary material for regional activities, a flowerpot in a plantation, or a vessel for children to drink water.

屋顶绿化 Roof Greening

Kindergarten
Grass on roof
Roof playground

It can enrich and improve the teaching environment, increase the chances of children's contact with nature, increase the area of urban green space, and have good ecological effects

It is not allowed to go to the roof greening to cool the interior and warm the greenhouse in a cheaper way. Herbaceous vegetation with strong resistance is planted on the roof greening structure layer, which is light in weight. Improve the quality of children's living and learning environment; Alleviate atmospheric dust and purify the air; Protect the top of buildings and prolong the service life of roof building materials.

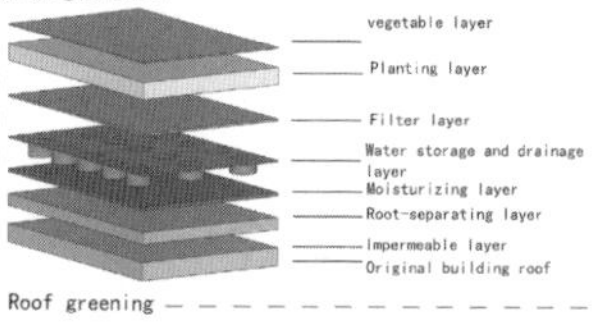
vegetable layer
Planting layer
Filter layer
Water storage and drainage layer
Moisturizing layer
Root-separating layer
Impermeable layer
Original building roof

Roof greening

North Elevation 1 : 250

East Elevation 1 : 250

达标率和评分标准　Achievement Rate and Scoring Standard

Building daylighting in

各层房间采光达标率

楼层	房间编号	房间类型	采光等级	采光类型	采光系数要求(%)	房间面积(m2)	达标面积(m2)	达标率(%)
1	1001	普通教室	III	侧面	3.30	104.89	68.72	66
	1002	普通教室	III	侧面	3.30	99.68	69.24	69
	1003	专用教室	III	侧面	3.30	230.56	145.25	63
	1004	普通教室	III	侧面	3.30	99.68	67.21	67
	1005	普通教室	III	侧面	3.30	99.68	69.24	69
	1006	办公室	III	侧面	3.30	66.65	41.32	62
	1007	办公室	III	侧面	3.30	34.19	28.13	82
2	2001	普通教室	III	侧面	3.30	104.89	69.63	66
	2002	普通教室	III	侧面	3.30	99.68	68.74	69
	2003	专用教室	III	侧面	3.30	230.56	149.86	65
	2004	普通教室	III	侧面	3.30	99.68	67.98	68
	2005	普通教室	III	侧面	3.30	99.68	69.80	70
	2006	专用教室	III	侧面	3.30	94.25	94.25	100
	2007	办公室	III	侧面	3.30	23.57	14.38	61
	2008	办公室	III	侧面	3.30	21.61	15.20	70
	2009	办公室	III	侧面	3.30	19.22	11.72	61
	2010	办公室	III	侧面	3.30	17.60	15.30	87
	2011	办公室	III	侧面	3.30	21.75	13.49	62
	2012	办公室	III	侧面	3.30	23.93	15.34	61
	2013	教师宿舍	IV	侧面	2.20	18.05	18.05	100
	2014	教师宿舍	IV	侧面	2.20	17.52	17.52	100
	2015	教师宿舍	IV	侧面	2.20	17.64	17.64	100
	2016	教师宿舍	IV	侧面	2.20	18.09	18.09	100
3	3001	普通教室	III	侧面	3.30	104.89	68.72	66
	3002	普通教室	III	侧面	3.30	99.68	68.74	69
	3003	普通教室	III	侧面	3.30	99.68	67.98	68

检查项	评价依据	结论	得分
室内采光	室内主要空间至少60%面积比例区域的采光照度值不低于采光要求的小时平均数不少于4h/d.	满足	3

Acoustic environment

Indoor

房间名称	室外传入室内噪声级 昼间	夜间	室内设备噪声级 昼间	夜间	相邻空间设备传入室内噪声级 昼间	夜间	室内噪声级 昼间	夜间
一层								
普通教室[1001]	27	--		-		-	27	-
普通教室[1002]	28	--		-		-	28	-
普通教室[1003]	29	--		-		-	29	-
普通教室[1004]	25	--		-		-	25	-
专用教室[1005]	30	--		-		-	30	-
办公室[1006]	27	--		-	<5	-	27	-
办公室[1007]	29	--		-	<5	-	29	-
二层								
普通教室[2001]	29	--		-		-	29	-
普通教室[2002]	29	--		-		-	29	-
普通教室[2003]	30	--		-		-	30	-
普通教室[2004]	27	--		-		-	27	-
专用教室[2005]	32	--		-		-	32	-
专用教室[2006]	30	--		-		-	30	-
办公室[2007]	10	--	37	-		-	37	-
办公室[2008]	29	--	37	-		-	38	-
办公室[2009]	10	--	37	-		-	37	-
办公室[2010]	29	--	37	-		-	38	-
办公室[2011]	14	--	37	-	<5	-	37	-
办公室[2012]	28	--	37	-		-	38	-
三层								
普通教室[3001]	32	--		-		-	33	-
普通教室[3002]	33	--		-		-	37	-
普通教室[3003]	31	--		-		-	31	-
普通教室[3004]	28	--		-		-	28	-
专用教室[3005]	25	--		-		-	25	-
办公室[3006]	29	--	37	-		-	38	-
办公室[3007]	28	--	37	-		-	38	-
办公室[3008]	28	--	37	-		-	38	-
办公室[3009]	15	--	37	-		-	37	-

注：普通教室（底限：≤45，高要求：≤40）；办公室（底限：≤45，高要求：≤40）；专用教室（底限：≤50，高要求：≤45）

检查项	评价依据	结论	得分
空气声隔声	满足平均要求	满足	3
撞击声隔声	本工程无此项评价	---	0
室内噪声级	满足高要求	满足	8

Outdoor

环境噪声综合得分表　　单位：dB(A)

	噪声最大值	2类噪声限值	3类噪声限值	得分情况
昼间	54	60	65	10
夜间	50	50	55	

经过软件模拟和结果统计分析，最终判定本建筑《绿色建筑评价标准》GB 50378-2019第8.2.6条，得10分。

After software simulation and statistical analysis of results, the project is finally decided <<Evaluation standard of green building>>GB 50378-2019 article 8.2.6.Get 10 points

Ventilation

Air changes in public buildings

公共建筑过渡季节典型工况下换气次数统计表

换气次数大于2次/h的面积比		
换气次数大于2次/h的面积	3178.85	m²
总面积	3397.45	m²
面积比例RR	93.57	%

主要功能房间换气次数统计表

分类	体积(m³)	面积(m²)	换气次数(次/h)
○建筑			
├○第1层			
│├○1001[普通教室]	394.98	104.69	31.61
│├○1002[普通教室]	373.82	103.84	71.96
│├○1003[专用教室]	853.70	237.16	11.29
│├○1004[普通教室]	373.82	103.84	64.11
│├○1005[普通教室]	373.82	103.84	46.54
│├○1006[办公室]	255.58	70.99	[illegible]
│├○1007[办公室]	132.07	36.82	120.44
├○第2层			
│├○2001[普通教室]	394.98	109.69	[illegible]
│├○2002[普通教室]	373.82	103.84	[illegible]
│├○2003[专用教室]	853.78	237.16	21.15
│├○2004[普通教室]	373.82	103.84	[illegible]
│├○2005[普通教室]	373.82	103.84	52.39
│├○2006[专用教室]	353.63	98.23	22.61
│├○2007[办公室]	92.51	25.70	64.96
│├○2008[办公室]	85.38	23.63	73.27
│├○2009[办公室]	76.32	21.17	85.36
│├○2010[办公室]	69.89	19.41	104.52
│├○2011[办公室]	85.15	23.71	42.48
│├○2012[办公室]	94.15	26.15	100.18
│├○2013[卧室]	71.60	[illegible]	[illegible]
│├○2014[卧室]	69.77	19.38	80.91
│├○2015[卧室]	69.77	19.38	[illegible]
│└○2016[卧室]	71.00	19.69	99.78
├○第3层			
│├○3001[普通教室]	394.98	109.69	42.78
│├○3002[普通教室]	373.82	103.84	43.87
│├○3003[普通教室]	373.82	103.84	75.31
│├○3004[普通教室]	373.82	103.84	80.39
│├○3005[专用教室]	198.48	55.13	121.90
│├○3006[办公室]	93.85	26.06	85.51
│├○3007[办公室]	72.36	20.10	193.36
│├○3008[办公室]	101.74	28.26	40.42
│└○3009[办公室]	91.47	25.46	84.36

检查项	评价依据	结论	得分
换气面积比例	该建筑主要功能房间换气次数大于2次/h的面积比例为93.57%，按照绿标5.2.10条款要求得7分。	满足	7

Summer

过渡季、夏季工况达标判断表

评价项目	标准要求	项目计算结果	达标判定	得分
无风区	场地内人活动区不出现涡旋或无风区，得3分	人行区无无风区	达标	3分
旋涡区		人行区无旋涡区		
外窗室内外表面的风压差	50%以上可开启外窗室内外表面的风压差大于0.5Pa，得2分。	可开启外窗室内外表面的风压差满足标准要求	达标	2分

过渡季、夏季典型风速和风向条件下，场地内人活动区不出现涡旋或无风区，得3分；50%以上可开启外窗室内外表面的风压差大于0.5Pa，得2分。

Under the conditions of typical wind speed and direction in transition season and summer, there is no vortex or in windless areas, score 3 points; 2 points if the wind pressure difference between indoor and outdoor surfaces of more than 50% openable windows is greater than 0.5Pa.

Winter

评价项目	标准要求	项目计算结果	达标判定	得分
风速	建筑物周围人行区距地高1.5m处风速小于5m/s，户外休息区、儿童娱乐区风速小于2m/s，且室外风速放大系数小于2，得3分；	人行区没有出现风速大于5m/s的区域，户外休息区、儿童娱乐区没有出现风速大于2m/s的区域	达标	3分
风速放大系数		人行区没有出现风速放大系数大于等于2的区域		
建筑迎风面背风面风压差	除迎风第一排建筑外，建筑迎风面与背风面表面风压差不超过5Pa，得2分	本项目没有出现建筑迎风面与背风面表面风压差大于5Pa的建筑	达标	2分

冬季典型风速和风向条件下，建筑物周围人行区距地高1.5m处风速低于5m/s，户外休息区、儿童娱乐区风速小于2m/s，且室外风速放大系数小于2，得3分；除迎风第一排建筑外，建筑迎风面与背风面表面风压差不超过5Pa，得2分。

Under the conditions of typical wind speed and direction in winter, the wind speed is low at the pedestrian zone around the building 1.5m above the ground At 5m/s, the wind speed in outdoor rest area and children's entertainment area is less than 2m/s, and the amplification factor of outdoor wind speed is less than 2. Get 3 points; Except for the windward row buildings, the wind pressure difference between windward and leeward surfaces of buildings shall not exceed 5Pa. Get 2 points.

太阳能系统分析　Solar Energy System Analysis

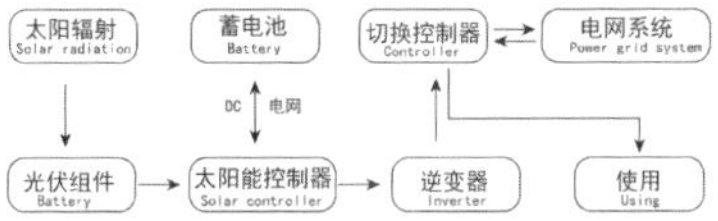

太阳辐射
Solar radiation
蓄电池
Battery
切换控制器
Controller
电网系统
Power grid system
DC
电网
光伏组件
Battery
太阳能控制器
Solar controller
逆变器
Inverter
使用
Using

（1）基于光伏板的转化效率，工艺以及电池的辐射形态，选择块状为单晶硅太阳能电池。

（2）以月为周期，用最大日照强度来确定电池型号，选取功率为150W/m²的光伏板。

（3）通过太阳能光伏板、太阳能停车棚、太阳能光伏树、太阳能座椅、以及太阳能路灯和太阳能垃圾桶集热，将太阳光能转化为建筑照明和供能使用。太阳能既是一次能源，又是可再生能源。它资源丰富，既可免费使用，又无需运输，对环境无任何污染。为儿童创造了一种新的生活学习形态。

(1) Based on the conversion efficiency of photovoltaic panel, technology and auxiliary emission form of the cell, the block is selected as monocrystalline silicon solar cell.

(2) On a monthly basis, the battery model is determined by the maximum sunshine intensity, and the photovoltaic panel with power of 150 W/m² is selected.

(3) Collect heat through solar photovoltaic panels, solar parking sheds, solar photovoltaic trees, solar seats, solar street lamps and solar garbage bins, Convert solar energy into architectural lighting and energy supply. Solar energy is both a primary energy source and a renewable energy source. It is rich in resources and can be used for free. Without transportation and without any pollution to the environment. It creates a new form of life and study for children.

无障碍设计　Barrier - free Design

(1) Stairs

楼梯设计时我们考虑幼儿园的踏步宽度大些，高度小些；楼梯踏板的材料应选用防滑材料，并且设无凸缘梯板，防止小孩玩耍时滑倒；楼梯两侧设置扶手，且扶手是连续的，以防止儿童攀爬时、脚步踏空时，可以用手稳住身体；扶手栏杆为为竖杆，间距符合幼儿的使用要求。

When designing stairs, we consider that the step width of kindergarten is larger and the height is smaller; The stair treads shall be made of non-slip materials and shall be provided with flangeless ladder boards. Prevent children from slipping while playing; Handrails are arranged on both sides of stairs, and the handrails are connected Continued, in order to prevent children from climbing and stepping empty, they can be stabilized by hand Body. Handrails are vertical bars, and the spacing meets the use requirements of children

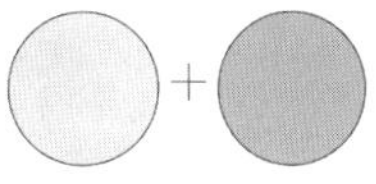

(2) Color

整体建筑风格偏黄+绿色。黄色搭配绿色最值得推崇。黄色能让人产生活力，诱发食欲，给人以温暖的感觉；绿色能使人感到舒适，消除疲劳，起到镇静的作用。两者搭配，可以使整个空间既有色彩的跳跃感，又不失稳重。

The overall architectural style is yellow+green. Yellow with green is the most respected. Yellow can make people energetic, induce appetite and give people a warm feeling; Green can make people feel comfortable, relieve fatigue and play a calming role. Thecombination of the two can make the whole space not only have a sense of color jumping, but also not Lose stability.

(3) Doors and windows

门窗是幼儿园无障碍设计时发生紧急事件时的逃生通道。我们在设计中留予足够的交通缓冲与回旋余地，以避免转变过急或迎面冲撞以保证儿童的安全。门的设计上使用平开门且要设置观察窗口方便老师。

Doors and windows are the escape routes when emergency happens in kindergarten barrier-free design. We leave enough traffic buffer and leeway in the design to avoid Change too quickly or collide head-on to ensure the safety of children. The design of the door makes Use flat doors and set observation windows for teachers' convenience.

(4) Toilet

门盥洗台的高宽度为0.5m*0.4m，且盥洗台下做成封闭的状态，防止幼儿钻到下面玩耍引起危险。在幼儿园的卫生间地面设计时考虑防滑问题，没有使用台。我们通大便蹲位下陷，来达到地面的平整性。盥洗间的洁具选择没有棱角、圆滑、白色的卫生洁具。卫生间和盥洗间满足通风良好和空间开阔的要求。卫生间门设置幼儿视觉适合的小观察窗与把手相配合。

The height and width of the door washstand are 0.5m*0.4m, and the washstand is closed State, to prevent children from drilling to play under the danger. Hygiene in kindergarten Anti-skid problem was considered in the design of the floor, and no table was used. We defecate Squat down to achieve the flatness of the ground. Did you choose the sanitary ware in the bathroom Angular, smooth and white sanitary ware. Toilets and washrooms meet Requirements for good ventilation and open space. The bathroom door is set with children's visual fitness The closed small observation window is matched with the handle.

6 记忆中的糖块

专项奖 · 技术奖
Special Prize Awarded · Technology Prize

注 册 号：7667

项目名称：稚梦飞扬（南平）
Childlike Dreams Flying (Nanping)

作 者：方心怡

参赛单位：北京建筑大学

指导教师：欧阳文

稚梦飞扬

Location analysis

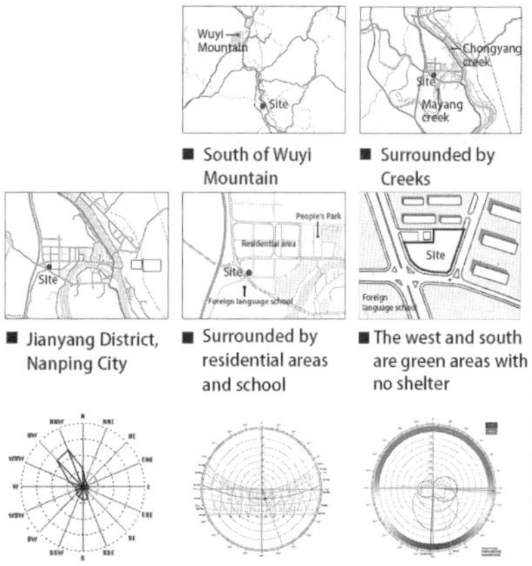
Wuyi Mountain
Site
Chongyang creek
Mayang creek
People's Park
Residential area
Foreign language school
Foreign language school

■ South of Wuyi Mountain
■ Surrounded by Creeks
■ Jianyang District, Nanping City
■ Surrounded by residential areas and school
■ The west and south are green areas with no shelter
■ Wind direction
■ Daylight trail
■ Best orientation

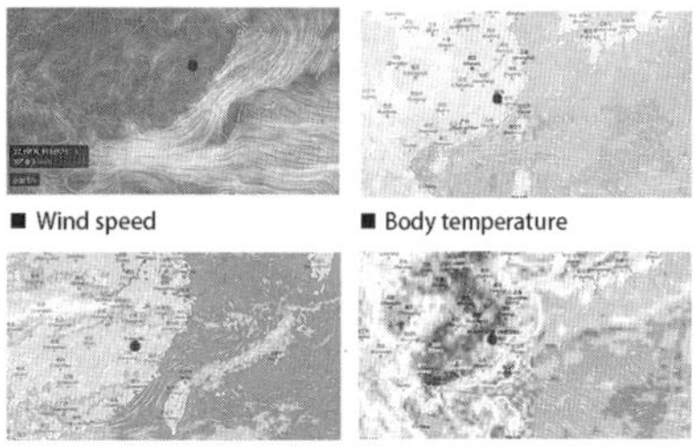

■ Wind speed
■ Body temperature
■ Precipitation
■ Air pollution

Site analysis

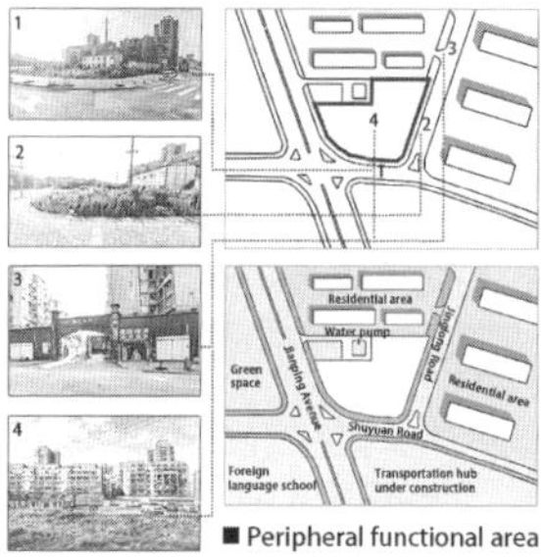
Residential area
Water pump
Green space
Residential area
Foreign language school
Transportation hub under construction

■ Peripheral functional area

Design explanation

本设计为低碳童趣的幼儿园设计，主要用庭院解决场地的高差，不同形式的庭院丰富了室外活动空间；室内活动充分考虑幼儿多元化的需求。建筑造型以交错的坡屋顶喻意“翻开的书本”和“展开的翅膀”，幼儿园不仅是教育场所，更是稚梦飞扬的地方，并附以节能高效的主被动式技术。同时，运用当地材料，从建造的角度使之更合理、可实施。

This design is a low-carbon childlike kindergarten design. The courtyard is mainly used to solve the height difference of the site, and different forms of courtyards also enrich the outdoor activity space; indoor activities fully consider the diverse needs of children. The architectural shape uses the staggered sloping roof as a metaphor for "opened books" and "spreading wings". The kindergarten is not only a place for education, but also a place for childish dreams. It is accompanied by more energy-saving and efficient active and passive technology. At the same time, local materials are used to make it more reasonable and feasible from the perspective of construction.

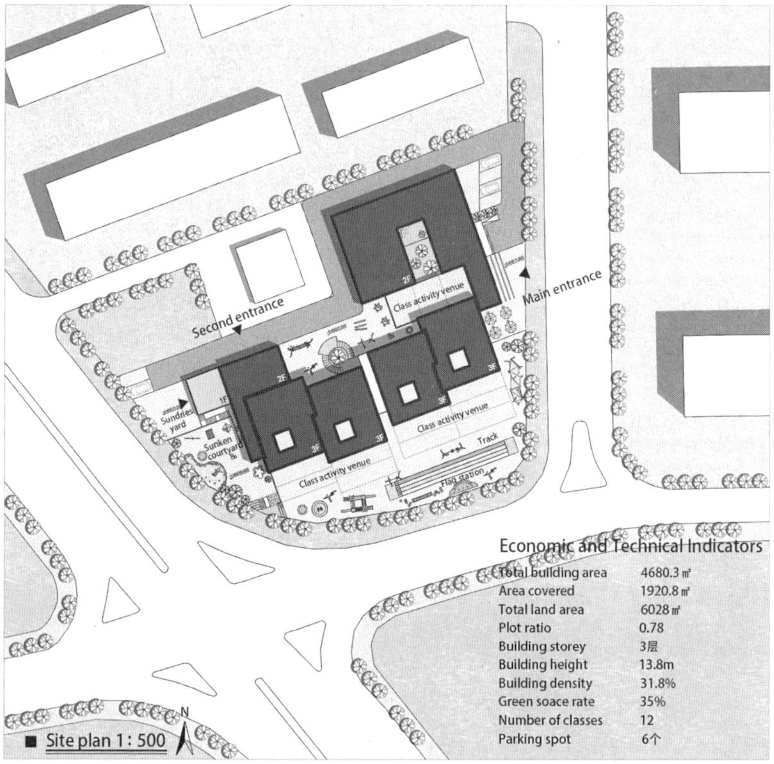
Second entrance
Main entrance
Class activity venue
Sundries yard
Sunken courtyard
Class activity venue
Class activity venue
Track
Flag station

■ Site plan 1 : 500

Economic and Technical Indicators

Total building area	4680.3㎡
Area covered	1920.8㎡
Total land area	6028㎡
Plot ratio	0.78
Building storey	3层
Building height	13.8m
Building density	31.8%
Green soace rate	35%
Number of classes	12
Parking spot	6个

Kindergarten building

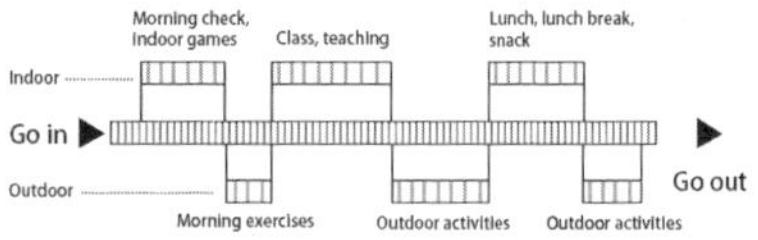
Morning check, indoor games
Class, teaching
Lunch, lunch break, snack
Indoor
Go in
Outdoor
Go out
Morning exercises
Outdoor activities
Outdoor activities

■ Toddler human body scale

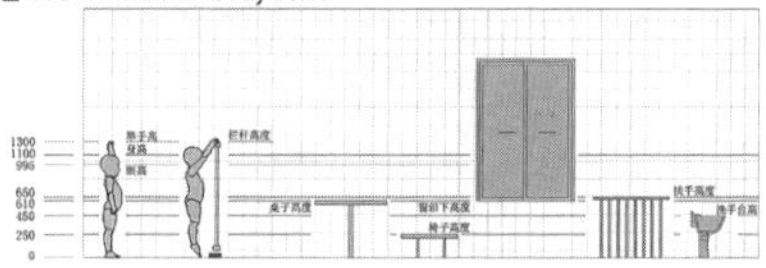

雏梦飞扬

Concept generation

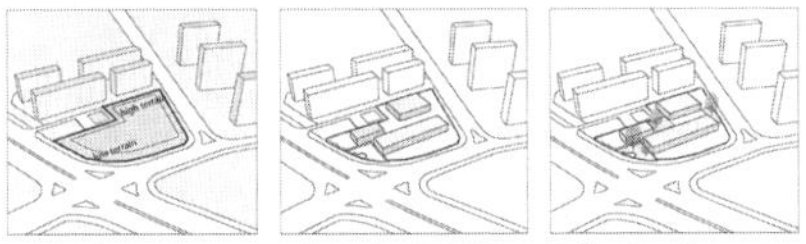

■ The site
The site is high in the northeast and low in the southwest, with convenient transportation.

■ Functional Division
Children' s acitivity unit
Office service part
Logistics supply part

■ Courtyards
The volumes are arranged to form the entrance courtyard, central courtyard and sunken courtyard.

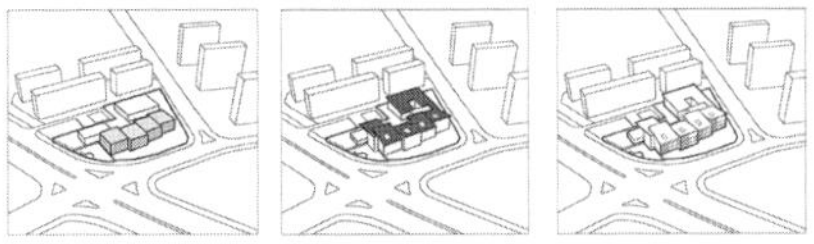

■ Staggered volume
The volume units are staggered to form self-shading, enriching the architectural form.

■ Solar sloping roof
The sloping roof is staggered and shaped like an open book and spreading wings.

■ Adjustable shutters
Adjustable sunshade shutters are used on the facade to form a unique light and shadow effect

Indoor and outdoor activities

■ Different layouts of the activity room

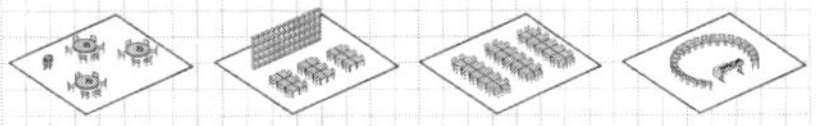

Hand-made　Reading　Dining　Teaching

■ Variety of outdoor activities

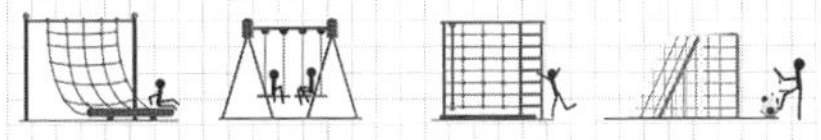

Climbing net　Swing　Combination　Sports

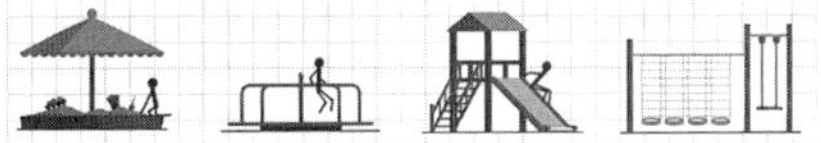

Sand pond　Rotating disc　Slide　Tire ring

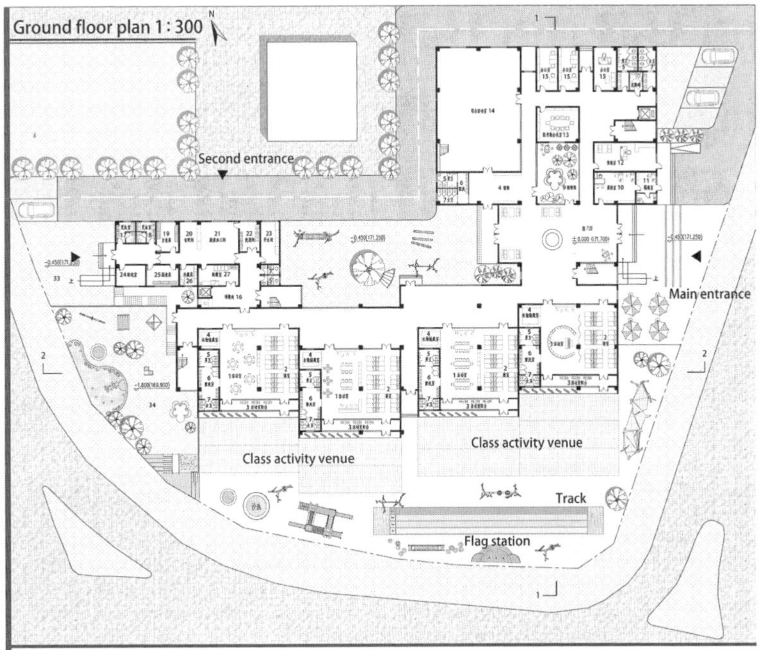
Ground floor plan 1 : 300
N
Second entrance
Main entrance
Class activity venue
Class activity venue
Track
Flag station

1 activity room
2 Bedroom
3 Sunshine room
4 Storage
5 Boy toilet
5' Male toilet
6 Bathroom
7 Girl toilet
7' Female toilet
8 hall
9 Botanical garden
10 Morning check room
11 Lsolation room
12 Duty room
13 Meeting room
14 Comprehensive activity room
15 Office
16 Get meal
17 Male locker room
18 Female locker room
19 Staple food bank
20 Cutting room
21 Kitchen processing room
22 Washing
23 Boiling water room
24 Distribution room
25 Non-staple food warehouse
26 Cold storage
27 Catering room
28 Outdoor platform
29 Small Restaurant
30 Laundry room
31 Teaching aid production room
32 Semi-outdoor platform
33 Sundries yard
34 Sunken courtyard

Plane function

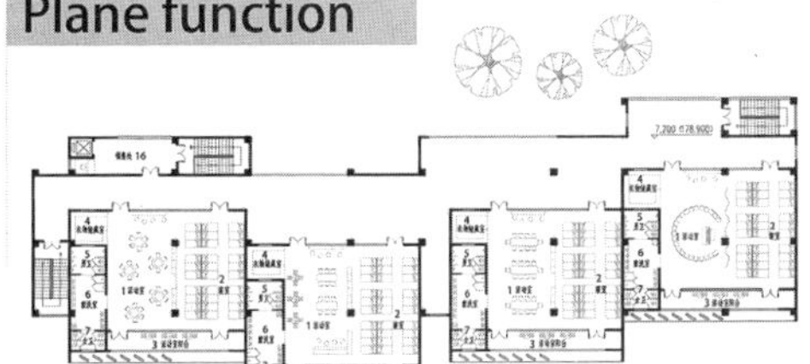

The third floor plan 1 : 300

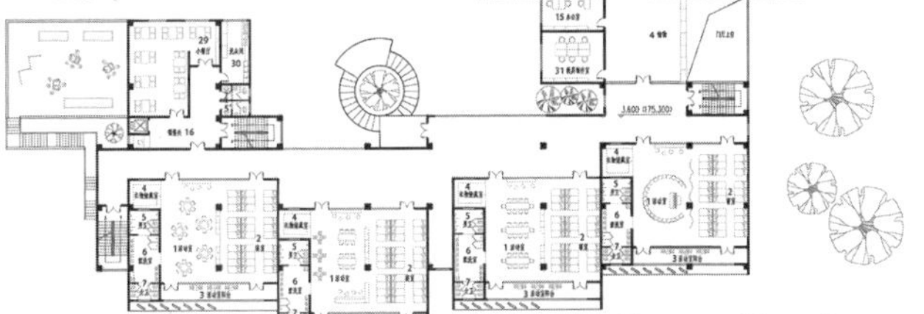

The second floor plan 1 : 300

稚梦飞扬
III

Overall plan

□ Chirdren' s activity unit

■ Windproof

In winter,the office part and logistics part can block northwest monsoon.

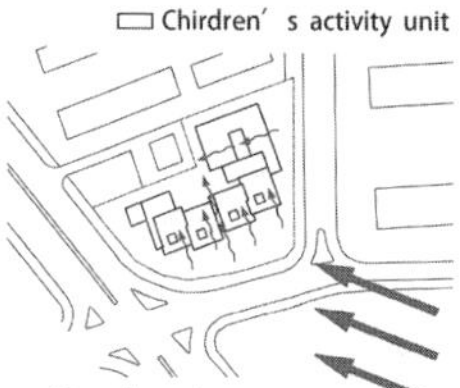

■ Ventilated

In summer,the southwest monsoon can directly blow through each activity unit.

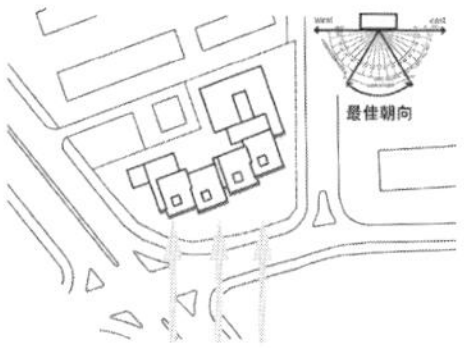
最佳朝向

■ Oriented

Each children' s activity unit is in the best orientation to ensure sufficient sunlight.

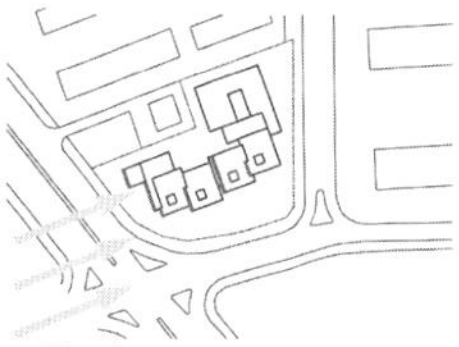

■ Shade

The northwest direction of the building is mainly solid walls, which can block part of the heat from entering the room.

Architectural design

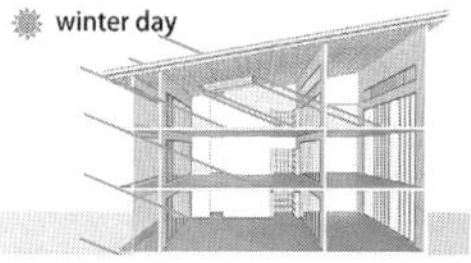
winter day

■ Heating

In winter, you can open the windows and the skylight to achieve direct indoor heating,the sunshine room can also be used for heating.

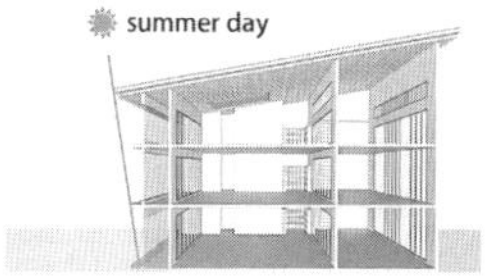
summer day

■ Shade

The eaves of the building at noon in summer form self-shading and cannot directly shine into the indoor space.

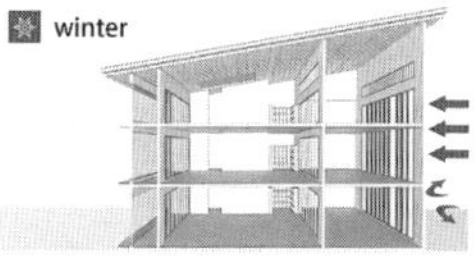
winter

■ Windproof

In winter, the movable shutters on the north side can be rotated and closed to block the winter monsoon

summer

■ Ventilated

In summer, you can open doors and windows to form a good ventilation environment.

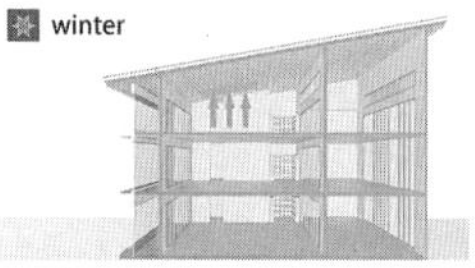
winter

■ Heat preservation

In winter, close doors and windows, Insulation structure of roof and wall.

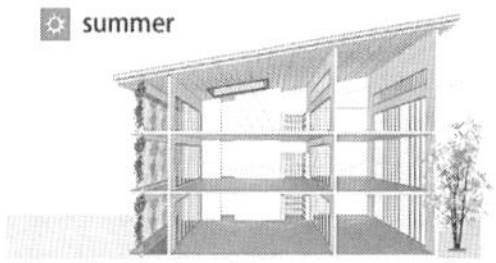
summer

■ Heat insulation

In summer, the indoor pool, vines and the landscape can provide cool indoors.

Sunshine room

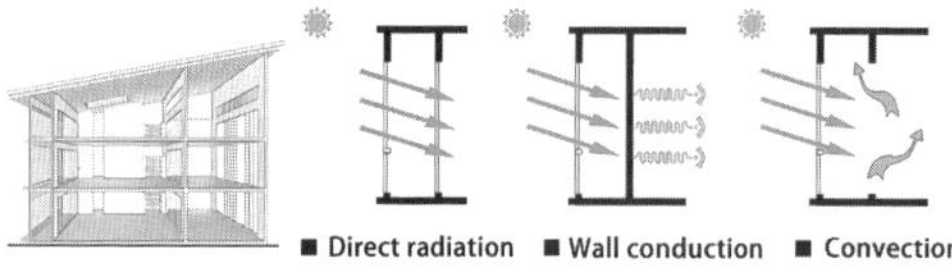

■ Direct radiation ■ Wall conduction ■ Convection

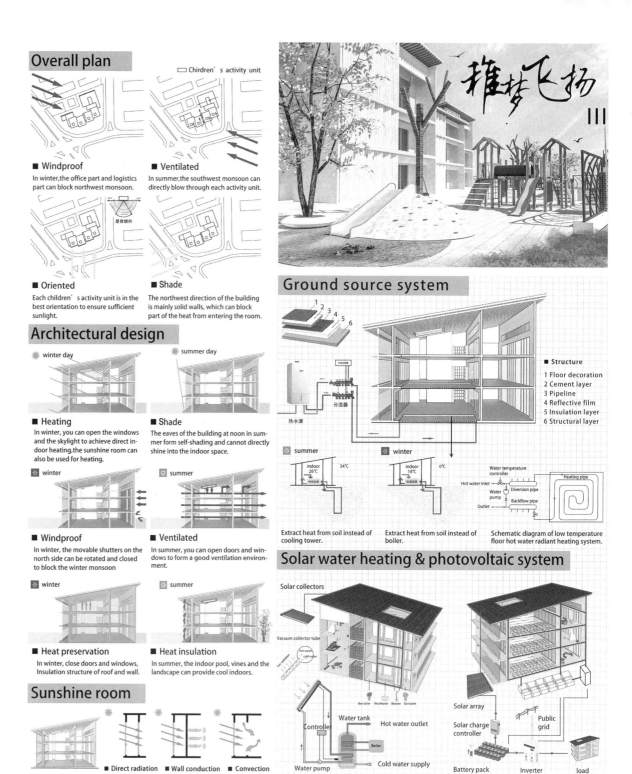

Ground source system

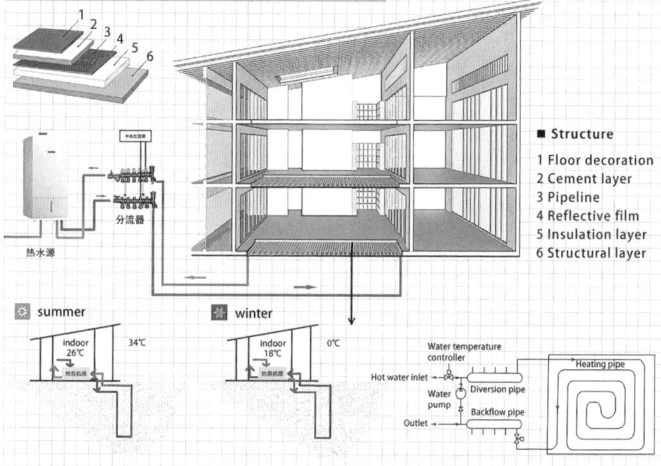
1
2
3
4
5
6
分流器
热水源
summer
indoor 26℃
34℃
winter
indoor 18℃
0℃
Water temperature controller
Hot water inlet
Water pump
Diversion pipe
Backflow pipe
Outlet
Heating pipe

■ Structure

1 Floor decoration
2 Cement layer
3 Pipeline
4 Reflective film
5 Insulation layer
6 Structural layer

Extract heat from soil instead of cooling tower.

Extract heat from soil instead of boiler.

Schematic diagram of low temperature floor hot water radiant heating system.

Solar water heating & photovoltaic system

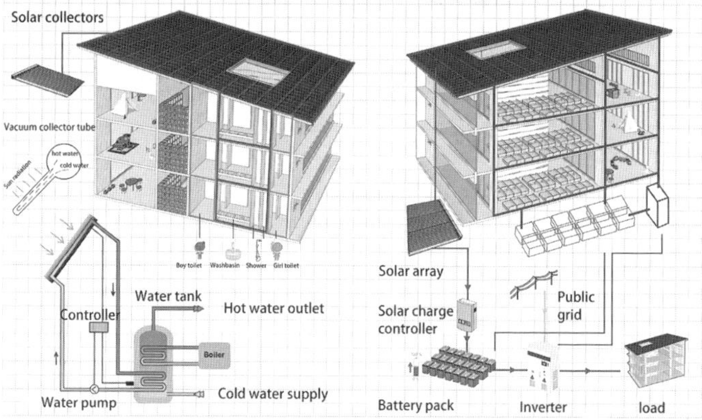
Solar collectors
Vacuum collector tube
Controller
Water tank
Hot water outlet
Boiler
Cold water supply
Water pump
Solar array
Solar charge controller
Public grid
Battery pack
Inverter
load

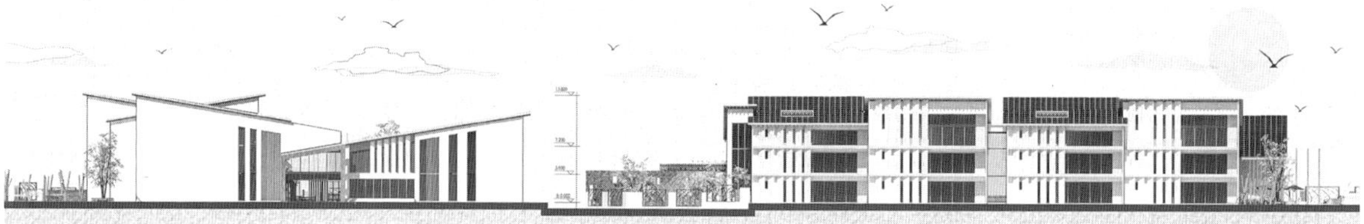

■ East elevation 1：300

■ South elevation 1：300

稚梦飞扬 IV

Green atrium

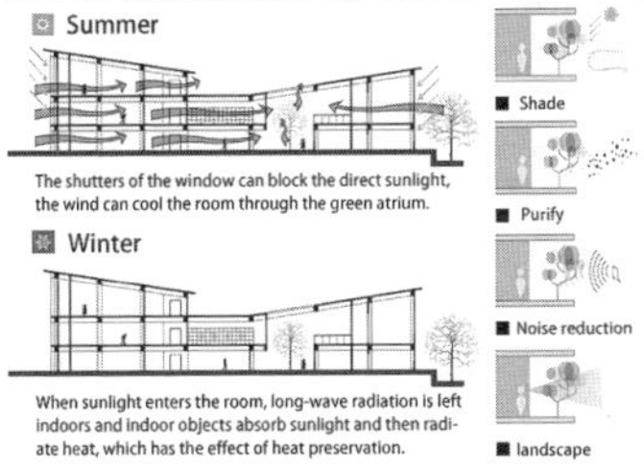

Summer
Winter
Shade
Purify
Noise reduction
landscape

The shutters of the window can block the direct sunlight, the wind can cool the room through the green atrium.

When sunlight enters the room, long-wave radiation is left indoors and indoor objects absorb sunlight and then radiate heat, which has the effect of heat preservation.

Noise reduction

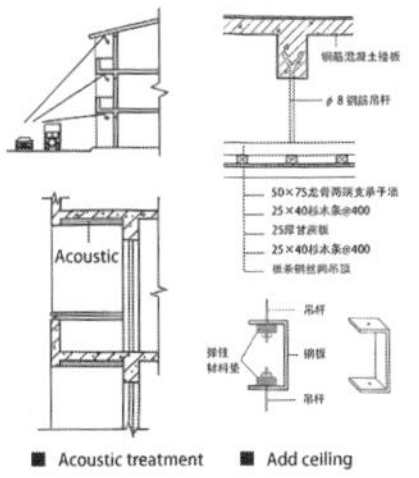

Acoustic
Acoustic treatment
Add ceiling

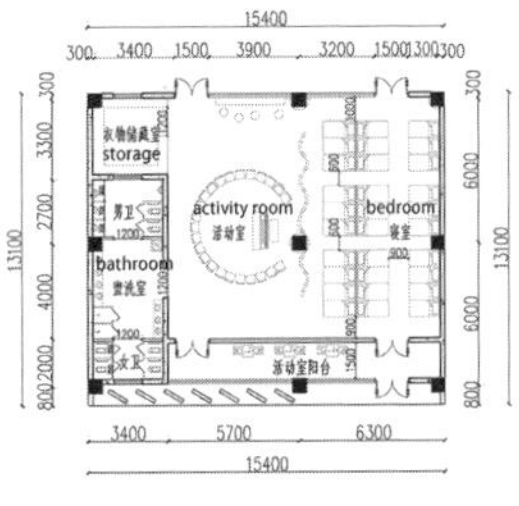

15400
storage
activity room
bedroom
bathroom
3400
5700
6300
13100

■ Detail drawing of children's activity unit

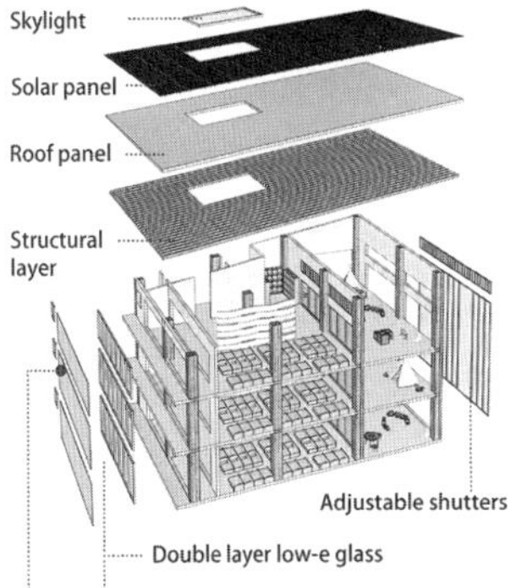

Skylight
Solar panel
Roof panel
Structural layer
Adjustable shutters
Double layer low-e glass

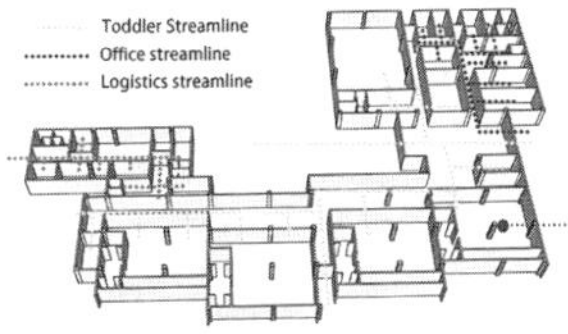

Toddler Streamline
Office streamline
Logistics streamline

■ Reinforced concrete frame structure

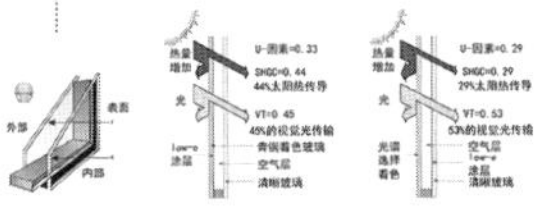

External adjustable shutters

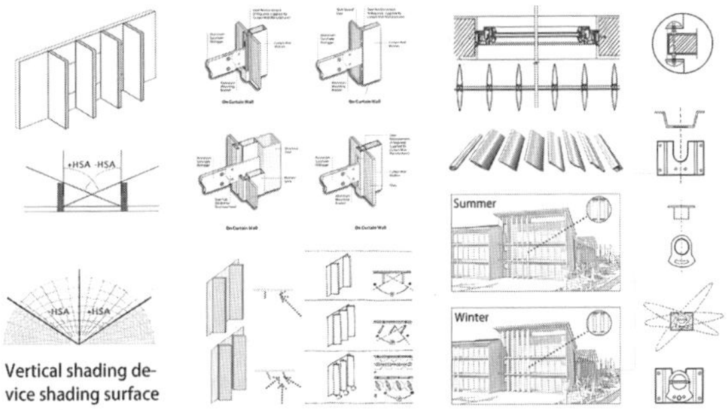

Summer
Winter

Vertical shading device shading surface

Use of local materials——Bamboo & Wood

■ The site is rich in bamboo and wood resources. Bamboo is used to divide the space indoors and outdoors to create a landscape. Wooden furniture and decorations are used indoors, and wooden shutters and wooden playground equipment are used outdoors.

Planted roof

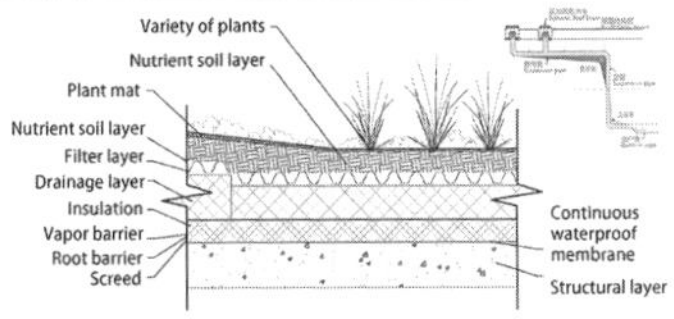

Variety of plants
Nutrient soil layer
Plant mat
Nutrient soil layer
Filter layer
Drainage layer
Insulation
Vapor barrier
Root barrier
Screed
Continuous waterproof membrane
Structural layer

Detail structures

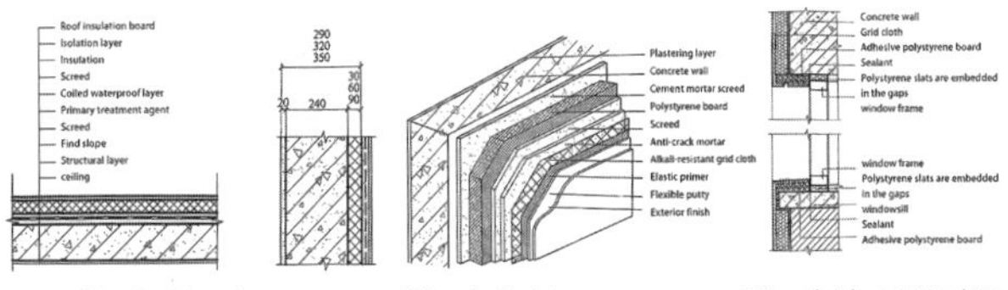

■ Inverted thermal insulation roof ■ External wall insulation ■ Door and window opening insulation

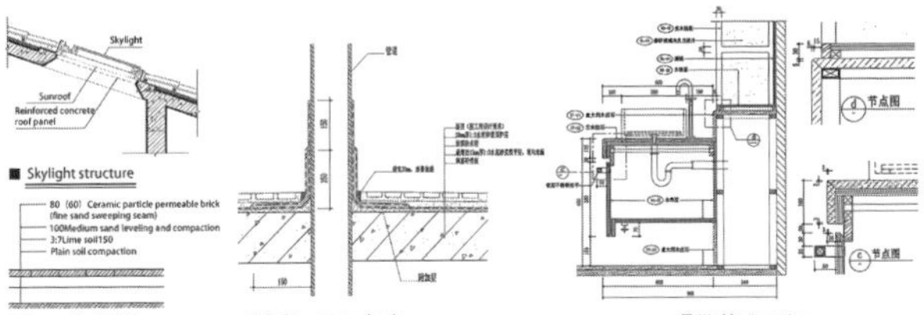

■ Skylight structure ■ Permeable brick floor ■ Toilet waterproof node ■ Washbasin section

Rain garden

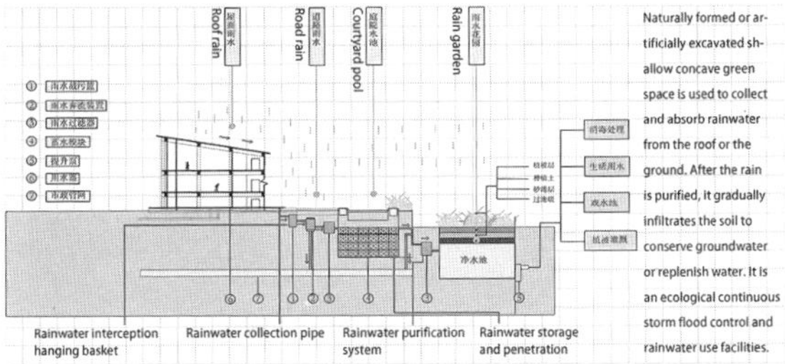

Roof rain
Road rain
Courtyard pool
Rain garden
Rainwater interception hanging basket
Rainwater collection pipe
Rainwater purification system
Rainwater storage and penetration

Naturally formed or artificially excavated shallow concave green space is used to collect and absorb rainwater from the roof or the ground. After the rain is purified, it gradually infiltrates the soil to conserve groundwater or replenish water. It is an ecological continuous storm flood control and rainwater use facilities.

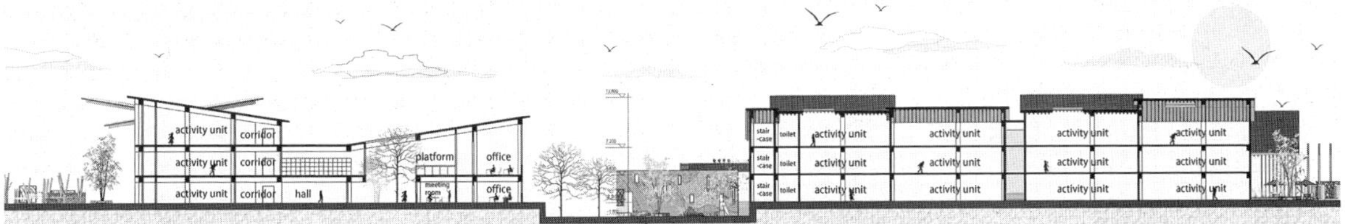

activity unit
corridor
hall
platform
office
stair case
toilet

■ 1-1 Section 1：300 ■ 2-2 Section 1：300

有效作品参赛团队名单

Name List of All Participants Submitting Valid Works

注册号	作者	单位名称	指导人	单位名称
5756	罗佳琦、叶桂明、刘艺蓉、陈东宇	华南理工大学	王静	华南理工大学
6667	王鹏举、李婷婷、盖婉婷	山西大学	沈涛	山西大学
6668	刘炳慧、林佳伟、王雨彤、李坤阳	烟台大学	孙佳媚	烟台大学
6673	蔡淑琪、赵子博、谢贤晟、吴明萱	南昌大学	叶雨辰	南昌大学
6678	张泽瑛、邹倩、王毅、郑庚锋	嘉兴学院	魏园园	嘉兴学院
6681	傅凯、刘志文、范佳溪、李睿琳	嘉兴学院	魏园园	嘉兴学院
6683	朱清尘、姚子安、温贺帆	北京交通大学	曾忠忠、陈泳全	北京交通大学
6686	王宇、黄云斌、左琪、梅亚岚	合肥工业大学	王旭	合肥工业大学
6687	洪亚聪、喻佩灵、黄逸凡、李俞锦	华东交通大学	彭小云	华东交通大学
6688	方鸣、刘晓者	广东博意建筑设计院有限公司		
6691	卢一迪、秦帅、姜雪梅、卡米然	新疆大学	斯依提艾力、樊辉	新疆大学
6692	张世超、别烨	昆明理工大学	李莉萍	昆明理工大学
6694	黄丹、谷智慧、吴靖、陈婷婷	武夷学院	林进益	武夷学院
6701	熊珊珊、吾米提江·吐逊江、孜木来提·吾布力哈斯木、麦尔旦·艾再孜	新疆大学	肉孜阿洪·帕尔哈提、阿迪力·赛买提	新疆大学
6703	钱超、赵念友、巴哈德尔·凯尤木、应缘炯	上海交通大学	何嘉伟	上海交通大学
6709	陈可嘉	厦门大学	林育欣、石峰	厦门大学
6716	李星林、叶江旺、黄友慧、陈方明	华东交通大学	彭小云	华东交通大学
6718	王雅楠、田佳瑶、冯 萌、孙雯雯、骆浩、曹傲彬	西安科技大学	孙倩倩	西安科技大学
6728	孙文龙、宋家辉、王树龙、叶葳蕤	青岛理工大学	高钰琛	青岛理工大学
6732	李磊、刘豪、刘合俊、李丽姝、开欣	合肥工业大学、苏州大学	郑先友、王旭	合肥工业大学
6734	夏小龙	上海马普建筑设计咨询有限公司		
6742	樊子钰、钱晓琪、王露婷、余建国、张峻旗	湖北工业大学	张辉、徐杨	湖北工业大学
6745	李金涛、马昊泓、陈思彤、钟旭	广州大学	席明波、李丽	广州大学
6762	黄赓、刘笑千	天津城建大学	任娟、刘芳	天津城建大学
6783	王奕文、杨文静、樊泽宇、尚诗越、李雯璐	西安科技大学	孙倩倩	西安科技大学
6800	陈斌、潘华威、蔡润栩、张文瑞	西安科技大学	孙倩倩	西安科技大学

续表

注册号	作者	单位名称	指导人	单位名称
6801	陈蕾、罗雅洁、陈伟东、邱妮波	武夷学院	林进益	武夷学院
6804	韩欣琰、李文珠、杨阳、谭雯	大连大学	黄世岩	大连大学
6805	侯凯怡、董璇	天津城建大学	万达	天津城建大学
6810	王润丰、姜昊、霍然、宋静怡、王心如、王新成、牛宁、孙百磊	天津大学、AA 建筑联盟、山东建筑大学、同圆设计集团有限公司	杨倩苗、汪江华	山东建筑大学、天津大学
6811	陈禹彤、于潇童	中国石油大学（华东）	李佐龙、陈瑞罡	中国石油大学（华东）
6817	郭嘉钰、张玉琪、王松瑞	天津大学、哈尔滨工业大学、上海同济规划院设计研究院	刘家韦华	华商国际工程有限公司
6834	邹佐、郝俊、黄金强、宋茁林	中国石油大学（华东）	王凌绪	中国石油大学（华东）
6836	张琴婷、李舒颖、张小雨、高天宇、李升瑞	西安科技大学	孙倩倩	西安科技大学
6838	梅丽宁、吴承龙、李小虎、高祚、张瑶、夏盛玉	沈阳工业大学	姜军	沈阳工业大学
6850	李欣冉、闵无非、邱乔、向以沫、杨妮	大连大学	黄世岩	大连大学
6854	王奕祺	厦门大学	林育欣	厦门大学
6866	徐启轩、何储悦、冯可心、白雨欣	西安科技大学	孙倩倩	西安科技大学
6874	王雨晴	厦门大学	林育欣	厦门大学
6875	张聪俐	厦门大学	林育欣、石峰	厦门大学
6877	柯少翔	厦门大学	林育欣	厦门大学
6880	于海玲	隆化县农业农村局		
6898	黄允、张伟聪、陈浩岚	惠州学院	黄汇雯	惠州学院
6899	罗臻昊、程悦、马腾宇	南京工业大学	薛洁	南京工业大学
6900	肖正天、王琦、王亚鑫、高鸣飞	天津大学、华北水利水电大学	冯刚、杨崴	天津大学
6902	黄东、陈一平、曾泽坤、曾志明	贵阳建筑勘察设计有限公司		
6907	Kgomotso Baloyi	武夷学院	林进益	武夷学院
6911	古丽哈娜・卡力甫汉、杨媛琨、杨航、华超美、徐婧	西安科技大学	石嘉怡	西安科技大学
6921	邹子琪、刘颖琪、徐婉琪、罗国辉	广州大学	李丽、周孝清	广州大学
6927	丁冉、黄洺月、吴双、周子谦	安徽建筑大学	杨雷鸣、吴华英、蔡迪、吴奇、蒋俊峰	安徽建筑大学

续表

注册号	作者	单位名称	指导人	单位名称
6932	杨宝奎、刘兴、刘威力、刘荣吉、沙永梅	昆明学院	陈虹羽	昆明学院
6938	林雨漩	厦门大学	林育欣	厦门大学
6943	鲁划、尹骏豪	天津城建大学	万达、郭莲	天津城建大学
6947	谢菁、张创	深圳市建筑设计研究总院有限公司、广州博厦建筑设计研究院有限公司深圳设计院		
6964	乔雨、何蕾	大连理工大学	金诚协	大连理工大学
6965	张家瑞、张嘉仪	长安大学	傅东雪、任娟	长安大学
6966	黄瑞龙、王子尧	重庆大学	周铁军、张海滨	重庆大学
6969	宋磊、张廷辉、邓兰、王彪、麻建钢、薛恺	昆明学院	陈虹羽	昆明学院
6975	游树华、王书帆 、巫相龙、张振贤	惠州学院	黄汇雯	惠州学院
6977	石婷霆、何惠姿、沈湘悦、郑煜川、任天漪、袁章拿、韦佳玲	中南林业科技大学	冯叶	中南林业科技大学
6979	王茜、晋英琪	西华大学	秦媛媛	西华大学
6984	丁海忠、潘兴清、施鹏仙、李松博、刘杰	昆明学院	陈虹羽	昆明学院
6986	周之恒、潘嘉豪、贺子琦、沈聪、叶慧瑶、沈立一、曹鑫媛	浙江理工大学	文强	浙江理工大学
6991	蔡智啟、吕泽钜、曾金安、叶伟浩	惠州学院	黄汇雯	惠州学院
7001	张缤月、袁蜀佳、袁传璞、芦乐平、张瑜珊	山东科技大学	冯巍	山东科技大学
7008	王悠然	北方工业大学	马欣	北方工业大学
7015	肖方扬、陈耀威、戴翊瑜、张智程	福州大学	林志森	福州大学
7016	师念、郭润妍、万佳丽	西安科技大学、鲍尔州立大学		
7017	张明远、程咏梅、崔帅杰、王诗雨、岳雅琦、李尚谦	北京工业大学、大连理工大学	戴俭	北京工业大学
7034	李航、陈峙宇、Yang Yang	南京工业大学、Syracuse University	杨亦陵	南京工业大学
7038	李起航、高雨凯、李思	中国矿业大学（北京）	贺丽洁、郑利军、曹颖	中国矿业大学（北京）
7042	郝上凯、杨潇静	西北工业大学	刘煜	西北工业大学
7046	郭靖、崔震宇、赛维杰	上海都市建筑设计有限公司		
7061	邹道圳、陈树强、黄宇敏、严淑园	惠州学院	黄汇雯	惠州学院
7063	吴昊天、徐建军、文冬霞	昆明理工大学	谭良斌	昆明理工大学

续表

注册号	作者	单位名称	指导人	单位名称
7068	古丽·玉素甫、阿依旦、冀雯思、刘静、刘梦君、刘帆、廖小苗	新疆大学	王万江、滕树勤、樊辉	新疆大学
7077	李弢、吴国亮	合肥工业大学	王旭	合肥工业大学
7078	蒋叡、文怀伟、杨易、蒋林晓	重庆大学	周铁军、张海滨	重庆大学
7080	艾晓雄、方振威、李成成、齐放、吴昊、张波	内蒙古工业大学	贾晓浒、常泽辉	内蒙古工业大学
7082	王宇、李婧、何玉龙、赵嘉诚	中国矿业大学（北京）	贺丽洁	中国矿业大学（北京）
7085	汤赵琳、李蕤泽、岳乐、白文琦	长安大学	夏博、任娟	长安大学
7087	郭志鹏、张闻祥、徐一涵、马源	西北工业大学、东南大学、华中科技大学	李静、毕景龙	西北工业大学
7088	马瑞、寇茜瑶、王梦迪	西安科技大学	孙倩倩	西安科技大学
7089	周梓彬、黄家威、黄嘉伟、李银标	惠州学院	黄汇雯	惠州学院
7097	王申、李鑫	北京工业大学	戴俭	北京工业大学
7104	夏雨诗、胡苏宁、孟轶彬、刘文洁、彭睿思、郝睿琪、艾立雨、李延欣	青岛理工大学	高钰琛、高艳娜	青岛理工大学
7106	苗成森、刘全超、普布桑珠	西藏大学	索朗白姆	西藏大学
7116	梁瑞、郑涛、陈兆哲	西安建筑科技大学	李涛、张斌	西安建筑科技大学、西耶建筑设计工作室
7120	李凌霄、王佩玉、孙逸达	中国矿业大学（北京）	贺丽洁	中国矿业大学（北京）
7124	贝莎莎、岳乐	长安大学	任娟	长安大学
7129	黄瑞、沈焜、熊旎颖、高青、卢惠阳、徐艺倩、杰布、李永晖、刘奎	湖南省建筑设计院有限公司、上迈新能源（北京）公司	蔡屹、Tim Mason	湖南省建筑设计院有限公司
7135	刘倩文、程璐瑶、李雪	中国矿业大学（北京）	贺丽洁	中国矿业大学（北京）
7144	孙佳攀、杨博、范冰冰、甄中豪	华北水利水电大学与俄罗斯乌拉尔联邦大学	张忆萌、卢梅珺	华北水利水电大学
7145	陈文静、董春朝、白一伟	西北工业大学	李静、毕景龙	西北工业大学
7153	李佳、袁晓霞、贾莹、赵力玮、余彬	成都理工大学	冯桢懿、肖瑜	成都理工大学
7161	郑曼如、刘世恒、陈佳怡、肖婉凝	福州大学	崔育新	福州大学
7162	肖婉凝、陈佳怡、郑曼如、刘世恒	福州大学	崔育新	福州大学
7164	陈睿、范孟辰、李乐恒、吕孟泽	广东工业大学	吉慧	广东工业大学
7165	明健、张蔚杰	华南理工大学	王静	华南理工大学

续表

注册号	作者	单位名称	指导人	单位名称
7166	孙镇郢、王琳杰、张铭洁、张秋瑶、居少捷、杨民阁	南京工业大学、华中农业大学	邵继中	南京工业大学、华中农业大学
7167	陈泉任、欧智凡、许子娴、阮朝锦、魏佳纯	重庆大学、华南理工大学、Royal College of Art	周铁军、张海滨	重庆大学
7171	李玉桢、胡然然	河北工程大学	连海涛	河北工程大学
7178	赵俊逸、许云鹏、史心怡	南京工业大学	郗皎如	南京工业大学
7182	张琳、黄和荣、张金锦	重庆大学	周铁军、张海滨	重庆大学
7193	罗岸霖、李浩杰、顾浩森、林子民	惠州学院	黄汇雯	惠州学院
7198	孟东漫、牛健壮、吴永涛	石家庄铁道大学	樊海彬	石家庄铁道大学
7200	谭若晨、付浩然、高力、王胜娟、刘宸溪、牛静仁	石家庄铁道大学、中铁建安工程设计有限公司	樊海彬、高力强	石家庄铁道大学
7203	宋函颖、田雨	重庆大学	周铁军、张海滨	重庆大学
7209	曹鑫磊、李婷、魏清、何瑷钰	石家庄铁道大学	樊海彬、高力强	石家庄铁道大学
7212	闫鑫懿、杨承超、高力、张清亮	石家庄铁道大学、中铁建安工程设计有限公司	樊海彬、高力强	石家庄铁道大学
7215	刘玺、周翔昀	石家庄铁道大学	高力强、樊海彬	石家庄铁道大学
7220	王佳捷、邱恭超、董紫阳、梁威豪	西安科技大学	孙倩倩	西安科技大学
7227	沈鹏远、黄鹤、黄丽梅、李霖、沈葛萍	湖州师范学院	张华、王丽晖	湖州师范学院
7231	曹钟月、金赫	大连理工大学	王丹、李国鹏	大连理工大学
7232	刘敏	石家庄铁道大学	高力强	石家庄铁道大学
7233	张尚书、吴凡诗、张伊婧、王汉斌	中国矿业大学（北京）	贺丽洁、李晓丹	中国矿业大学（北京）
7237	龙宇翔、王尧、宋航宇、李鹏、韩坤	西南交通大学	张樱子	西南交通大学
7239	严文杰、谢畅、徐荣归、张尼燕、高阳	湖州师范学院	张华	湖州师范学院
7243	林云菲、叶锦超、费姝慧	长安大学	任娟、夏博	长安大学
7253	程灿华、李宇锋	重庆大学	周铁军、张海滨	重庆大学
7259	刘利军、姚晨辉、荆萱、王云潇、颜实、杨泽宇	烟台大学、清华大学		
7260	张欣宇、汪紫涵	大连理工大学	高德宏	大连理工大学
7267	杨志祥、岳诗文、孙舒宜、胡畔、王阳	天津大学	杨崴	天津大学

续表

注册号	作者	单位名称	指导人	单位名称
7271	马磊	原乡（天津）景观设计有限公司		
7277	蔡双、胡诗雨、王叶涵	重庆大学	周铁军、张海滨	重庆大学
7278	李源鑫、陈嘉卉、包彦琨	北方工业大学、东南大学	马欣	北方工业大学
7279	刘静怡、许嘉诺、李之	南京工业大学	薛春霖、董凌	南京工业大学
7281	王瑞阳、吴启明	重庆大学	周铁军、张海滨	重庆大学
7285	朱郁松、金宇航、朱碧滢、朱可、曲恺辰、卢勇东、陈雨蒙、胡心怡、周超、顾瑜雯	东南大学、苏州科技大学	张宏、周欣、王伟、孔哲、塞尔江	东南大学、新疆大学
7292	邓冠中、费若雯、刘敏	石家庄铁道大学	高力强、樊海彬	石家庄铁道大学
7294	喀斯木・牙森、苏菲娅、麦如甫江・塞迪尔丁、苏巴提・珀拉提	新疆大学	肉孜阿洪・帕尔哈提、阿迪力・赛买提	新疆大学
7295	赵有亮、赵帅、肖靖宇	石家庄铁道大学	樊海彬	石家庄铁道大学
7299	叶文慧、尤畅、王资涵、陈梅旭	南京工业大学	叶起瑾	南京工业大学
7307	黄承佳、江逸妍	华南理工大学	王静、肖毅强	华南理工大学
7308	谢一、闫晨晨、周童、李俊良	石河子大学	王蒙、额热艾汗、任玉成	石河子大学
7313	冷依阳、张炎、夏闻典	辽宁工业大学	田波	辽宁工业大学
7319	罗懿鹭、王梅竹	北京交通大学	杜晓辉、胡映东	北京交通大学
7321	刘静怡、马宝裕、刘晓俐、王邑心、俞欣妤	华南理工大学、浙江大学、南加州建筑学院	史劲松、赵立华	西南交通大学、华南理工大学
7323	张淑娴、邓盛炜、曾小婷	福州大学	吴木生、林志森	福州大学
7324	侯尚哲	青岛理工大学	高钰琛	青岛理工大学
7325	冯美歌			
7327	宋恒之、谢苑仪、苗子健、成昱霖	华南理工大学	熊璐	华南理工大学
7332	冯美歌			
7335	周继发、刘小双	大连理工大学	高德宏	大连理工大学
7343	吴静文、谢逸冰、曹赓、韩旭、林文强、成恩宏	安徽建筑大学	桂汪洋	安徽建筑大学
7361	韩世翔、牛一凡、郭佳琪、赵晨林	北方工业大学	马欣、张宏然	北方工业大学
7363	张晟、单倩茹、李欣、郝宥恩	沈阳建筑大学、吉林建筑大学	王常伟	沈阳建筑大学

续表

注册号	作者	单位名称	指导人	单位名称
7371	王清妍、金童	北方工业大学	马欣、温芳	北方工业大学
7379	沈鹏、王凯强、王欣	浙江大学	吴放	浙江大学
7387	苏瑞雪、周浩、卢琰、孙文君、赵另志	西京学院、山东工艺美术学院	齐达、张创业	西京学院
7388	郭佳琪、韩世翔	北方工业大学	马欣、张宏然	北方工业大学
7392	成浩源、王凌曦	大连理工大学	高德宏	大连理工大学
7396	赵韦、孙沁郁、李玉、滕雨欣、刘瑶瑶	西华大学	丁玎、郭坤	西华大学
7400	彭露瑶、谢紫欣、粟媛媛、蒲美廷、严永灏	西华大学	郑澍奎	西华大学
7404	林邦汛、李康、自明娟、徐天娇	嘉兴学院	王中锋	嘉兴学院
7406	周扬空、李景秀、苗振轩、赵晓彤、金勇运、许仁杰、李正洁、甄成	天津大学	郭娟利	天津大学
7408	崔林森、薛荣骅、张丛媛	西安科技大学	孙倩倩	西安科技大学
7410	黄锐聪、杨嘉瑶、王兰、刘昊坤	西安科技大学	孙倩倩	西安科技大学
7420	陈栩欣、邓瑾茹、殷悦	南京工业大学	薛春霖	南京工业大学
7422	梁钟琪、朱游学、秦瑞炜	南京工业大学	叶起瑾、郗皎如	南京工业大学
7430	陈悦怡、郭玉琛、李雨茜、王嘉序	南京工业大学	董凌、薛春霖	南京工业大学
7433	陈璐、陈子菲、贾子玥、朱丹妮	南京工业大学	罗靖	南京工业大学
7437	李姝颖、孙烯、李仪	西华大学	谭璐薇	西华大学
7441	黄和谷、李崇玮、潘文涛	南京工业大学	罗靖	南京工业大学
7444	李诗蕾、袁智海、廖梦琦	江西理工大学	彭峰、蔡丽蓉	江西理工大学
7450	李宁、王昊天、张清亮、王胜娟	石家庄铁道大学、中铁建安工程设计有限公司	樊海彬、高力强	石家庄铁道大学
7451	刘时羽、陈香合、覃泷莹	北京交通大学	杜晓辉、胡映东	北京交通大学
7463	周笑、周宇弘、嵇晨阳、顾涛	南京工业大学	薛春霖	南京工业大学
7471	斯依提艾力·艾麦提、普拉提江·依麻木、王文举、陆胤京、丁英杰、艾麦提江	新疆大学、新疆九木建筑设计事务所		
7472	丁英杰、王文举、柏红喜、雷倩、伊敏江	新疆大学	斯依提艾力·艾麦提、王健	新疆大学
7478	古丽·玉素甫、阿依旦、冀雯思、刘静、刘梦君、刘帆、廖小苗	新疆大学	王万江、滕树勤、樊辉	新疆大学

续表

注册号	作者	单位名称	指导人	单位名称
7485	窦雨薇、邢艺馨、袁佳慧	北京交通大学	杜晓辉、胡映东	北京交通大学
7489	王笑涵、沈一飞、段昭丞、陈晴	重庆大学	张海滨、周铁军	重庆大学
7494	冯玥、温静	吉林建筑大学	宋义坤	重庆大学
7500	谢冲、彭榆棋、冉铖李、侯盈竹、韦小小	重庆大学	张海滨	重庆大学
7501	陈紫嫣、王琪瑶、田颖影	南京工业大学	罗靖	南京工业大学
7508	杨秋缘、冯宇欢、徐秋	南京工业大学		
7512	杨振宇、金弘毅、张赋祺、成燕、荀宇骁、吴欣桐、李彦一	上海大学	章国琴	上海大学
7517	金在纯、施忆梅、程舒婷、李金洋、顾瑜雯	苏州科技大学	刘长春、冯进、杨绍亮	苏州科技大学
7525	徐欣荣、王朝红、刘德利、王明琦、高家乐、陈志磊、田帅、刘佳辉、孙波、张溱旼、喀普兰巴依·艾来提江、张子安、金宇航、石紫宸	东南大学、新疆大学、吉林建筑大学	张宏、塞尔江·哈力克、周欣、孔哲、王伟、张玫英	东南大学、新疆大学
7529	成力、田翊翔	西华大学	郑澍奎、郭坤	西华大学
7536	陈博文、肖子一、唐夏旭、王重邦	西安建筑科技大学	何泉、范征宇	西安建筑科技大学
7545	郑仲意、徐衍新、相楠、李迪萌、刘权仪、张鹏娜	山东建筑大学、北京师范大学珠海分校	薛一冰、房涛、何文晶	山东建筑大学
7551	张问楚、黄昱豪、黄浩、林宇栋	华南理工大学	王静、赖文波	华南理工大学
7553	邓扬、王振宇、姚菲、孙辰、韩佳勤	西南交通大学	张樱子、王俊	西南交通大学
7562	李子逸、冯琳欢、毛敏、张艺婷、张新倩、宋家辉、王树龙	青岛理工大学	沈源	青岛理工大学
7568	陈俊安、杨豪广、顾祎霏、徐芊卉、刘旭	三江学院		
7574	耿进、杨桂玲、赵思嘉、郑鑫	三江学院	王加鑫	三江学院
7583	仲文、冯苗苗	大连理工大学	李国鹏	大连理工大学
7587	肖新怡、禹思奇、孙畅、尹鑫睿	青岛理工大学	高钰琛、高艳娜	青岛理工大学
7594	常逸凡、陆雨瑶、曹志昊	南京工业大学	胡振宇、薛春霖、董凌	南京工业大学
7598	费若雯	石家庄铁道大学	高力强、樊海彬	石家庄铁道大学
7600	范蕊、邢昊晟、苏忠豪、李晓伟	山东科技大学	冯巍	山东科技大学
7601	余泽明、田耕育	重庆大学	周铁军	重庆大学
7605	乔巧、陈妮、汪淼、王茜、庄韡荭	西华大学	秦媛媛、谭露薇	西华大学

续表

注册号	作者	单位名称	指导人	单位名称
7608	牛静仁、刘宸溪、张清亮、高力、付浩然、谭若晨	石家庄铁道大学、中铁建安工程设计有限公司	樊海彬、高力强	石家庄铁道大学
7611	喀普兰巴依·艾来提江、张龄之、陈炳合、史静毅、朱紫悦、马丽芳、张耀春、张雪娇、代彩琼	新疆大学	塞尔江·哈力克、齐典韦、范涛、陈菊香	新疆大学
7612	李奇芫、于汉泽、李洁、经翔宇、高皓然、陈柯欣	天津大学、天津大学建筑设计规划研究总院有限公司	杨崴、刘刚、贡小雷	天津大学
7614	赵哲、郑秋晨、王琪琪、仇诗语	中南大学	宋盈	中南大学
7616	张仲奇、蒋宜芳、邵奇	清华大学、哈尔滨工业大学、太平洋国立大学		
7627	裴让让、李雪萌	中国石油大学（华东）	王凌旭	中国石油大学（华东）
7630	徐衍新、郑仲意、李迪萌、相楠、刘权仪、张鹏娜	山东建筑大学、北京师范大学珠海分校	薛一冰、房涛、何文晶	山东建筑大学
7631	彭宜洁、王梦琪、丁洁、张润萌	北京建筑大学	张大玉、欧阳文	北京建筑大学
7638	边新元、张瀚文、高欣宇、刘欢、唐耀晟、廖为胤、曾祥有、程禹哲	华北理工大学	檀文迪、唐晨辉	华北理工大学
7653	陈雨欣、王云芸、尹科、李瑞琪、刘珂伶	西华大学	丁玎	西华大学
7654	王乐群、陈晓敏、刘纯铭、赵昱	青岛理工大学	高钰琛、高艳娜	青岛理工大学
7660	曹瑞、李晓萱、王艺洋、朱宇昂	西北工业大学	李静、毕景龙	西北工业大学
7663	聂甜玉、刁永怡、牛家渺、张艺文	山东建筑大学	魏瑞涵	山东建筑大学
7667	方心怡	北京建筑大学	欧阳文	北京建筑大学
7673	高煜喆、王奇、王若虹、陈晓超、孙仁文	天津大学	朱丽、尹宝泉	天津大学
7684	钱诗雨、王卓雅	西北工业大学	李静、毕景龙	西北工业大学
7701	徐从淮、董星瑶、程良浩、金玲玲、彭泓宇、陶靖煊、左艳琳、任志翔	武昌理工学院	刘玉曦、原菊蒲	武昌理工学院
7707	喻虎、周斌	北京凯盛建材工程有限公司		
7715	刘学奎、宋昊明、慕璟云、许嘉琪、徐珈璐、张文豪、孙成鹏	青岛理工大学	耿雪川、程然、解旭东	青岛理工大学
7724	李浩楠、李婷清、郭希晞、王冰璞	西安科技大学	孙倩倩	西安科技大学
7747	吕晓、王丹	大连理工大学	周博	大连理工大学

续表

注册号	作者	单位名称	指导人	单位名称
7761	陈铁、冯恩童、张龄予、刘学齐、易庄仪	武汉科技大学	梁宇鸣	武汉科技大学
7763	田晶、王家轩、牟朝群、王鑫汝、张忠瑞	山东科技大学		
7768	朱文婧、金怡婕、冒戎威	福州大学	林涛	福州大学
7769	裘昕、张昊月、黄顺发	福州大学	吴木生、王炜、崔育新	福州大学
7775	陈佳琪、肖志伟、李雅娴、龚永强、谷晓娜、张文莉	河北工程大学	莫菁洁、席晖	河北工程大学
7788	张迅、王懿	山东农业大学	董文	山东农业大学
7796	李尚锋、周从越、张鲁鹏	麦锐设计工作室		
7798	曹赓、王禹、刘益翔	安徽建筑大学	吴运法	安徽建筑大学
7799	孟泽、李一童、江明慧、胡家浩	山东建筑大学	魏瑞涵	山东建筑大学
7809	单后壮、高慧、王妍、张煜培	山东科技大学	冯巍	山东科技大学
7811	张溱旼、张子安、喀普兰巴依·艾来提江、金宇航、王明琦、王朝红、石紫宸、田帅、徐欣荣、刘家辉	东南大学、新疆大学	张宏、塞尔江·哈力克、周欣、孔哲、王伟	东南大学、新疆大学
7819	张喜庚、张旭、孙叶文、张晓梅、庞滨	山东农业大学	王舒敏	山东农业大学
7838	储栎杨、张子涵、沈翔宇、吴湉钰、周正、郭雨祺	安徽建筑大学	解玉琪、陈萨如拉	安徽建筑大学
7840	姚晓宇、余赛尔、姚乾列、谢成芸、陈奕璇、高澜、莫涵晞	浙江理工大学	李雯、毛万红	浙江理工大学
7848	耿子涵、吴和平	大连理工大学	李国鹏、张宇	大连理工大学
7851	谢星杰、高嘉婧、徐涵	重庆大学、华侨大学	黄海静、周铁军、张海滨、欧达毅	重庆大学、华侨大学
7857	杭明远、杨威海、查素琪、张絮、李一然、项欣、陆春华	安徽建筑大学	解玉琪、韩玲、陈萨如拉、沈念俊	安徽建筑大学
7860	国萧、罗頔、陈欣雨、李志强、周玥、王子腾	安徽建筑大学	解玉琪、韩玲、陈萨如拉、沈念俊	安徽建筑大学
7872	潘俊杰、程祖文、孟文哲、张艳红、陈明琼	昆明学院	陈虹羽	昆明学院

2020台达杯国际太阳能建筑设计竞赛办法

Competition Brief on International Solar Building Design Competition 2020

竞赛宗旨：

贯彻落实国家推动城镇居住区幼儿园建设、着力补齐农村学前教育短板的政策，以幼儿园为平台，将生态、绿色的理念融入学前教育中。

竞赛主题：**阳光·稚梦**

竞赛题目：**福建省南平市建阳区景龙幼儿园**

新疆巴音郭楞州和静县建设兵团牧场幼儿园及服务中心

主办单位：**国际太阳能学会**

中国可再生能源学会

中国建筑设计研究院有限公司

承办单位：**国家住宅与居住环境工程技术研究中心**

中国可再生能源学会太阳能建筑专业委员会

冠名单位：**台达集团**

技术指导：**全国高等学校建筑学学科指导委员会**

媒体支持：**《建筑技艺》(AT) 杂志**

评委会专家：**杨经文：马来西亚汉沙杨建筑师事务所创始人、2016年梁思成建筑奖获得者。**

Deo Prasad: 澳大利亚科技与工程院院士、澳大利亚勋章获得者、澳大利亚新南威尔士大学教授。

崔愷：中国工程院院士、全国工程勘察设计大师、中国建筑设计研究院有限公司总建筑师。

王建国：中国工程院院士、教育部高等学校建筑类专业指导委员会主任委员、东南大学建筑学院教授。

GOAL OF COMPETITION:

To implement the national policy of advancing the construction of kindergartens in urban residential areas and improving the drawbacks of preschool education in rural areas, ecological and green concepts are integrated into preschool education with kindergartens as the platform.

THEME OF COMPETITION:

Sunshine & Childlike Dream

SUBJECT OF COMPETITION:

1. Jinglong Kindergarten in Jianyang District, Nanping City, Fujian Province
2. Construction Corps Pasture Kindergarten and Service Center in Hejing County of Bayingol Mongolian Autonomous Prefecture, Xinjiang

HOST:

International Solar Energy Society (ISES)
China Renewable Energy Society (CRES)
China Architecture Design & Research Group (CAG)

ORGANIZER:

China National Engineering Research Center for Human Settlements (CNERCHS)
Special Committee of Solar Buildings, CRES

TITLE SPONSOR:

Delta Group

庄惟敏：中国工程院院士、全国工程勘察设计大师 、2019 梁思成建筑奖获得者、清华大学建筑学院院长。

林宪德：台湾绿色建筑发展协会理事长、台湾成功大学建筑系教授。

仲继寿：中国可再生能源学会太阳能建筑专委会主任委员、中国建筑设计研究院副总建筑师。

黄秋平：华东建筑设计研究总院总建筑师。

冯雅：中国建筑西南设计研究院顾问总工程师。

组委会成员：由主办单位、承办单位及冠名单位相关人员组成。办事机构设在中国可再生能源学会太阳能建筑专业委员会。

评比办法：

1. 由组委会审查参赛资格，并确定入围作品。
2. 由评委会评选出竞赛获奖作品。

评比标准：

1. 参赛作品须符合本竞赛“作品要求”的内容。

2. 作品应具有原创性和前瞻性，鼓励创新。

3. 作品应满足使用功能、绿色低碳、安全健康的要求，建筑技术与太阳能利用技术具有适配性。

4. 作品应充分体现太阳能利用技术对降低建筑使用能耗的作用，在经济、技术层面具有可实施性。

5. 由于两个场地的设计内容差别较大，所以分别设置评分办法。作品评定采用百分制，分项分值见下表：

福建省南平市建阳区景龙幼儿园

评比指标	指标说明	分值
规划与建筑设计	规划布局、建筑空间组合、功能流线组织、建筑艺术	35
被动太阳能利用技术	通过专门建筑设计与建筑构造实现隔热、降温和通风的技术	30
主动太阳能利用技术	通过专门系统设备收集、转换、传输、利用太阳能的技术	10
采用的其他技术	其他绿色、低碳、安全、健康技术	15
可操作性	作品的可实施性，技术的经济性和普适性	10

TECHNICAL INSTRUCTOR:

National Supervision Board of Architectural Education (China)

MEDIA SUPPORT:

Architecture Technique (AT)

EXPERTS OF JUDGING PANEL:

Mr. King Mun YEANG: President of T. R. Hamzah & Yeang Sdn. Bhd(Malaysia), 2016 Liang Sicheng Architecture Prize Winner

Mr. Deo Prasad: Academician of Academy of Technological Sciences and Engineering, Winner of the Order of Australia, and Professor of University of New South Wales, Sydney, Australia

Mr. Cui Kai: Academician of China Academy of Engineering, National Engineering Survey and Design Master and Chief Architect of China Architecture Design & Research Group (CAG)

Mr. Wang Jianguo: Academician of China Academy of Engineering, Director of Academic Council of School of Architecture, Professor of School of Architecture, Southeast University

Mr. Zhuang Weimin: Academician of Chinese Academy of Engineering, National Engineering Survey and Design Master, 2019 Liang Sicheng Architecture Prize Winner and Dean of School of Architecture, Tsinghua University

Mr. Lin Xiande: Chairman of Taiwan Green Building Committee, and Professor of Faculty of Architecture of Cheng Kung University, Taiwan

Mr. Zhong Jishou: Chief Commissioner of Special Committee of Solar Buildings, CRES,and Deputy Chief Architect of CAG

Mr. Huang Qiuping: Chief Architect of East China Architectural Design & Research Institute (ECADI)

Mr. Feng Ya: Chief Engineer of China Southwest Architectural Design and Research Institute Corp. Ltd.

MEMBERS OF THE ORGANIZING COMMITTEE:

It is composed of the competition host, organizer and title sponsor. The administration office is a standing body in Special Committee of Solar Buildings, CRES.

APPRAISAL METHODS:

1. Organizing Committee will check up eligible entries and confirm shortlist entries.
2. Judging Panel will appraise and select out the awarded works.

APPRAISAL STANDARD:

1. The entries must meet the demands of the Competition Requirement.
2. The entries should be original and forward-looking to encourage innovation.

新疆巴音郭楞州和静县建设兵团牧场幼儿园及服务中心

评比指标	指标说明	分值
规划与建筑设计	规划布局、环境利用与融入、功能流线组织、建筑空间组合、建筑艺术	40
被动太阳能利用技术	通过专门建筑设计与建筑构造实现采暖、隔热和通风的技术	30
主动太阳能利用技术	通过专门系统设备收集、转换、传输、利用太阳能的技术	10
采用的其他技术	其他环保、低碳、健康技术	10
可操作性	作品的建造成本控制、可实施性	10

设计任务书及专业术语等附件：（见资料下载）

附件 1：福建省南平市建阳区景龙幼儿园任务书

附件 2：新疆巴音郭楞州和静县建设兵团牧场幼儿园及服务中心项目任务书

附件 3：专业术语

附件 4：参赛者信息表

奖项设置及奖励形式：

1. 综合奖：

一等奖作品：2 名　颁发奖杯、证书及人民币 100000 元奖金（税前）；

二等奖作品：4 名　颁发奖杯、证书及人民币 20000 元奖金（税前）；

三等奖作品：6 名　颁发奖杯、证书及人民币 5000 元奖金（税前）；

奖金按偶然所得税缴纳，由组委会统一缴费。

优秀奖作品：30 名　颁发证书。

2. 优秀设计方法奖：10 名，颁发证书。

作品设计方法报告内容丰富充实，设计过程记录完整，设计方法新颖。

3. 技术专项奖：名额不限，颁发证书。

作品采用的技术或设计方面具有创新，实用性强。

4. 建筑创意奖：名额不限，颁发证书。

作品规划及建筑设计方面具有独特创意和先导性。

3. The submitted works should meet the demands for use, green and low-carbon concept, health and coziness, the building technology and solar energy utilization technology should be compatible.

4. The submitted works should play the role of reducing building energy consumption by using solar energy technology and have feasibility in the aspect of economy and technology.

5. Scoring methods are set respectively due to vast difference between the two sites. A percentile score system is adopted for the appraisal as follows:

Jinglong Kindergarten in Jianyang District, Nanping City, Fujian Province

Appraisal Indicator	Explanation	Scores
Planning and architecture design	Planning design, architectural space combination, functional division, streamline organization, architectural art	35
Utilization of passive solar energy technology	Use of solar cooling and heating technology by specific architecture and construction design	30
Utilization of active solar energy technology	Use of solar energy though collecting, transforming, and transmitting energy by specific equipment	10
Other technologies	Other technologies such as: green, low-carbon, safe and healthy technologies	15
Operability of the technology	Feasibility, economy, and popularity of relevant technology demands	10

Construction Corps Pasture Kindergarten and Service Center in Hejing County of Bayingol Mongolian Autonomous Prefecture, Xinjiang

Appraisal Indicator	Explanation	Scores
Planning and architecture design	Planning design, use of environmental resource and integrating into the surroundings, functional division and streamline organization, architectural space combination, architectural art	40
Utilization of passive solar energy technology	Use of solar cooling and heating technology by specific architecture and construction design	30
Utilization of active solar energy technology	Use of solar energy though collecting, transforming, and transmitting energy by specific equipment	10
Other technologies	Other technologies such as: green, low carbon, safe and healthy technologies	10
Operability of the technology	Feasibility, economy, and popularity of relevant technology demands	10

THE TASK BOOK OF DESIGN AND PROFESSIONAL GLOSSARY (FOUND IN ANNEX)

Annex 1: Jinglong Kindergarten in Jianyang District, Nanping City, Fujian Province (Draft for comments)

作品要求：

1. 建筑设计方面应达到方案设计深度，技术应用方面应有相关的技术图纸和指标。

2. 作品图面、文字表达清楚，数据准确。

3. 作品基本内容包括：

3.1 简要建筑方案设计说明（限 200 字以内），包括方案构思、太阳能综合应用技术与设计创新、技术经济指标表等。

3.2 项目的竞赛作品需进行竞赛用地范围内的规划设计，总平面图比例为 1 ：500~1 ：1000（含活动场地及环境设计）。

3.3 单体设计：

能充分表达建筑与室内外环境关系的各层平面图、外立面图、剖面图，比例为 1 ：200 或 1 ：250；

能表现出技术与建筑结合的重点部位、局部详图及节点大样，比例自定；其他相关的技术图、分析图、海峡两岸绿色建筑评价指标得分表等。

3.4 建筑效果表现图 1~4 个。

3.5 参赛者须将作品文件编排在 840 mm × 590 mm 的展板区域内（统一采用竖向构图），作品张数应为 4 或 6 张。中英文统一使用黑体字。字体大小应符合下列要求：标题字高 25 mm；一级标题字高 20 mm；二级标题字高 15 mm；图名字高 10 mm；中文设计说明字高 8 mm；英文设计说明字高 6 mm；尺寸及标注字高 6 mm。文件分辨率 100 dpi，格式为 JPG 或 PDF 文件。

4. 参赛者通过竞赛网页上传功能将作品递交竞赛组委会，入围作品由组委会统一编辑板眉、出图、制作展板。

5. 作品文字要求：除 3.1 "建筑方案设计说明" 采用中英文外，其他为英文；建议使用附件 3 中提供的专业术语。

参赛要求：

1. 欢迎建筑设计院、高等院校、研究机构、绿色建筑部品研发生产企业等单位，组织专业人员组成竞赛小组参加竞赛。

2. 请参赛者访问 www.isbdc.cn，按照规定步骤填写注册表，提交后会得到唯一的注册号，即为作品编号，一个作品对应一个注册号。提交作品时把注册号标注在每幅作品的左上角，字高 6mm。注册时间 2020 年 3 月 1 日～ 2020 年 6 月 30 日。

3. 参赛者同意组委会公开刊登、出版、展览、应用其作品。

Annex 2: Construction Corps Pasture Kindergarten and Service Center in Hejing County of Bayingol Mongolian Autonomous Prefecture, Xinjiang (Draft for comments)

Annex 3: Professional Glossary

Annex 4: Information Table

Award Setting and Award Form:

1. General Prize:

First Prize: 2 winners

The Trophy Cup, Certificate and Bonus RMB 100,000 (before tax) will be awarded.

Second Prize: 4 winners

The Trophy Cup, Certificate and Bonus RMB 20,000 (before tax) will be awarded.

Third Prize: 6 winners

The Trophy Cup, Certificate and Bonus RMB 5,000 (before tax) will be awarded.

The tax of bonuses which shall be levied based on accidental income will be paid by the Organizing Committee.

Honorable Mention Prize: 30 winners

The Certificate will be awarded.

2. Prize for Excellent Design Method: 10 winners

The Certificate will be awarded.

The report on works' design method is rich and full, the design process is complete, and the design method is novel.

3. Prize for Technical Excellence Works:

The quota is open-ended. The Certificate will be awarded.

Prize works must be innovative with practicability in aspect of technology adopted or design.

4. Prize for Architectural Originality:

The quota is open-ended. The Certificate will be awarded.

Prize works must be originally creative and advanced.

Requirements of Works:

1. The submitted drawing sheets should meet the requirements of scheme design level and should be accompanied with relevant technical drawings and technology data.

2. Drawings and text should be expressed in a clear and readable way. Mentioned data should be accurate.

3. The submitted work should include:

3.1 A project description (not exceeding 200 words) including the following elements: Schematic design concept; Integration of solar energy technology; Innovative design; Technical and economic indicators.

3.2 Participants should provide a planning design within the outline of the competition site, and provide a site plan (including the venue / environment design) with the scale of 1 ：500 or 1 ：1000.

3.3 Monomer Design:

Participants should provide floor plans, elevations and sections with the scale of

4. 被编入获奖作品集的作者，应配合组委会，按照出版要求对作品进行相应调整。

注意事项：

1. 参赛作品电子文档和作品设计方法报告须在 2020 年 9 月 15 日前提交组委会，请参赛者访问 www.isbdc.cn，并上传文件，不接受其他递交方式。

2. 作品中不能出现任何与作者信息有关的标记内容，否则将视其为无效作品。

3. 组委会将及时在网上公布入选结果及评比情况，将获奖作品整理出版，并对获奖者予以表彰和奖励。

4. 获奖作品集首次出版后 30 日内，组委会向获奖作品的创作团队赠样书 2 册。

5. 竞赛活动消息发布、竞赛问题解答均可登陆竞赛网站查询。

所有权及版权声明：

参赛者提交作品之前，请详细阅读以下条款，充分理解并表示同意。

依据中国有关法律法规，凡主动提交作品的“参赛者”或“作者”，主办方认为其已经对所提交的作品版权归属作如下不可撤销声明：

1. 原创声明

参赛作品是参赛者原创作品，未侵犯任何他人的任何专利、著作权、商标权及其他知识产权；该作品未在报纸、杂志、网站及其他媒体公开发表，未申请专利或进行版权登记，未参加过其他比赛，未以任何形式进入商业渠道。参赛者保证参赛作品终身不以同一作品形式参加其他的设计比赛或转让给他方。否则，主办单位将取消其参赛、入围与获奖资格，收回奖金、奖品及并保留追究法律责任的权利。

2. 参赛作品知识产权归属

为了更广泛推广竞赛成果，所有参赛作品除作者署名权以外的全部著作权归竞赛承办单位及冠名单位所有，包括但不限于以下方式行使著作权：享有对所属竞赛作品方案进行再设计、生产、销售、展示、出版和宣传的权利；享有自行使用、授权他人使用参赛作品用于实地建设的权利。竞赛主办方对所有参赛作品拥有展示和宣传等权利。其他任何单位和个人（包括参赛者本人）未经授权不得以任何形式对作品转让、复制、转载、传播、摘编、出版、发行、许可使用等。参赛者同意竞赛承办单位及冠名单位在使用参赛作品时将对其作者予以署名，同时对作

1 : 200 or 1 : 250, which can fully express the relationship between architecture and indoor and outdoor environment.

Participants should provide detailed drawings (without limitation of scale) that illustrate the integration of technology in the architectural project, as well as any other relevant elements, such as technical charts, analysis diagram, and scoring tables of Assessment Standard for Green Building between both sides of Taiwan Strait.

3.4 Rendering perspective drawing (1-4).

3.5 Participants should arrange the submission into four or six exhibition panels, each 840mm × 590mm in size (arranged vertically). Chinese and English font type should be both in boldface. Word height is required as follows: title: 25mm; first subtitle: 20mm; second subtitle: 15mm; figure title: 10mm; design description in Chinese: 8mm; design description in English: 6mm; dimensions and labels: 6mm. File resolution: 100dpi in JPEG or PDF format.

4. Participants should send (upload) a digital version of submission via FTP to the Organizing Committee, who will compile, print and make exhibition panels for shortlist works.

5. Text requirement: The submission should be in English, in addition to 3.1 "architectural design description" in English and Chinese. Participants should use the words from the Professional Glossary in Annex 3.

PARTICIPATION REQUIREMENTS:

1. Professionals from institutes of architectural design, colleges and universities, research institutions and green building product development and manufacturing enterprises are welcomed to take part in the competition in the form of competition groups.

2. Please visit www.isbdc.cn. You may fill in the registry according to the instruction and gain an ID of your work after submitting the registry. That's the number of your work, and one work only has one ID. The number should be indicated in the top left corner of each submitted work with word height of 6mm. Registration time: March 1, 2020 - June 30, 2020.

3. Participants must agree that the Organizing Committee may publish, print, exhibit and apply their works in public.

4. The authors whose works are edited into the publication should cooperate with the Organizing Committee to adjust their works according to the requirements of press.

IMPORTANT CONSIDERATION:

1. Participant's digital file and report on works' design method must be uploaded to the Organizing Committee's FTP site (www.isbdc.cn) before September 15, 2020. Other ways will not be accepted.

2. Any mark, sign or name related to participant's identity should not appear in, on or included within the submitted files, otherwise the submission will be deemed invalid.

3. The Organizing Committee will publicize the process and result of the appraisal online in a timely manner, compile and publish the awarded works. The winners will be honored and awarded.

4. In 30 days after the collection of works being published, two books of award works will be freely presented by the Organizing Committee to the competition teams who are awarded.

品将按出版或建设的要求做技术性处理。参赛作品均不退还。

3. 参赛者应对所提交作品的著作权承担责任，凡由于参赛作品而引发的著作权属纠纷均应由作者本人负责。

声明：

1. 参与本次竞赛的活动各方（包括参赛者、评委和组委），即表明已接受上述要求。

2. 本次竞赛的参赛者，须接受评委会的评审决定作为最终竞赛结果。

3. 组委会对竞赛活动具有最终的解释权。

4. 为维护参赛者的合法权益，主办方特提请参赛者对本办法的全部条款、特别是〝所有权及版权〞声明部分予以充分注意。

5. The information concerning the competition as well as explanation about all activities may be checked and inquired in the website of the competition.

ANNOUNCEMENT ABOUT OWNERSHIP AND COPYRIGHT:

Before submitting the works, participants should carefully read following clauses, fully understand and agree with them.

According to relevant national laws and regulations, it is confirmed by the competition hosts that all "participants" or "authors" who have submitted their works on their own initiative have received following irrevocable announcement concerning the ownership and copyright of their works submitted:

1. Announcement of originality

The entry work of the participant is original, which does not infringe any patent, copyright, trademark and other intellectual property; it has not been published in any newspapers, periodicals, magazines, webs or other media, has not been applied for any patent or copyright, not been involved in any other competition, and not been put in any commercial channels. The participant should assure that the work has not been put in any other competition by the same work form in its whole life or legally transferred to others, otherwise, the competition hosts will cancel the qualification of participation, being shortlisted and being awarded of the participant, call back the prize and award and reserve the right of legal liability.

2. The ownership of intellectual property of the works

In order to promote competition results, the participants should relinquish copyright of all works to competition organizers and titled sponsors except authorship. It includes but is not limited to the exercise of copyright as follows: benefit from the right of the works on redesigning, production, selling, exhibition, publishing and publicity; benefit from the right of the works on construction for self use or accrediting to others for use. Hosts of the competition have such rights to display and publicize all the works. Without accreditation, any organizations and individuals (including authors themselves) cannot transfer, copy, reprint, promulgate, extract and edit, publish and admit others to use the works by any way. Participants have to agree that competition organizers and titled sponsors will sign the name of authors when their works are used and the works will be treated for technical processing according to the requirements of publication and construction. All works are not returned to the author.

3. All authors must take responsibility for their copyrights of the works including all disputes of copyright caused by the works.

ANNOUNCEMENT:

1. It implies that everybody who has attended the competition activities including participants, jury members and members of the Organizing Committee has accepted all requirements mentioned above.

2. All participants must accept the appraisal of the jury as the final result of the competition.

3. The Organizing Committee reserves final rights to interpret the competition activities.

4. In order to safeguard the legitimate rights and interests of the participants, the organizers ask participants to fully pay attention to all clauses in this document, especially the Announcement about Ownership and Copyright.

附件 1：

福建省南平市建阳区景龙幼儿园任务书

1. 项目背景

本项目用地位于福建省南平市建阳区，北纬 27°19′58″，东经 118°05′47″，海拔 168m。项目处于城市中心地带，交通便利，用地东北高西南低。用地内无既有建筑和需要保留的树木，场地较为空旷。本项目为 12 班全日制幼儿园，为城市区域配套的公共服务设施。

图1 项目所在区位图
Figure 1 Project Location Map

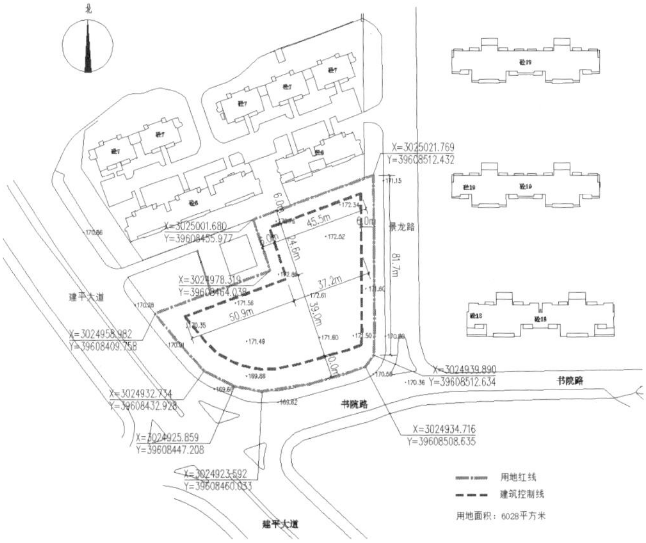

X=3025021.769
Y=39608512.432
X=3025001.680
Y=39608455.977
X=3024978.310
Y=39608464.039
X=3024958.982
Y=39608409.758
X=3024932.734
Y=39608432.928
X=3024925.859
Y=39608447.208
X=3024923.692
Y=39608460.033
X=3024939.890
Y=39608512.634
X=3024934.716
Y=39608508.635
45.5m
37.2m
50.9m
81.7m
景龙路
书院路
建平大道
用地红线
建筑控制线
用地面积：6028平方米

图2 项目用地平面图
Figure 2 Floor Plan of Project Site

Annex1：

Task Book of Design of Jinglong Kindergarten in Jianyang District, Nanping City, Fujian Province

1. Project background

The project site, higher in the northeast and lower in the southwest, is located in downtown areas of Jianyang District, Nanping City, Fujian Province, with the coordinate of latitude and longitude of 27°19′58″N and 118°05′47″E, an elevation of 168m, as well as convenient transportation. The site is relatively empty as there are no existing buildings and no trees to be reserved. This project is planned to be built into a full-time kindergarten with 12 classes as public service facilities in urban regions.

2. Natural condition

Jianyang District of Nanping City is located on the southeast side of the north section of Wuyi Mountain, with a subtropical monsoon climate which is genial and rich resources in light and heat. The region is characterized by short winters and long summers, calm wind and concentrated rainy seasons. There are also following features annually: the average temperature of 18.1℃ , 322 frost-free days, average precipitation of 1,742mm, average sunshine duration of 1,802 hours, and average relative humidity of 82%.

2. 自然条件

南平市建阳区位于武夷山脉北段东南侧，属亚热带季风气候，光热资源丰富，冬短夏长，气候宜人，静风多，雨季集中；年平均气温 18.1℃，无霜期 322 天，年平均降雨量 1742mm，年日照平均 1802h，年相对湿度平均值为 82%。

当地森林资源丰富，有“林海竹乡”之称。

3. 基础设施

场地周边基础设施完备，已有市政自来水、排水、雨水、天然气、供电及通信系统。

4. 竞赛场地

本项目用地，东侧与绿欧璟园小区隔景龙路相邻，南邻书院南路，西邻城市级干道建平大道，北侧紧靠融汇山水居住小区，用地面积 6028m^2。场地西北侧为水泵房和空地，有出入口通往场地，场地及周边道路标高见用地现状图。

5. 设计要求

（1）在给定的竞赛用地范围内设计 12 班共 360 人规模的全日制幼儿园一所，层数不超过 3 层，总建筑面积不超过 4700m^2，停车位不少于 4 个。

（2）场地包括室外游戏场地、集中绿地两部分，室外游戏场地人均面积不应低于 4m^2，其中分班游戏场地人均面积不应低于 2m^2，公用游戏场地人均面积不应低于 2m^2；幼儿园绿地率不应低于 30%，集中绿地人均面积不应低于 2m^2。

（3）充分结合当地的气候特征、资源条件、建筑特色以及建筑的使用特点，针对本幼儿园的建设条件和使用特点，合理选择和应用相关绿色技术，解决夏季通风、降温、除湿等问题，兼顾冬季保温的需求，达到降低建筑能耗、提高室内环境健康品质的目的。采用的技术应具有可实施性。宜考虑的技术问题包括但不限于以下内容：

①建筑节能。主要包括围护结构、空调系统、照明系统等方面的节能。

②可再生能源利用。主要包括光伏、光热，以及其他可再生能源等方面。

Local forests are known as "Seas of Forests and Bamboos" due to the abundant resources.

3. Infrastructure

The surrounding infrastructures are self-contained, with municipal tap water, drainage, rainwater, natural gas, power supply and communication systems equipped.

4. Competition venue

The project site, covering an area of 6,028m^2, is adjacent to Jinglong Road across Lv'ou Jingyuan Community in the east, South Shuyuan Road in the south, Jianping Avenue, namely a city-level artery, in the west, and Ronghui Shanshui residential area in the north. The pump room and open space lie on the northwest side of the site, where there are entrances and exits to the site. Moreover, the elevation of the site and surrounding roads is shown in the floor plan of project site.

5. Requirements for design

A. A full-time kindergarten with 360 people in 12 classes will be designed within the venue, with no more than 3 floors covering a construction area of no exceeding 4,700m^2, and no less than 4 parking spaces.

B. The venue includes the outdoor game field and the concentrated greenbelt. The per capita area of the former should not be less than 4m^2, of which the per capita area of the game field for divided class should not be less than 2m^2, and that of the public game field should not be less than 2m^2. The greening rate should not be less than 30%, with the per capita area of concentrated greenbelt should not be less than 2m^2.

C. Concerning local climates, resources, architectural features and buildings' characteristics in using, and based on the kindergarten's construction conditions and applied characteristics, relevant green technologies will be reasonably selected and applied to solve such problems as ventilation, cooling and dehumidification in summer, as well as thermal insulation in winter, thus reducing building energy consumption and improving healthy quality of indoor environment. Besides, the technology should be operable. Technical issues that should be considered include, but are not limited to:

a. Energy saving of buildings. It mainly includes energy saving in building envelope, air conditioning system and lighting system.

b. Renewable energy utilization. It mainly includes photovoltaic, solar thermal, and other renewable energy.

c. Indoor comfort. It mainly includes indoor temperature and humidity, natural light adjustment and natural ventilation.

d. Physical environment of the site. It mainly includes acoustic and wind environment.

e. Rainwater collection and utilization.

f. Site virescence and tridimensional virescence.

D. The building functions in three ways, namely children's living room, room for service management and room for logistics supply. See Table 1 for room composition and requirements for usable areas of each function.

③室内舒适度。主要包括室内温湿度、自然光线调节、自然通风等方面。

④场地物理环境。主要包括场地声环境、风环境。

⑤雨水收集与利用。

⑥场地绿化与立体绿化。

（4）建筑功能包括幼儿生活用房、服务管理用房、后勤供应用房三部分。各部分的房间组成及使用面积要求见表 1。

幼儿园各类用房与要求　　表 1

房间名称		数量	最小使用面积（m^2）	备注
幼儿生活用房	班级活动单元	12	1584 或 1908	包括活动室、寝室、卫生间、储藏间。班级活动单元使用面积，当活动室和寝室合并设置时不小于 1584m^2，当分开设置时不小于 1908m^2
	综合活动室	1	230	幼儿园共用，用于开展大型活动
服务管理用房	办公室	9	180	含园长、教师、财务
	图书兼会议室	1	30	
	保健室	1	30	含晨检、保健
	教具制作室	1	25	
	门卫室	1	30	兼收发、值班、监控
	储藏室	—	60	分设用品、玩具、杂物
	教工厕所	—	30	分设男、女、无障碍厕所
后勤供应用房	厨房	1	248	包括加工间、切配间、备餐间、洗消间、食库、更衣室
	开水间	1	12	
	洗衣房	1	18	
	配电室	1	12	

（5）其他未尽事宜符合以下标准要求

《民用建筑设计统一标准》GB 50352-2019

《建筑设计防火规范》GB 50016-2014（2018 年版）

《托儿所、幼儿园建筑设计规范》JGJ 39-2016（2019 年版）

《幼儿园建设标准》建标 175-2016

《公共建筑节能设计标准》GB 50189-2015

《福建省公共建筑节能设计标准》DBJ 13-305-2019

Various Types of Rooms and Requirements of the Kindergarten　　Table 1

Room		Quantity	Minimum Area (m^2)	Note
Children's Living Room	Space for class activities	12	1,584 or 1,908	Including activity room, dormitory, bathroom and storeroom. The usable area of the space for class activities is not less than 1,584m^2 when the activity room and dormitory are shared for use and not less than 1,908m^2 when they are separate for use
	Comprehensive activity room	1	230	Shared by the kindergarten for large-scale activities
Room for Service Management	Office	9	180	Set for the principal, teacher, financial staff, etc
	Library also used for conference	1	30	
	Fitness room	1	30	For morning inspection and health care, etc
	Teaching aid production room	1	25	
	Guard room	1	30	Also used for sending and receiving, on duty and monitoring
	Storeroom	—	60	Storing supplies, toys and sundries separately
	Faculty's toilet	—	30	Set men's, women's and accessible toilet separately
Room for Logistics Supply	Kitchen	1	248	Including the processing room, food cutting and matching room, preparation room, decontamination room, food storeroom and dressing room
	Water heater room	1	12	
	Laundry	1	18	
	Switch room	1	12	

E.Other unfinished matters should meet the following standards:

Uniform Standard for Design of Civil Buildings　GB 50352-2019

Code for Fire Protection Design of Buildings　GB 50016-2014 (2018 Edition)

Code for Design of Nursery and Kindergarten Buildings　JGJ 39-2016 (2019 Edition)

Construction Standard for Kindergartens　Construction Standard 175-2016

Design Standard for Energy Efficiency of Public Buildings　GB 50189-2015

Design Standard for Energy Efficiency of Public Buildings in Fujian　DBJ 13-305-2019

附件 2：

新疆巴音郭楞州和静县建设兵团牧场幼儿园及服务中心任务书

1. 项目背景

本项目用地位于新疆维吾尔自治区巴音郭楞州和静县巴伦台镇布热斯台地区，北纬 42°58′，东经 86°33′，海拔 2200m。项目处于山夹谷之中的小平原地区，一条溪流流经。场地距 G216 高速与 S201 省道 11.6km，距和静县直线距离 75km，距乌鲁木齐市直线距离 130km。

Annex2：

Task Book of Design of Construction Corps Pasture Kindergarten and Service Center in Hejing County of Bayingol Mongolian Autonomous Prefecture, Xinjiang

1. Project background

The project site is located in Buresitai Region, Baluntai Town, Hejing County, Bayingol Mongolian Autonomous Prefecture, Xinjiang Uygur Autonomous Region, with the coordinate of latitude and longitude of 42° 58′ N and 86° 33′ E, and an elevation of 2,200m. The project lies in a small plain area in a valley among

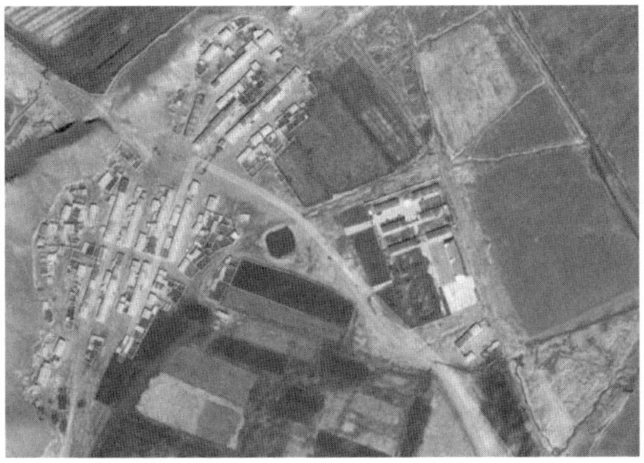

图1 项目所在区位图
Figure 1 Project Location Map

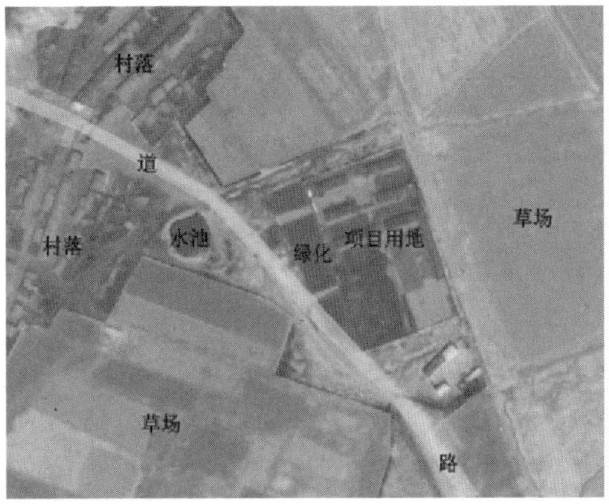
村落
道
水池
村落
绿化
项目用地
草场
草场
路

图2 项目用地周边功能图
Figure 2 Surroundings of Project Site

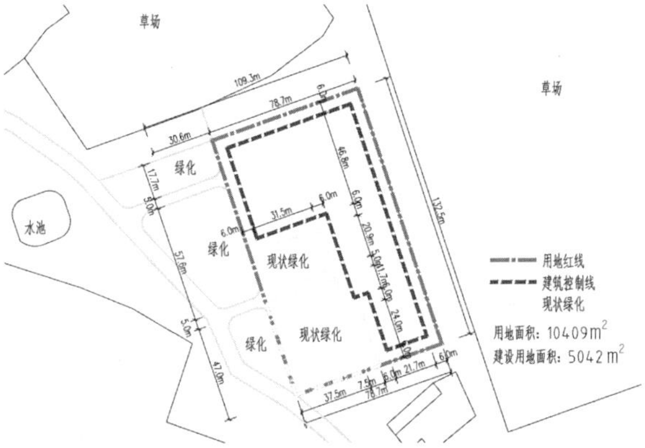
草场
草场
水池
绿化
绿化
绿化
现状绿化
现状绿化
用地红线
建筑控制线
现状绿化
用地面积：10409m²
建设用地面积：5042m²

图3 项目用地平面图
Figure 3 Floor Plan of Project Site

图4 项目用地入口现状
Figure 4 Entrance to Project Site

图5 项目用地内现状
Figure 5 Inside of Project Site

图6　村落民居
Figure 6　Village Houses

图7　项目用地外道路全景
Figure 7 Panorama of Roads Outside Project Site

该项目为新疆建设兵团某师某团牧三场连部所在地，连部原有房屋破旧，已成为危房，需要重新建设。新建连部除具备连部办公功能外，还作为附近牧场的服务中心，建设幼儿园、卫生室、图书室、活动室等。

2. 自然条件

和静县巴伦台镇位于新疆维吾尔自治区南疆地区，属温带大陆性气候，海拔较高，昼夜温差大，气候特点是冬季寒冷漫长，春季解冻慢，风多，夏秋天气凉爽，年平均气温 5.7℃，无霜期 131 天，年降水量在 111~289mm 之间，降水主要集中在 6、7、8 月间，季节分布不均匀，境内水资源丰富。项目用地场地平整，场

图8　项目用地周边照片
Figure 8 Photos Shot around Project Site

mountains, through which a stream flows. The site is 11.6km away from the G216 highway and S201 provincial highway, 75km from Hejing County in a straight line, and 130km from Urumqi in a straight line.

This site is home to a Company office of Musanchang of a certain regiment, a certain division, Construction Corps, Xinjiang, where original houses need rebuilding because they have become dilapidated. The newly-established Company functions as a service center for nearby pastures, and is put into use to build kindergartens, medical rooms, libraries and activity rooms except for its role of an office.

2. Natural condition

Baluntai Town of Hejing County is located in the south of Xinjiang Uygur Autonomous Region, with a temperate continental climate which features cold and long winters, slow speed in defrosting and windiness in spring, as well as cool weather in summer and autumn. Annually, there are also following features: the average temperature of 5.7 ° C, 131 frost-free days, concentrated precipitation in June, July and August between 111mm and 289mm, which is uneven in distribution in different seasons, as well as abundant resources in water. The project site is flat with many soil slopes around. Generally, it's cold in winter, while it's suitable in summer.

3. Infrastructure

The infrastructures inside the site are backward, with poor stability in power supply, no measures for drainage and unstable communications, merely a tap water system equipped.

4. Competition venue

This project, covering an area of 10,408.6m^2, is a service center of Musanchang of a certain regiment, a certain division, Construction Corps, Xinjiang, the site of which is 1m higher than surrounding sites. There are pastures on the east side and south side. The site is adjacent to the angled greenbelt on the southwest side, which connects the road to enter the village. And the site is near the pasture on the north side with a stream setting apart. The village is located on the northwest side of the site, about 100m away.

5. Requirements for design

A. The boarding kindergarten, pasture service center and supporting house are designed within the competition venue, with a construction area of no more than 2,200m^2;

B. The boarding kindergarten provides preschool education and accommodation

地周边多土坡。冬季寒冷、夏季温度适宜。

3. 基础设施

基地内基础设施较差，有自来水系统，供电稳定性不佳、无排水措施，通信不稳定。

4. 竞赛场地

本项目为新疆建设兵团某师某团牧三场服务中心，项目用地比周边场地高1m。东侧、南侧为牧场，西南侧紧邻夹角型绿地与进村道路相接，北侧隔一条小溪与牧场相邻，村庄在用地西北侧，距离约100m。用地面积10408.6m^2。

5. 设计要求

（1）在给定的竞赛用地范围内设计寄宿幼儿园、牧场服务中心、配套用房，总建筑面积不超过2200m^2。

（2）寄宿幼儿园为牧场周边村落儿童提供幼儿教育和住宿，建筑面积不大于1000m^2，独立设置，建筑为1层，两个班，每个班20个儿童，需满足《托儿所、幼儿园建筑设计规范》JGJ 39—2016（2019年版）的要求。室外活动场地、农业体验园和动物饲养体验区不计入建筑面积。

（3）牧场服务用房为牧场提供办公场所，并为牧场周边村落提供基础社区服务和集中交流活动空间，建筑面积不大于700m^2，建筑不高于2层。室外应设置不少于8个停车位，不小于500m^2的活动广场。

（4）配套用房是为办公人员、幼儿园教职员工等人员提供居住生活服务的场所，建筑面积不大于500m^2，建筑不高于2层。

（5）现状绿化范围内不得建设建筑，且不得设置活动广场及停车位，可设置农业体验园和动物饲养体验区。

（6）考虑当地的气候特点和自然环境，结合当地的建筑特色和材料，以及不同建筑的用能特征，解决建筑的冬季保温采暖和夏季遮阳通风等问题，合理应用主动、被动太阳能技术及其他可再生能源技术，并考虑技术的可实施性。

（7）考虑建筑的低成本建造的经济性和可推广性。

（8）合理组织场地内不同功能空间之间分区与流线的关系，保证联系便捷，减少相互干扰。

（9）项目用地内市政配套措施缺乏，应设置相应的垃圾处理、污水无害化处理等应对措施。

（10）建筑用房设置如下表所示：

for children in villages around the pasture, with a construction area of no more than 1,000m^2. Besides, the kindergarten is set independently with 1 floor and 2 classes, and 20 children are arranged in each class. The project should also meet the requirements of *Code for Design of Nursery and Kindergarten Buildings* JGJ 39—2016 (2019 Edition). Outdoor activity venues, agricultural experience parks and animal breeding experience areas are excluded from the construction area.

C. The pasture service center provides office space for pastures, and offers basic community service and communication space to surrounding villages. The building is not more than 700m^2 in construction area, and not higher than 2 floors. There should be at least 8 parking spaces outdoors and an activity square of no less than 500m^2.

D. Supporting houses are provided to office staff, faculty and other personnel for living. The building is not more than 500m^2 in construction area, and not higher than 2 floors.

E. No buildings can be built within the current greenbelt, and no activity squares and parking spaces can be set. However, agricultural experience parks and animal feeding experience areas can be built.

F. Considering local climate characteristics and natural environment, and based on local architectural features and materials, as well as the energy consumption of different buildings, efforts should be made to solve such problems as thermal insulation and heating in winter, as well as sun-shading and ventilation in summer. In addition, both the active and passive solar technology and other renewable energy technology should be rationally applied, and the feasibility of technology should be considered.

G. The economical performance in low-cost construction and possibility of popularization should be considered.

H. The relationship between subareas and streamlines among different functional spaces within the site should be reasonably organized to ensure convenient contact and less interference.

I. Corresponding measures such as garbage treatment and harmless treatment of sewage should be adopted due to a lack of municipal supporting measures in the project site.

J. Rooms with functions set are shown in the following table:

功能空间	功能要求	数量	总使用面积（m^2）	备注
寄宿幼儿园	幼儿活动用房	2 套	320	包括活动室、寝室、卫生间、盥洗室、衣帽间等
	综合活动用房	1	80	兼作专用活动教室
	管理用房	—	100	包括办公室、门厅、保健室、隔离室等
	配套用房	—	120	厨房、消毒室、门卫、教师值班室、储藏室等
	活动场地	—	400	室外不计入建筑面积
	农业体验园	—	≥ 400	包括室外种植区和设施农业种植区，不计入建筑面积
	动物饲养体验区	—	≥ 400	羊圈、牛棚等配套设施，不计入建筑面积
牧场服务中心	办公室	4 间	100	
	多功能厅	1 间	150	含设备用房
	医务室	1 间	80	含诊室、处置室及观察室
	图书室	1 间	50	
	活动室	2 间	60	
配套用房	宿舍	6 间	120	牧场工作人员宿舍
	厨房及餐厅	1 间	60	
	储藏室	1 间	20	
	卫生间等用房	—	40	按需设置

（11）其他未竟事宜符合以下标准要求

《民用建筑设计统一标准》GB 50352—2019

《建筑设计防火规范》GB 50016—2014（2018 年版）

《托儿所、幼儿园建筑设计规范》JGJ 39—2016（2019 年版）

《幼儿园建设标准》建标 175—2016

《公共建筑节能设计标准》GB 50189—2015

Functional Space	Function	Quantity	Total Usable Area (m^2)	Note
Boarding Kindergarten	Room for children's activity	2 sets	320	Including activity room, dormitory, bathroom, washroom and dressing room
	Room for comprehensive activity	1	80	Also used as the exclusive classroom for activities
	Room for management	—	100	Including the office, lobby, fitness room and isolation room
	Supporting room	—	120	Kitchen, disinfection room, guard room, faculty duty room, storeroom, etc
	Activity venue	—	400	As outdoor space, excluded from the construction area
	Agricultural experience park	—	≥ 400	Including outdoor planting areas and protected agriculture planting areas, excluded from the construction area
	Animal feeding experience area	—	≥ 400	Such supporting facilities as sheepfolds, cowsheds, excluded from the construction area
Pasture Service Center	Office	4 rooms	100	
	Multi-function room	1 room	150	Including rooms for equipment
	Medical room	1 room	80	Including consulting room, treatment room and observation room
	Library	1 room	50	
	Activity room	2 rooms	60	
Supporting Room	Dormitory	6 rooms	120	For staff of pastures
	Kitchen and dining room	1 room	60	
	Storeroom	1 room	20	
	Bathroom and other rooms	—	40	Set based on demands

K. Other unfinished matters should meet the following standards:

Uniform Standard for Design of Civil Buildings GB 50352-2019

Code for Fire Protection Design of Buildings GB 50016-2014 (2018 Edition)

Code for Design of Nursery and Kindergarten Buildings JGJ 39-2016 (2019 Edition)

Construction Standard for Kindergartens Construction Standard 175-2016

Design Standard for Energy Efficiency of Public Buildings GB 50189-2015

附件 3：／Annex 3：

专业术语 Professional Glossary

百叶通风 – shutter ventilation
保温 – thermal insulation
被动太阳能利用 – passive solar energy utilization
敞开系统 – open system
除湿系统 – dehumidification system
储热器 – thermal storage
储水量 – water storage capacity
穿堂风 – through-draught
窗墙面积比 – area ratio of window to wall
次入口 – secondary entrance
导热系数 – thermal conductivity
低能耗 – lower energy consumption
低温热水地板辐射供暖 – low temperature hot water floor radiant heating
地板辐射采暖 – floor panel heating
地面层 – ground layer
额定工作压力 – nominal working pressure
防潮层 – wetproof layer
防冻 – freeze protection
防水层 – waterproof layer
分户热计量 – household-based heat metering
分离式系统 – remote storage system
风速分布 – wind speed distribution
封闭系统 – closed system
辅助热源 – auxiliary thermal source
辅助入口 – accessory entrance
隔热层 – heat insulating layer
隔热窗户 – heat insulation window
跟踪集热器 – tracking collector
光伏发电系统 – photovoltaic system
光伏幕墙 – PV façade
回流系统 – drainback system
回收年限 – payback time
集热器瞬时效率 – instantaneous collector efficiency
集热器阵列 – collector array
集中供暖 – central heating
间接系统 – indirect system
建筑节能率 – building energy saving rate
建筑密度 – building density
建筑面积 – building area
建筑物耗热量指标 – index of building heat loss
节能措施 – energy saving method
节能量 – quantity of energy saving
紧凑式太阳热水器 – close-coupled solar water heater
经济分析 – economic analysis
卷帘外遮阳系统 – roller shutter sun shading system
空气集热器 – air collector
空气质量检测 – air quality test (AQT)
立体绿化 – tridimensional virescence
绿地率 – greening rate
毛细管辐射 – capillary radiation
木工修理室 – repairing room for woodworker
耐用指标 – permanent index
能量储存和回收系统 – energy storage & heat recovery system
平屋面 – plane roof
坡屋面 – sloping roof
强制循环系统 – forced circulation system
热泵供暖 – heat pump heat supply
热量计量装置 – heat metering device
热稳定性 – thermal stability
热效率曲线 – thermal efficiency curve
热压 – thermal pressure
人工湿地效应 – artificial marsh effect
日照标准 – insolation standard
容积率 – floor area ratio
三联供 – triple co-generation
设计使用年限 – design working life
使用面积 – usable area
室内舒适度 – indoor comfort level
双层幕墙 – double facade building

太阳方位角 — solar azimuth
太阳房 — solar house
太阳辐射热 — solar radiant heat
太阳辐射热吸收系数 — absorptance for solar radiation
太阳高度角 — solar altitude
太阳能保证率 — solar fraction
太阳能带辅助热源系统 — solar plus supplementary system
太阳能电池 — solar cell
太阳能集热器 — solar collector
太阳能驱动吸附式制冷 — solar driven desiccant evaporative cooling
太阳能驱动吸收式制冷 — solar driven absorption cooling
太阳能热水器 — solar water heating
太阳能烟囱 — solar chimney
太阳能预热系统 — solar preheat system
太阳墙 — solar wall
填充层 — fill up layer
通风模拟 — ventilation simulation
外窗隔热系统 — external windows insulation system
温差控制器 — differential temperature controller
屋顶植被 — roof planting
屋面隔热系统 — roof insulation system
相变材料 — phase change material (PCM)
相变太阳能系统 — phase change solar system
相变蓄热 — phase change thermal storage
蓄热特性 — thermal storage characteristic
雨水收集 — rain water collection
运动场地 — schoolyard
遮阳系数 — sunshading coefficient
直接系统 — direct system
值班室 — duty room
智能建筑控制系统 — building intelligent control system
中庭采光 — atrium lighting
主入口 — main entrance
贮热水箱 — heat storage tank
准备室 — preparation room
准稳态 — quasi-steady state
自然通风 — natural ventilation
自然循环系统 — natural circulation system
自行车棚 — bike parking